VOLUME 3

A HANDBOOK SERIES ON
ELECTROMAGNETIC INTERFERENCE AND COMPATIBILITY
(EMI CONTROL METHODS AND TECHNIQUES)

By Donald R. J. White, MSEE/PE

© Copyright 1973
Third Edition 1981

DON WHITE CONSULTANTS, INC.
State Route 625
P. O. Box D
Gainesville, Virginia 22065 USA
Telephone: (703) 347-0030
Telex: 89-9165 DWCI GAIV

Library of Congress Catalog Card No. 72-138444

Printed in the United States of America

ACKNOWLEDGMENT

The author wishes to thank the many people who encouraged him to write the EMI/EMC handbook series in general, and this volume in particular. He expresses his appreciation to the individuals and companies who have furnished several of the illustrative figures, for which acknowledgments in this handbook have been made.

The author expresses his appreciation to Marge Lovenberg and to his wife, Colleen for their assistance in typing, and in the many facets of logistics involved in preparation of the manuscript; and to Sylvester Mathis for his drafting, and the many others who have helped produce this publication.

VOLUME 3

A HANDBOOK ON

EMI CONTROL METHODS AND TECHNIQUES

DON WHITE CONSULTANTS, INC.
14800 Springfield Road
Germantown, Maryland 20767
301-948-0028

(c)

Copyright 1973
First Edition

Library of Congress Catalog Card No. 72-138444
Printed in United States of America

OTHER BOOKS PUBLISHED BY DWCI

(1) White, Donald R. J., *Electrical Filters - Synthesis, Design & Applications;* published 1963. Second Edition 1980.

(2) White, Donald R. J., *A Handbook on Methods & Procedures for Automating RFI/EMI;* published 1966.

(3) White, Donald R. J., Volume 1, *Electrical Noise and EMI Specifications;* published 1971.

(4) White, Donald R. J., Volume 2, *Electromagnetic Interference Test Methods and Procedures;* published 1974. Second Edition 1980.

(5) White, Donald R. J., Volume 4, *Electromagnetic Interference Test Instrumentation Systems;* published 1971. Second Edition 1980.

(6) Duff, William G. and White, Donald R. J., Volume 5, *Electromagnetic Interference Prediction & Analysis Techniques;* published 1972.

(7) Hill, James S. and White, Donald R. J., Volume 6, *Electromagnetic Interference Specifications, Standards & Regulations;* published 1975.

(8) White, Donald R. J., *A Glossary of Acronyms, Abbreviations and Symbols;* published 1971.

(9) White, Donald R. J., *A Handbook on Electromagnetic Shielding Materials and Performance;* published 1975. Second Edition 1980.

(10) Duff, William G., *A Handbook on Mobile Communications;* published 1976. Second Edition 1980.

(11) Mertel, Herbert K., *International and National Radio Frequency Interference Regulations;* Volume I, of the Multi-Volume EMC Encyclopedia Series, published 1978.

(12) Jansky, Donald M., *Spectrum Management Techniques;* Volume II of the Multi-Volume EMC Encyclopedia Series, published 1977.

(13) Herman, John R., *Electromagnetic Ambients and Man-Made Noise;* Volume III, of the Multi-Volume EMC Encyclopedia Series, published 1979.

OTHER BOOKS PUBLISHED BY DWCI (CONT'D)

(14) Hart, William C. and Malone, Edgar W., *Lightning and Lightning Protection*, Volume IV of the Multi-Volume EMC Encyclopedia Series, published 1979.

(15) Keiser, Bernhard E., *EMI Control in Aerospace Systems*, Volume V of the Multi-Volume EMC Encyclopedia Series, published 1979.

(16) Feher, Kamilo, *Digital Modulation Techniques in an Interference Environment*, Volume IX of the Multi-Volume EMC Encyclopedia Series, published 1977.

(17) Gard, Michael F., *Electromagnetic Interference Control in Medical Electronics*, Volume X of the Multi-Volume EMC Encyclopedia Series, published 1979.

(18) Carstensen, Russell V., *EMI Control in Boats and Ships*, Volume XXIV of the Multi-Volume EMC Encyclopedia Series, published 1980.

(19) White, Donald R. J., *EMI Control Methodology and Procedures (EMC Design Synthesis)*, published 1978.

(20) White, Donald R. J., *EMI Control in the Design of Printed Circuit Boards and Backplanes*, published 1981.

NOTICE

All of the books listed above are available for purchase from Don White Consultants, Inc., State Route 625, P. O. Box D, Gainesville, Virginia 22065, U.S.A. Telephone: (703) 347-0030; Telex: 89-9165 DWCI GAIV.

For details on ordering and price listing, please refer to a copy of our *Handbook Order Form* which is enclosed at the back of this book.

SUMMARY
OF THE HANDBOOK SERIES ON
ELECTROMAGNETIC INTERFERENCE AND COMPATIBILITY

PREFACE

The control or reduction of Electromagnetic Interference (EMI) or its predecessor names, Radio Noise, Electrical Noise, or Radio-Frequency-Interference (RFI), is a rapidly expanding technology. It covers the frequency spectrum from dc to about 40 GHz. EMI is the culprit which does not allow radio, TV, radar, navigation, and the myriad of electrical, electromechanical, and electronic and communication devices, apparatus and systems to operate compatibly in a common frequency spectrum environment. EMI can result in a jammed radio, heart pacer failures, navigation errors and many other either nuisance or catastrophic events. Therefore, it follows that this spectrum pollution problem has reached international levels of concern and must be dealt with in proportion to the safety and economic impact involved.

There is much written material on EMI which is generally available in trade journals, symposium records, and the like. With certain exceptions, in their total this material represents an anthology of miscellaneous subjects and topics which do not interrelate very well. Certainly, to either a newcomer to the EMI or Electromagnetic Compatibility (EMC) disciplines or to others already in these disciplines seeking tutorial or *how-to-do-it* knowledge, it is very frustrating. Accordingly, these considerations are one of the primary reasons for bringing together under a handbook series a number of interrelated topics so that the reader will not be penalized with often conflicting, confusing, and missing material. Thus, much of this handbook information has never appeared in print before.

This handbook series is captioned *Electromagnetic Interference and Compatibility*. Volume 1 covers the topics of Electrical Noise and EMI Specifications. It surveys the EMI Community and Committees, and presents a number of EMI specifications, control and test plans and reports. Volume 2, *EMI Test Methods and Procedures*, covers calibration, test specimen preparation, and the family of conducted and radiated, emission and susceptibility test procedures. Both component and system level testing are covered. Volume 3, *EMI Control Methods and Techniques*, reviews in depth the topics of missions and applications, EMI prediction, EMI control techniques, and EMC control devices and materials. Volume 4 covers the topics of *EMI Test Instruments and Systems*. It presents the test environment, EMI sensors and instruments, and automatic EMI measuring systems. Volume 5, *EMI Prediction and Analysis Techniques*, covers mathematical modeling of transmitters, receivers, and the intervening propagation media. It presents the data base, scoring criteria, and computer simulation together with many illustrative examples. Volume 6 covers *EMI Specifications, Standards, and Regulations*.

vii

This handbook, Volume 3 on *EMI Control Methods and Techniques,* is novel in a number of ways. It underscores the tutorial throughout, emphasizing fundamentals wherever possible so that the reader can understand the rationale. Secondly, the *how-to-do-it* is illustrated wherever possible so that much of the heretofore *black magic* is removed by the many figures, tables and examples.

Volume 3 covers the following topics: EMI sources and receptors, applications and performance criteria, and inter-system prediction and control. Since the latter is emphasized in Vol. 5, this volume emphasizes more intra-system prediction and control, via grounding and bonding, shielding; cabling, wiring, and harnessing, connectors and fittings, packaging and gasketing, filters and filtering and suppression devices. Many illustrative examples are given including applications to radiation hazard control.

The author invites readers of these volumes and the users of the information contained therein to communicate with him. He especially invites their comments, questions, or request for further elucidation where necessary. Since it is planned to publish subsequent editions of these volumes and to add others to the series, any additional useful information that can be passed along to the engineering profession and other users will further advance the degree of EMI/EMC understanding and proficiency.

Germantown, Maryland   Donald R. J. White
February, 1973

Gainesville, Virginia   Donald R. J. White
May 1980

TABLE OF CONTENTS
VOLUME 3 - EMI CONTROL METHODS AND TECHNIQUES

CHAPTER 1 AN INTRODUCTION TO EMI CONTROL

CHAPTER 2 SOURCES OF ELECTROMAGNETIC INTERFERENCE

TABLE OF CONTENTS

CHAPTER 5   INTER-SYSTEM EMI PREDICTION AND CONTROL

CHAPTER 6   CABLE WIRING AND HARNESSING

Table of Contents

TABLE OF CONTENTS

CHAPTER 9   ARCHITECTURAL GROUNDING, WIRING, AND SHIELDING

CHAPTER 10   SHIELDING THEORY AND MATERIALS

Table of Contents

TABLE OF CONTENTS

CHAPTER 14   POWER-LINE FILTERS

CHAPTER 15   POWER-LINE ISOLATION DEVICES

TABLE OF CONTENTS

CHAPTER 16 EMI CONTROL IN COMPONENTS

CHAPTER 17   EMI CONTROL IN CIRCUITS AND EQUIPMENTS

TABLE OF CONTENTS

LIST OF ILLUSTRATIONS

CHAPTER 1   AN INTRODUCTION TO EMI CONTROL

CHAPTER 2   SOURCES OF ELECTROMAGNETIC INTERFERENCE

ILLUSTRATIONS

CHAPTER 3   EMI RECEPTORS AND SUSCEPTIBILITY CRITERIA

CHAPTER 4   INTRA-SYSTEM EMI PREDICTION AND ANALYSIS

ILLUSTRATIONS

CHAPTER 5   INTER-SYSTEM EMI PREDICTION AND CONTROL

CHAPTER 6   CABLE WIRING AND HARNESSING

ILLUSTRATIONS

CHAPTER 7   CONNECTORS

Illustrations

CHAPTER 10   SHIELDING THEORY AND MATERIALS

ILLUSTRATIONS

CHAPTER 13   COMMUNICATION FILTERS

CHAPTER 14   POWER-LINE FILTERS

CHAPTER 15   POWER-LINE ISOLATION DEVICES

CHAPTER 16   EMI CONTROL IN COMPONENTS

CHAPTER 17   EMI CONTROL IN CIRCUITS AND EQUIPMENTS

ILLUSTRATIONS

LIST OF TABLES

LIST OF TABLES

LIST OF TABLES

CHAPTER 17   EMI CONDUCTORS AND CIRCUIT EQUIPMENT SYSTEMS

ABBREVIATIONS AND SYMBOLS - TECHNICAL & RELATED

A*	ampere (unit of electrical current flow)
A	Angstrom, unit of length (= 10^{-4} microns = 39.37 x 10^{-10} in.)
A	area (effective) of antenna
A-Band	0-250 MHz (see A-1 to A-10 Band)
A/D	analog-to-digital converter
ADC	analog-to-digital converter
ADP	Automatic data processing
AF	audio frequency ($\simeq$ 0 to 150 kHz)
AFC	automatic frequency control
AF/PC	automatic frequency/phase controlled (loops)
AGC	Automatic gain control
AJ	anti-jamming
Al	aluminum (metal)
AM	amplitude modulation
A/M*	ampere per meter (unit of magnetic field strength)
AMTI	airborne moving target indicator
AN	Army-Navy (nomenclature (prefix on equipment, viz., AN/UVW-XYZ)
Ant	antenna
AOC	automatic overload control
APC	Auto-Plot Controller
APD	amplitude probability distribution
ASDSRS	Automatic Spectrum Display & Signal Recognition System
ASK	amplitude-shift keying (see AM and OOK)
ATR	anti-transmit-receive (tube)
AUTO	automatic
AVE	Aerospace Vehicle Electronics
Avg	average
AWG	American Wire Gage
Az	azimuth (angles)
A-1 Band	0-25 MHz
A-2 Band	25-50 MHz
A-3 Band	50-75 MHz
A-4 Band	75-100 MHz
A-5 Band	100-125 MHz
A-6 Band	125-150 MHz
A-7 Band	150-175 MHz
A-8 Band	175-200 MHz
A-9 Band	200-225 MHz
A-10 Band	225-250 MHz
B	bandwidth (see BW)
B	magnetic flux density (webers/m^2 = gauss)
B	susceptance (imaginary portion of admittance)
BB	broadband
B-Band	250-500 MHz(see B-1 to B-10 Band)

---

* Abbreviations and Symbols followed by an asterisk (*) appear in MIL-
STD-463.

CW	continuous wave
C-1 Band	500 - 550 MHz
C-2 Band	500 - 600 MHz
C-3 Band	600 - 650 MHz
C-4 Band	650 - 700 MHz
C-5 Band	700 - 750 MHz
C-6 Band	750 - 800 MHz
C-7 Band	800 - 850 MHz
C-8 Band	850 - 900 MHz
C-9 Band	900 - 950 MHz
C-10 Band	950 - 1000 MHz
d*	deci (10.1 multiplier)
DA	demand assignment
D/A	digital-to-analog converter
DAC	digital-to-analog converter
DASS	demand assignment signaling and switching
dB	decibel
D-Band	new designation: 1-2 GHz; replaces obsolete L-Band (see D-1 to D-10 Band)
DBB	detector balanced bias
dBm	dB above 1 milliwatt
dBm/m^2	dB above 1 milliwatt per square meter
$dBm/m^2/MHz$	dB above 1 milliwatt per square meter per Megahertz
dBμV	dB above 1 microvolt
dBμV/m/MHz	dB above 1 microvolt per meter per Megahertz
dBμV/MHz	dB above 1 microvolt per Megahertz
deg	degree (unit of phase or angular measurement)
DF	direction finding
DINA	direct noise amplification
DM	delta modulation (cf. PCM)
DPDT	double-pole double throw (switch)
DPSK	differential phase-shift keying
DPST	double-pole single throw (switch)
DSB	double sideband (modulation)
DSB-SC	double sideband suppressed carrier (modulation)
DTL	(modified) diode-transistor logic
DVM	digital voltmeter
D-1 Band	1.0 - 1.1 GHz
D-2 Band	1.1 - 1.2 GHz
D-3 Band	1.2 - 1.3 GHz
D-4 Band	1.3 - 1.4 GHz
D-5 Band	1.4 - 1.5 GHz
D-6 Band	1.5 - 1.6 GHz
D-7 Band	1.6 - 1.7 GHz
D-8 Band	1.7 - 1.8 GHz
D-9 Band	1.8 - 1.9 GHz
D-10 Band	1.9 - 2.0 GHz
E	volts per meter, V/m (unit of electric-field strength)
E-Band	new designation: 2-3 GHz (see E-1 to E-10 Band; replaces old S-Band

```
ECCM        electronic counter-counter measures
ECL         emitter-coupled logic
ECM         electronic countermeasures
ECN         engineering change notice
ECR         engineering change request
ECU         electronic conversion unit
EDP         electronic data processing
EED         electro-explosive device
EHF         extremely-high frequency (30 GHz to 300 GHz)
El          elevation (angle)
EM*         electromagnetic
EMC*        electromagnetic compatibility
EME         electromagnetic energy
EMI*        electromagnetic interference
EMICE       electromagnetic interference control engineer
EMP         electromagnetic pulse
EMR         electromagnetic radiation
EMS         electromagnetic susceptibility
EMW         electromagnetic warfare
ESAR        electronically scanned array radar
ESC         electrostatic compatibility
eV          electron volt
EW          electronic warfare
E-1 Band    2.0 - 2.1 GHz
E-2 Band    2.1 - 2.2 GHz
E-3 Band    2.2 - 2.3 GHz
E-4 Band    2.3 - 2.4 GHz
E-5 Band    2.4 - 2.5 GHz
E-6 Band    2.5 - 2.6 GHz
E-7 Band    2.6 - 2.7 GHz
E-8 Band    2.7 - 2.8 GHz
E-9 Band    2.8 - 2.9 GHz
E-10 Band   2.9 - 3.0 GHz


f*          farad (unit of capacitance)
f           frequency (Hz)
fath        fathom (unit of length; 1 fath = 6 feet = 1.828 m)
F-Band      new designation: 3-4 GHz (see F-1 to F-10 Band; replaces old
              S-Band
f/D         focal (length) to diameter ratio
F           noise figure  (or factor)
F           Farenheidt (unit of temperature; 1° F = 0.556° C)
FDM         frequency-division multiplex
FDMA        frequency division multiple access
FDR         frequency-domain reflectometry
FET         field-effect transistor
F-F         flip flop
FI          field intensity
FIM         field intensity meter
FM          frequency modulation
FSD         full-scale deflection
```

```
FSK          frequency-shift keying
FSVM         frequency selective voltmeter
ft           foot (unit of length;  ft = 12 in. = 0.3048 m)
FTC          fast time constant
F-1 Band     3.0 - 3.1 GHz
F-2 Band     3.1 - 3.2 GHz
F-3 Band     3.2 - 3.3 GHz
F-4 Band     3.3 - 3.4 GHz
F-5 Band     3.4 - 3.5 GHz
F-6 Band     3.5 - 3.6 GHz
F-7 Band     3.6 - 3.7 GHz
F-8 Band     3.7 - 3.8 GHz
F-9 Band     3.8 - 3.9 GHz
F-10 Band    3.9 - 4.0 GHz


g            gram (unit of weight; 1 gr = 0.353 ounce)
G*           giga (10^9 multiplier)
G            Conductance
G            conductivity of material
G-Band       new designation 4-6 GHz (see G-1 to G-10 Band; replaces old
                C-Band)
Ga           gain of antenna
GaAr         gallium arsenide (semiconductor material)
Ge           germanium (semiconductor material)
Gc           Gigacycles (per second); now obsolete; use GHz
GHz          Gigahertz (1,000 MHz or 10^9 Hz)
GSE          ground support equipment
G-1 Band     4.0 - 4.2 GHz
G-2 Band     4.2 - 4.4 GHz
G-3 Band     4.4 - 4.6 GHz
G-4 Band     4.6 - 4.8 GHz
G-5 Band     4.8 - 5.0 GHz
G-6 Band     5.0 - 5.2 GHz
G-7 Band     5.2 - 5.4 GHz
G-8 Band     5.4 - 5.6 GHz
G-9 Band     5.6 - 5.8 GHz
G-10 Band    5.8 - 6.0 GHz


h*           henry (unit of inductance)
H            amperes per meter, A/m (unit of magnetic field strength)
H-Band       new designation: 6-8 GHz (see H-1 to H-10 Band)
HERO         Hazards of Electromagnetic Radiation to Ordnance
HF           high frequency (3 MHz to 30 MHz)
HN           connector type
hr           hour (3600 seconds)
HTL          high-threshold logic
Hz*          hertz (unit of frequency)
H-1 Band     6.0 - 6.2 GHz
H-2 Band     6.2 - 6.4 GHz
H-3 Band     6.4 - 6.6 GHz
H-4 Band     6.6 - 6.8 GHz
```

```
H-5 Band   6.8 - 7.0 GHz
H-6 Band   7.0 - 7.2 GHz
H-7 Band   7.2 - 7.4 GHz
H-8 Band   7.4 - 7.6 GHz
H-9 Band   7.6 - 7.8 GHz
H-10 Band  7.8 - 8.0 GHz
```

I	symbol for current
IAGC	instantaneous automatic gain control
I-Band	new designation: 8-10 GHz (see I-1 to I-10 Band; replaces old x-Band)
IBW	impulse bandwidth
IC	integrated circuits
ID	inside dimension
IF	intermediate frequency
IMP	integrated microwave products
in	inch (unit of length = 25.4 mm)
I/N	interference-to-noise ratio
INL	internal noise level
INR	interference-to-noise ratio
I/O	input-output (devices)
ips	inches per second
IR	infra-red
ISM	industrial, scientific, and medical (equipment)

```
I-1 Band   8.0 - 8.2 GHz
I-2 Band   8.2 - 8.4 GHz
I-3 Band   8.4 - 8.6 GHz
I-4 Band   8.6 - 8.8 GHz
I-5 Band   8.8 - 9.0 GHz
I-6 Band   9.0 - 9.2 GHz
I-7 Band   9.2 - 9.4 GHz
I-8 Band   9.4 - 9.6 GHz
I-9 Band   9.6 - 9.8 GHz
I-10 Band  9.8 - 10.0 GHz
```

J	vector operator at 90°, $\sqrt{-1}$
J	Joule (one watt-second or 10^7 ergs)
J-Band	new designation: 10-20 GHz (see J-1 to J-10 Band; replaces old K_u - Band

k*	kilo (10^3 multiplier)
k	Boltzmann's constant = 1.38×10^{-23} W/°C/Hz
K-Band	new designation: 10-20 GHz (see K-1 to K-10 Band)
K-Band	18-26.5 GHz (old designation; see new J and K-Bands)
K_a-Band	26.5-40 GHz (old designation; see new K-Band)
K_u-Band	12.5-18 GHz (old designation; see new J-Band)
kc	Kilocycle (per second); now obsolete; use kHz
kg	Kilogram (10^3 grams; 1 kg = 2.205 pounds)
kHz	kilohertz (10^3 Hz)
km	kilometer (10^3 meters = 0.624 statute miles)
kV	kilovolt (10^3 volts)
kVA	kilo-volt-amperes

kVDC	kilovolts DC
kΩ	kilohm (10^3 ohms)
kW	kilowatt (10^3 watts)
K-1 Band	20 - 22 GHz
K-2 Band	22 - 24 GHz
K-3 Band	24 - 26 GHz
K-4 Band	26 - 28 GHz
K-5 Band	28 - 30 GHz
K-6 Band	30 - 32 GHz
K-7 Band	32 - 34 GHz
K-8 Band	34 - 36 GHz
K-9 Band	36 - 38 GHz
K-10 Band	38 - 40 GHz
L	conversion loss
L	henry (unit of inductance)
L-Band	new designation; 40-60 GHz;
L-Band	1.1 - 1.7 GHz (sometimes 1 - 2 GHz; old designation; see D-Band)
lb	pound (unit of weight; 1 lb = 16 ounces = 453.59 grams)
LC	connector type
le*	antenna effective length for electric-field antennas
lem*	antenna effective length for magnetic-field antennas
LF	low frequency (30 kHz to 300 kHz)
LHC	left-hand circular (polarization)
LISN	line impedance stabilization network
LO	local oscillator
LBS	lower sideband
LSB	least significant bit
LS-Band	1.7 - 2.6 GHz (obsolete; see D and E-Bands)
LSI	large scale integration
LSN	line stabilization network
m*	meter
m*	milli (10^{-3} multiplier)
M*	Mega (10^6 multiplier)
mA	milliamperes (10^{-3} amperes)
M-Band	60 - 100 GHz (new designation)
Mbps	Megabits per second
Mc	Megacycles (per second); now obsolete; use MHz
MDS	minimum discernible (or detectable) signal
MF	medium frequency (300 kHz to 3 MHz)
MGC	manual gain control
MHD	Magneto-HydroDynamics
MHz	Megahertz (10^6 Hz)
MΩ	Megohm (10^6 ohms)
mi	mile (unit length; 1 mi = 5,280 feet = 1.6093 km)
MIL-STD	military standard
mm	millimeter (10^{-3} meter)
mo	month (unit of time; 1 mo = 30.44 days = 734.6 hours)
MOPA	master oscillator power amplifier
mm-Band	millimeter band, 40-300 GHz

MOS	metal-oxide semiconductor
msec	milliseconds (10^{-3} seconds)
MTI	moving target indicator
MTBF	mean time between failure
MTTF	mean time to failure
MTTR	mean time to repair
MUSA	multiple-unit steerable antenna
mV	millivolts (10^{-3} volts)
MVS	minimum visible signal
mW	milliwatts (10^{-3} watts)
MW	microwave
MW	Megawatt (10^6 watts)
n*	nano (10^{-9} multiplier)
n	connector type
N	noise
NB	narrowband
NBC	narrowband conducted
NBR	narrowband radiated
NF	noise figure
nh	nanohenry (10^{-9} henry)
n mi	nautical mile (1 n mi = 6076 feet = 1.151 statute miles)
NPM	connector type
NRZ	non-return to zero
nsec	nanosecond (10^{-9} seconds)
NTC	negative temperature coefficient
OCI	optically-coupled isolators (fiber-optics)
OCR	optical character recognition
OD	outside dimension
OOK	(binary) on-off keying
OSM	connector type
oz	ounce (unit of weight; 1 oz = 0.0625 pounds = 28.35 grams)
p*	pico (10^{-12} multiplier)
PA	public address
PABFA	post-amplification beam forming array
P-Band	225 - 400 MHz (old designation; see B-Band)
PAM	pulse amplitude modulation
PC	printed circuits
PCM	pulse-code modulation
PCW	pulsed continuous wave
PDM	pulse duration modulation
pf	picofarad (10^{-12} farad)
Pk	peak
PLL	phase-lock loop
PM	phase modulation
PPI	plan position indicator
PPM	parts per million
PPM	pulse position modulation
PRF	pulse repetition frequency

PRR	pulse repetition rate (same as PRF)
psi	pounds per square inch
PTC	positive temperature coefficient
PVC	Polyvinyl chloride
PWM	pulse width modulation (same as PDM)
Q	Q-Factor (f_o/BW at 3 dB points)
QP	quasi-peak
R	range or distance between transmitter and receiver (target)
R	resistance in ohms
rad	radian (unit of angular measurement; 1 rad = 57.296 deg)
RAD-Haz	radiation hazard
RAM	random access memory
RC	resistor-capacitor (combination)
RE*	radiated emission
RF	radio frequency
RFI*	radio-frequency -interference
RHC	right-hand circular (polarization)
RHI	range-height indicator
RI-FI	radio interference and field intensity
RLC	resistor-capacitor-inductance (combination)
rms	root mean square
ROM	read only memory
rpm	revolutions per minute
RS*	radiated susceptibility
RTL	resistor-transistor logic
RX	receiver
RZ	return to zero
s*	second (unit of time)
S	signal
S-Band	2.6 - 4.0 GHz (sometimes 2-4 GHz; old designation; see E and F-Bands
SB	sideband
SC	connector type
SC	silvered copper (wire)
SCO	subcarrier oscillator
SCR	silicon-controled rectifier
SCW	silvered copper-clad steel wire
SGP	structure ground point
SHF	super-high- frequency (3 GHz to 30 GHz)
Si	silicon (semiconductor material)
sig gen	signal generator
SLF	super-low frequency (30 Hz to 300 Hz)
S/N	signal-to-noise ratio
SNR	signal-to-noise ratio
SOM	start of message
SPDT	single pole double throw (switch)
SPEC	specification
SPST	single pole single throw (switch)

```
sq          square
SSB         single sideband (modulation); same as SSB-SC
SSB-SC      single sideband, suppressed carrier
SSG         small signal gain
STC         sensitivity time control
STD         standard
SWR         standing wave ratio (cf. VSWR)
SAE         Society of Automotive Engineers
t           temperature
T           ton (unit of weight; 1 T = 2,000 pounds = 907.2 kg)
T*          Tesla (unit of magnetic flux density)
TC          tantallum capacitor
TDA         tunnel diode amplifier
TDM         time-division multiplex
TDMA        time-division multiple access
TDR         time-domain reflectometry
TE          transverse electric (field)
TEM         transverse electromagnetic (dominant field)
TM          transverse magnetic (field)
TM          telemetry
TNC         connector type
TTL         transistor-transistor logic
TR          transmit-receive
Tu*         tuning unit number
TV          television
TWA         traveling wave amplifier
TWS         track-while-scan
TWT         traveling-wave tube
TWTA        traveling wave tube amplifier
TX          transmitter

UHF         ultra-high frequency (300 MHz to 3 GHz)
ULF         ultra-low frequency (300 Hz to 3 kHz)
USB         upper sideband

V*          volt (unit of electric potential)
V-Band      26.5 - 40 GHz (old designation;  see K-Band)
VAC         volts AC
VCO         voltage-controlled oscillator
VDC         Volts DC
VF          video frequency (0-5 MHz)
VGP         vehicle ground point
VHF         very-high frequency (30 MHz to 300 MHz)
VLF         very-low frequency (3 kHz to 30 kHz)
V/m*        voit per meter (unit of electric field strength, E)
vol         volume
VSB         vestigial sideband
VSWR        voltage standing wave ratio
VTM         voltage tunable magnetron
VTO         voltage-tuned oscillator
VTVM        vacuum-tube voltmeter
```

W*	watt (unit of power)
Wb*	Weber (unit of magnetic flux)
wpm	words per minute
X	reactance
X_c	capacitive reactance in ohms
X_1	inductive reactance in ohms
X-Band	8.2 - 12.5 GHz (old designation; see I and J-Band)
X_n-Band	5.3 - 8.2 GHz (old designation; see H-Band)
X_b-Band	7 - 10 GHz (old designation; see I-Band)
X-Y	X-Y (cartesian coordinate system)
Y	admittance
yd	yard (unit of length; 1 yd = B feet = 0.9144 m)
YIG	yittrium-iron garnet
yr	year (unit of time; 1 yr = 365.24 days)
Z	impedance in ohms
Z_s	impedance of free space in ohms (cf. η)

SYMBOLS - Technical & Related

α	attenuation or attenuation constant
β	phase constant
δ	duty cycle (PRF x τ)
δ	skin depth
Δ	an increment or small change
ϵ	electric permitivity of medium (for air, $\epsilon = 8.85 \times 10^{-12}$ f/m)
ϵ_0	dielectric constant (relative permitivity)
η	Impedance of free space ($120\pi = 377\Omega$)
θ	angle in degrees
λ	wavelength
μ*	micro (10^{-6} multiplier)
μ	magnetic permeability of medium
μ	micron (unit of length; 10^{-4} cm)
μA	microamperes (10^{-6} amperes)
μh	microhenry (10^{-6} henry)
μj	microjoule (10^{-6} watt-seconds)
μ_0	relative permeability
μsec	microseconds (10^{-6} seconds)
μV	microvolts (10^{-6} volts)
$\mu V/m$	microvolts per meter (10^{-6} V/m)
$\mu V/m/MHz$	microvolts per meter per Megahertz)
π	pi (3.1416 - ratio of circumference to diameter of circle)
Π	product of indicated terms
σ	resistivity of medium
σ	standard deviation
σ	target (radar) cross sectional area in m^2
Σ	summation of indicated terms
τ	pulsewidth
ϕ	angle in degrees
ψ	magnetic flux
ω	angular frequency ($\omega = 2\pi f$ radians/sec)
Ω	ohm (unit of electric resistance)

---

*   Abbreviations and symbols follwed by an asterisk (*) appear in
    MIL-STD-463, "Definitions and System of Units", Electromagnetic
    Interference Technology, 9 June 1966.

CHAPTER 1
AN INTRODUCTION TO EMI CONTROL

1.1   WHAT IS ELECTROMAGNETIC INTERFERENCE?

1.2   FREQUENCY SPECTRUM USE
      1.2.1   Frequency and Wavelength
      1.2.2   Nomenclature
      1.2.3   Band Designation
      1.2.4   Frequency Allocations by International Treaty
      1.2.5   Frequency Allocations in the United States

1.3   BASIC ELEMENTS OF ALL EMI SITUATIONS

1.4   INTRA-SYSTEM VS INTER-SYSTEM INTERFERENCE
      1.4.1   Intra-system EMI
      1.4.2   Inter-system EMI

1.5   AN OVERVIEW OF EMI-CONTROL TECHNIQUES
      1.5.1   Intra-system EMI-Control Techniques
      1.5.2   Inter-system EMI-Control Techniques

1.6   HOW TO USE THIS HANDBOOK
      1.6.1   Inter-system EMI Problem Identification
      1.6.2   Intra-system EMI Problem

1.7   BIBLIOGRAPHY

CHAPTER 1
AN INTRODUCTION TO EMI CONTROL

The required compliance of electrical and electronic equipment with electromagnetic interference (EMI) specifications is intended to assure electromagnetic compatibility (EMC). EMC is the ability of either equipments or systems to function as designed without degradation or malfunction in their intended operational electromagnetic environment. Further, the equipments or systems should not adversely affect the operation of, or be adversely affected by any other equipment or system.

This chapter reviews what EMI and EMC are, uses of the frequency spectrum, basics of all EMI situations, and examples of EMI, including intra and inter-system interference*. It summarizes EMI control techniques for both forms of interference. The chapter ends with an overview of how to use this handbook and a comprehensive bibliography. Subsequent chapters treat the topics of EMI intra-system prediction and control in the design, operational and retrofit stages. Inter-system prediction and control is summarized in one chapter but is the essence of Vol. 5.

1.1 WHAT IS ELECTROMAGNETIC INTERFERENCE?

EMI is one kind of environmental pollution together with its associated reactions and damage. With the national emphasis today on reducing environmental pollution, the layman readily recognizes and understands water, air, noise, and other forms of contamination. However, he probably has heard very little about spectrum pollution (sometimes called electromagnetic or electrical-noise pollution) because it is more esoteric, i.e., it cannot be directly seen, tasted, smelled, or felt. Therefore, he asks, how can it be a problem? As explained

---

* Intra-System EMI pertains to electromagnetic emissions generated and coupled to a victim within an equipment or system. Inter-System EMI involves electromagnetic emissions from one or more equipments or systems coupling into one or more other victim equipments or systems.

below, it is just as damaging as the other forms of pollution.

Actually, the layman is familiar with some simple forms of spectrum pollution - at least some of those that are nuisances. For
example, it is recognized that certain types of electric shavers can
jam a nearby radio. The resulting *buzzing* or *crackling noise* denies
the ability to listen to the radio while a shaver is in use. Here,
either conducted or radiated electrical noise jams the radio picking
up the broadcast stations. Another example is the ignition of an
automobile idling outside of a house or an apartment. It causes interference to one's TV picture by developing intermittent dash lines or
bars, or even causes complete flipping (loss of synch) of the picture.
These situations are only a nuisance. They are not catastrophic as
are some situations pointed out below.

More serious examples of spectrum pollution are evident if a person having a heart pacer uses electrical appliances, shop equipment,
automobiles, or other RF energy-emitting sources which cause his pacer
to operate improperly, such as trigger it into a different mode. Here,
results could manifest themselves in fainting or near death. Another
example of more serious EMI consequences includes a sudden loss of
telephone conversation by jamming due to high electrical or interfering noise level backgrounds. If one were in the middle of giving his
stockbroker an important sell or buy order, or if an important business
deal were being negotiated, the EMI problem is no longer a nuisance
only. It can be economically damaging or catastrophic.

The spectrum pollution problem has and can be far more damaging
than the loss of a single life or of one's fortune. It can involve
many people and substantial property values. For example, if two air-
planes collide during inclement weather due to either navigation errors resulting from EMI or computer loss of memory because of electrical transients during a storm at an air traffic control center, the
loss of life and property become substantial. If police mobile radios
are jammed during a riot at *rush hour* due to the composite radiation
effects of many automobile ignition noises, the consequences can be
enormous. Alternatively, if a field army during war finds that its
communications, radar, and other modes of combat effectiveness are
jammed by its own spectrum pollution noise makers, a battle can be
lost together with many lives and much property. Some major electrical power blackouts can result from a pyramiding of fault-sensing
transient devices during an electrical storm. Here the impact can involve millions of lives and have enormous economic consequences.

Thus, it develops that EMI or electrical noise resulting in spectrum pollution is indeed of national concern even though its more
esoteric properties are not readily perceived by the layman. Fortunately, the Government and certain industry bodies have issued specifications with which all electrical, electromechanical, and electronic
equipment must comply. Unfortunately, policing and enforcing these

specifications are another matter and that is why problems exist. This is explained more fully in Vol. 1 of this handbook series, *Electrical Noise and EMI Specifications*.

1.2   FREQUENCY SPECTRUM USE

Having presented examples of spectrum pollution and EMI, the frequency spectrum is next examined. The frequency spectrum covers from *DC to light* (and beyond). As a practical matter, there rarely exists a pure DC EMI problem, and frequencies in the infrared, optical and above spectrum regions, while also electromagnetic in nature, are outside the definition of EMI. Thus, EMI is somewhat arbitrarily defined to cover the frequency spectrum from about 10 Hz to 100 GHz*. For radiated emissions a lower frequency limit of 10 kHz is often used, although infra-system EMI can exist in many equipments and systems below this frequency.

1.2.1   FREQUENCY AND WAVELENGTH

Sometimes the term wavelength instead of frequency is used. To convert from frequency, f, in Hz to wavelength, λ, (length of one cycle of frequency) in meters, the velocity of propagation in air** is used:

$$\lambda f = C; \quad \lambda = C/f \text{ or } f = C/\lambda \qquad (1.1)$$

where: $C \approx 3 \times 10^8 \text{m/sec}$ in air

Thus: $\lambda_m = 3 \times 10^8/f_{Hz}$ meters

$$= 300/f_{MHz} \qquad (1.2)$$

where: f_{Hz} = frequency in Hz

f_{MHz} = frequency in MHz

Eq. (1.1) is plotted in Fig. 1.1 from 10 Hz to 100 GHz and for λ in units of centimeters, inches, meters, kilometers and miles. The

---

* The FCC defines the upper frequency in its Rules and Regulations (Part 15.4) as 3,000 GHz which is also in the far infrared spectrum.

** For propagation in other than air, the velocity is $C^1 = C/\sqrt{\mu\varepsilon}$, where ε is the dielectric constant (relative permeability) and μ is the relative permittivity, i.e., relative to air.

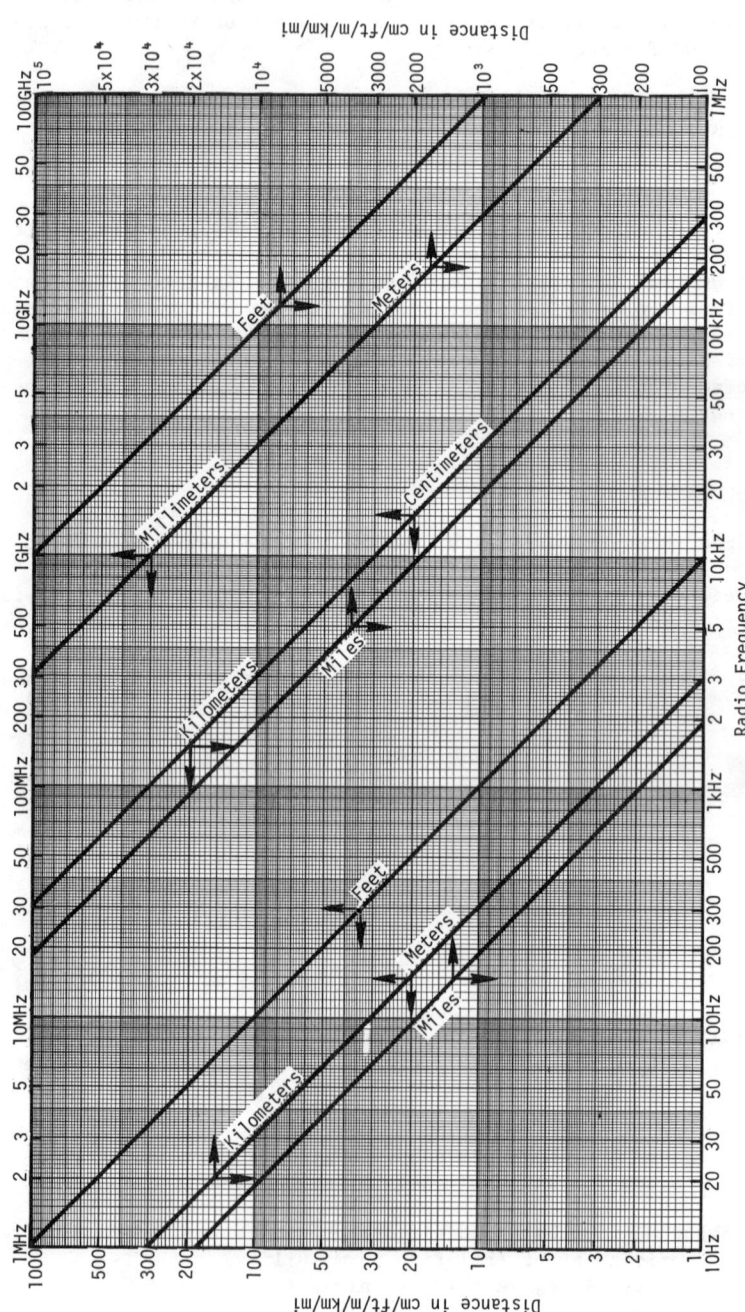

Figure 1.1 - Wavelength in Air vs. Radio Frequency

arrows on the graphic lines instruct which axis to read from (right or left distance axis, or top or bottom frequency axis).

1.2.2 NOMENCLATURE

From Article 2, Section 11, *Radio Regulations of the Internation-al Telecommunications Union (ITU)*, Geneva, 1959, the following band nomenclature is defined:

Table 1.1 - Nomenclature of Frequency Bands

Band No.*	Freq. Range	Metric Div. (Wavelength)	Abbrev.	Meaning
1	3-30 Hz		SAF	Sub-Audio Freq.
2	30-300 Hz	Megameter waves	ELF	Extremely-low Freq.
3	300-3000 Hz		VF	Voice Frequency
4	3-30 kHz	Myriameter waves	VLF	Very-low Freq.
5	30-300 kHz	Kilometer waves	LF	Low Frequency
6	300-3000 kHz	Hectometer waves	MF	Medium Frequency
7	3-30 MHz	Decameter waves	HF	High Frequency
8	30-300 MHz	Meter waves	VHF	Very-high Freq.
9	300-3000 MHz	Decimeter waves	UHF	Ultra-high Freq.
10	3-30 GHz	Centimeter waves	SHF	Super-high Freq.
11	30-300 GHz	Millimeter waves	EHF	Extremely-high Freq.

*   "Band No. η" extends from $0.3 \times 10^{\eta}$ to $3 \times 10^{\eta}$ Hz (this is one decade or 3.32 octaves). The upper limit is included in each band; the lower limit is excluded.

1.2.3 Band Designation

During and subsequent to World War II, it was common practice to use a letter designation for a portion of the microwave band (above 200 MHz). While there exist no official international acceptance, these designations are frequently used.

In 1966, an official re-classification of the band designation was made but its application has been slow to catch on. Both the old and new band designations are listed in Table 1.2.

1.2.4 Frequency Allocations By International Treaty

The following information is adapted from *Radio Regulations of the ITU,* Geneva, 1959. Copies of these regulations may be obtained from the Secretary General, International Telecommunicatio Union, Palais Wilson, Geneva, Switzerland.

For pusposes of frequency allocations, the world is divided into three regions. The Western Hemisphere including the United States is located in Region No. 2. See Article 5, Section 1 of the ITU Radio Regulations for definitions of the regions.

Frequency bands are allocated to services defined as follows:

Fixed: Radio communication between specified fixed points. Examples: point-to-point high-frequency circuits and microwave relay links.

Mobile: Radio communication between stations intended to be used while in motion or during stops at unspecified points or between such stations and fixed stations.

Aeronautical Mobile: Radio communication between a land station and an aircraft or between aircraft.

Maritime Mobile: Radio Communication between a coast station and a ship or between ships.

Land Mobile: Radio communication between a base station and a mobile station or between mobile stations. Examples: radio communication with taxicabs and police squad cars.

Radio Navigation: The determination of position or location for purposes of navigation by means of the propagation properties of radio waves. This includes obstruction warning. Example: Loran and Omega.

Aeronautical Radio Navigation: A radio navigation service intended for the benefit of locating or positioning aircraft. Example: VOR and Tacan, aeronautical radio beacons, instrument landing systems (ILS), radio altimeters, and airborne obstruction-indicating (terrain avoidance) radar.

Table 1.2 Old and New Band Designations

Freq. Range	New Designatn	Old Designatn	Freq. Range	New Designatn	Old Designatn
0-25 MHz	A-1 Band		3.0-3.1 GHz	F-1 Band	
25-50 MHz	A-2 "		3.1-3.2 GHz	F-2 "	S-Band
50-75 MHz	A-3 "		3.2-3.3 GHz	F-3 "	
75-100 MHz	A-4 "		3.3-3.4 GHz	F-4 "	
100-125 MHz	A-5 "		3.4-3.5 GHz	F-5 "	
125-150 MHz	A-6		3.5-3.6 GHz	F-6 "	
150-175 MHz	A-7 "		3.6-3.7 GHz	F-7 "	
175-200 MHz	A-8 "		3.7-3.8 GHz	F-8 "	
200-225 MHz	A-9 "		3.8-3.9 GHz	F-9 "	
225-250 MHz	A-10 "		3.9-4.0 GHz	F-10 "	
250-275 MHz	B-1 Band		4.0-4.2 GHz	G-1 Band	
275-300 MHz	B-2 "		4.2-4.4 GHz	G-2 "	
300-325 MHz	B-3 "	P-Band	4.4-4.6 GHz	G-3 "	
325-350 MHz	B-4 "		4.6-4.8 GHz	G-4 "	
350-375 MHz	B-5 "		4.8-5.0 GHz	G-5 "	
375-400 MHz	B-6 "		5.0-5.2 GHz	G-6 "	C-Band
400-425 MHz	B-7 "		5.2-5.4 GHz	G-7 "	
425-450 MHz	B-8 "		5.4-5.6 GHz	G-8 "	
450-475 MHz	B-9 "		5.6-5.8 GHz	G-9 "	
475-500 MHz	B-10 "		5.8-6.0 GHz	G-10 "	
500-500 MHz	C-1 Band		6.0-6.2 GHz	H-1 Band	
550-600 MHz	C-2 "		6.2-6.4 GHz	H-2 "	
600-650 MHz	C-3 "		6.4-6.6 GHz	H-3 "	
650-700 MHz	C-4 "		6.6-6.8 GHz	H-4 "	
700-750 MHz	C-5 "		6.8-7.0 GHz	H-5 "	
750-800 MHz	C-6 "		7.0-7.2 GHz	H-6 "	
800-850 MHz	C-7 "		7.2-7.4 GHz	H-7 "	
850-900 MHz	C-8 "		7.4-7.6 GHz	H-8 "	
900-950 MHz	C-9 "		7.6-7.8 GHz	H-9 "	
950-1000 MHz	C-10 "		7.8-8.0 GHz	H-10 "	
1.0-1.1 GHz	D-1 Band		8.0-8.2 GHz	I-1 Band	
1.1-1.2 GHz	D-2 "		8.2-8.4 GHz	I-2 "	
1.2-1.3 GHz	D-3 "		8.4-8.6 GHz	I-3 "	
1.3-1.4 GHz	D-4 "		8.6-8.8 GHz	I-4 "	
1.4-1.5 GHz	D-5 "	L-Band	8.8-9.0 GHz	I-5 "	
1.5-1.6 GHz	D-6 "		9.0-9.2 GHz	I-6 "	X-Band
1.6-1.7 GHz	D-7 "		9.2-9.4 GHz	I-7 "	
1.7-1.8 GHz	D-8 "		9.4-9.6 GHz	I-8 "	
1.8-1.9 GHz	D-9 "		9.6-9.8 GHz	I-9 "	
1.9-2.0 GHz	D-10 "		9.8-10.0 GHz	I-10 "	
2.0-2.1 GHz	E-1 Band		10 - 11 GHz	J-1 Band	
2.1-2.2 GHz	E-2 "		11 - 12 GHz	J-2 "	
2.2-2.3 GHz	E-3 "		12 - 13 GHz	J-3 "	
2.3-2.4 GHz	E-4 "		13 - 14 GHz	J-4 "	
2.4-2.5 GHz	E-5 "		14 - 15 GHz	J-5 "	
2.5-2.6 GHz	E-6 "		15 - 16 GHz	J-6 "	
2.6-2.7 GHz	E-7 "		16 - 17 GHz	J-7 "	
2.7-2.8 GHz	E-8 "		17 - 18 GHz	J-8 "	
2.8-2.9 GHz	E-9 "		18 - 19 GHz	J-9 "	
2.9-3.0 GHz	E-10 "		19 - 20 GHz	J-10 "	

Maritime Radio Navigation: A radio navigation service intended for the benefit of ships. Examples: coastal radio beacons, direction-finding stations, and shipboard radar.

Radio Location: The determination of position for purposes other than those of navigation by means of the propagation properties of radio waves. Examples: land-based radars, coastal radars, and aircraft-tracking systems.

Broadcasting: Radio communication intended for direct reception by the general public. Examples: amplitude-modulation (AM) broadcasting on medium and high frequencies, frequency-modulation (FM) broadcasting, and television (TV).

Amateur: Radio communication carried on by persons interested in the radio technique solely with a personal aim and without financial interest.

Space: Radio communication or information relay between space stations.

Earth-Space: Radio communication between earth and space stations. Example: between the earth and a satellite.

Radio Astronomy: Astronomy based on the reception of radio waves of cosmic origin.

Standard Frequency: Radio transmission of specified frequencies of stated high precision, intended for general reception for scientific, technical, and other purposes.

Fig. 1.2 is a chart of the frequency spectrum from .001 Hz to 10^{22} Hz showing the above band allocations together with other pertinent information.

1.2.5 FREQUENCY ALLOCATIONS IN THE UNITED STATES

The following listings are obtained from Part 2 - *Frequency Allocations and Radio Treaty Matters; General Rules and Regulations (R&R)* of the Federal Communications Commission as revised to January 1, 1970. A copy of Part 2 (actually Parts 0 to 19) Title 47, Code of Federal Regulations (CFR-47) may be obtained from the Superintendent of Documents, U.S. Government Printing Office for $1.75.

(1) Aeronautical Mobile (Ground-air-ground; air-air)

a. General: Many - see Part 2, FCC R&R
b. Calling and Distress: 490-510 kHz; 2170-2194 kHz; 156-725-156-875 MHz; 243 MHz.
c. Airdome Control: 117.975-121.425 MHz
d. Aero Search and Rescue: 121.575-121.625 MHz
e. Aero Utility: 121.625-121.975 MHz
f. Private Aircraft: 121.975-123.075 MHz

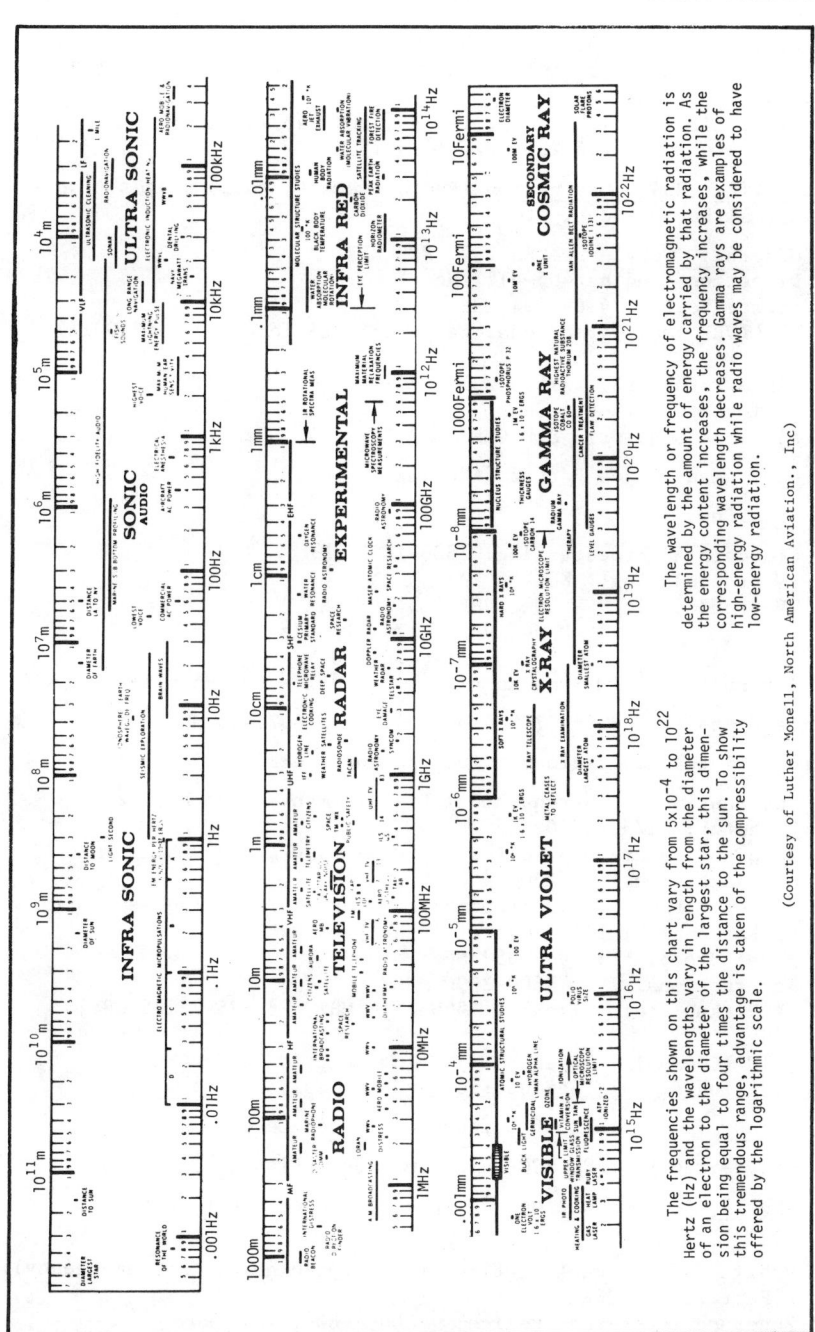

The frequencies shown on this chart vary from 5×10^{-4} to 10^{22} Hertz (Hz) and the wavelengths vary in length from the diameter of an electron to the diameter of the largest star, this dimension being equal to four times the distance to the sun. To show this tremendous range, advantage is taken of the compressibility offered by the logarithmic scale.

The wavelength or frequency of electromagnetic radiation is determined by the amount of energy carried by that radiation. As the energy content increases, the frequency increases, while the corresponding wavelength decreases. Gamma rays are examples of high-energy radiation while radio waves may be considered to have low-energy radiation.

(Courtesy of Luther Monell, North American Aviation., Inc)

Figure 1.2 - The Frequency Spectrum Chart

    g. Flight Test: 123.075–123.575 MHz
    h. Aviation Instructional: 123.075–123.125 MHz: 123.275–123.-
325; 123.475–123.525 MHz
    i. Civil Air Patrol: 26.62, 143.90, and 148.15 MHz
    j. Telemetering: 216.220 MHz; 1435–1535 MHz

(2) Aeronautical Radio Navigation (Radio beacons, radio ranges, landing systems, airborne radar, etc.):

    a. General: Many – see Part 2, FCC R&R
    b. Direction Finding 405–415 kHz
    c. Marker Beacon 74.6–75.4 MHz
    d. VOR (VHF omni directional range and ILS localizer 108.0–117.-
975 MHz
    e. Glide Path, ILS; 328.6–335.4 MHz
    f. Altimeter: 4.2–4.4 GHz
    g. Airborne Doppler: 8.8 GHz; 13.25–13.40 GHz

(3) Amateur: Many – see Part 2, FCC R&R

(4) Broadcasting:

    a. Standard AM Broadcasting: 536–1605 kHz
    b. FM Broadcasting: 88–108 MHz
    c. TV Broadcasting: Lower VHF: 54–88 MHz
                                  Upper VHF: 174–216 MHz
                                    UHF: 470–890 MHz
    d. International AM Broadcasting: Several – See Part 2

(5) Citizens Radio: 26.96–27.23 MHz; 462.525–467.475 MHz

(6) Fixed (Point-Point radio service; neither terminal mobile)

    a. International & Aeronautical Fixed: Many – see Part 2
    b. Disaster: 1750–1800 kHz
    c. Zone and Inter-zone Police; 5.005–5.450 & 7.300–8.195 MHz
    d. Omnidirectional: 2150–2160 MHz
    e. Domestic Fixed Public (common carrier): Several – see Part 2
    f. Operational Fixed: Several – see Part 2
    g. Aural Broadcast: 942–952 MHz
    h. Instructional TV: 2500–2690 MHz
    i. TV Pickup (relay/link): 1990–2110 MHz; 6.875–7.125 GHz;
12.7–13.2 GHz
    j. CATV (Community Antenna TV) Relay: 12.70–12.95 GHz
    k. ISM (Industrial, Scientific, and Medical): Several

(7) Government (Department of Defense and Other Government):

    a. Many – see Part 2, FCC Rules and Regulations

(8) Land Mobile (Communications in which at least one terminal is
mobile:

    a. Public Safety (Public, fire, forestry, highway, and emergency):
Many – see Part 2, FCC R&R
    b. Zone and Inter-zone Police: 2804, 2808, 2812 kHz

    c. Disaster 1750–1800 kHz
    d. Industrial (Power, petroleum, pipeline, forest products, factories, builders, ranchers, motion picture, press relay, etc.) Many – see Part 2, FCC R&R
    e. Land Transportation (taxi, trucks, busses, railroads) several – see Part 2, FCC R&R
    f. Land & Mobile: Several – see Part 2
    g. Domestic Public: Several – see Part 2
    h. Broadcast Remote Pickup: Several – see Part 2
    i. TV Pickup: 1990–2110 MHz; 6.875–7.125 GHz; 12.70–13.20 GHz
    j. Common Carrier: 6.425–6.525 and 11.70–12 GHz

(9) <u>Maritime Mobile</u> (Communications in which at least one terminal is a ship):

    a. General: Several – see Part 2, FCC R&R
    b. Calling, Safety, and Distress:
    c. Coast Stations (Telegraphy & Facsimile):
    d. Ship Stations (Telegraphy & Facsimile): R&R
    e. Coast Stations (Telephony) Facsimile
    f. Ship Stations (Telephony and Facsimile)
    g. Ship Stations (SSB-Telephony): Several
    h. Ship Calling (Telegraphy): Several
    i. Ship Calling (DSB-Telegraphy): Several
    j. Intership Telephony: 2638 and 2738 kHz

(10) <u>Meterological Aids</u>

    a. Radiosondes: 400.05–406.0 MHz; 1660–1700 MHz
    b. Ground-Based Weather Radars: 5.60–5.65 and 9.3–9.5 GHz

(11) <u>Radio Astronomy</u>: Many – see Part 2, FCC R&R

(12) <u>Radio Location</u> (Coastal radar, tracking systems): several

(13) <u>Radio Navigation</u> (Radio Beacons, ship-board radar, navigation and DF systems, etc.

    a. General: Several – see Part 2, FCC R&R
    b. Maritime Radio Navigation: 285–325 kHz; 2900–3100 MHz; 5.47–5.65 GHz
    c. Maritime Direction Finding: 405–415 MHz
    d. Loran: Loran C: 90–110 kHz; Loran A: 1.8–2.0 MHz
    e. Land Radio Navigation: 1638 and 1708 kHz

(14) <u>Earth-Space and Space Research</u>:

    a. General: 20.00–20.01 MHz; 8.4–8.5, 15.25–15.35, and 31.5–31.8 GHz
    b. Telemetering and Tracking: 136–137, 500.05–401.0 MHz; 1.70–1.71 and 2.29–2.30 GHz
    c. Telecommand: 148.25 and 154.2 MHz; 1.427–1.429 GHz

(15) <u>Space Communications Between Stations</u>: Telemetering & Tracking: 137–138, 401–402 MHz; 1.525–1.540 GHz

(16)   <u>Satellites</u>

    a. Communications Satellites: 3.7–4.2 GHz; 5.925–6.425; 7.25–7.75, and 7.9–8.4 GHz

    b. Meterological Satellites: 137–138 MHz; 1.66–1.67, 1.69–1.70, and 7.30–7.75 GHz

    c. Radio Navigation Satellites: 149.9–150.05 MHz; 399.9–400.05 MHz; 14.3–14.4 GHz.

1.3   BASIC ELEMENTS OF ALL EMI SITUATIONS

Fig. 1.3 illustrates the three basic elements required to produce EMI. They consist of electrical noise emitters, propagation media, and receptors, as the necessary but not sufficient conditions required to produce either degradation or malfunction responses in receptors. The method of coupling between emitters and receptors of electrical noise shown in the figure are divided between radiation (space separation with no hard-line connection) and conduction such as through wires or cables.

Many equipments listed in Fig. 1.3 act as both emitters and receptors of EMI. This is quite obvious for transmitters and receivers which are part of the same equipment. However, it may not be too apparent for equipments, such as computers. Here, computer peripherals, such as line or page printers, exhibit high broadband transient noise while at the same time computer memory sense amplifiers are rather susceptible to higher level emissions.

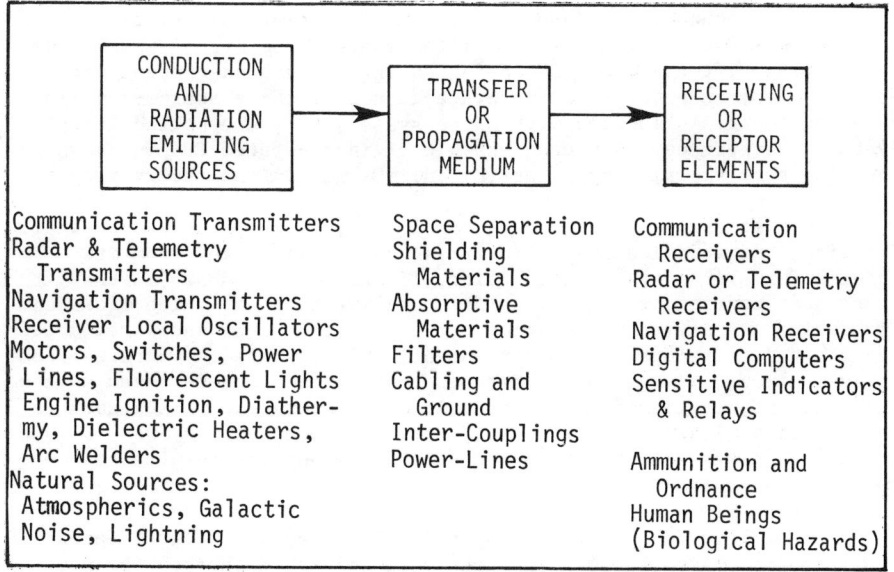

Figure 1.3 - Three Basic Elements of an Emitting-Susceptibility Situation

1.4   INTRA-SYSTEM VS INTER-SYSTEM INTERFERENCE

Electromagnetic Compatibility (EMC) is defined as the gainful operation of electric, electromechanical, and electronic devices, equipments,
and systems in a common environment such that no degradation of performance exists due to either internally or externally conducted or
radiated electromagnetic emissions.

EMC, then, covers the technological discipline of controlling the
otherwise damaging effects of electromagnetic interference (EMI). This
is best accomplished in the planning stage, i.e., the conception and
early design stages of devices, equipments, and systems. The ability
to implement this pragmatically is predicated in part on whether or
not EMI develops within a system or equipment (intra-system interference) or between two or more removed and discrete systems (inter-
system interference). Fig. 1.4 shows a conceptual block diagram involving both forms of EMI. The transmitter on the left is attempting to
communicate with the receiver in the center of the figure. The receiver,
however, is also subject to a number of other electromagnetic emissions including those internally developed and those from both intentional and unintentional emitters shown on the right.

Examining Fig. 1.4 in further detail reveals that the information
to be conveyed starts at the *information source* associated with the
transmitter shown in the lower left. For example, this may be voice,
pictures, data, or a radar pulse train. The information is then converted through a source transducer to the proper electronic format for
direct electromagnetic transmission. At this point EMI may first
develop as shown by either (E) radiated energy from higher power levels
of the transmitter and/or (G) conducted through ground-current loops
or from a common power supply. Since either of these source-coupling-
receptor routes is located *within* the system, they are called intra-
system interference. A second example of intra-system EMI is shown in
the center of Fig. 1.4 by paths (E) and (G) within the receiver system.

After modulation, frequency translation, if applicable, and amplification, the modified source information is transmitted from its
antenna (or over a transmission line, if applicable) shown in the
upper left portion of Fig. 1.4. The receiver antenna, which picks up
the desired transmitter radiations, may also intercept undesired anten-
na-radiated emissions from other intentional emitters (A) such as from
communications, navigation, and radar equipments. The receiver antenna
may also intercept emissions from unintentional radiators (B) such as
from incidental man-made devices. Interference resulting from (A) and/
or (B) is called inter-system interference because it results *between*
two or more systems.

To complete the situation shown in Fig. 1.4, intercepted radiations are processed through the receiver, converted through the user
transducer, and presented in the form of output voice, teleprint,
pictures, meters, or other graphic displays or outputs required by
the user. Here is where the above latent form of EMI finally manifests

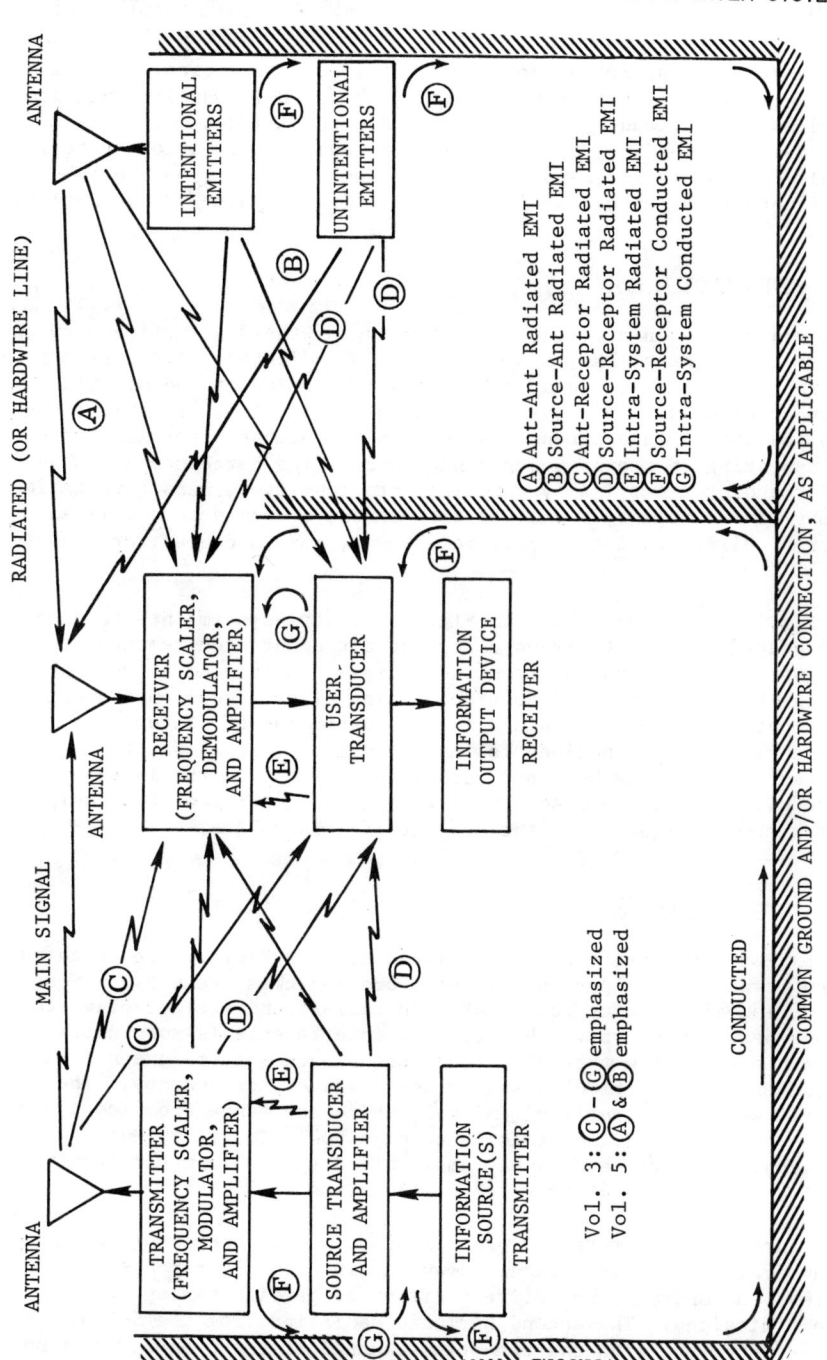

Figure 1.4 – Example of Both Intra-System and Inter-System   Interference

itself by either distortion, erroneous, or lost information. Fig. 1.4
also shows other alternate inter-system paths of EMI coupling including
radiated routes (C) and (D), and conduced routes (F) such as by a
common co-site grounding, power supply, or cable distribution system.
This volume presents intra-system EMI problems and solutions in substantial detail while Vol. 5 emphasized the inter-system EMI situation.

1.4.1   Intra-system EMI

In order to gain a better understanding of the distinction between
the above two forms of EMI, a closer look is indicated. Intra-system
interference comes about as a result of self-jamming or undesirable
emission coupling within a system as shown in Fig. 1.5. In this illustration, undesired interference is developing because electrical noise
spikes appearing on nearby power cables and wiring harnesses are magnetically and/or electrically coupling into low-level, sensitive cables.
Noise may also couple by a voltage drop across a common ground impedance, or by direct radiation from box-to-box, box-to-cables, or cables-
to-box.

In addition to that shown in Fig. 1.5, intra-system interference
control practices almost always carry the attendant requirements for
control of overall emissions from the equipment or system as well as
hardening it to emissions of outside-origin. These requirements exist
in anticipation of the possible simultaneous intra- and inter-system
interference once the specimen is installed in its real operational
environment. Design methods and techniques to contain the intra-system
form of EMI, which are almost always within the control of the equipment or system designer, are the subject of this volume.

1.4.2   Inter-system EMI

Inter-system interference, is illustrated in Fig. 1.6 in which many
different types of equipments, systems, and vehicles are shown. Most
communication-electronics equipment both radiate and are receptive to
electromagnetic emissions. This form of interference is more difficult
to contain because the totality of equipments and systems are *not* generally under the control of a single user, agency, or company. The
figure illustrates intentional emitters which can range from low power
(e.g., 3 watts) from handy talkies to high power (e.g., 10 Mwatts)
from radars. The frequency spectrum involved typically may range from
power-line frequencies of 60 Hz to field radars at 35 GHz.

To further illustrate the problem of *no single user control*, a
microwave relay link and a radar shown in Fig. 1.6 are purchased,
installed, and operated by entirely different parties having totally
different missions. The second harmonic radiation from the S-Band
radar may interfere with the reception channel of the C-Band microwave

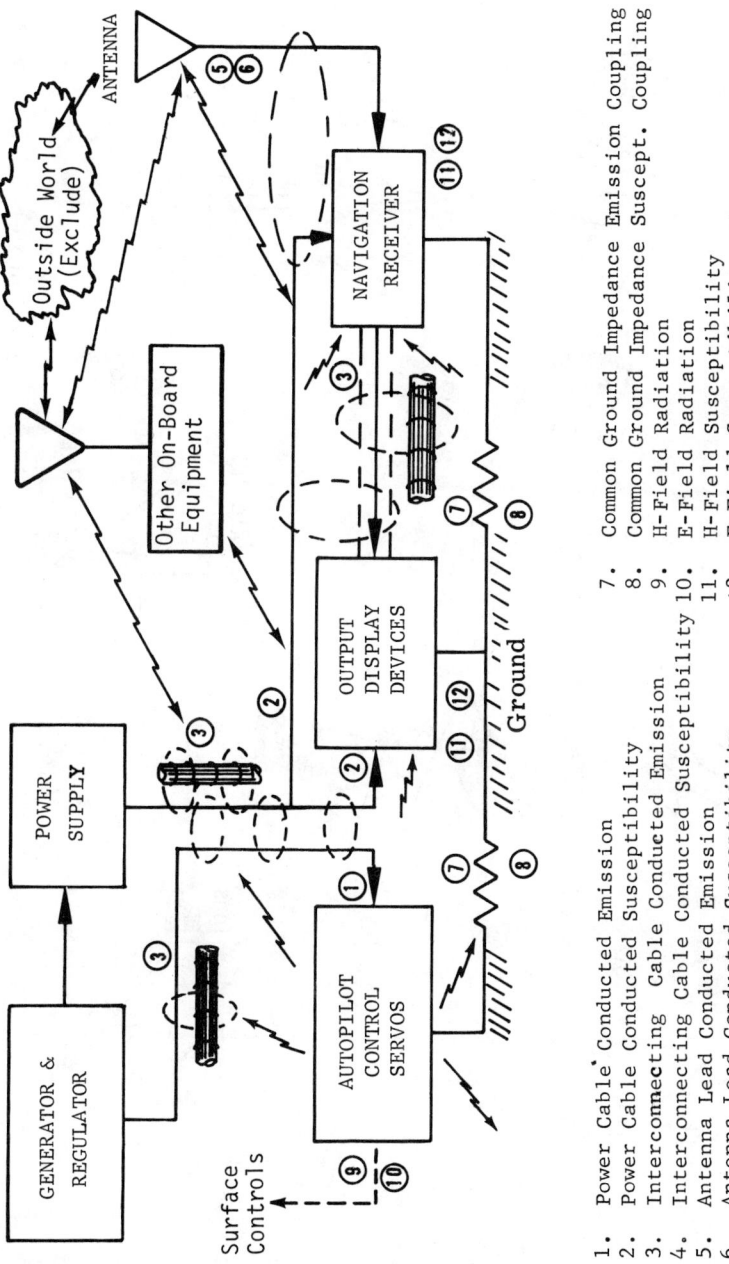

1. Power Cable Conducted Emission
2. Power Cable Conducted Susceptibility
3. Interconnecting Cable Conducted Emission
4. Interconnecting Cable Conducted Susceptibility
5. Antenna Lead Conducted Emission
6. Antenna Lead Conducted Susceptibility
7. Common Ground Impedance Emission Coupling
8. Common Ground Impedance Suscept. Coupling
9. H-Field Radiation
10. E-Field Radiation
11. H-Field Susceptibility
12. E-Field Susceptibility

Figure 1.5 - Examples of Intra-System EMI

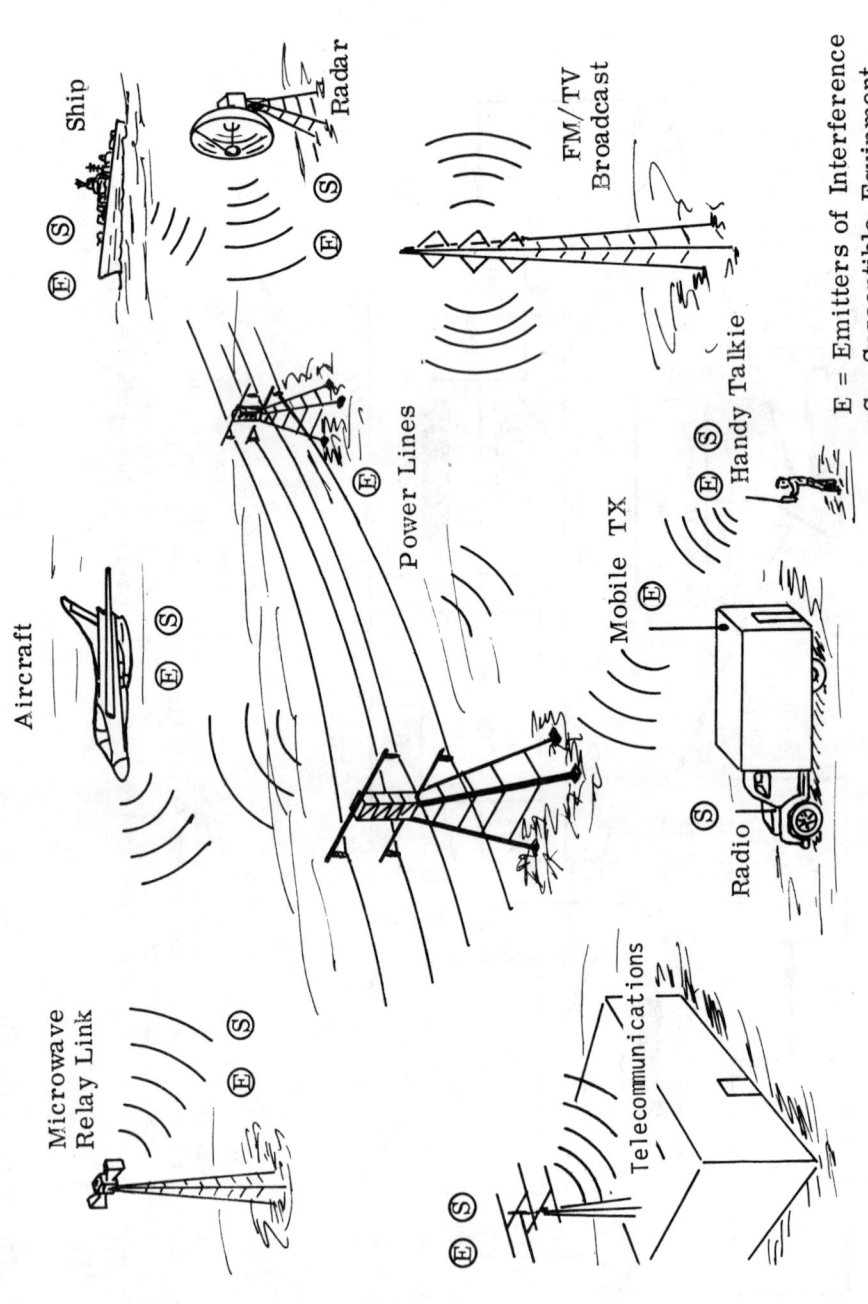

Ship

Radar

FM/TV Broadcast

E = Emitters of Interference
S = Susceptible Equipment

Power Lines

Handy Talkie

Aircraft

Mobile TX

Microwave Relay Link

Radio

Telecommunications

Figure 1.6 - Examples of Inter-System EMI

relay link. In this case, the radar is the culprit. The EMI solution
may involve using another relay link channel, but that poses problems
of frequency assignment. The radar user may change frequency or install
a harmonic filter, but that takes cooperation between the two users.
Since inter-system interference involves many equipments and users,
EMI problems and control are formidable. Chapter 5 of this volume and
Vol. 5 of this handbook series are dedicated to the inter-system EMI
situation.

1.5 AN OVERVIEW OF EMI-CONTROL TECHNIQUES

EMI-control techniques involve both hardware and methods and procedures. They may also be divided into *intra-system* and *inter-system*
EMI control. Engineers and technicians may become knowledgeable and
even accomplished in intra-system EMI-control techniques and yet remain
relatively naive in the other, or vice versa. Accordingly, this volume
emphasizes intra-system EMI control while Vol. 5 of this handbook series
stresses inter-system control. This section, however, presents an
overview of both techniques.

1.5.1 INTRA-SYSTEM EMI-CONTROL TECHNIQUES

Fig. 1.7 illustrates the basic elements of concern in an intra-
system EMI problem. The test specimen may be a single box, an equipment, subsystem, or system (an ensemble of boxes with interconnecting
cables). From a strictly near-sighted or selfish point-of-view, the
only EMI concern would appear to be degradation of performance due to
self jamming such as suggested at the top of Fig. 1.7. While this is
the primary emphasis (cf. MIL-E-6051D; Chap. 5, Vol. 1), latent problems associated with either (1) susceptibility to outside conducted
and/or radiated emissions or (2) tendency to pollute the outside world
from its own undesired emissions (cf. MIL-STD-461A, Vol. 1), come under
the primary classification of *intra-system* EMI. Corresponding EMI-
control techniques, however, address themselves to both self-jamming
and emission/susceptibility in accordance with applicable EMI specifications.

Fig. 1.8 presents one organization tree which groups intra-
system EMI-control techniques by five fundamental categories often
appearing in the literature: circuits and components, filtering,
shielding, wiring, and grounding. Bonding, connectors, gasketing, and
other topics appear as sub-categories as shown in the figure. These
categories together with some sub-categories are the topics of separate
chapters in this volume.

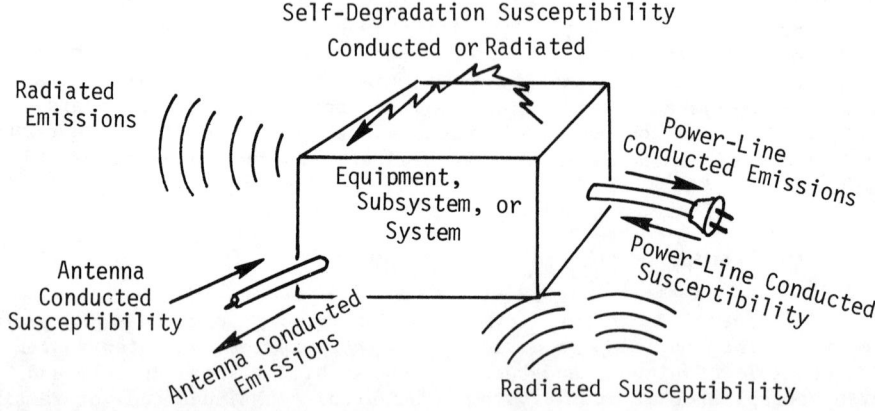

Figure 1.7 - Intra-System EMI   Manifestations

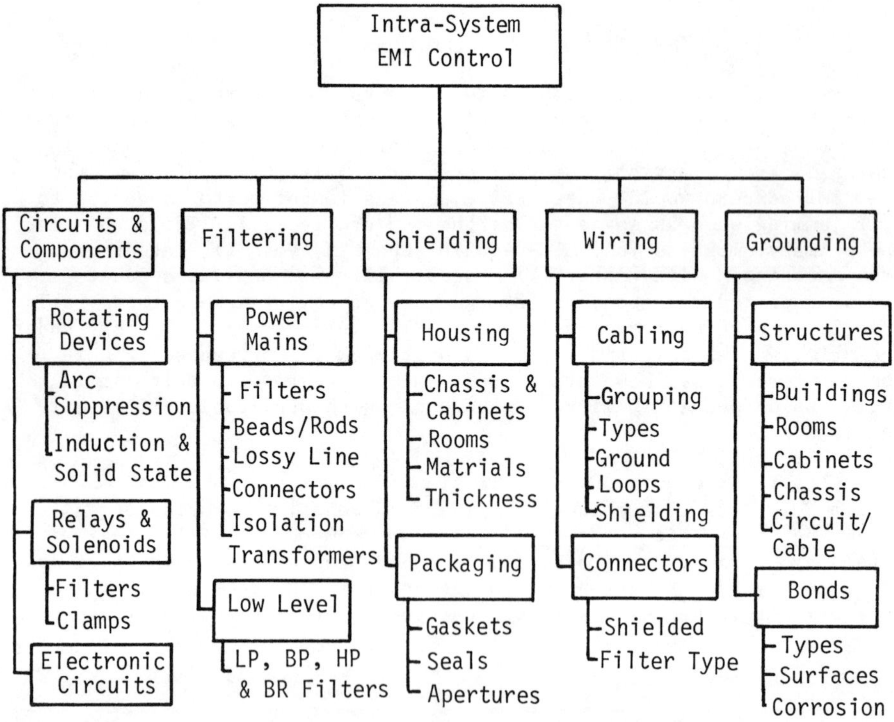

Figure 1.8 - Intra-System EMI-Control Organization Tree

1.5.2   INTER-SYSTEM EMI-CONTROL TECHNIQUES

Fig. 1.9 illustrates the basic elements of concern in an inter-system EMI-control problem. This type of EMI distinguishes itself by interference between two or more discrete systems or platforms which are frequently under separate user control. Culprit emissions and/or susceptibility situations are divided into two classes: (1) antenna entry/exit and (2) *back-door** entry/exit. More than 95% of inter-system EMI problems involve the antenna entry/exit route of EMI. The relatively few back-door EMI situations use either intra-system or inter-system techniques, as applicable, for EMI control. The parameters measured in MIL-STD-449D and antenna conducted receiver parameters in MIL-STD-461A (see Chap. 5, Vol. 1) constitute the principle characteristics to be controlled in inter-system EMI problems.

Fig. 1.9 presents one organization tree which attempts to group inter-system EMI-control techniques by four fundamental categories: frequency management, time management, location management, and direction management. System isolation/separation, antenna polarization, and other control parameters appear in the figure as sub-categories. These inter-system EMI control topics are the subject of Chapter 5 of this volume and Vol. 5 in this handbook series.

---

* *Back-door* inter-system EMI is a colloquialism used in the trade to denote EMI existing between a communications-electronic transmitter-receiver pair in which either one or no antenna is involved. For example, emissions from case penetration of one source modulator may enter the antenna of a receiver, or a transmitter antenna radiation may penetrate a receiver housing and couple into the I-F. Another example includes R-F transmitter cable radiation coupling into the I-F cable of a receiver belonging to a different system.

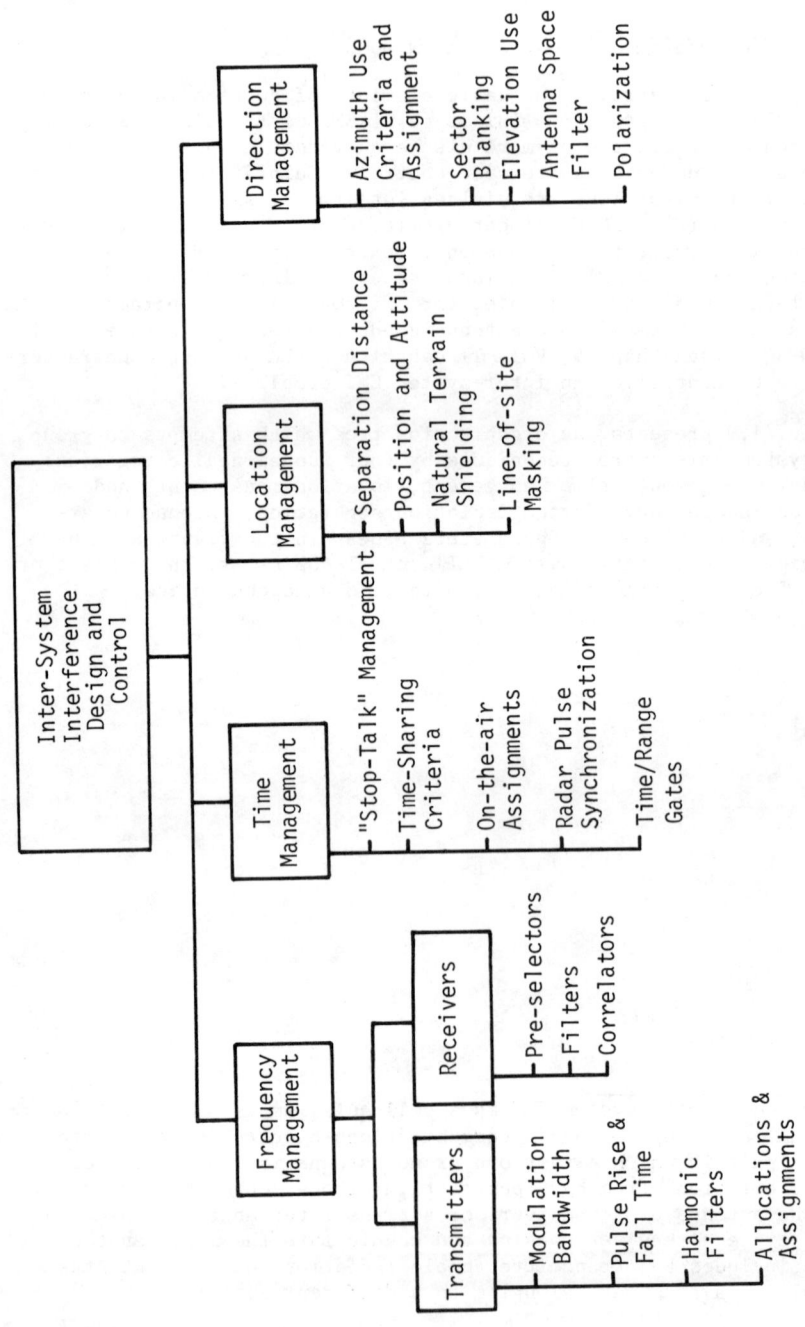

Figure 1.9 - Inter-System Electromagnetic Interference Control Techniques

1.6 HOW TO USE THIS HANDBOOK

This handbook is written in a style which introduces theory first; develops design data, charts, and tables next; then applies the information through illustrative examples. Theoretical background is offered for the curious mind, i.e., it is intended to satisfy how or why certain relations exist and the supporting rationale. If they choose, application engineers or technicians may skip the interlaced theoretical parts on each topic and move directly to design data graphs and tables.

While the table of contents and index are the classical ways of locating information sought, they are not always the most efficient. An engineer or technician, less practiced in EMI, may waste much time on either false starts or improperly diagnosing systems. It would be helpful if he could be steered to the material he needs which is not necessarily that which he is seeking. Alternatively, he should be able to find the material required even if he doesn't know the proper subject topic. Such is the objective of this section which hopefully is achieved in part through two flow diagrams and some accompanying discussion.

1.6.1 INTER-SYSTEM EMI PROBLEM IDENTIFICATION

The first step in locating a solution is to identify the problem as either an inter-system or intra-system EMI situation. Generally, if the specimen has an antenna and the problem develops from what exits or enters the antenna from *another* specimen or ambient, then the problem is identified as an *inter-system* EMI one. For this case, Chap. 5 of this book or Vol. 5 of this handbook series will offer the solution. This is the substance of what is illustrated in Fig. 1.10.

If, on the other hand the problem is diagnosed as an *intra-system* EMI problem, then the solution lies in one or more chapters in this volume. In this event the following discussion and Fig. 1.11 will help.

1.6.2 INTRA-SYSTEM EMI PROBLEM

Having determined that it is not an inter-system EMI problem, Fig. 1.11 is then used. Note that emphasis here is to use the test specimen as its own EMI-degradation performance monitor. This is only one-half the problem (susceptibility) since the specimen may perform well and still create a potentially disturbing electrical noise to the outside ambient by conducted or radiated emissions.

For susceptibility situations, the culprit can often be attributed to EMI appearing on AC or DC power-line mains as shown in Fig. 1.11. Another frequent cause of EMI involves an R-F potential difference existing between a chassis, frame, or housing and reference ground. When placing the test specimen inside a shielded enclosure, if the problem disappears it was due to either inadequate power-line filtering or outer case or cable shielding deficiencies. Fig. 1.11 takes the reader through a series of the more likely causes of EMI and references pertinent chapters for detail reading.

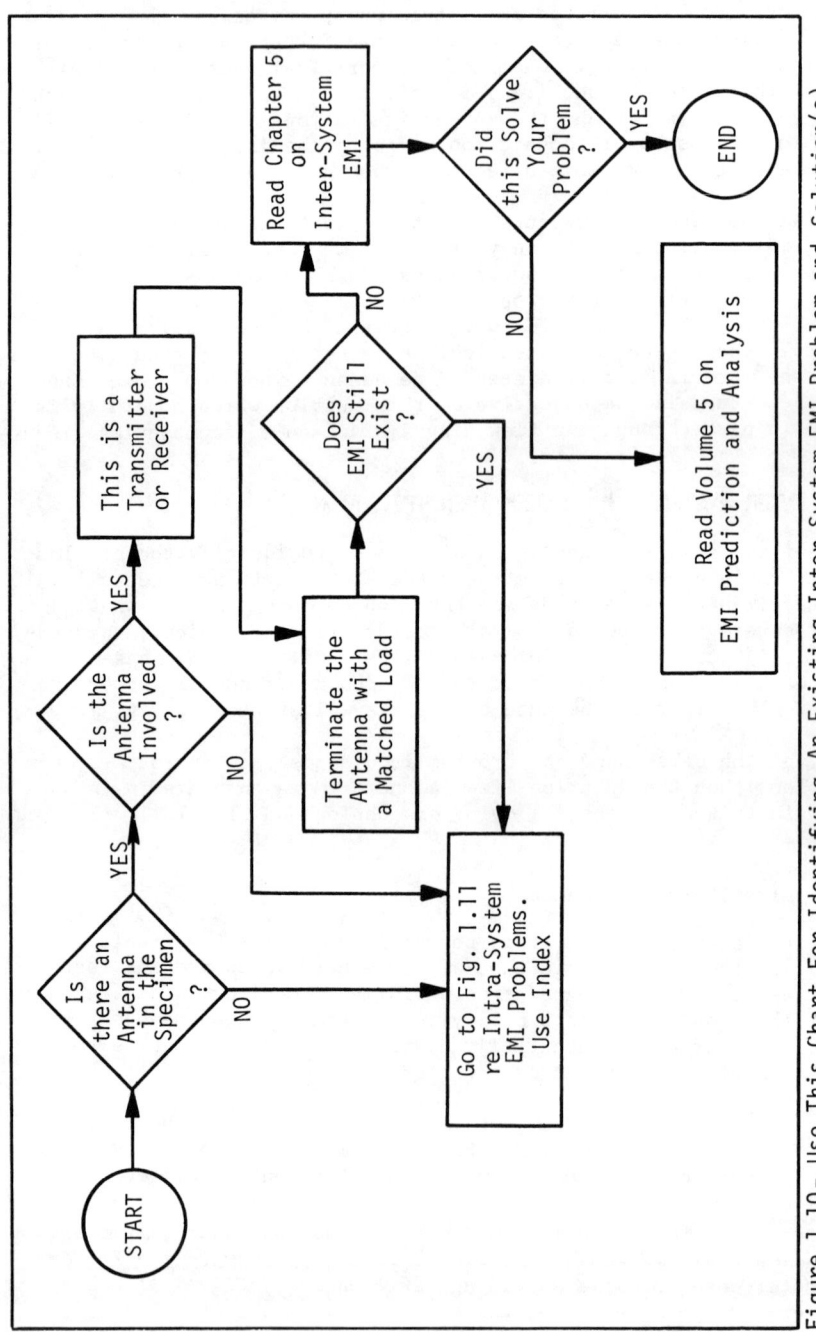

Figure 1.10- Use This Chart For Identifying An Existing Inter-System EMI Problem and Solution(s).

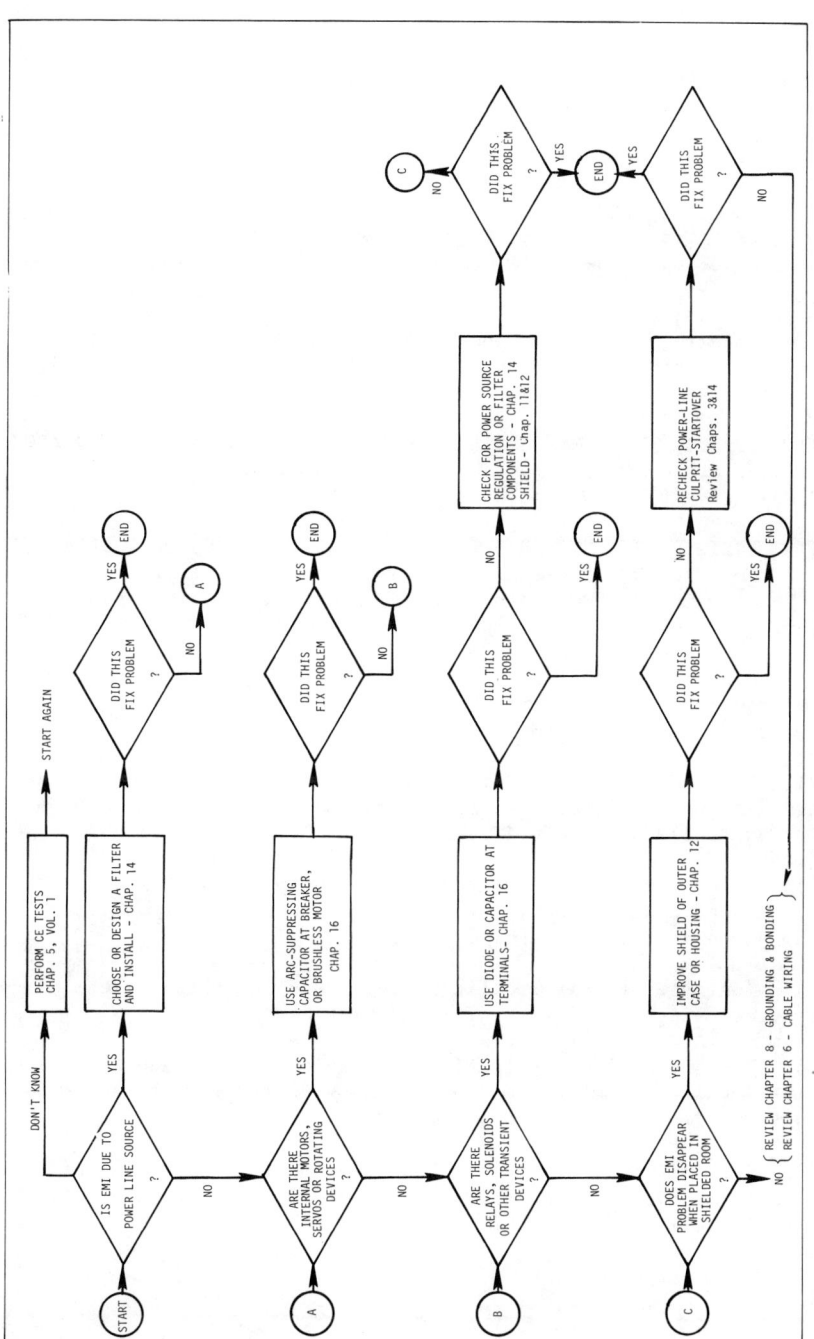

Figure 1.11 - Fault Isolation Chart for Intra-System EMI Problem

1.7 BIBLIOGRAPHY

(1)   Bridges, J. E. and Brueschke, Dr. E. E., "Hazardous Electromagnetic Interaction with Medical Electronics," *Record of the 1970 IEEE International Symposium on Electromagnetic Compatibility,* Vol. 70 C28-EMC, July 14-16, 1970; Anaheim, pp. 173-182.

(2)   Buehler, W. E., and Lunden, C. D., "Signature of Man-Made HF Noise," *IEEE Transactions on Electromagnetic Compatibility,* September, 1966; pp. 143-152.

(3)   Clarke, C. M. and Hoffort, H., "Electromagnetic Compatibility Awareness Seminars," *Record of the 1970 IEEE International Symposium on Electromagnetic Compatibility,* Vol. 70 C28-EMC, July 14-16, 1970; Anaheim; pp. 314-319.

(4)   Daniels, R., "New Horizons in EMC," *Record of the 1970 IEEE International Symposium on Electromagnetic Compatibility,* Vol. 70 C28-EMC, July 14-16, 1970; Anaheim; pp. 160-167.

(5)   Daniels, R., "The Key to Interdisciplinary Communications," *Record of the 1970 IEEE Regional Electromagnetic Compatibility Symposium,* Vol. 70 C64-REGEMC, October 6-8, 1970; San Antonio, pp. II-C-1 to II-C-9.

(6)   Dean, W., "Jogging for Improved EMC," Keynote Address, *Record of the 1970 IEEE International Symposium on Electromagnetic Compatibility,* Vol. 70 C28-EMC, July 14-16, 1970; Anaheim; pp. 1-4.

(7)   Ecom, U. S. Army, "Interference Reduction Guide for Design Engineers," August 1964; *Defense Documentation Center, No. AD 619 666 (Vol. I) and AD 619 667 (Vol. II).*

(8)   "Electromagnetic Compatibility Principles and Practices," Office of Manned Space Flight, Apollo Program, *National Aeronautics and Space Administration,* October, 1965.

(9)   "Enroute IFR Air Traffic Survey, Peak Day, Fiscal Year, 1964," Federal Aviation Agency, Office of Management Services.

(10)   FCC (Federal Communication Commission), "Man-Made Noise," Report to Technical Committee of the Advisory Committee for Land Mobil Radio Services from Working Groups 3.

(11)   Federal Register, "Control of Electronic Product Radiation," *Federal Register,* Vol. 35, No. 100, May 22, 1970, Washington, D. C.; pp. 7851-7954.

(12)   Fraser, D., "Spectrum Management and Interference Control," *Record of the 1969 IEEE Symposium on Electromagnetic Compatibility,* Vol. 69 C3-EMC, June 17-19, 1969, Asbury Park; pp. 298-303.

(13)   Frazier, R. A. and Freeman, E. R., "EMC in Air Traffic Control," *Record of the 1970 International IEEE Symposium on Electromagnetic Compatibility,* Vol. 70 C28-EMC, July 14-16, 1970; Anaheim; pp. 71-80.

(14)   Freeman, E. R. and Sachs, H. M., "Compatibility Management-A Situation Report," *Record of the 1969 IEEE Symposium on Electromagnetic Compatibility,* Vol. 69 C3-EMC, June 17-19, 1969; Asbury Park; pp. 138-142.

(15)   "General Secretariat of the International Telecommunication Union, Radio Regualtions," Geneva, Switzerland, 1959.

(16)   Haber, F., Celebiler, M., Ho, E., Lefferts, R., and Fass, L., "Frequency Assignment Guideline for Satellite Radio Links," NASA Contract NAS5-14923 (Apr. 22, 1968, to July 31, 1969), Moore School Report No. 70-01, University of Pennsylvania. Issued July, 1969.

(17)   Heisler, K. G. and Hewitt, H. J., "Interference Notebook," RADC-TR-66-1.

(18)   Henkel, R. and Mealey, D., "Electromagnetic Compatibility Operational Problems Aboard the Apollo Spacecraft Tracking Ship," *Record of the 1967 IEEE Symposium on Electromagnetic Compatibility,* Vol. 27 C80, July 18-20, Washington, D. C.; pp. 70-81.

(19)   Inglis, L. P., "Why the Double Standard? A Critical Review of Russian Work on the Hazards of Microwave Radiation," *Record of the 1970 IEEE International Symposium on Electromagnetic Compatibility,* Vol. 70 C28-EMC, July 14-16, 1970; Anaheim; pp. 168-172.

(20)   *International Frequency List, International Telecommunication Union,* 3rd ed., Vol. 4, Parts (b), (c), and (d), Feb. 1, 1965.

(21)   Interference Notebook, Volume I, Rome Air Development Center, U. S. Air Force, RADC TR-66-1, 1966.

(22)   Interference Notebook, Volume II, Rome Air Development Center, U. S. Air Force, RADC TR-66-1, 1966.

(23)   JTAC (Joint Technical Advisory Committee) of the *IEEE and EIA,* "Spectrum Engineering-The Key to Progress," *Institute of Electrical and Electronics Engineers,* 1968; New York.

(24)   Kunkel, George M., "The Role of University Extension in the Developments of the Professional EMC Engineer," *Record of the 1970 IEEE International Symposium on Electromagnetic Compatibility,* Vol. 70 C28-EMC, July 14-16, 1970; Anaheim; pp. 320-327.

(25)   McKay, H. D., "Current Status Electromagnetic Pollution Management," *Record of the 1970 IEEE Regional Electromagnetic Compatibility Symposium,* Vol. 70 C64-REGEMC, October 6-8, 1970; San Antonio; pp. II-D-11.

(26)   Metzger, S. D., and Burrus, B. R., "Radio Frequency Allocation in the Public Interest: Federal Government and Civilian Use," *Duquesne University Law Review, 4;* pp. 1-96, 1965-1966.

(27)   Mills, A. H., "Measurement and Analysis of Radio Frequency Noise in Urban, Suburban and Rural Areas," *General/Dynamics/Convair Report GDO-AWV68-001,* NASA CR-72490, February 1969.

(28)   Morris, M., "A New Look at the EMC Problem" *Record of* the *1969 IEEE Symposium on Electromagnetic Compatibility,* Vol. 69 C3-EMC; June 17-19, 1969; Asbury Park; pp. 325.

(29)   Mumford, W. W., "Some Technical Aspects of Microwave Radiation Hazards," *Proceedings of the IRE;* February, 1961; pp. 427-447.

(30)   Myers, H., "Industrial Equipment Spectrum Signatures," *IEEE Transaction on Radio Frequency Interference,* March 1963; pp. 30-42.

(31)   Nichols, F. J., "EMC Reaches an Every-Day Problem Level," *Record of the 1970 IEEE International Symposium on Electromagnetic Compatibility,* Vol. 70 C28-EMC, July 14-16, 1970; Anaheim; pp. 40-44.

(32)   "Partial Revision of the Radio Regulations, Geneva, 1959, and Additional Protocol," General Secretariat of the International Telecommunication Union, adopted at the Extraordinary Administrative Radio Conference, Geneva, Switzerland, 1963.

(33)   Plantz, V. C., "The Structures EMC Engineer: Something New on the Horizon," *Record of the 1969 IEEE Symposium on Electromagnetic Compatibility,* Vol. 69 C3-EMC, June 17-19, 1969, Asbury Park; pp. 312-318.

(34)   "Radio Frequency Allocations for Space and Satellite Requirements in Accordance with EARC Geneva, 1963," National Aeronautics and Space Administration, Washington, D. C., Nov., 1964.

(35)   *Radio Spectrum Utilization, IEEE/EIA* Joint Technical Advisory Committee, IEEE, New York, 1969.

(36)   Skomal, E. N., "The Dimensions of Radio Noise," *Record of the 1969 IEEE Symposium on Electromagnetic Compatibility,* Vol. 69 C3-EMC, June 17-19, 1969; Asbury Park; pp. 18-28.

(37)   Space Radio Facing Frequency Shortage by End of Decade," N. Y. Times, May 4, 1966.

(38)   "The 6th National Resources," A Pamphlet on Two-Way Radio, *Electronics Industries Association;* Washington, D. C., 1965.

(39)   U. S. Senate, "Hearings Before the Committee on Commerce," *Radiation Control for Health and Safety Act of 1967,* U. S. Senate, Ninetieth Congress on S. 2067, S.3211, and H. R. 10790, Part 2, May, 1968; U. S. Government Printing Office.

(40)   Walter, C., "Safe Electrical Environment in Hospitals," *Proceedings of First National Conference on Electronics in Medicine,* Electronic/Management Center, pp. 144; McGraw-Hill, New York.

(41)   White, D. R. J., et al, "Spectrum Pollution in Metropolitan Washington, D. C.," *Record of the 1970 IEEE Regional Electromagnetic Compatibility Symposium,* Vol. 70 C64-REGEMC, October 6-8, 1970; San Antonio; pp. II-A-1 to II-A-9.

CHAPTER 2

SOURCES OF ELECTROMAGNETIC INTERFERENCE

CHAPTER 2
SOURCES OF ELECTROMAGNETIC INTERFERENCE

This chapter surveys sources of electromagnetic interference (EMI) emissions - both natural and man made - in the 10 Hz to 30 GHz frequency spectrum. These sources are summarized in Fig. 2.1. Natural sources include terrestrial atmospheric noise and precipitation static and extraterrestrial emissions originate from the sun, cosmos, and radio stars.

Man-made EMI sources include both intentional and unintentional radiations. The former emphasizes fundamental emissions from communications electronics (C-E) equipments, such as radar, navigation, and tele-communications. Unintentional radiations include emissions at non-fundamental frequencies from the same C-E equipments. They also include incidental emissions from automotive-ignition systems; power lines; electric tools, machines, and appliances; industrial devices and certain consumer products. Table 2.1 is a further amplification of Fig. 2.1 by listing many specific EMI sources.

Whenever practical, this chapter presents typical radiated emissions in terms of either power density or field intensities, generally at stipulated distances from identified EMI sources. Most such data are presented in broadband radiated units of either $dBm/m^2/kHz$ or $dB\mu V/m/kHz$ (see Vol. 1, Chap. 5, or Vol. 4 - Terms and Definitions). Some C-E spectrum signatures, however, are reported in terms of power relative to the fundamental in the transmitter output.

A summary example of the above is shown in Fig. 2.2 in which field intensity of several emission sources are plotted vs radio frequency. For reference, typical receiver sensitivity (S=N) is also shown in the figure corresponding to an isotropic, pick-up antenna (antenna gain, G = 0 dB). This permits converting receiver noise to an equivalent field intensity. Note that all natural and man-made emissions above receiver noise may be expected to mask-typical receiver sensitivity by the vertical displacement shown in dB. Conversely, extra-terrestrial sources that are below the receiver sensitivity will pose no EMI problem, provided an associated high-gain antenna does not *look-at* these sources.

Readers who may not be particularly interested in details of man-

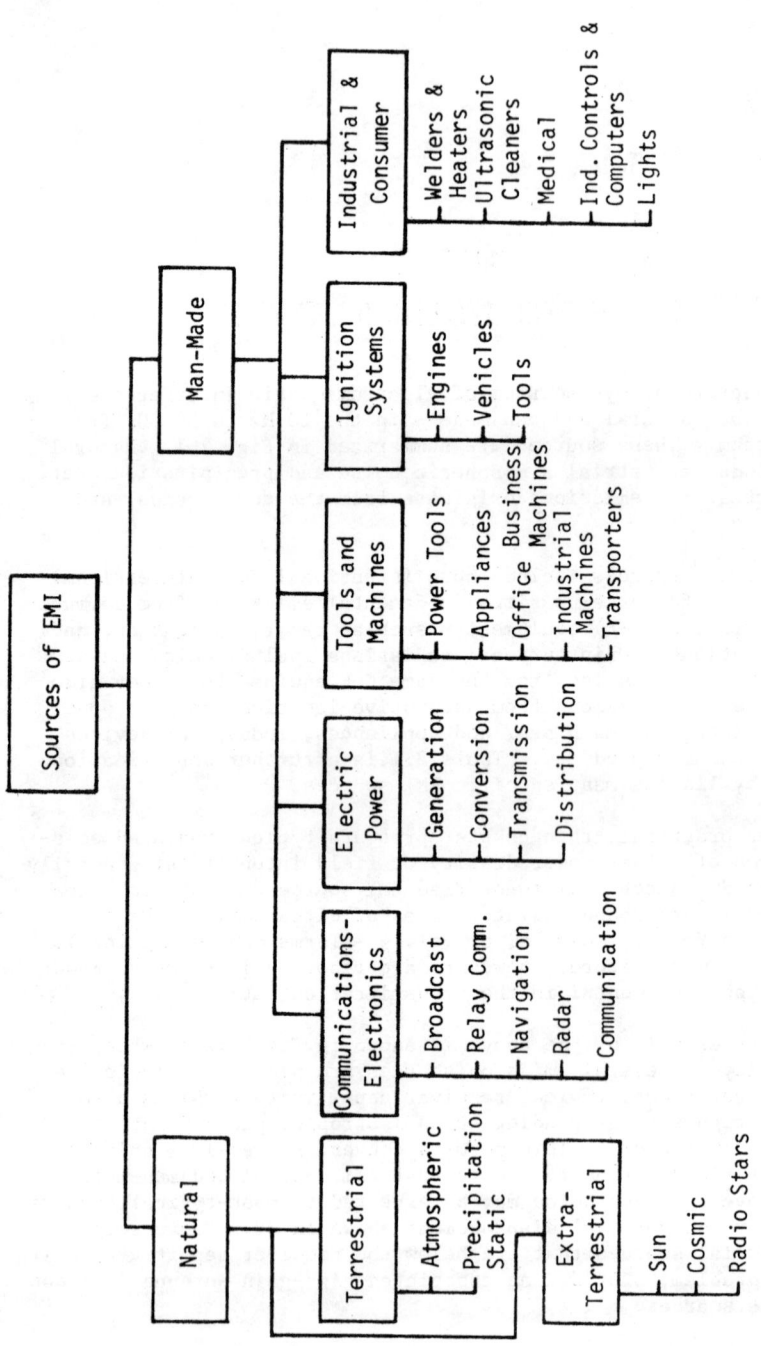

Figure 2.1 - Sources of Electromagnetic Interference

made noise, but are concerned only about the impact it may have upon the performance of their communications-electronics receiver, may wish to go directly to Sec. 2.3.3, *Overall Mathematical Models of Incidental Emitters*.

Table 2.1 - Examples of Sources of Electromagnetic Interference

The following is a list of representative sources of electric, magnetic or electromagnetic interference. While far from complete, the list is fairly typical:

1. NATURAL SOURCES	VHF/FM
	VHF/UHF TV
1.1 Terrestrial Sources:	b. Communications (non relay)
a. Atmospherics (thunderstorms around the world)	Aeronautical Mobile
	Amateurs (Hams)
b. Lightning discharges (local storms)	Citizens radio
	Facsimile
c. Precipitation Static	HF Telegraphy
d. Whistlers	HF Telephony
	Land Mobile
1.2 Extra-Terrestrial Sources:	Maritime mobile
a. Cosmic Noise	Radio-Control Devices
b. Radio Stars	Telemetry
c. Sun:	Telephone Circuits
Disturbed	Wireless Microphone
Quiet	c. Navigation (non radar)
	Aircraft Beacons
2. MAN-MADE SOURCES	Instrument Landing Systems
	Loran
2.1 Electric-Power:	Marker Beacons
a. Conversion (Step Up/Down)	Omega
Faulty/Dirty Insulators	VOR/TACAN/VORTAC
Faulty Transformers	d. Radar
b. Distribution	Air Search
Faulty/Dirty Insulators	Air Surface Detection
Faulty Transformers	Air Traffic Control
Faulty Wiring	Harbor
Pick-up and Re-radiation	Mapping
Poor Grounding	Police Speed Monitor
c. Generators	Surface Search
d. Transmission Lines	Tracking/Fire Control
Faulty/Dirty Insulators	Weather
Pick-up and Re-radiation	e. Relay Communications
2.2 Communications-Electronics (C-E):	Ionospheric Scatter
a. Broadcast	Microwave Relay Links
MF Amplitude Modulation	Satellite Relay
VHF/FM	Tropospheric Scatter

Table 2.1 - (Continued)

2.3 Tools and Machines	Lawn Mowers
a. Appliances	Portable Saws
Air Conditioners	c. Vehicles
Blenders	Aircraft
Deep Freezes	Automobiles
Fans	Farm Machinery
Lawn Mowers, electric	In-board Motors
Mix Masters	Mini-bikes
Ovens, electric	Motorcycles
Ovens, Microwave	Out-board Motors
Refrigerators	Tanks
Sewing Machines	Tractors
Vacuum Cleaners	Trucks
Water Pumps	2.5 Industrial & Consumer
b. Industrial Machines	(non motor/engines)
Electric Cranes	a. Heaters and Gluers
Fork-lift Trucks	Diaelectric Heaters
Lathes	Plastic Pre-Heaters
Milling Machines	Wood Gluers
Printing Presses	b. Industrial Controls &
Punch Presses	Computers
Rotary Punches	Card Punches
Screw Machines	Card Readers
c. Office/Business Machines	Computers
Adding Machines	Machine Controllers
Calculators	Peripheral Equipment
Cash Registers	Process Controllers
Electric Typewriters	Silicon-Controlled
Reproduction Equipment	Rectifiers
d. Power Tools	Teletypewriters
Band Saws	c. Lights
Drill Press	Faulty Incadescent
Electric Drills	Fluorescent Lamps
Electric Hand Saws	Light Dimmers
Electric Grinders	Neon Lights
Electric Sanders	R-F Excited, Gas-Display
Hobby Tools	Signs
Routers/Joiners	R-F Excited, Gas Laser
Table Saws	d. Medical Equipment
e. Transporters	Defibrillators
Converyor Belts	Diathermy
Elevators	X-Ray Machines
Escalators	e. Ultra-sonic Cleaners
Moving Sidewalks	f. Welders & Heaters
	Arc Welders
2.4 Ignition Systems	Heli-Arc Welders
a. Engines	Induction Heaters
b. Tools	Plastic Welders
Auxiliary Generators	R-F Stabilized Welders

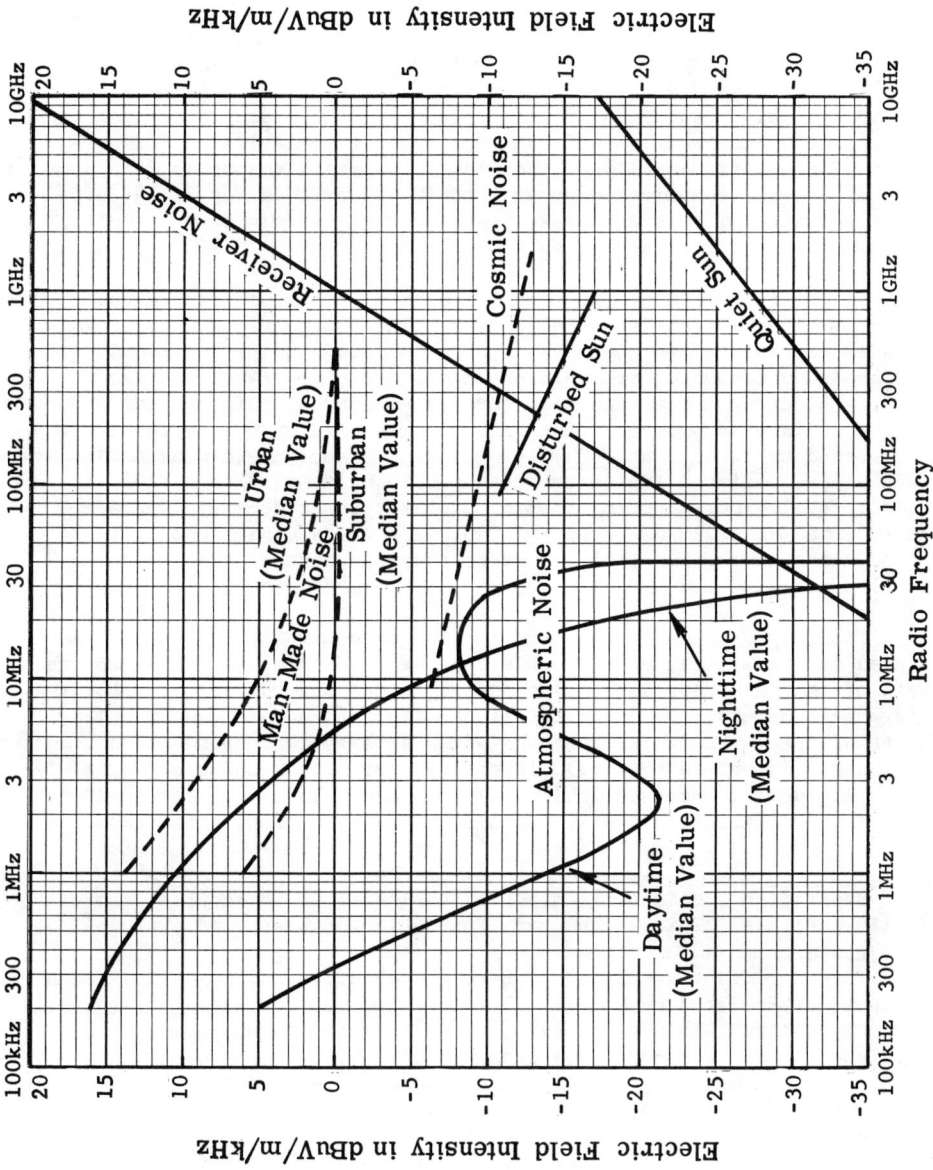

Figure 2.2 – Summary of Electromagnetic Noise Sources and Levels

2.1 DATA-REPORTING FORMATS

In any technical topic involving relationship of parameters, the question presents itself regarding useful ways to format data. The matter is especially important when empirical manifestations of phenomena involve many variables which are often theoretically inseparable.* In this chapter, only principle effects are displayed and hopefully some of the more significant independent variables are identified. For spectrum-amplitude profiles from emitting sources, the problem is further complicated by such topics as near vs far field, distance between emitters and measuring receiver, terrain effects, directional properties, polarization, and the like.

A spectrum signature is identified as an amplitude vs frequency portrayal of emissions from one or more electromagnetic sources. This immediately raises two questions: what amplitude units are involved and are they conducted or radiated emissions? The latter is easy to specify since most spectrum signatures discussed in this chapter are of the radiated type, viz., they are not measured on wire or cable, but sensed by an antenna or a field pick-up probe. Regarding amplitude units, radiated fields are measured in units of either field intensity (Volts/meter, $\mu V/m$, $dB\mu V/m$, etc.) or power density (Watts/m^2, dBW/m^2, dBW/cm^2, dBm/cm^2, etc.). Table 2.2 lists equivalency between these units in the far field,** where the impedance of free spece (377 ohms) applies.

The most frequently reported (or converted amplitude unit of radiated spectrum signatures in this chapter is $dBm/m^2/kHz$. The capture area of the intercept antenna in square meters and the receiver bandwidth must be known to convert this amplitude to a received signal in units of dBm. When this broadband power density corresponds to an equivalent incoherent*** field intensity, conversion units of 0 $dBm/m^2/kHz$ = 116 $dB\mu V/m/kHz$ (see Table 2.2) are used. This applies provided far-field conditions prevail. Displayed as a function of frequency,

---

* Separation and identification of variables is, of course theoretically possible. This is the very foundation of basic and exploratory research.
** Far-Field is that distance where the electric and magnetic fields are uniquely related by $Z=377\Omega$. For emitting sources having dimensions much less than a wavelength, this distance is apporximately a wavelength. For sources whose greatest dimension, D, is equal to or greater than a wavelength (λ), the distance = D^2/λ is the crossover region between near and far fields. This topic is discussed in further detail in Chap. 3 of Vol. 4.
*** An incoherent emission source is one in which there exists no amplitude and phase relations across the measuring receiver bandwidth. When coherency exists, such as from an impulsive source, power density is expressed in units of dBm/m^2kHz^2.

Table 2.2 - Field Intensity and Power Density Relationships (Related by free space impedance = 377 ohms)

Volts/m	dBµV/m	Watts/m²	dBW/m²	Watts/cm²	dBW/cm²	mW/cm²	dBm/cm²
10,000	200	265,000	+54	27	+14	26,500	+44
7,000	197	130,000	+51	13	+11	13,000	+41
5,000	194	66,300	+48	6.6	+8	6,630	+38
3,000	190	23,900	+44	2.4	+4	2,390	+34
2,000	186	10,600	+40	1.1	0	1,060	+30
1,000	180	2,650	+34	.27	-6	265	+24
700	177	1,300	+31	.13	-9	130	+21
500	174	663	+28	.066	-12	66	+18
300	170	239	+24	.024	-16	24	+14
200	166	106	+20	.011	-20	11	+10
100	160	27	+14	27×10^{-4}	-26	2.7	+4
70	157	13	+11	13×10^{-4}	-29	1.3	+1
50	154	6.6	+8	6.6×10^{-4}	-32	.66	-2
30	150	2.4	+4	2.4×10^{-4}	-36	.24	-6
20	146	1.1	0	1.1×10^{-4}	-40	.11	-10
10	140	.27	-6	27×10^{-5}	-46	.027	-16
7	137	.13	-9	13×10^{-6}	-49	.013	-19
5	134	.066	-12	6.6×10^{-6}	-52	66×10^{-4}	-22
3	130	.024	-16	2.4×10^{-6}	-56	24×10^{-4}	-26
2	126	.011	-20	1.1×10^{-6}	-60	11×10^{-4}	-30
1	120	27×10^{-4}	-26	27×10^{-8}	-66	2.7×10^{-4}	-36
0.7	117	13×10^{-4}	-29	13×10^{-8}	-69	1.3×10^{-4}	-39
0.5	114	6.6×10^{-4}	-32	6.6×10^{-8}	-72	66×10^{-6}	-42
0.3	110	2.4×10^{-4}	-36	2.4×10^{-8}	-76	24×10^{-6}	-46
0.2	106	1.1×10^{-4}	-40	1.1×10^{-8}	-80	11×10^{-6}	-50
0.1	100	27×10^{-6}	-46	27×10^{-10}	-86	2.7×10^{-6}	-56
70×10^{-3}	97	13×10^{-6}	-49	13×10^{-10}	-89	1.3×10^{-6}	-59
50×10^{-3}	94	6.6×10^{-6}	-52	6.6×10^{-10}	-92	66×10^{-8}	-62
30×10^{-3}	90	2.4×10^{-6}	-56	2.4×10^{-10}	-96	24×10^{-8}	-66
20×10^{-3}	86	1.1×10^{-6}	-60	1.1×10^{-10}	-100	11×10^{-8}	-70
10×10^{-3}	80	27×10^{-8}	-66	27×10^{-12}	-106	2.7×10^{-8}	-76
7×10^{-3}	77	13×10^{-8}	-69	13×10^{-12}	-109	1.3×10^{-8}	-79
5×10^{-3}	74	6.6×10^{-8}	-72	6.6×10^{-12}	-112	66×10^{-10}	-82
3×10^{-3}	70	2.4×10^{-8}	-76	2.4×10^{-12}	-116	24×10^{-10}	-86
2×10^{-3}	66	1.1×10^{-8}	-80	1.1×10^{-12}	-120	11×10^{-10}	-90
1×10^{-3}	60	27×10^{-10}	-86	27×10^{-14}	-126	2.7×10^{-10}	-96
700×10^{-6}	57	13×10^{-10}	-89	13×10^{-14}	-129	1.3×10^{-10}	-99
500×10^{-6}	54	6.6×10^{-10}	-92	6.6×10^{-14}	-132	66×10^{-12}	-102
300×10^{-6}	50	2.4×10^{-10}	-96	2.4×10^{-14}	-136	24×10^{-12}	-106
200×10^{-6}	46	1.1×10^{-10}	-100	1.1×10^{-14}	-140	11×10^{-12}	-110
100×10^{-6}	40	27×10^{-12}	-106	27×10^{-16}	-146	2.7×10^{-12}	-116
70×10^{-6}	37	13×10^{-12}	-109	13×10^{-16}	-149	1.3×10^{-12}	-119
50×10^{-6}	34	6.6×10^{-12}	-112	6.6×10^{-16}	-152	66×10^{-14}	-122
30×10^{-6}	30	2.4×10^{-12}	-116	2.4×10^{-16}	-156	24×10^{-14}	-126
20×10^{-6}	26	1.1×10^{-12}	-120	1.1×10^{-16}	-160	11×10^{-14}	-130
10×10^{-6}	20	27×10^{-14}	-126	27×10^{-18}	-166	2.7×10^{-14}	-136
7×10^{-6}	17	13×10^{-14}	-129	13×10^{-18}	-169	1.3×10^{-14}	-139
5×10^{-6}	14	6.6×10^{-14}	-132	6.6×10^{-18}	-172	66×10^{-16}	-142
3×10^{-6}	10	2.4×10^{-14}	-136	2.4×10^{-18}	-176	24×10^{-16}	-146
2×10^{-6}	6	1.1×10^{-14}	-140	1.1×10^{-18}	-180	11×10^{-16}	-150
1×10^{-6}	0	27×10^{-16}	-146	27×10^{-20}	-186	2.7×10^{-16}	-156

amplitude defines a typical spectrum signature such as shown in Fig. 2.3 for a fluorescent lamp. Note that a *best* straight line fit to this curve (the math model) is $40 - 40 \log_{10} f_{MHz}$ dBµV/m/kHz.

Sometimes an emission is formatted as a function of time (or numbers of events) corresponding to many quantitized subincrements of the cumulative look interval (or total events). It is then identified as an amplitude probability distribution (APD). This is a measure of the percentage of the total look interval (or events) in which any particular amplitude is exceeded. Such an example is shown in Fig. 2.4, for 100 fluorescent lamps of the same type measured at 1 MHz. This example indicates that the most probable (50%) radiation level is 40 dBµV/m/kHz and that 10% of the time a level of 51 dBµV/m/kHz may be exceeded. APD's are of special value, but many are needed to define different radio frequencies, bandwidths, time of day, or other variables, as applicable.

Perhaps the best of all worlds would be to develop composite spectrum signature – APD combinations. Here, the basic plot is obtained from many repeated spectrum signatures in which radio frequency is either quantitized or sampled in defined increments. For each such frequency increment, certain APD points are calculated such as the 1%, 10%, 50%, 90%, and 99% levels, or perhaps, simply the mean value, μ, and standard deviation, σ. Fig. 2.5 illustrates this concept for the same hypothetical fluorescent lamp group. It is likely that radio-noise maps of metropolitan areas in years to come will use a format similar to this.

Data representing radiated electromagnetic emissions reported in this chapter follow the first two formats illustrated in Figs. 2.3 and 2.4. Relatively little data of the form shown in Fig. 2.5 are available.

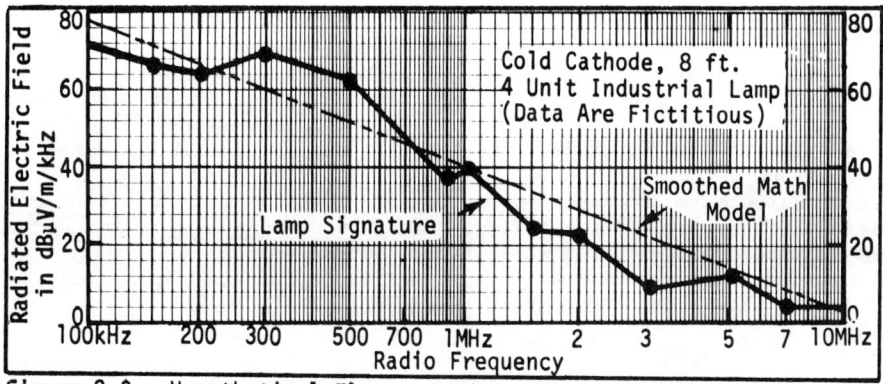

Figure 2.3 - Hypothetical Fluorescent Lamp Spectrum Signature

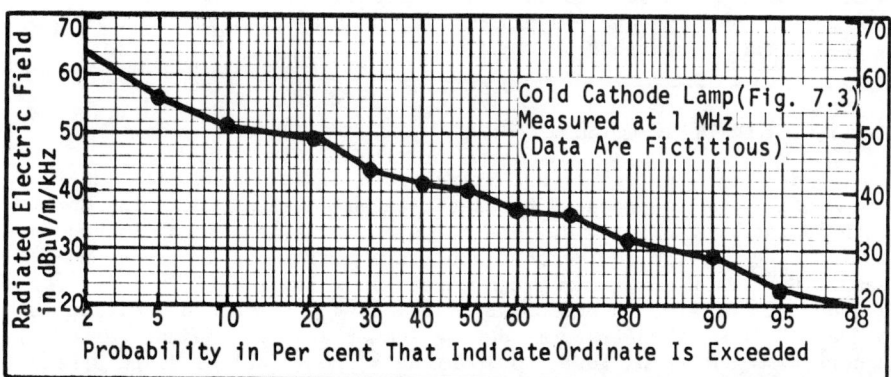

Figure 2.4 - Amplitude Probability Distribution (APD) of Lamp

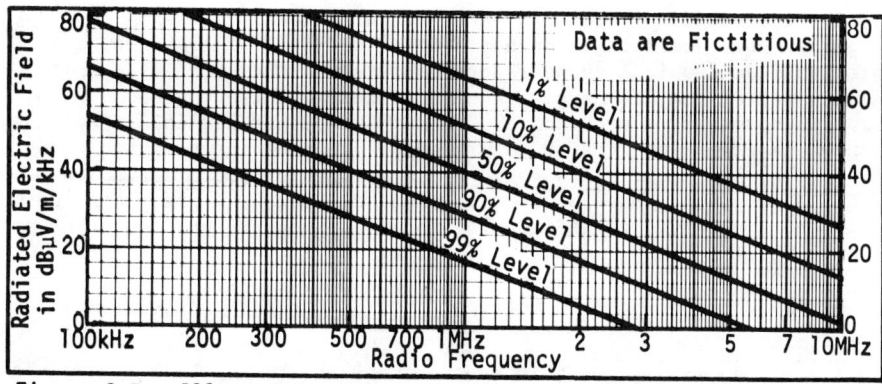

Figure 2.5 - Illustrative Spectrum Signature - APD Curve

2.2 NATURAL SOURCES OF EMI*

For convenience of discussion and because of different physical properties, natural sources of EMI are divided into: (1) terrestrial sources (sources eminating from the earth's atmosphere) and (2) extra-terrestrial sources (sources eminating from regions beyond the earth). As explained below, terrestrial sources of interference tend to be more transient-like in nature whereas, with certain exceptions, noise originating from stars, the galaxy, and other regions tend to behave more like bandwidth-limited white noise. These natural sources of noise often limit the maximum useful sensitivity obtainable by most receivers below about 300 MHz, as shown in Fig. 2.2.

2.2.1 TERRESTRIAL SOURCES OF NATURAL INTERFERENCE

Terrestrial EMI sources discussed here include those natural emissions coming from the earth's atmosphere, such as atmospheric noise and precipitation static.

2.2.1.1 ATMOSPHERIC NOISE

The dominant, naturally-occurring radio noise source below 30 MHz is atmospheric noise produced by electrical discharges occurring during thunderstorms. Electrical-discharge noise has a moderately broad emission spectra with the largest amplitude components occurring between 2 kHz and 30 kHz. For frequencies below ionospheric cutoff, the preponderence of atmospheric noise detected at a temperate location is produced by local thunderstorms during summer and by tropical-region thunderstorms during winter. Radio signals from tropical disturbances are propagated by an ionospheric sky wave over distances of several thousand kilometers.

For frequencies above ionospheric cutoff, local lightning discharges are the dominant atmospheric-noise sources. The resulting isolation from tropical thunderstorm centers and the decreasing spectral density of the discharges cause atmospheric noise to become an insignificant noise source above 30 MHz in temperate and polar regions as shown in Fig. 2.6. Four CCIR grades of atmospheric noise are shown (20, 50, 70, and 100). Each noise grade curve displays a variation of average atmospheric noise power with frequency for a geographical location whose observed relative noise level is proportional to the noise grade number.

Within the 300 kHz - 30 MHz spectrum, atmospheric noise is not

---

* Substantial portions of this section were obtained from "*A Review of Incidental Radio Noise in Metropolitan Areas*", by E.N. Skomol, Aerospace Corporation.

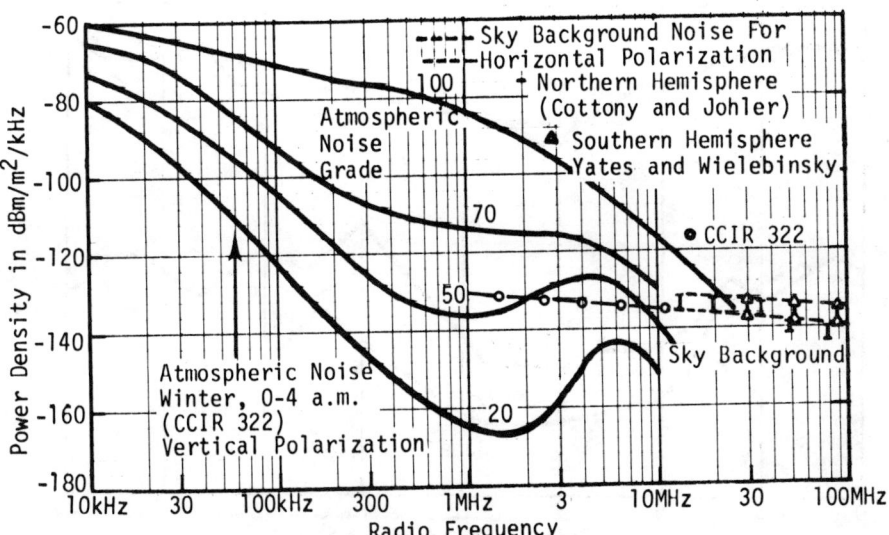

Figure 2.6 - Spectral Distribution of Atmospheric Noise

always the dominant natural noise source. Sky-background or galactic noise as discussed later, often dominates atmospheric noise in temperate and tropical regions and may be dominant in the polar regions. Over sub-polar continents and oceans, atmospheric noise clearly dominates sky-background noise only during the daily 8 PM to 4 AM period. This interval is extended somewhat beyond either limit during the summer and autumn periods but for the remainder of the day, atmospheric noise levels are commonly less than sky-background noise.

The diurnal increase in atmospheric noise begins in the afternoon and rises in level throughout the evening as a consequence of reduced ionospheric D-region absorption. Electrical discharge activity is maximum during the afternoon within the sunlit tropical regions of the continents and decreases after sunset attaining a minimum level sometime after midnight. Because of long range, moderate-loss sky-wave propagation during the evening, transpolar and transequatorial propagation paths from sunlit tropical regions contribute large atmospheric noise signals at sites in the northern hemisphere.

In Fig. 2.7 the RMS atmospheric-noise amplitude diurnal cycle is shown for two frequencies and two seasons at a Canadian observation station. Evening and midnight noise maxima are observed. The summer double maxima indicates emissions from diametrically opposite tropical thunderstorm centers peaking during the local afternoon periods. Morning noise minima are noticeable during all seasons.

Electric fields from atmospheric-noise are impulsive containing

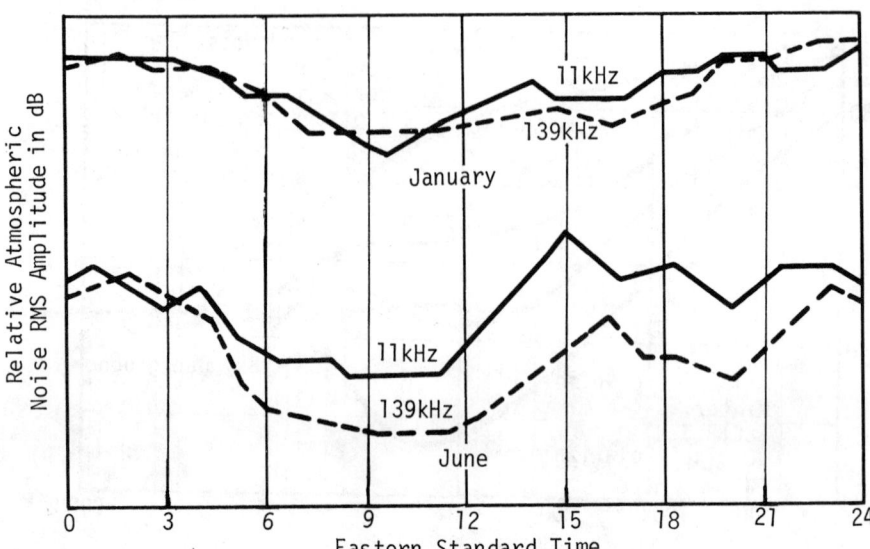

Figure 2.7 - Relative Atmospheric Noise Showing The Diurnal Variation

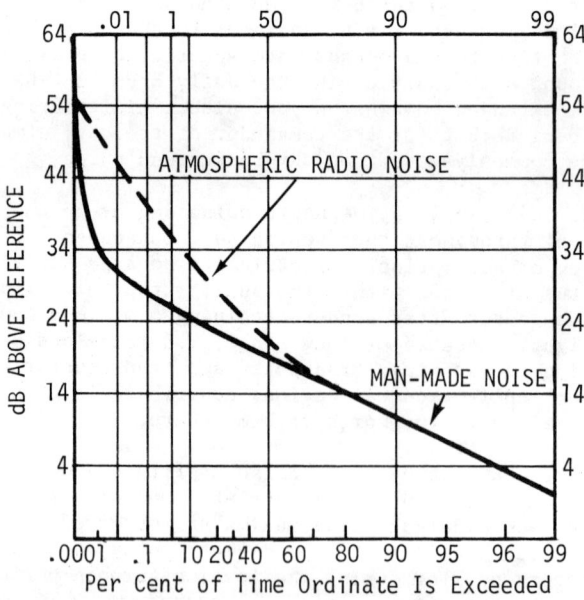

Figure 2.8 - Amplitude Probability Distributions of Atmospheric and Man-Made Noise

significant numbers of very large amplitude pulses arising from light-
ning-discharge return strokes. Intense pulses are superimposed upon
a background of smaller pulses, the voltage envelop of which is
Rayleigh distributed. Resulting cumulative distribution functions of
atmospheric-noise amplitude are typically as shown in Fig. 2.8. Surface man-made noise is also shown for comparison.

Illustrative Example 2.1

An automobile AM broadcast radio has a 1 meter whip antenna
(electrical length = 0.5 meters). Because of the inefficiency of it's
antenna*, the coupling loss is about 30 dB. The receiver sensitivity
for a 10-kHz bandwidth at 1 MHz is -122 dBm. Representative broadcast signal strengths in the area are 60 dBμV/m. Determine if atmospheric noise for a number 100 noise grade will compromise reception.

A sensitivity, N, of -122 dBm corresponds to:

$$N = -122 \text{ dBm} - (-107 \text{ dBm/dB}\mu V) = -15 \text{ dB}\mu V$$

A 60 dBμV/m broadcast signal, S_B, corresponds to a receiver input signal, S_R, of:

$$S_{RdB\mu V} = S_{BdB\mu V/m} - AF^*_{dB/m}$$

$$= 60 \text{ dB}\mu V/m - (6 \text{ dB} + 30 \text{ dB}) = 24 \text{ dB}\mu V$$

Thus, the S/N ratio (without atmospherics) is 24 dBμV - (-15 dBμV)
    = 30 dB.

For a 100-grade atmospheric noise, N_a, at 1 MHz, Fig. 2.6 shows
that the power density is -83 dBm/m^2/kHz = -73 dBm/m^2 for a 10 kHz
receiver bandwidth. Since: 0 dBμV/m = -116 dBm/m^2, the atmospheric
noise, N_a = -73 dBm/m^2 - (-116 dB) = 43 dBμV/m. Using the 36 dB/m
antenna factor, the induced atmospheric noise is: N_a - AF = 43 dBμV/m
= 7 dBμV. From the above, this corresponds to an I/N ratio of 7 dBμV
- (-15 dBμV) = 22 dB.

While the atmospherics are below the particular broadcast station
level by 17 dB, the effective 22 dB increase in receiver noise will
mask weaker stations.

---

* See Chap. 3, Vol. 4, *EMI Test Instrumentation and Systems*, Antenna
Factor (AF) in dB/meter = effective height in dB/meter + conversion
loss in dB.

2.2.1.2   PRECIPITATION STATIC

Precipitation static is a source of radio noise which arises from
an electric charge accumulation on a surface followed by corona discharge and/or dielectric breakdown on elements of an antenna, ground
plane, or surroundings. The frequency spectrum, the intensity, and
the occurrance rate of precipitation static are governed by structural
features of the antenna environment and the conditions of operation.
Aircraft-receiving systems are those most often compromised by this
noise source. Remedies which have been successfully employed to reduce or eliminate precipitation interference consists of static dis-
chargers, loop rather than linear antennas, and high-dielectric
strength materials.

2.2.2   EXTRA-TERRESTRIAL SOURCES OF NATURAL INTERFERENCE

Extra-Terrestrial EMI sources discussed here include those natural emissions coming from beyond the earth's atmosphere, such as sky-
background noise, solar noise, and secondary cosmic noise sources.

2.2.2.1   SKY-BACKGROUND NOISE

The sky background noise shown earlier in Fig. 2.6 for frequencies
above 1 MHz is generated within the galaxy by a combination of unre-
solved discrete sources and an approximately continuous spatial distribution of emission, a portion of which has a noticeable concentration in the galactic plane. Continuously distributed noise arises
from two mechanisms one of which is ionized hydrogen displaying a
blackbody spectrum and the other electron-synchroton radiation presenting a non-thermal spectra with a linearly polarized radiation field.
The non-thermal sky background component is not concentrated in the
galactic plane. Composite background noise sources produce significant radio emission above 300 kHz throughout both hemispheres. Sky-
background noise undergoes a daily variation which is determined by
the relative location of the closest approach of the receiving antenna beam to the galactic equator.

2.2.2.2   SOLAR NOISE

During periods of high solar activity, intense, visible bursts
of energy are emitted in the vicinity of sun spots, often as frequently
as 12 times per day. These intense flares are accompanied by a sharp
rise in solar radio-noise output. Solar-flare noise varies in spectral content, duration, and polarization depending upon which of five
types of flares predominates. For example, the power density of Type-
IV flares is shown in Fig. 2.9. Noise emission of a Type-IV flare is
not the largest of the five classes (Type I, II, and III may emit

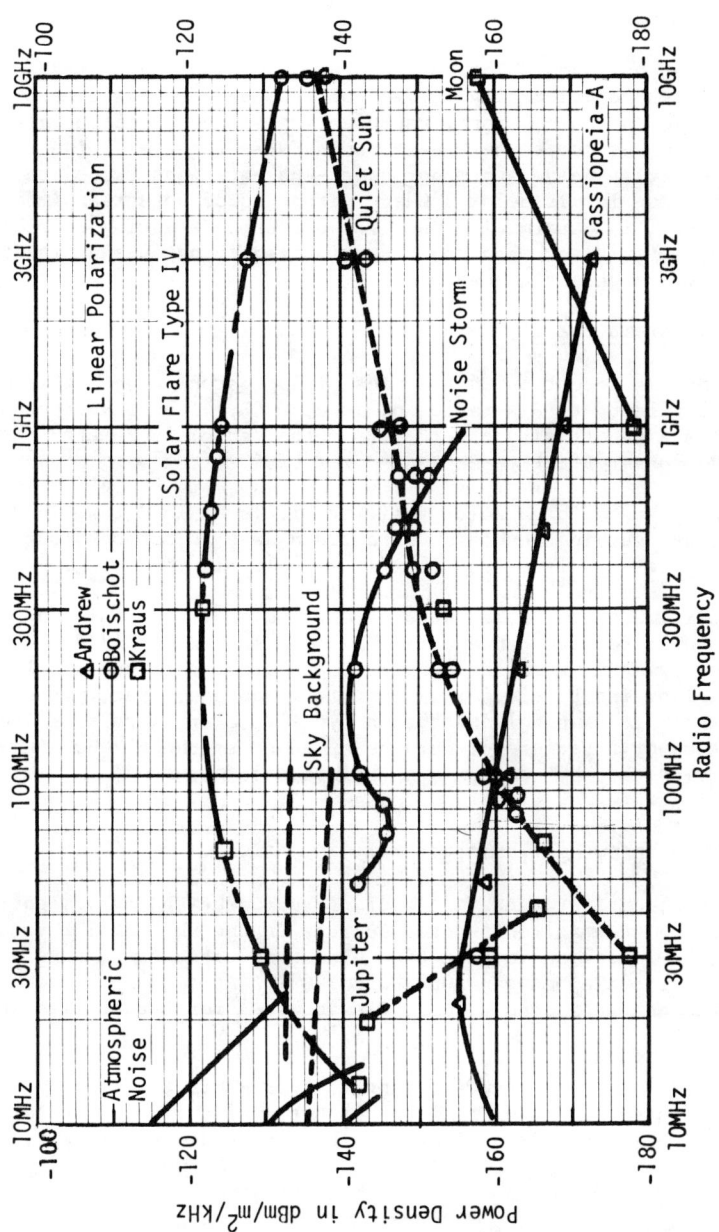

Figure 2.9 - Solar, Planetary and Stellar Noise

higher noise power densities), but represents the class which emits a stable, very intense broadband noise output which may endure for days. It is evident from Fig. 2.9 that Type-IV solar flares exceed other natural radio sources above 20 MHz.

Quiet-sun noise emission represents the minimum solar radiation condition which occurs during 11-year periods of very-low sunspot activity. A solar-noise storm, representing an intermediate-noise intensity case, produces short-duration bursts of narrow-noise impulses. Neither solar-noise storm nor quiet-sun emissions exceed the sky background noise levels below 3 GHz. Either is of potential importance, however, in determining receiver performance during periods of low-sunspot activity or when the beam of a receiving antenna is directed away from the galactic equator.

Illustrative Example 2.2

Land-mobile reception in a typical fringe area at 40 MHz is marginal corresponding to S/N ratios of about 0 dB. The 25-kHz bandwidth receiver has a sensitivity of -120 dBm. The vehicle's 2-meter whip corresponds to an effective antenna* area of $6.7m^2$.

During years of high solar activity, it is noticed that land-mobile, fringe service reception on certain days is not possible at all. Determine if a Type-IV solar flare is likely the cause.

From Fig. 2.9, the power density of a solar flare impinging upon the land-mobile receiver antenna is -127 dBm/m^2/kHz = -113 dBm/m^2 for a 25 kHz bandwidth. From the above antenna area, this equals -105 dBm at the receiver input terminals. This is 15 dB above receiver sensitivity. Thus, it is concluded that a Type-IV solar flare corresponds to I/N ratios or interference margins of 15 dB to the 40 MHz land-mobile service thereby compromising fringe-area service.

2.2.2.3   Secondary Cosmic Noise

The moon, Jupiter, and Cassiopeia-A represent the only significant additional cosmic radio-noise sources in the VHF/UHF/SHF bands. SHF-band lunar emission is that of a blackbody whose temperture varies with the lunar cycle between 100°K and 300°K. As a radio-noise source, the moon exceeds in intensity only Cassiopeia-A for frequencies above 2 GHz.

Cassiopeia-A, a supernova remanent, represents the most intense extra-solar noise source. It emits a non-thermal, linearly polarized, spectrum which at HF and the lower VHF band exceeds in intensity the quiet sun.

---

* $A = G\lambda^2/4\pi = 1.5(7.5m)^2/4\pi = 6.7m^2$.

HF/VHF band radiation from Jupiter is non-thermal, linearly polarized noise. At frequencies below the earth's ionospheric cutoff, this noise does not penetrate to the earth's surface. However, in the frequency interval 20 to 40 MHz, when the planet is in a receiving antenna beam and at the peak of its noise emission cycle, the noise emission is exceeded only by the disturbed sun and sky background.

Natural noise sources of lesser intensity also exist and include: the universal thermal background radiation (3°K); hydrogen emissions from ionized clouds; line emissions from neutral hydrogen, the OH radical and most recently observed from ammonia. Also emitting radio noise of very-low intensity at the earth's surface are flare stars and radio galaxies.

2.3   MAN-MADE SOURCES OF EMI

Unlike natural sources of EMI, man-made sources originate from devices, equipments, and machines created by humans. Their emissions may come from either terrestrial or extra-terrestrial (earth satellite, space ships, etc.) locations. With some exceptions, extra-terrestrial man-made sources are of modest concern today since they represent but a tiny fraction of the total electromagnetic noise-making populations. However, to such elevated platforms vast sections of the earth are in the field-of-view of either transmitters or receivers and the numbers are likely to increase exponentially in the future.

Sometimes man-made electromagnetic emissions are classified as either intentional or unintentional. However, this division proves to be more academic than useful since many radiators simultaneously emit both types or emissions. Another man-made classification is by a mixture of both applications and users. Fig. 2.1 previously illustrated such a classification. In that example, man-made sources are divided into (1) communications-electronics, electric power, tools and machines (all applications) and (2) industrial and consumer (users). Actually, a division between applications and users, per se, is not particularly useful since each also contains many of the other.

Man-made radio noise also may be classified as either: (1) coherent interference such as harmonic output of radar transmitters, emission sidebands of communication and navigational transmitters or normal transmission from these and most other electronic transmitters, or (2) incoherent random signals which are emitted incidentally to the normal functioning of any electronic equipment.

The following sections review the above sources of man-made EMI. Typical radiation intensities of many such sources are presented together with a bibliography.

2.3.1   COMMUNICATIONS-ELECTRONICS EMITTERS

This section reviews a class of man-made radiaters called communications-electronics (C-E) emitters. This class is further divided into the following five categories:

(1)   Broadcast (MF/AM, VHF/FM, VHF/UHF TV)
(2)   Communications (amateur, aeronautical mobile, citizens, facsimile, HF telephony/telegraphy, land mobile, maritime mobile, radio-control devices, telemetry, telephone circuits, etc.).
(3)   Relay communications (microwave relay link, satellite relay, ionospheric and tropospheric scatter, etc.).
(4)   Navigation (VOR/TACAN/VORTAC, aircraft and marker beacons, instrument landing systems, loran, omega, etc.).
(5)   Radars (air-traffic control, air search, harbor surveillance, mapping, surface search, tracking and fire control, weather, etc.).

One or more examples of spectrum signatures, of each of the five C-E categories is presented in the following sections.

2.3.1.1 Broadcast C-E Emitters

Broadcast bands cover:

> HF Amplitude Modulation (535-1605 kHz)
> VHF Frequency Modulation (88-108 MHz)
> VHF Television: Lower Bands (54-88 MHz),
>   Upper Bands (174-216 MHz)
> UHF Television (470-890 MHz)
> International AM broadcasting

Two examples of broadcast spectrum signatures will serve to illustrate spectrum-amplitude profiles. Fig. 2.10 illustrates a typical signature of an F-M broadcast transmitter. Note the nominal 300 kHz emission bandwidth around the fundamental frequency and the relatively low (-90 dB) out-of-band levels as controlled by FCC Rules and Regulations.

Fig. 2.11 illustrates the spectrum signature of a typical VHF TV emission. The nominal 6-MHz transmission bandwidth levels are readily observed. For both these examples the measurement receiver bandwidth is necessarily much less than the transmitter bandwidth* and the amplitude is presented in relative units of dBµV/m/kHz for a pre-scribed distance from the transmitter and receiver bandwidth.

2.3.1.2 Communications C-E Emitters

Communications equipments are the greatest in number and most varied of all C-E types. They occupy portions of the spectrum inter-laced between other activities from 20 kHz to about 1 GHz. Above 1 GHz, point-to-point communications is generally of the relay type.

Fig. 2.12 shows the spectrum signature of a VHF voice-communica-tion, land-mobile transmitter. The 25-kHz transmitter bandwidth is evident. The spectrum emission shows that the levels are below 100 dB down at 150 kHz on either side of carrier.

---

* See Chap. 2 of Vol. 5, "EMI Prediction and Analysis," regarding system of units for receiver bandwidth greater than and less than the transmitter bandwidth.

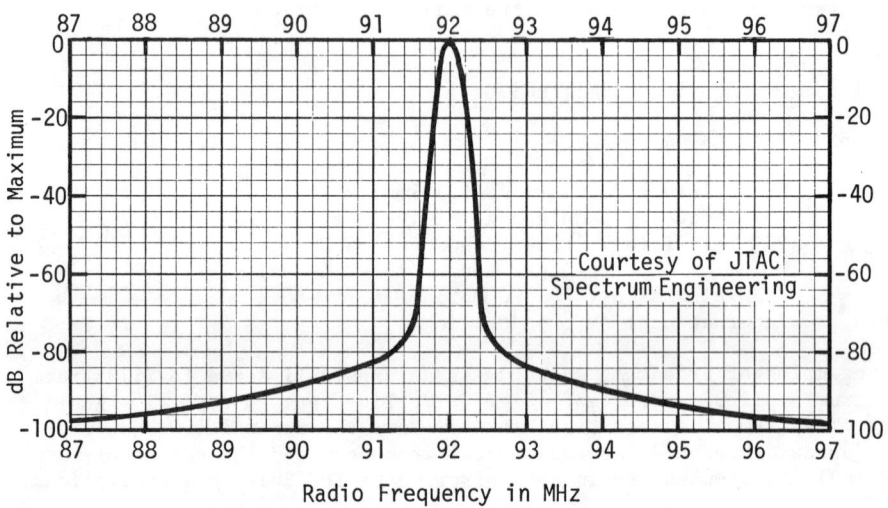

Figure 2.10 - Typical Spectrum Signature of an F-M Broadcast Transmitter

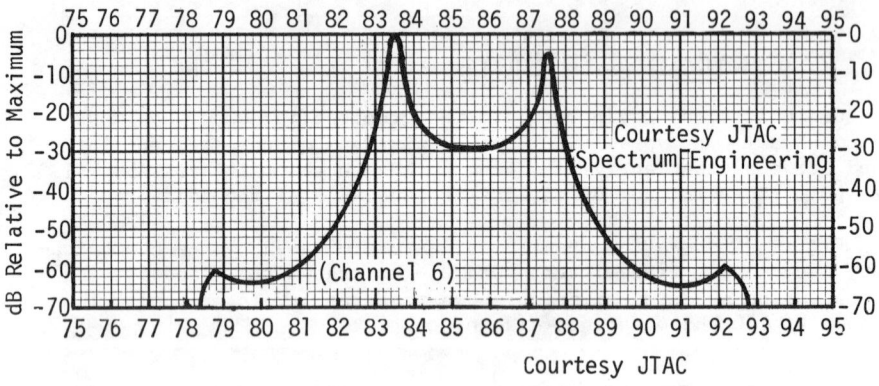

Courtesy JTAC

Radio Frequency in MHz

Figure 2.11 - Typical Spectrum Signature of a VHF T-V Broadcast Transmitter

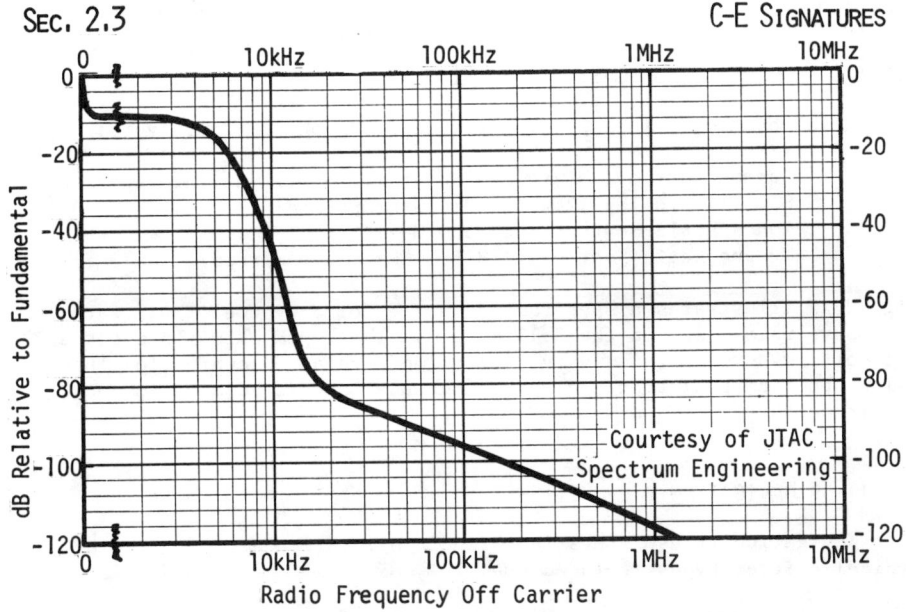

Figure 2.12 Typical Spectrum Signature of VHF Land-Mobile Transmitter

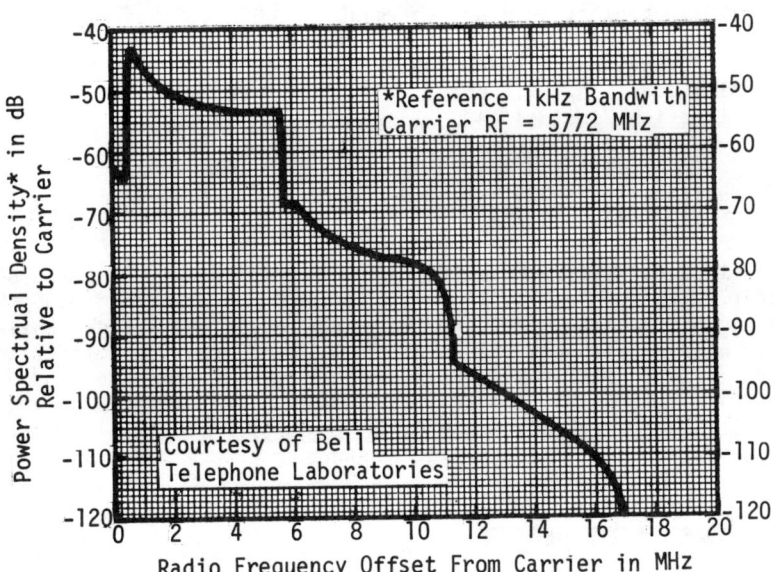

Figure 2.13 - Typical Spectrum Signature of a C-Band Microwave Relay Transmitter

2.3.1.3 RELAY COMMUNICATION C-E EMITTERS

Relay communications generally consist of one of four types:

> Common Carrier, Microwave Relay (2.1–11.7 GHz intersperced)
> Satellite Relay (2.4–16 GHz intersperced)
> Ionospheric Scatter (400–500 MHz)
> Tropospheric Scatter (1.8–5.6 GHz intersperced)

Fig. 2.13 illustrates a spectrum signature for a typical TM, 1 watt microwave relay used in the Bell System. The rapid emission spectrum fall-off is noted. This is typical of most relay communications having a low-spectrum pollution and operating under severe adjacent-channel requirements.

2.3.1.4 NAVIGATION C-E EMITTERS

Navigation transmitters in this classification exclude radars. Typical emitter types included are:

> VOR(VHF Omni Range): 108–118 MHz
> TACAN(Tactical Air Navigation)
> Marker Beacons: 74.6–75.4 MHz
> ILS(Instrument Landing System):
>   ILS Localizer: 108–118 MHz
>   Glide Path: 328.6–355.4 MHz
> Altimeter: 4.2–4.4 GHz
> Direction Finding: 405–415 kHz
> Loran C: 90–110 kHz
>   A: 1.8–2.0 MHz
> Maritime: 285–325 kHz; 2.9–3.1 GHz; 5.47–5.65 GHz
> Land: 1638–1708 kHz

Fig. 2.14 illustrates a spectrum signature for a typical, LF Loran C transmitter. Its 20-kHz, 99.5% power bandwidth centered about 100 kHz, the sideband emission fall-off, and harmonic levels are noted. While relatively high emission levels exist out-of-band, the transmitter is also operated in a relatively uncongested portion of the spectrum.

Fig. 2.15 shows the signature of a VHF doppler VOR transmitter. The 10 kHz variable-phase subcarrier components are readily observed.

2.3.1.5 RADAR C-E EMITTERS

Radars are perhaps the greatest offenders of EMI from C-E emitters because of their large peak pulse powers (up to about 5 MWatts) and attendent spectrum spread due to short pulses occupying broad basebands. They are also offensive because of their relatively high harmonic radiations.

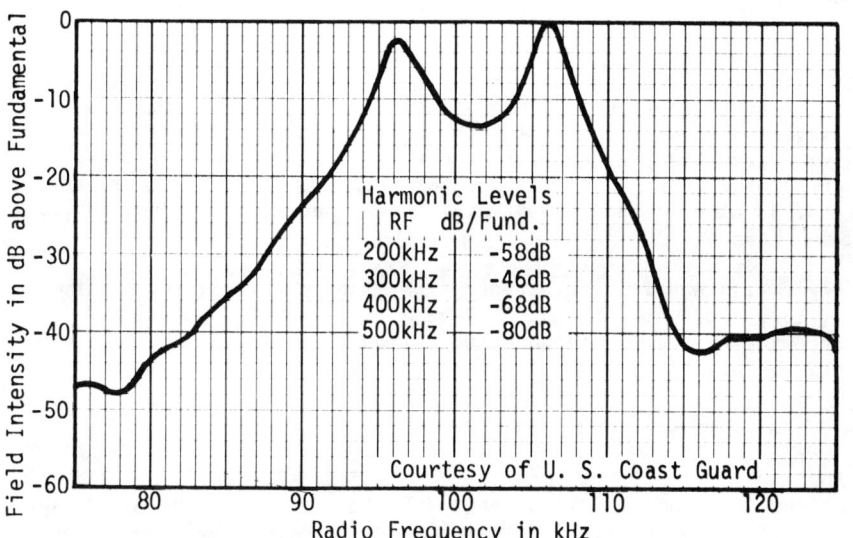

Figure 2.14 - Typical Spectrum Signature of a 100kHz
Loran-C Transmitter

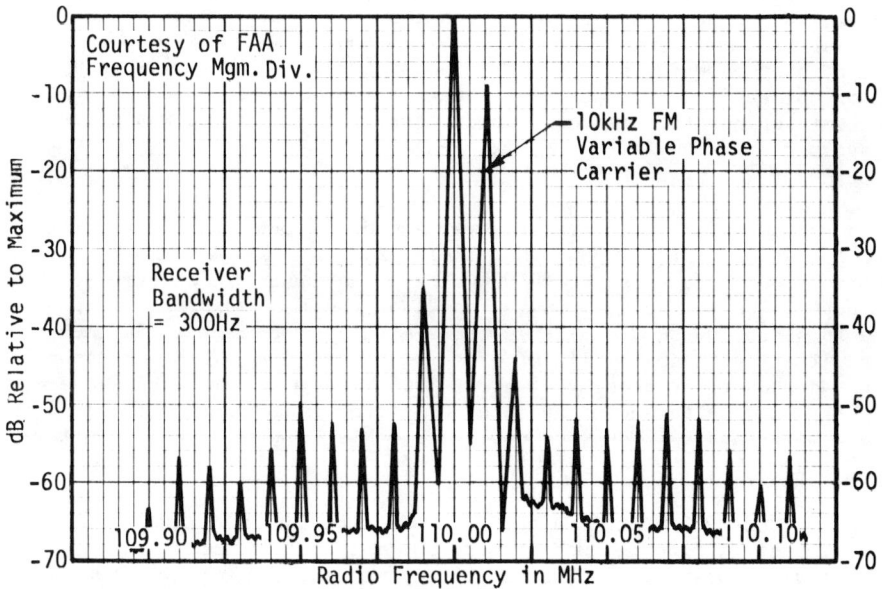

Figure 2.15 - Spectrum Signature of FAA, VHF Doppler VOR

Radars are used in intermittent portions of the spectrum from about 225 MHz to 35 GHz. They are employed in many capacities (the Department of Defense is the largest user of the more powerful radars) including: air traffic control, air and surface search, harbor surveillance, mapping, tracking and fire control, police-speed monitoring, and weather. There is believed to be upwards of 100,000 such radars used in the U. S. alone although most are of the innocuous, low-power police-speed monitors.

Figs. 2.16 and 2.17 illustrate typical spectrum signatures from P-band, long-range search and FAA S-band, air-traffic control radars. Note the wide emission sidebands and harmonic levels.

2.3.1.6 C-E POPULATION STATISTICS

Some measure of intentional communications-electronics emitter populations is beneficial in order to obtain a feel for the magnitude of EMI predictions required, especially in the more populated bands and areas. Only a few emitter types having power outputs above 100 watts are reviewed.

● Broadcast

As of 1972, there are about 18,000 broadcast stations in the country including AM and FM radio and commercial and educational TV. The radiated powers range from a low of 500 watts for some AM Stations to a high of 5 MW for some UHF TV transmitters. There exists an average of one such station per four square miles of urban/suburban area. This corresponds to a radiated power density of about 400 watts per octave per square mile.

● Land Mobile

As of 1972, there exists over 4,000,000 land-mobile transmitters operating in the 30-50 MHz, 150-160 MHz, and 450-460 MHz bands. Average power output is about 100 watts. Based on more than 50% of the population in the U.S. living on 2% of the land-mass area (U.S. area is 3,615,000 square miles), there exists an average of 50 such transmitters per square mile of urban/suburban area. This corresponds to a radiated power density of about 6500 watts per octave per square mile if all were turned on at once. In terms of average power per square meter and a 1% duty or use cycle, this equals -15 dBm/m^2/octave.

● Radar

As of 1972 there are about 3,000 fixed radar systems in the U.S. excluding airborne, mobile, waterborne, and classified military C-E installations. Typical peak-pulse power radiated ranges from 100 kW to 5MW. This corresponds to an average such radar per 24 square miles of urban/surburban area and a radiated power density of about 80,000 watts (ppp) per octave per square mile. This equals a peak pulse power of +16 dBm/m^2/octave or an average power of about

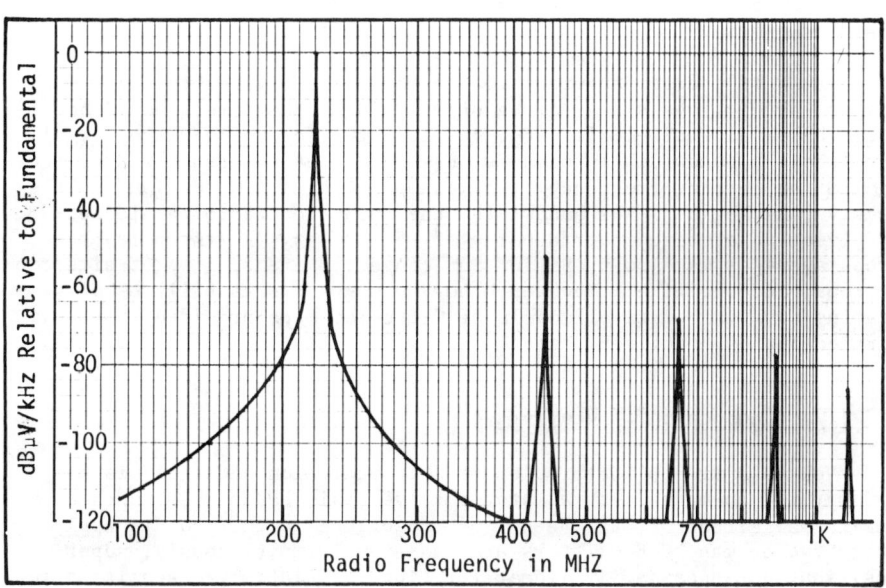

Figure 2.16 - Typical Spectrum Signatures of a P-Band (VHF) Long-Range Search Radar

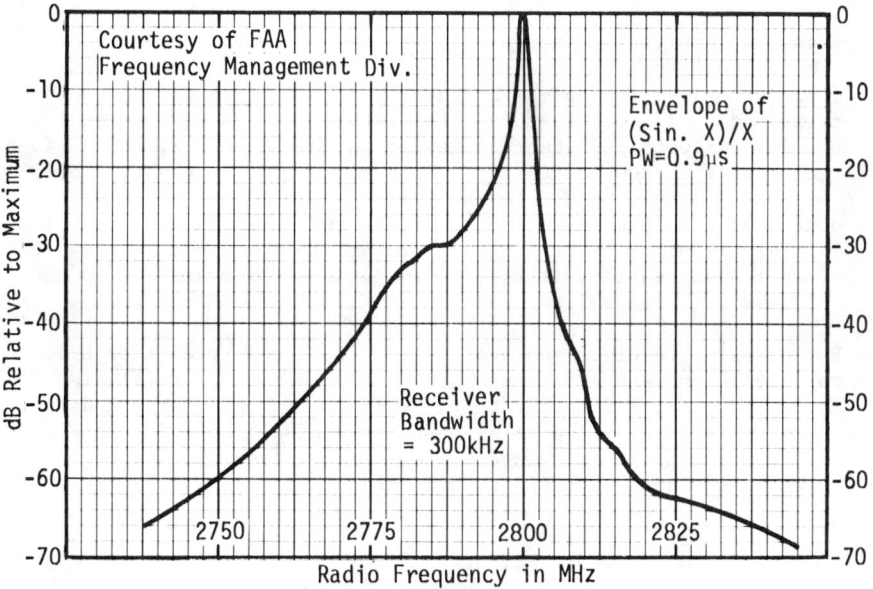

Figure 2.17 - Typical Spectrum Signature of an FAA ASR, S-Band Radar

2.25

-14 dBm/m^2/octave.

It is of course dangerous to use the above figures in performing any EMI predictions since, among other things, the levels would be reduced by a value of $-10 \log_{10} B_{RXO}* = 10 \log_{10}(f_{CO}/B_{RX})$, where B_{RXO} is the receiver bandwidth in octaves, B_{RX} is the receiver bandwidth, and f_{CO} is the carrier frequency. For example, for a 150 MHz land-mobile receiver having a 25-kHz bandwidth, the reduction factor would be -38 dB giving an average power density of -53 dBm/m^2. Furthermore, these gross averages make no provision for congested areas, line-of-site conditions, and the like. However, they do suggest possibilities of generating some mathematical models of C-E noise maps provided they were carried out on a refined basis.

2.3.1.7   COMPOSITE C-E RADIATIONS

Volume 5 treats the problem of EMI prediction and analysis on a discrete man-made, C-E radiator-receiver pair basis. For such a one-to-one situation, the techniques described in that volume apply. Where two or many C-E emitters are radiating simultaneously, superposition, not necessarily linear (cf. intermodulation, desensitization and cross-modulation), is used. However, a problem develops in that the temporal characteristics of a multi-emitter environment are not generally known.

When the receiver's vantage point takes in a substantial amount of earth real estate, many emitters are involved. For example, at ground elevations, a line-of-sight path becomes hard to define because of terrain and building structure irregularities. However, something of the order of 10 to 100 square miles is indicated in Table 2.3. To an aircraft at a 10,000 ft. altitude, the earth area increases about three orders of magnitude and to an orbiting satellite the area is five orders of magnitude greater. Thus, to sufficiently elevated platforms, literally millions of C-E emitters are in the field of view of a receiver. A somewhat similar situation exists for receivers in the field of view of a transmitter.

For elevated receiver vantage points, it is concluded that C-E electromagnetic ambient environments should be composed of (1) discrete emitters of particular interest to the EMI prediction problem and (2) statistical, multi-dimension noise maps for translation and immersion of the receiver's antenna. These noise maps are basically amplitude probability distributions (APD) as a function of:

- Location
- Platform height

---

* Multiplier is 20 for radar and impulsive sources. See Vol. 5.

Table 2.3 - Earth Area vs. Receiver Height Above MSL*

Receiver Height	Earth Area Field-of-View	Percentage of Total Earth Area
100 ft. (30m)	630 mi^2	0.0003%
1,000 ft. (300m)	6.3k mi^2	0.003%
10k ft. (3km)	63k mi^2	0.03%
10 mi (6.2km)	300k mi^2	0.15%
100 mi (62km)	2.2M mi^2	1.1%
1,000 mi (620km)	20M mi^2	10%
Synchronous height	83M mi^2	43%

* The earth has a total geographical area of about 195,000,000 sq. mi or 505,000,000 sq. km. To a ground-based transmitter, an elevated receiver is in the field-of-view for the ground areas or percentages of earth area listed in the table.

- Time of day, week, and season
- Look-time intervals
- Radio Frequency
- Bandwidth
- Beam width or sector angle
- Polarization

For the most part such C-E APD's are not available today. There exists, however, some long-range plans in the U.S. to develop noise maps of more limited scope including the FCC's Chicago Land-Mobile Program and subsequent follow-through mapping of the U.S.

While not in any sense a composite C-E population spectrum signature, Fig. 2.18 illustrates the maximum allowable spurious radiation from FCC licensed and other devices in terms of transmitter power output. For the most part, transmitter spurious radiation need only be 60 dB down from fundamental. For example, Channels 14-83 are limited to fundamental emissions of 5 MW (+97 dBm); Fig. 2.18 shows their spurious radiation levels 60 dB down or at 5 watts (+37 dBm). Since this is not defined as a power spectral intensity (dBm/kHz), the level could exist everywhere or as a broad-band jammer.

Illustrative Example 2.3

An aircraft TACAN receiver operating at 1090 MHz is 100 miles from its destination transmitter which emits 1 kW (+60 dBm) of power. Flying near Washington, D.C. at the moment, TV Channel 26 (542-548; 4.5 MW = 97 dBm) is located three miles away. Might an EMI problem exist?

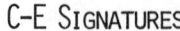

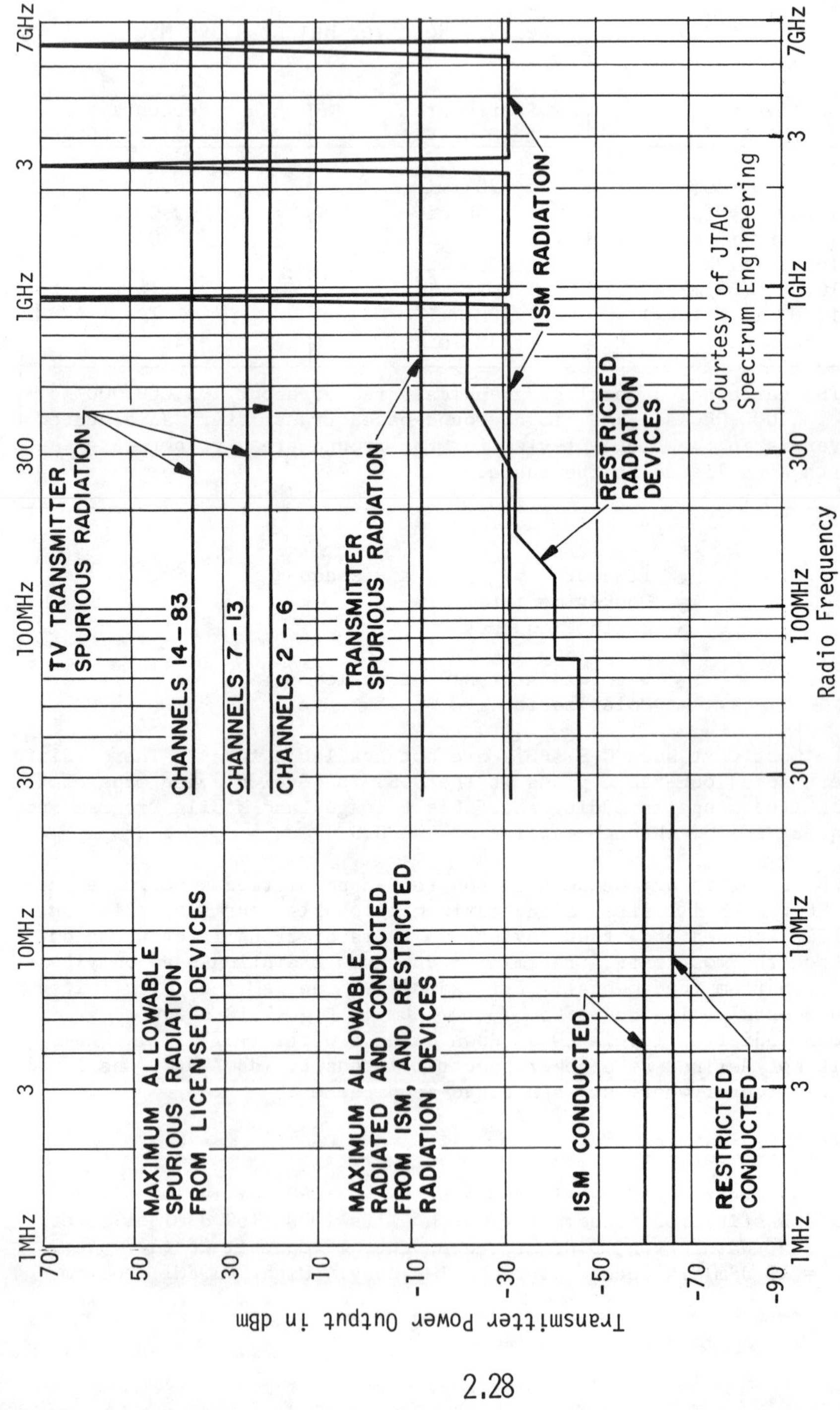
1MHz
3
10MHz
30
100MHz
300
1GHz
3
7GHz
70
50
30
10
-10
-30
-50
-70
-90
Transmitter Power Output in dBm
Radio Frequency
MAXIMUM ALLOWABLE SPURIOUS RADIATION FROM LICENSED DEVICES
TV TRANSMITTER SPURIOUS RADIATION
CHANNELS 14-83
CHANNELS 7-13
CHANNELS 2-6
MAXIMUM ALLOWABLE RADIATED AND CONDUCTED FROM ISM, AND RESTRICTED RADIATION DEVICES
TRANSMITTER SPURIOUS RADIATION
ISM CONDUCTED
RESTRICTED CONDUCTED
RESTRICTED RADIATION DEVICES
ISM RADIATION
Courtesy of JTAC Spectrum Engineering

Figure 2.18 - Maximum Allowable Spurious Radiation From FCC Licensed And Other Devices

The ratio of the TACAN transmitter power output to the allowable
TV Channel 26 second harmonic output (FCC regulations) is 60 dBm –
(97 dBm – 60 dB) = 23 dB. Since both transmitter antennas are omni-
directional, the power densities at equal distances to a receiver are
in the same 23 dB ratio. However, the stated distance ratio is 100
mi/3 mi = 15 dB or, for inverse square propagation loss, the ratio
squared is 30 dB. Thus, the arriving S/I ratio = $P_{dB} - D_{dB}^2$ = 23 dB –
30 dB = –7 dB. Hence EMI may exist since the TV Channel 26 receiver
second harmonic power input is greater than that from the TACAN transmitter. Frequency culling and detailed prediction (see Chaps 2 and 9)
will determine if EMI may really exist.

2.3.2   Incidental Electromagnetic Emitters[*]

This class of radio-noise sources are incidental in that they
are not C-E emitters. They are shown in Fig. 2.1 and include listings
2.1, and 2.3 through 2.5 in Table 2.1.

Unintentionally generated radio noise consists of radiated interference emitted by electrical and electromechanical equipment incidental to its intended operation. Such radiated noise may be either random in time and representable by a stochastic process or deterministic
and display a line spectrum. Within a metropolitan area many types of
unintentional noise generators exist. Among the many noise sources,
several dominant examples have been examined in sufficient detail to
permit formation of quantitative descriptions and parametric depen-
dences. These dominant types include automotive-ignition noise, power
transmission and distribution lines, fluorescent lamps, and RF stabilized arc welders.

Other urban incidental radio noise sources have received lesser
attention either because of their lower emission levels or their less
frequent encounter. Thus, quantitative information on the behavior of
this group with varying conditions is significantly incomplete. For
example, electric trains and busses, dielectric heaters, and plastic
and wood-gluing equipment are typical. Finally, in industrial areas
there exists millions of incidental noise sources for which quantitative information on the radiated output is either of little validity
or virtually nonexistent. Among these are electrical switching equipment. Brush-type electric drive motors, household appliances, and
electric-powered commercial and consumer vehicles.

Interest in determining radio noise produced by industrial, commercial, or consumer equipment results from responses by manufacturers

---

[*]   Substantial portions of this section were obtained from "A Review
of Incidental Radio Noise in Metropolitan Areas," by E.N. Skomal,
Aerospace Corporation.

or customers to pressures from regulatory agencies or from public ir-
ritation. Those noise sources which have been the least conspicuous
have largely escaped systematic examination.

Efforts directed towards achieving a general understanding of intensity and geographical extent of incidental radio noise levels in a
metropolitan area have consisted of area noise surveys performed in and
above several cities and towns of the world. Each survey has been appreciably restricted in scope. Only one has examined the variation of
metropolitan area noise over a period as long as one year. Most surveys have covered only a small portion of the radio spectrum. All
have been geographically restricted.*

Surface incidental radio noise surveys dominate the area noise
investigations which have been performed and reported. Aerial surveys
have only been systematically undertaken within the past few years.
The reported data represent a small body of information. Surface and
airborne area incidental-noise survey work has employed radio propagation observations to develop a description of the noise dissemination
mechanism and stochastic theory to describe spectral characteristics
of the area noise sources. Upon these foundations, analytical models
have been developed which accurately represent several aspects of the
range and frequency dependence of metropolitan area incidental radio
noise on and above the surface.

2.3.2.1 AUTOMOTIVE-IGNITION NOISE

Automotive-ignition noise is the composite of radiation from metropolitan area vehicular traffic. Individual vehicles radiate pulse
trains consiting of periodic, narrow pulses of one to five nanoseconds
duration. Peak-pulse amplitudes from single vehicles vary with auto
ignition type, vehicle speed, and mechanical loading during daily operation. Over longer periods of time, single vehicle radiations change
with aging and wear. Varying engine and drive-train designs, occurring
in the vehicle mix comprising urban traffic, add to the diversity of
radiated ignition-system wave forms which in total compose the impulsive noise of automotive traffic.

A selection of automotive traffic noise is shown in Fig. 2.19. It
is based on measurements from four investigators working on three continents. The ordinate of the data are presented in radiated electric-
field strength measured in units of dBμV/m/kHz. Measurements by
Crichlow and George were performed in Washington, D.C. and on Long
Island, New York respectively. These measurements represent automotive
traffic noise levels during the business day in the vicinity of major
roadways. Measurements by George characterize automotive interference

---

* FCC, Chicago Land-Mobile Project, loc. cit.

10kHz
100kHz
1MHz
10MHz
100MHz
1GHz
60
20
-20
-60
Field Intensity in dBμV/m/kHz
Peak of 21 cars
Average of Peaks
George & Ellis
WEI
Crichlow
30 Auto=/min
10 Autos/min
Suzuki
Radio Frequency
Emissions Measured in Vicinity of Roads(≈10meters)

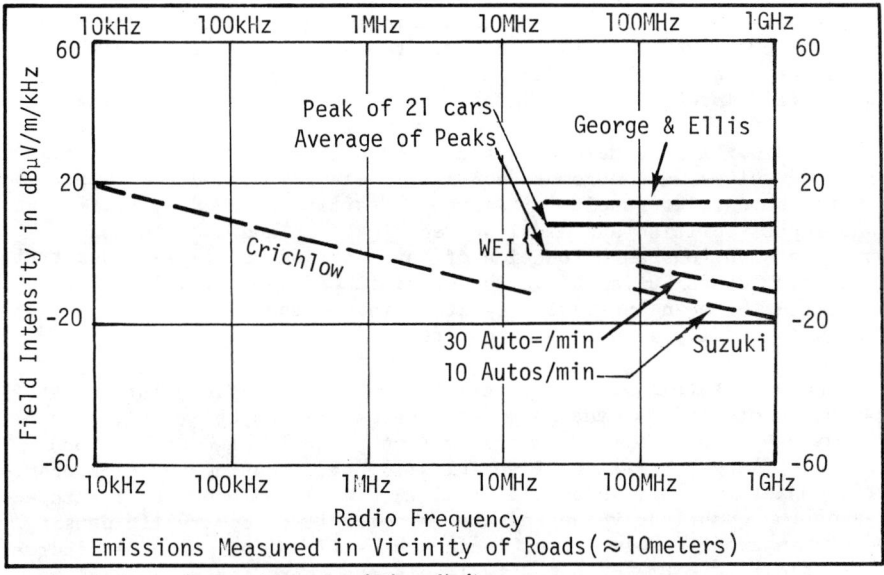

Figure 2.19 - Automotive Ignition Noise

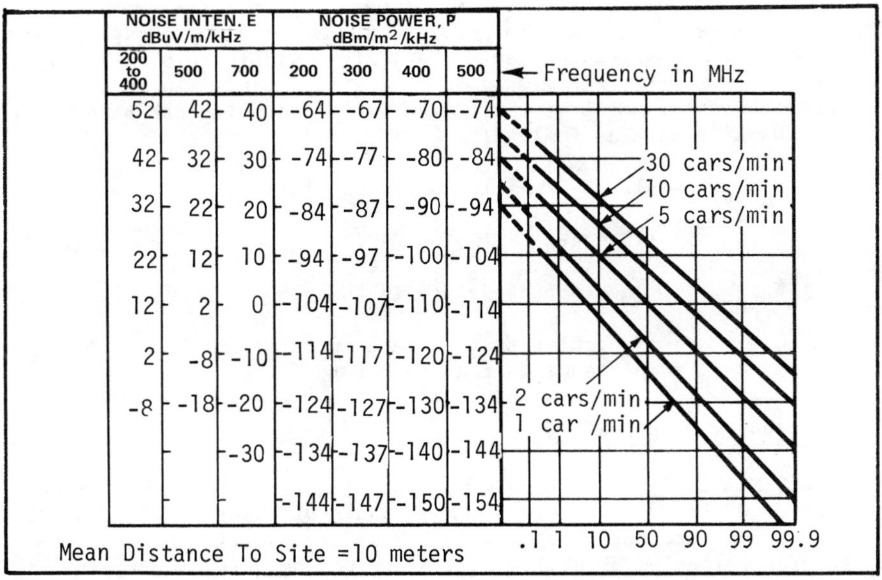

NOISE INTEN. E dBuV/m/kHz
NOISE POWER, P dBm/m2/kHz
200 to 400
500
700
200
300
400
500
Frequency in MHz
52 42 40 -64 -67 -70 -74
42 32 30 -74 -77 -80 -84
32 22 20 -84 -87 -90 -94
22 12 10 -94 -97 -100 -104
12 2 0 -104 -107 -110 -114
2 -8 -10 -114 -117 -120 -124
-8 -18 -20 -124 -127 -130 -134
-30 -134 -137 -140 -144
-144 -147 -150 -154
30 cars/min
10 cars/min
5 cars/min
2 cars/min
1 car /min
.1 1 10 50 90 99 99.9
Mean Distance To Site =10 meters

Figure 2.20 - Percent of Total Time Electric Field or Power Density Is Exceeded

from vehicles of pre-world War II manufacture. Ignition-noise suppression techniques used in the vehicles comprising the traffic observed by Crichlow in 1960 were advanced over those encountered by George. This difference in vehicle design at least partially accounts for the higher field intensities of radiated noise observed on Long Island.

Ellis working in Melbourne and other cities of eastern Australia observed business day automotive noise levels at VHF and UHF comparable to the earlier results of George. Traffic observed by Ellis were predominantly British and American vehicles. In addition to noise-power distributions as a function of daily period, Ellis recorded traffic density contributing to the observed interference levels. A strong correlation between automotive-ignition interference level and traffic density is readily seen in his results.

The most extensive statistical accumulation of field intensity of automotive origin was made in three Japanese cities by Suzuki. His data were accumulated from an observation point located at the road side. The large variation of vehicle flow rate passing Suzuki's observation point, which is shown in total in Fig. 2.20, provides a perspective on the quantitative dependence of noise level on traffic density. The median traffic noise level increases by 17 dB for a 10-fold increase in traffic flow. Field intensity levels are approximately normally distributed for all but the smallest exceedance values where data are insufficient to establish the form of the distribution function. Multiple ordinate scales of Fig. 2.20 are included to aid in the comparison of noise levels for several frequencies of measurement. The combined noise field-intensity scales for 200 and 400 MHz reflect a negligible variation of the observed median noise voltage within this interval.

A mathematical model of the median field intensity can be developed from Suzuki's data as follows:

$$E_{dB\mu V/m} \approx -11 \text{ dB}\mu V/m + 10 \log_{10} B_{RkHz} + 17 \log_{10} C$$

$$-20 \log_{10} (R/10m) \qquad\qquad (2.1)$$

where, B_{RkHz} = receiver bandwidth in kHz

C = vehicular traffic rate in cars/min.

R = distance to road in meters

Illustrative Example 2.4

A 450 MHz land-mobile receiver, operating at a dispatch station, is located 100 feet (30 meters) from a road. Off-hour traffic corresponds to a few cars per minute, while rush-hour traffic reaches sustained proportions of about 100 cars/min. During quiet hours, the dispatcher can barely hear communications from fringe-area cabs corresponding to received field intensities of about 10 dBμV/m (3μV/m).

Determine the effect of nearby auto traffic upon reception quality of
the 50-kHz bandwidth, dispatch-station receiver.

The median (50% probable) field intensities from the nearby road
traffic (5 cars/min) during off hours is obtained from Fig. 2.20
or Eq. (2.1).

$$E_{dB\mu V/m} = -11 \text{ dB}\mu V/m + 10 \log_{10} B_{RkHz} + 17 \log_{10} C - 20 \log_{10}(R/10m)$$

$$= -11 \text{ dB}\mu V/m + 10 \log_{10} 50 + 17 \log_{10} 5 - 20 \log_{10}(30/10m)$$

$$= -11 \text{ dB}\mu V/m + 17 \text{ dB} + 12 \text{ dB} - 10 \text{ dB} = 8 \text{ dB}\mu V/m$$

This approximates the normal fringe reception and may explain why performance is marginal or range is limited (Note that for 90% probable
service the received signal field intensity would have to be 10dB
higher).

During rush hour, a 100 car/minute traffic rate increases the
median automotive ignition noise by: $17 (\log_{10} 100 - \log_{10} 5) = 22$ dB.
Thus, degradation (median noise now = 30 dBμV/m) would either (1) service only 1% of the fringe area cabs, or (2) the median service range
would be reduced to 10% (22 dB space loss).

More recent ignition radiation measurement data on American automobiles raises the question of the credibility of some previously
measured and reported data. During 1970 White Electromagnetics
reports the results of a survey performed for the Automobile Manufac-
tures Association on various 1969-1970 model cars. All automobiles,
representing several manufacturers, were previously tested to comply
with SAE J55/a specification limit. Tests were performed from 20MHz
to 1 GHz, at a distance of 33 feet with the measuring antennas located
10 feet above the ground. The receiver impulse bandwidth was 16 kHz
below 30 MHz and 350 kHz above, both feeding a peak detector.

Figure 2.19 shows the general results corresponding to 21 individual vehicles and the same vehicles taken as a simultaneous operating
groups. The following general conclusions can be made from the WEI
tests when reference is made to their plotted results.

• Radiations below about 100 MHz show a greater tendency to be
  vertically polarized. The effect is more pronounced (0-20 dB
  variations with a 10 dB average) for single automobiles, than
  as a simultaneous groups.

• Peak emission levels from automobiles as a group (0 to 10 dB
  μV/m/kHz) shows no enhancement over that of the most offending
  within the group (5 to 10 dBμV/m/kHz).

- The average of peak emissions from automobiles as a simultaneous group (-10dBµV/m/kHz) is about equal to that of the
group taken individually.

- Above 1 GHz, radiation levels decrease at a rate of about
20 dB per octane.

It is significant to note that the total number of pulses coming
from all automobiles never exceeded the instrument bandwidth. Thus,
while both the average signal level and duty cycle would rise as more
automobiles are simultaneously operated, there may be no change in
peak emissions unless an offending automobile with high peaks is operated. All this raises the topic of the impact on a C-E receiver which
is discussed in Chapter 3.

2.3.2.2   POWER TRANSMISSION-LINE NOISE

Although automotive-ignition noise is generally the largest con-
tributor to incidental radio-noise level in industrialized areas above
20 MHz, other noise sources produce measurable contributions. High-
voltage generation equipment and transmission lines produce detectable
R-F noise which for transmission lines reaches maximum intensity during
conditions of rain, snow, fog, and high relative humidity. Conditions
of high air turbulences and solar radiation are also believed to stim-
ulate increased transmission-line noise in desert and arid regions. When
power transmission lines and components have deteriorated or have been
damaged, their impulsive-noise emissions above 50 MHz are observed to
rise appreciably. They are readily detectable in the absence of ignition noise.

The waveform of power transmission-line noise is random and impulsive, but of greater pulse width and less frequent occurrence than
commonly found for automotive-ignition noise. Bursts of interference
are observed, the typical duration of which is several milliseconds.
Fine structure of the burst pulses consists of short duration, fast
rise time, distorted square waves occurring often at high repetition
rates. Very wideband receivers have recorded power-line pulse durations and rise times of approximately 10 nsec and 2 nsec, respectively.
The origin of this was identified to be actuations of inductive loads
on power lines.

Shorter duration transients have been associated with corona
discharges on lines and support elements arising from line imperfections due to component aging and damage. Burst repetition periods
are comparable to burst durations. Resulting noise spectra have been
analyzed and found to extend from SLF to VHF.

Power lines are capable of supporting long distance, low attenuation propagation of H-F transients by virtue of their ability to function as either coaxial waveguides (the power flow being confined between the inner conductor and sheath) or as a single line above ground. In addition, transmission lines may reasonably enhance peaks of the noise spectra, the resonances arising from periodicities in the mechanical construction of the power line, such as directional changes in runs or support spacings of one-half wavelength separation for particular harmonic components.

A composite presentation of radiated interference from high-voltage transmission lines as a function of frequency, line voltage, and distance from the lines is shown in Fig. 2.21. The ordinate is electric-field strength presented in units of dBμV/m/kHz. Measurements of power-line noise are presented for varying line voltages and lateral distances of the observation point from the line. Line voltage values are identified as (4), (16), etc., in units of kV. Distances from the outer conductor of power lines are measured in feet and indicated by the notation <μ> beneath the conductor, viz., <25> ft, <50 ft>, etc.

Field-intensity data representing lateral distances do not clearly reveal the $1/d^2$ dependence of the field strength upon distance from the outer conductor that has been observed for the near-field case. Controlled experimental conditions existing on "test" transmission lines used for the confirmation of the $1/d^2$ dependence do not exist for measurements of Fig. 2.21, where all data were obtained on service lines of widely varying conditions of repair. "Test" transmission-line noise data were intentionally excluded from the summary as being unrepresentative of typical service-line noise encountered in industrial regions. All power-line data of Fig. 2.21 represent dry-weather conditions. Transmission-line noise levels may increase 10 to 20 dB if the relative humidity is high, or if fog, mist or rain is present.

2.3.2.3 ARC WELDERS, HEATERS, AND GLUERS

Several primary types of industrial material processing and assembly equipment are known to be major sources of incidental-radio noise. Among equipment for which there exist extensive radio-noise intensity data are R-F stabilized arc welders, plastic heaters, and wood-gluing equipment. Within this group only arc welders emit a spark spectrum. For the others radiation occurs as a line spectrum at harmonics of the radio-excitation frequency.

The most extensive compilation of R-F stabilized arc welder radiated interference was published by Garlan on more than one hundred operationally-installed units. These data provide sufficient detail to determine both the frequencies of dominant noise emission and the

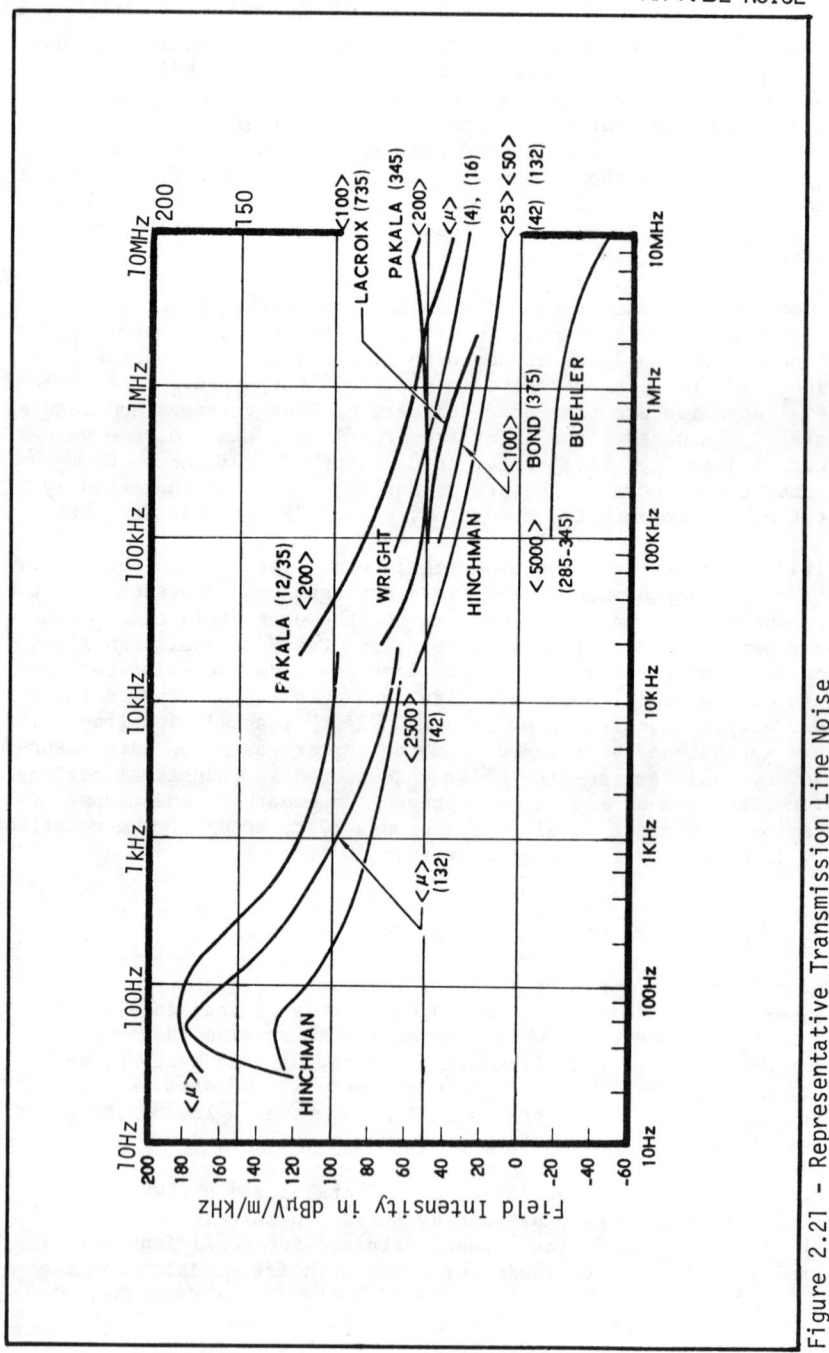

Figure 2.21 – Representative Transmission Line Noise

range dependence of interference in a typical industrial environment. The range variation of the dominant noise field emissions in the MF and HF bands is proportional to $d^{-1.5}$ for the separation distance of 1000 feet to one mile.

By compiling data supplied by Garlan on 72 extensively investigated welder units at a separation distance of 1000 feet, dominant radiation bands for the spark-emission spectrum are identified in Fig. 2.22 at 750 kHz, 3 MHz and 20 MHz. The field intensity is in broadband units of dBµV/m/kHz (left ordinal). Individual welders usually do not radiate equally in each frequency interval. Often radiation is concentrated in one of three bands.

Pearce has examined field intensities radiated by several types of industrial processing and fabricating equipment including plastic welders, plastic pre-heaters and wood gluers. Equipment tested was powered by R-F sources which radiated appreciable energy in the form of a line spectrum via coupling from leads connecting the power supplies to the working head or unit or by leakage from power supply enclosures.

Selected radiated fields from plastic welders are also presented in Figs. 2.22. Radiated interference from a plastic welder operating at 3 kW output is compared with radio-noise from an R-F stabilized arc welder. For both emissions the observer is separated by 1000 feet in an urban area. The right-ordinal of Fig. 2.22 (dBµV/m) is to be used with plastic welder data which, being a line spectrum, exhibit a field intensity that is independent of receiver bandwidth. The plastic welder emits a fundamental interference frequency at 35 MHz and most of the harmonics up to the limit of Pearce's measurements which was about 1 GHz. Spectral-line intensity from the 3 kW plastic welder is comparable or greater than interference from the arc welder.

Fig. 2.23 presents a similar plot for a 3-kW wood gluer and a 12-kW plastic pre-heater both compared to R-F stabilized arc welders at equal observer distances of 1000 feet in an urban area. Fundamental frequency of the wood-gluer noise is 10 MHz and for the plastic pre-heater it is 70 MHz. Most harmonics of each are generated and radiated with intensities comparable or exceeding the spectrum of R-F stabilized arc welders.

2.3.2.4 FLUORESCENT LAMPS

Several configurations of fluorescent lighting fixtures have been studied by Clark to determine the intensity of radiated noise. Reported data include unmodified fluorescent units in the form received from the manufacturer and units operated after special modifications had been

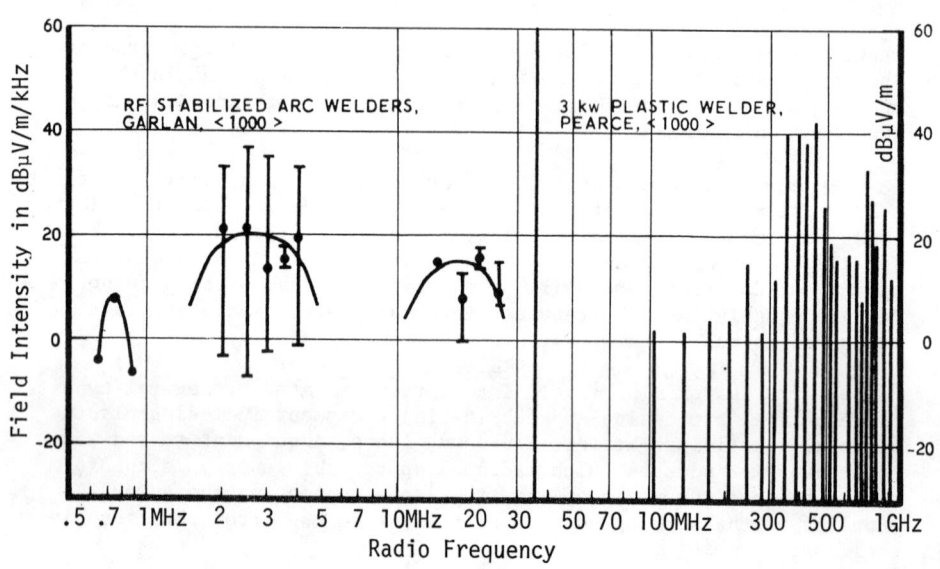

Figure 2.22 - R-F Stabilized Arc Welder And Plastic Welder Noise

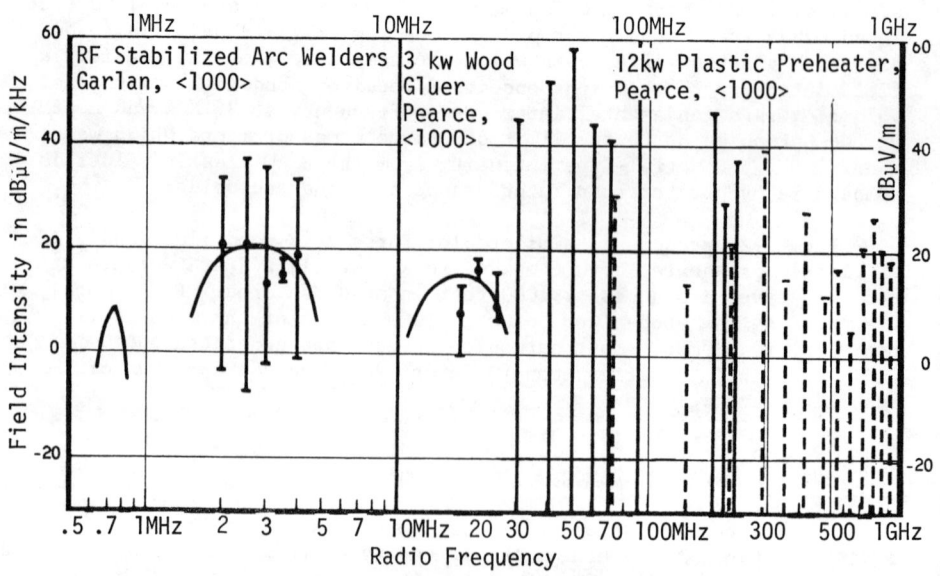

Figure 2.23 - R-F Stabilized Arc Welder, Wood-Heater, And Plastic Preheater

made to reduce their broad spectrum noise emissions. Included in the study were both hot and cold-cathode fluorescent light assemblies, the latter yielding much lower radiated noise in and above the HF band.

Maximum values of radiated noise in dBμV/m/kHz for two cold and one hot cathode, unmodified fluorescent assemblies are shown in Fig. 2.24. The legend indicates the length and number of fluorescent bulbs used in each unit. For each assembly studied the intercept antenna was located three feet from the installation. No measurements were reported at greater distances. The distinctive high-frequency radiation of the hot-cathode fluorescent unit is clearly evident in the VHF and UHF bands.

2.3.2.5 OTHER INCIDENTAL EMITTERS

Electric trains and busses are additional noise sources found in metropolitan areas. For electric trains, arc discharge noise is generated by sporadic contact between the electrified bus and the pantograph. A radiated-spark pulse width of a fraction of a microsecond is commonly found occurring with an average repetition frequency of one or fewer pulses per second. The noise emission spectrum is usually confined to below 30 MHz. However, there exist evidence that high-speed electric trains produce a radiated noise spectrum which extends into the VHF band.

The drive motors of both electric trains and busses can be significant noise sources as can on-board equipment. Both groups of noise sources radiate energy concentrated in the lower frequency bands. For electric trains and busses very little quantitative radiated noise have been reported. Amamiya has provided one of the few references in the field; yet it is of little general value since measuring instrument characteristics were unreported and the emphasis was placed on conducted interference.

D-C electric motors and some types of AC motors of both multiple and fractional horsepower are noise sources of importance. Heavy D-C drive motors used in older elevator installations are noisy. Some small appliance A-C motors produce intense broad spectrum noise radiations extending up to the UHF band.

2.3.3 OVERALL MATHEMATICAL MODELS OF INCIDENTAL EMITTERS

Composite surface, incidental radio-noise environment of a metropolitan area is produced by distributions of individual noise sources discussed in preceding sections. Composite metropolitan area noise has been the subject of several experimental investigations during the past twenty years. Motivation for experimental investigations

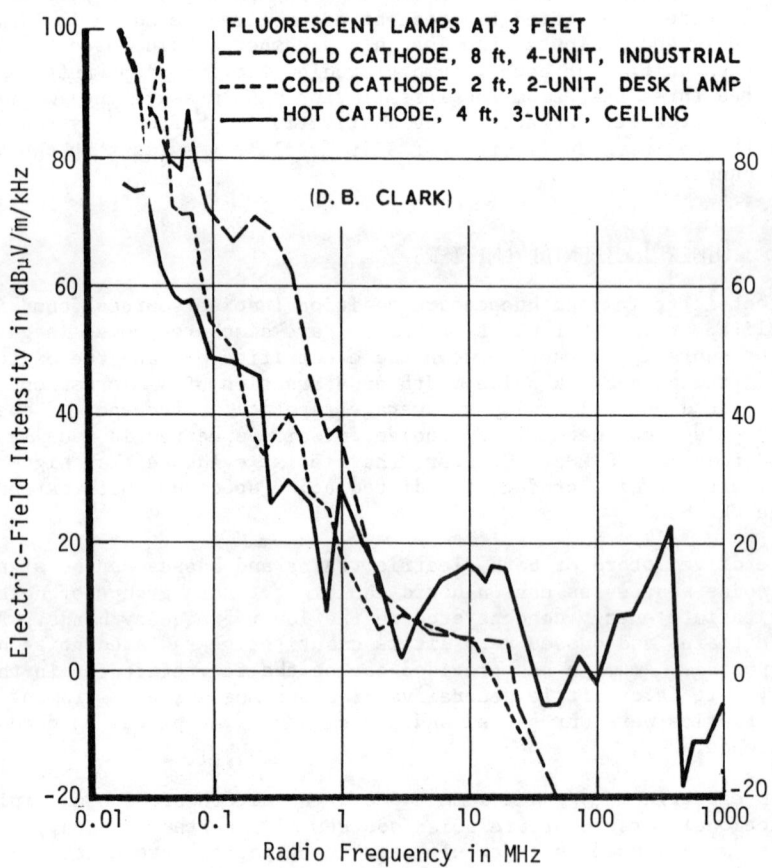

Figure 2.24 - Radiated Fluorescent Lamp Noise

and the more recently undertaken theoretical studies has been provided
by radio communication and information-handling system requirements.
This is primarily because performance of all urban radio telephone,
aircraft navigation, radio broadcast and television systems is adversely affected by excessive surface incidental radio-noise levels.

Wide geographic and short-term time variations in man-made radio
noise, apparent from limited observations, make it difficult to suggest
specific interference levels for different times and geographic locations. Although specific levels are unknown, it is possible to express
typical levels of unintended radiation. The convention has been
adopted of segregating the levels of unintended man-made radiation
into categories based on the level of urbanization in the area of measurement or by contour mapping of noise levels. It is recognized that
this categorization may ultimately be replaced by classifications
based on local industrial or commercial activity as more data becomes
available and improved data correlation methods are introduced. Effort
has been made in the collation of data to include only incidental
noise and to exclude signals that eminate coherently from C-E radiators.

Fig. 2.25 shows typical levels of unintended man-made radio noise
to be found at three arbitrarily chosen types of locations - urban
(typically 0-10 miles from city center), suburban (10 to 30 miles),
and rural (beyond 30 miles). Since sources contributing most to
general level of man-made radio noise apparently change in the neighborhood of 10 to 20 MHz, and since no measurements used to obtain
Fig. 2.25 were made between these frequencies, all values measured
at 10 MHz and below were considered separately from those at 20 MHz
and above*. For frequencies 20 MHz and above, curves were obtained
from data given by Young, Simpson, Hamer, Ellis, FCC ACLMRS, and
the results of some recent measurements by ITSA/ESSA. At frequencies
of 10 MHz and below, the curves were derived from data supplied by
ITSA/ESSA.

The division of areas, where measurements were taken, into urban,
suburban, and rural were made primarily on the basis of types of
locations such as the business areas of New York City, New York;

---

*   The reason for the disparity of data across the 10-20 MHz interface is not clear other than below 10MHz all data were measured with
a vertically polarized whip or rod antenna and most data above 20 MHz
(especially above 100 MHz) were measured using a tuned dipole, hori-
zonally polarized. Yet polarization, per se, does not explain the
disparity since measurement which were made with both polarizations of
the same antenna evidenced an average of only 2 dB difference. Accordingly, Fig. 2.25 suggested an interpolation model from about 1 to 200
MHz.

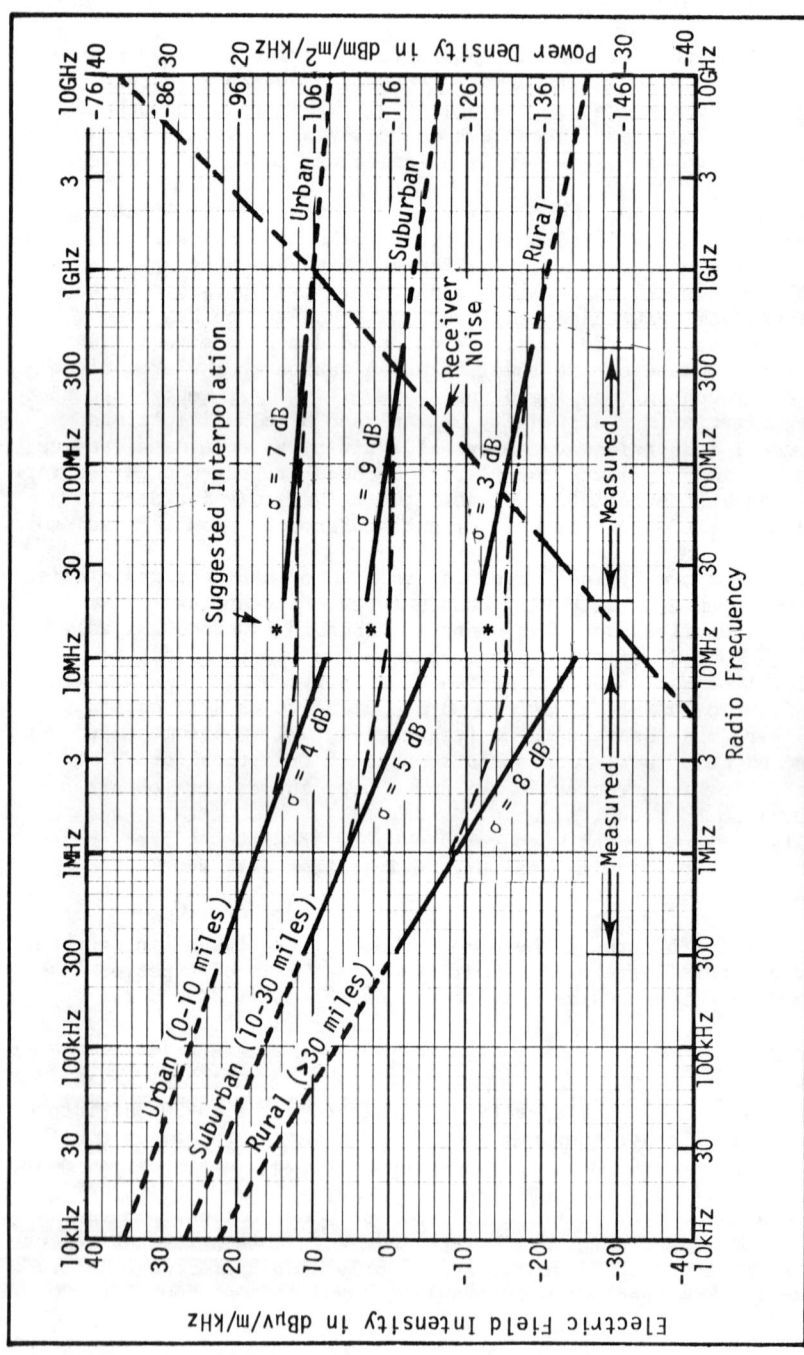

Figure 2.25 - Math Models of Median Incidental Man-Made Noise Based on Lossless Omni-directional Antenna Near Surface

Baltimore, Maryland; Washington, D.C.; Denver, Colorado; Melbourne, Australia; and Tel Aviv, Haifa, and Jerusalem, Israel.

The "suburban" curve was obtained from data taken at Boulder, Colorado; near Washington, D.C.; Melbourne, Tel Aviv, Haifa, and in suburban areas of England.

Those measurements which were considered rural were made at locations chosen to be as free as possible of man-made noise (the stations given in CCIR Report 322 and experimental radio facilities near Boulder, Colorado.

Since measurements at 20 MHz and above were made at only two or three frequencies at each location, frequencies used at other locations did not generally correspond, measurements at the upper end of the spectrum usually were not made near 20 MHz, and vice versa. In order to eliminate a frequency-location bias, the best fit for each set of observations was taken, and equal frequency points were then used from these curves to obtain a best fit to all data.

To indicate the variability of the average noise power, the RMS deviation of the median values used to obtain each segment of the curves is indicated as a σ in units of dB. Note that this deviation is found from the average power values for each location and frequency and is not the instantaneous (peak) variation which is considerably larger than the σ values shown.

In estimating noise level at the receiver due to external sources, the gain, polarization, and orientation of the receiving antenna should be considered. Fig. 2.25 shows the median operating field intensities and equivalent power densities from various noise sources. These values are the levels expected from an omni-directional, short, loss-less vertical antenna near the surface of the earth. For convenient reference, typical receiver noise is shown in the figure based on a rising 10 dB noise figure from 1 MHz to 15 dB at 1GHz to 20 dB at 10 GHz. Many government and commerical receivers would be better than this and many consumer receivers would be poorer.

Illustrative Example 2.5

TV reception in fringe areas is always a problem for millions of viewers. Typical modern VHF log-periodic TV antennas exhibit gains of about 12 dB. To Channel 9 (186-192 MHz), for example, this antenna exhibits an effective area of 2.8 m^2, and the receiver sensitivity (S-N) for a 6 MHz bandwidth is about -93 dBm. This corresponds to an arriving power density of -93 dBm - 5 dB meter^2 = -98 dBm/m^2 per 6 MHz bandwidth or -132 dBm/m^2/kHz. Determine the effect of incidental man-made noise on reception.

Fig. 2.25 indicates that man-made noise in a typical suburban/
rural area at 200 MHz may average between -117 and -133 dBm/m^2/kHz.
If the emphasis is on the rural side, the I/N ratio is -133 dBm/m^2/kHz
-(-132 dBm/m^2/kHz) = -1 dBm. On the other hand, if emphasis is more
on suburban areas, typical I/N ratios would be more like -117 dBm/m^2/kHz
-(-132 dBm/m^2/kHz) or 15 dB - thereby compromising reception. Note
that if the TV transmitter were in the center of the city, the signal
level would be reduced by 6 dB in doubling the location distance from,
say, 15 to 30 miles. Since the man-made noise would be reduced by 15
dB, and the signal by 6 dB, the S/I ratio actual improves in the rural
area. This explains why reception is sometimes better in rural areas
than in the closer-in suburban areas, especially those located near
highways or industrial centers.

2.4 COMPONENT SOURCES OF EMI

This section summarizes other sources of EMI which involve components that may conduct and/or radiate electromagnetic energy. Details are not presented here since they appear in later chapters and references are given throughout the text. Component emitters are sources of EMI which emanate from a single element rather than an ensemble of components such as equipments or devices discussed in the previous section. Examples of component noise sources includes wires and cables, connectors, motors and other rotating devices, switches, relays and solenoids, vacuum tubes, transistors, diodes, gas lamps, and the like units. Actually, this distinction is somewhat academic since these components require energy and connecting wires from other sources to perform. Thus, they are not true sources of EMI, but are *EMI transducers* in that they convert electrical energy to electrical noise.

2.4.1 CABLES AND CONNECTORS

Wires and cables, while not generating EMI by themselves, provide an induction or radiating physics media to couple undesired energy into or out of other wires, circuits, or equipments. Magnetic and/or electric field coupling and EMI control techniques are discussed in Chapter 6.

While not a direct source of electrical noise, connectors may develop EMI indirectly due to poor contacts. In effect this acts like a variable-impedance switch which may be environmental sensitive, such as to shock or vibration. The result is to impedance modulate a current or voltage source which can then emit EMI. VSWR, inadequate circumferential shielding (poor back shells), and/or contact potential are other attributes of connectors contributing to interference. Chap. 7 and Sec. 16.3 discusses this subject in detail along with EMI-control techniques.

2.4.2 MOTORS AND GENERATORS

Motors and generators, which use brushes and commutators in order to perform, are inherent sources of broadband-transient EMI noise. Transients develop as a result of an arc discharge upon separation of the rotating brush-commutator interface. An arc is generated because of the rapidly collapsing magnetic field ($e = -d\phi/dt$) in the motor or generator armature windings. The coupling mode of EMI may be either conducted or radiated and significant transients may exist up to 100 MHz or higher. Interference due to magnetic induction may be significant below 100 kHz. Sec. 16.7.4 discusses these components in detail including techniques for EMI suppression.

2.4.3   RELAYS AND SOLENOIDS

Electromagnetic relays and solenoids are also capable of producing EMI in sensitive equipment up to 300 MHz or higher. Upon de-
energizing a relay, viz., opening a switch to the source of relay
power, the stored magnetic energy develops a reverse voltage typically
10-20 times the supply voltage. Thus, arcing will develop at the
switch contacts which may conduct and/or radiate broadband transient
EMI. Sec. 16.7.3 discusses these components in detail including
techniques for EMI suppression.

2.4.4   ACTIVE DEVICES

Vacuum tubes, diodes, transistors, and gas lamps can also generate
EMI. Vacuum tubes generate Johnson (thermal) and shot noise, may
develop microphonics at low frequency, or produce parasitic oscillations at high frequency. Diodes may produce transients as a result
of reverse recovery periods which develop transient spikes from A-C
supplied sources. Transistors generate thermal, shot, and flicker
noise which limits their sensitivity for low-level amplification;
parasitic oscillation may also take place at high frequencies. Secs.
16.5 and 16.6 review these active component sources of EMI in detail.

2.4.5   ENERGY AND SIGNAL SOURCES

Virtually any device consuming power or generating electromagnetic
energy or control signals represents a potential EMI culprit. Generally, the higher energy level sources constitute the greater EMI
threat. Typical examples include A-C power mains sources, modulators,
servo-control circuits, and base-band digital data transmitters. These
and many other topics and their associated EMI-control techniques are
discussed in other chapters throughout this handbook (see Index).

2.5 BIBLIOGRAPHY

(1) ACLMRS, Technical Committee, "Man-Made Noise," Working Group 3 report, June 1966.

(2) Akima, H., "Speeding-up of HF Radio-Transmissions and Broadcasts," (World Weather Watch Study T.21), ESSA Tech. Memo. IERTM-ITSA 42, March 1967.

(3) Akima, H. (1972), "A Method of Numerical Representation for the Amplitude-Probability Distribution of Atmospheric Radio Noise," Report OT/TRER 27, Office of Telecommunications (U.S. Department of Commerce, Washington, D.C.).

(4) Amamiya, Y., "A Research on Radio Noise in Medium and High Frequency Regions Generated by Electric Cars," *Bull. I.R.C.A.*, pp. 192-203, April 1962.

(5) Arnold, J.G., "Predicting Spurious Transmitter Signals," *Electronics*, Vol. 34, No. 16, p. 68, Apr. 21, 1961.

(6) Barghausen, A.F., J.W. Finney, L.L. Proctor and L.D. Schultz (1969), "Predicting Long-Term Operational Parameters of High-Frequency Sky-Wave Telecommunications Systems," Report ERL 110-ITS 78, ESSA (U.S. Department of Commerce, Washington, D.C.).

(7) Barsis, A.P., and Miles, M.J., "Cumulative Distributions of VHF Field Strength Over Irregular Terrain Using Low Antenna Heights," NBS Report 8891, October 1965.

(8) Barsis, A.P., K.A. Norton, P.L. Rice and P.H. Elder (1961), "Performance Predictions for Single Tropospheric Communication Links and for Several Links in Tandem," Technical Note 102, NBS (U.S. Department of Commerce, Washington, D.C.).

(9) Bean, B.R., and Dutton, E.J., "Radio Meteorology," NBS Monograph 92, 1966.

(10) Beckmann, P., "Amplitude-Probability Distribution of Atmospheric Radio Noise," *Radio Sci. J. Res. NBS,* Vol. 68D, No. 6, pp. 723-736, June 1964.

(11) Bello, P.A., "Error Probabilities Due to Atmospheric Noise and Flat Fading in HF Ionospheric Communications Systems," *IEEE Trans. Communications Technology,* Vol. 13, No. 3, pp. 266-279, September 1965.

(12) Bernard, C.R.W., "VHF Noise Levels Over Large Towns," Royal Aircraft Establishment, Tech. Rpt. 67213, August 1967.

(13) Boischot, A., "Bruits Cosmiques en Micro-ondes," *L. 'Onde Electrique,* pp. 252-257, February 1967.

(14) Bond, C.R., Pakala, W.E., Graham, R.E., and O'Neil, J.E.,

"Experimental Comparisons of Radio Influence Fields from Short and Long Transmission Lines," *IEEE Trans. on Power Apparatus and Systems*, pp. 175-185, Vol. 82, April 1963.

(15) Bowen, B.A., "Some Analytical Techniques for a Class of Non-Gaussian Processes," Queens University Research Report No. 63-3, June 1963.

(16) Buehler, W.E., King, C.H., Lunden, C.D., "VHF City Noise," *1968 IEEE Electromagnetic Compatibility Symp. Rec.,* pp. 113-118.

(17) Buehler, W.E., and Lunden, C.D., "Signatures of Man-Made High-Frequency Radio Noise," *IEEE Trans. Electromagnetic Compatibility,* Vol. EMC-8, pp. 143-152, No. 8, September 1966.

(18) Burgess, R.E. (1941), "Noise in Receiving Aerial Systems," Proc. Phys. Soc., 53, 293-304.

(19) C.C.I.R. Report 322, "World Distribution and Characteristics of Atmospheric Radio Noise," Geneva, 1964.

(20) C.C.I.R.,"Bandwidths and Signal-to-Noise Ratios in Complete Systems," Rec. 339-1, Documents of the XIth Plenary Assembly, Oslo, 1966, Vol. III, 28-30.

(21) C.C.I.R. (1967), "Operating-Noise Threshold of a Radio Receiving System," CCIR Report 413 (International Telecommunication Union, Geneva, Switzerland).

(22) C.C.I.R. (1957), "Revision of Atmospheric-Radio Noise Data," CCIR Report 65 (International Telecommunication Union, Geneva, Switzerland)

(23) C.C.I.R. (1970a), "Revision of Atmospheric Radio Noise Data," Resolution 8-2, Documents of the XIIth Plenary Assembly, New Delhi 1970, Vol. II, Part 2, Ionospheric Propagation (International Telecommunication Union, Geneva, Switzerland).

(24) C.C.I.R. (1970b), "Use of Atmospheric Radio-Noise Data," Recommendation 372, Documents of the XIIth Plenary Assembly, New Delhi 1970, Vol II, Part 2, Ionospheric Propagation (International Telecommunication Union, Geneva, Switzerland).

(25) Clack, C., Bradley, P.A., and Mortimer, D.C., "Characteristics of Atmospheric Radio Noise Observed at Singapore," *Proceedings IEEE,* Vol. 112, No. 5, pp. 849-860, May 1965.

(26) Clark, D.B., "Evaluation of Interference Suppression of Florescent Lamps," U.S.Naval Civil Engineering Lab., Tech. Rpt. 166, Oct. 1961.

(27) Cochrane, R.C. (1966), "Measures for Progress: A History of the National Bureau of Standards," Misc. Publ. 275, NBS (U.S. Department of Commerce, Washington, D.C.)

(28) Conda, A.M., "The Effect of Atmospheric Noise on the Probability of Error for an NCFSK System," *IEEE Trans. Communications Technology,* Vol. 13, No. 3, pp. 280-283, September 1965.

(29) Cottany, H.V., Johler, J.R., "Cosmic Radio Noise Intensities in the VHD Band," *Proc. IRE,* pp. 1053-1060, September 1952.

(30) Crichlow, W.Q., and Disney, R.T., "Predictions of Radio and Systems Application," Lecture No. 31, NBS Course in Radio Propagation, 1961.

(31) Crichlow, W.Q., and Disney, R.T., "Man-made Radio Noise Measurement at the NBS Gaithersburg Site, and NBS, Washington, D.C.," Memorandum Report, Boulder Laboratories, NBS, February 1961.

(32) Crichlow, W.Q., Robique, C.J., Spaulding, A.D., and Berry, W.M. "Determination of the Amplitude-Probability Distribution of Atmospheric Radio Noise from Statistical Moments," *J. Res. NBS,* Vol. 64D, No. 1, pp. 49-56, January-February 1960.

(33) Crichlow, W.Q., Smith, D.F., Morton, R.N., and Corliss, W.R., "World-wide Radio Noise Levels Expected in the Frequency Band 10 Kilocycles to 100 Megacycles," NBS Circular 557A, August 1955.

(34) Crichlow, W.Q., Spaulding, A.D., Robique, D.J., and Disney, R.T., "Amplitude-Probability Distribution for Atmospheric Radio Noise," NBS Monograph 23, November 1960.

(35) Crichlow, W.Q. (1948), "The Effects of Antenna Circuit Loss and External Noise on Radio Reception in the Frequency Band from 50 to 5,000 Kilocycles," NBS Report CRPL-4-4, NBS (U.S. Department of Commerce, Washington, D.C.).

(36) Crichlow, W.Q. and R.T. Disney (1964) "Terrestrial Radio Noise," Radio Science, NBS/USNC-URSI, 68D, 5.

(37) Diamessis, J.E., "Investigation of Corona Noise in a Three-Phase Transmission Line," *Proc. Fifth Conference on Radio Interference Reduction and Electronic Compatibility,* Chicago, Ill.: Armour Research Foundation, Oct. 1959.

(38) Diamond, H., K.A. Norton and E.G. Lapham (1938), "On the Accuracy of Radio Field-Intensity Measurements at Broadcasting Frequencies," *NBS Journ. of Res.,* Vol. 21, pp. 795-818.

(39) Disney, R.T. (1972), "Estimates of Man-Made Radio Noise Levels Based on the Office of Telecommunications," ITS Data Base, Proc. IEEE International Conference on Communications, 72 CHO 622-1-COM, 20-13 to 20-19.

(40) Disney, R.T. and A.D. Spaulding (1970), "Amplitude and Time Statistics of Atmospheric and Man-Made Radio Noise," Report ERL 150-ITS

98, ESSA (U.S. Department of Commerce, Washington, D.C.).

(41) Eaglesfield, C.C., "Motorcar Interference," *Wireless Engineer,* Vol. 23, No. 277, pp. 265-272, Oct. 1946.

(42) Eaglesfield, C.C., "Car Ignition Radiation," *Wireless Engineer,* Vol. 28, No. 382, pp. 17-22, Jan. 1951.

(43) Ellis, A.G., "Site Noise and Its Correlation with Vehicular Traffic Density," *Proc. IRE* (Australia), pp. 45-52, January 1963.

(44) Final Report to Space and Missile Systems Organization, Communications Task-Project 672A, prepared by ESSA under contract AF D/0(04-694)-67-3, October 1967.

(45) Friis, H.T. (1944), "Noise Figures of Radio Receivers," Proc. IRE, 32, 7, 419-422.

(46) Furutsu, K., and T. Ishida, "On the Theory of Amplitude Distribution of Impulsive Random Noise and its Application to the Atmospheric Noise," *Journal of the Radio Research Laboratories of Japan,* Vol. 7, No. 32, pp. 279-307, July 1960.

(47) Galejs, J., "Amplitude Distributions of Radio Noise at ELF and VLF," *J. Geophys. Res.,* 71, pp. 201-216, January 1966.

(48) Garlan, H., and Whipple, G.L., "Field Measurements of Electromagnetic Energy Radiated by RF Stabilized Arc Welders," FCC Tech. Div. Rpt. T-6401, February 1964.

(49) Gauper, H.A., "Final Report, Radio Influence Noise Test on PASNY Niagara-Adirondack 345 kV Tie-line," General Electric Co., General Engineering Lab., Feb. 1962.

(50) George, R.W., "Field Strength of Motor Car Ignition Between 40 and 450 Megacycles," *Proc. IRE,* Vol. 28, pp. 409-413, September 1940.

(51) Greene, F.M. (1951), "Calibration of Commercial Radio Field-Strength Meters at the NBS, NBS Circ. 517.

(52) Haber, F. and J.E. Diamessis, "Corona Noise Models Based on Modulated Gaussian Noise," *Proc. Sixth Conference on Radio Interference Reduction and Electronic Compatibility,* Chicago, Ill.: Armour Research Foundation, Oct. 1960.

(53) Hall, H.M., "A New Model for 'Impulsive' Phenomena: Applicatio to Atmospheric-Noise Communication Channels," Stanford Electronic Laboratories, Tech. Rpt. 3412-8 and 7050-7, August 1966.

(54) Halton, J.H., and Spaulding, A.D., "Error Rates in Differential

Coherent Phase Systems in Non-Gaussian Noise," *IEEE Trans. Communications Technology,* Vol. COM-14, No. 5, pp. 594-601, October 1966.

(55) Hamer, E.G., "Noise Performance of VHF Receivers," *Electronics Engineering,* pp. 68-70, February 1953.

(56) Heisler, K.G., and H.J. Hewitt, "Interference Notebook," Rome Air Development Center Report No. RADC-TR-66-1, June 1966.

(57) Herbays, Ing. E.H. (1962), "The Measurement of Characteristics of Terrestrial Radio Noise," URSI Special Report 7 (Elsevier Publ. Co., Amsterdam).

(58) Hinchman Corp. Engineers, Detroit, Mich., "Report and Criteria of Methods of Measurement and Recommended Allowable Limits of Electromagnetic Interference Voltages from High Voltage Transmission Lines," October 1957.

(59) Hogg, D.C., "Effective Antenna Temperatures Due to Oxygen and Water Vapor in the Atmosphere," *J. Appl. Phys.,* pp. 1417-1419, Sept. 1959.

(60) Hogg, D.C., and W.W. Mumford, "Effective Noise Temperature of the Sky," *The Microwave Journal,* 3, Issue 3, pp. 80-84, March 1960.

(61) Horner, F., and J. Harwood, "An Investigation of Atmospheric Noise at Very Low Frequencies," Proc. IEEE, Vol. 103D, pp. 743-751, 1956.

(62) Ibukun, O., "Structural Aspects of Atmospheric Radio Noise in the Tropics," Proc. of IEEE, 54, pp. 361-67, March 1966.

(63) "Interference From Fluorescent Tubes," *Wireless World,* Vol. 56, No. 3, pp. 90-93, March 1950.

(64) IRE (1942), "Standards on Radio Wave Propagation Measuring Methods," Supplement to Proc. IRE, Vol. 30, pp. 7, Pt. II.

(65) IRE (June 1960), "Report of the Television Allocations Study Organization," Proc. IRE, 48, No. 6, Part I. Fredendall, G.L., and W.L. Behrend, Picture Quality-Procedures for Evaluating Subjective Effects of Interference, 1030-1034; Dean C.E., Measurements of the Subjective Effects of Interference in Television Reception, 1035-1050.

(66) IRPL (1943), Radio Propagation Handbook, NBS (U.S. Department of Commerce, Washington, D.C.).

(67) JTAC (1968), "Unintended Radiation," Suppl. 9 to Spectrum Engineering - The Key to Progress, IEEE.

(68) Karr, B.M., Private Communication, A Detailed Documentation of the Pertinent Data is Available in E.N. Skomal, "Distribution and Frequency

Dependence of Unintentionally Generated Man-made VHF/UHF Noise in Metro-politan Areas," Aerospace Corp. Tech. Rpt. TDR 469 (S5805-45)-1, 1965 (Available from DDC by AD 452 953).

(69) Kaya, P.A., "Noise Generated by Fluorescent Lamps," master's thesis, Moore School of Electrical Engineering, Univ. of Pennsylvania, August 22, 1951.

(70) Kraue, J.D., "Radio Astronomy," McGraw Hill Book Co., 1966.

(71) Krstansky, J.J., and R. F. Elsner, "Environment-generated Intermodulation in Communication Complexes," *Proc. Tenth Tri-Service Conference on Electromagnetic Compatibility*, Chicago, Ill.: ITT Research Institute, pp. 77-79, Nov. 1964.

(72) Lacroiz, R., Charbonneau, H., "Radio Interference from the First 735 kV Line of Hydro-Quebec," *IEEE Trans. on Power Apparatus and Systems*, pp. 932-939, April 1968.

(73) Lerner, R.M., "Design of Signals," Lectures on Communication System Theory, E. J. Baghdady, ed., McGraw-Hill Book Co., Inc., New York, pp. 213-277, 1961.

(74) Lindenlaub, J.C., and K.A. Chen, "Performance of Matched Filter Receivers in Non-Gaussian Noise Environments," IEEE Trans. on Communications Technology, Vol. 13, No. 4, pp. 545-547, December 1965.

(75) Linfield, R.F., "High Performance Reliable VLF Component of the Naval Advanced Communication System - VLF System Design, Specification and Evaluation Process:" DECO Electronics, Inc., Final Report No. 34-F, pre-pared for U.S. Navy Bureau of Ships.

(76) Lucas, D.L., and J.D. Harper (1965), "A Numerical Representation of CCIR Report 322 High Frequency (3-30 Mc/s) Atmospheric Radio Noise Data," NBS Technical Note 318 (U.S. Department of Commerce, Washington, D.C.).

(77) Lucas, D.L., and G.W. Haydon (1966), "Predicting Statistical Performance Indexes for High Frequency Ionospheric Telecommunications Sys-tems," Report IER 1-ITSA 1, ESSA (U.S. Department of Commerce, Washington, D.C.).

(78) "Man-Made Noise," *FCC Advisory Committee for Land Mobile Radio Services Report*, Appendix A, Vol. 2 Part 2, November 1967.

(79) "Man-Made Radio Noise," Report of Joint Technical Advisory Comm. Sub-committee 63.1.3 (Unintended Radiation), February 1968.

(80) Marcus, R. B., "The Analysis and Synthesis of Radar Emission Spectrums by Digital Computer Techniques," *Proc. Ninth Tri-Service Confer-ence on Electromagnetic Compatibility*, Chicago: Armour Research Foundation, pp. 231-260, Oct. 1963.

(81) Maxwell, E.L. (1967), "Atmospheric Noise From 20 Hz to 30 kHz," *Radio Science,* 2, 637-644.

(82) McMillan, F.O., "Radio Interference From Insulator Corona," *AIEE Trans.,* Vol. 51, pp. 385-391, 1932.

(83) McPetrie, J.S., and J.A. Saxton (1941), "Theory and Experimental Confirmation of Calibration of Field-Strength Measuring Sets by Radiation," *Journ. B.I.E.E.,* Vol. 88, Pt III, pp. 11.

(84) Mertz, P., "Impulse Noise and Error Performance in Data Transmission," *The Rand Corporation,* Memorandum RM-4526-PR, April 1965.

(85) Middleton, D., Introduction to Statistical Communications Theory, McGraw-Hill Book Co., Inc., New York, 1961.

(86) Montgomery, G.F., "Comparison of Amplitude and Angle Modulation for Narrowband Communication of Binary-Coded Messages in Fluctuation Noise," *Proc. IRE,* Vol. 42, pp. 447-454, February 1954.

(87) Mumford, W.W., and E.H. Schelbe (1968), "Noise Performance Factors in Communication Systems," Horizon House, Dedham, Mass.

(88) NBS (1948), "Radio Propagation," Circular 462, NBS (U.S. Department of Commerce, Washington, D.C.).

(89) Nethercot, W., "Car-Ignition Interference," *Wireless Engineer,* Vol. 26, No. 311, pp. 251-255, Aug. 1949.

(90) Nicholson, A.J., and M.J. Kay, "Reduction of Impulse Interference in Voice Channels," *IEEE Trans. on Communications Technology,* December 1964.

(91) North, D.O. (1942), "The Absolute Sensitivity of Radio Receivers," *RCA Review,* 6, 332-343.

(92) North, D.O. (1945), "Discussion on Noise Figures of Radio Receivers," by H.T. Friis, *Proc. IRE,* 33, 2, 125-127.

(93) Norton, K.A., "Operating Noise-Threshold of a Radio Receiving System," to be published.

(94) Norton, K.A., and A.C. Omberg (1947), "The Maximum Range of a Radar Set," *Proc. IRE,* 35, 1.

(95) Norton, K.A. (1953), "Transmission Loss in Radio Propagation," *Proc. IRE,* 41, 146.

(96) Norton, K.A. (1959), "System Loss in Radio-Wave Propagation," *Proc. IRE,* 47, 9, 1661.

(97) Norton, K.A. (1962), "Efficient Use of the Radio Spectrum," *Technical Note 158,* NBS (U.S. Department of Commerce, Washington, D.C.).

(98) "Operating Noise-Threshold of a Radio Receiving System," *CCIR Report 413,* Documents of the XIth Plenary Assembly, Oslo, 1966. Reports by the International Working Party III/1, Geneva, 1966.

(99) Otteson, H., "Electromagnetic Compatibility of Randomly Scattered Man-Made Noise Sources," Ph.D.Dissertation, University of Colorado, August 1968.

(100) Pakala, W.E., Raylor, E.R., Jr., and Harold, R.T., "Radio Noise Measurements on High Voltage Lines," *1968 IEEE Electromagnetic Compatibility Symp. Rec.,* pp. 96-107.

(101) Pearce, S.F., and J.H. Bull, "Interference from Industrial, Scientific and Medical Sea Radio-Frequency Equipment," *The Electrical Research Association,* Leatherheed, Surrey, England, Rpt. 5033, Oct. 1964.

(102) Pressey, B.G., and G.E. Ashwell, "Radiation From Car Ignition Systems," *Wireless Engineer,* Vol. 26, No. 304, pp. 31-36, Jan. 1949.

(103) "Report of UHF Propagation and Ignition Interference Tests," Sylvania Electronic Systems, Central Engineering Labs., Buffalo, N.Y., Rpt. TR 08-63.7, July 1963.

(104) Rorden, H.L., "Radio Noise Influence of 230 kV Lines," *AIEE Trans.,* Vol. 66, pp. 677-681, 1947.

(105) Rosa, A.J., "HF and VHF Automobile Ignition Measurements," *IEEE Electromagnetic Compatibility Regional Symposium,* San Antonio, Texas, Oct. 6-8, 1970.

(106) RPU (1945), "Minimum Required Field Intensities for Intelligible Reception of Radio Telephony in Presence of Atmospherics or Receiving Set Noise," *RPU Report 5* (Holabird Signal Depot, Baltimore, Md.).

(107) Schildknecht, R.O., "Ignition Interference to UHF Communication Systems," *IRE Trans. on Radio Frequency Interference,* Vol. RFI-4, No. 3, pp. 63-66, Oct. 1962.

(108) Schnelleng, J.C., Burrows, C.R., and Ferrell, E.B., "Ultra-short-wave Propagation," *Proc. IRE,* Vol. 21, No. 3, pp. 427-463, March 1933.

(109) Schultz, L.D., Spaulding, A.D., and A.F. Barghausen (1972), "Radio Spectrum Occupancy - Signals and Noise," International EMC Symposium Record, *IEEE,* 72CH 0638-7E-MC, pp. 42-49.

(110) Schwartz, M., Bennett, W.R., and S. Stein, Communication Systems and Techniques, McGraw-Hill, New York, 1966.

(111) Shepelavey, B., "Non-Gaussian Atmospheric Noise in Binary-Data, Phase-coherent Communication Systems," *IEEE Trans. Communications Systems*, Vol. CS-11, pp. 280-284, September, 1963.

(112) Shrhukin, A. N., On a Method of Combating Impulse Interference in Radio Reception, Ann. USSR Phys. Ser., 19 (trans. by P. E. Green and W. Dolye).

(113) Simpson, L. C., "Israel Intercity VHF Telecommunication System," *RCA Rev.*, pp. 100-124, March, 1953.

(114) Simpson, L. C., "System Parameters Using Tropospheric Scatter Propagation," *RCA Rev.*, pp. 432-457, 1955.

(115) Sisco, W. B., "Bit and Message Error Rates in Atmospheric Noise With and Without Peak Limiting," TRW Space Technology Laboratories Report No. 7020-6323-RU-000, November, 1964.

(116) Skomal, E. N., "Comparative Radio Noise Levels of Transmission Lines, Automotive Traffic and RF Stabilized Arc Welders," *IEEE Trans. on Electromagnetic Compatibility*, Vol. EMC-7, pp. 73-76, September, 1967.

(117) Skomal, E. N., "Analysis of Airborne VHF/UHF Incidental Noise Over Metropolitan Areas," *IEEE Trans. on Electromagnetic Compatibility*, Vol. EMC-11, No. 2, pp. 75-83, May, 1969.

(118) Skomal, E. N., "Analysis of the Frequency Dependence of Man-made Radio Noise," *1966 IEEE International Convention Record*; Part 2, pp. 125-129, March, 1966.

(119) Skomal, E. N., "Distribution and Frequency Dependence of Incidental Man-made HF/VHF Noise in Metropolitan Areas," *IEEE Transactions on Electromagnetic Compatibility*, Vol. EMC-11, No. 2; pp. 66-75, May, 1969.

(120) Skomal, E. N., "Distribution and Frequency Dependence of Unintentionally Generated Man-made VHF/UHF Noise in Metropolitan Areas," *IEEE Transactions on Electromagnetic Compatibility*, Vol. EMC-7; pp. 263-278; September, 1965.

(121) Skomal, E. N., "Distribution and Frequency Dependence of Unintentionally Generated Man-made VHF/UHF Noise in Metropolitan Areas, Part 2, Theory," *IEEE Transactions on Electromagnetic Compatibility*, Vol. EMC-7; pp. 420-428; December, 1965.

(122) Slemen, G. R., "Radio Influence from High Voltage Corona," *AIEE Trans.*, Vol. 68, Part 1, pp. 198-205, 1949.

(123) Smith, J. S. and Shepherd, N. H., "The Gaussian Curve-transmitter Noise Limits Spectrum Utilization," *Proc. of the Unclassified Sessions of the Symposium on Electromagnetic Interference,* U. S. Army Signal Corps Research and Development Laboratory, Fort Monmouth, N. J., pp. 138-144, June 15, 1958.

(124) Spaulding, A. D., Determination of Error Rates for Narrowband Communication of Binary-coded Messages in Atmospheric Radio Noise, *Proc. IEEE,* Vol. 52, No. 2, pp. 220-221, February, 1964.

(125) Spaulding, A. D., The Characteristics of Atmospheric Radio Noise and Its Effects on Digital Communication Systems, *1966 IEEE International Communications Conference,* Paper No. CP-1126.

(126) Spaulding, A. D., Ahlbeck, W. H. and Espeland, L. R., Urban Residential Man-made Radio Noise Analysis and Predictions, Report OT/TRER 14, Office of Telecommunications (U. S. Department of Commerce, Washington, D. C.), 1971.

(127) Spaulding, D. C., Robique, C. J. and Crichlow, W. Q., "Conversion of the Amplitude-probability Distribution Function for Atmospheric Radio Noise from one Bandwidth to Another," NBS, Vol. 66D, No. 6; pp. 713-720, November-December, 1962.

(128) Steele, H. L. R., Jr., "Physical Processes in the Fluorescent Lamp Which Cause Radio Noise," *Illuminating Engineering,* Vol. 47, No. 7, pp. 349-356, July, 1954.

(129) Sylvania Electronics, Waltham, Massachusetts, Design Review Report for Project 124-221, September, 1963. Contained in Report on Hole Puncher Performance Investigation of Radio Subsystems of Minuteman Ground Electronic System.

(130) Systems Development Corporation, Decision Information Distribution System; Intelligibility Study Report, January, 1968.

(131) Thomas, H. A., "Some Measurements of Atmospheric Noise at High Frequencies," *Proc. BIEE,* Pt. III, <u>97</u>, pp. 335-343, 1950.

(132) Thomas, H. A., "A Subjective Method of Measuring Radio Noise," *Proc. BIEE,* Pt. III, <u>97</u>, pp. 329-334, 1950.

(133) Thomas, H. A. and Burgess, R. E., "Survey of Existing Information and Data on Radio Noise Over the Frequency Range 1-30 mc/s," Special Report 15 (Dept. of Scientific and Industrial Research, H.M.S.S., London), 1947.

(134) Tomiyasu, K., "On Spurious Outputs from High Power Pulsed Microwave Tubes and Their Control," *IRE Trans. on Microwave Theory and Techniques,* Vol. MTT9, No. 6, pp. 480-484, Nov., 1961.

(135) URSI Special Report No. 7 on The Measurement of Characteristics of Terrestrial Radio Noise, Elsevier Publishing Company, 1962.

(136) Watt, A. D., Coon, R. M., Maxwell, E. L., and Plush, R. W., "Performance of Some Radio Systems in the Presence of Thermal and Atmospheric Noise," *Proc. IRE*, Vol. 46, pp. 1914-1923, December, 1958.

(137) Watt, A. D. and Maxwell, E. L., "Measured Statistical Characteristics of VLF Atmospheric Noise," *Proc. IRE*, Vol. 45, No. 1, pp. 55-62, Jan., 1957.

(138) White, H. E., "Atmospheric Noise FSK Error Probabilities for an Envelope-detection Receiver," *IEEE Trans. on Communications Technology*, pp. 288-289, April, 1966.

(139) Young, W. R., "Comparison of Mobile Radio Transmission at 150, 450, 900, and 3700 Mcps," *Bell System Technical Journal*, 1068-1085, November, 1952.

(140) Zacharisen, D. H., and Jones, W. B., "World Maps of Atmospheric Radio Noise in Universal Time by Numerical Mapping," Report OT/ITSRR 2, Office of Telecommunications (U. S. Department of Commerce, Washington, D. C.), 1970.

CHAPTER 3

EMI RECEPTORS AND SUSCEPTIBILITY CRITERIA

CHAPTER 3

EMI RECEPTORS AND SUSCEPTIBILITY CRITERIA

The previous chapter surveyed sources of electromagnetic interference (EMI). Since it takes both an emission source and a susceptible receptor to make EMI possible, the latter topic is the subject of
this chapter. The term *receptor* here refers to the generic class of
devices, equipments, and/or systems which when exposed to conducted
and/or radiated electromagnetic energy from emitting sources will
either degrade or malfunction in performance. Thus, receptors include
those summarized in Fig. 3.1. It is noted that many devices, equipments and systems can serve as both emission sources and susceptible
receptors. Examples include most communications-electronics (C-E)
equipments since they contain both transmitters and receivers, and
computers including their peripherals.

This chapter surveys EMI receptors such as C-E receivers, low-
level sensors, I-F amplifiers, video and audio amplifiers, computers,
and status monitors and indicators. Associated susceptibility criteria
of these receptors are also summarized. Among others, these criteria
include voice intelligibility, digital error acceptance, radar and
visual displays and other outputs. The chapter is concluded with a
bibliography.

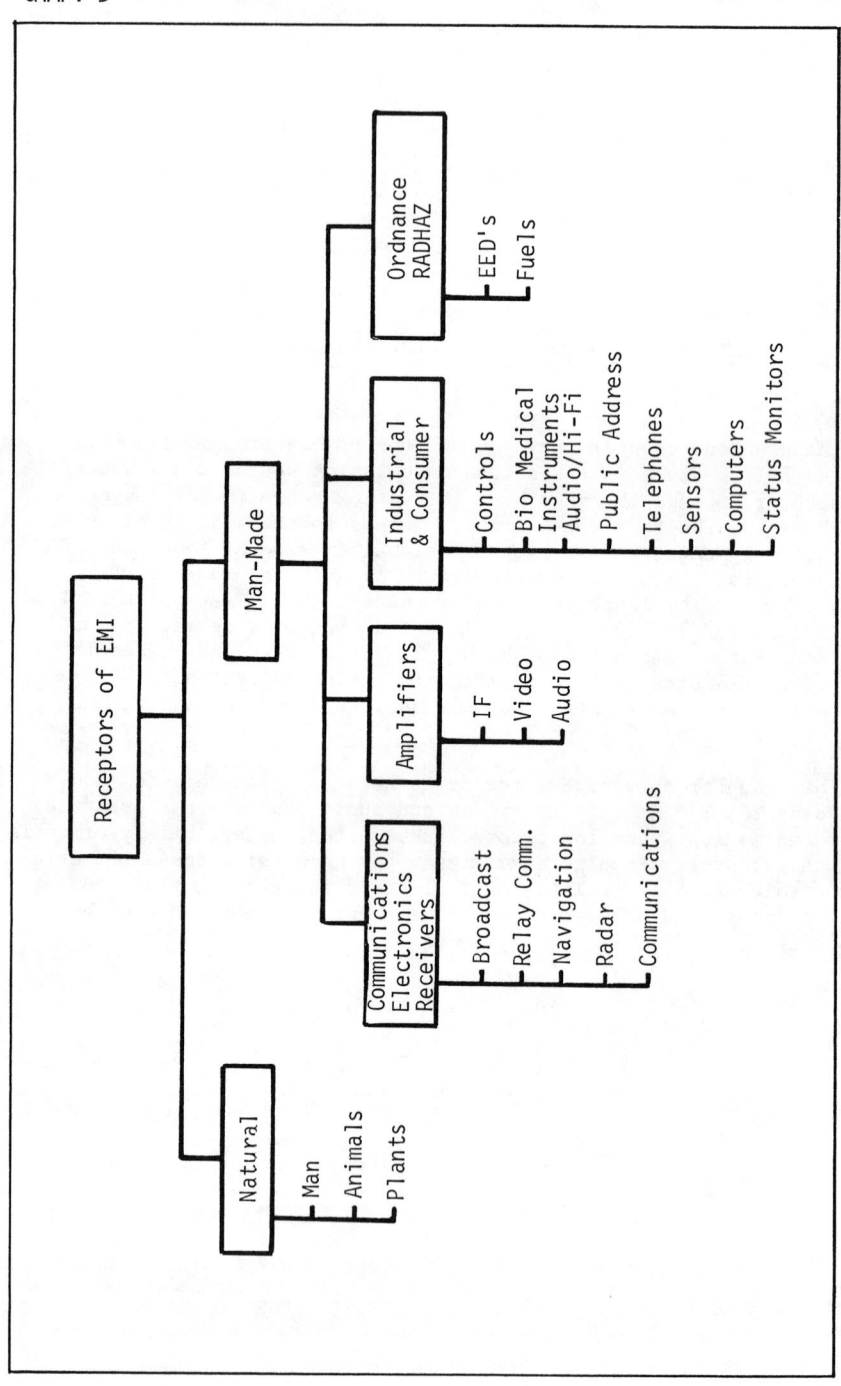

Figure 3.1 - Receptors of Electromagnetic Interference

3.1 NATURAL RECEPTORS OF EMI

Fig. 3.1 shows that receptors of EMI may be divided into natural and man-made receptors. This parallels the dichotomy for EMI emitters previously shown in Fig. 2.1. The far greater EMI problem, insofar as is known today, is interference to man-made devices. Thus, this chapter so weights this division. Natural receptors of EMI include biological and psychological effects to humans and animals.

3.1.1 RADIATION HAZARDS TO HUMANS

This section discusses electromagnetic radiation levels to humans which are regarded as both safe and hazardous. Controversial aspects of these limits are presented together with protective measures to be taken when humans must be exposed to high R-F fields.

3.1.1.1 EXPOSURE LEVELS AND REACTIONS

Several government agencies have stipulated that an average R-F power density level of greater than 10 mw/cm^2 (194 V/m in the far field) is regarded as hazardous to human life.* An exposure to a power density of 10 mw/cm^2 should not be allowed for more than 6 minutes and 1 mw/cm^2 is considered safe indefinitely. Specifically, damaging results may develop to reproductive organs resulting in sterility or to the eyeball resulting in cataracts. Relatively little is known, however, about the possible cumulative biological effects of lower-level R-F field exposure over an extended period of time. It is also not known whether high peak power density levels of short durations and low duty cycle have an effect and if lower levels play a role in adversely affecting human behavior.

One approach used to determine tolerable levels of electromagnetic radiation by the military in U. S. during the 1960's was to quantify the environmental conditions. Of special importance are climatic environmental conditions, R-F environmental conditions, and the physical condition of an individual exposed to radiation.

Climatic environmental conditions include ambient temperature, humidity, and wind velocity. Important R-F environmental conditions are power density, radio frequency, duty cycle and type of modulation. Physical condition of an individual includes health (especially circulatory condition), the physical labor engaged, and the apparel worn at the time of exposure. These factors define, in essence, heat input and heat removal from the body.

---

* Earlier reports indicated that 155 mw/cm^2 was the threshold for cataract formation by a single exposure to the eye. 100 mw/cm^2 was estimated to be tolerable under favorable conditions for total immersion of the human body (except for the testes). 40 mw/cm^2 was lethal to dogs, and 5 mw/cm^2 was considered to be the maximum exposure for no observable change in the testes of dogs.

One guide for determining tolerable levels of R-F radiation from 10 MHz to 30 GHz, under whole or partial body irradiation, is shown in Fig. 3.2. This applies to R-F fields that radiate from radio, television, radar, and high-power telecommunications. The figure applies to situations where personnel may be exposed unintentionally, in contrast to deliberate and supervised medical exposure. The line identified as *normal-conditions* designates a 10 mW/cm^2 level for periods of 6 min. or more and a 1 mW/cm^2 for exposure periods up to 6 min. These levels hold for both continuous and intermittent radiation.

It is believed that deviation from normal-conditions can take place with caution. The limits of deviation for the *caution zone* are shown in Fig. 3.2 as a dashed line. The extent of possible deviation is a function of heat-input and heat-removal of a particular situation. Either a person suffering from a circulatory condition or a laborer working in high-ambient temperature should observe the limits of the *safe zone*. By contrast, a person in excellent health working in cold and windy conditions could approach the boundary between the *caution* and *hazard zones.*

In recent years, the Russians have indicated that the thermal biological effects of exposure of humans to average R-F fields is not the only significant reaction. They report that a much more significant impact exists to the central nervous system of humans even when exposed to R-F power density levels over an extended time basis from two to three orders of magnitude below the 10 mw/cm^2 level used in USA. This in part results from the peak-pulse power densities involved, the much shorter time constants associated with the human central nervous system contrasted with the thermal heating effects of organs, and the cumulative effects of extended time of exposure. Reaction, according to the Russians, is evident in headaches, listlessness, sexual behavior, loss of memory, indecisiveness, and other responses.

Table 3.1 summarizes various standards for RF and microwave exposure throughout the world. In most cases, including the U. S. A., standards have not been set in lower-frequency bands. The Admiralty Surface Weapons Establishment in England uses, as an unofficial guide, the value of 1000 V/m for continuous human exposure below 30 MHz. This operating guide is based on experimental and analytical studies of RF-induced heating in tissue-equivalent dielectric absorbers of various geometries.

Static and fluctuating fields at very low frequencies were reported to have effects upon biological materials including man. Properly oriented static fields in the 3-5 kgauss (230-234 dBpT) range were discovered to produce profound alterations in the behavior of amphibians. Low-strength fields (6-12 gauss or 175-181 dBpT) with 25% modulation at frequencies of 0.1-0.2 Hz were found to produce measurable alterations in human reaction time. Strong statistical evidence has also been presented indicating that naturally occurring magnetic

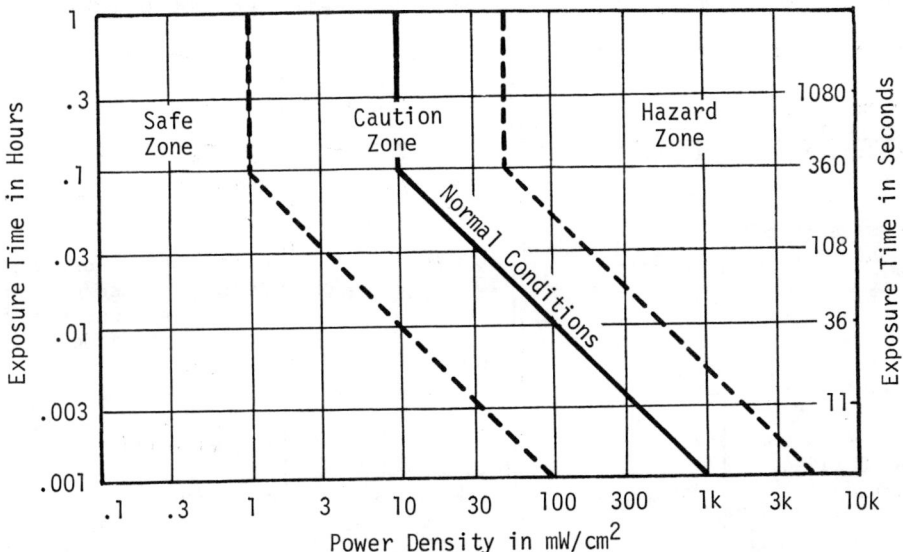

Figure 3.2 - Electromagnetic Ambient Exposure Levels to Humans

storms influence the behavior of the susceptible portion of the human population.

Disparities between Soviet bloc and Western standards reflect a basic difference in experimental approach to related biological research. As previously remarked, Russia and certain Eastern European countries centers on changes exhibited by the functional state of animal subjects; i.e., behavioral or psychological changes. Concepts that have been developed to explain many of these effects, have not been substantiated by sufficient experimental evidence to ensure widespread acceptance of their applicability. Conversely, Western investigators tend to rely upon observations of physiological and/or biochemical changes.

The fact of the matter is that the whole topic of biological effects of R-F radiation is quite contraversial today. It is felt in some circles that the Russian conclusions are more fanciful which at times are not apparently supported by reported measurements and results. Nevertheless, little is known about the effect on humans when exposed to lower-power densities of various frequencies, peak vs average powers of different duty cycles, and the cumulative time of exposure.

By way of postscript to this section, it is noteworthy to remark about the ERMAC report. A coordinated intergovernmental program for control of electromagnetic pollution of the environment has been recommended to the Director of Telecommunications, Office of Telecommunications Policy (OTP) by the Electromagnetic Radiation Management

Table 3.1 - Summary of Maximum Recommended Levels of Human Exposure

Country and Source	Radiation Frequency	Maximum Recommended Level	Condition or Remarks
U.S.A.(ANSI)	10 MHz–100 GHz	10 mW/cm^2	Periods of 0.1 hr
		1 mW hr/cm^2	Averaged over any 0.1-hr period
U.S. Army and Air Force	...	10 mW/cm^2	Continuous exposure
		10 to 100 mW/cm^2	Maximum exposure time in minutes at $W(mW/cm^2)=6000\ W^{-2}$
		100 mW/cm^2	No occupancy
Great Britain (Post Office Regulation)	30 MHz–30 GHz	10 mW/cm^2	Continuous 8-hr exposure, average power density
NATO (1956)	...	0.5 mW/cm^2	
Canada	10 MHz–100 GHz	1 mW hr/cm^2	Averaged over any 0.1-hr period
		10 mW/cm^2	Periods of 0.1 hr
Poland	300 MHz	10 µW/cm^2	8-hr exposure/day
		100 µW/cm^2	2 to 3 hr/day
		1 mW/cm^2	15 to 20 min/day
German Soc. Republic	...	10 mW/cm^2	
U.S.S.R.	0.1–1.5 MHz	20 V/m	Alternating magnetic fields
		5 amp/m	
	1.5–30 MHz	20 V/m	
	30–300 MHz	5 V/m	
	300 MHz	10 µW/cm^2	6 hr/day
		100 µW/cm^2	2 hr/day
		1 mW/cm^2	15 min/day
Czech. Soc. Rep.	0.01–300 MHz	10 V/m	8 hr/day
	300 MHz	25 µW/cm^2	8 hr/day, CW operation
		10 µW/cm^2	8 hr/day, pulsed

Advisory Council (ERMAC). This council was formed by OTP to advise and recommend constructive measures for the investigation and mitigation of undesirable side effects of radiation arising from telecommunications activities. This is a program for survey, testing, and research to establish a rational scientific basis for determining potential hazards to man and his environment of radio-frequency and other nonionizing electromagnetic radiation.

The principle objective is to establish meaningful safety criteria for the protection of man and his environment while assuring optimal use of radiation equipment and avoiding unnecessary limitation or withdrawal of equipment. Specifically, ERMAC has recommended a comprehensive integrated and sustained research effort estimated to cost $63 million for fiscal years 1974-1978 as contrasted with the current spending by the federal government, which approximates $4 million per year.

3.1.1.2 High-Field Protective Measures

Protection from hazardous R-F energy can be accomplished by measuring and/or predicting the R-F environment and then enforcing safety precautions when the electromagnetic energy level is too high. In potentially hazardous areas, safety procedures and practices should become part of normal working habits.

Persons responsible for operating and maintaining transmitters and other sources of high R-F radiation, who are subjected to R-F fields in excess of the allowable limits shown in Fig. 3.2 should exercise special precautions. One type of precaution is abstention but this is often not a practical solution. A second precaution involves shielding by either reflecting or absorbing incident R-F energy, or both. Shielding is used to protect both operation and maintenance personnel and the general public.

Wearing apparel in the form of radiation protective coveralls, fabricated from a silverized nylon mesh material (heavy marquisette, leno having a maximum D-C resistance of 2 ohms/sq.) is used in special situations. Fig. 3.3 shows a suit developed by the U. S. Navy to protect personnel from R-F fields up to 200 mW/cm^2. It provides a minimum power attenuation (mostly reflection loss) of 20 dB in radiated fields over the radar frequency range of 200 MHz to 10 GHz. A hard hat is worn under the hood to keep the fabric away from the face. To preclude arcing amongst different parts of the body as a result of induced potential difference, an arc-preventative overgarment is worn.

Other protective measures which should be taken include:

- Make measurements on a periodic basis of R-F radiation levels in areas where personnel will be working.
- Place warning signs or symbols in suitable places to alert both operation personnel and the general public.
- Install fences around R-F hazardous areas.
- Turn off R-F sources when doing maintenance work on transmitters or when it is necessary to work in hazardous areas.
- Wear R-F protective apparel, if it is not possible to turn off a transmitter, or reduce the power output to a safe level.

Figure 3.3 - Protective Suit for Intense Electromagnetic Fields

- Perform tests and experimental programs in shielded enclosed rooms when possible.
- Give medical attention to personnel who have been exposed to high intensity fields. Measure the radiation level and record the time and circumstances of measurement.

3.1.2 RADIATION HAZARDS TO ANIMALS

Most of the data obtained for drawing conclusions regarding either the physiological/biochemical or psychological/behavioral changes discussed in the previous section, involved controlled experiments on animals. Among the many results reported in the literature are those listed in Table 3.2.

Table 3.2 - Thermally Important Values of RF Intensities
in Various Animals

Radio Frequency	Animal	Intensity*	Duration of Exposure	Remarks
500 kHz	Rats and rabbits	8,000 V/m 160 A/m	...	Threshold for increase in rectal temperature
50-500 Hz	Mouse	650,000 V/m	1-2 hours	70-90% mortality
50 Hz	Mouse	650,000 V/m	4-1/2 hrs.	50% mortality
14,88 MHz	Rats and rabbits	2,500 V/m	...	Threshold for increase in rectal temperature
69.7 MHz	Rats and rabbits	200 V/m	...	
14,88 MHz	Rat	9,000 V/m	10 min.	100% mortality
	Rat	5,000 V/m	100 min.	80% mortality
	Rat	4,000 V/m	100 min.	25% mortality
69.7 MHz	Rat	5,000 V/m	5 min.	100% mortality
	Rat	2,000 V/m	100 min.	83% mortality
200 MHz	Dog	300 mW/cm^2	15 min.	50% mortality
	Dog	200 mW/cm^2	21 min.	25% mortality
200 MHz	Guinea pig	590 mW/cm^2	20 min.	67% mortality
	Guinea pig	410 mW/cm^2	20 min.	100% mortality
	Guinea pig	330 mW/cm^2	20 min.	100% mortality
200 MHz	Rabbit	165 mW/cm^2	30 min.	100% mortality

Gordon of USSR has written a chapter in a handbook dealing with
various occupationally related hazards. He presents the results of
tests-to-the-death on laboratory rats by exposure of strong fields of
various frequencies. The results reported are somewhat startling.

Wavelength	Energy Intensity ergs/cm^3	Mean Time to Death
Medium Waves	2830 x 10^{-6}	Non-Lethal
Short Waves	1100 x 10^{-6}	100 Minutes
Ultra-Short Waves	1100 x 10^{-6}	5 Minutes
Decimeter Waves	33 x 10^{-6}	60 Minutes
Centimeter Waves	33 x 10^{-6}	15 Minutes
Millimeter Waves	33 x 10^{-6}	180 Minutes

Gordon has also studied the thermal effects from several months
exposure to fields of low enough intensity (<10 mw/cm^2). No thermal
effects were observed. Gordon notes that after two months, they observed severe functional changes in the central nervous system, in
particular a change in the reactivity to stimuli such as light, a loss
of conditioned reflexes, and even the appearance in some animals of a

---

* 1 mW/cm^2 = 0 dBm/cm^2 = 61 V/m = 156 dBµV/m in the far field

pre-disposition to some of the symptoms of epilepsy. He concludes that
the action of radio waves upon the central nervous system is achieved
both directly on the nerve cells of the brain as well as reflexly, by
the transmission of impulses from receptors.

The above reactions of animals to electromagnetic fields is only
a small sample already reported in the literature. It is expected that
animal (and human) experiments will increase significantly in the latter
half of the 1970 decade as an aide in understanding the natural effects
of electromagnetic exposure. Acceptable exposure standards of man and
animal to his environment and meaningful hazard levels will be only
one of many useful outputs. Insofar as U.S.A. is concerned, among the
many sources of support for future experiments will be that set forth
in the ERMAC report discussed at the end of Sec. 3.1.1 and the National
Cancer Institute.

3.2 MAN-MADE RECEPTORS OF EMI

Fig. 3.1 showed that receptors of EMI may be divided into natural and man-made receptors. This section discusses the latter. As shown in Fig. 3.1, man-made receptors are classified into communications-electronics receivers, amplifiers, industrial and consumer devices, and ordnance including fuels.

It is helpful to establish some measure of latent susceptibility of receptors to electromagnetic exposure in order to rate them for classification purposes. This then requires that a scoring system be used to calculate and assess relative vulnerability to EMI environments. Such a scoring technique has been developed and will now be discussed.

Sensitivity, N, and bandwidth, B, appear to be the two most pertinent parameters required for a rating system. The greater the sensitivity (a lower number) and/or bandwidth, the greater the tendency that a receptor has to EMI. In terms of sensitivity equal internal noise, this means that the bandwidth term will be in the numerator and the noise term will be in the denominator. If an interfering source is coherent* (e.g. a broadband transient or pulse emission), then a receptor susceptibility *voltage-to-noise* ratio, RS_c (subscript c stands for coherent), is proportional to bandwidth (see Sec. 16.4):

$$RS_c = \frac{kB}{N_v} = \frac{kB}{\sqrt{4RFKTB}} = \sqrt{\frac{Bk^2}{4RFKT}} \qquad (3.1)$$

where, B = bandwidth in Hz

N_v = internal noise voltage

R = resistive component of the equivalent input impedance in ohms

F = equivalent noise figure of receptor

$KT = 4 \times 10^{-21}$ watts/Hz bandwidth

If the EMI source is non-coherent (e.g. bandwidth-limited white noise, such as an unmodulated arc discharge), the RS voltage-to-noise ratio is proportional to the square root of bandwidth:

$$RS_n = \frac{k\sqrt{B}}{N_v} = \frac{k\sqrt{B}}{\sqrt{4RFKTB}} = \frac{k}{\sqrt{4RFKT}} \qquad (3.2)$$

Higher values of RS correspond to greater tendencies for EMI susceptibility. Since potential EMI sources may be either coherent or non-coherent, Eq. (3.1) is chosen as the more damaging of the two equations, it becomes the basis for latent receptor susceptibility. Setting $k = 1$ in Eq. (3.1) yields:

---

*   As explained in Chap. 16, a coherent signal or EMI emission has a specified frequency and phase relation between its environmental frequency components, whereas a non-coherent emission has a random phase and often a random amplitude between increments.

$$RS_v = \sqrt{\frac{B}{4R \times 4 \times 10^{-21}F}} = 0.8 \times 10^{10}\sqrt{B/RF}$$

$$= 198 + 10\,\log_{10}(B/RF)\ dB \tag{3.3}$$

Sometimes it is more useful to calculate RS directly from the rated (power) sensitivity, N_p, whereupon Eq. (3.1) becomes for k = 1:

$$RS_p = RS_v^2 = \frac{B^2}{N_v^2} = \frac{B^2}{4R \cdot N_p} = 20\,\log_{10}B - N_{dBW} - 10\,\log_{10}4R$$

$$= 20\,\log_{10}B - (N_{dBm} - 30\ dB) - 10\,\log_{10}4R$$

$$= 24\ dB + 20\,\log_{10}B - 10\,\log_{10}R - N_{dBm} \tag{3.4}$$

where, N_{dBW} = sensitivity in units of dBW

             N_{dBm} = sensitivity in units of dBm = N_{dBW} + 30 dB

Eq. (3.4) is plotted in Fig. 3.4 with resistance and sensitivity as parameters. To select the pertinent curve in the figure, first use Table 3.3.

Illustrative Example 3.1

Determine the RS receptor susceptibility rating of a receiver having a sensitivity of −104 dBm, an input impedance of 50 ohms, and a bandwidth of 1 MHz. For this situation, Eq. (3.4) or Fig. 3.4 is used:

$$RS = 24\ dB + 20\,\log_{10}10^6 Hz - 10\,\log_{10}50 -(-104\ dBm)$$

$$= 24\ dB + 120\ dB - 17\ dB + 104\ dBm = 231\ dB$$

As shown in Chap. 5, this same receiver exhibits a noise figure of 10 dB or a noise factor of 10. For this, Eq. (3.3) would yield:

$$RS = 198 + 10\,\log_{10}(10^6/50 \times 10)\ dB$$

$$= 198 + 33 = 231\ dB$$

In order to obtain a *feel* for the ranges of RS, a few different ones will now be computed.

Illustrative Example 3.2

Determine the RS receptor susceptibility rating of the following amplifiers:

Amplifier Type	$N_{dB\mu V}$	Z Level	N_{dBm}	Bandwidth	RS
TWT Amplifier	(+37)	50Ω	−70	26Hz	263 dB
I-F Amplifier	(+37)	50Ω	−70	1MHz	197 dB
Crystal-Video RX	(+75)	1kΩ	−45	3MHz	169 dB
Video Amplifier	+106	50Ω	(−1)	1MHz	128 dB

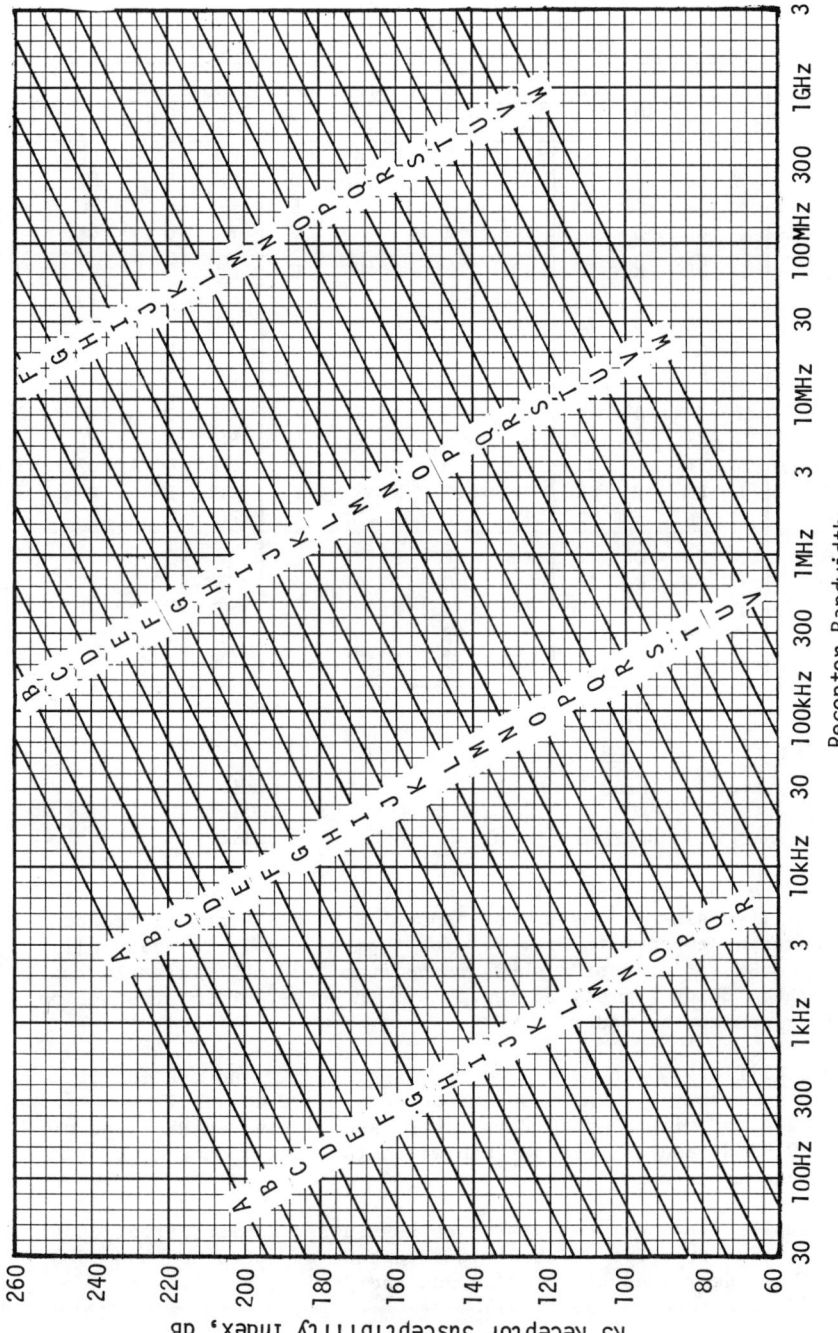

Figure 3.4 - Receptor Susceptibility Index vs Receptor Bandwidth (see Table 3.3)

Table 3.3 - Sensitivity and Impedance Levels for
Determining Curve Used in Fig. 3.4

Sensitivity N_{dBm}	Equivalent Source Impedance in Ohms												
	10	30	50	100	300	600	1k	3k	10k	30k	100k	300k	1M
-160	-	--	A	A	AB	B	B	BC	C	CD	D	DE	E
-150	A	AB	B	B	BC	C	C	CD	D	DE	E	EF	F
-140	B	BC	C	C	CD	D	D	DE	E	EF	F	FG	G
-130	C	CD	D	D	DE	E	E	EF	F	FG	G	GH	H
-120	D	DE	E	E	EF	F	F	FG	G	GH	H	HI	I
-110	E	EF	F	F	FG	G	G	GH	H	HI	I	IJ	J
-100	F	FG	G	G	GH	H	H	HI	I	IJ	J	JK	K
-90	G	GH	H	H	HI	I	I	IJ	J	JK	K	KL	L
-80	H	HI	I	I	IJ	J	J	JK	K	KL	L	LM	M
-70	I	IJ	J	J	JK	K	K	KL	L	LM	M	MN	N
-60	J	JK	K	K	KL	L	L	LM	M	MN	N	NO	O
-50	K	KL	L	L	LM	M	M	MN	N	NO	O	OP	P
-40	L	LM	M	M	MN	N	N	NO	O	OP	P	PQ	Q
-30	M	MN	N	N	NO	O	O	OP	P	PQ	Q	QR	R
-20	N	NO	O	O	OP	P	P	PQ	Q	QR	R	RS	S
-10	O	OP	P	P	PQ	Q	Q	QR	R	RS	S	ST	T
0	P	PQ	Q	Q	QR	R	R	RS	S	ST	T	TU	U
10	Q	QR	R	R	RS	S	S	ST	T	TU	U	UV	V
20	R	RS	S	S	ST	T	T	TU	U	UV	V	VW	W

Amplifier Type	$N_{dB\mu V}$	Z Level	N_{dBm}	Bandwidth	RS
Audio Amplifier	+100	600Ω	(-18)	10kHz	94 dB
Sensor Amplifier	+60	1kΩ	(-60)	100Hz	94 dB
Digital Data RX	+126	150Ω	(+14)	1MHz	108 dB

From the above, an RS level ranking can be created in terms of
susceptibility value judgments corresponding to all types of receptors.
One such rating appears in Fig. 3.5 for typical receivers and amplifiers. The RS rating scores shown range from >230 dB (extremely susceptible) to <80 dB (rarely susceptible). A number of interesting observations can now be made:

● Amplifiers which have a voltage sensitivity independent of
  bandwidth (e.g. base-band video amplifiers driven from a second
  detector with about 0.2 volts threshold) have an RS slope of
  20 dB/decade of receptor bandwidth (N_{dBm} is independent of
  bandwidth in Eq. (3.4)).

● Amplifiers and receivers which have noise-limited front ends
  have an RS slope of about 10 dB/decade of receptor bandwidth
  (N_{dBm} decreases at 10 dB/decade and $20 \log_{10} B$ increases at
  20 dB/decade in Eq. (3.4)).

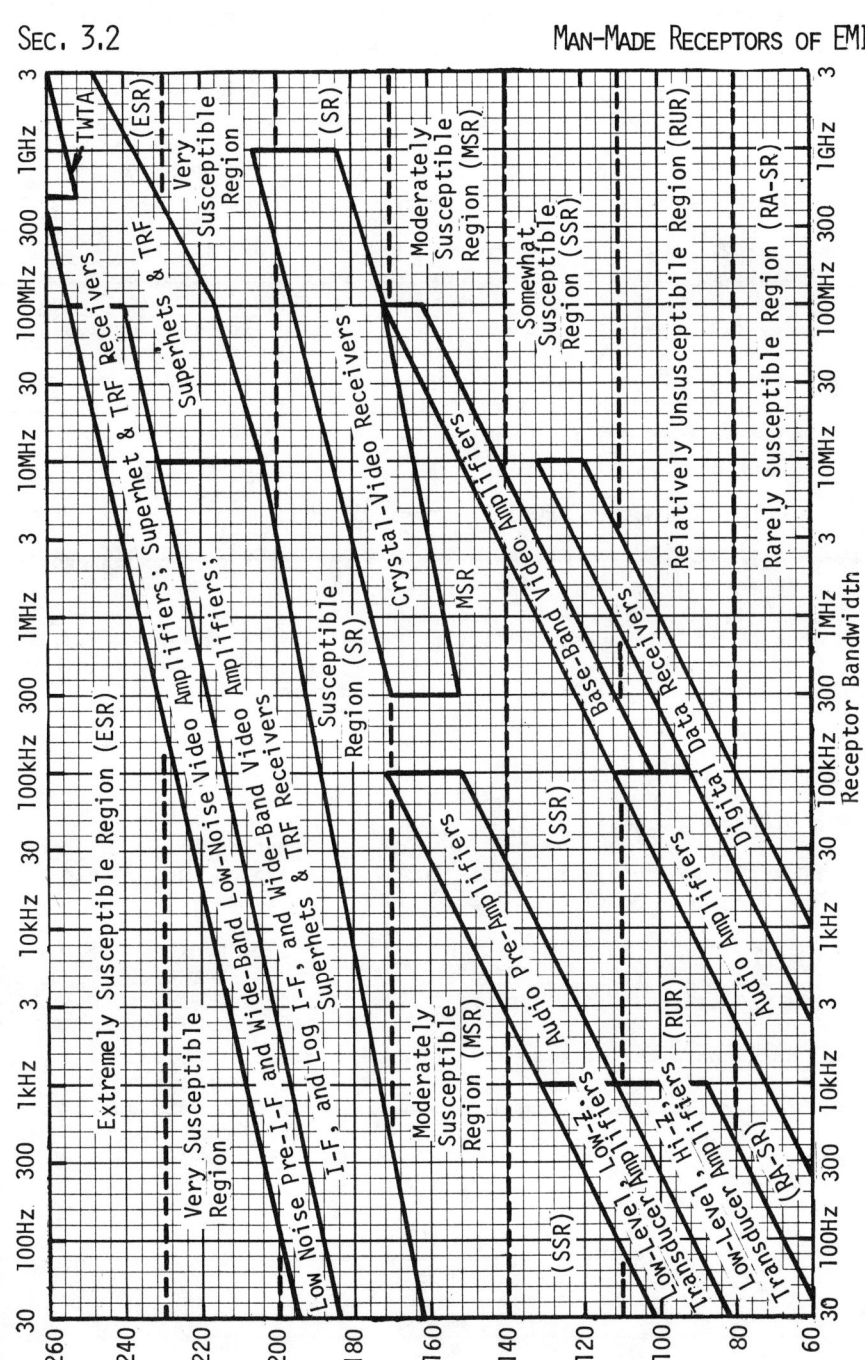

Figure 3.5 - Typical Receptor Susceptibility Scores of Receivers and Amplifiers

- The spread in RS susceptibility of receivers and amplifiers is in excess of 200 dB - an enormous range!

- Broadband amplifiers and receivers are more susceptible than their narrowband counterparts.

- Low-noise (2-20 dB) receivers, pre-IF, and wide-band video amplifiers are very susceptible receptors.

- Low-level (0.1-1mV), narrowband transducer amplifiers are relatively unsusceptible. However, a low-impedance, baseband transducer amplifier is more susceptible (see magnetic fields, Sec. 6.2) than high-impedance amplifiers at low frequencies.

- Assuming all amplifiers and receivers exhibit a 100 dB rejection to out-of-band EMI emission, wide-band, low noise devices still evidence moderately susceptible characteristics.

3.2.1 COMMUNICATIONS-ELECTRONICS (C-E) RECEIVERS*

Fig. 3.2 showed that C-E receivers are the first of four categories of man-made receptors. This class is generally the most susceptible to EMI since their RS rating varies from 160 dB to over 260 dB (see Fig. 3.5). Thus, most EMI susceptibility problems exist here. Sec. 5.3.2 and 18.2 discuss receivers in detail. Hence, they are only summarized here with regard to their RS susceptibility scores.

3.2.1.1 BROADCAST C-E RECEIVERS

The broadcast bands cover:

	RS Rating	Value Judgment
HF Amplitude Modulation (535-1605 kHz)	≈ 195 dB	Susceptible
VHF Frequency Modulation (88-108 MHz)	≈ 215 dB	Very Susceptible
VHF Television:		
Lower Bands (54-88 MHz)	≈ 230 dB	Extremely Suscep.
Upper Bands (174-216 MHz)	≈ 230 dB	Extremely Suscep.
UHF Television: (470-890 MHz)	≈ 225 dB	Very Susceptible

In addition to man-made sources of electromagnetic emissions, A-M broadcast receivers are susceptible to broadband atmospheric noise. F-M and T-V receivers, while immune to atmospheric noise are quite susceptible to automobile ignition noise if their associated antennas are located near roads and are not located well above the ground (e.g. less than 10 meters.

---

*   This Section matches Sec. 2.3.1, Communications-Electronics Emitters.

3.2.1.2 Communication C-E Receivers

Communications equipments are the greatest in number and most varied of all C-E types. They occupy portions of the spectrum interlaced between other activities from about 10 kHz to about 2 GHz. Many of the C-E receivers here (about 4,000,000) are of the land-mobile type (see Sec. 1.2). Above 2 GHz, point-to-point communications is generally of the relay type. RS ratings for communications receivers vary from 150 dB (moderately susceptible) at VLF to about 235 dB (extremely susceptible) at UHF.

Since many C-E receivers are of the land-mobile type, they are susceptible to automobile ignition and nearby industrial-area noise. This is a lesser problem for the military since all vehicles employ ignition suppression devices. There, co-channel and adjacent channel interference is more of a problem because of the perishability of combat frequency assignments.

3.2.1.3 Relay Communications C-E Receivers

Point-to-point communications generally consist of one or more of the following four types:

	RS Rating	Value Judgment
Common Carrier, Microwave Relay (2.1-11.7 GHz interspersed)	≈ 245 dB	Extremely Suscept.
Satellite Relay (2.4-16 GHz interspersed)	≈ 225 dB	Very Susceptible
Ionospheric Scatter (400-500 MHz)	≈ 220 dB	Very Susceptible
Tropospheric Scatter (1.8-5.6 GHz interspersed)	≈ 230 dB	Extremely Suscept.

It is interesting to note that microwave relay receivers, per se, are the most susceptible (RS rating of 245 dB). Yet, pragmatically, microwave relay links are relatively immune to EMI. This is due to both the enormous off-axis interference rejection of their antennas (typically 60-70 dB) and high S/N ratios (typically about 50 dB) of the links. This illustrates what can be done in good EMI-control in system equipment design.

3.2.1.4 Navigation C-E Receivers

Navigation receivers in this classification exclude radars since radars are covered in the next section. The RS rating of navigation receivers varies from about 170 dB (moderately susceptible for certain VLF receivers to about 230 dB (extremely susceptible) for certain UHF types. Typical navigation receiver types included are:

VOR (VHF Omni Range): 108–118 MHz
TACAN (Tactical Air Navigation)
Marker Beacons: 74.6–75.4 MHz
ILS (Instrument Landing System):
   ILS Localizer: 108–118 MHz
   Glide Path: 328.6–355.4 MHz
Altimeter: 4.2–4.4 GHz
Direction Finding: 405–415 kHz
Loran C: 90–110 kHz
   A: 1.8–2.0 MHz
Maritime: 285–325 kHz; 2.9–3.1 GHz; 5.47–5.65 GHz
Land: 1638–1708 kHz.

3.2.1.5 Radar C-E Receivers

Radars are used in intermittent portions of the spectrum from about 225 MHz to 35 GHz. They are employed in many capacities (the Department of Defense is the largest user of the more powerful radars) including: air-traffic control, air and surface search, harbor surveillance, mapping, tracking and fire control, police-speed monitoring, and weather.

Except for narrowband doppler police and CW radars, radar receivers are usually broadband and operate at higher frequencies. Hence their RS rating typically is about 230–240 dB (extremely susceptible). This is a bit ironic since radars are significant offending sources of EMI because of their high effective radiated powers (+95 dBm to +135 dBm).

3.2.2 Amplifiers

The preceding section summarized C-E receivers and stated that they are generally the most susceptible of the four man-made receptor categories shown in Fig. 3.1. There exists a few exceptions such as when low-noise, wide-band video amplifiers exhibit higher RS susceptibility ratings than receivers. Many applications of such amplifiers, however, are to improve the sensitivity of receiver front-ends, and therefore become a part of the C-E receiver, per se. On the other hand, low-noise, wide-band video amplifiers are used in other applications.

Intermediate-frequency (I-F) amplifiers are always part of C-E receivers by definition since they are the natural devices to the superheterodyne process. Either video (not low-noise, wide-band types) and/or audio amplifiers are also always found in superheterodyne receivers. In contrast to I-F amplifiers, however, video and audio amplifiers are used in many applications other than receivers. Since all these amplifiers constitute a separate class of potential EMI victims, whether or not they appear in C-E receivers, they are shown as a distinct category of man-made receptors in Fig. 3.1.

3.2.2.1 I-F Amplifiers

I-F amplifiers constitute another input terminal pair to which a receiver may be susceptible. Culprit emissions ordinarily enter a receiver input terminal by the antenna and sometimes by the lead-in R-F transmission line pick-up (leakage). If the input signals exist at intermediate frequencies, the receiver front-end rejection may or may not be adequate to prevent a susceptible situation. Assuming such rejection to I-F emissions were adequate, it may still be possible for these emissions to by-pass the receiver front-end and be picked up by the I-F input cable. One example is when the receiver front-end is located up on a mast with an antenna and the I-F is piped down to the deck where the remainder of the receiver is located. Here, the long I-F coaxial cable acts as a pick-up *antenna* and the I-F amplifier (and remainder of the receiver) now may be susceptible.

An I-F amplifier may have a greater or less sensitivity than its associated overall receiver sensitivity. This depends upon the loss of the R-F input cable, the loss or gain, noise figure and bandwidth of the pre-mixer circuitry, the conversion loss or gain of the converter, and the noise figure and bandwidth of the I-F amplifier. This topic is discussed in further detail in Chap. 5.

Fig. 3.5 shows that low noise, pre-I-F amplifiers may have RS ratings comparable to receivers. These ratings typically vary from about 190 dB (susceptible) for small bandwidths to about 240 dB (extremely susceptible) for very large bandwidths.

Fig. 3.5 also shows less sensitive I-F amplifiers which typically accommodate receivers having either pre-mixer R-F gain, first-converter gain, and/or pre-I-F amplifiers. Such I-F amplifiers may have RS ratings which are 10 to 30 dB less than low-noise pre-I-F amplifiers. Their susceptibilities, however, are still significant especially when operated under conditions in which they are physically removed from their receiver front-ends by a long interconnecting cable as discussed above.

3.2.2.2 Video Amplifiers

There are many types of video amplifiers. One simple classification of such amplifiers is (1) low-level video amplifiers and (2) baseband, high-level video amplifiers. A few examples of each classification are:

Low-Level Video Amplifiers	High-Level Video Amplifiers
Receiver Front-End Noise Figure Improvement	Receiver Post-Detector Amplifiers
Crystal-Video Receiver Pre-Amplifiers	Digital Data Receivers
Non-Tunable, Low-Noise Receivers	Telemetry Data Amplifiers
Low-Noise Oscilloscope Amplifiers	A/D and D/A Converters

The distinguishing features of the above video amplifiers classification are that low-level types are generally noise limited by either typical 50-ohm (72-ohm or other) input impedances or high input impedances (100 kΩ, 1 MΩ or other). On the other hand, high-level video amplifiers are generally characterized by requiring a minimum voltage to perform. For example, receiver post-detector video amplifiers typically need a level of about 0.2 volts because of the second detector characteristics. A second example is digital data receivers which may require a minimum level of 2 volts to sense a mark or "1-bit."

Fig. 3.5 shows that low-noise, video amplifiers have RS ratings comparable to receivers and pre-I-F amplifiers (e.g., from about 190 dB to 240 dB). Unless protected by a low and/or high-pass filter, the wide-band, low-noise video amplifier is highly susceptible to EMI emissions and intermodulation (see Sec. 5.3.2). This comes about because they generally operate over a broad-frequency base band having a low-frequency cut-off of perhaps 100 Hz to 1 kHz, where many broadband EMI emissions are the highest (see Sec. 16.4).

Low-noise video amplifiers can even have pass bands of 1 GHz or more due to recent advances in solid-state circuitry, where gain-bandwidth products of active devices may be of the order of 3 GHz or more. Here, Fig. 3.5 can be extrapolated to yield R-S ratings of 250 to 260 dB - an extremely susceptible situation.

Finally, when the low-noise, wide-band video amplifier has a high
low-frequency cutoff, it takes on other names such as traveling-
wave tube (TWT) amplifiers shown in the figure. Low-noise TWTAs have
the highest RS ratings of all (250 to 270 dB). Fortunately, broadband
EMI noise and/or C-E transmitter spectrum activity in the 10 GHz and
higher frequency regions render the TWATs less susceptible than their
latent RS ratings would indicate. From 500 MHz to 10 GHz, however,
they are extremely susceptible to EMI.

Another EMI problem associated with low-level video amplifiers is
their out-of-band rejection performance. Unless protected by additional
low-pass and/or high-pass filters, such amplifiers may exhibit 60 –
80 dB rejection to out-of-band emissions at one decade or more beyond
the amplifier cut-off frequencies. Thus, such emissions indicate that
video amplifiers have out-of-band RS ratings of perhaps 110 dB (190
dB – 80 dB = somewhat susceptible) to 180 dB (240 dB – 60 dB = susceptible). Consequently, a 30 MHz, low-level video amplifier, for example,
has been and can be very susceptible to a high-level UHF TV transmission or to a nearby VHF land-mobile transmitter.

The high-level video amplifiers pose an altogether different
susceptibility situation. Fig. 3.5 indicates that RS ratings may vary
from about 90 dB (relatively unsusceptible) for 100 kHz base-band bandwidths to about 160 dB (moderately susceptible) for 30 MHz bandwidths
(high data rates and/or short pulse widths). For similar bandwidths
this corresponds to 80 to 130 dB lesser susceptibilities than their
low-level video amplifier counterparts. This notwithstanding, the high-
level, wide-band video amplifiers such as used in high-clock rate digital receivers and computers can be susceptible to EMI.

3.2.2.3 AUDIO AMPLIFIERS

There are also many types of audio amplifiers. One simple classification of such amplifiers somewhat parallels that for video amplifiers, except that 100 kHz is arbitrarily defined here as the upper
frequency limit for audio amplifiers. The classification suggested is
(1) low-level audio amplifiers and (2) high-level audio amplifiers. A
few examples of each classification are:

Low-Level Audio Amplifiers	High-Level Audio Amplifiers
Audio Pre-Amplifiers	Telephone-Line Repeaters
Low-Noise Oscilloscope Amplifiers	Public Address Amplifiers
Low-Level Sensor Amplifiers	Hi-Fi Power Amplifiers
Bio-Medical Instruments	Chart Recorder Drivers

The distinguishing feature of the above audio amplifier classification is that low-level types are often noise limited with either
low input impedances (approximately 10-ohms) or high input impedances
(10 kΩ, 100 kΩ or other). If not noise limited, they have input sensitivities less than about 10 mV. On the other hand, high-level audio

amplifiers are generally characterized by requiring a minimum voltage
to perform such as about 100 mV as in the case of many Hi-Fi audio
power amplifiers.

Fig. 3.5 shows that audio pre-amplifiers have RS ratings ranging
from about 100 dB (somewhat susceptible) to 170 dB (susceptible). If
the pre-amplifier is used for either a low-level, high impedance transducer or oscilloscope applications, Fig. 3.5 shows that it may be 30
to 40 dB more susceptible than its lower impedance counterpart.

Another EMI problem associated with low-level video amplifiers
is their out-of-band rejection to broadcast, FM, TV, radar and other
high-level emissions. Typical rejections vary from about 60 to 100 dB.
Thus, it is not uncommon for EKG bio-medical instruments having a 100
μV sensitivity and 100 Hz bandwidth, for example, to be susceptible to
a 100 watt, 30 MHz physicians' paging system in a hospital. Another
better known example is either the Hi-Fi amplifier or public address
system which is susceptible to the nearby FAA or military radar.

The high-level video amplifiers pose an altogether different
susceptibility situation. Fig. 3.5 indicates that RS ratings may vary
from about 60 dB (rarely susceptible) to 110 dB (somewhat susceptible).
Thus, audio power amplifiers and audio signal boosters are not often
susceptible to EMI.

3.2.3 INDUSTRIAL AND CONSUMER RECEPTORS

The preceding two sections discussed communications-electronics
receivers and amplifiers as two of the four man-made classes of receptors shown in Fig. 3.1. This section presents a third class, industrial and consumer receptors. The concept of an RS (receptor susceptibility) rating, introduced in Sec. 3.2, will be continued here to establish some measure of relative susceptibility to EMI.

Industrial and consumer receptors are electrical, electromechanical,
and electronic systems, equipments, and/or products which may mal-per-
form or degrade in the presence of electromagnetic ambient environments indigenous to these items. Most of these items require amplifiers
of one form or another in order to accomplish their intended performance;
as such, they sense lower-level signals and deliver higher-level outputs (see Sec. 3.2.2, Amplifiers). This accounts for their relative
susceptibility. Some examples of such systems, equipments, and products
are:

Industrial Receptors	Consumer Receptors
Computers	Radio and TV Broadcasting*
Industrial Process Controls	Hi-Fi Equipment
Electronic Test Instruments	Intercoms
Bio-Medical Instruments	Electronic Musical Instruments
Public-Address Systems	Climate Control
Telephones and Teletype	Heart Pacers

---

*   Previously discussed in Sec. 3.2.1, C-E Receivers

Industrial Receptors   Consumer Receptors (Cont.)
Electronic Security Alarms   Automobiles, Boats and Aircraft
Electrical Recorders
Land-Mobile Receivers*

The following sections outline a few of the industrial and consumer receptors. It is noteworthy that relatively little is reported
in the technical literature about the susceptibility of many of these
receptors. Consequently, relatively little quantative results are
reported here.

3.2.3.1 Digital Computers

Chap. 2 indicated that computers and peripherals both conduct
and radiate electromagnetic energy. Most such emissions emanate from
the higher level peripherals including, card punches and readers,
character and line pointers, and the like. Computers, on the other
hand, may be susceptible to EMI because they generally operate at
lower levels than their peripherals. For example, the computer areas
most sensitive to pulse and transient emissions are the amplifier circuits whose direct inputs are from low-level output storage devices,
such as magnetic tape and disk (most susceptible) and drum and ferrite
core (less susceptible).

Other aspects of computers, which make them susceptible to EMI,
include power-line conducted transients, ground shifts, and electrostatic discharge. All are characterized by short duration emissions
which are sensed as a mark (1-bit) or space (0-bit) resulting in
character or word-error readouts. Line transients are superimposed
on the A-C power system and may couple to the computer logic buss,
sense amplifiers, or logic circuits. Ground shifts are transient
potential differences between two or more portions of the computer
ground reference. Thus common-mode impedance drops result in sensed
logic errors. Personnel-induced electrostatic discharge where carpets
are located near computers or static discharge associated with moving
belts, paper and tape can also common-mode impedance couple into logic
circuits. Chap. 17 discusses some of the EMI-control techniques which
may be used in computers and peripherals.

3.2.3.2 Industrial Process Controls

Many manufacturing quality-control distribution, and the like
processes are automated today. This includes a myriad of operations,
a few of which are: automatic machinery operations, in-plant climate
control, oil refining and processing, item weighing and counting, steel
fabrication parameter control, etc. Nearly all are characterized by
a closed-loop servo system in which intended controls are injected
(often from mini computers), results sampled, departures or errors
sensed, and an adjusted control is re-inserted. Thus, these devices

contain amplifiers if not computers in addition. Consequently they
are susceptible to the same type of EMI sources as amplifiers and
computers.

Because servo amplifiers may have a frequency response from DC
to perhaps only a few Hz, it does not follow that they are not responsive to higher level AM broadcast, FM, TV, radar, and like MF-UHF
ambients. While offering 60 - 120 dB rejection to such radiated out-
of-band concessions, servo amplifiers and associated control processes
will often malfunction when the electric-field intensity exceeds about
5 V/m. Not infrequently, this results from the pick-up of the sense
amplifier harness or cable. Thus, cable shielding and shield grounding
becomes an EMI-control requirement.

3.2.3.3 ELECTRONIC TEST INSTRUMENTS

This class of devices is also very numerous in type and variety.
A few examples include, oscilloscopes, frequency and time-event counters,
wave and spectrum analyzers, recorders and X-Y platters, AF and RF
millivoltmeters, impedance bridges, etc. Again, nearly all these
instruments contain amplifiers of one form. Thus, they may be susceptible in the same sense as the amplifiers discussed in Sec. 3.2.2. The
mode of susceptibility entry may be either through power-line conducted
or radiated input cable pick-up and case leakage.

3.2.3.4 BIO-MEDICAL INSTRUMENTS

This class of devices is similar to electronic test instruments
except that upper 3-dB frequency responses are often limited to about
100 Hz or less. Consequently, their susceptibility on average is less.
A few typical bio-medical instruments are EKG and ECG recorders, arterial and venous-pressure monitors, and respiration indicators. Because
they all use amplifiers to boost lower-level transducer sensors for
display purposes they may be susceptible to both 60 Hz power line
emissions and R-F ambients. Regarding the latter, it is not uncommon
in a hospital for an EKG monitor to pick up a 30 MHz physician's
paging signal through its sensor leads.

3.2.3.5 PUBLIC ADDRESS SYSTEMS AND INTERCOMS

Public address systems are occasionally susceptible to EMI emissions because their microphone lead-in cable to the amplifier may
represent a length of many feet. As such, the cable can act as a
pick-up antenna to electromagnetic ambients and be amplified along
with a voice message. Intercoms (intercommunication system amongst
different rooms of an office or private residence) may pose an even
greater susceptibility problem. This results from the substantial
amount of multi-conductor wiring which is routed back and forth between

station speakers and a central control unit and amplifier. Here, the cable *pick-up antenna* to the master station can run in the order of 100 feet or more.

3.2.3.6   Automobiles, Boats, and Aircraft

Another class of receptors involves those existing inside of consumer automobiles, boats, and aircraft. These vehicular platforms, per se, merely house a family of individual receptors, most of which include an AM radio and many of which have several electronic devices. Except for some boats, they all contain an on-board gasoline engine ignition system which can compromise receptor susceptibility.

Modern automobiles of the more expensive type contain several receptor devices such as AM and FM radio, stereo-tape players, climate control, cruise control, and headlight dusk initiators. This is only the beginning of what is in the offing for future automobiles. They may be susceptible to outside-world (inter-system) interference such as to the AM and FM radio or from within (intra-system) because of engine ignition noise or electrical-system transients resulting from activation of automatic internal controls.

Consumer in-board motor boats of the cruiser and yacht class may contain some of the same receptors as used in automobiles. They also are likely to contain inexpensive passive sonars, fathometers, ship-to-shore communications and, in some cases, harbor radars or anti-collision alarms. Thus, depending upon the complement of actual receptors used, such motor craft may be susceptible to electromagnetic emissions - both self generated and from the outside world.

The more expensive pleasure aircraft usually contain the largest array of receptors (and emitters) and constitute the largest electro-magnetic susceptibility problem of the three consumer vehicle types. This results from the extra receptors required or desired such as auto-pilot, navigation receivers, communications, collision-avoidance system and in some cases weather radar, beacons, and minicomputers. Some cockpit instruments are also susceptible to on-board aircraft emissions even though no amplifiers may be involved.

3.2.4   Radiation Hazards to Ordnance

The fourth class of man-made receptors shown in Fig. 3.1 involves radiation hazards to electro-explosive devices (EEDs) and fuel. EEDs are the devices used to electrically ignite explosives by the application of a specified current. Their input levels act as a pick-up antenna by induction or radiation thereby delivering electromagnetic energy to the susceptible igniter. Ignition of fuels, on the other hand, generally comes about as a result of sparks from either structural members immersed in a high R-F field or from brush-commutator interfaces, inductive switching transients, and the like.

The U. S. Department of Defense has been actively engaged in determining the extent of radiation hazards and methods for controlling them. The problem investigated has come to be known as radiation hazards (RADHAZ). The Navy has been involved in the RADHAZ problem in connection with ordnance programs known as Hazards of Electromagnetic Radiation to Ordnance (HERO). Initiated in 1958, HERO covers research and development, with special attention to EEDs. Primary research by the Air Force has been related to the biological effects (see Sec. 3.1.1) and some studies have covered the effects of R-F radiation on volatile fuels (SPARKS).

3.2.4.1   Radiation Hazards to EEDs

R-F power dissipated in an EED depends on the characteristic of a signal appearing across its leads, the impedance characteristics of both the EED and the leads, and on other factors. Electric initiators are classified under seven types: high-resistance wire, low-resistance wire, carbon (graphite) bridge, conductive film, semiconductor, and spark gap. The resistance wire and the carbon bridge are the most commonly used EEDs. Combinations of more than one of these types is used to obtain special characteristics for a particular application.

The wire-bridge detonator uses a fine wire of tungsten or other noble metal between two electrodes to form the bridge. Electricity flowing in the bridge heats the wire, igniting a spot charge, which in turn sets off the detonator base charge. A typical low-energy bridge detonator is shown in Fig. 3.6. The bridge resistance ranges from 2 to 5 ohms and is made of tungsten wire. The detonator is rated at a nominal 5000 ergs at less than 10 μsec, with a capacitor charged to 50 V. The bridge wire is coated with an explosive *spot charge*.

The carbon-bridge initiator shown in Fig. 3.7 uses a colloidal-graphite charge to form the electrical detonation circuit path. It functions like the wire bridge, but is more sensitive. The lead wires inside the plug are coated with insulating varnish and twisted to attain the small separation at the face of the plug needed for graphite-bridge detonators. This detonator will function in 10 μsec., with 300 volts applied from a 2200 pF capacitor.

The explosive mixture contains conductive material which forms the electric circuit. The heating effect of the current in this mixture causes detonation. The conductive material is a mixture of metals and graphite and is mixed with explosives. Conductive mixtures vary in resistance from approximately 1 to 8000 ohms.

Because of manufacturing difficulties with the wire-type bridge and its relatively low sensitivity, a deposited-metal film initiator is sometimes used. Titanium smears on a glass base have detonated the explosive with an average energy of 50 ergs at 30 to 40 ohms resistance.

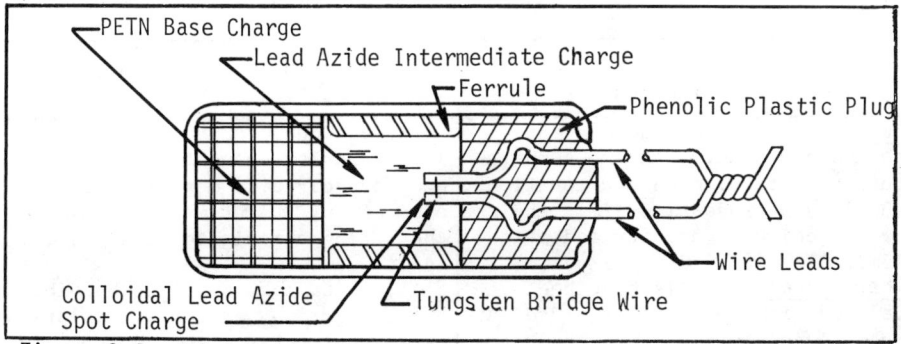

Figure 3.6 - Wire Bridge Type Electric Initiator

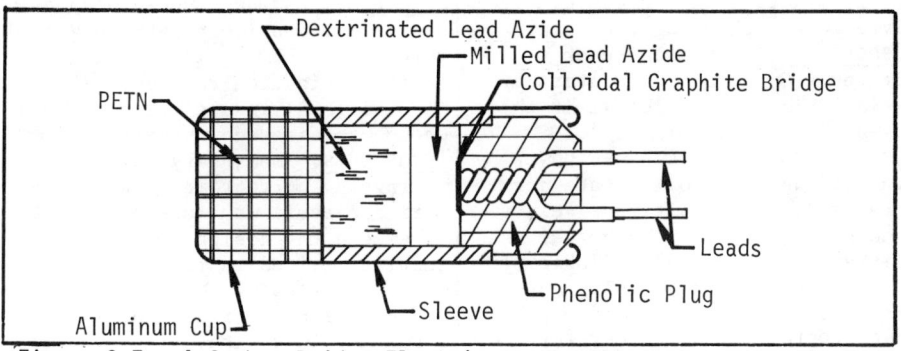

Figure 3.7 - A Carbon Bridge Electric Initiator

Three principle modes of electrical initiation of EEDs are: arcing, heating and shock wave. Arcing exists at levels greater than 25 volts and usually occurs in carbon bridge and thin-film initiators. Hot-wire initiators as well as conductive film and conductive-mix initiators operate by the heating mode. The third mode of initiation occurs in the exploding bridge wire type of initiator. The shock-wave mode requires greater than 300 volts to cause the formation of an intense shock wave which initiates a secondary high-explosive directly. Most initiators can operate under more than one mode, depending on the magnitude of the electrical stimulus.

EEDs are initiated by intentional sources; they also may be activated by unintentional sources. The conventional sources are those designed into a firing circuit to supply a controlled amount of energy to initiate the EED at the time of firing. Some circuits use a capacitor band which is charged slowly some time prior to firing time. Unintentional activating sources are those that couple sufficient EMI energy to the EED to result in inadvertent ignition. Both sources are listed in Table 3.4

Table 3.4 - Intentional and Unintentional Sources of Energy
Which May Initiate EED Firing

Intentional EED Activating Sources	Unintentional EED EMI Activating Sources
Battery. Both wet and dry batteries are used. Extreme temperature variation, high accelerations, and long storage time limit the use of wet cells. Capacity and weight limit the use of dry cells. **Thermal cell**. Cells have been developed with solid electrolytes at room temperature. When required, a small thermite charge inside the cell is activated and the electrolyte melts, charging the cell. **Generator**. Mechanical-energy driven A-C and D-C generators are used. Size and weight limit their uses. **Converter**. Converted A-C power by vacuum tubes or solid-state devices are used. The additional weight and space can often be better used either by using the available A-C power directly or by using batteries. **Electrostatic (dust generator)**. Whirling dust generators are used to develop potentials up to 5000 volts. Generation occurs only when the dust whirls with sufficient speed. High whirling speeds are obtained in most projectiles fired from a gun. Large amounts of energy may be produced with relatively small generators.	**Stray voltages**. Ground current loops, faulty connections, shorts and open circuits may produce voltages of sufficient magnitude in the firing circuit to cause ignition. **Static**. Charges built up by vibration, friction, or the accumulation of a static charge on a person's body may actuate sensitive initiators. **Lightning**. An ungrounded weapons system may provide a path to ground through the EED or its firing circuit. Ionization voltages from lightning may induce currents great enough to fire an initiator. **Transients**. Momentary surges of voltage or current in or near a firing circuit may induce currents or exceed the design limits of protective devices in firing circuits. **Magnetic effects**. Magnetic fields of sufficient strength may induce a voltage in a wire. **Thermoelectric effects**. When two dissimilar metals at different temperatures are joined, a small thermoelectric voltage is generated, such as between a copper EED lead and an aluminum ground. Since such voltages are so small, ignition would be highly improbable. **Test equipment**. Some circuit test equipment used for checkout purposes inject voltages and/or currents into a weapons system that could possibly activate an initiator. **Electrostatic**. Whirling clouds of dust or steam may produce electrostatic charges to fire a sensitive detonator.

When R-F radiated energy is the cause for unintentional firing of EED's the transmission path of a firing circuit becomes an important factor. Parameters which define the R-F source and thereby provide information for the design of protective measures are field intensity or power density, type of modulation and duty cycle, radio frequency, and polarization. Other important factors include electromagnetic coupling and thermal parameters. Here, R-F waves propagating in free space become guided waves in a coupled firing circuit. The thermal parameter represents the mechanism by which R-F initiation of EEDs take place by heating.

There exists a number of mathematical models which have been used to predict the probability of firing an EED when the EMI sources are radars and/or nearby communications transmitters. Rather simplified models assumed that the source transmitter antennas were boresighting the EED leads. Using prediction models similar to those discussed in Chap. 5, the effective area of the leads convert the arriving power densities to power available to ignite the EEDs. When superimposed on the EED known firing characteristics, the probability of initiation is predicted.

Nearly all of the above math models yield pessimistic results for radar EMI sources and optimistic results for some communication transmitter sources. This occurs for radar models because the simultaneous superposition of a boresight condition, matched polarization, ignoring the effect of EED lead twist and/or lossy material, and under-estimating the shielding effect of an intervening barrier (e.g. missile skin), necessarily results in a dangerous cumulatively probability which, in effect, may exist arbitrarily close to 0% of the time. Thus, the recommended measures are to shut-down radars to the chagrin of a ship's captain or a range safety officer. Some communication models, on the other hand, fail to recognize that an entire missile skin or aircraft can act as a pick-up antenna and capacitively couple intercepted energy to the EED leads.

Among the additional parameters complimenting the prediction process are personnel proximity or contact with the EED or its housing, structure of the EED container and ground plane, openings in the EED container, configuration of lead wires from control stations to EED or container, and R-F impedance of the firing circuit. As a result some attempt has to be made to establish safe distances of separation between explosive and EMI emitter source. While not particularly useful, Table 3.5 has been used as one measure of RADHAZ protection.

It is complicated to protect EED's from R-F radiated energy and yet allow the devices to be operable in the firing circuit. Basically,

Table 3.5 - Explosive Safe Distances

Explosive	Safe Distance in in Feet	Source	Power in Watts
Ammunition Blasting Caps	>50 ft.		
	100 ft.	Radar	3
	1000 ft.	Radar	2 kw
	1 mile	Radar	100 kw
	100 ft.	AM Radio	10 watts
	1000 ft.	AM Radio	1 kw
	1 mile	AM Radio	25 kw
	5 ft.	FM Mobile	5
	10 ft.	FM Mobile	25
	30 ft.	FM Mobile	100

factors concerning the R-F generators, the transmission media, the coupling parameters to the EED, and the EED firing characteristics should be known, as well as the type of susceptible material, its geometry and its location with a weapon system. Ambient conditions, antenna configuration, polarization, type of modulation, and such factors as frequency, power output and field intensity are pertinent factors in the problem of protecting a sensitive EED from unwanted ignition.

Protective devices that have been developed for different EED's include:

1. Attenuators - R-F attenuating materials such as carbonyl iron powder, long-chain polymers, tantalum peroxide, magnesium dioxide and ferrite materials.

2. Shielding - *Scotch* type metallic tape and twisted braid, complete metallic enclosures, etc.

3. Thermoelectric Attenuator - With a Peltier junction, DC will cause both Peltier and Joule heating, but AC will cause only Joule heating. The results are a 2 to 1 ratio of alternating current to direct current for the same heating effect: for the same current, DC has a heating ratio of 4 to 1.

4. Transmission Lines - Lossy coaxial, parallel-wire, shielded parallel-wire and parallel strip lines.

5. Bypass Capacitors - A certain amount of R-F attenuation can be achieved by shunting the leads of the EED with a high dielectric constant bypass capacitor. The size is determined by the resistance of the squib to be protected and the frequency to be bypassed.

6. Relays - Various methods have been devised to use a relay in the firing circuit in a manner that requires a fairly heavy current for the relay to operate. Experimental work has been done on special coaxial relays that pass only DC and very low-frequency AC.

7. Filters - RL and RLC dissipative filters have been used. The size, weight and mounting, and effect on the firing time are limiting factors in the use of this type of protective device.

8. Shorting Straps - Usually a temporary *shorter* is applied during manufacture to protect the initiator during loading, handling and shipping, and remains in place until the initiator is armed. When the initiator is in the circuit the temporary *shorter* is removed and a second one installed. The effectiveness of this device falls off at higher frequencies.

9. Connectors - A shielded cable connector to make contact before and break contact after the two power contacts are made.

10. Fuses - A fuse in parallel with the EED. A limitation is the one-time nature of its protection.

3.3 SUSCEPTIBILITY CRITERIA

The preceding sections discussed man-made receptors of EMI. The receptor susceptibility rating concept, introduced in Sec. 3.2, was based on the notion of signal equals noise as part of the bases for scoring. As it develops, S = N may not be an acceptable criteria for susceptibility threshold because:

● Different detection systems are affected differently by the same interference or noise spectral intensity levels.

● Different read-out and display systems may respond differently to the same detection systems.

● Different user requirements impose different interpretations to any given level of receptor performance.

Thus, it is necessary to introduce the concept of receptor susceptibility criteria in order to relate any given interference level referenced to the input of a receptor to the impact upon the user's satisfaction with the degree of performance.

Another way to introduce the receptor susceptibility criteria is to recognize that some receptors either perform correctly or malfunction entirely. A few examples of these *black and white* situations are:

● An electroexplosive device (EED) either detonates or it does not (e.g., for explosive belts, an associated fuel wing tank is dropped; a canopy and pilot are jettisoned, a missile is launched, etc.).

● A carrier-operated relay is closed or it is not.

● Any two-state output device from the receptor is either in state 0 or 1.

The problem, however, develops since most receptor output devices are not two-state output devices. In fact, they are not three-state or n-state, but rather they represent a continuum of degradation (a very large or infinite state) from *white to black*, i.e., from perfect performance to complete malfunction. Consequently, it is necessary to relate the status of performance vs S/(N+I) referred to either the input of the receptor or to it's output. The next sections summarize this for voice intelligibility, digital-error acceptance, TV-picture displays, radar displays, and other outputs.

3.3.1   Voice Intelligibility

The problem of specifying an operational performance measure for
voice communication systems is complicated by the random nature of
a received voice signal; variations in message content; and differences in hearing and understanding abilities from one receiver operator to another.

3.3.1.1   Articulation Score

One performance measure used for voice systems is *articulation
score* which is obtained by using trained talkers and listeners to
determine the percentage of words scored correctly by the listener
out of the total number of words contained in the test.

The procedure used in an articulation test consists of a talker
(or a standardized voice generator such as a tape-recorded voice)
reading a set of selected words or syllables over a communication
system (which may be subjected to interference). The listener panel
interprets what it hears. Various levels of interference may be introduced into the receiver system along with the selected words.
The percentage of words interpreted correctly by the listener indicates the intelligibility level or articulation score for the particular set of conditions tested.

Resulting empirical data can then be translated into suitable
electrical characteristics (such as signal-to-interference ratio)
which in turn can be used in an EMI prediction process to determine
voice system performance.

Fig. 3.8 shows the relationship between signal-to-interference
ratio and articulation score for different combinations of desired
and interfering signal conditions. All of the cases illustrated are
for co-channel interference conditions. One very significant factor
that is evident in the figure is that there is a fairly rapid transition from good to poor performance vs S/I ratio.

3.3.1.2   Articulation Index

Another method for specifying performance of voice communications systems is the articulation index.* Fig. 3.9 shows the rela-

---

*   See Sec. 8.2.1 of Vol. 5, EMI Prediction and Analysis

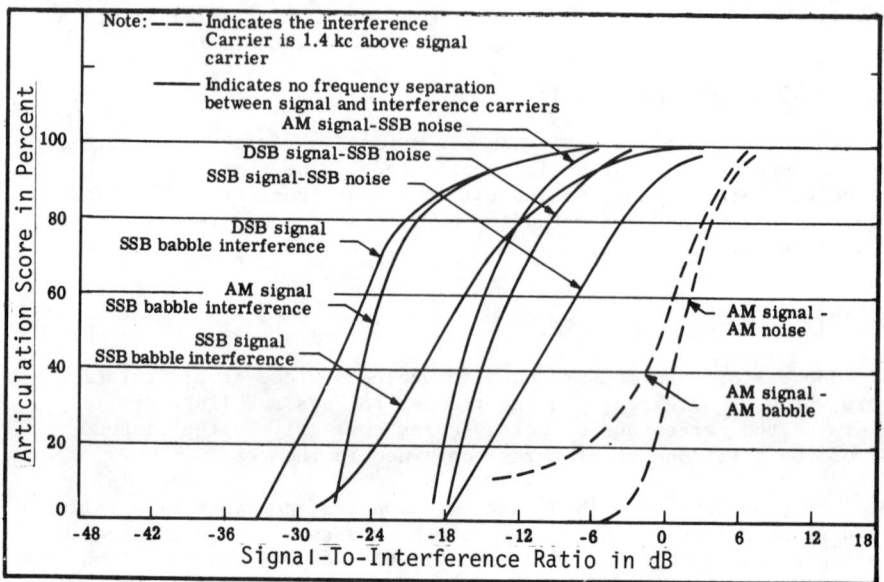

Figure 3.8 - Voice System Performance

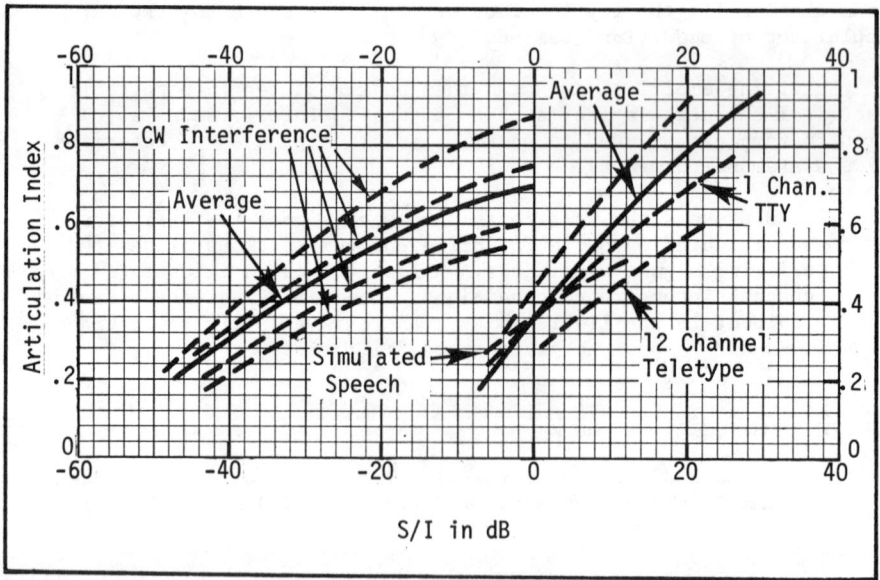

Figure 3.9 - Measured Data, On-Tune Interference

tionship between articulation index and signal-to-interference ratio for a number of different types of interfering signals.

The curves may be grouped into one of two classes: C-W interference and modulated interference. Considering the many uncer-
tainties that exist in an EMI prediction, the variations between the
different curves within a group are relatively insignificant and all
of the curves within a group may be approximated by the average curve
shown in the figures.

Figs. 3.10 through 3.15 show the relationship between articulation index, signal-to-interference, and signal-to-noise ratio for
different types of interference. Except for single frequency tones,
which do not severely degrade performance, the articulation index
relationships shown in the figures exhibit similar trends.

3.3.2 DIGITAL COMMUNICATION SYSTEMS

The evaluation of performance for a digital system consists of
calculating the probability of error. Two basic types of errors are
false acceptance (i.e., mistaking interference or noise for the signal) and false dismissal (i.e., not recognizing the presence of the
signal). For on-off binary transmission, false acceptance is equal
to the probability of false alarm and false dismissal is equal to
one minus the probability of detection.* The relative occurence of
false acceptance and/or dismissal can be determined from the probability densities for signal, interference, and noise at the receiver output.

The relationship between the basic decision process and the two
types of errors is illustrated in Fig. 3.16. The density function
designated IN(x) refers to the output probability distribution
density when interference and noise are present while SIN(x) is the
output distribution density when a signal, interference, and noise,
S+I+N, are present. Decision regions are defined such that when the
output exceeds a certain threshold, T, the decision is *signal present*
whereas if the output is less than T the decision is *no signal
present.*

---

* Probability of false alarm is the conditional probability of
deciding that a signal is present when no signal was transmitted.
Probability of detection is the conditional probability of deciding
that a signal is present, given that a signal was transmitted.

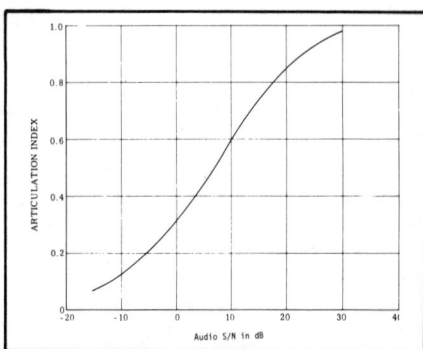

Figure 3.10- Audio Performance,
White Noise Interference

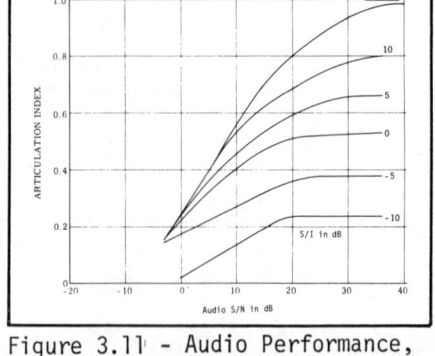

Figure 3.11 - Audio Performance,
White Noise-Simulated Speech
Interference

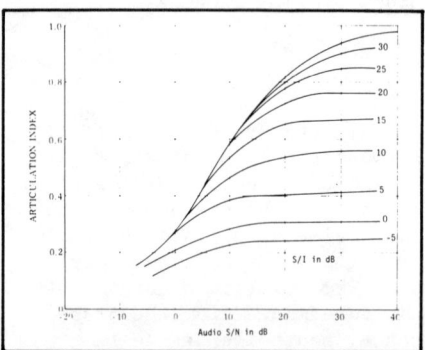

Figure 3.12 - Audio Performance,
White Noise-12 Channel Multiplex

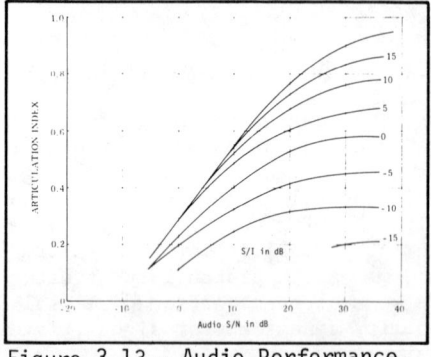

Figure 3.13 - Audio Performance,
White Noise Single Channel Teletype
Interference

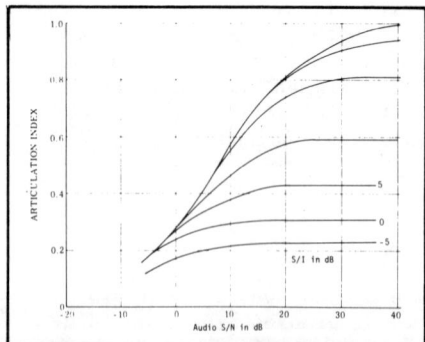

Figure 3.14 - Audio Performance,
White Noise-Pulse Interference,
200 pps

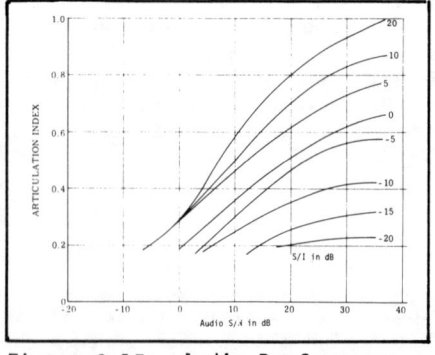

Figure 3.15 - Audio Performance,
White Noise-500 Hz Tone Interference

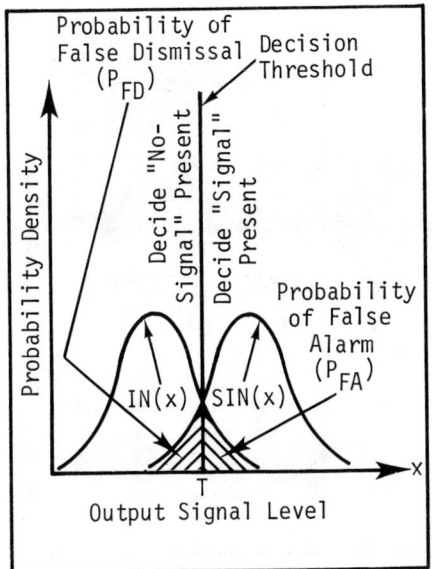

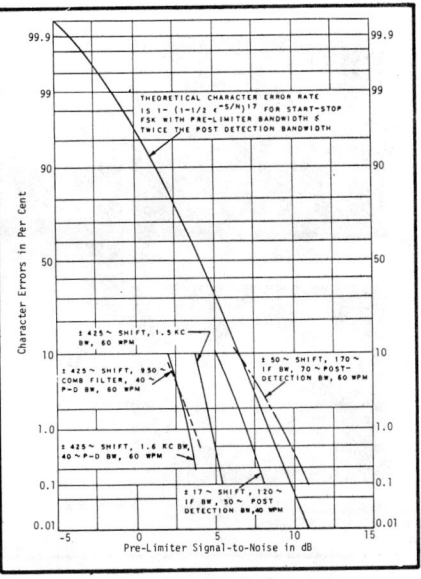

Figure 3.16 - Binary Decision
Process

Figure 3.17 - Digital System
Performance

Fig. 3.17 illustrates the relationship between signal-to-noise
ratio and error rate. Many types of interference result in noise-
like signals at the receiver output and for these situations Fig.
3.17 provides a good approximation of the error rate. One very
significant factor that is evident in the figure is that there is a
rapid transition from good to poor performance.

3.3.3 Picture Communication Systems

Television and facsimile systems transmit information that is
eventually displayed in the form of a picture, an important form of
communication. Aside from the many TV sets currently in use, this
form of communication is experiencing increasing use by law-enforce-
ment and criminal justice agencies for transmitting *mug-shots* and
line-ups to neighboring agencies. The picture phone is another example of picture communication which is expected to become widely
used in the future.

Interference can degrade picture transmission by introducing
dots, lines or bars, causing the picture to be blurred, or causing
the receiver to lose sync and roll. Fig. 3.18 shows typical effects
of different types of interference to TV. Television receivers (particularly color TV receivers) are relatively sensitive to interference.
For example, with pulse interference such as would be produced by a

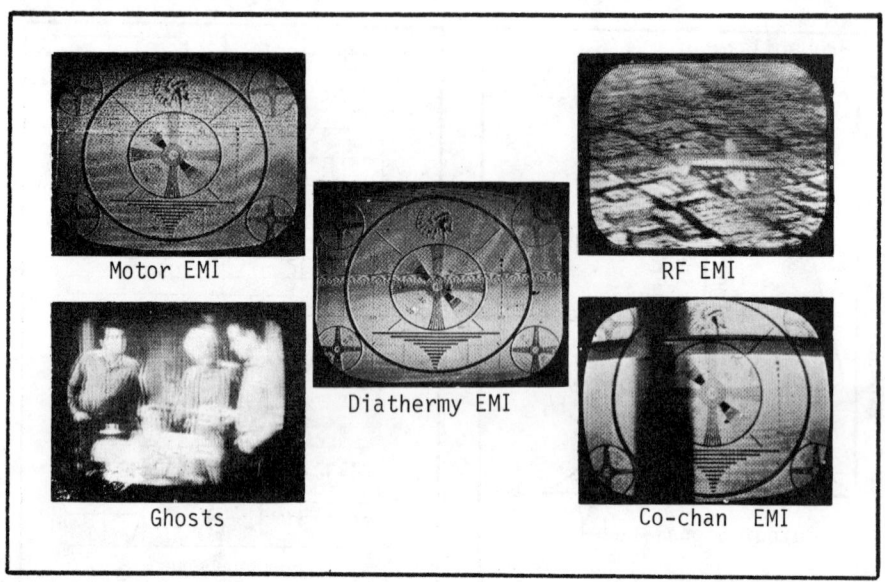

Figure 3.18 - Interference to TV Pictures

radar, a 15 dB ratio of peak signal to peak interference is required
to avoid picture degradation in the form of *snow* in the picture.

The effects of interference on the performance of picture communication systems are somewhat subjective. One technique for rating
television (TV) performance in the presence of interference is based
on establishing six rating grades as follows: (1) excellent, (2)
fine, (3) passable, (4) marginal, (5) inferior, and (6) unusable.
An extensive measurement program was conducted using the above rating
scheme and approximately 38,000 ratings were obtained on color and
monochrome TV pictures having different injected interference. Nearly
200 observers participated in these experiments.*

Representative results are shown in Figs. 3.19 and 3.20 for co-
channel interference from another TV station and for random noise
interference. For co-channel interference the signal-to-interference
ratios required for a *passable score* or better by 50 percent of the
observers are tabulated in Table 3.6.

_____

* Dean, Charles E., Measurements of the "Subjective Effects of
Interference in Television Receivers," *Proceedings of the IRE*, pp.
1035-1049, 1960.

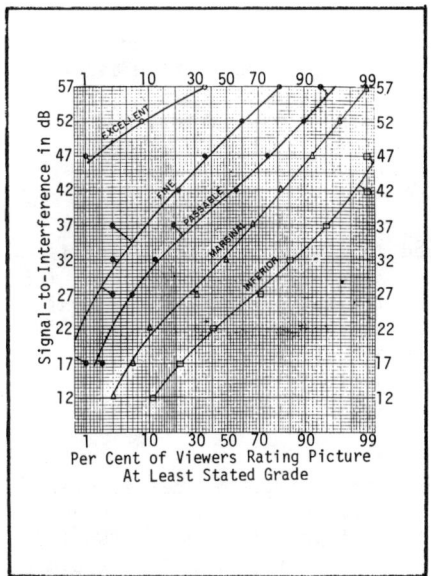

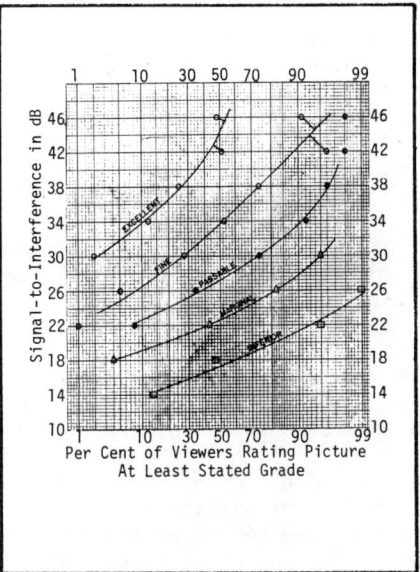

Figure 3.19 - Co-channel
Interference 604 Hz Carrier
Separation

Figure 3.20 - Random Noise
Interference

Table 3.6 - S/I Required for at Least a Passable
Score by 50% of Observers

Interfering Signal Offset in Hz	Required S/I in dB
604	41
9,985	24
10,010	17
19,995	29
20,020	17

For random noise interference the S/I requirement for at least
a passable rating by 50 percent of the observers was +27 dB on the
basis of root-mean-square (RMS) sync amplitude to RMS noise over the
6-MHz TV channel.

3.3.4   RADAR SYSTEMS

It has been estimated that over 12,000 high-power radars are
currently operating in the continental United States. Many of these
radars are situated in congested areas and because of real estate

considerations are frequently co-located at specific sites such as
airports, military bases, and missile-launching sites. In addition,
there has been and will be in the future, an increasing use of radar
in airborne systems, navigational aids, weather observation, and
satellites and space probes. This trend toward greater spectrum
usage and congestion has resulted in a mutual interference problem
which is becoming increasingly complex.

Surveillance radar systems are used in a wide variety of applications where it is necessary to monitor a relatively large area. The
primary use of surveillance radars is to monitor marine and air traffic for the purpose of national defense and traffic control. A typical surveillance radar provides an operator with a scope display of
both angular position and range of objects within the radar coverage
area. The scope used for this type of display is referred to as a
Plan Position Indicator (PPI).

The most common form of interference is the appearance of interfering dots or spirals on the radar scope presentation caused by pulse
interference from other radars. This type of interference is usually
moving continuously (called rabbits) and may cover a large portion of
the scope face, making targets difficult to detect. Interference of
this type is annoying to the operator and over a period of time causes
fatigue which reduces effectiveness. If the interference sector contains a target, delayed detection is likely to result. If the interference is extreme, false target reports become likely.

One measure of operational performance which is used for this
type of interference to surveillance radars is *scope condition*. Interference effects on radar PPI scopes have been classified into five
scope conditions which are illustrated in Fig. 3.21.

3.3.5 OTHER RECEPTOR OUTPUTS

The foregoing emphasized communications-electronics (C-E) equipments susceptibility criteria. However, there exists a myriad of
non-C-E receptors. For example, what is the impact of different EMI
levels upon:

● A digital computer intervening between monitoring and displaying the status of many patients in a hospital intensive care ward?

● A digital computer time sharing the debiting and crediting of
bank customers' accounts?

● The control systems of an urban rapid transit rail system?

● The control system of a computer-fed automatic milling machine?

● The monitoring and control system of an electrical utility
substation?

3.40

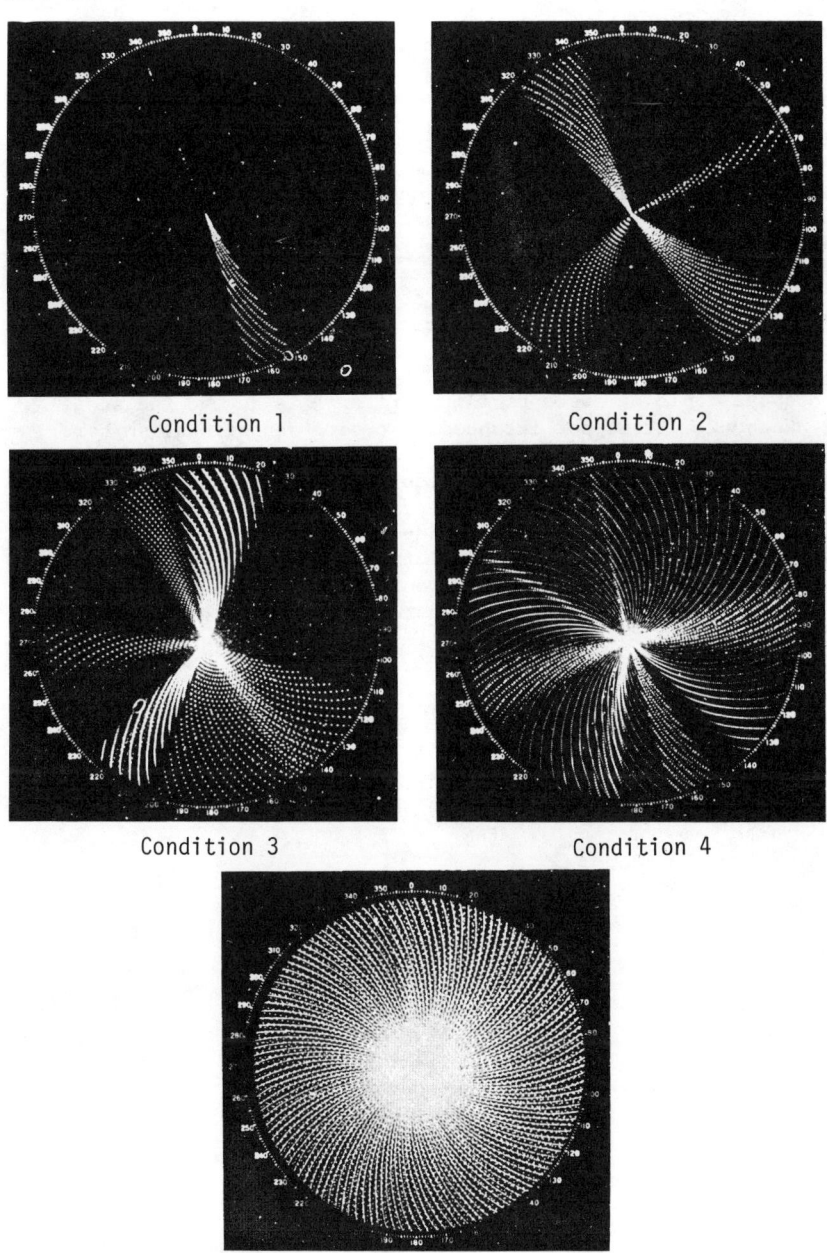

Condition 1   Condition 2

Condition 3   Condition 4

Condition 5

Figure 3.21- Typical Scope Conditions

● The transmission and status display of stock-market transactions?

● The *squiggles* on the output recorder of a patient's EKG?

● The *squiggles* on a polygraphic recorder?

Some of the above, depending upon the degree of manifested EMI, might
only flag the operator that the data are not reliable or repeatable.
Some might cause medical doctors to improperly diagnose a patient's
condition and result in wrong corrective measures. Some might cause
a customer to lose many dollars. But some could result in death to
several people or cause a power blackout over wide areas affecting
millions.

The above topic of susceptibility criteria is beyond the scope
of this handbook. It is not intended to create sensationalism here.
Rather, since each of the above EMI situations has precedent, it is
intended to aware the reader that susceptibility criteria are necessary for other than inconsequential receptors. Thus, unlike an emitter which may be radiating EMI (see Chap. 2) and not penalized, a
receptor (this chapter) will be the victim of whatever consequences
EMI may herald. Unfortunately, the proof of cause is more often
borne by the victim rather than the source even though the victim is
frequently the cause.

3.4 BIBLIOGRAPHY

(1) Abromavage, M. M., Merchant, C. C., and de Pian, L., "Electromagnetic Coupling to Ordnance Systems," Final Report Phases I and II, NWL Contract N178-7604, Alexandria, Virginia: Jansky and Bailey, Inc. (August 25, 1961).

(2) American Machine and Foundry Company, Alexandria, Virginia, "Study and Testing of Radio Frequency Insensitive Electroexplosive Matches (Squibs)," Final Report, AFSWC TR-61-88, Contract AF 29 (601)-2769, AD 266454 (October, 1961).

(3) American National Standards Institute, *Methods for Calculation of the Articulation Index*, ANSI Standard 53.5-1969.

(4) Association of Home Appliance Manufacturers, Chicago, "Safety and Measurement Standard; Microwave Ovens," February, 1970.

(5) Barron, C., Love, A., and Baraff, A., "Physical Evaluation of Personnel Exposed to Microwave Emanations," *Inst. Radio Engineers, Trans. on Medical Electronics*, p. 144, 1958.

(6) Bauner, E. J., Duffy, R., Haber, F., Polk, C., Ricci, V. C., and Salati, O. M., "Handbook for Calculating Pulsed Radar and CW Interference to AM Communications Receivers," Contract No. AF 30(602)-583; Moore School of Engineering.

(7) Bridges, J. E., and Brueschke, E. E., "Hazardous Electromagnetic Interaction with Medical Electronics," *1970 IEEE EMC Symposium Proceedings*, Anaheim, Calif., pp. 173-182, July 14-16, 1970.

(8) Brody, S. L., "Military Aspects of the Biological Effects of Microwave Radiation," *Inst. Radio Engineers, Trans. on Medical Electronics*, PGME-4, pp. 8-9, February, 1956.

(9) Communications Designer's Digest, "Receiver Design Study Identifies Interference Sources and Antidotes," Vol. 4, No. 3; March-April, 1970; pp. 48-51.

(10) Constant, P. C., Jr., Rhodes, B. L., and Chambers, G. E., "Investigation of Premature Explosions of Electroexplosive Devices and Systems by Electromagnetic Radiation Energy," Midwest Research Institute, Final Report, Vol. II, Bibliography, Contract AF 42(600)-22447, AD 275302, April, 1962.

(11) Constant, P. C., Jr., et al., "Survey of Radio Frequency Radiation Hazards," Midwest Research Institute, Final Report, U. S. Navy, Bureau of Ships, Contract No. bs-77142, June 30, 1962.

(12) Constant, P. C., Jr. and Martin, E. J., Jr., "The Radiation Hazards (RAD HAZ) Program on the Formulation of Standards," *IEE Trans. on RFI*, Vol. RFI-5, No. 1, pp. 56-77, March, 1963.

(13) "Control of Hazards to Health from Microwave Radiation,"
Army Technical Bulletin TB Med. 270, Air Force Manual AFM 161-7,
Dec., 1965.

(14) Daily, L. E., "A Clinical Study of the Results of Exposure
of Laboratory Personnel to Radar and High Frequency Radio," *U. S. Naval
Med. Bull.*, Vol. 41, pp. 1052-1065, 1943.

(15) Dickinson, W. T., "A Method to Determine Electromagnetic
Coupling to Ordnance Devices Aboard Ships," Technical Report No. 5456,
Contract N178-7604, AD 258914, Jansky and Bailey, Inc., June 15, 1961.

(16) Duff, W. G., "EMC Figure of Merit for Receivers," *Presented
at the 1969 IEEE Symposium on EMC*, June, 1969.

(17) Duff, W. G., et al., "Adjacent Signal Interference," *Communication Designer's Digest*, December, 1968.

(18) Duff, W. G., et al., "Determination of Receiver Susceptibility Parameters," *Presented at the IEEE Symposium on EMI*, July, 1970.

(19) Duff, W. G., et al., "Receiver Susceptibility Criteria,"
*presented at a joint meeting of the Boston Section EMI and Aerospace
Groups*, November, 1969.

(20) Duff, W. G., Heisler, K. G., Jr., et al., *Voice Communication
Degradation Study*, RADC-TR-67-556, February, 1968.

(21) Ebstein, B., et al., "Third-Order Intermodulation Study,"
Interference Notebook, RADC-TR-66-1, RADC-TR-67-344, July, 1967.

(22) Evans, G. R., "Effects of Electromagnetic Fields upon
Instrumentation Components and Systems," A Bibliography, Report SB-60-
27, AD 243538, Sunnyvale, California: Lockheed Aircraft Corporation,
July 27, 1960.

(23) Fitts, R. E., *Electronic Evaluation of Voice Communications
Systems*, RADC-TDR-63-355, August, 1963.

(24) Fleming, H., "Effect of High-Frequency Fields on Micro-
Organisms," *Elec. Eng.*, Vol. 63, pp. 18-21, 1944.

(25) French, N. R., and Steinberg, J. E., "Factors Governing the
Intelligibility of Speech Sounds," *Journal of the Acoustical Society of
America*, Vol. 19, pp. 90-119, January, 1949.

(26) Frequency Energy: A Potential Hazard in the Use of Electric
Blasting Caps," Safety Library Publication No. 20 (1968) - also under
preparation as ANSI Guide.

(27) Frey, A. H., "Biological Function as Influenced by Low-power Modulated RF Energy," *IEEE Trans. on Microwave Theory and Techniques*, vol. MTT-19, pp. 153-164, February, 1971.

(28) Frey, A. H., "Brain Stem Evoked Responses Associated with Low Intensity Pulsed UHF Energy," *J. Appl. Physiol.*, Vol. 23, pp. 984-988, 1967.

(29) Goldman, J., "Multiple Error Performance of PSK Systems with Cochannel Interference and Noise," *IEEE Transaction of Communication Technology*, COM-19, pp. 420-430, August, 1971.

(30) Hines, H. and Randall, E., "Possible Industrial Hazards in the Use of Microwave Radiation," *Elec. Eng.*, Vol. 71, pp. 879-881, 1952.

(31) Huenemann, R. G., "Whistle Generation in Superheterodyne Receivers," *IEEE Transactions on Communication Technology*, ITT Research Institute, Chicago, Illinois, COM-18, No. 2, April, 1970, pp. 158.

(32) King, G. R., Hamburger, A. C., Parsa, F., Heller, S. J., and Carleton, R. A., "Effect of Microwave Oven on Implanted Cardiac Pacemaker," JAMA 212:7, 1213, May 18, 1970.

(33) Korbel, S. F., "Behavioral Effects of Low Intensity UHF Radiation," Proceedings of Symposium on the Biological Effects and Health Implications of Microwave Radiation, Office of Information, Bureau of Radiological Health, Rockville, Md., pp. 436-448, 1970.

(34) Kryter, K. D., "Methods for the Calculation and Use of the Articulation Index," *Journal of the Acoustical Society of America*, Vol. 34, No. 11, pp. 1689, November, 1962.

(35) LaFond, Charles D., "Microwave Hazards are Exaggerated," *Astrionics, Missiles, and Rockets*, December 14, 1959.

(36) Lebowitz, S., "AM Desensitization Effects Due to Desired and Undesired Modulated Signals," ECAC-TN-003-266.

(37) Lustgarten, N. M., "An Approach to a Simplified Adjacent Signal Interference Model," ECAC-TN-002-19, October, 1968.

(38) Lustgarten, N. M., "An Approach to a Receiver Intermodulation (IM) Model," ECAC-TN-002-18, August, 1968.

(39) Lustgarten, N. M., "An Approach to a Simplified Spurious Response Prediction Model (UHF In-Band Response)," ECAC-TN-002-21, February, 1969.

(40) Lustgarten, N. M., "COSAM (Co-Site Analysis Model)," ECAC-TN-002-25, September, 1969.

(41) Marha, K., "Microwave Radiation Safety Standards in Eastern Europe," *IEEE Trans. on Microwave Theory and Techniques*, Vol. MTT-19, pp. 165-168, February, 1971.

(42) Mayher, R., Basic Performance Thresholds, ESD-TR-66-9, December, 1966.

(43) Meahl, H. R., "Critical Survey of Literature Pertinent to Microwave Hazards," Report No. R55GL373, General Electric Laboratory, October, 1955.

(44) Meahl, H. R., "Protective Measures for Microwave Radiation Hazards, 750-30,000 mc.," *Inst. Radio Engineers, Trans. on Med. Electronics*, p. 16, 1956.

(45) Mumford, W. W., "Heat Stress due to RF Radiation," *Proc. IEEE*, 57, pp. 171-178, February, 1969.

(46) "New Biological Effects of R-F," *Proc. 12th Annual Conf. on Electrical Technology in Medicine and Biology*, 1959.

(47) Performance Standard, "Microwave Ovens," Control of Electronic Product Radiation; PHS, HEW, Federal Register, Oct. 6, 1970.

(48) *Proc. Tri-Service Conf. on Biological Hazards of Microwave Radiation*, Rome Air Force Base, Rome, New York, ARDC-TR-58-51, ASTIA AD 115-603, July 15-16, 1957.

(49) *Proc. 2nd Annual Tri-Service Conf. on Biological Effects of Microwave Energy*, Rome Air Force Base, Rome, New York, Pattishall, E. G., and Banghart, F. W. (eds.), University of Virginia, ASTIA AD 131-477, July 8-10, 1958.

(50) *Proc. 3rd Annual Tri-Service Conf. on Biological Hazards of Microwave Radiating Equipments*, Rome Air Force Base, Rome, New York, August 25-27, 1959.

(51) *Proc. 4th Annual Tri-Service Conf. on Biological Effects of Microwave Radiation*, RADC-TR-60-180, August, 1960.

(52) Ruggera, P. S., and Elder, R. L., "Electromagnetic Radiation Interference with Cardiac Pacemakers," U. S. Dept. of Health, Education and Welfare, Public Health Service, Publication BRH/DEP 71-5, April, 1971.

(53) Sachs, H. M., "An Analysis of the Effects of CW Interference on Receiver Desensitization," ECAC-TN-002-2A; September, 1966.

(54) Schwan, H., and Li, L., "Capacity and Conductivity of Body Tissue at Ultrahigh Frequencies," *Proc. IRE.*, Vol. 41, pp. 1835-1840, 1953.

(55) Schwan, H. P., and Li, K., "Hazards Due to Total Body Irradiation by Radar," *Proc. IRE*, 44, p. 1572, 1956.

(56) Schwarzlander, H., "Intelligibility Evaluation of Voice Communications," *Electronics*, May, 1959.

(57) Steiner, J. W., "An Analysis of Radio Frequency Interference Due to Mixer Intermodulation Products," *IEEE Transactions on Electromagnetic Compatibility*, January, 1964, pp. 62-68.

(58) "Supplement to the Information Bulletin for the Discipline of Industrial Hygiene and Occupational Diseases and for Radiation Hygiene," Prague, June, 1968.

(59) "Technical Manual for RF Radiation Hazards," Dept. of Navy, Naval Ships System Command, NAVSHIPS 0900-005-8000, July, 1966.

(60) "Temporary Safety Regulations for Personnel in the Presence of Microwave Generators," USSR Ministry of Hygiene, Publication No. 273-58, Nov. 26, 1958.

(61) Thompson, A. S., *The Application of the Voice Analysis Interference System to the Prediction of Voice Intelligibility*, U. S. Army Electronic Proving Ground Pub. No. USAEPG-DR-433.

(62) Trammell, R. D., Jr., Donaldson, E. E., Jr., and Spence, P. T., "Mixer Interference Characteristics," Contract DA-36-039-AMC-02294(E), August 15, 1964.

(63) U. S. Air Force, "Radio Frequency Hazards," *Handbook*, T. O. 31-1-80, April 15, 1958.

(64) USA Standards Institute (now the American National Standards Institute, New York), "Safety Level of Electromagnetic Radiation with Respect to Personnel," USAS C95.1, 1966.

(65) U. S. Naval Weapons Laboratory, *Proc. 2nd HERO Congress on Hazards of Electromagnetic Radiation to Ordnance*, The Franklin Institute, U. S. Navy Contract, Bureau of Weapons, Contract N178-8083, April, 1963.

(66) U. S. Public Law 90-602, "Radiation Control for Health and Safety Act of 1968," Oct. 18, 1968.

(67) Wass, C. A. A., "A Table of Intermodulation Products," *IEE Journal*, 95, Part 3, January, 1949, pp. 31-39.

(68) Yatteau, R. F., "Radar-Induced Failure of Demand Pacemaker," *N. E. Jour. Med.*, 283, 26, 1447, Dec. 24, 1970.

CHAPTER 4

INTRA-SYSTEM EMI PREDICTION AND ANALYSIS

CHAPTER 4

INTRA-SYSTEM EMI PREDICTION AND ANALYSIS

Chap. 1 discussed the distinction between inter-system and intra-system EMI. This is further illustrated in Fig. 4.1 where systems A, B, C, and X each emits from it's antenna(s) to the antennas of the others. When EMI results, it is due to *inter-system* coupling and is the subject of Vol. 5 of this EMI/EMC handbook series. Inter-system EMI prediction and analysis is also discussed in Chap. 5 of this volume. When EMI results between an emitter and receptor located *within the same* package, housing, platform, vehicle, etc., it is due to *intra-system* coupling. This situation is shown in Fig. 4.1 for system X in which an antenna may or may not be involved. This chapter discusses prediction and analysis due to intra-system EMI, as illustrated in system X.

In discussing intra-system EMI, one of two prediction approaches may be used: (1) programmed mathematical models for digital computer prediction and analysis, or (2) math models in which the solution is carried out manually. It does not necessarily follow that the former approach is the more creditable simply because computer simulation is used. Certainly, computer modeling is a more powerful tool that lends itself to a more comprehensive approach and removes the drugery of calculations so that the analyst can concentrate on EMI-control trade offs and sound solutions. Manual prediction can be just as creditable and, in certain situations, more accurate provided the basic mathematical models are validated and are more applicable to the problem at hand. Thus, the emphasis here is the technique of prediction, not the machine tools, especially when many emitters and receptors are involved.

This chapter surveys two existing intra-system prediction and analysis models using computer simulation: (1) the McDonnell Douglas program and (2) the TRW program. Most engineers, tasked with the responsibility of performing an intra-system EMI prediction, will not have any of the above two programs available to them. Because they often need a fast feedback or because the problem is less complex, a manual prediction technique is also discussed in this chapter. It consists of two parts: (1) identifying possible EMI emitter-receptor

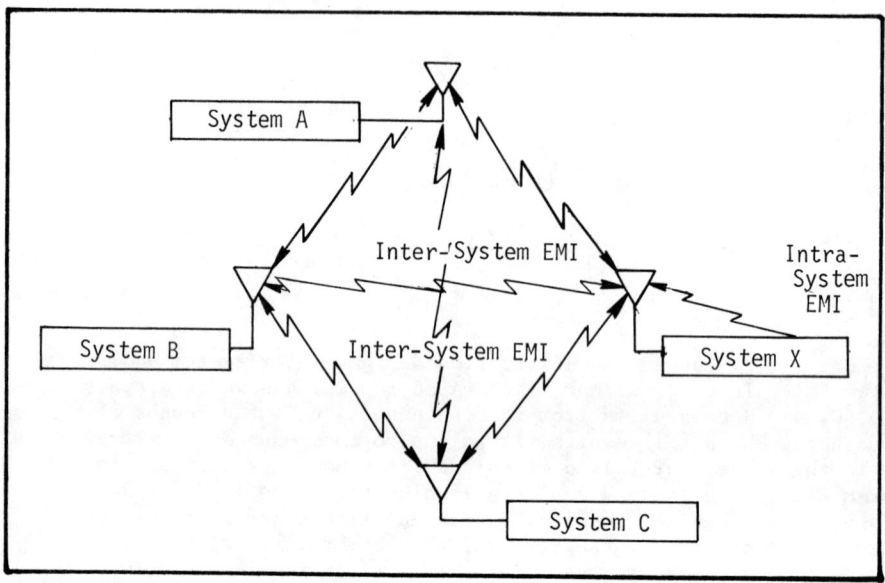

Figure 4.1 - Illustrating Inter-System and Intra-System EMI

combinations and culling out unlikely EMI situations to the point
where only a few probable ones survive, and (2) employing a detail
prediction on these surviving emitter-receptor combinations.   An
EMI prediction form is included to facilitate calculations and assure
comprehensiveness in considering modes of coupling.

4.1 EMI PREDICTION PREPARATION

Before any EMI prediction can be carried out, whether computer assisted or manual, it is necessary to perform certain preliminary data-collection steps. This section discusses:

(1) Identifying possible emitter-receptor combinations, viz:

- Identify all potential or actual emitters
- Identify all potential or actual receptors

(2) Determining spectrum signatures, i.e.:

- Determine spectrum emission signature for each emitter
- Determine spectrum susceptibility signature for each receptor

(3) Identifying possible coupling paths, i.e.:

- Antenna-to-antenna radiated
- Antenna-to-box radiated
- Antenna-to-wire radiated
- Box-to-antenna radiated
- Box-to-box radiated
- Box-to-wire radiated
- Wire-to-antenna radiated
- Wire-to-box radiated
- Wire-to-wire radiated
- Wire-to-wire conducted
- Common-mode conducted
- Complex combinations

(4) Compiling repertoire of potential EMI control techniques:

- Cabling criteria
- Grounding and bonding
- Component and circuit selection
- Shielding
- Filtering
- Other criteria

The first two steps must be performed before prediction can start. If a rapid-culling prediction process is used to narrow down the emitter-receptor combinations, as explained in Sec. 4.3, the third step can be deferred. In that way gainful effort is conserved for application to a manageable number of the more likely surviving EMI situations. The fourth step outlined above can be applied after the initial prediction in order to budget the necessary EMI controls or fixes to firm up the system design. In practice, this step is or should be included in the initial design stage.

4.1.1 IDENTIFYING EMITTER-RECEPTOR COMBINATIONS

The problem starts with identifying all possible system electrical-noise makers including: (1) intentional communications-electronics emitters and (2) internal-noise sources such as generators and power suppliers, modulators, digital-data and control signals,

4.3

brush-commutator type motors, relays, solenoids, SCR's and the like. This is shown as *identify all E_n Emitters* in box #1 of Fig. 4.2. Associated with this is collecting, defining or estimating both a spectrum and temporal signature (see Chap. 2) of each E_n emitter (box #2). Temporal signatures or electromagnetic scenarios, as they are sometimes called, are important since they impact the probability of EMI.

The next step is to identify all receptors, i.e., responding devices, displays, indicators, and the like which can unintentionally exhibit EMC performance degradation, malfunction, or end in mission abort. As applicable, this includes output display devices, cockpit instruments, mechanical surface controls in vehicles,* electro-explosive devices (EED), etc. This is shown in Fig. 4.2, box 3 as *Identify Output Devices and Indicators*. Associated with this is either defining a go/no-go threshold of undesirable response** (box 4; also see Sec. 3.3, Susceptibility Criteria) or establishing a gray-scale scoring scheme to weight degrees of performance degradation.

Since output devices, per se, generally are not *directly* affected by culprit emitter(s), the next step is to identify more sensitive receptors which drive these output devices. This is carried out at one or two of the most sensitive levels. For example, in a superheterodyne receiver, this would correspond to the antenna/R-F input terminals and the I-F input. For a digital computer, it would be the memory sense amplifiers. This is shown in Fig. 4.2, box 5 as *Identify Lower-Level Sensors for Output Devices*. Concurrent with this, the associated susceptibility levels of these receptors used to drive or activate the output devices (box 6), is defined or estimated.

The foregoing discussion (exclusive of spectrum signatures) is illustrated in the form of a graph of principle emission and susceptibility amplitude vs frequency. Fig. 4.3 shows such an example for a hypothetical aircraft system.*** The frequency span of all significant emitter-receptor situations ranges from about 300 Hz to 10 GHz, representing a spectrum coverage of 7-1/2 decades (about 24 octaves). The left ordinate of Fig. 4.3 is power in units of dBm. For a 50-ohm impedance level, the right ordinate in units of dBμV may be used.

---

* Note that not all output manifestations of EMI are electrical or electronic displays, indicators, etc. A hunting-servo system, for example, can cause unnecessary bearing wear or reduction in mechanical life of aircraft control surfaces. A second example is one in which a wing-tip tank could be dropped if an EED were inadvertently triggered.
** See MIL-STD-461A, MIL-E-6051D and Chap. 5 of Vol. 1, Electrical Noise and EMI Specifications.
*** To facilitate the example, all identifiable emitters and receptors are not shown.

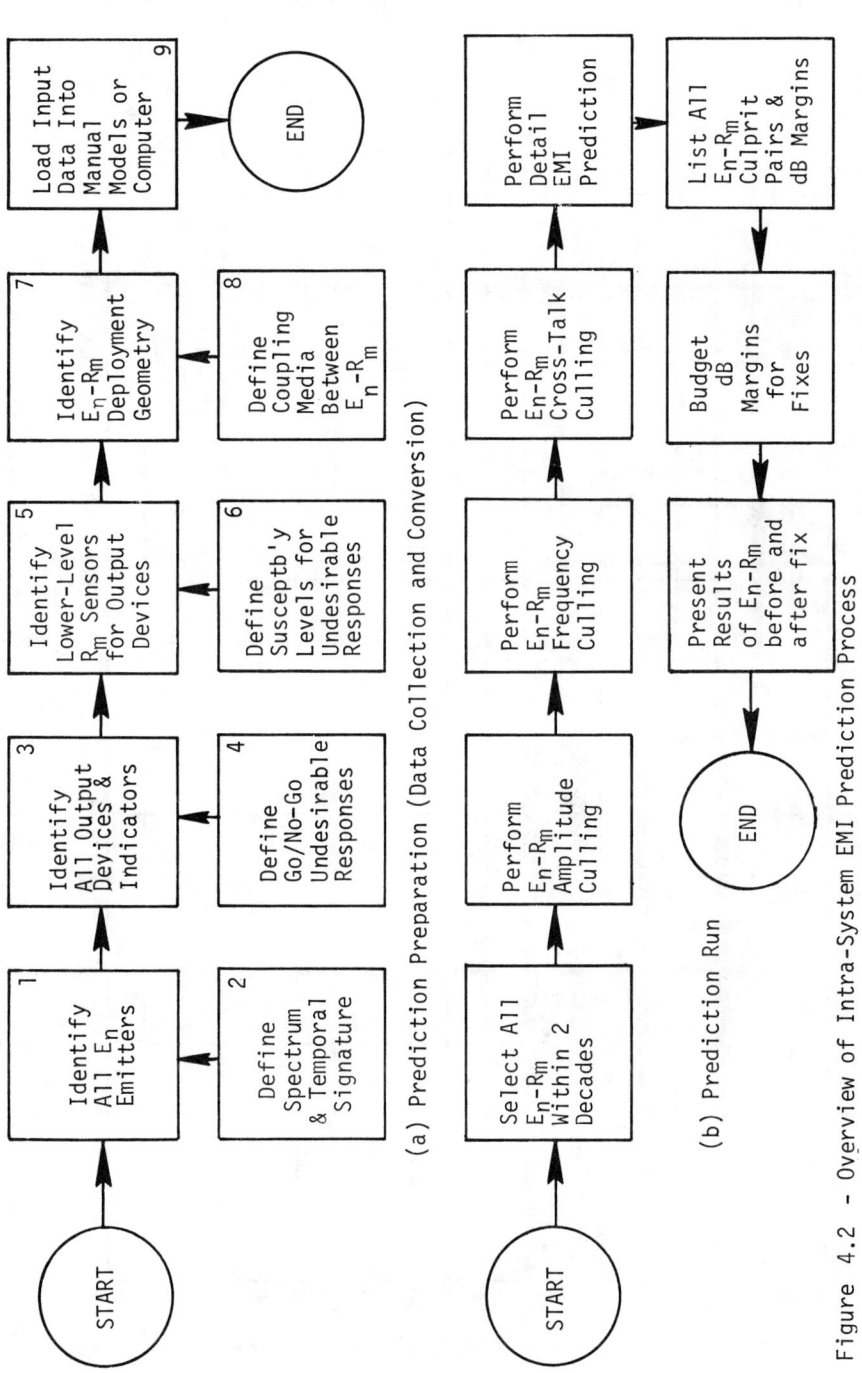

(a) Prediction Preparation (Data Collection and Conversion)

(b) Prediction Run

Figure 4.2 - Overview of Intra-System EMI Prediction Process

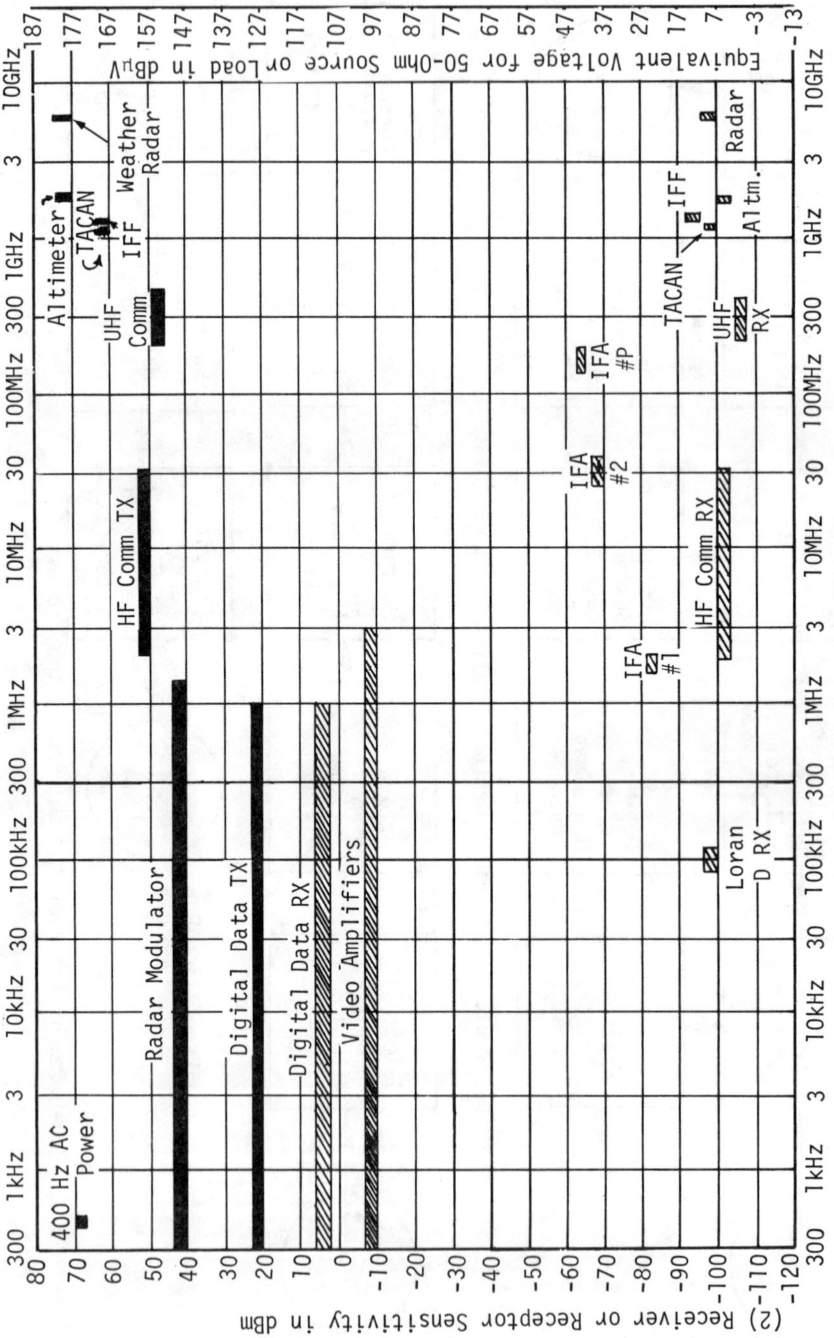

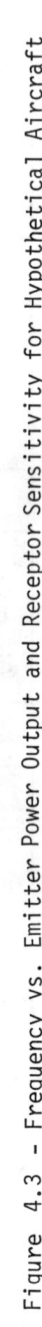

Figure 4.3 – Frequency vs. Emitter Power Output and Receptor Sensitivity for Hypothetical Aircraft

The heavy horizontal lines shown in Fig. 4.3 represent the principle emitters and the hatched-lines are the receptors. Note that
some emitters are shown occupying a significant spectrum width. This
is because: (1) they either are tunable or are allocated this spectrum,
such as communications emitters, or (2) their fundamental emission
is pulsed, such as base-band digital data or modulator drivers. An
analogous situation applies to most of the receptors which are either
allocated a spectrum or are operated at base-band frequencies. A
further examination of Fig. 4.3 shows that some emitters are fixed
in frequency (e.g., 400-Hz AC power mains) and some receptors occupy
a narrowband (e.g., the I-F amplifiers, viz., IFA, and the Loran D
receiver).

In practice, it is not necessary to actually draw such a graph
as illustrated in Fig. 4.3. Equivalent data can be listed in tabular
format, put into a matrix, or put on punched cards for computer use.
However, four points become immediately apparent in examining Fig.
4.3:

(1) All* emitters are rated at higher power levels than are
the susceptibility levels of all receptors.

(2) Some emitters are in vertical (frequency) alignment or
overlap with either their intentional receivers or potentially-victim
receptors.

(3) Some emitters, while not overlapping in frequency with
potential victim receptors, may constitute either an *adjacent-channel*
or *out-of-band* coupling threat.

(4) The vertical displacement (dB level difference) between an
emitter and receptor may be small (e.g., 30 dB between the digital
data transmitter and video amplifier of Fig. 4.3) so that EMI is unlikely. On the other hand, the displacement may be enormous (e.g.,
140 dB between the radar modulator and the Loran D receiver) so that
EMI is considerably more likely.

There exists 136 emitter-receptor pairs shown in Fig. 4.3, only
a few of which will end up as likely EMI problems. The technique for
narrowing down the emitter-receptor pairs to these problem situations
is discussed in Sec. 4.3, on culling models.

4.1.2 DETERMINING SPECTRUM SIGNATURES

The horizontal emitter and receptor lines shown in Fig. 4.3
might tend to imply that emitters do not emit outside the indicated

---

*   This may not always be true for every significant EMI situation.

frequency limits and that receptors are not susceptible outside of their indicated limits. However, this is not only *not* the case, it is the principle reason for many EMI situations, viz. an emitter's out-of-band emissions may jam a receptor's in-band response or an emitter's in-band emissions will jam a receptor's spurious responses.

Chap. 5 presents typical spectrum signature data for transmitter out-of-band emissions and receiver spurious responses. It is always best to use known or measured data. If this is not available, other math models, similar to the class of emitter/receptors in question, can be used. While it is sometimes possible to predict such spectrum signatures, parasitic resonances and higher-order effects make this dangerous (e.g., errors of 30 to 50 dB may develop) when an emitter-receptor fundamental frequency alignment exceeds about one decade.

Spectrum-signatures for several man-made noise emitting devices are presented in Chap. 2. Spectrum signatures in the form of susceptibility data are presented in Chap. 3 for many receptors. However, one of the largest problems confronting the EMI community, is that the spectrum signature data bank for non C-E emitters and receptors is grossly inadequate.

4.1.3   Identifying Possible Coupling Paths

The mode(s) of coupling an emitter to a receptor can become very complicated. In general, the coupling paths are more extensive than those for inter-system EMI and are not as well defined since considerably less money has been spent in developing coupling-path models. Coupling can also result from a combination of paths, such as conducted from an emitter to a point of radiation, then picked up by induction, and re-conducted to the victim.

Some idea of EMI coupling paths is suggested in Fig. 4.4, where both radiation and conducted paths are illustrated. While not all inclusive, these paths account for, perhaps, 98% of all intra-system EMI situations. The object is to classify each potential EMI situation into one or more of the coupling paths illustrated. The radiation paths are:

- Antenna-to-antenna
- Antenna-to-box
- Antenna-to-wire
- Box-to-antenna

- Box-to-wire
- Wire-to-antenna
- Wire-to-box
- Wire-to-wire

- Box-to-box

The conduction paths are:

- Wire-to-wire
- Filters

- Common-ground impedance
- Common-source impedance

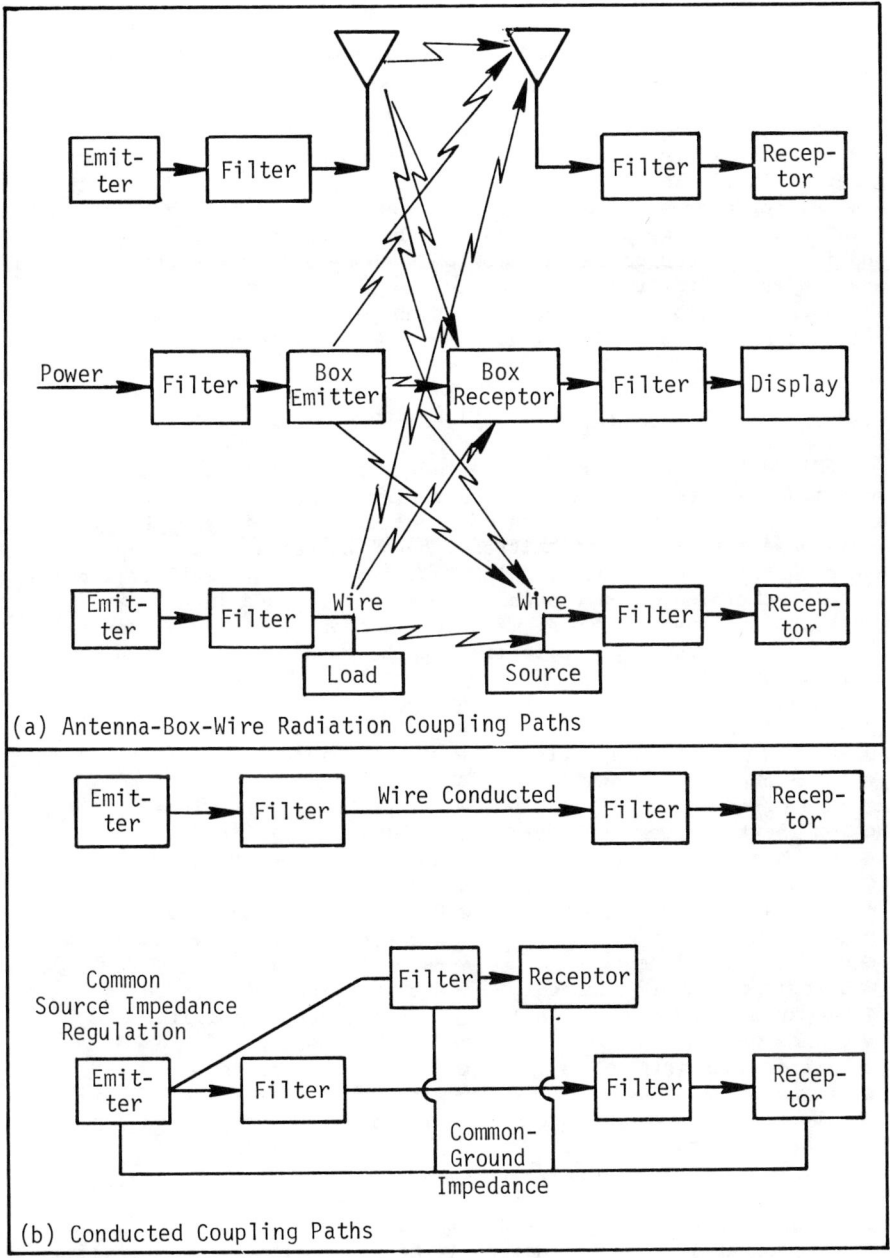

(a) Antenna-Box-Wire Radiation Coupling Paths

(b) Conducted Coupling Paths

Figure 4.4 - Principle EMI Coupling Paths

The above coupling paths are discussed in Sec. 4.2, Computer Models and in Chaps. 6 through 16 in this handbook.

Where many emitter-receptor pairs are involved in a prediction, a rapid-culling process is used to narrow down candidates to a smaller number of likely EMI situations (see Sec. 4.3). The third step in this culling process involves elimination by coupling-paths sorting in which rather simplified models for the above are used. Thus, coupling paths need be defined for only those emitter-receptor pairs surviving the first two stages of culling. This avoids much unnecessary coupling computation on pairs which are first eliminated through amplitude and frequency culling.

4.1.4   COMPILING REPERTOIRE OF EMI CONTROLS

EMI-control can be applied in the design, installation, and/or operational stages of equipment and systems. A best practice is to apply EMI control in all stages of the life cycle of systems. Thus, the mathematical models of emitters, coupling paths, and receptors should contain the essence (explicit or implicit) of nearly all EMI-control techniques, even though many may not be used in any given problem. To illustrate this, math models should contain as applicable, the following alphabetic listings as a function of frequency or other variables:

- Antenna beamwidth
- Antenna gain, hemispheric
- Antenna polarization
- Bond impedance
- Box aperture leakage
- Box shield grounding
- Box shielding effectiveness
- Cable shield grounding
- Cable shielding
- Circuit load impedance
- Circuit source impedance
- Compartment shielding
- Connector leakage
- Emitter bandwidth
- Emitter harmonics
- Emitter spurious emissions
- Filter transmission loss

- Free space loss
- Gasket shielding
- Ground impedance
- Harness cable groupings
- Harness cable shields
- Non-RLOS propagation
- Pulse rise and fall times
- Pulse duration and rep rate
- Receptor selectivity
- Receptor sensitivity
- Receptor spurious responses
- Receptor susceptibility
- Terrain masking
- Twisted-wire pairs
- Wire/cable capacitive coupling
- Wire/cable inductive coupling
- Wire/cable shields

Wire orientation

Many of the above will be included in any particular model, but many also may not apply. By including all at the beginning, the process of EMI prediction and control then becomes one of trade-off and manipulating the variables.

4.2   COMPUTER PREDICTION

This section reviews intra-system EMI prediction and analysis using computer techniques. Two such computer models have been developed: (1) the McDonnell Douglas program and (2) the TRW program.

4.2.1   THE MCDONNELL DOUGLAS PROGRAM

This intra-vehicle program was developed under USAF Contract No. F33615-70-C-133 with the AFSC Avionics Laboratory to predict and analyze EMI between avionics systems on aerospace vehicles. Four principle EMI coupling paths are mathematically modeled: (1) antenna-to-antenna, (2) wire-to-wire, (3) electromagnetic field-to-wire, and (4) box-to-box. Each of these models was assembled into a CDC 6600 computer program using FORTRAN IV language. The programs are modular and can be run separately or combined into one internal program.

4.2.1.1   ANTENNA-TO-ANTENNA MODELS

Mathematical models based on UHF-blade antennas were developed for describing coupling loss in the 225-800 MHz spectrum. Three geometric aircraft situations were modeled and validated: (1) ground plane only to simulate line-of-sight coupling between two antennas on a common metallic plane, (2) ground plane obscured by a half cylinder to simulate coupling around the fuselage, and (3) ground plane obscured by both a cylinder and wing to simulate coupling around the wing portions of aerospace vehicles. The ground plane only was tested over a 30 to 40-dB isolation geometry and validation tests indicated that coupling loss predictability averaged 2 dB in error. The ground plane and cylinder situation was tested over a 40 to 55-dB isolation geometry and tests indicated average errors of 5 dB. Finally, the ground plane, cylinder, and wing situation was tested over a 40 to 60-dB isolation geometry and tests showed that errors averaged 7 dB.

The limitations of this program are based on far-field, UHF models. Thus, errors would become significant for H-F communications antennas, especially those close to each other. Further, the model is based on co-planar antenna locations and an obscuring cylinder of 2-1/2 foot radius. The effect of these constraints is unknown.

4.2.1.2   WIRE-TO-WIRE MODELS

Mathematical models based on 60 wires inside a single harness (40 wires and 20 shields) and 40 signal sources were developed in which all wire geometry are adjacent. Thus, the model calculates EMI at wire loads within the harness due to cross talk and interaction of all other wires. However, capacitance between wires inside and outside a shield is assumed zero. Twisted groups of wires are modeled so that common-mode rejection is included.

Validation tests were performed on a 16-foot, unshielded harness located two inches above a ground plane. Four sources of EMI and loads were used. Coupling of a 115 VAC, 400-Hz source to different wires at the fundamental and third harmonic resulted in a 40-90 dB cross talk.

Errors between calculated and measured data at 400 Hz averaged 3 dB
while the difference averaged 8 dB at 1200 Hz. For 28 VDC switching
tests, errors below 16 kHz ranged to 20 dB and approached 45 dB between 10 kHz and 10 MHz. Average errors of both, however, were 8 dB.
A 10V square-wave test from 10 kHz to 10 MHz resulted in peak model
errors of 40 dB with average errors of 11 dB. In addition to the
above, some limitations of the wire-to-wire models include:

- Performance was restricted to wires in a single harness only.
- No outer harness shield was used and no harness-to-harness
  coupling was modeled.
- Wire separation was based on each wire adjacent to every
  other wire.
- Coupling in coaxial and triaxial lines was not evaluated.
- Harness break-outs (branched bundles) was not accommodated.

4.2.1.3   Electromagnetic Field-to-Wire Models

Mathematical models were developed to predict EMI induced on
wires from on-board transmitting antennas. This form of EMI occurs
when harness routing is close to openings in aircraft skin such as
radomes, landing-gear wells, pylons, plastic composite panels, and
canopies. The model is based in part on portions of the preceding
two models. The coupling is formed by the ratio of the induced wire
voltage at the load to the antenna output voltage.

Validation tests were performed from 100 MHz to 10 GHz with an
excitation antenna located 20 feet away illuminating an 8 inch x
16 inch slot. The harness was exposed two inches behind the slot.
Wires in the bundle contained both shielded and unshielded wires and
twisted pairs. Overall results indicated errors of 9 dB from 100 MHz
to 1 GHz and 11 dB from 1 GHz to 10 GHz.

The principle limitations of this model are the very large peak-
to-peak errors resulting primarily from incorrect resonances in the
models. Since EMI receivers were used to measure data, the loading
effect of these receivers raises questions regarding the accuracy of
the validation tests.

4.2.1.4   Box-to-Box Models

The mathematical model used is based on L-F magnetic coupling
between devices contained in aluminum enclosures located in aerospace
vehicles. Electric-field and plane-wave coupling is not included
since most equipment enclosures provide good shielding effectiveness.
It was rationalized that most H-F coupling is traceable to wire
coupling, covered by a previous model.

Validation tests consisted of using a 1 kHz source to drive a transformer inside a box and to sample the outside magnetic field over a 10 to 100 cm. distance for radiation and to reverse the procedure for susceptibility tests. Predictability was within 1 dB of measured results. Model limitations are restricted to transformers at 400 Hz or other ELF. Models are needed for broadband magnetic sources such as solenoids, motors, servo transformers, and for magnetic receptors such as CRT, flight-control system actuators, magnetic-tape heads, and sensors containing coils.

4.2.2   The TRW Program

This program was developed by the TRW Systems Company under NASA Contract #9-7305 entitled *Development of a Space Vehicle Electromagnetic Interference/Compatibility Specification*, dated 28 June 1968. The program consists of a number of generator, transfer, and receptor models. The generator models include voltage and current density spectrum sources, E- and H-field density spectrum conversions from antennas and wires, and filters. The transfer models include a close-coupled inductive and capacitive transfer for wires and on H-field transfer models. The receptor models are limited to filters only.

EMI may be transferred from any generator terminal to any receptor terminal via a number of paths. These paths are characterized by three parameters which can be unique for each generator-receptor terminal pier: close-coupled, common-run wire; close-coupled separa-distance; and a field-coupled separation distance. Close-coupled capacitive, inductive, and resistive wire-to-wire transfer of EMI takes place over a specified length and distance and impedance. Each generator may produce EMI as either a wire or antenna and is picked up by the receptor after the fields have decayed through a specified distance. Each wire may respond to fields as a wire and/or an antenna. Wire-to-antenna and wire-fields can be reduced by a bulkhead factor.

4.2.2.1   Generator Models

Commensurate with EMI paths of the system model, each generator terminal is modeled as having four kinds of interference emanations: a wire-voltage and current density spectrum, and an E- and H-field density spectrum at a one meter distance. The four spectral emanations are described directly by a frequency-domain spectral profile. One of four-voltage source models may be modified by one or two filters to produce the wire-voltage spectrum. One of three E-field conversion models is then used to generate the E-field at one meter from an antenna or wire in terms of the terminal voltage output. The equivalent process is used to generate the wire-current spectrum and the corresponding H-field output of the generator.

4.2.2.2 TRANSFER MODELS

There are four kinds of transfer models corresponding to the four
kinds of generated EMI. The close-coupled, a capacitive transfer
function is the amount of spectral voltage induced in the various
types of receptor wires per unit voltage on various types of generator
wires. Similarly, the close-coupled, inductive and resistive function
is the amount of spectral voltage induced in receptor wires per unit
current flowing on generator wires. The E-field transfer function
is the amount of spectral voltage induced in either receptor wires or
antennas per unit E-field at one meter from the generator wire or
antennas. Similarly the H-field transfer function is the amount of
spectral voltage induced in receptor wires and antennas per unit H-
field at one meter from the generator wire and antennas. A field
distance function model translates this one-meter reference to the
actual separation distance. The purpose for the one-meter is to
relate all this to MIL-STD-461A, as desirable for certain side benefit applications.

The close-coupled transfer models calculate the voltage induced
in a load connected to a wire in terms of spectral voltage and currents present on an EMI generating wire. The models apply to many
wire configurations including shielding and twisting effects over a
frequency range of 10 Hz to 1 GHz. Predictability of the models is
about $\pm$ 3 dB below 10 kHz, $\pm$ 5 dB from 10 kHz to 10 MHz and about
$\pm$ 20 dB from 10 MHz to 1 GHz.

4.2.2.3 RECEPTOR MODELS

The receptor model configuration represents a voltage detector
with a specifiable frequency response described by the product of
two filter frequency-response curves. Each of these may have low-
pass, high-pass, or band-pass characteristics. The filter models
are used with both generators and filters. The equations are limited to the Butterworth maximally-flat response. A parameter is
added to limit the out-of-band attenuation.

4.3 EMITTER-RECEPTOR CULLING PROCEDURES

This section outlines the procedure for taking many emitter-receptor potential EMI situations and, by a culling process, narrow down the number of candidates to a few likely EMI situations. Thereafter, a detailed EMI prediction can be carried out on the surviving few. This avoids wasted effort on performing EMI predictions between emitters and receptors which would never cause mutual interference. The technique is useful for both computer and manual prediction.

The process involves:

• Listing in a matrix all emitter-receptor combinations in which the matrix entry corresponds to dB displacement regardless of lack of frequency alignment.

• Performing an amplitude culling which accounts in a rough way for lack of frequency alignment.

• Performing a frequency culling which accounts for receptor bandwidth correction of the emitter bandwidth emissions and for certain adjacent-channel situations.

• Performing a coupling-path culling which accounts in a rough way for spatial separation.

The best way to explain the above process is to carry out an example. Thus, Fig. 4.3 will be used as the starting point in which potential emitters and receptors already have been identified (see Sec. 4.1).

4.3.1 MATRIX COMPILATION

The first step preceding culling is to:

(1) List all identifiable emitters in the heading columns of a matrix.

(2) List all identifiable receptors in the rows of a matrix.

(3) Enter dB displacement (emitter dBm - receptor sensitivity dBm) in each matrix box without regard to frequency alignment. Do not enter *intentional* emitter-receptor pairs.

When the above three steps are carried out for Fig. 4.3, Table 4.1 results. Note that 136 potential emitter-receptor EMI situations exist. The dB entries vary from 15 dB (digital data transmitter to computer) where EMI is highly unlikely, to 175 dB (altimeter and radar emitter to UHF communications receiver) where EMI may be a distinct possibility.

Table 4.1 - Intra-System, Emitter-Receptor Matrix (1st Cull)

Receptors \ Emitters →	400 Hz AC	Digital Data	HF Comm	UHF Comm	TACAN	IFF	Altimeter	Radar	Radar Modul.	Receptors
Digital Data	65	▨	45	40	55	55	65	65	35	Digital data
Video Ampl #1	80	30	60	55	70	70	80	80	50	Video Ampl #1
Video Ampl #2	80	30	60	55	70	70	80	80	50	Video Ampl #2
Computer	65	15	45	40	55	55	65	65	35	Computer
IFA #1	155	105	135	130	145	145	155	155	125	IFA #1
IFA #2	140	90	120	115	130	130	140	140	110	IFA#2
IFA #P	135	85	115	110	125	125	135	135	105	IFA#P
Loran D	170	120	150	145	160	160	170	170	140	Loran D
HF Comm	170	120	▨	145	160	160	170	170	140	HF Comm
UHF Comm	175	125	155	▨	165	165	175	175	145	UHF Comm
TACAN	170	120	150	145	▨	160	170	170	140	TACAN
IFF	165	115	145	140	155	▨	165	165	135	IFF
Altimeter	170	120	150	145	160	160	▨	170	140	Altimeter
Radar	170	120	150	145	160	160	170	▨	▨	Radar
EED #1	80	30	60	55	70	70	80	80	50	EED #1
EED #Q	80	30	60	55	70	70	80	80	50	EED#Q
E_n-R_m Cases :	136									

4.3.2   Amplitude-Culling

In the parlance of Chap. 5, four cases are examined for each pair of E_n-R_m situations shown in Table 4.1, viz:

FIM-Fundamental Interference Margin:

FIM corresponds to cases where there exists a vertical frequency alignment in Fig. 4.3 of an E_n emitter with an unintentional R_m receptor including any edge overlap. For FIM situations involving receivers, the band edges are extended by 20% of the edge frequency shown in Fig. 4.3. For baseband amplifiers and receptors, the band edges are extended by one octave. Where such modified E_n-R_m frequency alignment exists, a 0-dB correction factor is applied to Table 4.1 and entered in the upper part of the corresponding box entry in Table 4.2. The lower part of the same box then equals Table 4.1 entry - 0 dB. If the case is not an FIM situation, then proceed to the next case for each E_n-R_m pair.

TIM - Emitter Interference Margin
RIM - Receptor Interference Margin

These cases are processed equally here* and are illustrated in Fig. 4.5. They correspond to coupling from: (1) an emitter fundamental emission to a receptor spurious response (TIM case), and (2) an emitter spurious level to a receptor fundamental response (RIM). The test for this is to identify if a given E_n-R_m pair in Table 4.1 exhibits vertical frequency alignment within 1 decade of the nearest frequencies of approach for each in Fig. 4.3. This is illustrated in Fig. 4.5. Where such a frequency separation is $\leq$ 1 decade, a -60 dB correction factor is applied to Table 4.1 and entered in the upper part of the corresponding box entry in Table 4.2. The lower part of the box then equals the Table 4.1 entry - 60 dB.

SIM - Spurious Interference Margin

This corresponds to cases where the closest approach of vertical frequency alignment, Δf, for an E_n-R_m pair in Fig. 4.3 is: 1 decade $\leq \Delta f$ 2 decades. A correction factor of -120 dB is applied in Table 4.1 and entered in the upper part of the corresponding box entry in Table 4.2. The lower part of the box then equals Table 4.1 entry - 120 dB.

BTIM - Break Through Interference Margin

This case corresponds to situations covering *high-field effects*, the electromagnetic pulse (EMP), and in general brute-force entry. A correction factor of -150 dB is applied to these Table 4.1 entries

---

*   cf. Chap. 5.

Table 4.2 - Intra-System, Emitter-Receptor Matrix (Amplitude Cull)

Receptors	400 Hz AC	Digital Data	HF Comm	UHF Comm	TACAN	IFF	Altimeter	Radar	Radar Modul.	Receptors
Digital Data	0 / 65	▨							0 / 35	Digital data
Video Ampl #1	0 / 80	0 / 30	0 / 60						0 / 50	Video Ampl #1
Video Ampl #2	0 / 80	0 / 30	0 / 60						0 / 50	Video Ampl #2
Computer	0 / 65	0 / 15							0 / 35	Computer
IFA #1	-150 / 5	-60 / 75	0 / 135				-150 / 5	-150 / 5	-60 / 65	IFA #1
IFA #2		0 / 120	-60 / 55	-120 / 10	-120 / 10	-120 / 20				IFA#2
IFA #P			-60 / 55	-60 / 50	-60 / 65	-60 / 65	-120 / 15	-120 / 15		IFA#P
Loran D	-150 / 20	0 / 120	-120 / 30		-150 / 10	-150 / 10	-150 / 20	-150 / 20	0 / 140	Loran D
HF Comm	-150 / 20	-60 / 60	▨	-60 / 85	-120 / 40	-120 / 40	-120 / 50	-150 / 20	-60 / 80	HF Comm
UHF Comm	-150 / 20		-60 / 95	▨	-60 / 105	-60 / 105	-60 / 115	-60 / 115		UHF Comm
TACAN	-150 / 20		-120 / 30	-60 / 85	▨	-60 / 100	-60 / 110	-60 / 110		TACAN
IFF	-150 / 20		-120 / 30	-60 / 80	0 / 155	▨	-60 / 105	-60 / 105		IFF
Altimeter	-150 / 20		-120 / 30	-60 / 85	-60 / 100	-60 / 100	▨	-60 / 110		Altimeter
Radar	-150 / 20			-120 / 25	-60 / 100	-60 / 100	-60 / 110	▨	▨	Radar
EED #1	0 / 80	0 / 30	0 / 60	0 / 55	0 / 70	0 / 70	0 / 80	0 / 80	0 / 50	EED #1
EED #Q	0 / 80	0 / 30	0 / 60	0 / 55	0 / 70	0 / 70	0 / 80	0 / 80	0 / 50	EED#Q
E_n-R_m Cases Eliminated:			43							
E_n-R_m Cases Reduced:			57							
E_n-R_m Cases Remaining:			93							

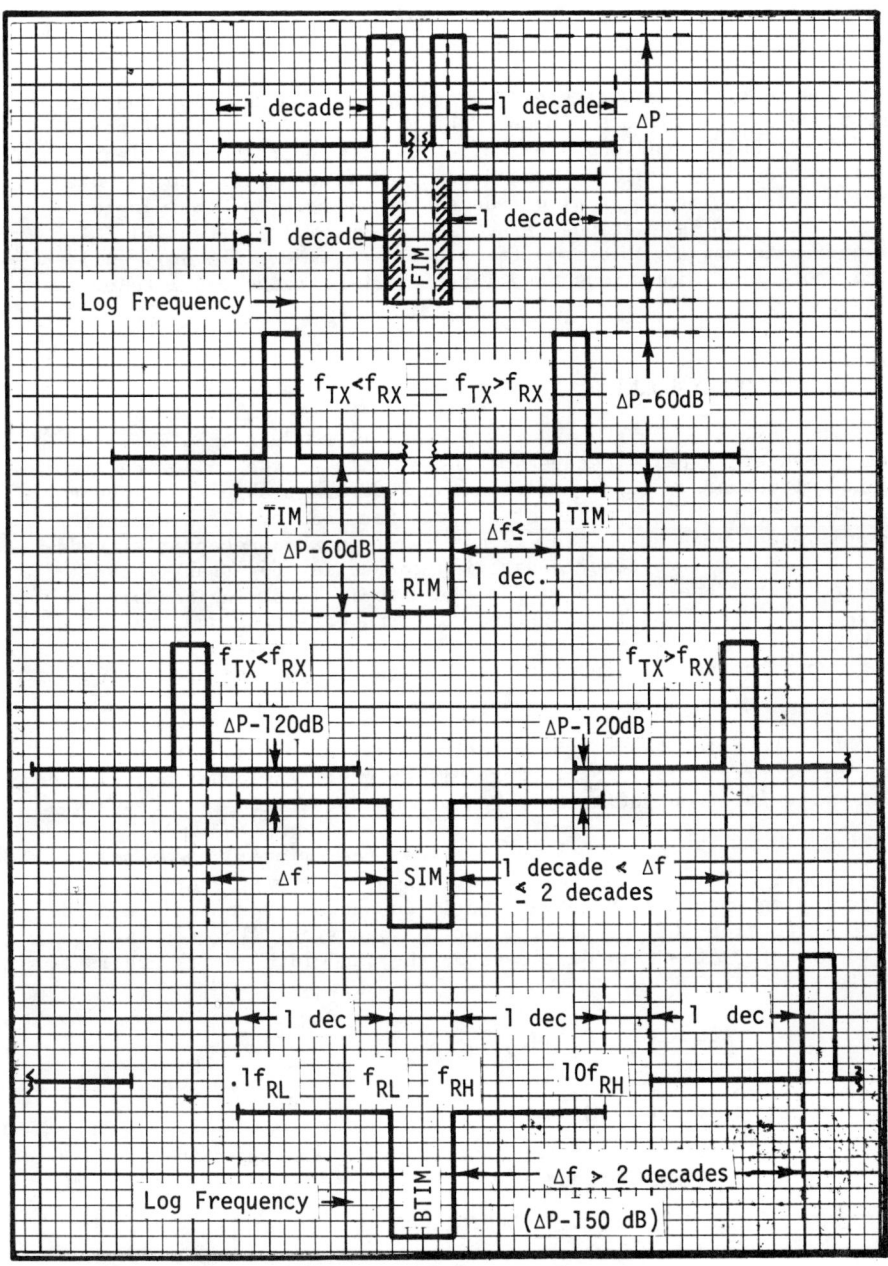

Figure 4.5 - The Five Amplitude-Culling Cases of EMI

for receptors which differ by more than 2 decades in frequency from
the bandwidth of the emission source as shown in Fig. 4.5.

The foregoing constitutes the second step in the amplitude-culling process. In the example used in Table 4.1, 136 pairs of potential E_n-R_m EMI situations are implied by row and column entries. The
amplitude-culling process, just discussed, reduced this to 93 potential EMI situations. It also reduced the interference margins of
57 potential EMI situations. The largest surviving potential EMI
situation is now TACAN transmitter to IFF receiver (155 dB). All
other potential situations have interference margins equal to or
less than 115 dB.

4.3.3   Frequency Culling

The next step in the culling process is to further eliminate
emitter-receptor pairs because of one or both of the following frequency situations: (1) convolution or superposition of the receptor's
bandwidth upon the bandwidth of the emitter, and/or (2) frequency-
mis-alignment when an emitter's emission is near a receptor's fundamental response (frequency separation correction for transmitters
and receivers; see Chap. 5). The first topic involving bandwidth
correction is divided into on-frequency and off-frequency cases.

Two subcases for bandwidth correction for on-frequency situations ($\Delta f = 0$) exist:

(1) *Receptor bandwidth is either equal to or greater than the
emitter bandwidth ($B_R \geq B_E$).* For this case all the power associated
with the emitter output is received and no correction is necessary.

(2) *Receptor bandwidth is less than the emitter bandwidth*
($B_R < B_E$). For this case only a portion of the power associated with
the emission output is received and it is necessary to apply a bandwidth correction, CF, to account for the bandwidth differences. This
correction for $\Delta f = 0$ is dependent on the bandwidth ratios and is of
the form:

$$CF(\Delta f = 0) = K \log_{10}(B_R/B_E) \text{ dB for } B_R < B_E \qquad (4.1)$$

where: B_R = receptor 3-dB bandwidth in HZ

B_E = emitter 3-dB bandwidth in Hz

K = a constant for a particular emission-response
combination*   (4.2)

= 0 for $B_R \geq B_E$ and co-channel frequency alignment

---

*   See Table 5.15.

> = 10 for noise-like (incoherent) emissions for
> which RMS levels apply and $B_R < B_E$
>
> = 20 for pulse signals or transients (coherent)
> emissions for which peak levels apply and
> $B_R < B_E$*

Illustrative Example 4.1 ($B_R < B_E$)

The Loran D receiver shown in Fig. 4.3 has a bandwidth, B_R = 10 kHz and the radar modulator has an emission bandwidth of 1.5 MHz. Eq. (4.1) applies in which K = 20 since the modulator emits coherent pulsed signals. Thus, the Loran receiver intercepts only a fraction of the available emitter's power:

$$CF(\Delta f = 0) = 20 \log_{10}(10kHz/1500kHz)$$
$$= -44 \text{ dB}$$

When applying the frequency-culling step to the radar modulator – Loran D receiver entry (140 dB) in Table 4.2, both the correction (–44 dB) and the resultant EMI level (140 dB –44dB = 96 dB) are posted in the corresponding box entry of new Table 4.3.

The above frequency-culling** process is carried out for each of the emitter-receptor entries of Table 4.2 to result in Table 4.3. This culling step further eliminated 37 emitter-receptor pairs from further consideration. It also reduced the levels of 45 pairs. Only two remaining emitter-receptor pairs (HF communications transmitter to IF-Amplifiers #1 and #2) have interference margins in excess of 100 dB. As illustrated in the next culling stage, coupling-path culling, both the final elimination and emitter-receptor pair reduction are very significant.

4.3.4 Coupling-Path Culling

The previous culling steps made no allowance for physical or spatial isolation of surviving potential emitter-receptor E_n-R_m situations. It assumes a 0-dB cross talk, i.e., all potential EMI margins reduced in levels by the above culling, is coupled directly into the victim receptor. Fortunately, this is never the case. Thus, coupling-path culling will both further eliminate the number of surviving situations and reduce the degree of EMI in these E_n-R_m pairs.

---

* For pulse signals, peak levels are applicable as long as B_R is greater than the pulse-repetition rate.
** See Chap. 5 for further description of off-frequency and frequency separation cases of frequency culling.

Table 4.3 - Intra-System, Emitter-Receptor Matrix (Frequency Cull)

Emitters →

Receptors	400 Hz AC	Digital Data	HF Comm	UHF Comm	TACAN	IFF	Altimeter	Radar	Radar Modul.	Receptors
Digital Data	0/65	▨							-8/27	Digital data
Video Ampl #1	0/80	0/30	-35/25						-4/46	Video Ampl #1
Video Ampl #2	0/80	0/30	-35/25						-4/46	Video Ampl #2
Computer	0/65								-8/27	Computer
IFA #1	0/5	-40/30	-30/105							IFA #1
IFA #2			0/120							IFA#2
IFA #P			-30/25	-10/40	-40/25					IFA#P
Loran D	0/20	-46/84							-44/96	Loran D
HF Comm	0/20	-40/20	▨	-35/50		-25/25			-58/22	HF Comm
UHF Comm		-30/65	-30/75	▨	-52/23	-20/95	-32/83	-32/48		UHF Comm
TACAN			-10/75		▨	-30/70	-30/80	-25/85		TACAN
IFF			-20/60	-55/100		▨	-25/80	-20/85		IFF
Altimeter			-30/55	-30/70	-25/75		▨	-30/80		Altimeter
Radar					-35/65	-35/65	-30/80	▨	▨	Radar
EED #1	0/80		-3/57	-3/52			-10/70	-66/12		EED #1
EED #Q	0/80		-3/57	-3/52			-10/70	-66/12		EED#Q
E_n-R_m Cases	Eliminated:		37							
E_n-R_m Cases	Reduced:		45							
E_n-R_m Cases	Remaining:		56							

This culling step commences with identifying the physical deployment geometry of the above E_n-R_m pairs. This involves using a plan and elevation view in which surviving pairs are located along with pertinent structures, housings, cabinets, bulkheads, partitions, interconnecting harnesses, cable raceways, etc. Coupling paths are then identified and defined for each surviving E_n-R_m combination. Principle coupling paths were discussed in Sec. 4.1.3 and may be grouped into five row and column headings as shown below:

Receptor \ Emitter	Ant.	Box	Wire	Cable	Com. Mode
Antenna Radiated	AA	BA	WA		
Box/Circuit Radiated	AB	BB	WB		
Wire/Cable Radiated	AW	BW	WW		
Wire/Cable Conducted				CC	MC
Common Mode-Conducted				CM	MM_2

Coupling paths are discussed in Chaps. 5, 6, 10 and 11. However, a few words of overview are presented now.

If the method of interference coupling is hard-wire (e.g., common-mode source impedance coupling) or E_n-R_m pairs share a common return path or ground plane (e.g., common-mode ground impedance coupling), then the mode of coupling is conducted. If it is radiated, it may be either near-field (induction field) or far-field (radiation field). However, the more conservative situation, far-field, is used in the culling operation.

The conservative models listed in Table 4.4 may be used for the coupling-path culling:

where: λ_R = wavelength corresponding to highest susceptible frequency of receptor

d_{ER} = distance from emitter antenna, box, or wires to receptor antenna, box, or wires

A_b = box area of surface facing antenna, box or wires

K_{dB} = minimum shielding effectiveness of metallic box

= 10 dB for H-field coupling

= 40 dB for E-field or electromagnetic coupling

a = 2 for metallic-box-to-metallic-box coupling

= 1 for non-box emitter

ℓ_R = length of receptor cable/wire

W_R = equivalent spacing of receptor outgoing and return wires

Table 4.4 - Radiated and Conducted Coupling-Path Culling Models

RADIATION COUPLING			
Emitter / Receptor	Antenna	Box	Wire
Antenna	$\longleftarrow$ 20 $\log_{10}(\lambda_R/4\pi d_{ER})$ $\longrightarrow$		
Box	$\longleftarrow$ 10 $\log_{10}(A_b/4\pi d_{ER}^2) + aK_{dB}$ $\longrightarrow$		
Wire	$\longleftarrow$ 10 $\log_{10}\left(\dfrac{\ell_R W_R}{4\pi d_{ER}^2}\right)$ $\longrightarrow$		Use Fig. 4.6

CONDUCTED COUPLING	
Direct Wire	No Correction + Filter (if applicable)
Common-Mode Grounding	20 $\log_{10}(R_g/2R_L)$ + Filter (if applicable)
Common-Mode Regulation	20 $\log_{10}(R_R/R_L)$ + Filter (if applicable)

Illustrative Example 4.2

The TACAN transmitter and Radar receiver (9.3 GHz, λ = .032m) in Table 4.3 represent a potential E_n-R_m EMI pair having 65 dB of interference margin. The separation distance of the two on the air-craft is 30 feet (10m). Determine if coupling-path culling will eliminate the pair as a potential EMI situation. From Table 4.4 for antenna-to-antenna coupling:

$$\text{Coupling} \geq 20 \log_{10}\left(\frac{.032m}{4\pi \times 10m}\right) = -72 \text{ dB}$$

Thus, the pair is eliminated since the coupling loss exceeds 65 dB.

Illustrative Example 4.3

A 115 VAC, 400 Hz power-line wire run of 10 feet (ℓ = 3 meters) shown in Fig. 4.3 is located one foot away (d_{ER} = 30 cm) from a parallel run of twisted-pair computer cable. Since the power mains are putting out 70 dBm (i.e., 10 kW), the load impedance, $Z \approx 1$ ohm. Table 4.4 indicates that for wire-to-wire coupling, Fig. 4.6 is to be used:

$$\text{Coupling} \geq 97 \text{ dB} - 20 \log_{10}(50 \times 3/30 \times 1) = 83 \text{ dB}$$

For the twisted-wire pair, Fig. 4.6 is conservative because it is based on parallel un-twisted wires. Since Table 4.3 showed this

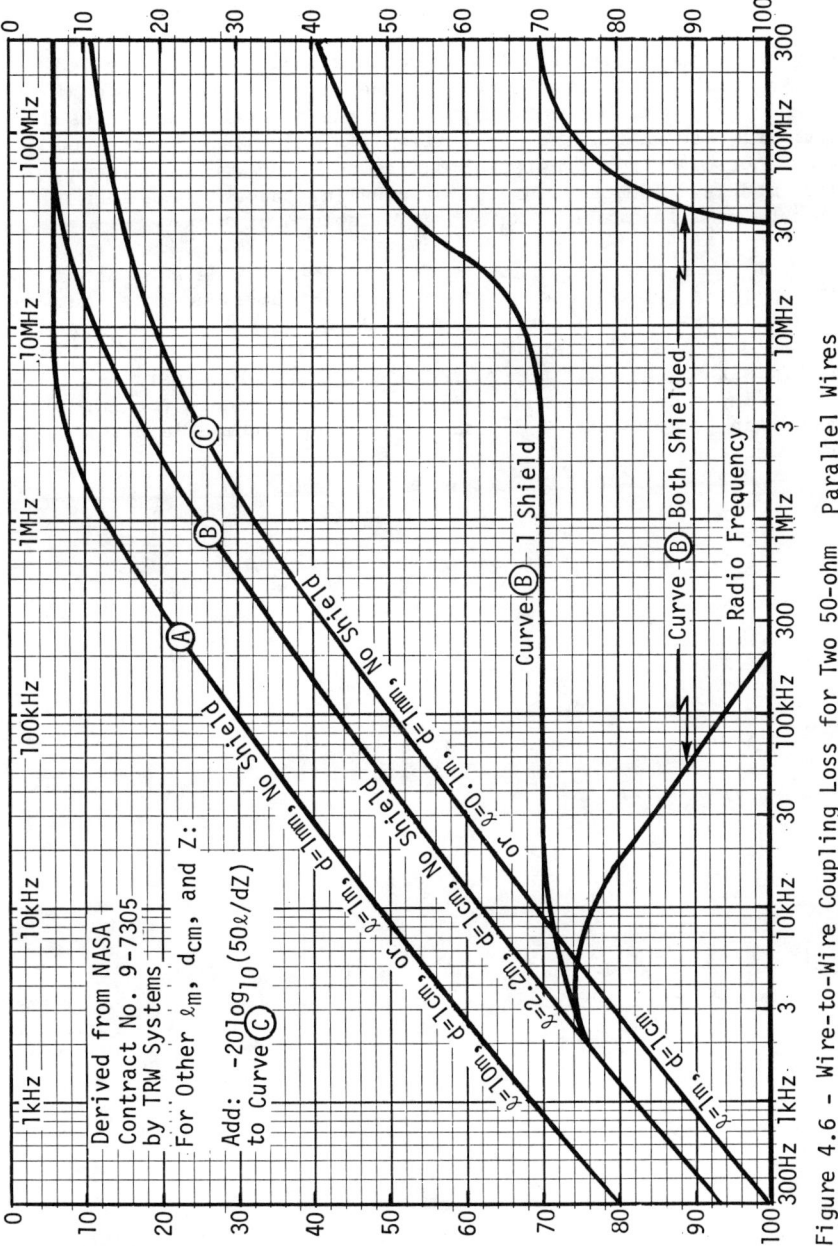

Figure 4.6 – Wire-to-Wire Coupling Loss for Two 50-ohm Parallel Wires

E_n-R_m pair to have an interference margin of 65 dB, this pair is also eliminated because the coupling loss exceeds 65 dB.

When the conservative coupling models of Table 4.4 are applied for coupling-path culling to Table 4.3, the remaining E_n-R_m situations are reduced from 56 to a relatively few in the aircraft example used. For surviving cases after the application of Table 4.4, the reduced levels are not used in a prediction operation, since they are conservative models. The only purpose of coupling-path culling then is to eliminate or retain the E_n-R_m pairs in Table 4.3

Many of the subsequent chapters discuss detail models of intra-system coupling and techniques that may be used to reduce this coupling. Thus, most of the chapters involve detail models to use for the prediction stage to determine which E_n-R_m pairs are indeed culprit-victim situations, and their predicted interference margins. After effecting a shield, ground, filter, etc., to reduce the interference margin to less than 0 dB, the prediction process is repeated for confirmation.

4.4 BIBLIOGRAPHY

(1)   "An Analysis Technique for Investigating Aerospace Data Link Susceptibility to Electromagnetic Compatibility," USAF Report No. Al-TDR-64-184, Sept. 14, 1964, Ad 447508.

(2)   "Antenna Models Based on Statistical Tendencies of Spectrum Signature Data," Electromagnetic Compatibility Analysis Center Report ECAC-TN003263, Sept. 1966.

(3)   B. L. Carlson, W. R. Marcelja, D. A. King, Computer Analysis of Cable Coupling for Intra-System Electromagnetic Compatibility, U. S. Air Force Report No. AFAL-TR-65-142, May 1965, AD 465040.

(4)   C. D. Taylor, S. Satterwhite, C. W. Harrison, Jr., "The Response of a Terminated Two-Wire Transmission Line Excited by a Nonuniform Electromagnetic Field," IEEE Transactions Antennas and Propagation, Vol AP-13, Nov. 1965, pp. 987-989.

(5)   C. W. Stuckey, J. C. Toler, and O. B. Francis, "Statistical Description of Near Zone Spurious Emissions," USAF Report AFAL-TR-67-37, AD 815146.

(6)   G. Hasserjian and A. Ishmirau, "Excitation of a Conducting Cylindrical Surface of Large Radius of Curvature," Transactions of IRE, Vol. AP-10, No. 3, May 1962, pp. 264-273.

(7)   J. A. M. Lyon, C. J. Dignis, et al, "Derivation of Aerospace Coupling - Factor Interference Prediction Techniques," USAF Report No. AFAL-TR-66-57, AD 483051.

(8)   J. A. M. Lyon, C. J. Dignis, et al, "Electromagnetic Coupling Reduction Techniques," USAF Report No. TR-68-132, June 1968, AD 834900.

(9)   John D. Kraus, Antennas, New York: McGraw-Hill, 1950, pp. 25, 54-55.

(10)   M. D. Siegel, "Aircraft Antenna-Coupled Interference Analysis," IEEE Electromagnetic Compatibility Symposium Record, June 17-19, 1969, pp. 85-90.

(11)   M. D. Siegel, "Near Field Antenna Coupling on Aerospace Vehicles," IEEE Electromagnetic Compatibility Symposium Record, July 14-16, 1970, pp. 211-216.

(12)   O. M. Salati, "Compatibility Studies," USAF Report No. RADC-TR-67-636, Vol. I, Feb. 1968.

(13)   R. D. Parlo, E. R. Freeman, H. M. Sachs, and A. M. Singer, "An Intra-System Compatibility Analysis Program (ISCAP)," USAF Report No. ESD-TR-70-261, June, 1970.

(14)   R. Goldman, A. Kalviste, and H. L. Rehkopf, "Application of Computer Techniques to System Interference Analysis, Electro-Interference, Class II Research," Boeing Company Document No. D2-23036, 1963.

(15)   S. Ramo, J. R. Whinnery, T. VanDuzer, Fields and Waves in Communication, Electronics, New York: John Wiley and Sons, Inc., 1967, pp. 291-321.

(16)   T. K. Foley and A. Rudzitis, "Feasibility Study of Computer Prediction of Broadband Near-Field Electromagnetic Interference in a Space Vehicle - Final Report," Contract NAS S-5608, Boeing Document No. D2-90642-1, Dec. 31, 1964, AD 463972.

(17)   William G. Duff, J. E. Balwin, Jr., F. F. Ferrante, et al, "Electromagnetic Compatibility Prediction Studies," USAF Report No. RADC-TR-66-560, Nov. 1966.

(18)   W. L. Weeks, Electromagnetic Theory for Engineering Applications, New York: John Wiley and Sons, Inc., 1964, Ch. 2.

(19)   W. R. Johnson, B. B. Cooperstein, and A. K. Thomas, "Development of a Space Vehicle Electromagnetic Interference/ Compatibility Specification," Final Report, NASA Contract No. 9-7305, Vols 1 & 2, TRW Document No. 08900-6001-1000, June 28, 1968.

CHAPTER 5

INTER-SYSTEM EMI PREDICTION AND CONTROL

CHAPTER 5

INTER-SYSTEM EMI PREDICTION AND CONTROL

This chapter summarizes interference prediction and control between one or more potential culprit transmitters and one or more
victim receivers emersed in a common electromagnetic ambient environment. The EMI prediction emphasizes the antenna-to-antenna mode of
coupling. Considerably more detail on this topic is presented in
Vol. 5 which is dedicated to the subject of this chapter. Thus, the
reader who desires a comprehensive treatment of *inter*-system EMI
prediction and analysis should review this reference. On the other
hand, *intra*-system EMI prediction and control is emphasized in this
volume, Vol. 3.

This chapter develops the basic EMI prediction equation, and
summarizes its more comprehensive version involving many parameters.
Separate mathematical models for transmitters, receivers, antennas,
propagation, and signal acceptability criteria are presented to support the prediction relations. Several illustrative examples show
how to make EMI predictions and how to effect an EMI control in the
presence of the signal of interest.

Most EMI inter-system predictions involve two or more potential
culprit emitters each having several spurious emissions. Additionally,
each receiver exhibits several spurious responses one or more of
which may result in interference. Thus, to facilitate selection and
computation of EMI, the multi-transmitter/receiver prediction problem
is based on a rapid culling process. This eliminates non-EMI situations early in the prediction to provide a more efficient process.
Illustrative examples include predicting EMI at sites where several
emitters and/or receptors are located.

The final section of this chapter reviews the topics of inter-
system EMI control based on frequency management, time management,
location management, direction management, and post-detector EMI control devices. The chapter is concluded with a bibliography.

5.1 INTRODUCTION

This section presents an overview of the EMI inter-system prediction and control process. Inter-system here involves antenna-to-antenna coupling in which at least one undesired transmitter emission and one victim receiver are involved. Sometimes many transmitters and/or receivers are involved with separations ranging anywhere from a few feet to thousands of miles distance.

This section discusses the topics of problem definition, input data collection, system performance requirements, and supporting data reduction and presentation. Subsequent sections then apply this input information in performing EMI predictions and control.

5.1.1 PROBLEM DEFINITION

The first step in EMI prediction is problem definition. Here, the prediction frequency range, geographical area, equipment involved, relative geometry, prediction detail, necessary sources of input data, output results required, and related considerations are all defined.

5.1.1.1 PREDICTION FREQUENCY RANGE

Although each EMI prediction problem is different and requires separate consideration, there are certain generalizations that apply. For example, regarding limitations on frequency range to be examined, the fundamental interference margin (FIM) is considered when the transmitter and receiver are separated in frequency by 20% or less.*

Significant transmitter spurious emissions and significant receiver spurious responses are generally limited to frequencies ranging from 0.1 to 10 times the respective fundamental operating frequency of each. Hence, transmitter-interference margin (TIM) and receiver-interference margin (RIM) are usually limited to cases for which the transmitter and receiver fundamental frequencies are mutually separated by more than 20% but less than one decade.** Finally, spurious interference margin (SIM) is usually considered in cases for which

---

* Fundamental interference margin (FIM) refers to EMI situations resulting from transmitter fundamental emissions interfering with receiver fundamental response in which fundamental frequency separation of both exists within about 20% (see Figs. 5.51 and 5.52).
** Transmitter interference margin (TIM) refers to a transmitter fundamental emission interfering with a receiver spurious response(s). Receiver interference margin (RIM), on the other hand, refers to transmitter spurious emission(s) interfering with a receiver fundamental response. For each of these situations, mutual frequency separation of the fundamentals ranges between 20% and one decade.

transmitter and receiver fundamental frequencies are mutually separated by more than one decade but less than two decades in frequency.*

5.1.1.2 PREDICTION GEOGRAPHICAL BOUNDS

The geographical area to be considered in an EMI inter-system prediction problem is a function of operating frequency, transmitter power output, antenna gains, and receiver susceptibility threshold. As a guidance, general criteria may be developed to bound the geographical area. For example, in planning a site survey it is required to determine the maximum geographical separation between off-site transmitters and site victim receiver(s). The first step is to calculate the Effective Power Margin, EPM, in dB for each potential transmitter-receiver EMI situation:

$$EPM = P_t + G_t + G_r - P_r \text{ dB} \qquad (5.1)$$

where,

P_t = culprit transmitter power output in dBm

G_t = culprit transmitter antenna gain in dB

G_r = victim receiver antenna gain in dB

P_r = victim receiver sensitivity in dBm

The second step is to determine what FIM, RIM, TIM and SIM emitters, if any, exist and their distance. Using Fig. 5.1, the maximum radial distance is obtained from the site or victim receiver to each class of emitter type to be surveyed. The relationships shown in Fig. 5.1 are based on worst cases: (1) co-channel EMI situations for FIM in which no off-frequency rejection levels are assumed, (2) 60-dB rejection corresponding to either separate transmitter fundamental emission to receiver spurious response (TIM) or transmitter spurious emission to receiver fundamental response (RIM), and (3) 100-dB rejection for SIM cases corresponding to transmitter-to-receiver spurious situations. In these preliminary sorting models, propagation loss up to radio-line-of-sight (RLOS) is assumed to be 20 dB per decade, and beyond RLOS a rate of attenuation of 60-dB per decade is used.

Illustrative Example 5.1

Assume a 1050 MHz Tacan, ground-based receiver has a sensitivity, P_r, of -100 dBm and an antenna gain of +3 dB. Determine the distance out to which all ground-based, L-Band (1300 MHz) radar transmitters (P_t = 1 MW = +90 dBm; G_t = 30 dB) and UHF TV (470-890 MHz) transmitters

---

* Spurious interference Margin (SIM) refers to a transmitter spurious emission interfering with a receiver spurious response in which fundamental emission and response are separated by more than one but less than two decades in frequency (see Figs. 5.51 and 5.52).

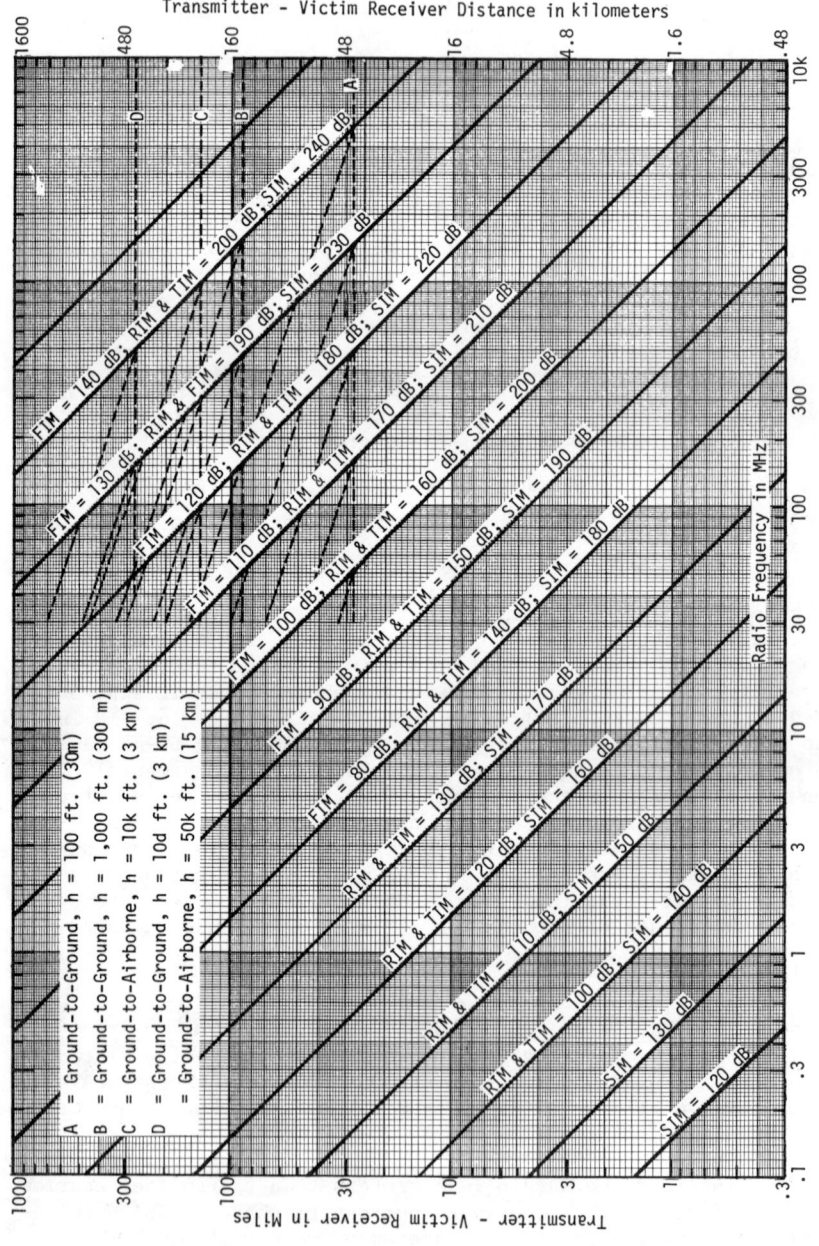

Figure 5.1 - Radial Distance from Transmitter to Receiver Site for FIM, RIM, TIM, and SIM Situations

(P_t = 5 MW = +97 dBm; G_t = 5 dB) should be surveyed as potential
sources of EMI. Maximum transmitter-receiver antenna height variation
is 1,000 feet. Note that both sources are TIM cases. From Eq. (5.1)
there results:

$$EPM_L = 90 \text{ dBm} + 30 \text{ dB} + 3 \text{ dB} - (-100 \text{ dBm}) = 223 \text{ dB}$$

$$EPM_{UHF} = 97 \text{ dBm} + 5 \text{ dB} + 3 \text{ dB} - (-100 \text{ dBm}) = 205 \text{ dB}$$

Based on a regional ground-to-ground altitude variation (including
antenna height) of 1,000 feet (300 meters), Fig. 5.1 shows that the
RLOS is 90 miles corresponding to a TIM EMP of 198 dB. This is less
than either of the above EPMs, so emission intercepts will go beyond
RLOS. For the more severe case (EPM_L radar = 223 dB), the maximum
distance for the site survey for all emitters from $0.1f_r$ (105 MHz)
to $10f_r$ (10.5 GHz) is:

$$198 \text{ dB} + 60 \log_{10}R = 223 \text{ dB}$$

$$60 \log_{10}R = 25 \text{ dB}$$

$$\log_{10}R = 0.42$$

$$R = 2.6 \times 90 \text{ miles} = 235 \text{ miles}$$

Thus, all transmitters* from about 105 MHz to 10.5 GHz should be
surveyed out to a distance of 235 miles.

5.1.1.3 TRANSMITTER-RECEIVER IDENTIFICATION

When the geographical area of consideration has been defined, such
as above, the next step is to identify specific emitters and receivers
within that area. This information may be obtained from various
sources such as the Federal Communication Commission files on licensees
and applicants for a license to use the spectrum, the Electromagnetic
Compatibility Analysis Center (ECAC)** environmental files, and/or
the Office of Telecommunications of the Department of Commerce. If
none of these are used, it may then be necessary to identify potential
emitting sources by an ambient site survey. This is discussed later.

---

* There exists a more refined planning quantization process than
shown here. However, when Eq. (5.1) is repeatedly applied to each
transmitter-receiver pair, some further delimiting in distance may
exist. For example, the 235 mile range applies to L-Band radars only;
100 miles for UHF TV stations; and lesser distances for emitters from
105 MHz to 470 MHz and above L-Band to 10.5 GHz.
** Access to ECAC files is difficult and/or slow for other than DOD
and certain other Government agencies and some of their contractors.

5.1.2   DATA COLLECTION

The next step in preparing for an EMI prediction is collecting
required data. There are two types of data used in EMI prediction:
problem input data and interference characteristic data.

The first type of data required consists of nominal equipment
characteristics and specific problem data. Nominal characteristics
data include transmitter bandwidth and fundamental power output, receiver bandwidth and sensitivity; and antenna gain, polarization and
beamwidth. Specific problem data are assigned frequencies, actual
equipment three-dimensional locations, and antenna direction of
transmission or reception. Table 5.1 is a sample data form which is
used in identifying required problem input data.

The second type of data used in EMI prediction is interference
characteristics. Although it is desirable to obtain these data for
all communications-electronics equipment considered, this is not always possible. In order to provide data for use in prediction for
which specific information is not available, statistical interference
characteristics summaries are available for various equipment types.
If specific interference characteristics data are not available, an
EMI prediction may still be performed based on the summary models.
Thus, specific interference data are not required to perform an EMI
prediction in the same sense as is the *required problem input data*.
Table 5.2 serves as a sample data form used in collecting interference
characteristics data.

Data required for performing an EMI prediction problem may be obtained from a number of different sources. For example, nominal characteristics data may be obtained from equipment operation manuals or
specifications, design specifications, ECAC's nominal characteristics
files, or direct solicitation of equipment users. Information on
specific equipment locations, antenna directions, frequencies, etc.,
can be obtained from operational plans, system deployment plans, frequency assignments and schedules, ECAC's environmental files, FCC applications for frequency assignments, and direct solicitation of
users.

Specific interference characteristics data are not in general as
readily available as nominal data and specific problem input data.
However, there are still a number of possible sources for these types
of data such as: MIL-STD-449C measurement results, equipment limits
provided in specifications and standards, FCC rules and regulations
governing spurious emissions and responses, and design, test, and
evaluation results. In the event that specific interference characteristics data are available from one of these sources it is used as
the basis for the input of the equipment models for EMI prediction.
If specific interference data are not available, summary models presented in Sec. 5.3 are used.

Table 5.1 - Required Problem Input Data

Characteristic ↓/#→	1	2	3	4	5
Transmitter					
Nomenclature					
Frequency-Upper					
(MHz) Lower					
Power (dBm)					
Type Modulation					
Bandwidth (kHz)					
Antenna (TX)					
Type					
Gain (dB)					
Polarization					
Location (X ft)					
(Y ft)					
(H ft)					
Beamwidths (°AZ)					
(°EL)					
Direction (°/North)					
Receiver					
Nomenclature					
Frequency-Upper					
(MHz) Lower					
Sensitivity (dBm)					
Type Modulation					
Bandwidth (kHz)					
LO Frequency (MHz)					
1st IF (MHz)					
Sign of IF					
Antenna (RX)					
Type					
Gain (dB)					
Polarization					
Location (X ft)					
(Y ft)					
(H ft)					
Beamwidths (°AZ)					
(°EL)					
Direction (°/North)					

Table 5.2 - Interference Characteristics Data*

(Amplitude & Frequency Cull)

Transmitter	1	2	3	4	5
Harmonic A (dB/decade)					
(f>f_OT) B (dB)					
σ (dB)					
Maximum Frequency (MHz)					
Spurious A' (dB/decade)					
(f<f_OT) B' (dB)					
σ' (dB)					
Minimum Frequency (MHz)					
Modulation Δf_0 (kHz)					
Envelope M_0 (dB/decade)					
Δf_1 (kHz)					
M_1 (dB/decade)					
Δf_2 (kHz)					
M_2 (dB/decade)					

Antenna	1	2	3	4	5
Main Beam					
Harmonic C (dB/decade)					
(f>f_0) D (dB)					
σ (dB)					
Spurious C' (dB/decade)					
(f<f_0) D' (dB)					
σ' (dB)					
Waveguide Cutoff (MHz)					
Side & Back Lobe					
Mean G[p(θ≠θ_0)]					
Gain (dB/isotrope)					
σ (dB)					

Receiver	1	2	3	4	5
Spurious I (dB/decade)					
(f>f_OR) J (dB)					
σ (dB)					
Maximum Frequency (MHz)					
Spurious I' (dB/decade)					
(f<f_OR) J' (dB)					
σ' (dB)					
Minimum Frequency (MHz)					
Selectivity BR20 (kHz)					
BR60 (kHz)					
Adjacent Δf_{max} (MHz)					
Signal					
Allowable FIM (dB)					
Antenna					

* See Glossary for definition of symbols.

5.1.3   ELECTROMAGNETIC AMBIENT SITE SURVEY

If either all significant data required to perform an EMI prediction are not obtainable or if nominal substitutive data leave un-
filled gaps, it may then be necessary to perform an ambient site survey by emission intercept. One method is to make a site survey at
the location of the potential victim receiver(s) from $0.01f_r$ to
either 100 f_r or about 12 GHz, whichever is lower. Fig. 5.2 shows
conceptual results of what such an ambient survey may look like. Note
that intercept signal levels, modulation type, and bearing angle of
all significant detected emissions are documented.

Since the potential victim receiver and it's antenna will differ
from the instrumentation used to make the site survey, an adjustment
in the measured levels, $V_{mdB\mu V}$, must be made to translate intercepts
to values more nearly representing the victim situation:

$$V_{rdB\mu V} = V_{mdB\mu V} + AF_{mdB} - AF_{rdB} \quad dB\mu V \qquad (5.2)$$

where, $V_{rdB\mu V}$ = adjusted intercept levels in units of dBμV,*
                     as potentially seen by the victim receiver

AF_{mdB} = antenna factor* of EMI measuring antenna
                (see EMI antenna manufacturer for listing of
                antenna factors; also see example in Fig. 5.2)

AF_{rdB} = antenna factor of actual potential victim
                receiver antenna to be used $\approx 20 \log_{10}(9.7/\lambda\sqrt{G_r})$

and, λ = receiver operating wavelength in meters

G_r = receiver antenna gain ratio for in-band
                conditions ≈ 0.1 (−10 dB) for out-of-band
                operation.

When this correction is applied to Fig. 5.2 for the significant
measured emissions, a result similar to Fig. 5.3 may develop.

It now remains to draw a ± 20% frequency band around the intended
frequency of the potential victim receiver at the level of its sensitivity as shown in Fig. 5.3. This is the FIM region discussed in
Sec. 5.1.1. The figure also shows the 60-dB step levels from $0.1f_r$
to $10f_r$ for the RIM and TIM cases, and the 100-dB levels from $0.01f_r$
to $0.1f_r$ and from $10f_r$ to $100f_r$ to accommodate the SIM cases. It is
concluded that all site intercepted emissions using the measurement
instruments existing above these levels may constitute potential
sources of EMI to the victim receiver to be installed at the site.
These emissions require further examination.

---

*   See Vol. 4 for definition of units or Abbreviations and Symbols
in front of this volume.

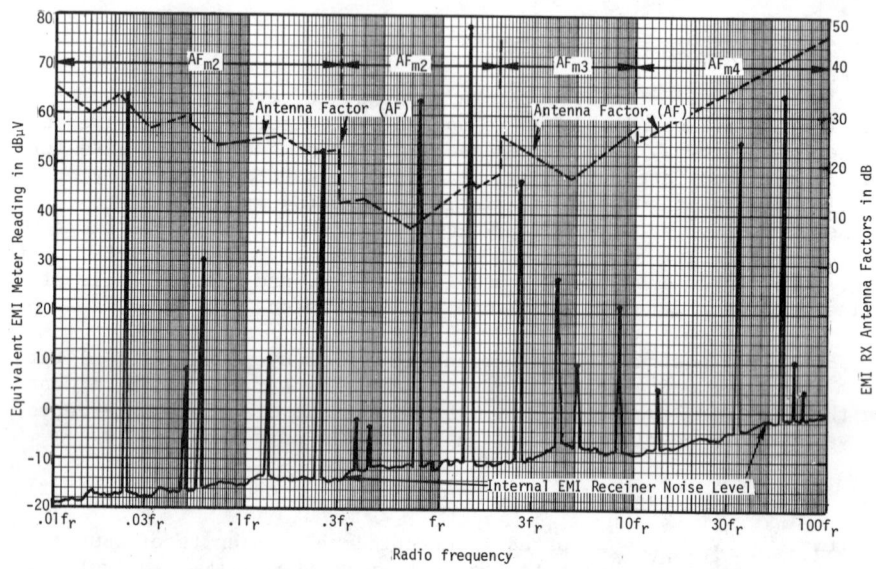

Figure 5.2 - Results of Electromagnetic Ambient Site Survey and Antenna Factors.

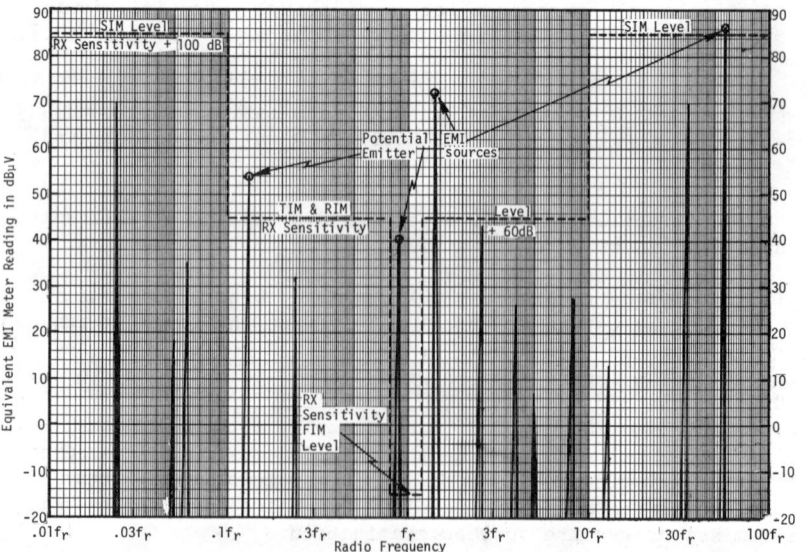

Figure 5.3 - Ambient Site Survey of Fig. 5.2 Corrected for Instrument and Specimen Antenna Factors.

5.1.4 Emitter Population Census

Based upon either the geographical distance and frequency coverage
discussed in Sec. 5.1.2 or the electromagnetic ambient site survey
performed in Sec. 5.1.3, a topographic map is used to locate potentially
interfering EMI emitters. It may be necessary to actually visit site
locations of such emitters in order to position them on the map and
to establish missing data not completable in Tables 5.1 and/or 5.2.
The results of such an emitter population census may yield a map similar to that shown in Fig. 5.4. Note the contour lines (height) and
location (latitude and longitude) of the potential EMI emitters.

Finally, by using 4/3 earth-radius graphs of line-of-site conditions between each emitter and the victim receiver(s), the propagation path physics to be used above 30 MHz can be determined. This
is illustrated in Fig. 5.5 for the path between the victim VHF receiver and the F-M transmitter in Fig. 5.4. Note that a clear path
RLOS condition exists. Thus, no propagation correction is required.
This is discussed in further length in Sec. 5.3.4 and is given comprehensive treatment in Vol. 5.

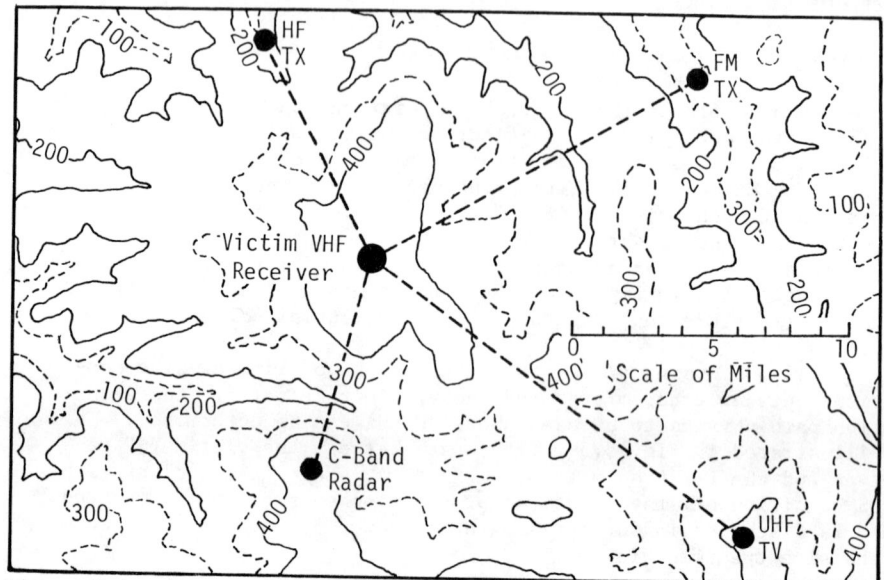

Figure 5.4 - Topographic Map of Surviving Potential EMI Emitters

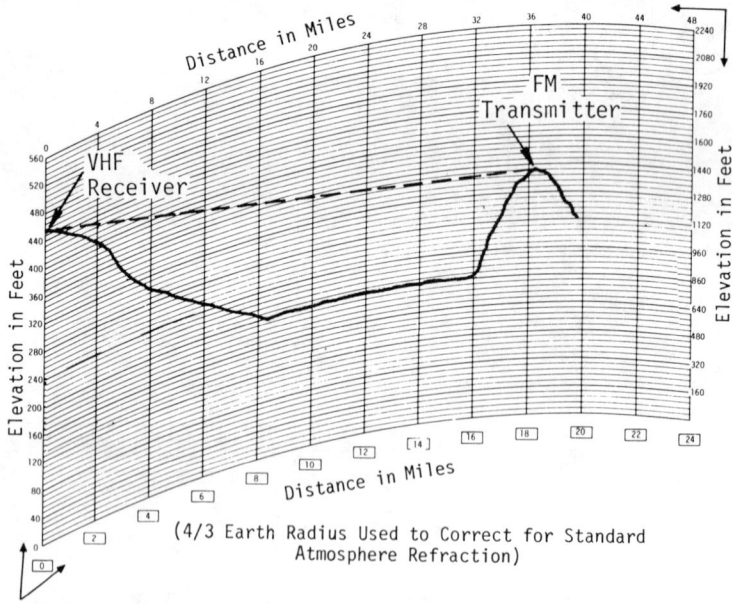

(4/3 Earth Radius Used to Correct for Standard
Atmosphere Refraction)

Figure 5.5 - Elevation Plot of Transmitter - Receiver Pair to
Determine if ROLS Conditions Exist.

5.2 THE BASIC ONE-WAY PREDICTION EQUATION

The following development results in what is known as the *one-way transmission-reception equation* or the *beacon equation*. While of limited use for either intentional signal of interest or EMI prediction, it constitutes the basic relation from which more comprehensive prediction equations (i.e., math models) may be developed.

5.2.1 ARRIVING POWER FLUX DENSITY

If a transmitter delivers P_t watts* to a loseless isotropic radiating antenna, the power will be radiated homogeneously in all directions, viz., with equal flux density over a sphere or 4π sterradians. Thus, the power density, P_D in watts per unit area exiting from a sphere of radius R, will be:

$$P_D = \frac{P_t}{4\pi R^2} \text{ watts/unit area} \tag{5.3}$$

where, R^2 and the unit area are in the same system of units.

Since it is wasteful to radiate the transmitter power in all directions, except for certain situations, a transmitting antenna is introduced which exhibits a forward gain, G_t, over the original isotropic radiating antenna (G_i =1=0dB as shown in Fig. 5.6(b). Thus, G_t is introduced into Eq. (5.3) to yield:

$$P_D = \frac{P_t G_t}{4\pi r^2} \text{ watts/unit area} \tag{5.4}$$

5.2.2 ANTENNA INTERCEPTED RECEIVED POWER

To make Eq. (5.4) useful in a communications-electronics transmitting-receiving system, it is necessary to provide a receiving antenna which exhibits an intercept or capture area, A_r, in the path of maximum power flux density as shown in Fig. 5.6(c). Here the normal to the intercepting antenna area is coincident with the main axis (boresight axis) of the transmitting antenna. Consequently, Eq. (5.4) will yield a net power flow, P_r, into the receiving antenna:

$$P_r = P_D A_r \text{ watts} \tag{5.5}$$

$$= \frac{P_t G_t A_r}{4\pi R^2} \text{ watts} \tag{5.6}$$

---

* Subscript t is used to denote *transmitter* and subscript r is used to denote *receiver*.

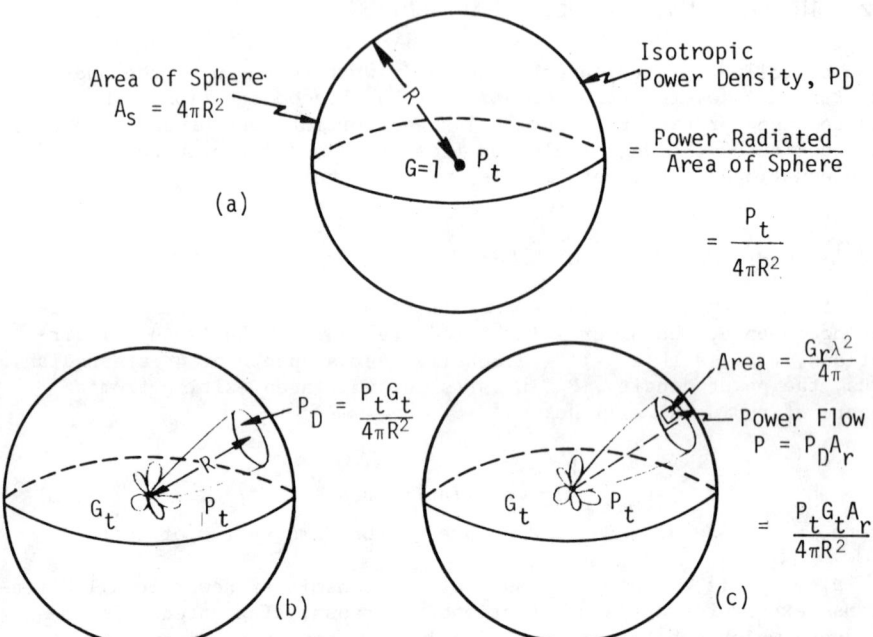

Figure 5.6 - Three-Dimensional Geometry Illustrating Gain of Antenna.

Eq. (5.6) is not particularly useful in the format presented since the antenna area is generally unknown or it is not even rated. Therefore, the relation between the antenna capture area and its gain is introduced:

$$G = \frac{4\pi A\varepsilon}{\lambda^2} \qquad (5.7)$$

where,

A = actual area of antenna

ε = antenna efficiency

$A\varepsilon$ = effective area of antenna

λ = wavelength in units of R

Table 5.3 lists the efficiencies and effective areas of a number of frequently used antennas:

Table 5.3 - Typical Antenna Efficiency and Effective Areas

Antenna Type	Antenna Efficiency	Effective Area
Isotropic Radiator	1	$\lambda^2/4\pi$
Small Dipole or Loop ($\ell<\lambda/10$)	1	$1.5\lambda^2/4\pi$
Half-wave Dipole	1	$1.64\lambda^2/4\pi$
Horn	0.45	0.45A
Parabola	0.5-0.6	0.5A-0.6A

Solving Eq. (5.7) for $A\varepsilon$, yields:

$$A\varepsilon = \frac{G\lambda^2}{4\pi} \tag{5.8}$$

When Eq. (5.8) is substituted into Eq. (5.6), there results:

$$P_r = \frac{P_t G_t G_r \lambda^2}{(4\pi R)^2} \text{ watts} \tag{5.9}$$

Eq. (5.9) is one form of the basic *one-way transmission-reception equation*. The equation is more useful if λ is expressed in equivalent operating frequency rather than wavelength:

$$\lambda f = C \text{ or } \lambda = C/f \tag{5.10}$$

where,   f = radio frequency in Hz

C = velocity of light (electromagnetic propagation in air)

= 3 x 10^5 km/sec (metric system of units)

= 1.86 x 10^5 miles/sec (English system of units)

Substituting Eq. (5.10) into (5.9) yields:

$$P_r = \frac{P_t G_t G_r C^2}{(4\pi R f)^2} \text{ watts} \tag{5.11}$$

5.2.3 THE BASIC PREDICTION EQUATION

To facilitate using Eq. (5.11), f is expressed in units of MHz, f_{MHz}; R is expressed in units of C; and all constants may be collected to form a single multiplier:

$$P_r = \left(\frac{3 \times 10^5}{4\pi \times 10^6}\right)^2 \times \frac{P_t G_t G_r}{\left(R_{km} f_{MHz}\right)^2}$$

$$= \frac{.00057 \ P_t G_t G_r}{\left(R_{km} f_{MHz}\right)^2} \qquad \text{watts (metric system)} \quad (5.12)$$

$$= \left(\frac{1.86 \times 10^5}{4\pi 10^6}\right)^2 \times \frac{P_t G_t G_r}{\left(R_{mi} f_{MHz}\right)^2}$$

$$= \frac{.00022 \ P_t G_t G_r}{\left(R_{mi} f_{MHz}\right)^2} \qquad \text{watts (English system)} \quad (5.13)$$

At this juncture, it is common practice to express either Eq (5.12) or Eq. (5.13) in decibel form, and to use the milliwatt as the basic unit of power reference rather than the watt. Carrying out this operation yields:

$$P_{rdBm} = -(32 \ dB + 20 \ \log_{10} R_{km} + 20 \ \log_{10} f_{MHz}) + P_{tdBm}$$
$$+ G_{tdB} + G_{rdB} \quad \text{(metric system)} \qquad\qquad (5.14)$$

$$= -(37 \ dB + 20 \ \log_{10} R_{mi} + 20 \ \log_{10} f_{MHz}) + P_{tdBm}$$
$$+ G_{tdB} + G_{rdB} \quad \text{(English system)} \qquad\qquad (5.15)$$

The terms in parentheses are called the propagation gain between transmitter and receiver, or more commonly, the propagation loss when the minus sign is absorbed. Fig. 5.7 is a plot of Eqs. (5.14) and (5.15) for propagation loss vs distance with frequency as a parameter. This figure will be used many times in subsequent prediction problems.

Illustrative Example 5.2

Determine the propagation loss between a line-of-site, transmitter-receiver pair operating at a frequency of 1 GHz and having a spatial separation distance of 10 km (6.2 miles). From Fig. 5.7 the propagation loss is 113 dB.

Illustrative Example 5.3

Determine the received power for the above TX-RX pair in which the transmitter radiates 10 kW (+70 dBm), its antenna gain is 30 dB, and the associated receiver antenna gain is 20 dB. From Example 5.2 and Eq. (5.14):

$$P_r = -113 \ dB + 70 \ dBm + 30 \ dB + 20 \ dB = + 7 \ dBm = 5 \ mW$$

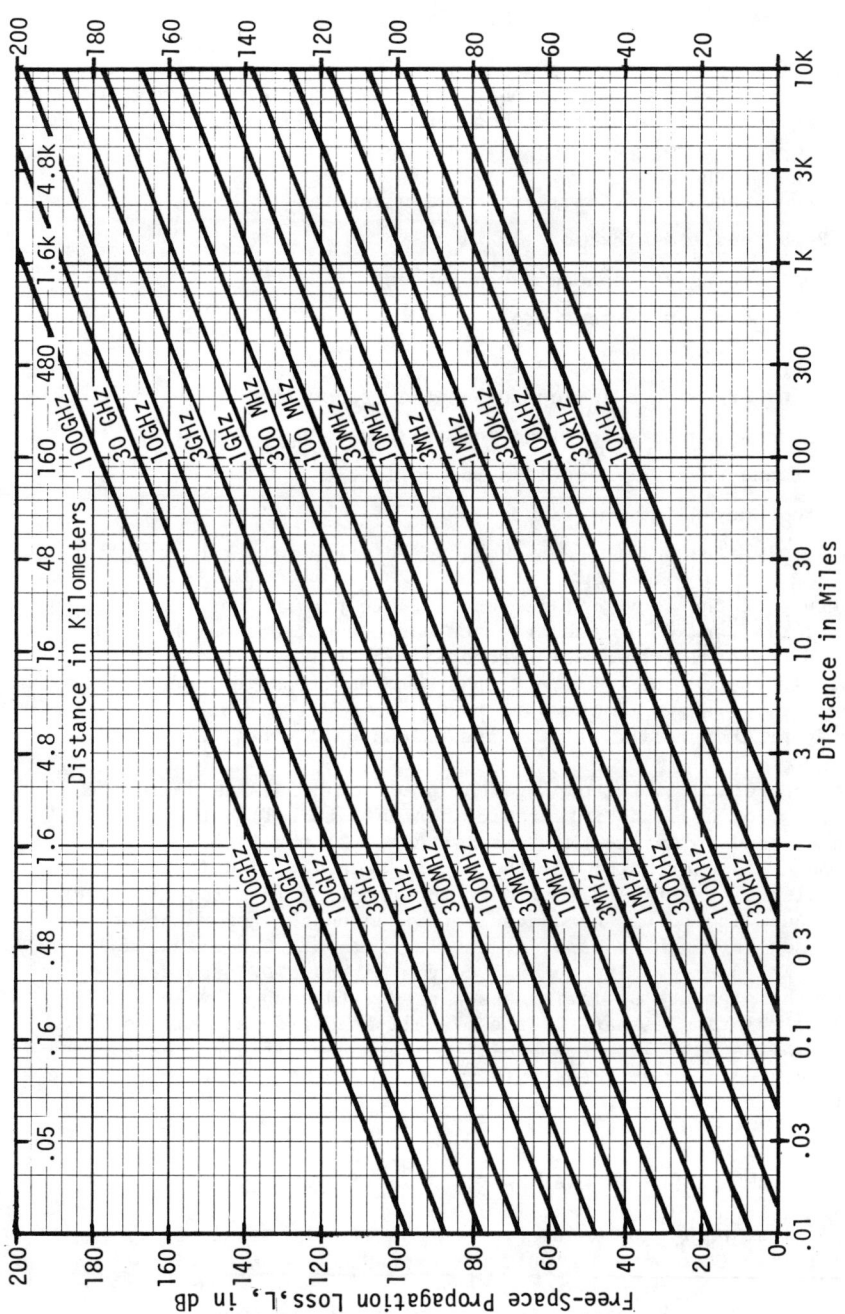

Figure 5.7   Free-Space Propagation Loss

5.2.4   A More Comprehensive Prediction Equation

Eqs. (5.14) and (5.15) are often useful for predicting the inter-cepted power between an *intentional* TX-RX pair. However, they are not often useful for predicting EMI between an *unintentional* TX-RX pair because so many conditions are assumed, including:

- Free-Space (far-field) propagation applies
- RLOS conditions exist
- Intervening atmosphere has no effect
- Antenna polarizations are matched
- Transmitting antenna is looking at the receiver
- Receiver antenna is looking at the transmitter
- Antenna transmission line loss is negligible
- Antenna receiver line loss is negligible
- Frequency alignment exists
- Receiver bandwidth equals or exceeds transmitter bandwidth

As a result of the above, it is necessary to provide correction terms to Eqs. (5.14) and (5.15) so that the basic equations may be used for either intentional transmission-reception or EMI prediction and analy-sis. To introduce these corrections, a composite signal and EMI pre-diction form is shown on the next page. Some details of how to determine or compute the entries are discussed in the next section.

For convenience of grouping, the prediction form is divided on the left row of headings into topics dealing with (1) both the signal-of-interest and potential EMI transmitters, (2) the intervening propa-gation path loss for both, (3) the victim receiver, and (4) the proces-sed signal and potential EMI through the receiver in terms of output performance. The column headings identified on the top right of the form provide for both *+dB* and *-dB* entries for both the signal and potential EMI source. The central column marked *STDV* provides for entering the standard deviation associated with each dB value. This permits calculating either (1) the overall standard deviation of the output S/(N+I)* or (2) the probability that a specific EMI damage may be obtained. All dB entries in the form are to be rounded off to the nearest integer dB value.

5.2.4.1   Transmitter-Antenna Effective Radiated Power

The first entry in the prediction form is the measured, rated, or nominal transmitter power output, P_t, expressed in units of dBm:

---

*   S/(N+I) = ratio of signal to noise-plus-interference

ELECTROMAGNETIC INTERFERENCE PREDICTION FORM

Signal-Source TX:_____MHz; _____Watts; Other:_____

Potential EMI TX:_____MHz; _____Watts; Other:_____

Victim Receiver:_____MHz; _____kHz Predetection BW; _____ kHz Baseband BW

	SIGNAL		STDV	INTERFERENCE	
TRANSMITTERS (Signal and Interfering Sources)	+dB	-dB	dB	+dB	-dB
1. Transmitter Power Available in dBm	30			93	
2. Power Reduction for Out-of-Band Emission					64
3. Antenna Transmission Line Loss		2			O
4. Antenna Gain	35			30	
5. Gain Reduction for Out-of-Band Frequency					O
6. Gain Reduction in Direction of Receiver					O
7. TOTAL MODIFIED ERP (Sum Lines 1 thru 6)	63 dBm			59 dBm	
PROPAGATION PATH (Signal and Interfering Sources)					
8. Free-Space Loss (Sig: mi; EMI: mi.)		140			110
9. Non-RLOS Correction Loss		O			O
10. Fade Margin Required (% Reliability)		30			
11. Rain, Water Vapor, and Oxygen Loss		O			O
12. TOTAL PROPAGATION LOSS (8+9+10+11)		170			110
RECEIVER (Signal and Interfering Victim)					
13. Receiver Antenna Gain	35			35	
14. Gain Reduction in Direction of EMI TX)					45
15. Gain Reduction due to Polarization					—
16. Antenna Transmission Line Loss		2			2
17. TOTAL ANTENNA GAIN OR LOSS (13+14+15+16)	33			12	
18. RECEIVER-ANTENNA INPUT POWER (7+12+17)	-74 dBm			-63 dBm	
19. Frequency Mis-Alignment Correction				12	
20. Bandwidth Correction (Channel Baseband)				O	
21. Receiver Sensitivity (RMS Noise Level)	-98 dBm			-98 dBm	
22. S/N & I/N Before Detection (18+19+20-21)	26			35	
23. Modulation Noise Improvement					
24. Video/Audio Discrimination and A-J Rejection					
25. S/N and I/N at Output (22+23+24)	26			35	
26. Combined S/(N+I) (Line 25: S/N-I/N if I/N>3)	-9				
27. Interference Margin: S/(N+I) - Std. Dev.					

CONCLUSIONS:_____

$$P_{tdBm} = 30 \text{ dB} + 10 \log_{10} P_t \text{(watts)} \qquad (5.16)$$

The reader may choose to use Table 5.4 to determine the nearest dBm
equivalent of the power output in watts.

The next entry provides for a potential EMI emitter power reduction correction for out-of-band emissions when the victim receiver
is operating more than 50% above the transmitter fundamental frequency.
This correction does not provide for harmonic/subharmonic frequency mis-
alignments, bandwidth corrections, and the like which are discussed
later. Thus, for such out-of-band radiations, Table 5.5 is used in
the absence of known or measured data. No correction is used when the
receiver is tuned below the transmitter fundamental emission since it
will remain to examine EMI between the transmitter fundamental to receiver spurious response (cf. TIM case). Sec. 5.3.1 discusses transmitter models in further detail.

The entry on line 3 in the prediction form involves the antenna
transmission line loss in dB attenuation at the prediction frequency.
Note that this may not correspond to the frequency of the fundamental
power output in an EMI prediction situation as explained later, but
rather at the frequency of a harmonic or other interfering emission.
Fig. 5.8 shows the attenuation in dB per 100 feet (30 meters) for
several coaxial lines. Fig. 5.9 shows the attenuation for several common waveguides.

The prediction form entry for the transmitter antenna gain (line
4) is the rated nominal gain in dB at the antennas' intended frequency
of operation. When the potential EMI transmitter antenna is emitting
out-of-frequency band (line 5) or its main beam is at off-axis angles
(unintentional region) to the victim receiver (line 6), gain correction
is required as shown in Table 5.6. Thus, the correction in dB should
be in an amount to result in the final gain shown in the table. Note,
if the antenna is an integral part of the transmitter such as for radars,
no gain correction is entered in line 5 when the transmitter data in
Table 5.5 is used since it already includes the out-of-band effects
of the antenna. Table 5.6 is used for any combinations of transmitter
antenna gain corrections including the composite effects of lines 5
and 6. For this situation either line 5 or line 6 is used but not
both. Two examples will serve to clarify this. Sec. 5.3.3 discusses
antenna models in further detail.

Illustrative Example 5.4

A potentially interfering L-Band (1335 MHz) radar has third-
harmonic emissions existing in the 4 GHz, common-carrier, microwave
relay spectrum. Calculate the antenna gain corrections for both lines
5 and 6 of the prediction form.

Table 5.4 Transmitter Power in Watts vs Equivalent dBm

dBm	Watts	dBm	Watts	dBm	Watts	dBm	Watts
110	100 MW	80	100 kW	50	100 W	20	100 mW
109	80 MW	79	80 kW	49	80 W	19	80 mW
108	63 MW	78	63 kW	48	63 W	18	63 mW
107	50 MW	77	50 kW	47	50 W	17	50 mW
106	40 MW	76	40 kW	46	40 W	16	40 mW
105	32 MW	75	32 kW	45	32 W	15	32 mW
104	25 MW	74	25 kW	44	25 W	14	25 mW
103	20 MW	73	20 kW	43	20 W	13	20 mW
102	16 MW	72	16 kW	42	16 W	12	16 mW
101	13 MW	71	13 kW	41	13 W	11	13 mW
100	10 MW	70	10 kW	40	10 W	10	10 mW
99	8 MW	69	8 kW	39	8 W	9	8 mW
98	6.3 MW	68	6.3 kW	38	6.3 W	8	6.3 mW
97	5 MW	67	5 kW	37	5 W	7	5 mW
96	4 MW	66	4 kW	36	4 W	6	4 mW
95	3.2 MW	65	3.2 kW	35	3.2 W	5	3.2 mW
94	2.5 MW	64	2.5 kW	34	2.5 W	4	2.5 mW
93	2 MW	63	2 kW	33	2 W	3	2 mW
92	1.6 MW	62	1.6 kW	32	1.6 W	2	1.6 mW
91	1.3 MW	61	1.3 kW	31	1.3 W	1	1.3 mW
90	1 MW	60	1 kW	30	1 W	0	1 mW
89	800 kW	59	800 W	29	800 mW	-1	800 µW
88	630 kW	58	630 W	28	630 mW	-2	630 µW
87	500 kW	57	500 W	27	500 mW	-3	500 µW
86	400 kW	56	400 W	26	400 mW	-4	400 µW
85	316 kW	55	316 W	25	316 mW	-5	316 µW
84	252 kW	54	252 W	24	252 mW	-6	252 µW
83	200 kW	53	200 W	23	200 mW	-7	200 µW
82	159 kW	52	159 W	22	159 mW	-8	159 µW
81	126 kW	51	126 W	21	126 mW	-9	126 µW
80	100 kW	50	100 W	20	100 mW	-10	100 µW

Table 5.5 - Summary of Transmitter (TXMR) Harmonic Average
Emission Levels in dB Above Fundamental*

TXMRS Harmonic Number	All TXMRS Combined (σ=15 dB)	Transmitters Categorized According to Radio Frequency		
		Below 30MHz (σ=10 dB)	30MHz–300MHz (σ=15 dB)	Above 300MHz (σ=20 dB)
2	-51	-41	-54	-55
3	-64	-53	-68	-64
4	-72	-62	-78	-70
5	-79	-69	-86	-75
6	-85	-74	-92	-79
7	-90	-79	-97	-82
8	-94	-83	-102	-85
9	-97	-87	-106	-88
10	-100	-90	-110	-90

* This includes the contributions from high-power emitters having
   integral transmitter antennas such as radar, troposcatter
   communications and TV broadcast.

Table 5.6 - Antenna Gain Corrections for Different Operation Conditions

Nominal Antenna Gain	Operating Conditions				Beamwidths		Mean Antenna Gain	Standard Deviation σ, in dB
	Intentional Region		Unintentional Region		Horiz.	Vert.		
	Frequency	Polariz.	Frequency	Polariz.				
High Gain G>25dB	Design	Design	----	----	α	β	G	2
	Design	Orthognal	----	----	10α	10β	G-20	3
	NonDesign	Any	----	----	4α	4β	G-13	3
	----	----	Design	Design	----	----	-10	14
	----	----	Design	Orthognal	----	----	-10	14
	----	----	NonDesign	Any	----	----	-10	14
Medium Gain Resonant Non-Resonant 10≤G≤25dB	Design	Design	----	----	α	β	G	2
	Design	Orthognal	----	----	10α	10β	G-20	3
	NonDesign	Any	----	----	3α	3β	G	3
	NonDesign	Any	----	----	α	β	G	3
	----	----	Design	Design	----	----	-10	11
	----	----	Design	Orthognal	----	----	-10	13
	----	----	NonDesign	Any	----	----	-10	10
Low Gain G<10dB	Design	Design	----	----	α	β	G	1
	Design	Orthognal	----	----	6α	6β	G-16	2
	NonDesign	Any	----	----	360°	180°	0	2
	----	----	Design	Design	----	----	0	6
	----	----	Design	Orthognal	----	----	-13	8
	----	----	NonDesign	Any	----	----	-3	6

Notes: α = 10-dB azimuth beamwidth corresponding to design frequency and polarization. If unknown, use 2 x 3-dB beamwidth.

β = 10-dB vertical (elevation) beamwidth corresponding to design frequency and polarization. If unknown, use 2 x 3-dB beamwidth.

Antenna gains expressed in either (1) for intentional region, dB relative to maximum gain (G) at design frequency and polarization (e.g. G-20dB), or (2) for unintentional region, dB relative to isotropic antenna (e.g. -10dB).

α, β, G, and σ are functions of both frequency and polarization.

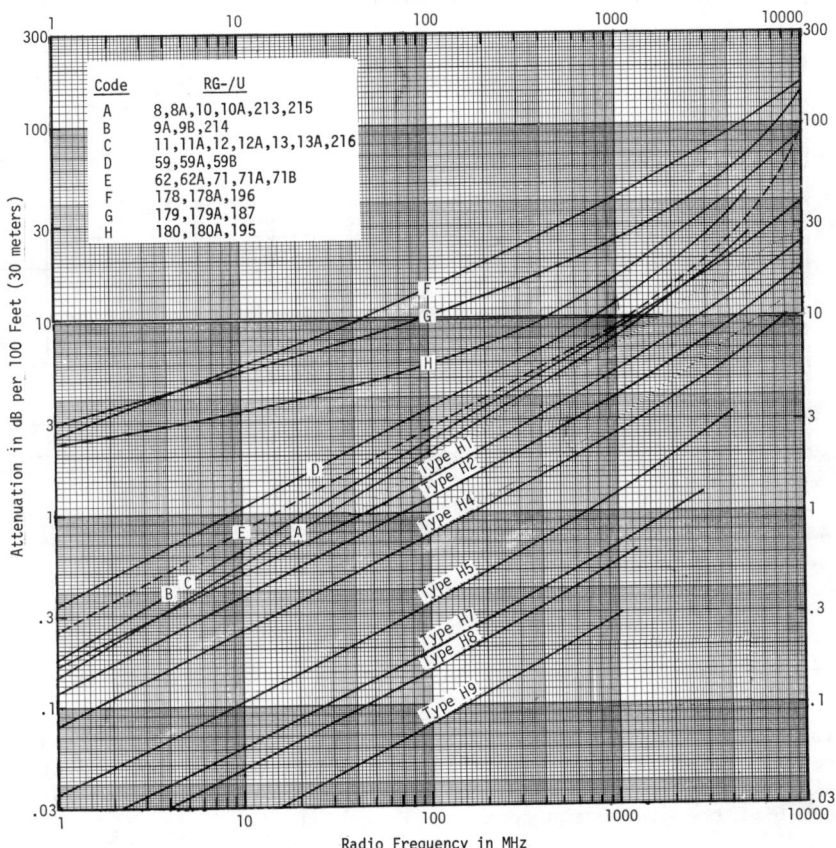

Figure 5.8 - Coaxial Cable Attenuation vs Frequency

Table 5.5 is used for calculating the third harmonic emission level to enter in line 2 (-64 dB). Since it already contains the contribution from the radar antenna, line 5 entry is 0 dB. Because the radar antenna rotates through 360°, there will always exist one situation in which its antenna is looking at the microwave receiver antenna. Thus, the correction for line 6 is also 0 dB.

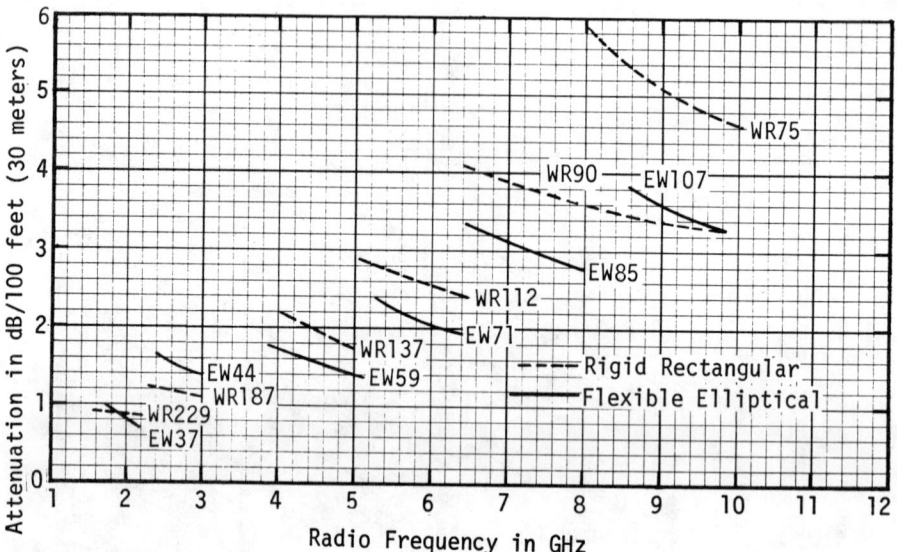

Figure 5.9 - Waveguide Attenuation vs Frequency

Illustrative Example 5.5

A 1 GHz, 40-dB gain troposcatter antenna is looking 60 degrees off-axis horizontally to the above (Example 5.4) microwave relay receiving antenna. Possible interference is expected from the tropo's fourth harmonic radiations. Calculate the antenna gain corrections for both lines 5 and 6 of the prediction form.

Again, since line 2 contains the on-axis entry from Table 5.5 for high power, integral transmitter-antenna combination, line 5 entry is 0 dB. Had this not been the case, Table 5.6 entry for line 5 would have been -13 dB corresponding to high gain, intentional region, non-design frequency. Since the tropo antenna is also looking away from the microwave relay antenna (unintentional region), Table 5.6 indicates that the equivalent absolute antenna gain is more nearly -10 dB, or -50 dB above (+ 50 dB below) the on-frequency, on axis situations of G = 40 dB entered on line 4. Thus, the additional correction for line 6 entry would be -37 dB, obtained from -50 dB -(-13 dB) absorbed in line 2. Stating this another way, line 6 entry would be the difference between the two non-design frequencies of Table 5.6 or -[(G-13) - (-10)] = -37 dB. Sec. 5.3.4 will further clarify this special situation.

5.2.4.2   PROPAGATION PATH CORRECTION

Line 8 of the prediction form corresponds to the free-space prop-agation loss in dB obtained from Fig. 5.7. The frequency parameter to be used in this figure corresponds to that of the victim receiver for an FIM or RIM case, while the frequency for a TIM case corresponds to the transmitter fundamental.

When non-RLOS conditions exist (line 9), such as would exist in Fig. 5.5 if blocked by an intervening hill or mountain, an additional propagation correction is required. This correction is obtained from Sec. 5.3.4 involving propagation corrections. Similarly, for trans-missions above about 5 GHz, rain, water vapor, and/or oxygen losses (line 11) should be computed from Sec. 5.3.4. Where transmission re-liability is important and fade margins must be provided for (line 10), Sec. 5.3.4 is used.

5.2.4.3   RECEIVER-ANTENNA INPUT POWER

Line 13 entry corresponds to the measured or nominal receiver an-tenna gain at the transmitter frequency. When the direction of arrival of the transmitted signal is off-axis (i.e., the interference signal) and/or out-of-band, line 14 entry is determined from Table 5.6 in lieu of using measured data. This entry also corrects for polarization so that line entry 15 would not be used. However, for both on-axis and in-band frequencies an option to use an explicit polarization cor-rection exists as listed in Table 5.7.

Table 5.7 - Antenna Polarization Correction in dB*

TX	Horizontal		Vertical		Cir
RX	G<10dB	G≥10dB	G<10dB	G≥10dB	
Hor G<10dB	0	0	-16	-16	-3
Hor G≥10dB	0	0	-16	-20	-3
Ver G<10dB	-16	-16	0	0	-3
Ver G≥10dB	-16	-20	0	0	-3
Circular	-3	-3	-3	-3	0

5.2.4.4   FREQUENCY ALIGNMENT AND BANDWIDTH CORRECTIONS

Most EMI problems result from other than co-channel interference. Thus, a correction (line 19 of the prediction form) is required for mis-alignment of (1) the transmitter fundamental or one of its spurious emissions with (2) the receiver fundamental or one of its spurious re-sponses. The correction to be used is shown in Fig. 5.10 if better

---

*   Applies only to intentional radiation region and design frequency.

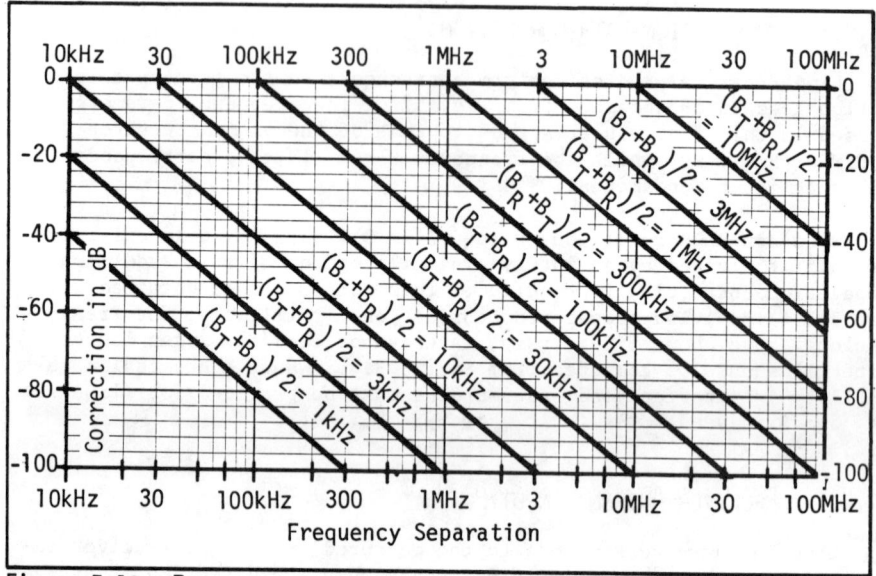

Figure 5.10 - Frequency Separation Correction

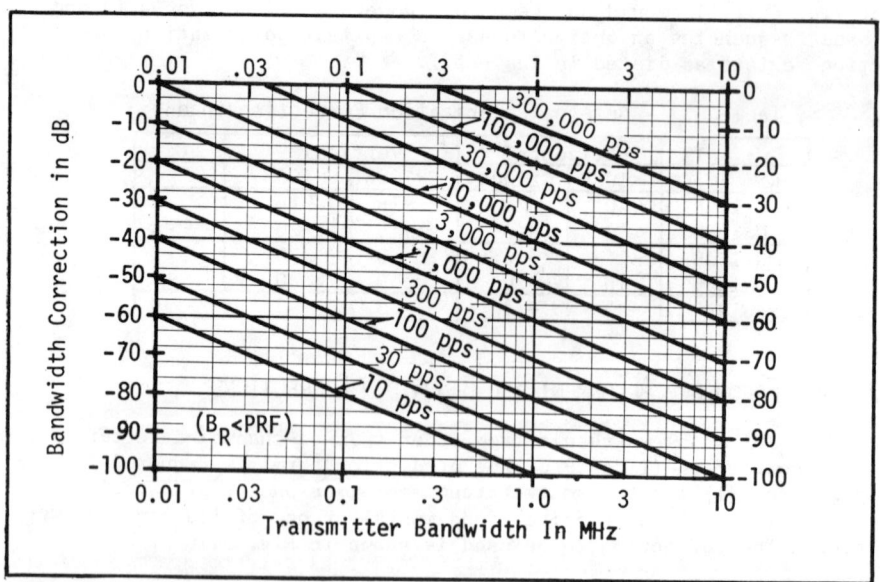

Figure 5.11 - Bandwidth Correction For Narrow Band Receiver and Pulses

off-frequency measured data are not available. The parameters, B_T (transmitter 3-dB bandwidth) and B_R (receiver bandwidth) must be known since this correction in dB is both a function of bandwidth and frequency separation.

A separate correction for dissimilar transmitter and receiver bandwidths is required in line 20. If the transmitter is pulse modulated and if the receiver bandwidth in Hz is less than the transmitter pulse repetition rate, then Fig. 5.11 is used. If the receiver bandwidth exceeds the pulse repetition rate, or if the transmitter is not pulse modulated, the bandwidth correction shown in Fig. 5.12 may be used. Sec. 5.3.1 presents more details on corrections to be used for both frequency separation and bandwidth differences, especially when known types of transmitter modulations are used.

The receiver sensitivity for signal equal noise (S=N) is entered on line 21. If the sensitivity is not known, but the noise figure and receiver bandwidth are known, Fig. 5.13 may be used to determine receiver sensitivity. If the noise figure is also unknown, it may be estimated from Table 5.8.

Table 5.8 - Typical Noise Figures of Operational C-E Receivers*

Radio Frequency	Old Equipment > 5 yrs. old	Std. Dev. σ in dB	New Equipment < 5 yrs. old	Equip. Type*	
				Expen.	Cheap
$f_r < 30$ MHz	7 dB	2	5 dB	−2dB	+2dB
30 MHz $\leq f_r < 300$ MHz	9 dB	3	7 dB	−3dB	+3dB
300 MHz $\leq f_r \leq 3$ GHz	12 dB	4	10 dB	−4dB	+5dB
$f_r > 3$ GHz	18 dB	5	15 dB	−5dB	----

5.2.4.5   OUTPUT S/(N+I) RATIOS AND INTERPRETATION

The previous S/N and I/N ratio calculations (see line 22 of prediction form) are based on performance up to but excluding the receiver post detector. When multiplied and certain non-AM demodulators are used, a modulation noise improvement (line 23) results. This and the impact of anti-jamming interference rejection (line 24) is discussed in Sec. 5.3.2 and Vol. 5 of this handbook series.

The final S/(N+I) ratio calculated in line 26 may be interpreted with the use of performance data discussed in Sec. 5.3.5. At the risk

---

*   If C-E receivers belong to expensive equipment (e.g., radio telescopes, high-power radars) then the noise figures will be less, and if the equipment is inexpensive (cheap, e.g., consumer broadcast receivers) the noise figure will be greater than indicated. For these cases, subtract or add the indicated amounts in lieu of known data.

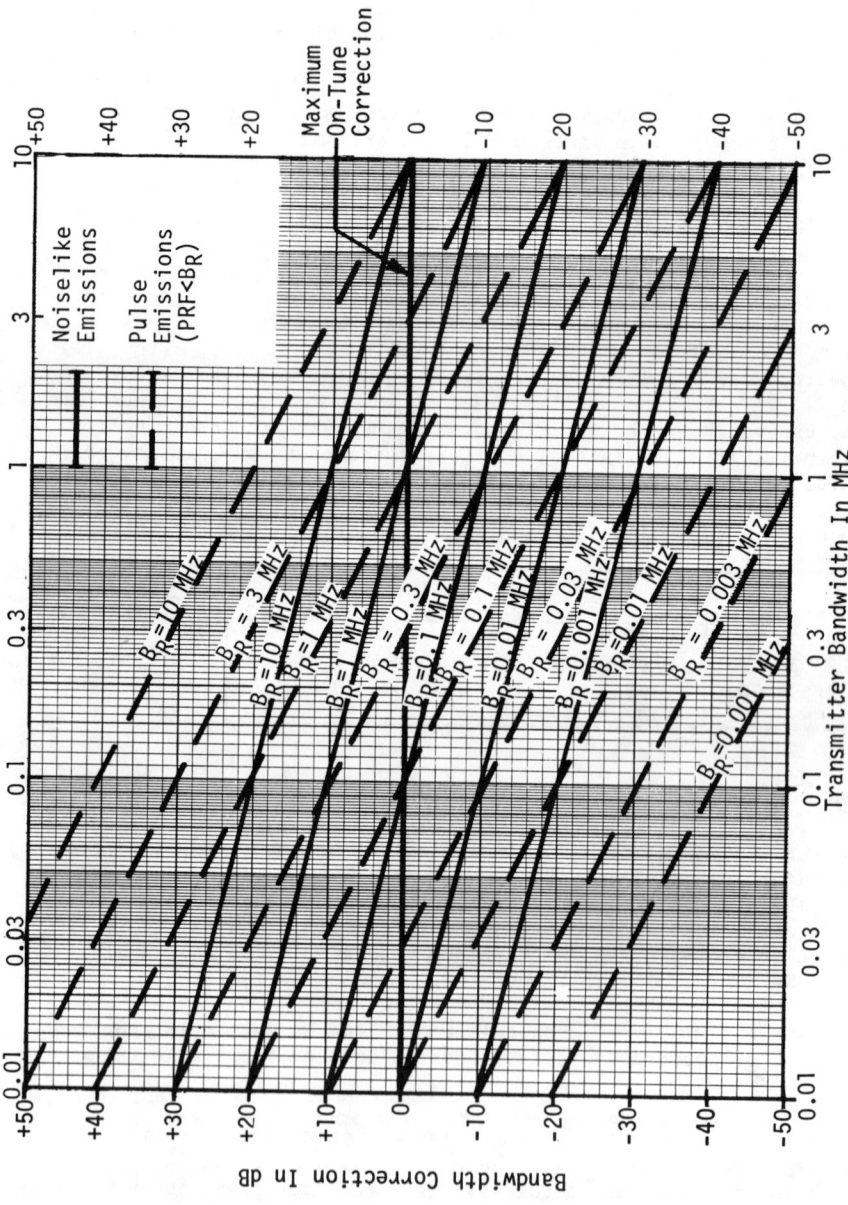

Figure 5.12 – Bandwidth Correction Factor

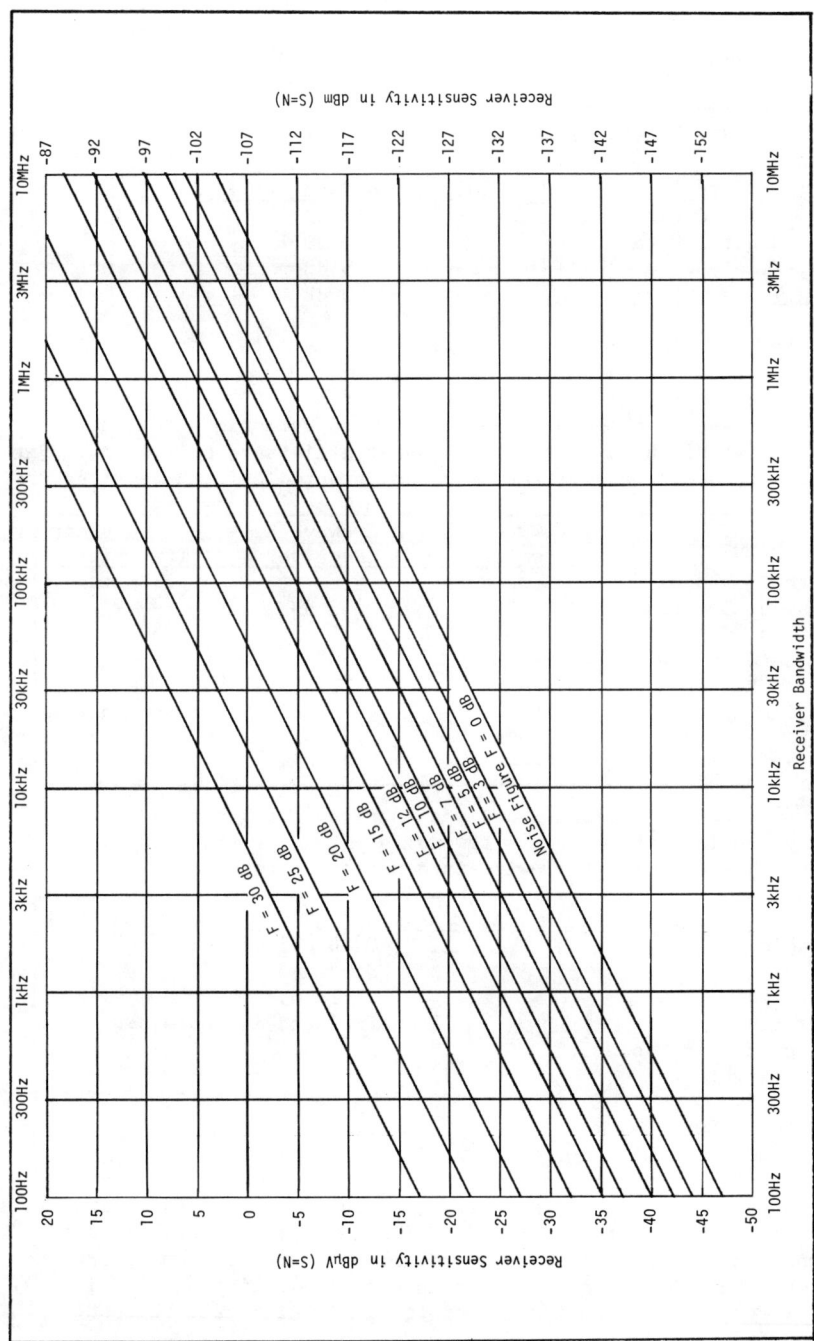

Figure 5.13- Receiver Narrowband Sensitivity vs Bandwidth and Noise Figure

of generalizing here, an interference problem may be considered to exist when S/(N+I) is less than 0 dB, or, more correctly as listed in
Table 5.9.

Table 5.9 - General Go/No-Go EMI Acceptability Criteria

Output	EMI Exists When S/(N+I)
Video/Optical	< 15 dB
Digital Data	< 10 dB
Teletype	< 5 dB
Voice	< 0 dB

Illustrative Example 5.6

A 4.0 GHz microwave relay receiver is located 30 miles from its
transmitting mate. An L-Band (1335 MHz; third harmonic = 4005 MHz)
radar exists within RLOS of the receiver at a distance of 7 miles. Pertinent data for both transmitters and the potential victim receiver are:

	Microwave Relay Transmitter	Microwave Relay Receiver	L-Band Radar Transmitter
Power Output	1 watt	----	2 MWatt
Transmission Line Loss	2 dB	2 dB	----
Antenna Gain	35 dB	35 dB	30 dB
TX-RX-TX Antenna Angle	----	20°	----
Fade Margin Required	----	30 dB for 99.9% Rel.	----
Polarization	Vertical	Vertical	Horizontal
Bandwidth	3 MHz	5 MHz	300 kHz
Sensitivity	----	-98 dBm	----

The EMI Prediction Form shows the entries made for this EMI illustrative problem. By way of a few words of explanation, the line
entries are based on the following information sources.

LINE	SOURCE	LINE	SOURCE
1	Table 5.4	11	0 dB below 5 GHz
2	Table 5.5	13	Stated
3	Stated	14	Table 5.6
4	Stated	15	Included in Table 5.6
5	Included in Table 5.5	16	Stated
6	Boresight	19	Fig. 5.10
8	Fig. 5.7	20	$B_R > B_T$, Fig. 5.12
9	Within RLOS	21	Stated
10	Stated		

The prediction form shows that the output S/I ratio is -9 dB. This
is considerably poorer than the +10 dB (Table 5.9) required for digital data. Thus, EMI would likely present a severe problem during the
moments when both the rotating radar antenna is looking at the microwave relay receiver and during a deep fade. To eliminate this EMI

situation, at least 19 dB (+10 dB – (–9 dB)) improvement in S/I is required. Thus, a somewhat greater objective (e.g., 25 dB improvement)
is sought. This may be accomplished in one or more of the following
ways:

EMI Control Technique	Comments
Change interfering transmitter (radar) frequency.	Effective, quick, and economical if user will cooperate.
Change microwave relay channel.	May be difficult to do because of other frequency assignment problems.
Install low-pass filter in radar transmitter.	Effective, but expensive and requires long lead time.
Sector blank radar antenna.	Effective, but creates a sector hole in air surveillance coverage.
Install *hood* around receiving antenna.	Expensive and awkward to do.
Install *billboard screen* line-of-sight mask to radar.	Effective but expensive and re-quires long lead time.

The first EMI control technique listed above is often the quick-
est and easiest to effect. To gain an additional 25 dB of interference
rejection, the 12 dB rejection in line 19 obtained from Fig. 5.10 would
have to be increased to 37 dB. Note, the $(B_T + B_R)/2 = (0.3$ MHz + 5
MHz$)/2 = 2.65$ MHz line intersects with –37 dB at about 20 MHz. If the
radar were tuned from 1335 MHz to 1340 MHz, its third harmonic would
change from 4005 MHz to 4020 MHz. Since the microwave receiver is
tuned to 4000 MHz, the frequency separation would then be 20 MHz.

5.3 COMPONENT MATHEMATICAL MODELS

This section presents more details on transmitter, receiver, antenna, propagation, and signal acceptability performance models than those introduced in the previous section in connection with the EMI prediction form. The reader should review Vol. 5, *EMI Prediction and Analysis,* for a comprehensive discussion of this subject.

5.3.1 TRANSMITTER MODELS

The primary function of a transmitter is to generate radio-frequency power containing direct or latent intelligence within a specified frequency band. In addition to the desired power, transmitters produce numerous unintentional emissions at spurious frequencies. A spurious emission is any radiated output that is not required for transmitting the desired information. The desired and/or undesired R-F power generated by transmitters may produce EMI in receivers or other receptors. Therefore, in evaluating EMC, it is necessary to consider all transmitters as potential sources of interference.

For EMI prediction, it is necessary to describe both the desired and undesired emission power spectrum of each transmitter. In general, the transmitter emission spectrum is a composite of several different transmitter interference functions. The next section briefly describes various transmitter functions, identifies the type of information required, and discusses problems involved in specifying each of these functions.

5.3.1.1 TRANSMITTER FUNCTIONS FOR EMI PREDICTION

Transmitter emissions are classified into one of four categories: (1) fundamental emissions, (2) harmonically-related emissions, (3) non-harmonic emissions, and (4) broadband noise. Examples of emissions in each of these categories are illustrated in Fig. 5.14. Although it is convenient for EMI prediction to consider discrete transmitter emissions, the power associated with any given emission is actually spread over a finite frequency range. Furthermore, broadband electrical noise is superimposed on a discrete emission spectrum. Thus, the complete transmitter spectral emissions would more nearly appear as shown in Fig. 5.15, where the overall spectrum-amplitude relations represent one example of a mathematical model.

For EMI prediction, the important parameters that are described for each type of emission are: (1) the output power used in amplitude culling, (2) a reference frequency and envelope that define the power content in the sidebands around the reference frequency for use in frequency culling, and (3) a description of the modulation for use in both detailed and performance prediction.

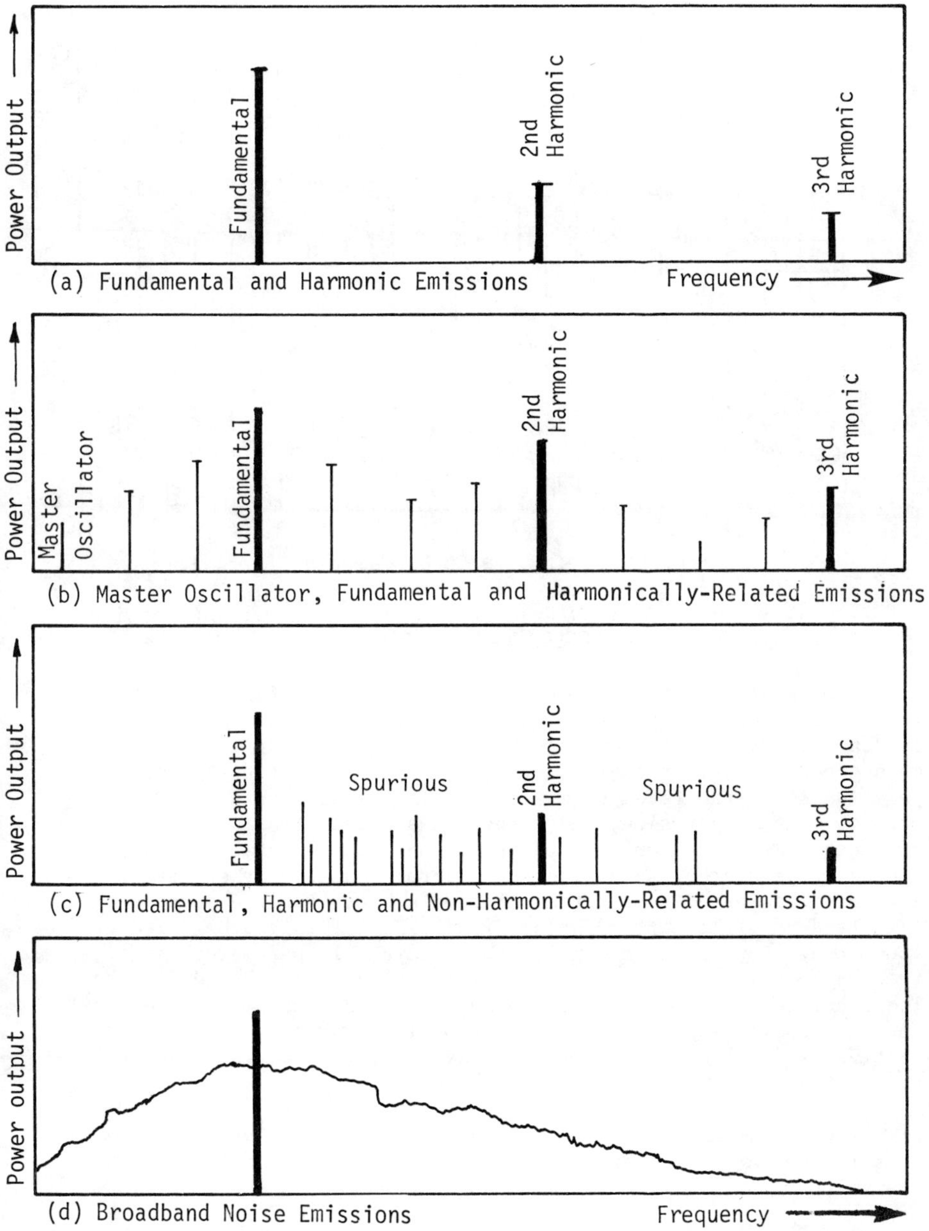

Power Output →

Fundamental

2nd Harmonic

3rd Harmonic

(a) Fundamental and Harmonic Emissions   Frequency →

Power Output →

Master Oscillator

Fundamental

2nd Harmonic

3rd Harmonic

(b) Master Oscillator, Fundamental and **Harmonically-Related** Emissions

Power Output →

Fundamental

Spurious

2nd Harmonic

Spurious

3rd Harmonic

(c) Fundamental, Harmonic and Non-Harmonically-Related Emissions

Power output →

(d) Broadband Noise Emissions   Frequency →

Figure 5.14 - Typical Transmitter Emission Spectra

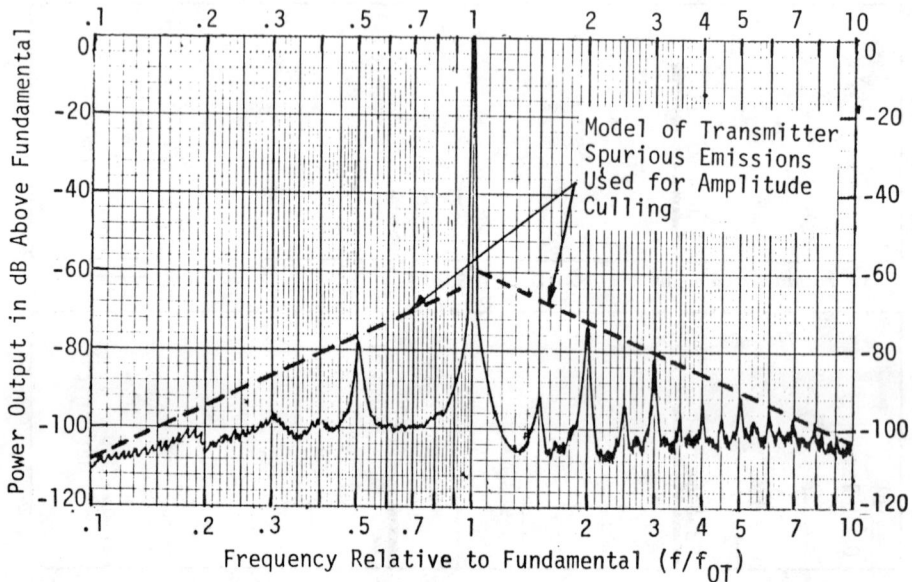

Figure 5.15 - Transmitter Output Spectrum Resulting from Composite of Broadband Noise and Discrete Emissions

EMI prediction may be performed at several different levels of varying complexity each giving different result. The term *amplitude culling* refers to the simplest type of EMI prediction which is based on a general consideration of transmitter power, antenna gains, propagation loss, and receiver susceptibility. Parameters such as frequency, time, separation, and direction are only considered in a limited sense. *Frequency culling* refers to an EMI prediction in which the frequency parameter is considered in detail but the other parameters are only considered in a rather limited sense. For the purpose of this handbook, the term *detailed prediction* refers to the third culling level of prediction complexity in which all parameters (i.e., frequency, time, separation, and direction) are considered in detail. The final culling level, *performance prediction,* considers the resulting impact of EMI problems on operational performance and mission effectiveness. These topics are discussed in Sec. 5.4 involving EMI prediction in multi-emitter environments.

The following paragraphs discuss some of the major considerations that influence the specification of these parameters for different transmitter emissions.

A. FUNDAMENTAL EMISSION

For amplitude culling, it is necessary to specify power output at the intended fundamental frequency. Typical sources used to determine the fundamental power are the transmitter technical manual, spectrum-signature measurements on the equipment, and design specifications. Although the nominal fundamental power is usually specified, there are variations in the actual output from different equipment serial-numbers of a particular transmitter nomenclature. Therefore, a statistical representation of the transmitter fundamental output level is recommended for use in EMI prediction.

For the second-stage of EMI prediction, viz frequency culling, it is necessary to specify the frequency associated with the fundamental emission. This frequency is the nominal operating or *carrier* frequency for each transmitter to be considered in a prediction. For problems in which different operating frequencies are to be considered for each transmitter (e.g., a frequency band or assignment problem) each fundamental frequency is separately considered.

The other transmitter parameter that is required for frequency culling, and must be specified for the fundamental emission, is the relative power in the sidebands around the carrier frequency. The sidebands on either side of the carrier result from the time-domain modulation process and from nonlinearities that exist in the transmitter. A base-band modulation envelope is used to describe relative power levels of the sidebands with respect to the carrier.

For detailed and performance predictions, it is necessary to consider information concerning the transmitter and antenna modulation characteristics in the time domain, the type of modulation used, and the type of information transmitted.

B. HARMONICALLY-RELATED EMISSIONS

Harmonically-related transmitter emissions include those undesired spurious emissions that are integer multiples of either the fundamental frequency or frequencies used to generate the fundamental (e.g., a master oscillator or crystal-controlled clock). Specification of harmonic-output power is complicated because of significant variations from serial number to serial number for a particular transmitter nomenclature. As a result, harmonic output power level must be represented in a statistical manner.

Spectrum signature measurements* provide one of the best sources of data on harmonic-emission output levels for use in EMI prediction. However, in the event that spectrum signature data are not available for a particular transmitting equipment, there are other means of obtaining information required. Transmitter specifications are one possible source of information regarding spurious emission levels. These specifications sometimes provide an indication of harmonic emission output levels.

* See MIL-STD-449D and Chap. 5, of Vol. 1, *Electrical Noise and EMI Specifications*.

For harmonic emissions, as for the fundamental emission, power is spread over a frequency range. Thus, for these emissions, it is also necessary to define relative content of power over the frequency band associated with the emission. The techniques for representing the modulation envelopes at the fundamental are also applied to harmonic emissions.

C. NON-HARMONIC SPURIOUS EMISSIONS

There are certain cases, such as in the operation of a magnetron or other high-peak pulse power source, where spurious emissions occur at frequencies that are not harmonically related to the fundamental or a frequency used in producing the fundamental. For these emissions, the relative frequency is subject to random variations. Thus, it is necessary to use statistical representations to define the probability that an output will occur in any given frequency increment. The output power level and the sideband envelope for the non-harmonic emissions are described in a manner similar to that for harmonically-related emissions.

D. TRANSMITTER BROADBAND NOISE

With the possible exception of high-power transmitters greater than about 1 kW, the transmitter noise level is relatively insignificant in comparison to the other interfering signals that are present in a given electromagnetic environment. For those cases where it is necessary to consider transmitter broadband noise, the level is specified in terms of available power per Hertz of bandwidth as a function of frequency, viz., dBm/Hz.*

E. OTHER TRANSMITTER FUNCTIONS

In addition to the specific transmitter outputs delineated above, there is another that is considered in EMI prediction, viz., transmitter intermodulation. For prediction purposes, those signals that are produced when the output of one transmitter mixes with the output of another in the non-linear circuits of a transmitter, are considered to be in a different category from those identified above.

5.3.1.2 TRANSMITTER MODELS FOR AMPLITUDE CULLING

In the first-stage EMI prediction process for multi-transmitter environments (see Sec. 5.4), amplitude culling is used. Transmitter mathematical models used to describe amplitude culling are divided into fundamental emission, harmonic emission, and non-harmonic emission models. Chap. 3 of Vol. 5 explains how these are obtained.

---

* Were this not random noise, but due to impulses, the term would be peak voltage intensity in V/Hz (or dBV/Hz) or peak power intensity in W/Hz^2 (or dBm/Hz^2).

For the purpose here, the fundamental emission levels are those of the rated nominal value with a 2 dB standard deviation. The transmitter harmonic amplitude models were summarized in Table 5.5 of Sec. 5.2. Here the standard deviation is more nearly 10-20 dB for the second to tenth harmonics. Finally, the non-harmonic emission amplitude models fall off with the logarithm of frequency departure in a manner similar to that of Fig. 5.15. Details are discussed in Vol. 5.

For the purpose of amplitude culling, when all else fails in collecting the data needed for an EMI prediction, a value of 60 dB down from fundamental emission may be used for out-of-band emissions out to one decade above and below fundamental frequency. On the upper side, however, it would be better to use Table 5.5. Beyond these frequencies (more than $\pm$ one decade) an amplitude value of 100 dB down may be used. These values are fairly conservative since emissions (excepting the second harmonic) are almost always less.

5.3.1.3 TRANSMITTER MODELS FOR FREQUENCY CULLING

This section presents methods and techniques for modeling transmitter characteristics for use in the second-stage frequency culling of the EMI prediction process. It identifies sources of information that may be used to determine characteristics of specific transmitters. It also describes general models that may be used to represent transmitter characteristics when specific information is not available.

For frequency culling, a reference frequency is associated with each transmitter emission and the distribution of power around this frequency is described. The reference frequency is specified in a simple deterministic manner for fundamental and harmonically related outputs. For non-harmonically-related outputs, however, it is necessary to use probabilistic techniques for describing these emission outputs. In general, the power distribution is described by defining a bandwidth (which is determined by the transmitter modulation characteristics) and an envelope (which describes the manner in which the power varies with frequency).

A. FUNDAMENTAL-EMISSION FREQUENCY MODELS

In performing an EMI prediction of a specific C-E equipment complex, it is necessary to specify the actual operating frequency or frequencies for each transmitter. This frequency is usually obtained from the equipment operational plans; it is simply specified by the nominal operating frequency of the transmitter. For other types of prediction which involve either (1) determining frequency-distance separation criteria for transmitter-receiver combinations, or (2) predicting EMI on the basis of frequency bands rather than discrete frequencies, the definition of a specific operating frequency is not required.

The transmitter fundamental output is not actually confined to a single frequency; it is distributed over a range of frequencies around the fundamental. The characteristics of the power distribution in the vicinity of the fundamental are determined primarily by the base-band modulation characteristics of the transmitter. The resulting spectral components are termed *modulation sidebands*. For frequency culling, the power distribution in the modulation sidebands is represented by a modulation envelope function. In general, the modulation envelopes are described by specifying bandwidths or frequency ranges and functional relationships which describe the variation of power with frequency, $M(\Delta f)$. The modulation envelope model is:

$$M(\Delta f) = M(\Delta f_i) + M_i \log_{10} \left(\frac{\Delta f}{\Delta f_i}\right), \qquad (5.17)$$

$$\text{for } \Delta f_i \leq \Delta f \leq \Delta f_{i+1}$$

where, Δf = separation from reference frequency

Δf_i = bandwidth of applicable region

M_i = slope of modulation envelope for applicable region

One example of the resulting functional relationship is shown in Fig. 5.16. The parameters that are required to specify the modulation envelope are the bandwidths of applicable regions of constant slope and the rate at which the envelope falls off over the frequency region of interest.

A.1 Bandwidth Models

One of the most important parameters associated with the modulation envelope is the transmitter nominal bandwidth (usually the 3 dB bandwidth). Most of the transmitter power is located within this region and the power decreases rapidly with frequency separation outside this region* Usually the transmitter nominal bandwidth may be determined from transmitter specifications. In the event that such information is not available, the bandwidth can then be determined from a consideration of the transmitter modulation characteristics.

A.1.a Amplitude Modulation (AM)

In an amplitude-modulated (AM) wave, the amplitude of a *carrier* is varied in accordance with the baseband intelligence being transmitted. This is shown in Fig. 5.17. When the amplitude of the R-F carrier is varied in this manner, sideband frequencies are generated which carry the intelligence. The nominal width of the frequency

---

*   For some situations to be discussed, less than half the power is contained within the transmitter bandwidth defined. This is done to establish the first break point for Δf, of Fig. 5.16.

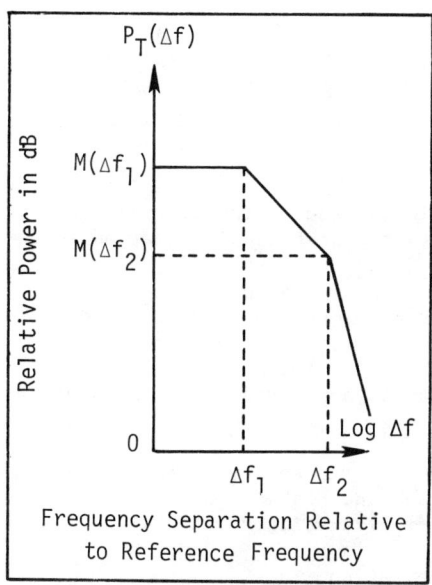

Figure 5.16 - Modulation Envelope
Representation

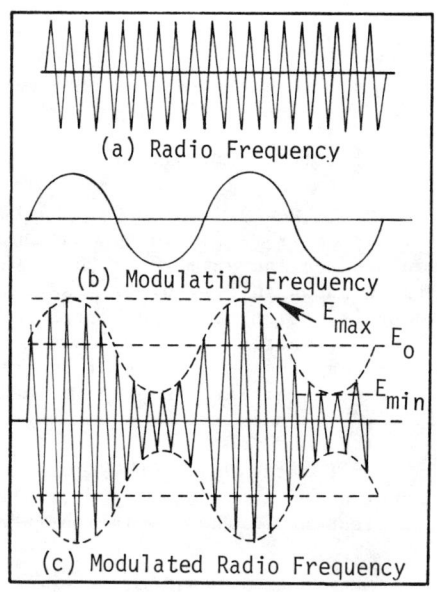

Figure 5.17 - Amplitude Modulation

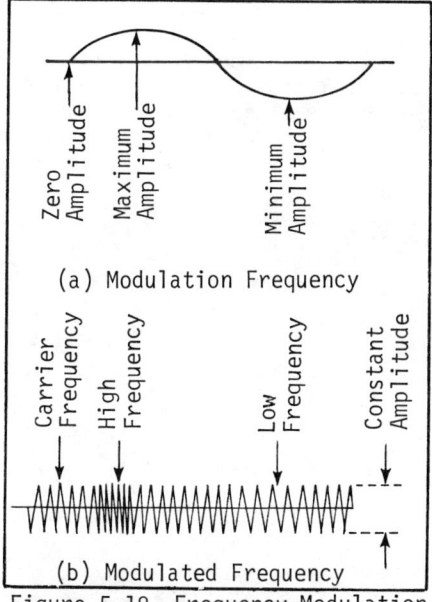

Figure 5.18 Frequency Modulation

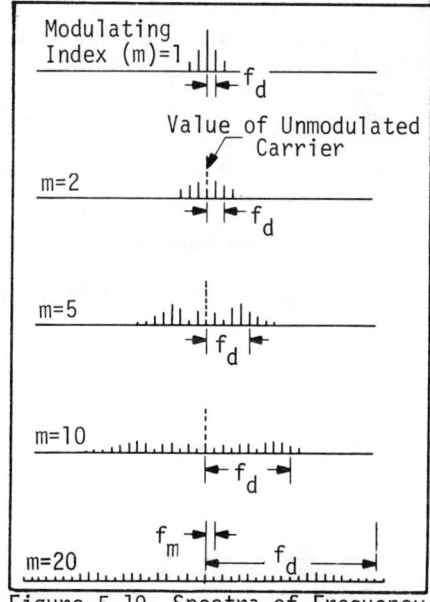

Figure 5.19 Spectra of Frequency
Modulated Waves (Constant
Modulating Frequency)

spectrum occupied by such an A-M wave is twice that of the highest
frequency contained in the baseband information. Thus, the bandwidth
required of the A-M wave is determined by the rate at which the amplitude is varied.

A.1.b Frequency Modulation (FM)

In frequency modulation (FM) the instantaneous frequency of an
R-F carrier is varied in accordance with the baseband information to be
modulated on the wave while the amplitude of the carrier is held constant.
This is shown in Fig. 5.18. The rate at which the instantaneous frequency is varied about the carrier is determined by the modulating
frequency, whereas the *frequency deviation* (i.e., the excursions or
amount that the excursions or amount that the frequency varies from
the central carrier) is proportional to the amplitude of the modulating
signal.*

For FM, the bandwidth in which most of the energy is contained is
a function of the frequency deviation and the highest modulating base-
band frequency. The modulation index (m_f) is defined as:

$$m_f = \frac{\text{frequency deviation}}{\text{modulation frequency}} = f_d/f_m \qquad (5.18)$$

Fig. 5.19 illustrates the relationship between modulation index
and bandwidth for a given modulating frequency. When the modulation
index is less than one-half (i.e., the frequency deviation is less than
half the modulating frequency), the power distribution is confined to
the bandwidth as with AM. On the other hand, if the modulation index
is greater than unity (i.e., the frequency deviation is greater than
the modulating frequency), the F-M wave contains significant sideband
components on both sides of the carrier. This exists over a frequency
interval approximating the sum of the frequency deviation and the modulating frequency. For EMI prediction, the transmitter bandwidth B_T
for an F-M signal is considered to be twice the sum of the frequency
deviation plus the modulating frequency, viz:

$$B_T = 2 \ (f_d + f_m), \text{ for } m_f > 1 \qquad (5.19)$$

where, f_d = frequency deviation

                    f_m = modulating frequency

Because the frequency deviation can be expressed as the product
of the modulating index and modulating frequency, the bandwidth can be
expressed as:

$$B_T = 2 \ (1 + m_f) \ f_m \qquad (5.20)$$

---

*   Frederick E. Terman, 4th ed., *Electronic Radio Engineering*,
McGraw-Hill Book Company, Inc., New York, New York, 1955.

When the modulating index (m_f) is one or more orders of magnitude greater than unity, the bandwidth is approximately twice the frequency deviation or twice the product of the modulation index and the modulating frequency. Thus,

$$B_T \simeq f_d \simeq 2m_f f_m, \text{ for } m_f \gg 1 \qquad (5.21)$$

Illustrative Example 5.7

Consider an FM transmitter modulated with a baseband signal having a 5 kHz bandwidth (i.e., f_m = 5 kHz). Assume that the frequency deviation is 45 kHz. Calculate the bandwidth.

From Eq. (5.18), the modulation index (m_f) is 9. The bandwidth for this transmitter is given by Eq. (5.20):

$$BW = 2 \ (1 + m_f) \ f_m$$
$$= 2 \ (1 + 9) \ 5 \text{ kHz}$$
$$= 100 \text{ kHz}$$

A.1.c. Phase Modulation (ΦM)

A phase-modulated (ΦM) wave is one in which the reference phase of the carrier is varied in proportion to the instantaneous amplitude of the modulating baseband signal. ΦM is similar to FM and is merely a way of obtaining an F-M wave in which the frequency deviation is proportional to the frequency instead of the amplitude of the modulating wave.

A.1.d. Pulse Modulation (PM)

For pulse-type emissions, such as those used with radar and navigation systems, the transmitter bandwidth, B_T, used in EMI prediction is given by:

$$B_T = 2/\pi\tau$$

where, τ = pulse width   (5.22)

The bandwidth is calculated from nominal data on the pulse width.

Illustrative Example 5.8

Consider that a radar transmitter has a pulse width of 1 μsec. The bandwidth is determined directly from Eq. (5.22):

$$B_T = \frac{2}{(\pi \times 10^{-6})} = 0.63 \times 10^6 \text{Hz} = 0.63 \text{ MHz}$$

A.2 Modulation Envelope Models

For performing an EMI prediction, it is necessary to evaluate the constants in the general mathematical model, Eq. (5.17) for the fundamental output modulation envelope. Often, the only available modulation information in the transmitter instruction manual will be the 3-dB frequencies and/or the 10-dB frequencies. Either measurements or theoretical prediction methods must be used to determine the modulation envelope. When measurements are available for a particular transmitter, the data can be utilized to determine the constants in the mathematical model for the fundamental output modulation envelope.

Mathematical techniques exist for converting a function in the time domain to an equivalent representation in the frequency domain. However, in many cases there is not sufficient information available to perform the conversion. Thus, general representations of modulation envelopes are required for those cases where detailed information required for the Fourier analysis is not available.

In the event that specific data on transmitter modulation envelopes are not available, generalized representations shown in Figs. 5.20 through 5.23 may be used to describe the transmitter modulation characteristics for use in EMI prediction. The slope, M, in dB/decade is always referenced with respect to the baseband notch frequency and not that of the same frequency with carrier inserted. However, the slope is negative on the upper carrier side and positive on the lower carrier side. Table 5.10 presents appropriate values for Δf_i, $M(\Delta f_i)$, and M_i that should be used in the modulation envelope model, Eq. (5.17), for the types of modulation indicated.

Illustrative Example 5.9

Consider an AM communication transmitter with an emission bandwidth of 10 kHz. Use Table 5.10 and Eq. (5.17) to model the modulation envelope for the transmitter.

For an AM transmitter, the modulation envelope shown in Fig. 5.23 is completely defined once the emission bandwidth is specified. Thus, for a 10 kHz bandwidth, the modulation envelope model, Eq. (5.17) is:

$$M(\Delta f) = 0, \text{ for } \Delta f \leq 5 \text{ kHz}$$

$$= -133 \log_{10} \frac{\Delta f_{kHz}}{5 \text{ kHz}} \text{ , for } 5 \text{ kHz} \leq \Delta f \leq 10 \text{ kHz}$$

$$= -40 - 67 \log_{10} \frac{\Delta f_{kHz}}{10 kHz} \text{ , for } 10 \text{ kHz} \leq \Delta f$$

For this example, the transmitter modulation envelope is -40 dB above the fundamental level at a frequency 10 kHz removed from the fundamental (i.e., Δf = 10 kHz), and is -80 dB above the fundamental at a frequency 40 kHz removed from the fundamental.

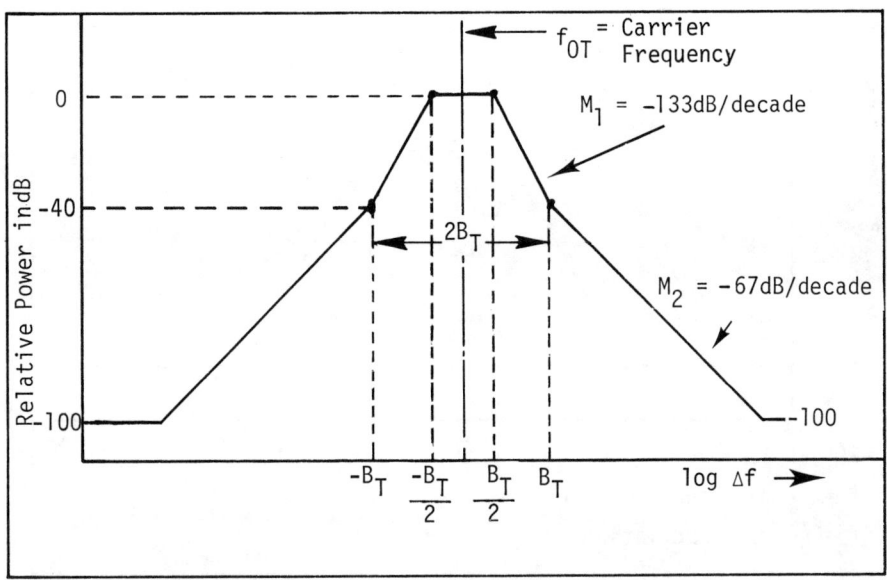

Figure 5.20 - Frequency Spectrum Envelope for AM Communication and
CW Radar

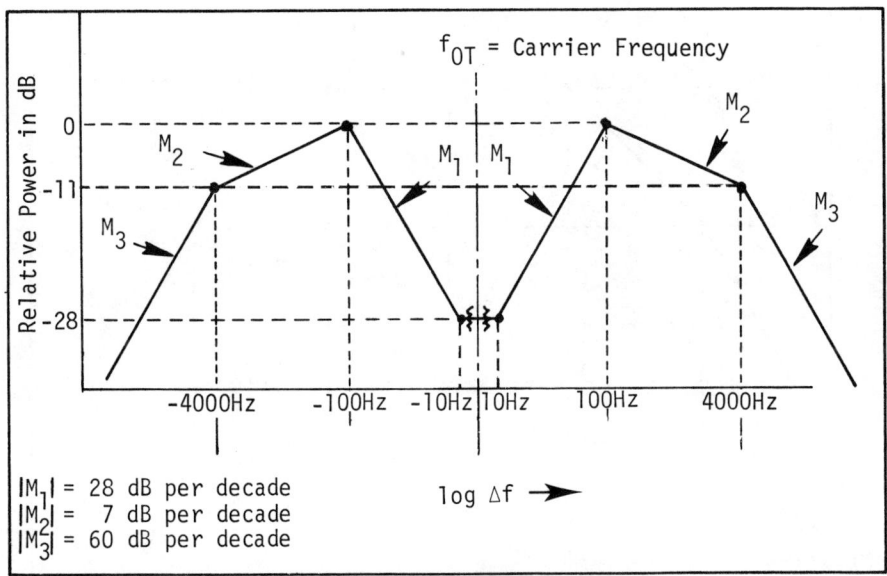

Figure 5.21 - Model for AM Voice Modulation

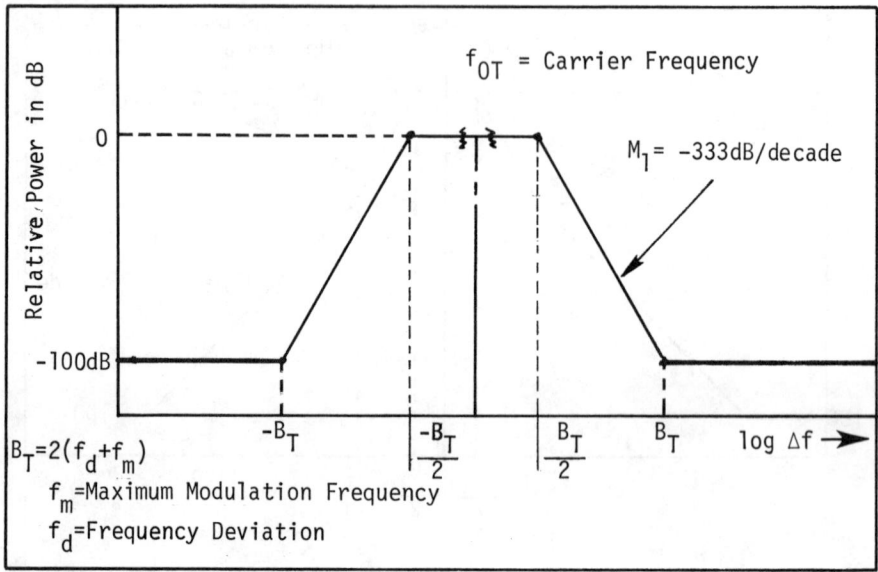

Figure 5.22 - FM Modulation Envelope

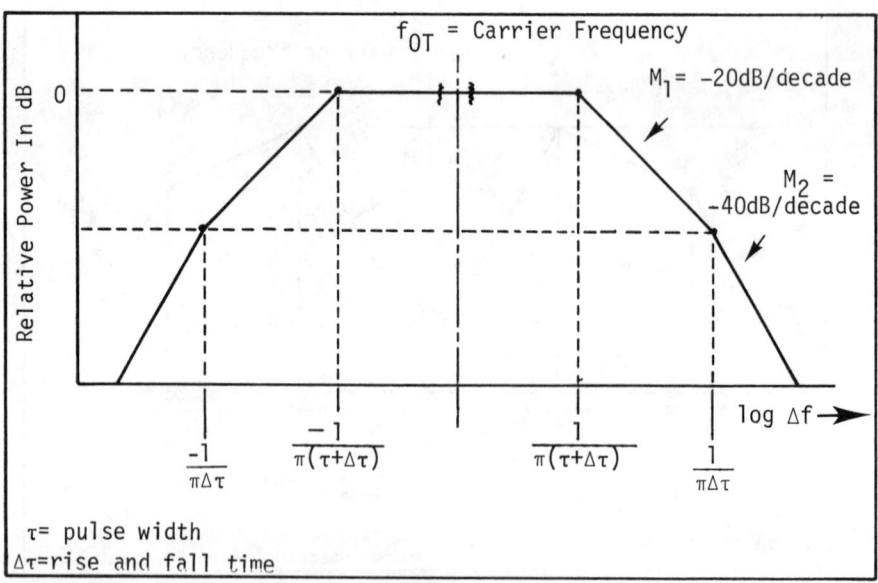

Figure 5.23 - Pulse Modulation Envelope

Table 5.10 – Constants for Modulation Envelope Model

Type of Modulation	i	Δf_i	$M(\Delta f_i)$ (dB above fundamental)	M_i* (dB/decade)
AM Communication and CW Radar	0	$0.1B_T$	0	0
	1	$0.5B_T$	0	133
	2	B_T	-40	67
AM Voice	0	1 Hz	-28	0
	1	10 Hz	-28	-28
	2	100 Hz	0	7
	3	1000 Hz	-11	60
FM	0	$0.1B_T$	0	0
	1	$0.5B_T$	0	333
	2	B_T	-100	0
Pulse**	0	$\dfrac{1}{10\tau}$	0	0
	1	$\dfrac{1}{\pi(\tau+\Delta\tau)}$	0	20
	2	$\dfrac{1}{\pi\Delta\tau}$	$-20 \log_{10} 1 + \dfrac{\tau}{\Delta\tau}$	40

A.3 Fourier Analysis Models

If measured data are not available, Fourier analysis methods may be used to determine the modulation envelope.

These methods are particularly useful and are widely used for pulse-type signals. Although Fourier analysis methods may also be applied to AM and FM, it is not as readily applied to the complex modulation waveforms associated with these signals. Such periodic and nonperiodic pulse modulation and AM and FM models are discussed in Chap. 3 of Vol. 5.

A.4 Harmonic-Emission Frequency Models

For all types of transmitters, the relationship between the frequencies of the harmonics and fundamental emission outputs are described by:

---

* M_i is negative for frequencies above the fundamental and positive for frequencies below the fundamental.
** Modulation envelope for trapezoidal pulse with half-voltage width, τ, and rise and fall times, $\Delta\tau$.

$$f_{NT} = Nf_{OT} \qquad (5.23)$$

where,

f_{NT} = harmonic frequency

N = harmonic number

f_{OT} = fundamental frequency

Outputs that are harmonically related to frequencies other than the transmitter fundamental (e.g., harmonics of a master oscillator or clock) are significantly lower in amplitude than harmonics of the fundamental. Therefore, they may be ignored except in situations where large interference margins are present. If emission outputs which are harmonically related to frequencies other than the fundamental are considered, then the interference margin obtained from the amplitude cull should be reduced by 20 dB for those outputs.

Some transmitters utilize a mixing scheme for frequency generation. The process by which these transmitters generate the fundamental frequency depends on nonlinear signal combination, such that the final output may contain signals at all frequencies corresponding to the sum and difference of the input signals and their harmonics. This relationship is:

$$f_{pq} = |pf_1 \pm qf_2| \qquad (5.24)$$

where,

f_{pq} = output emission frequencies

p, q = positive integers

f_1, f_2 = input signals

One of these outputs is the fundamental frequency of the transmitter. Even though the other undesired outputs may be attenuated by filters or band-limiting amplifiers within the transmitter, some are significant in amplitude. Thus, they are considered in performing an EMI prediction.

The modulation envelopes associated with harmonic outputs are directly related to the base-band modulation envelopes used for developing the fundamental output. However, for some pulse-modulated transmitters, the modulation envelope associated with the Nth harmonic is assumed to have the same shape as the envelope used for the fundamental but the bandwidths are assumed to be proportional to the harmonic number.

5.3.2   RECEIVER MODELS

Receivers are designed to respond to certain types of electromagnetic signals within a predetermined frequency band(s). However,

receivers also respond to undesired signals having various modulation
and frequency characteristics. Thus, it is necessary to treat a receiver as potentially susceptible to all transmitter emissions considered in a prediction.

In performing an EMI prediction it is necessary to describe receiver susceptibility to different types of emission sources. Basically, what is required is a composite susceptibility function which
represents the maximum interfering signal level that can be tolerated
as a function of both frequency and type of the interfering signal.
However, there are many problems involved in determining susceptibility
of a particular receiver to a specific type of interference. The next
section briefly describes receiver susceptibility effects that are
considered. It discusses the type of information required and problems involved in specifying each of these effects.

5.3.2.1 Receiver Functions for EMI Prediction

There are a number of interference effects that an undesired signal can produce in a receiver. In order to represent receiver composite susceptibility, it is necessary to consider these effects and to
determine which effect(s) dominate within a given range of frequencies.

Fig. 5.24 is a functional diagram useful in discussing various
receiver EMI effects. A superheterodyne receiver generally employs
R-F stages which provide frequency selectivity and/or amplification
and one or more mixers which translate the R-F signal to intermediate
frequencies (IF). It also contains I-F stages which provide further
frequency selectivity and amplification; a detector which recovers
the modulation; and post-detection stages that process the signal and
drive one or more output displays. Since tuned-radio-frequency (TRF)
and crystal-video receivers do not use the superheterodyne principle,
they do not contain mixers and I-F amplifiers. In specifying receiver
susceptibility it is necessary to consider the effects of an interfering signal on each of these stages. The resulting susceptibility
function then represents a composite of the most significant effects.

For the purpose of EMI prediction, potentially interfering signals
are considered to be in one of three basic categories: (1) co-channel,
(2) adjacent channel, (3) and out-of-band. These three categories are
defined with reference to Fig. 5.25.

A. CO-CHANNEL INTERFERENCE

Co-Channel Interference refers to signals having frequencies that
exist within the narrowest pre-detection passband of the receiver. For
superheterodyne receivers, the culprit frequency of co-channel interference must be such that the interference is translated to the I-F
passband in the same manner as the desired signal. This requires that

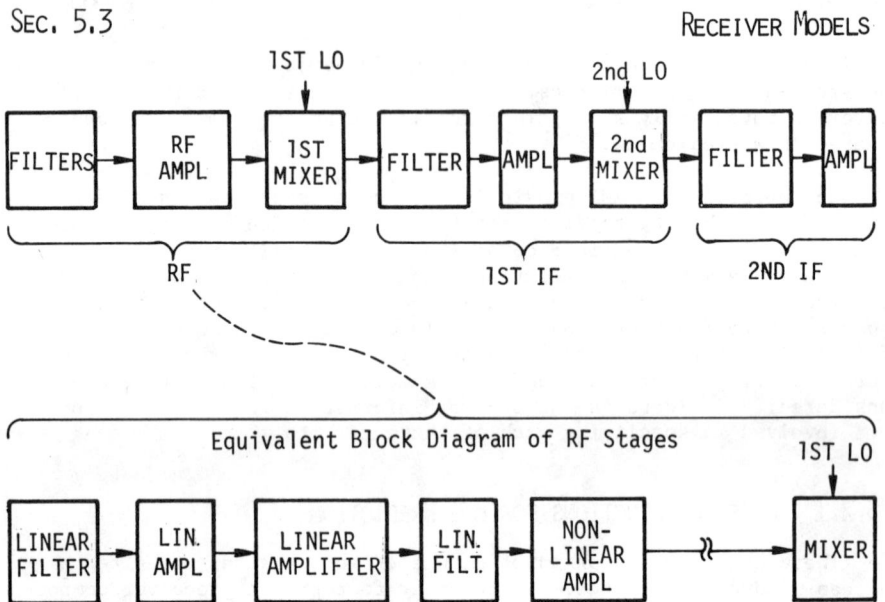

Figure 5.24 - General Pre-detection Representation for Superheterodyne Receiver.

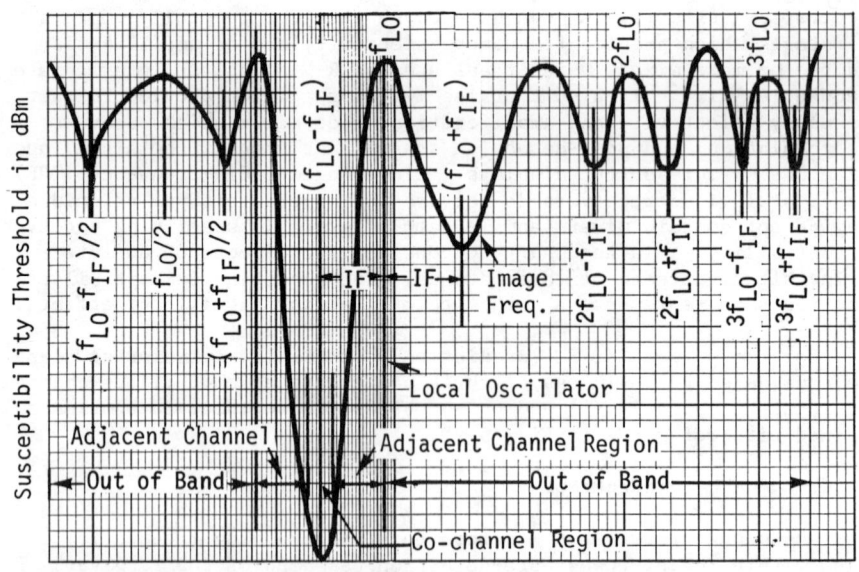

Figure5.25 -Receiver Susceptibility Characteristics

the frequency of co-channel interfering signals equal the tuned radio
frequency ± one-half the narrowest I-F bandwidth for superheterodyne
receivers. In the case of either TRF or crystal-video receivers,
selectivity is provided at RF, and the R-F passband determines the applicable frequency range for co-channel signals.

Because co-channel interfering signals are amplified, processed,
and detected in the same manner as the desired signal, the receiver is
particularly vulnerable to these emissions. Thus, co-channel EMI may
either desensitize the receiver or override or mask the desired signal.
It may also combine with the desired signal to cause serious distortion
in the detected output, or cause the automatic-frequency control circuitry to retune to the frequency of the interference, if this is
applicable.

In Sec. 5.4 it will be shown that the first phase of EMI prediction in a multi transmitter-receiver environment should be based
primarily on the amplitude of the various potential interference
functions. These include transmitter emission output, receiver susceptibility threshold, antenna gains, and propagation loss. The purpose of *amplitude culling* is to permit making a *quick look* at the
total EMI prediction problem in order to eliminate obvious non-interfering cases from further consideration. This defines the range of
parameters that must be considered for surviving EMI cases. In performing *amplitude culling*, relatively conservative assumptions are used
for various interference functions. The variables of frequency, time,
and direction are considered only in a rather general manner.

For amplitude culling, it is necessary to specify receiver susceptibility threshold to co-channel signals. The receiver noise level
(S=N) is used for co-channel susceptibility threshold. Receiver noise
level is related to the sensitivity, which may be obtained from receiver technical manual, spectrum signature data,* or design specifications. For example, receiver sensitivity is often specified for a
3-dB signal-plus-noise-to-noise ratio; this corresponds to a 0-dB signal-to-noise ratio (S=N). Therefore, receiver noise (and thus the
co-channel susceptibility threshold) is equal to the receiver sensitivity. This is shown in Fig. 5.7 as a function of noise figure and
bandwidth.

The second phase of the EMI prediction process in multi transmitter-
receiver environments is *frequency culling*. During this phase, surviving potential EMI situations obtained from amplitude culling are
refined by a more complete consideration of the frequency variable.
Amplitude culling results are modified to account for transmitter and
receiver bandwidths and frequency misalignment.

---

* Vol. 1, Chap. 5 for a discussion of MIL-STD-449C which identifies
measurement procedures and techniques used in the spectrum signature
data collection program.

For the frequency cull, receiver tuned frequency and I-F bandwidth are specified. Receiver tuned frequency is usually defined for each receiver to be included in the prediction, except for special situations such as those encountered in frequency assignment problems. For these situations, one important consideration is whether interference will exist if a receiver is tuned exactly to a transmitter output and this information is available as an output from the amplitude cull. Information on the I-F bandwidth is obtained from the receiver technical manual, specification, or spectrum signature data.

Two final phases of the EMI prediction process described in Sec. 5.4 are the detailed prediction and the performance prediction. In the former, results obtained from the frequency cull are further modified to include effects of time and direction variables. Also, the combined and simultaneous effects of multiple interfering signals are considered. Performance prediction involves a consideration of operational performance and effectiveness factors to determine the resulting impact on overall system operational requirements.

Co-channel interfering signals do not require any special consideration in the detailed prediction phase of multi transmitter-receiver environments. However, in the performance prediction it is necessary to consider modulation characteristics of both the desired and interfering signals, signal-to-noise and signal-to-interference ratios, and detector characteristics as they affect these ratios.

B. ADJACENT-CHANNEL INTERFERENCE

Adjacent-Channel Interference refers to emissions having frequency components that exist within or near the *widest* receiver passband. They may be sufficiently separated from the receiver tuned frequency such that they do not fall within the narrowest receiver passband as shown in Fig. 5.25. For example, in the case of a superheterodyne receiver, adjacent-channel interference falls within or near the R-F passband but after conversion will fall outside of the I-F passband. For TRF receivers, adjacent-channel interference refers to signals that are outside of the R-F passband but are still close enough to the receiver tuned frequency so that they are attenuated by less than 60 dB.

Adjacent-channel interference can produce any one of several effects in a receiver. For example, interference may be translated through the receiver together with the desired signal and both appear at the input to an I-F stage. In this case, the I-F selectivity and the adjacent-channel emission spectrum will both influence the relative level of the interfering signal appearing at the input to the detector. Alternately, one or more adjacent interfering emissions may produce nonlinear effects in the R-F amplifier or mixer. Major nonlinear effects include desensitization, cross modulation, and intermodulation.

Desensitization is a reduction in the receiver gain to the desired signal as a result of an interfering emission producing automatic-gain control (AGC) action or causing one or more stages of the receiver to operate nonlinearily due to saturation. *Cross modulation* is the transfer of the modulation from an undesired emission to the desired signal as a result of the former causing one or more stages of the receiver to operate nonlinearily. *Intermodulation* is the generation of undesired signals from the nonlinear combination of two or more input signals which produce frequencies existing at the sum or difference of the input frequencies or their harmonics.

Adjacent-channel signals are not considered separately in the initial amplitude cull of a multi transmitter-receiver environment (see Sec. 5.4). Results of the co-channel amplitude cull form the basis for the adjacent-channel EMI prediction.

For the second-step frequency cull, it is necessary to consider frequency separation between receiver tuned frequency and interfering emissions. This includes combined effects of signal and receiver bandwidths, emission spectrum, and receiver selectivity. It identifies emissions that are within the receiver R-F passband which have sufficient amplitude to produce nonlinear effects.

In the third-step detailed prediction, it is necessary to consider strong, adjacent-channel interfering emissions both singly and collectively to evaluate effects of nonlinear interactions in the receiver *front end*. For the final-stage performance prediction, considerations for adjacent-channel interfering signals are the same as those that apply for co-channel interference.

C. OUT-OF-BAND INTERFERENCE

Out-Of-Band Interference refers to signals having frequency components which are outside of the widest receiver passband. These are identified above and below the adjacent channel regions in Fig. 5.25.

Strong out-of-band interference may produce spurious responses in a receiver. The superheterodyne receiver is most susceptible to those out-of-band signals that mix with L-0 harmonics to produce a signal at the IF. Spurious responses in such a receiver usually occur at specific frequencies and other out-of-band frequencies are attenuated by the receiver I-F selectivity. For a tuned-radio-frequency or crystal-video receiver, the receiver will be susceptible to those out-of-band interfering signals that are not adequately rejected by the R-F selectivity.

For amplitude culling, it is necessary to describe the receiver susceptibility threshold to spurious responses. In general, the comments in Sec. 5.3.1 regarding transmitter harmonic emission amplitudes apply to receiver spurious responses amplitudes. For example, spurious

responses are subject to random serial-number-to-serial-number variations. Thus, receiver spurious response susceptibility is represented
statistically.

Frequencies associated with spurious responses are calculated in
the second-step frequency cull, and the separation between an interfering emission and a spurious response frequency is considered in the
same manner as for co-channel interference. Similarly, considerations
that apply to the detailed and performance predictions are the same as
those described for co-channel signals.

5.3.2.2   Receiver Models for Amplitude Culling

In the first-stage EMI prediction process for multi transmitter
and/or receiver environments (see Sec. 5.4), amplitude culling is used.
Receiver mathematical models used to describe amplitude culling are
divided into co-channel, adjacent-channel, and out-of-band susceptibility models. Chap. 4 of Vol. 5 explains how these are obtained.

In order for a spurious response to be generated in a superheterodyne receiver, it is necessary that the interfering signal or one
of its harmonics mix with the LO or one of its harmonics to produce
an output in the receiver I-F passband.* Frequencies, f_{SR} which are
capable of producing spurious responses for a single conversion superheterodyne, are:

$$f_{SR} = \frac{pf_{LO} \pm f_{IF}}{q} \qquad (5.25)$$

or

$$f_{IF} = pf_{LO} \pm qf_{SR} \qquad (5.26)$$

where:

p = harmonic number of local-oscillator

q = harmonic number of interfering signal

f_{LO} = L-O frequency

f_{IF} = intermediate frequency

The majority of spurious responses for a superheterodyne receiver
are defined in Eq. (5.25). Tests have shown that in multi-conversion
type receivers, spurious responses created by the second and/or higher
order conversion processes occurring in more than one stage may be
present but are not a serious source of interference susceptibility.

Signal levels required to produce q=1 responses (i.e., responses
due to the fundamental of the interfering emission) are generally

---

* The signal and local oscillator harmonics that contribute to spurious responses are produced in the mixer as a result of its non-linear
operation.

lower in amplitude than the signal levels required to produce high order q responses. Those responses for which q=2 are less significant than q=1, and q=3 responses are lower than q=2 responses.

For a TRF or crystal-video receiver, the susceptibility to out-of-band signals is not as frequency dependent as it is for the super-heterodyne receiver. The out-of-band susceptibility threshold is a relatively smooth function of frequency for TRF and crystal-video receivers.

The amplitude-culling process considers the frequency variable only in a general sense. Thus, discrete frequency response character-istics of superheterodyne receivers as shown in Fig. 5.25 are not con-sidered during this phase of EMI prediction. Instead, for amplitude culling it is assumed that the receiver is capable of responding to out-of-band signals at any frequency and the susceptibility, $P_R(f)$, is represented as a continuous function of frequency, f.

The amplitude-culling receiver model used for superheterodyne, TRF, and crystal-video receivers tuned to any frequency, f_{OR}, is:

$$P_R(f) = P_R(f_{OR}) + I \log_{10}(f/f_{OR}) + J \qquad (5.27)$$

where, $P_R(f_{OR})$ = receiver sensitivity (co-channel susceptibility threshold) at tuned frequency, f_{OR}

In order to apply Eq. (5.27) to a specific problem, it is neces-sary to determine the parameters I and J for the particular receiver or receiver type being considered. Also it is necessary to determine the standard deviation $\sigma_R(f_{SR})$ associated with variations about the average susceptibility level. These parameters are obtained from statistical summaries of available data, receiver specifications, or analysis of specific measured data. Each of these means of determin-ing out-of-band receiver susceptibility models is discussed in the following sections.

A. STATISTICAL SUMMARY OF SPURIOUS RESPONSE

When specific measured receiver data are not available, one al-ternative for obtaining an out-of-band susceptibility model for use in the amplitude-culling process is to derive statistical summaries from data for groups of similar receivers. From these summaries, mathematical models are developed which are representative of a group or class of receivers. In the absence of measured data, pertinent group models should be used for the receiver under study. Statistical summary models have been evaluated from available spectrum signature data and the applicable constants are summarized in Table 5.11.

Table 5.11 - Constants for Receiver Spurious Response Models
Obtained from Statistical Summary of Available Data*

Receiver Category Based on Fundamental Frequency	Summary Values for Constants In Spurious Response Amplitude Models		
	I dB/decade	J dB Above Fundamental	$\sigma_R(f_{SR})$ dB
All receivers combined	35	75	20
Below 30 MHz (MF & HF)	25	85	15
30 MHz to 300 MHz (VHF)	35	85	15
Above 300 MHz (UHF & SHF)	40	60	15

The first row entry presents values for I, J, and $\sigma_R(f_{SR})$ derived from the data for all receivers combined. The second, third, and fourth row entries provide values for I, J, and $\sigma_R(f_{SR})$ obtained by grouping receiver data on the basis of fundamental frequency range of operation. The appropriate set of values for I, J, and $\sigma_R(f_{SR})$ may be used in Eq. (5.27) to model receiver spurious response suscepti-bility levels for frequencies above the fundamental. The resulting average spurious response susceptibility level for receivers within each of the indicated frequency ranges are presented in Table 5.12.

Table 5.12 - Summary of Spurious Response Average Susceptibility

Local Oscillator Harmonic Number, (p)	Spurious Response Average Susceptibility Threshold (dB Above Fundamental Sensitivity; q=1)			
	All Receivers Combined	Receivers Categorized According to Tuned Frequency		
	$\sigma_R(f_{SR})$=20 dB	Below 30MHz $\sigma_R(f_{SR})$=15dB	30MHz – 300MHz $\sigma_R(f_{SR})$=15 dB	Above 300MHz $\sigma_R(f_{SR})$=15dB
1 (image)	75	85	85	60
2	85	93	95	72
3	92	97	102	79
4	96	100	106	84
5	99	102	109	88
6	102	104	112	91
7	105	106	115	94
8	107	107	117	96
9	108	109	118	98
10	110	110	120	100

---

* These constants apply to q=1 responses (see Eq. (5.25)). For q=2 responses and frequencies above the receiver fundamental add 15 dB to J and for q=3 responses add 20 dB to J.

5.3.2.3 RECEIVER MODELS FOR FREQUENCY CULLING

Superheterodyne receivers are susceptible to *out-of-band* signals that can generate a spurious response in the receiver. In order to predict the effect of mixer-generated spurious responses on receiver interference, it is necessary to utilize mathematical models that define frequencies at which spurious responses occur.

A spurious response may be generated if the frequency of an interfering signal is such that the signal or one of its harmonics can mix with the local oscillator or one of its harmonics to produce an output in the receiver IF passband. The frequencies which are capable of mixing with the local oscillator to produce spurious responses can be determined from the p,q relationship as specified earlier in Eq. (5.25). As shown in the equation, there are two response frequencies for each value of p and q. The response pairs result from mixing which produces outputs at frequencies corresponding to both the sum and difference of harmonics of the local oscillator and the input signal.

Illustrative Example 5.10

To illustrate the use of Eq. (5.25) in calculating receiver spurious response frequencies, consider a radar receiver used for aeronautical navigation with a fundamental frequency of 1500 MHz, a 60 MHz IF, and a local-oscillator frequency of 1560 MHz. The spurious response frequencies may be calculated directly from Eq. (5.25) by substituting 1560 MHz for f_{LO} and 60 MHz for f_{IF}, and letting p and q assume integer values over the frequency range of interest. The results are tabulated in Table 5.13.

Table 5.13 - Spurious Response Frequencies for Radar
Tuned to 1500 MHz

P	Sign of IF	Spurious q = 1	Response q = 2	Frequencies q = 3
1	+	1620	810	540
	−	Tuned Freq.	750	540
2	+	3180	1590	1060
	−	3060	1530	1020
3	+	4740	2370	1580
	−	4620	2310	1540
4	+	6300	3150	2100
	−	6180	3090	2060
5	+	7860	3930	2620
	−	7740	3870	2580

The tabulation shows that spurious responses may exist below the radar tuned frequency due to higher q's and low p's in addition to above the radar frequency for all q=1 responses and higher q's and p's.

5.3.2.4   RECEIVER MODELS FOR DETAILED PREDICTION

This section describes receiver models that are used in the detailed prediction phase, viz., the third stage of the EMI prediction
process for multi-transmitter-receiver environments. Receiver EMI
effects evaluated during this phase of prediction are the adjacent-
channel effects resulting from nonlinearities in the receiver front
end. This section presents some models for intermodulation, desensitization and cross-modulation. Detailed data of intermodulation models
and both desensitization and cross-modulation models are presented in
Chap. 4 of Vol. 5.

A.   INTERMODULATION MODELS

Because of nonlinearities within a superheterodyne receiver, two
or more interfering signals may mix, viz., intermodulate to produce
signals at other frequencies. If these new frequencies are close
enough to the receiver tuned frequency, these signals may be amplified and detected by the same mechanism which processes the desired
signal. Thus, possible degradation of performance may result.

The purpose for performing an intermodulation prediction is to
identify pairs of transmitters within the electromagnetic environment
which are emitting at frequencies that may cause such disturbance
in a particular receiver. The resulting signal-to-interference ratios
are then computed. Intermodulation mixes which are considered in this
section are second, third, and fifth orders.*

In order for an intermodulation product to cause interference, it
must be transformed to a frequency within the I-F passband for detection
to occur. The method considered here is intermodulation in the R-F
amplifiers and first mixer which results in an intermodulation frequency at or near the receiver tuned frequency, f_{OR}.

Signals which are capable of producing intermodulation interference in a receiver must satisfy the following:

$$|mf_1 \pm nf_2| = |f_{OR} \pm B_R| \qquad (5.28)$$

where, f_1 & f_2 = frequencies of two interfering emissions

f_{OR} = receiver tuned frequency

B_R = IF bandwidth in which intermodulation
products are significant

m & n = integers

---

* The order of an intermodulation product mix is determined by adding
m and n of Eq. (5.28). Other than second order (m = n = 1), the even
order products result in insignificant levels of EMI probabilities.

Eq. (5.28) may be normalized to the receiver fundamental frequency, f_{OR}, and plotted to show the intermodulation relationship between two culprit signals:

$$m \frac{f_1}{f_{OR}} \pm n \frac{f_2}{f_{OR}} = 1 \pm \frac{B_R}{f_{OR}} \qquad (5.29)$$

Fig. 5.26, obtained from Eq. (5.29), shows the resulting graph for second and third-order intermodulation products. Intermodulation signal combinations falling on or within one of the dashed-line boxes are capable of generating an intermodulation product in the vicinity of the receiver tuned frequency. Note that the third-order products are more significant sources of EMI than second-order products.

For the purpose of EMI prediction, the only signals that are considered potentially serious sources of intermodulation interference are those that are in the vicinity of the receiver tuned frequency and produce intermodulation products which fall within the receiver overall 60-dB bandwidth around f_{OR}. The following relations present the frequency criteria which two interfering signals must meet to offer potentially serious intermodulation problems:

$$
\begin{aligned}
f_N \pm f_F - f_{OR} &\leq B_{R60} \quad \text{(second order)} \\
2f_N - f_F - f_{OR} &\leq B_{R60} \quad \text{(third order)} \\
3f_N - 2f_F - f_{OR} &\leq B_{R60} \quad \text{(fifth order)}
\end{aligned}
\qquad (5.30)
$$

where, f_N = frequency of interfering emission Nearest to f_{OR}

f_F = frequency of interfering emission Farthest from f_{OR}

B_{R60} = receiver overall 60 dB I-F bandwidth

The area on the chart marked *region of major significance* is particularly important because of the proximity of the signals to the receiver tuned frequency.* Signals within this region will in general experience less R-F selectivity (rejection) than will signals outside of the region. Thus, they are more likely to produce significant intermodulation products. The extent of this region is in general a function of R-F selectivity, but the area indicated is representative of typical receivers.

Illustrative Example 5.11

Assume that an air-traffic control communication transmitter is tuned to 360 MHz. Determine other transmitter emission frequencies

---

* R-F bandwidths of superheterodyne receiver front ends are generally of low Q-factors (of order of 10) whereas Q-factors of selectivity due to the I-F transferred to the RF are typically of the order of 1,000.

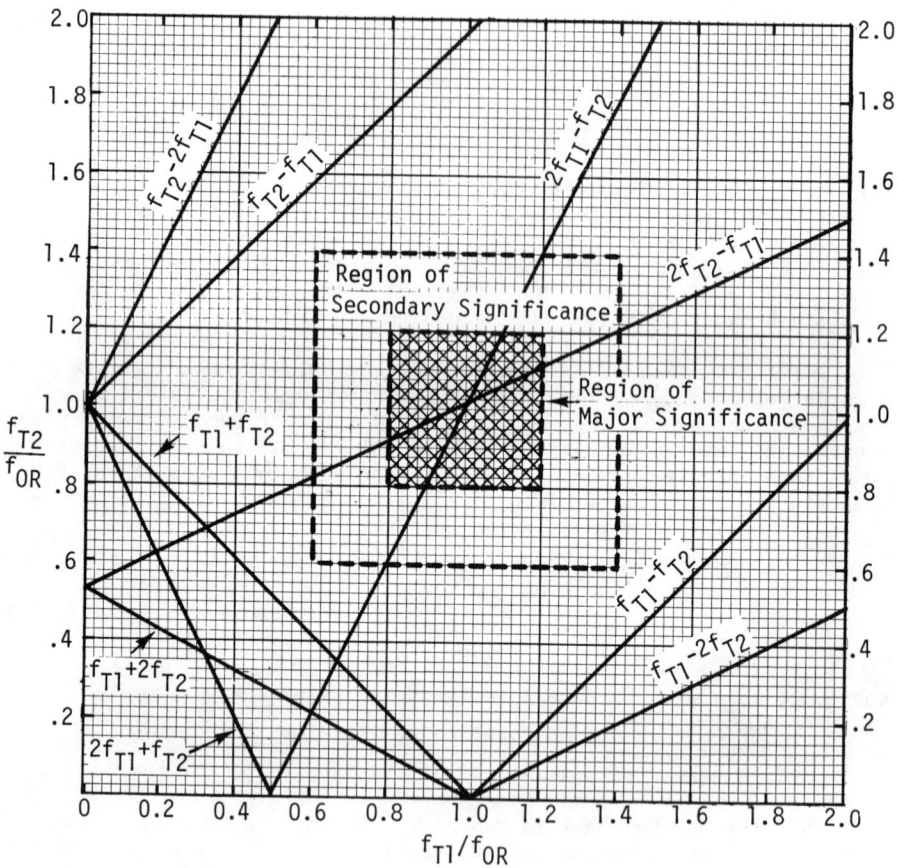

Figure 5.26 - Second and Third Order Intermodulation Chart

which may combine with the 360 MHz emission to produce second and third order intermodulation frequencies in a receiver used by the air-traffic controllers to intercept transmissions from pilots. A receiver is tuned to 300 MHz and co-located with the transmitter. For this situation, the transmitter frequency is 1.2 times the receiver frequency (i.e. $f_1/f_{OR} = 1.2$).

From Fig. 5.26, potential second and third-order intermodulation frequencies within the region $0 \leq f_2/f_{OR} \leq 2f_{OR}$ are:

$$\frac{f_2}{f_{OR}} = \begin{array}{ll} \text{0.1 third order} & (f_1 - 2f_2) \\ \text{0.2 second order} & (f_1 - f_2) \\ \text{1.1 third order} & (2f_2 - f_1) \\ \text{1.4 third order} & (2f_1 - f_2) \end{array}$$

The most significant frequencies from the standpoint of potential inter-modulation interference for the specified situation are $f_2/f_{OR} = 1.1$ and 1.4. Thus, f_2 = 330 MHz and 420 MHz emission sources could be po-tentially hazardous to the air-to-ground receiver tuned to 300 MHz. This is especially true of the former emission. The frequencies cor-responding to f_2/f_{OR} = 0.1 and 0.2 would be 30 MHz and 60 MHz respec-tively and these frequencies should be sufficiently rejected by the R-F front-end selectivity.

5.3.3 ANTENNA MODELS

The preceding two sections presented mathematical models of trans-mitters and receivers, exclusive of their antennas. This section pre-sents mathematical models that represent antenna radiation characteris-tics for use in EMI prediction. Although the models presented in this section are directed specifically toward communications-electronics equipment that are intended to radiate or receive electromagnetic energy, the general techniques described for representing antennas may be ap-plied to any source which acts as a radiator or receptor of emissions. Thus, the term *antenna* as used in this section may be considered in the broadest sense to be any device that either carries an electric current and thereby produces an electromagnetic field, or that responds to an electromagnetic field. Chap. 5 of Vol. 5 presents considerably more detail on the topic of antennas.

Antennas are designed to radiate and/or receive signals over a specific solid angle and within a specified frequency range. Some antennas, such as those used in land-mobile and many broadcast appli-cations, are designed to radiate or receive uniformly over all sectors surrounding the antenna. In other cases, such as fixed point-to-point communications, radar, and certain telemetry systems, it is desirable to confine the functional radiated or received signals to certain limited sectors. In practice, however, it is not possible to accomplish perfect discrimination with antennas in either the spatial or frequency domain. Thus, antennas that are intended to restrict the radiation to specific regions also radiate into or receive signals from other unin-tentional regions. In addition, undesired signals at non-design fre-quencies are inadvertently radiated or received by antennas, and the spatial characteristics of an antenna for spurious frequencies are significantly different from characteristics at the design frequency.

5.3.3.1   Antenna Functions for EMI Prediction

For interference prediction, emission levels received at a given antenna as a result of radiations from another must be determined. In order to provide this information it is necessary to specify antenna radiation characteristics: (1) in both intended and unintended directions, (2) at both design and non-design frequencies for different polarizations, and (3) for situations in which either near-field or far-field conditions may prevail.

The antenna representation must be consistent with the over-all objectives of the total EMI prediction and analysis, and must be sufficiently general so that it is applicable to many different types of antennas that may be encountered. Because of the lack of applicable data for every problem, antenna representations must be such that it is possible to extrapolate existing information and prepare generalized mathematical antenna models for use in cases where specific data are not available.

In a given prediction problem, a wide variety of antenna types may exist, ranging from those that exhibit essentially omnidirectional radiation to those having a highly directional radiation characteristic. Although omnidirectional type of antennas must be considered in interference prediction it does not present serious problems because antenna gains are relatively low and are essentially independent of direction.*

For directional antennas, problems involved in specifying antenna radiation characteristics are considerably more complex because of the wide range of variation that exists in the spatial domain. Furthermore, directional antennas are often utilized with both high-power transmitters and sensitive receivers in applications such as radar, satellite communications, and troposcatter. This combination of high-gain antennas with either high-power transmitters or sensitive receivers increases the propensity for EMI under certain conditions.

Because of the potential seriousness and complexity of EMI problems associated with directional antennas, the discussion presented in this section is primarily directed toward high-gain antennas. However, the methods presented are general and are also applicable to low or medium-gain antennas.

There are several criteria that are used for classifying antennas. However, for EMI prediction purposes, one of the most important characteristics of an antenna is its relative gain above isotropic levels.

---

*   The term omnidirectional antenna as used in this volume refers to a practical *omni* which generally is *omni* azimuth only and may have some gain relative to an ideal *omni* or isotropic antenna due to vertical angle selectivity.

Accordingly, three antenna classifications are used: high-gain antennas (greater than 25 dB gain), medium-gain antennas (10-25 dB gain), and low-gain antennas (less than 10 dB gain). Each class may be further subdivided according to the analysis required for the antenna within its basic class. Table 5.14 presents the classification of some typical antennas.

Table 5.14 - Antenna Classification by Gain*

Low Gain (G < 10 dB)	Medium Gain (10 dB ≤ G ≤ 25 dB)	High Gain (G > 25 dB)
1. Linear Cylindrical Biconical Dipoles Folded Dipoles Asymmetrical Dipoles Sleeve Dipoles Monopole Discone Quadrant Colinear Array 2. Traveling Wave Long Wire 3. Loop 4. Aperture Slot 5. Helix (Omnidirectional Mode)	1. Array Yagi Broadside Curtain End-Fire Curtain 2. Traveling Wave Rhombic Surface and Leaky Wave 3. Aperture Horn Corner Reflector 4. Equiangular Log Periodic Conical Log Spiral 5. Helix (Axial Mode)	1. Array Matress Electronic Steerable 2. Aperture Horns Reflector Antennas Lens Antennas

A. FAR-FIELD RADIATION REPRESENTATION

The far-field radiation characteristics** of directional antennas are often represented by a polar plot of an antenna pattern as shown in Fig. 5.27. This pattern represents the radiation characteristics in one plane (horizontal shown) which may be contrasted with radiation which occurs in all directions. Thus, in order to completely describe an antenna with a pattern representation, it is necessary to use a three-dimensional pattern. From the antenna pattern shown in Fig. 5.27, there are two sectors (i.e., intentional and unintentional radiation regions) that must be considered.

---

*   Gain in units of dB above isotropic.
** See Fig. 5.28 for development of the far-field distance or Vol. 4, *EMI Test Instruments and Systems.*

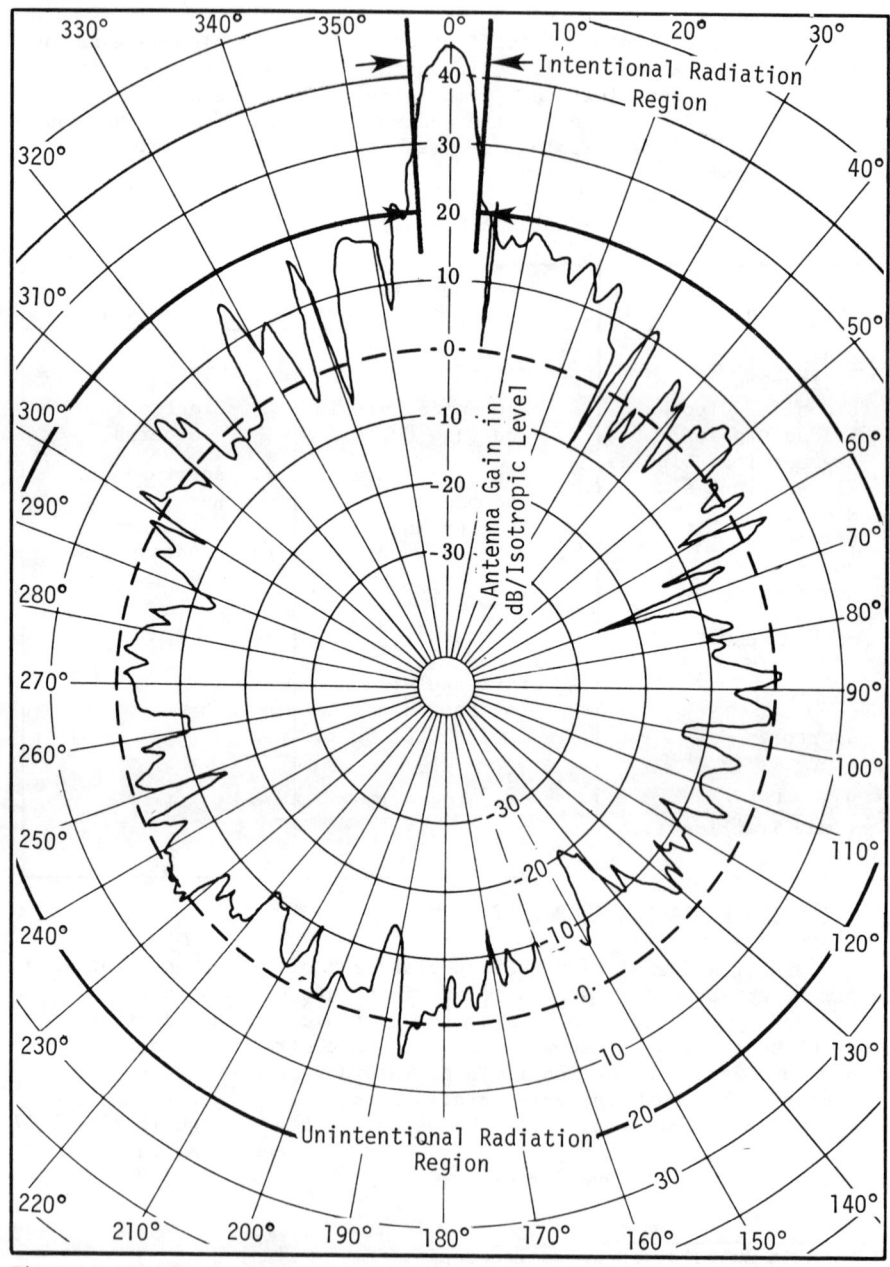

Figure 5.27 - Typical Directional Antenna Pattern

A.1 Intentional Radiation Region

The first sector consists of the intentional radiation region,
i.e., the region of space in which an antenna is designed to radiate.
For the directional type of radiation illustrated, this region is relatively limited in solid angle. On the other hand, for a true omni-
directional type of antenna, viz., an isotrope, the intentional-
radiation region would encompass all space (4π steradians) around the
antenna.

Within the design frequency range of the antenna, the extent of
the intentional radiation region may be represented by azimuth and
elevation beamwidths. The relative level of radiation may be represented by antenna gain over an isotropic reference. These parameters
are usually obtained directly or calculated. For non-design frequencies, the beamwidths and gains are more difficult to obtain or
calculate.

A.2 Unintentional Radiation Region

The second sector consists of the unintentional radiation region,
i.e., the region of space outside of the intentional radiation region.
Representation of antenna radiation in this region is also required
for performing an EMI prediction. Although a specific antenna pattern
provides a method for representing radiation characteristics of an antenna, it will be shown that such representation is not usually practical
or adequate for interference prediction. The primary problem results
from the fact that patterns measured on different serial numbers of a
given antenna nomenclature, or at different frequencies within the design bandwidth, or at different geographical sites do not exhibit corresponding gains for the same off-axis angles. They may differ significantly in a particular direction.

B. VARIATIONS IN PATTERN CHARACTERISTICS

Antenna patterns for two antennas differing only in serial number
when measured at the same location, utilizing the same equipment and
measurement techniques, do not exhibit a point-by-point correspondence.
They may in fact differ significantly at particular angular orientations.
However, the fluctuations in the patterns are generally confined to the
same range of values. For different frequencies within the design band
and for different sites, if the main-beam and principle side lobe region
are excluded, the fluctuations in the pattern occur at random. Therefore, at a given angular orientation, it is possible for a particular
measured pattern to exhibit any level within a permissible range of
values.

It is not practical to attempt to describe these random variations precisely. Furthermore, once the existence of these random variations is recognized, the extreme amount of detailed information which is represented by an antenna pattern is neither justified nor necessary for performing an EMI prediction.

One possibility for limiting the amount of useful antenna information that must be stored and handled in EMI prediction is to simply specify the maximum level of all sidelobe (off-axis) radiation. Although this approach has some merit, in many cases it leads to a gross over-estimation of the potential interference. A more realistic approach is to represent the antenna by a pattern distribution function which specifies the probability that various radiation levels are exceeded in the unintentional radiation region. When this pattern-distribution representation is used, radiation from an antenna may be expressed in terms of mean radiation level, standard deviation about this mean level, and the type of statistical distribution involved. This representation has proved to be very useful for providing a statistical model of the random variables associated with antenna radiation characteristics.

Another advantage of using the pattern-distribution function representation for the unintentional radiation region is that a single pattern distribution function may be used to represent a particular antenna type over a wide range of frequencies and polarizations in both the near and far-field. Pattern distribution functions are described in detail in Chap. 5 of Vol. 5.

5.3.3.2   ANTENNA MODELS FOR AMPLITUDE CULLING

For the amplitude-culling stage of EMI prediction in multi transmitter-receiver environments, it is necessary to specify beamwidths
and radiation levels associated with antennas. Beamwidths are used
to identify the directional region of interest in a particular problem.
Radiation levels are specified in terms of pattern-distribution functions containing mean gains and standard deviations. They are defined
as functions of frequency, polarization, and direction.

The following sections describe mathematical models that may be
used in the amplitude-culling stage to represent antenna radiation
levels in both the intentional and unintentional radiation regions.
Methods are presented for determining applicable antenna regions that
must be considered in a particular problem.

A.   INTENTIONAL RADIATION REGION

For EMI prediction, it is necessary to specify antenna radiation
characteristics within the solid angle for which the antenna is designed to radiate or receive energy. It is necessary to define the
gain and beamwidths of the antenna for both design and non-design frequencies and polarizations. This section describes EMI prediction
models for radiation within this region which is often referred to as
the main-beam region.

For a given set of frequencies and polarizations, statistical
techniques are used to represent small variations in gain that occur
between antennas of the same type, between different frequencies, for
different sites, etc. Antenna gain is represented by a normal distribution and the resulting antenna models for the intentional radiation region are based on the following principles.

A.1   beamwidth

The intentional radiation region is defined by the 10-dB azimuth
and elevation beamwidths, α and β.*

A.2   Mean Gain

The mean gain levels $G(f,p)$, are 3 dB below the maximum gain (i.e.,
the gain at the 3-dB beamwidth limits).

---

* It is recommended that 10-dB beamwidths be used for EMI prediction.
If specific data on the 10-dB beamwidths are not available, it may be
estimated by using two times the 3-dB beamwidths for high-gain antennas.

A.3 Standard Deviation

The standard deviation, $\sigma(f,p)$ accounts for gain variations that occur within the intentional radiation region and between antennas and EMI prediction conditions. It is assumed that those gain variations are random and may be described by a normal distribution.

If specific data are available on antenna gain and beamwidths, they should be used as the basis for the antenna models. On the other hand, if specific data are not available (which is usually the case), the generalized antenna models presented earlier in Table 5.6 may be used to represent antenna radiation characteristics in the intentional radiation region. These generalized models were derived from available antenna data, and they define the relationship between design and non-design conditions. The following sections discuss some of the major considerations that relate to the antenna intentional radiation region.

Illustrative Example 5.12

Consider a ten foot parabolic dish designed to operate at 10 GHz with a nominal gain of 44 dB and 3-dB azimuth and elevation beamwidths of 1.1°. Determine the amplitude cull models for the gain and beamwidths of the main-beam region at the design frequency and polarization.

For EMI prediction, the gain of the main beam (intentional radiation region) at the design frequency and polarization is represented by a normal distribution with a mean gain of 41 dB (i.e., 3 dB below the nominal gain) and a standard deviation of 2 dB (see Table 5.6). The beamwidth of the region is defined by the 10-dB beamwidths which are approximately two times the 3-dB beamwidths, or 2.2°.

B. UNINTENTIONAL RADIATION REGION

For EMI prediction, it is also necessary to specify antenna characteristics in the unintentional, off-axis radiation region. It was previously shown that the most logical, practical, and adequate means for such a representation is the pattern-distribution function which describes mean gains and standard deviations. In order to specify the antenna, it is necessary to define the effect of frequency, polarization, and site location on the resulting pattern distribution functions for the unintentional radiation region. When measured data are not available the generalized antenna models presented in Table 5.6 are used for EMI prediction.

The mean gain levels shown in Table 5.6 are given with respect to dB above an isotropic level. An example is presented below to illustrate how this table may be used to model antenna radiation characteristics in the unintentional radiation region.

Illustrative Example 5.13

Consider a high-gain parabolic dish with a six foot diameter and a 10-GHz design frequency. It was previously determined that the main beam gain for this antenna at the design frequency and polarization would be 44 dB/isotropic with a 2-dB standard deviation. Suppose that it is necessary to include this antenna in an EMI prediction involving the *unintentional* radiation region and a non-design frequency and polarization condition. For this situation, Table 5.6 shows the mean gain is -10 dB relative to an isotropic level and that there exists a 14-dB standard deviation.

5.3.3.3 ANTENNA MODELS FOR DETAILED PREDICTION

Once the amplitude and frequency-culling processes have been performed for multi transmitter-receiver environments, the surviving cases have a significant probability of causing interference. Thus, they must be examined in considerably more detail. The term *detailed prediction* applies specifically to consideration of the variables involving distance, direction, and time. For antennas, the primary considerations that apply during the detailed prediction are limited to: (1) an evaluation of antenna effects for situations involving near-field conditions, and (2) an evaluation of time-dependent considerations. The latter applies for situations involving antennas which move to rotate in a manner such that both main beam and side and back lobe considerations can occur for portions of the time interval over which EMI prediction considerations apply.

A. NEAR-FIELD MODELS

For EMI prediction, it is often necessary to specify radiation characteristics of an antenna for near-field conditions. This is particularly necessary in the case of high-gain, microwave antennas. Near-field conditions for these antennas may extend out to one mile or more. Because of the gains associated with these antennas they are potentially serious from an EMI viewpoint as well as possibly constituting a radiation hazard.* In many cases, operational requirements are such that several potentially interfering systems may be co-located in a mutual near-field region.

In the near-field region, the representation of characteristics presents a much more complicated problem than in the far-field region. Under idealized conditions, near-field characteristics cannot be

---

* In the main-beam of a high-gain antenna which is also radiating high power, the resulting power density may be a threat to human life or ordnances and fuel. See Vol. 1, Sec. 6.3, *BRH/PHS Rules and Regulations* for more details.

represented by a single pattern because radiation characteristics are functions of both angular position and distance with respect to the antenna. Also, in the near-field region, there are complex relation-ships between the electric and magnetic fields and there may be little correspondence between resulting field-intensity patterns.

For the detailed prediction, it is necessary to examine each po-tentially interfering situation remaining after amplitude and frequency culls to determine whether a near-field situation exists. For those cases, antenna gain models used in the amplitude and frequency cull must be modified to account for near-field effects. The following sections present techniques for determining the near-field to far-field transition distances and calculating the gain reduction resulting from near-field effects.

A.1 Transition Distance

The distance for which far-field approximations are no longer valid and for which near-field considerations must be applied is called the transition distance. Basically the transition from near-field to far-field conditions is a gradual one. However, by specifying the error in the far-field pattern as one moves closer to the antenna, a specific transition distance relationship can be obtained.

One criterion used to define transition distance is to limit the phase error to 1/8 of a wavelength.* This corresponds to about a 1 dB error in gain obtainable at an arbitrarily large distance. Now consider the antenna configuration of Fig. 5.28. The distance to a field point along the normal axis of the antenna is different from the distance between the edge of the antenna and the field point. This difference is denoted the space-phase error.

Assuming that the antenna dimension, ℓ, is large compared to the wave-length (i.e., $\ell \gg \lambda$), in order to limit this error to $\lambda/8$, the dis-tance, R, to the field point from the antenna must satisfy:

$$R > \ell^2/\lambda, \text{ for } \ell \gg \lambda \qquad (5.31)$$

Eq. (5.31) applies for high and medium-gain antennas. When this re-quirement is not met (i.e., ℓ is not $\gg\lambda$, for low-gain antennas), the equation is no longer valid, and to assume far-field conditions it is necessary to adopt the criterion:

$$R > 3\lambda, \text{ for } \ell \leq \lambda \qquad (5.32)$$

---

* See Sec. 3.1, Vol. 4, *EMI Instrumentation and System* for derivation of the near/far-field distance.

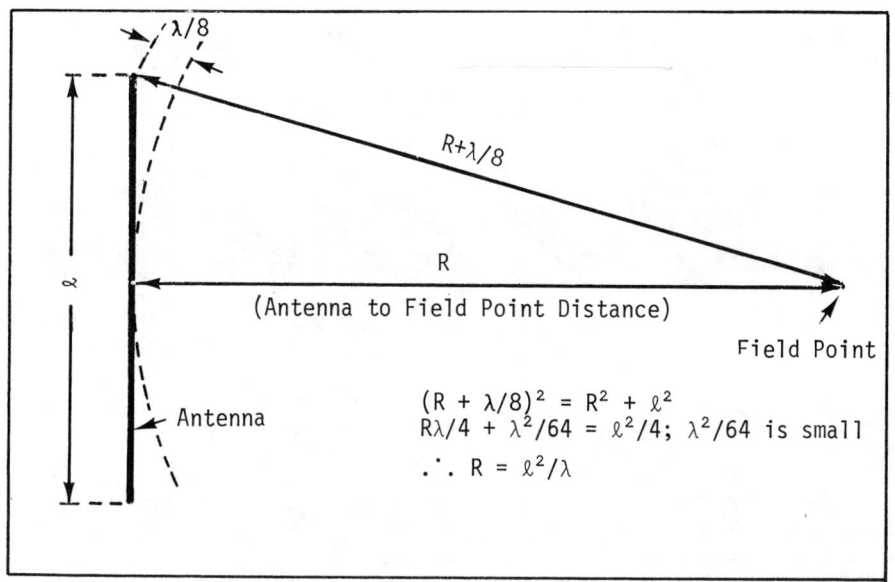

Figure 5.28 - Illustration of Space Phase Error

Eq. (5.32) is in agreement with the definitions given in MIL-STD-449C.*
Thus, in order to insure that acceptable far-field conditions exist
for EMI prediction, it is required that:

$$R > \ell^2/\lambda \text{ and } R > 3\lambda$$

Both conditions must be satisfied at all times. These transition-dis-
tance criteria are illustrated graphically in Fig. 5.29.

As one moves off main-beam axis, the near-field to far-field transition distance is reduced considerably. For high-gain antennas, the
transition distance for this off-axis condition is determined from Fig.
5.30 in the following manner. First, locate the solid curve that corresponds to the aperture dimension, ℓ, for the antenna of interest.
Next, locate the intersection of this solid curve with the dashed curve
that corresponds to the frequency of interest. The valid region of the
solid curve lies above and to the left of the intersection with the
frequency curve. For this region, the relationship between transition
distance and angular displacement can be obtained directly from the
solid curve. If the angular displacement is such that it lies in the
invalid region of the solid curve, the near-field criteria given in Eq.
(5.31) should be used to determine the near-field transition distance.

---

*See Chap. 5, Vol. 1, *Electrical Noise and EMI Specifications* re.
MIL-STD-449C.

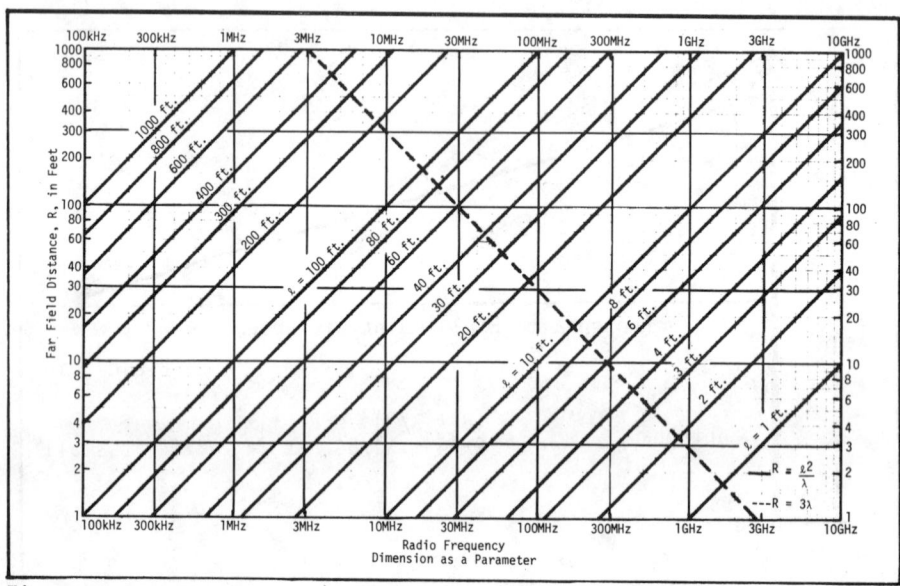

Figure 5.29 - Far-Field Distance vs Frequency with Aperture Dimension
As A Parameter.

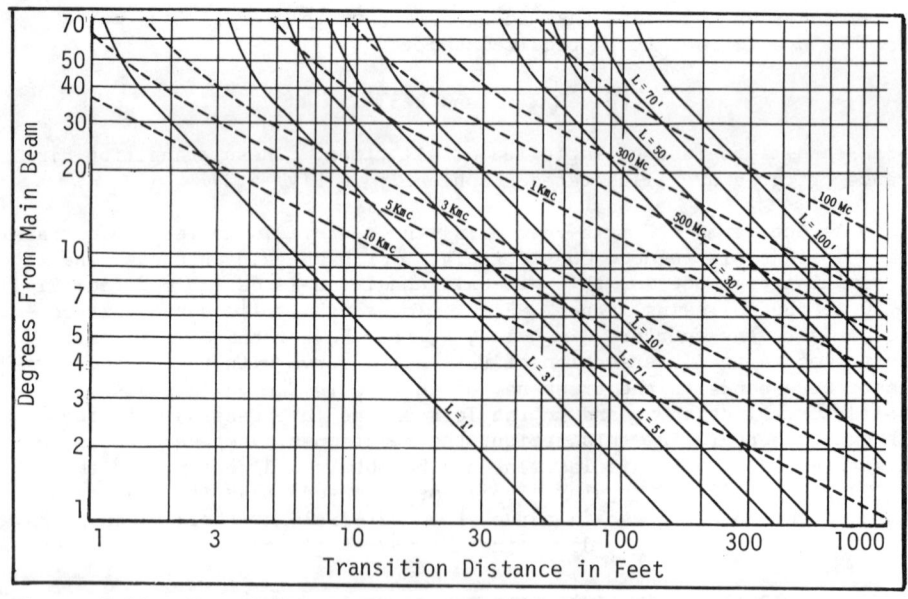

Figure 5.30 - Transition Distance for Off - Axis Pointing of Antenna

From Fig. 5.30, it is apparent that the transition distances decrease significantly as one moves off-axis. Thus, many EMI prediction
problems that would require near-field analysis considerations for an
on-axis condition can actually use far-field approximations for off-
axis conditions. Furthermore, when moving into the near-field, one of
the first effects is that antenna-gain nulls tend to become less pronounced. However, from the standpoint of EMI prediction, the nulls
are least important because they represent the lowest levels of radiation from the antenna.

Illustrative Example 5.14

Consider a parabolic antenna having an aperture diameter of 30
feet operating at a frequency of 1 GHz. Determine the near-field
transition distance for an angular main-beam offset of 10°. From
Fig. 5.30, the intersection between the solid curve representing the
aperture dimension and the dashed curve representing frequency, occurs
at an angular offset of 7°. The corresponding transition distance is
270 ft. Therefore, the solid curve representing the 30 foot aperture
dimension may be used directly to determine transition distance for
angular offsets *greater* than 7°. Since the specific value of 10° in
this example satisfies this requirement, the actual transition distance
is determined to be 190 ft. by the intersection of ℓ = 30 ft. and the
10° off-axis value.

A.2 Collimated Beam Approximation For Near-Field Gain

For the infrequent case where it is necessary to consider the on-
axis, near-field situation, it may be assumed that all of the transmitted power is contained in a cylindrical volume around the antenna
axis with a cross-sectional area, A, equal to the antenna aperture.
This is the collimated-beam effect. When this conservative approximation is used, the resulting antenna *near-field gain* at a distance R
from the antenna is:

$$G = \frac{4\pi R^2}{\text{Area}} \quad \text{for } G < G_{FF} \tag{5.33}$$

or

$$G_{dB} = 11 + 20\log_{10}R - 10\log_{10}A \tag{5.34}$$

where

$$G_{FF} = \text{far-field gain}$$

Illustrative Example 5.15

Consider another parabolic antenna having a circular aperture 10
ft. in diameter, operating at 10 GHz and exhibiting a far-field gain
of 48 dB. Determine whether a near-field condition exists on-axis at
100 ft., and if so, calculate the near-field gain to be used for EMI
prediction.

Eq. (5.31) or Fig. 5.29 may be used to determine if a near-field condition exists. From Fig. 5.29 the transition near/far-field distance is about 1,000 ft. For on-axis distances less than 1,000 ft., near-field conditions must be considered together with the attendent reduction in gain.

Eq. (5.33) is now used to calculate the antenna gain at 100' distance from the 48-dB nominal gain:

$$\frac{4\pi R^2}{Area} = \frac{4\pi R^2}{\pi \ell^2 / 4} = 16 \left[\frac{R}{\ell} \right]^2$$

$$= 16 \left[\frac{100^2}{10} \right] = 1600 = 32 \text{ dB}$$

This corresponds to a 16-dB loss (48 dB - 32 dB) in gain at the specified near-field distance.

A.3 Near-Field Gain Correction

Within the near-field or Fresnel region, the antenna on-axis gain is always less than the far-field gain. For EMI prediction purposes, the near-field gain can be determined by subtracting the appropriate gain correction from the far-field gain. Figs. 5.31 through 5.34 give resulting gain corrections for rectangular apertures with various aperture illumination functions. The gain correction that must be applied to the far-field gain can be read directly from the graphs for specified distances in wavelengths from the antenna, aperture dimension in wavelengths, and aperture-distribution functions.

Illustrative Example 5.16

To illustrate how Figs. 5.31 through 5.34 are used to correct antenna gains for near-field conditions, consider a microwave communication system operating at approximately 10 GHz and using an antenna with dimensions of 10 ft. by 5 ft. Assume that the antenna has a cosine aperture distribution and provides a nominal gain of 45 dB above isotropic. Calculate the effective near-field gain on-axis at 100 ft. from the antenna. The antenna dimension of 10 ft. is approximately 100 wavelengths and 5 ft. is approximately 50 wavelengths. The 100 ft. distance from the antenna is 1,000 wavelengths. Using these dimensions and distances in wavelengths, the gain correction (see Fig. 5.32) for an aperture dimension of 100 wavelengths and a distance of 1,000 wavelengths is -6 dB. Similarly, the gain correction for an aperture dimension of 50 wavelengths and a distance of 1,000 wavelengths is approximately -1 dB. Hence the total gain correction that must be applied to the antenna for the situation considered is the greater of the two or -6 dB, and the resulting effective gain on-axis 100 ft. away is (45 dB - 6 dB) or 39 dB.

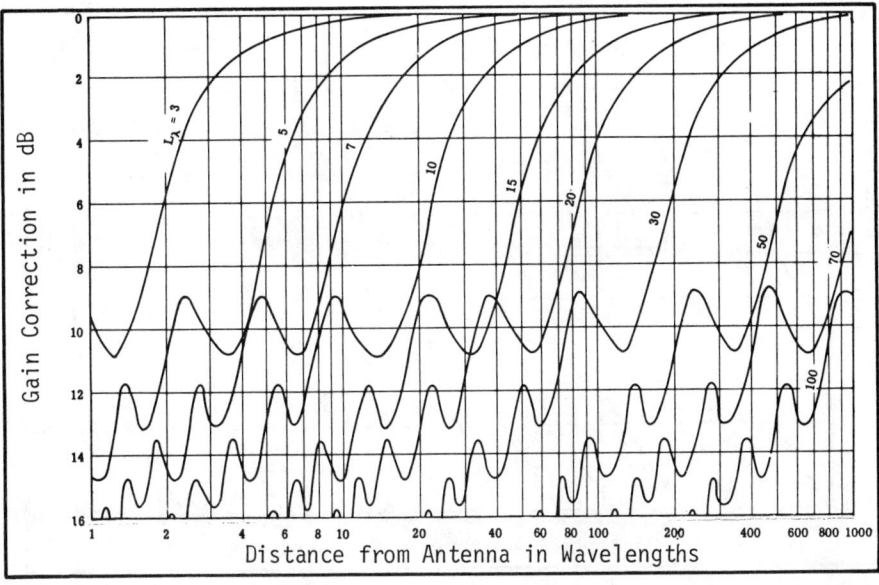

Figure 5.31   Fresnel Region Gain Correction for Uniform Illumination

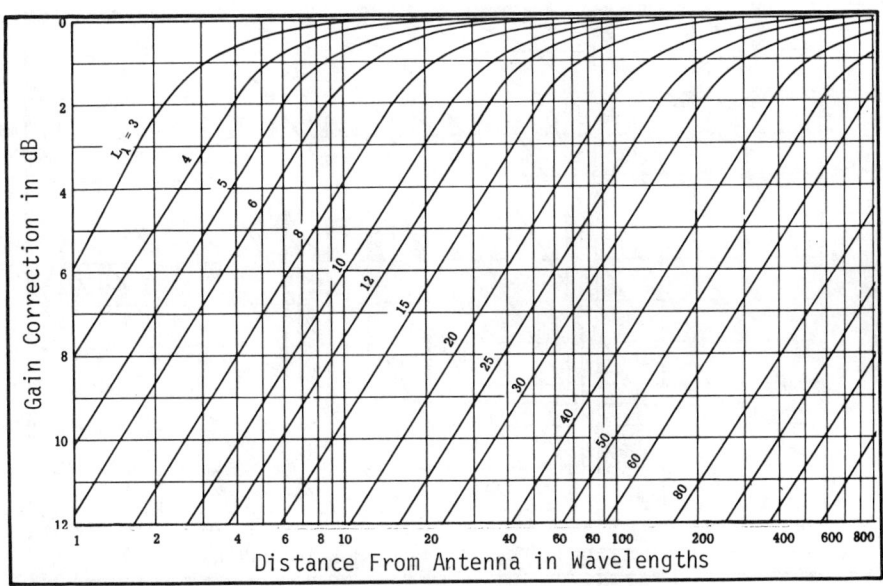

Figure 5.32 - Fresnel Region Gain Correction for Cosine Illumination

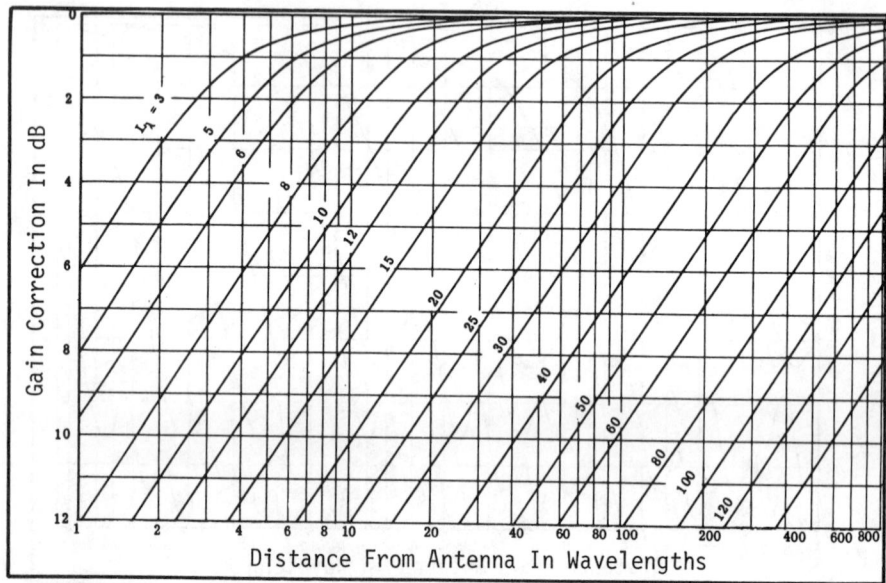

Figure 5.33 - Fresnel Region Gain Correction for Cosine Squared
Illumination

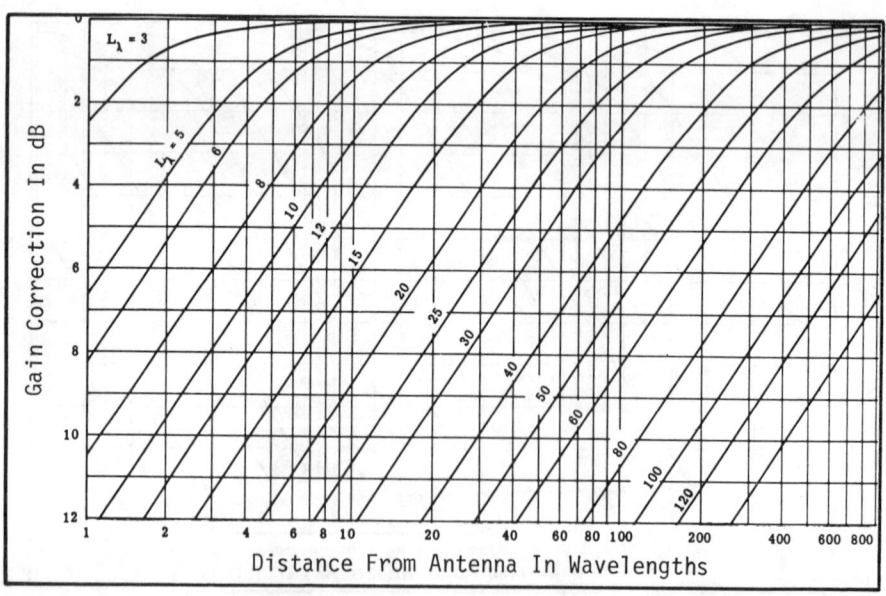

Figure 5.34 - Fresnel Region Gain Correction for Cosine Cubed
Illumination

5.3.4   PROPAGATION MODELS

Previous sections discussed prediction models for transmitters, receivers and antennas. This chapter continues the approach by presenting propagation models. It describes various modes by which an interfering signal may be propagated to a victim receiver, provides criteria that may be used to identify the most significant mode(s) of propagation, and presents mathematical models which describe the propagation loss resulting from these modes. The section also describes how these models may be used to calculate propagation loss for use in an EMI prediction.

Most communication-electronic systems depend on electromagnetic-wave propagation for the transfer of signals between the transmitter and receiver. In addition to being propagated to the intended point of reception, signals are also radiated into and received from unintended directions and locations. Thus, strong signals, intended for reception at a distant point, may produce an undesired electromagnetic field at a nearby receiver and result in EMI. In evaluating EMC, it is necessary to examine various propagation modes that may exist between emission sources and susceptible devices to determine the amount of isolation (or attenuation) provided by the physical separation of equipments.

Interference often results from undesired modes of propagation and thus each significant mode must be considered. The prediction must include any mode of propagation which yields a signal of sufficient strength to cause interference, even with low probability. There is a definite contrast between the system-design criteria of one mode with high reliability and the EMI criteria of any mode with low reliability. In order that a basic system design has the high degree of reliability demanded of present circuits, it is necessary to provide a high degree of protection against interfering emissions.

The following identifies the parameters that are important in analyzing propagation loss and briefly describes various propagation modes that must be considered.

5.3.4.1   PROPAGATION CONSIDERATIONS

There are a number of parameters that affect the mode of propagation and the propagation loss between two given points. Some of the more important factors which must be considered include frequency, distance, polarization, antenna heights, terrain, ground constants, time, season, weather, and radiated power. Specific factors that must be considered in a given circumstance and the degree to which they influence the resulting propagation loss depend on the frequency, distance, and antenna heights. The following paragraphs discuss some of the major propagation modes that are important in EMI prediction.

A.   FREE-SPACE PROPAGATION

In discussing concepts regarding various propagation modes it is helpful to begin with a discussion of free-space propagation between antennas. This was done in Sec. 5.2.3. Once the principles governing propagation under these conditions are understood, it is easier to follow the concept of propagation between either omnidirectional or directional antennas in the presence of earth and atmospheric discontinuities such as the ionosphere and the troposphere.

The term in parenthesis of either Eq. (5.14) or Eq. (5.15) is the propagation loss relation. It was plotted in Fig. 5.7. By rearranging these equations, the free space transmission coupling, C_{TR}, is obtained:

$$C_{TR} = P_r - P_t = -(32 \text{ dB} + 20 \log_{10}R_{km} + 20 \log_{10}f_{MHz})$$

$$+ G_t + G_r \text{ dB for metric system} \qquad (5.35)$$

or,
$$= -(37 \text{ dB} + 20 \log_{10}R_{mi} + 20 \log_{10}f_{MHz})$$

$$+ G_t + G_r \text{ dB for English System} \qquad (5.36)$$

Illustrative Example 5.17 - Free-Space Propagation Loss

Determine the free-space propagation loss and transmission coupling between a UHF transmitter operating at 400 MHz and a receiver operating in the same frequency range. Consider that the transmitter and receiver are separated by 528 ft., i.e., 0.1 mile and the antenna gains for both transmitter and receiver are 6 dB. From Fig. 5.7, the propagation loss is 69 dB. From Eq. (5.36), the transmission coupling is:

$$C_{TR} = -69 \text{ dB} + 6 \text{ dB} + 6 \text{ dB} = -57 \text{ dB}$$

B.   GROUND-WAVE PROPAGATION

The previous section summarized a propagation mode termed *free space*, i.e., where there are no objects and/or surfaces in the propagating path (other than the two antennas). This section discusses the more frequent situation where there are surfaces between the transmitter and receiver.

The ground wave is a radio wave which is propagated over the earth and is affected by the presence of the ground and the troposphere but is not affected by the ionosphere. The ground wave includes all components of a radio wave over the earth except ionospheric and tropospheric waves, and is the sum of their components. For example, in the surface-wave region, the ground wave is the sum of the direct, the ground-reflected, and the surface wave.

The following components are included under the general heading of ground wave and are illustrated in Figs. 5.35 and 5.36.

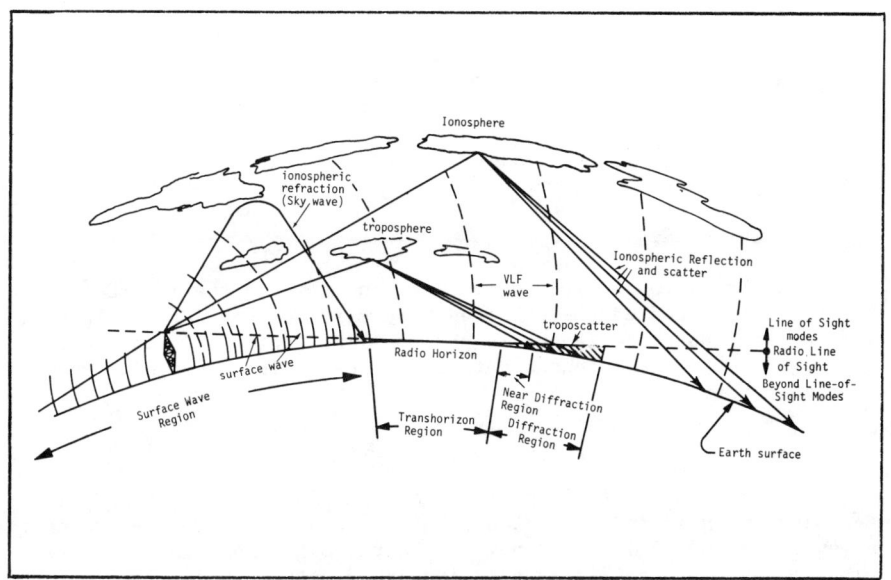

Figure 5.35- Illustration of Primary Propagating Modes

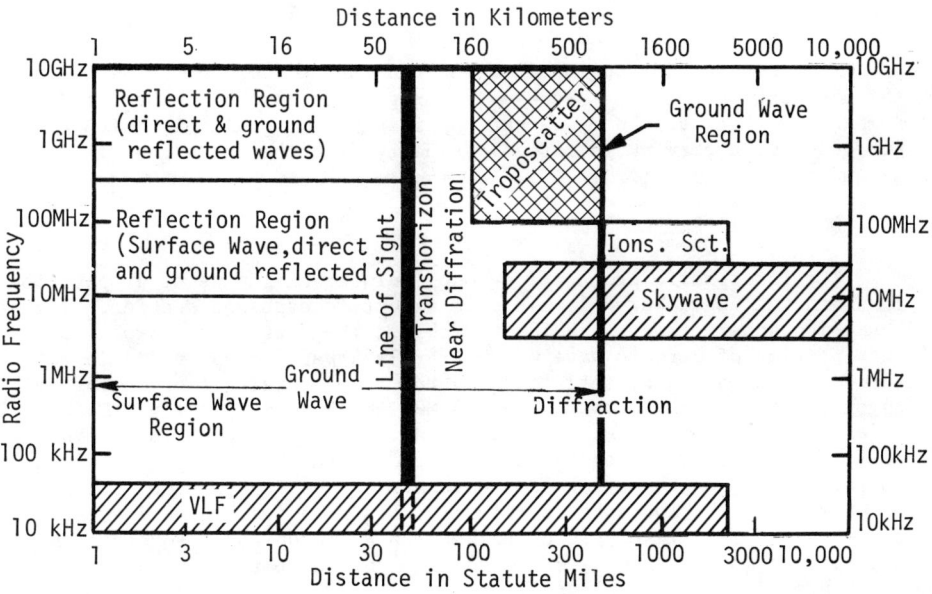

Figure 5.36- Generalized Frequency and Distance Limits of Primary Propagation Modes

B.1 Surface Wave

The surface wave is that portion of the ground wave which travels
over the surface of the earth, and is considered only out to the radio
horizon (line of sight). It predominates between 30 kHz and 3 MHz.
Beyond the radio horizon, diffraction is considered to be the predominate ground-wave component.

B.2 Reflection Region

The reflection region is that region in which the interaction between a direct and earth-reflected wave must be considered. This region
lies within and above the radio horizon. Reflection-region techniques
apply only when the antennas have significant elevations with respect
to a wavelength.

B.3 Diffraction

Well beyond the radio horizon, diffraction over rough or irregular
terrain is considered to be the principle source of ground-wave energy.
However, in the region well beyond the radio horizon two other principle modes, ionospheric propagation and scatter propagation, must be
considered. The diffraction region applies to frequencies greater
than 30 kHz.

B.4 Transhorizon

The transhorizon region lies between the reflection region and
the diffraction region. Because this region is also beyond the radio
horizon, reflection theory is not applicable. In addition, neither
surface-wave theory nor diffraction theory describe the signal levels
adequately.

C. IONOSPHERIC PROPAGATION

The ionospheric propagation mode includes all means of propagation
in which the ionosphere plays a role except ionospheric scatter. Two
basic ionospheric modes are identifiable. The first mode, *skywave*, is
that portion of a radio wave which travels upward in space and is returned to the earth by refraction in the ionosphere. It predominates
beyond line of sight between about 3 MHz to 30 MHz. The lower bound-
aries of the ionosphere play an important part in propagation of energy
at VLF. VLF depends upon the existence of both the ionosphere and the
surface of the earth acting as a concentric spherical waveguide for
propagation. The ionospheric modes involve relatively high losses,
and thus are less likely to contribute to EMI problems. For this reason
ionospheric modes are not emphasized in this chapter.

D. SCATTER PROPAGATION

Propagation of electromagnetic energy by scattering in either the ionosphere or the troposphere represents the fourth basic form of propagation to be considered. Propagation by these modes is characterized by extremely high losses. However, these modes have become significant with the advent of high power and highly directive high gain antenna systems. Three propagation modes are included under the heading of scatter propagation: ionospheric scatter, tropospheric scatter and meteor-burst scatter.

D.1 Ionospheric Scatter

Ionospheric scatter involves propagation by scattering of energy which takes place in an ionospheric volume of space. Ionospheric scatter is significant in the frequency range from about 30 MHz to 100 MHz and over distances from about 500 miles to 1500 miles.

Tropospheric scatter involves the scattering of energy within an illuminated volume of space in the troposphere. It is significant over the frequency range from 100 MHz to 10 GHz and over distances from 100 miles to 500 miles.

Meteor scatter represents a rather exotic mode of propagation which has attracted some attention in recent years. This mode is supported by scattering from the ionized trails left by meteors as they pass through the outer atmosphere. This mode will support intermittent propagation over the frequency range from about 50 MHz to 150 MHz and over distance ranges from about 500 miles to 1500 miles.

In general, scatter propagation modes involve large losses, and thus they are not usually primary modes from the standpoint of EMI prediction considerations. For this reason, these modes are not emphasized in this chapter.

For the reader desiring more information on propagation, the bibliography at the end of this chapter is presented. Additionally, Vol. 5 of this EMI handbook series, *EMI Prediction and Analysis*, contains considerably more on propagation models.

5.3.4.2 PROPAGATION MODELS FOR AMPLITUDE CULLING

As described in Sec. 5.4, amplitude culling consists of a quick look at candidate transmitter-receiver pairs utilizing free space propagation loss techniques. This first level of cull attempts to minimize the prediction problem by eliminating from further consideration those environmental emitters which obviously do not represent a potential EMI situation.

The mathematical models used in amplitude culling are consistent with the overall requirements of this phase of analysis and must be

compatible with the models used for other functions and other phases
of analysis. The basic requirement for the model is that it provides
a method for calculating propagation loss such that non-interfering
cases are eliminated but potentially-interfering cases are retained.
Additionally, the models must be such that propagation loss may be obtained with a minimum of input data and calculation time.

Free-space propagation is generally assumed for the amplitude
cull. This assumption provides a propagation model that is relatively
simple, can be implemented easily, requires minimum input data, minimizes calculations, and provides a reasonable estimate of the propagation loss for many situations of interest but does not eliminate potentially-interfering situations. Some important considerations concerning free-space propagation were presented in Sec. 5.2.3.

5.3.4.3 PROPAGATION MODELS FOR FREQUENCY CULLING

During frequency culling, the results of the amplitude cull are
modified to account for frequency separation between transmitter outputs and receiver responses. The primary factors that must be considered during this phase of EMI prediction are the transmitter modulation envelope and the receiver selectivity characteristics. Because these factors are relatively independent of propagation conditions, additional propagation considerations are not required for this
culling phase.

5.3.4.4 PROPAGATION MODELS FOR DETAILED PREDICTION

This section describes propagation models that are used in the
detailed prediction phase and presents summary graphics of propagation
loss for various conditions. Before performing propagation calculations for a detailed prediction, it is necessary to select the specific
propagation mode or modes that predominate for a given EMI prediction
situation. Sec. 5.3.4.1 discussed various propagation regions and the
primary frequency and distance limits for which each is applicable.
Fig. 5.36 should be used to select the modes of interest. If two or
more modes apply to a particular prediction situation, calculations
should be performed for each mode separately and the mode providing
the least propagation loss should be used.

In performing amplitude culling, free-space propagation models
were used. If during the preliminary phases of the detailed prediction
it is determined that other propagation modes prevail, it is necessary
to apply a correction factor to the amplitude cull results to account
for additional propagation loss. Propagation models used to determine
the average basic transmission loss for free-space, ground-wave, ionospheric propagation, and scatter propagation are presented in this
section.

A. FREE-SPACE PROPAGATION MODELS

Free-space propagation models, which formed the basis for the amplitude-cull propagation loss calculation, are applicable to many EMI prediction situations. Fig. 5.7 is the mathematical model for basic transmission loss where free-space conditions apply. This loss model is used for EMI prediction in two cases: (1) where high-gain antennas are used, e.g. radar antennas where there are no reflections from the ground or other conducting surfaces, and only the direct ray exists; and (2) where the transmitter and/or receiver systems are air or space borne and the reflected waves, if any, are minimal. If free-space conditions prevail, it is not necessary to perform any additional propagation calculation during the detailed prediction.

B. GROUND-WAVE PROPAGATION MODELS

When a radio wave is propagated over the earth, there are several different phenomena that occur and the selection of an applicable model depends on the particular phenomena that are most significant for the specific conditions being considered. Important factors which influence the selection of an appropriate model are the frequency range of interest and whether line-of-sight or beyond line-of-sight conditions prevail as suggested in Figs. 5.35 and 5.36.

B.1 Surface Wave

The surface wave is one of the three ground-wave components. It applies to line-of-sight situations involving vertically-polarized antennas and predominates from about 30 kHz to about 10 MHz. Above this frequency the direct and reflected-wave components of the ground wave generally predominate.

Fig. 5.37 illustrates basic transmission loss for the surface wave over average-land conditions. It corresponds to zero-height (or low elevation, say less than 10 meters) antennas and vertical polarization. For distances beyond line-of-sight conditions, transhorizon or diffraction propagation models must be used. Many graphs of the type shown in Fig. 5.37, corresponding to different land (and salt water) conditions and antenna heights, appear in some of the references in the bibliography at the end of this chapter.

B.2 Reflection Region

Other components of ground-wave propagation include the direct and reflected-wave. These components generally predominate over the surface wave above about 10 MHz in what is called the *reflection region* as shown in Fig. 5.36. Actually, the crossover frequency may occur anywhere between 3 MHz and 300 MHz depending upon terrain conditions. Propagation in this region is limited to nominal line-of-site conditions, local land undulations notwithstanding.

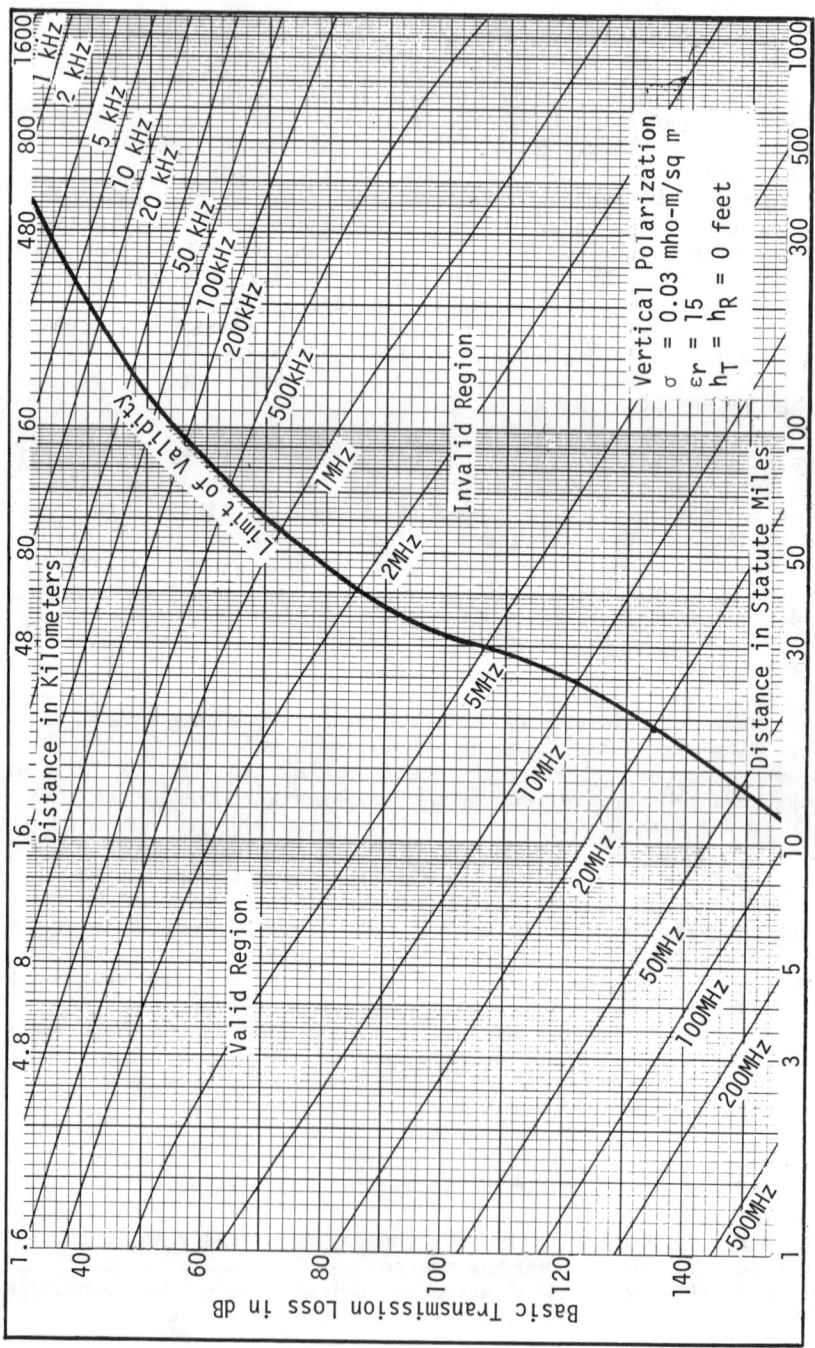

Figure 5.37 - Surface Wave Region, Average Land

For frequencies between about 40 MHz and 400 MHz and transmitter-receiver distances less than about 40 miles, an empirical model may be used to define mean propagation loss, L_{dB}, over irregular terrain:

$$L_{dB} = 117 + 20 \log_{10}f_{MHz} - 20 \log_{10}h_t h_r + 40 \log_{10}d_{mi} \quad (5.37)$$

where, f_{MHz} = radio frequency in MHz

h_t = transmitting antenna height above ground in feet

h_r = receiving antenna height above ground in feet

d_{mi} = transmitter-receiver separation in miles

To facilitate use of Eq. (5.37), it is plotted in Fig. 5.38 as a multiple parameter graph. The product of the transmitter and receiver antenna height (in ft^2 or m^2) is determined and the closest value is selected in the table insert column heading. The nearest frequency is then selected from the row heading. The intersection of the column and row indicate a letter from A to R. The resulting letter is the parameter value line to be used. Where this line intersects with the transmitter-receiver separation distance in miles or kilometers (X-axis), is the transmission loss in dB (Y-axis).

Illustrative Example 5.18

Determine the propagation loss at 100 MHz over irregular terrain between a base station antenna whose height is 100 feet and a vehicular-receiving antenna, 10 miles away, whose height is 10 feet. Also determine which propagation model to use.

While free-space conditions do not apply, free-space propagation will be determined for reference purposes. From Fig. 5.7, free-space loss is 96 dB.

It remains to determine whether the surface wave or reflection-region wave has the lesser loss. Fig. 5.37 indicates that the surface-wave transmission loss for zero-height antennas (not the situation here) is 156 dB. For the reflection region, the intersection of the 100 ft X 10 ft = 1,000 ft^2 column heading and 100 MHz row heading indicates parameter line "E" is to be used. The intersection of line "E" with 10 miles distance indicates that the propagation loss is 137 dB. Since this is less than that for the surface wave, the mean transmission loss is 137 dB. By contrast, this is 41 dB poorer (more loss) than that for free-space conditions.

For reflection region conditions exceeding those illustrated in Fig. 5.38, Figs. 5.39 or 5.40 may be used. More detailed situations are presented in Vol. 5 of the handbook series.

B.3 Diffraction Region

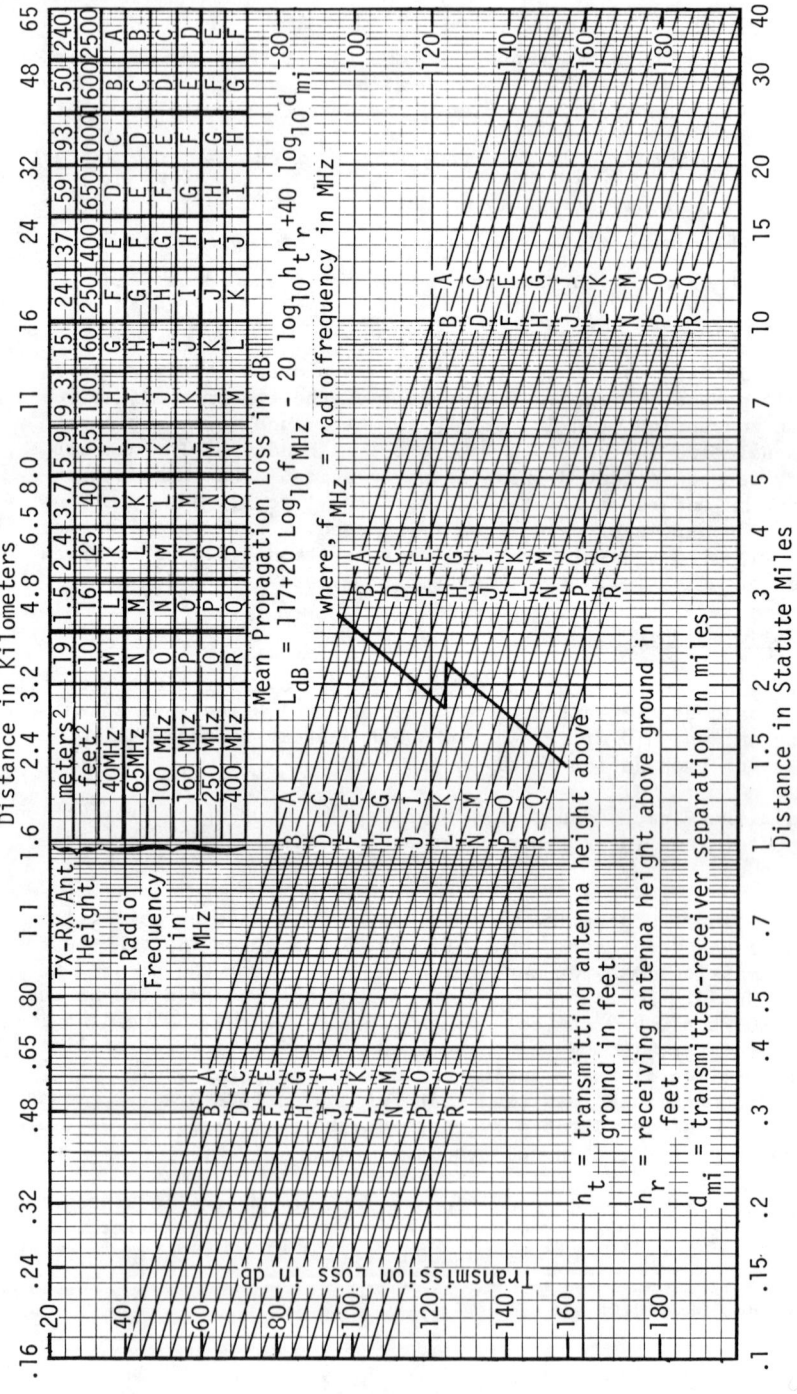

Figure 5.38 - Transmission Loss of Reflection Region of VHF over Irregular Terrain

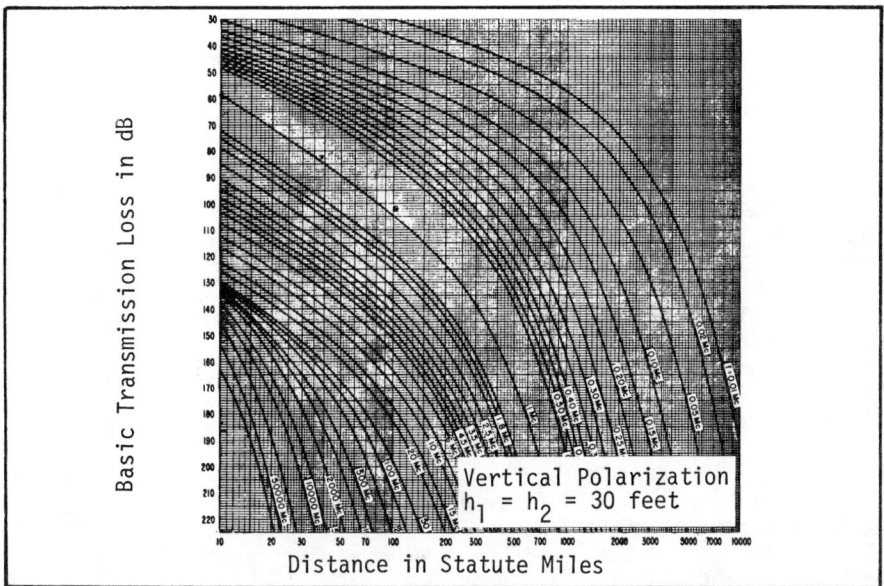

Figure 5.39 - Reflection Region, Smooth Spherical Earth, Land

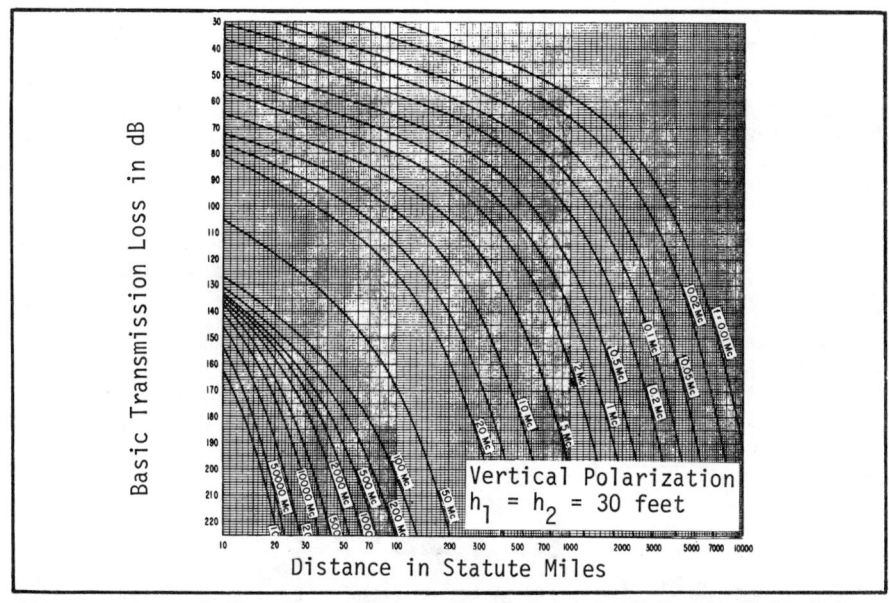

Figure 5.40 - Reflection Region, Smooth Spherical Earth, Salt Water

At distances just beyond the radio horizon, the diffracted surface wave (and in the case of elevated antennas, the diffracted ground
wave) is the predominant source of propagation as shown in Fig. 5.35.
As distance is increased still further beyond the radio horizon, the
level of the tropospheric scatter signal approaches the level of the
diffracted surface wave. The present discussion is limited to that
region where the effect of the tropospheric scatter signal need not
be considered. For convenience, this region is called the near-dif-
fraction region as shown in Fig. 5.36.

For smooth-earth conditions, the diffraction region model is based
on the zero-height surface wave discussed earlier. The calculation requires determining the zero-height, surface-wave field strength and
then modifying this value by appropriate antenna-height gain functions.
The curves given in Figs. 5.41 through 5.46 may be used directly to
determine diffraction-region basic transmission loss. The procedure
is:

1) Use Fig. 5.41 or 5.42 to determine the basic transmission loss
for the zero-height situation,
2) Use Figs. 5.43 through 5.46 to determine the height gain of
the transmitting and receiving antennas,
3) Calculate the basic transmission loss for the elevated antenna
by subtracting the two height gains from the zero height basic transmission loss.

Illustrative Example 5.19 - Application of Diffraction Region Curves

Determine the basic transmission loss between a 27 MHz potentially-
interfering signal transmitted from a vertically polarized antenna with
a height of 300 ft. and received by a vertically polarized land-mobile
antenna with a height of 10 ft. located 50 miles from the transmitter.
Assume that the intervening path is over land.

The first step is to use Fig. 5.41 to determine the basic transmission loss over a 50 mile path for zero-height antennas. The basic
transmission loss is 164 dB.

The next step is to use Figs. 5.43 and 5.44 to determine the
height gain for the transmitting and receiving antennas. It is necessary to examine the relationship between the antenna heights and frequency in order to determine which of the two figures apply (see
equation inserts in the figures). For the transmitting antenna height
of 300 ft:

$$h_T > \frac{882}{(f_{MHz})2/3} = \frac{882}{(27)^{2/3}} = 98 \text{ ft.} \qquad (5.38)$$

Therefore, Fig. 5.43 is used to determine the antenna-height gain.
From the figure, this gain is approximately 16 dB.

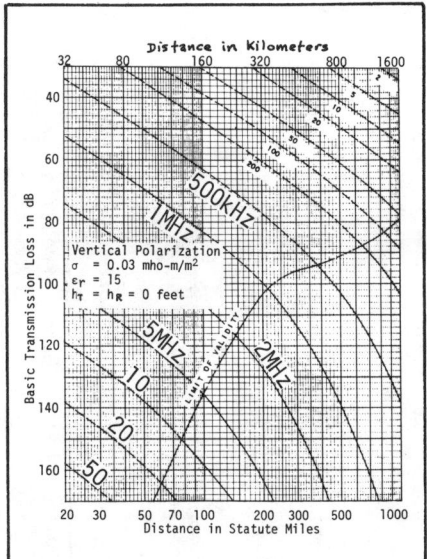

Figure 5.41 - Diffraction Region, Land

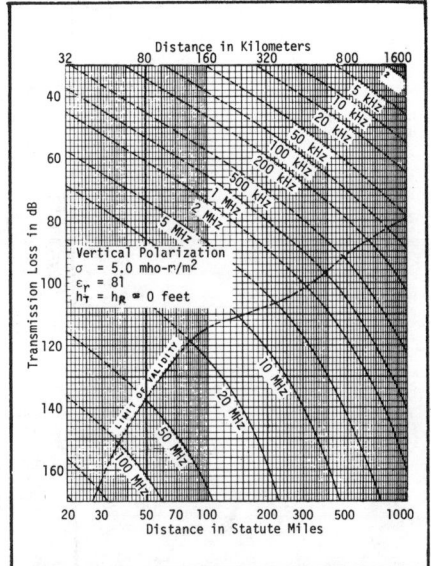

Figure 5.42 - Diffraction Region, Salt Water

For the receiving antenna height of 10 ft:

$$h_R < \frac{2000}{(f_{MHz})^{2/3}} = \frac{2000}{(27)^{2/3}} = 222 \text{ ft.} \tag{5.39}$$

Therefore, Fig. 5.44 is used to determine the antenna height gain, or approximately −1 dB.

The last step is to subtract the transmitting and receiving antenna height gains from the zero height basic transmission loss to determine the diffraction-region loss for the elevated antenna. Thus, for the illustrative example:

Corrected Loss = 164 dB − 16 dB − (−1 dB) = 149 dB

While not within the region of validity, this result may be contrasted with an extrapolation of curve A with 50 MHz in Fig. 5.38 to give about 148 dB. While somewhat fortuitous, it illustrates that mean differences may not be significant in transitional regions.

C. IONOSPHERIC PROPAGATION MODELS

Ionospheric propagation refers to all modes of propagation dependent upon one or more ionized layers of the atmosphere with the

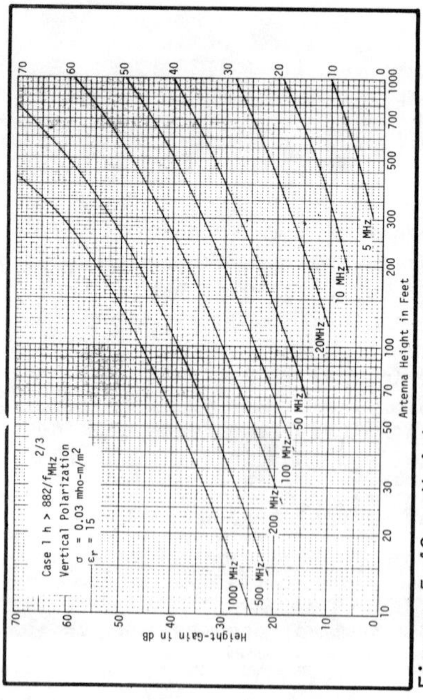

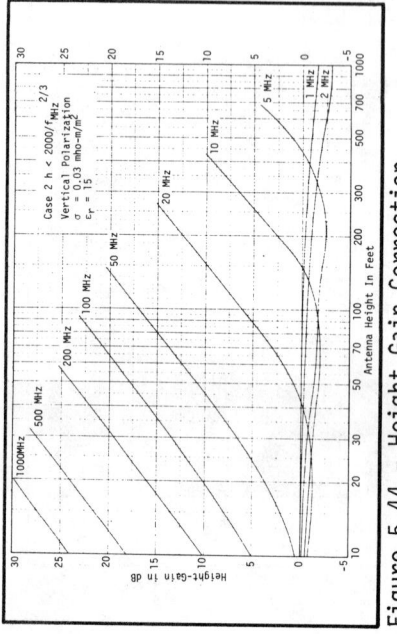

Figure 5.44 – Height Gain Correction,
Case 2, Over Land

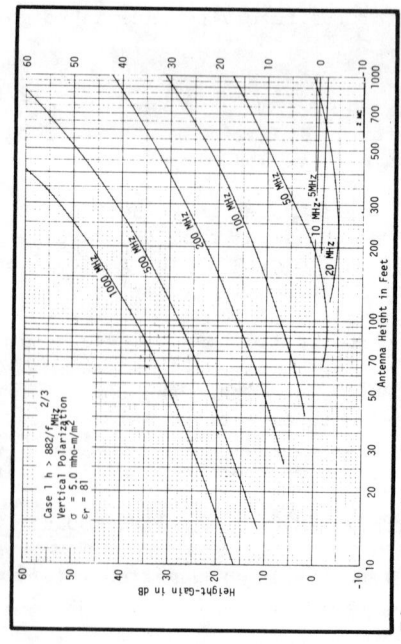

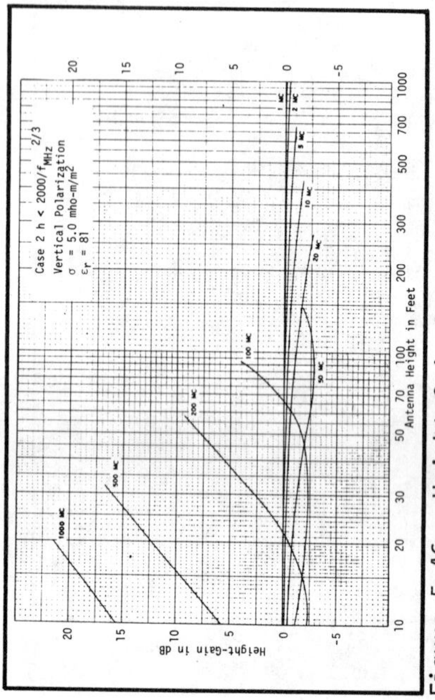

Figure 5.46 – Height Gain Correction,
Case 2, Over Sea Water

Figure 5.43 – Height Gain Correction,
Case 1, Over Land

Figure 5.45 – Height Gain Correction,
Case 1, Over Sea Water

exception of ionospheric scattering. The ionosphere is a complex, variable and sometimes unpredictable result of the periodic and anomalous characteristics of the sun. The absorption of polychromatic ultraviolet radiation from the sun produces ionized layers in the upper atmosphere with the most strongly absorbed radiation forming the highest layers. The height of the maximum ionization level for a particular layer varies with the zenith angle of the sun while the lower limit is a function of the rate of absorption of the radiant energy. The layers from the lowest to the highest are the D, E, sporadic-E, F_1 and F_2 layers. The layers responsible for most H-F communication are the E and F_2 layers.

Propagation which involves reflection from the ionosphere layers is termed *skywave* and falls into two cases, VLF and LF-HF frequency ranges. In the VLF range, the lower boundary of the D-layer and the surface of the earth form a waveguide and propagation in this frequency range is accounted for by waveguide theory. EMI is seldom associated with VLF propagation because of the low frequencies involved. For this reason, only a brief summary of the resulting propagation losses in the H-F spectrum is presented in Figs. 5.47 and 5.48. More detailed models are provided in the bibliography.

In the LF-HF band, two basic approaches are available for calculating the propagation loss on a skywave path. Either the median transmission loss can be computed directly by a relatively simple equation or a more cumbersome process using many graphs, charts, and overlays may be employed. Because skywave propagation is not likely to result in large interfering signals, specific techniques for calculating the propagation loss are not presented in this section. The median transmission loss equation and techniques for the field strength calculation are provided in the references.

D. SCATTER PROPAGATION MODELS

Just beyond the radio horizon the dominant mode of propagation is diffraction. However, the diffraction losses increase rapidly with increasing distance until a new mechanism, tropospheric scatter, becomes significant. As distance increases still further, i.e., beyond the limits of troposcatter, ionospheric scattering becomes significant. R-F signals scattered off the lower regions of the ionospheric E-layer, for frequencies between 3 and 75 MHz, attenuate very slowly with increasing distance. Fig. 5.49 is a plot of median basic transmission loss curves versus distance for the ground wave, troposcatter and ionospheric scatter modes. This curve serves to illustrate the relation between the three modes.

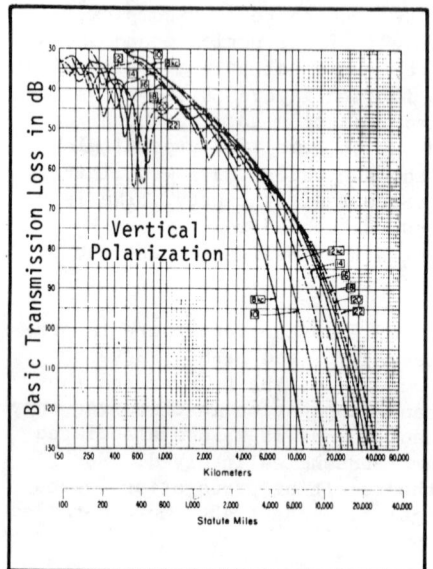

Figure 5.47 - Ionospheric
Propagation, Day-Over Land

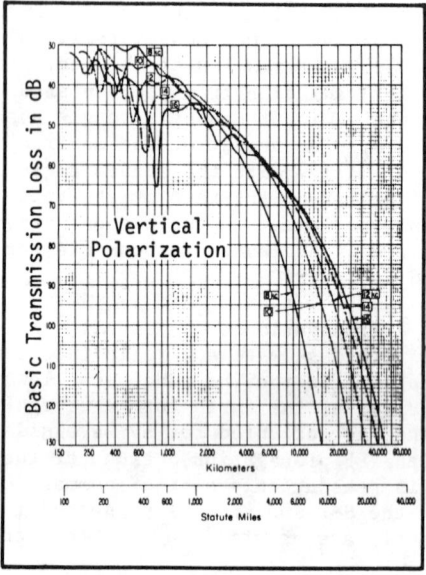

Figure 5.48 - Ionospheric
Propagation, Night - Over Land

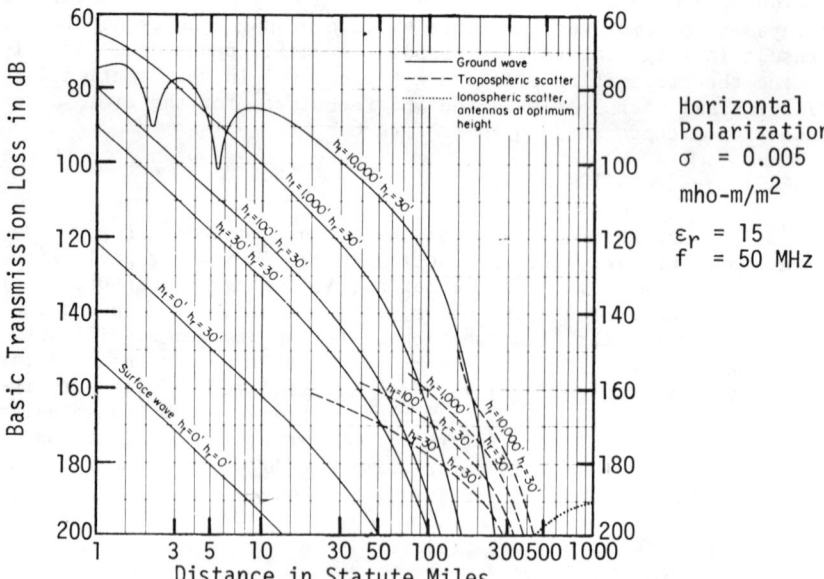

Figure 5.49 - Troposcatter Ground Wave, and Ionospheric Scatter
Propagation

5.4 MULTI-TRANSMITTER/RECEIVER PREDICTION TECHNIQUES

In executing the EMI prediction form introduced in Sec. 5.2.4, a prediction may be carried out in which it is concluded that no EMI exists. While this is good in the sense that EMI is not wanted, it can be wasteful of time and resources if many such EMI predictions were executed only to find no EMI problems exist. These situations may result when there exists requirements for determining where EMI exists in a multi-transmitter and/or receiver population.

To illustrate the foregoing, a transmitter will radiate 10 harmonics including its fundamental over a decade and a superheterodyne receiver may exhibit 40 spurious responses including its fundamental over $\pm$ 1 decade of frequency. Thus 400 transmitter-receiver combinations for a single pair alone exist over $f_R \pm$ 1 decade. Further, if there exists 10 transmitters and 10 potentially victim receivers, 100 fundamental emission-reception combinations exist and about 40,000 spurious situations may be possible. Accordingly, a technique is required to permit prediction efficiency by rapidly eliminating most situations at the onset for which EMI is not possible. This technique, to be explained in this section, is called the *Multi-Level Culling Process*.

5.4.1 MULTI-LEVEL PREDICTION PROCESS

For large transmitter and/or receiver populations, it is possible to take a *quick look* at the total EMI prediction problem using rather simple prediction models. This eliminates from further consideration many output-response pairs which would likely not represent EMI situations. This *culling* process can be extended to several levels of prediction detail. At each subsequent level additional factors are introduced and more non-interfering cases are eliminated from further consideration.

When the prediction is performed in the above manner it is only necessary to carry out a complete prediction on those surviving cases that exhibit a significant potential for creating EMI. The number of culling levels that gainfully should be used and the specific assumptions and calculations that should be performed at each level depend on the particular EMI prediction problem. As a general objective, each culling level of prediction should remove about 90% of the non-interference situations.

Four culling prediction levels are considered. They are based on amplitude, frequency, parameter detail, and performance. These prediction levels and the factors that are considered in each as well as the results that are obtained are illustrated in Fig. 5.50.

The most fundamental level of EMI prediction is based on emission-response amplitude. Frequency, time, distance, and direction effects are considered only in a relatively gross sense. The *amplitude cull*

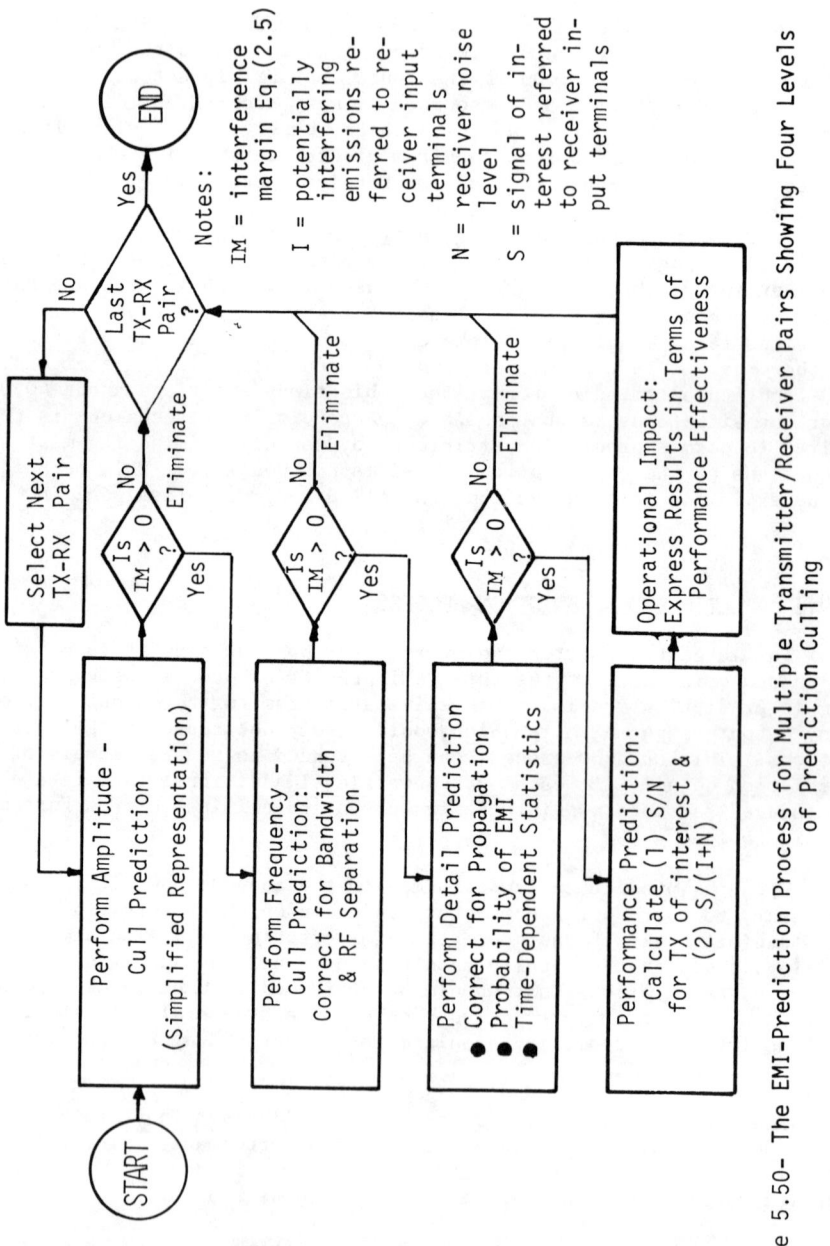

Figure 5.50- The EMI-Prediction Process for Multiple Transmitter/Receiver Pairs Showing Four Levels of Prediction Culling

uses simple, conservative approximations for each of the input functions. This separates a large number of unimportant interference possibilities from the relatively few cases which represent significant interference threats. For those cases that survive the amplitude prediction cull, there remains at least a small probability that interference will occur.

The amplitude cull is more than just a *weeding-out process* because the basic functions computed during this stage remain meaningful throughout the remainder of the finer-level predictions. It forms a basic nucleus to which a number of adjustment models are applied.

The frequency cull shown in Fig. 5.50 is the second prediction sorting level to be considered. During this stage those cases which remain as potential interference culprits are analyzed further using the results of the amplitude cull as a base. Frequency culling treats the frequency variable in more detail by considering additional interference rejection that is provided as a result of frequency separation (lack of alignment) between a susceptible response and a potentially interfering source and bandwidths.

The third stage of culling includes *detailed* consideration of time, distance and directional variables. Here potential interference amplitudes resulting from previous stages or culling are translated into probabilities of interference and time-dependent terms.

The final stage of prediction includes a consideration of factors such as transmitter and receiver modulation characteristics and operational response and *performance* analysis. At this stage prediction results are translated into terms that are more meaningful to the user from an operational standpoint. For example, interference conditions are interpreted in quantities such as intelligibility of voice systems, digital error-rates, scope presentation conditions, false alarms, missed targets, reduction in range of radar systems, and other measures that are related to overall system effectiveness and performance. They are based upon signal-to-noise and/or signal-to-interference-plus-noise ratios.

5.4.1.1   Amplitude-Culling Process

The amplitude-culling process is a first step in EMI prediction involving a relatively large number of transmitter-receiver main and spurious emission and response pairs. It seeks to separate those combinations of signal amplitudes and frequencies which under worst circumstances will not result in interference, from those situations which exhibit at least a small probability of producing interference. The primary purpose of the amplitude cull is (1) to examine meaningful combinations of emitter outputs and susceptible receptor responses and (2) to eliminate from further consideration as many noninterfering

cases as possible. The basic philosophy used to accomplish this goal
is to calculate potential interference margins assuming that transmission
losses are minimized. It also assumes emission outputs and receptor responses are aligned in frequency in a manner such that the
susceptible device provides minimum rejection to the potential interfering signal.

Specific operations performed during amplitude culling are:

$$IM(f,t,d,p) = I/N = P_T(f_E) + G_T(f_E,t,d,p) - L(f_E,t,d,p) + G_R(f_E,t,d,p)$$
$$- P_R(f_R) + CF(B_T,B_R,\Delta f) \qquad (5.40)$$

where, $P_T(f_E)$ = Power transmitted in dBm at emission frequency, f_E

   $G_T(f_E,t,d,p)$ = Transmitter antenna gain in dB at emission frequency, f_E, in direction of receiver.

   $L(f_E,t,d,p)$ = Propagation loss in dB at emission frequency, f_E, between transmitter and receiver.

   $G_R(f_E,t,d,p)$ = Receiver antenna gain in dB at emission frequency, f_E, in direction of transmitter.

   $P_R(f_R)$ = Receiver susceptibility threshold in dBm at response frequency, f_R.

   $CF(B_T,B_R,\Delta f)$ = Factor in dB that accounts for transmitter and receiver bandwidths, B_T and B_R, respectively and the
frequency separation, Δf, between transmitter emission and receiver response.

The amplitude-cull stage considers (1) transmitter fundamental
and spurious emission power levels $P_T(f)$, and (2) receiver fundamental
and spurious response susceptibility threshold levels $P_R(f)$. Amplitude
culling considers antenna gains and propagation loss, but uses simple
conservative approximations to represent the effects of time, distance
separation, and direction of these parameters. The correction factor
$CF(B_T,B_R,\Delta f)$ is not considered during amplitude culling, i.e.,
$CF(B_T,B_R,\Delta f)$ is assumed to be zero.

If the resulting interference margin exceeds a preselected culling
level, the emission-response combination is retained for the next finer
level of prediction. On the other hand, if the interference margin is
less than the preselected cull level, the emission-response combination is eliminated from further prediction considerations. When the
cull level is selected correctly, the cases that are eliminated will
have an insignificant probability of interference. Problems in selecting a suitable cull level will be discussed later.

With the above basic assumptions for amplitude culling, the problem of predicting EMI between transmitters and receivers is then
grouped into four significant and different cases of transmitter-out-
put and receiver-response pairs. Fig. 5.51 illustrates each of the

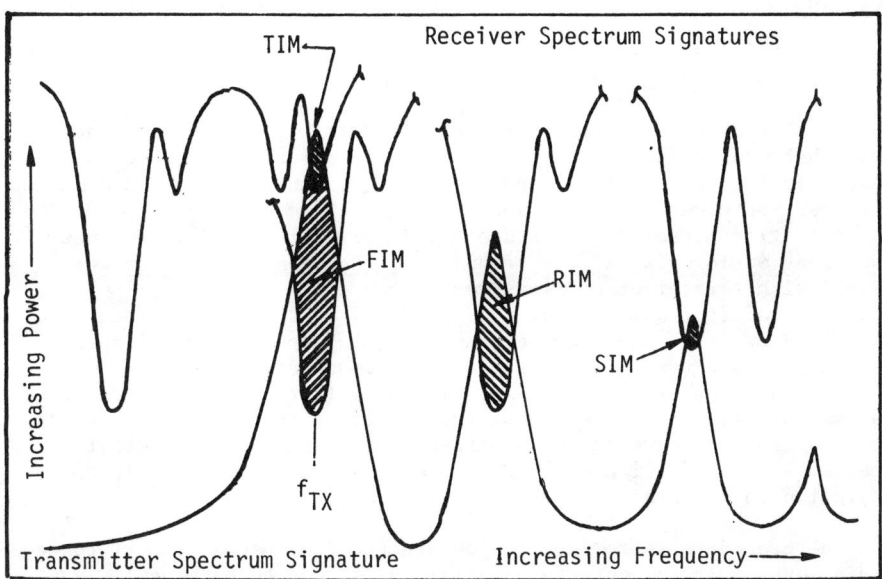

Figure 5.51 - Illustrating the Four Classes of Transmitter-to-Receiver Interference Margins

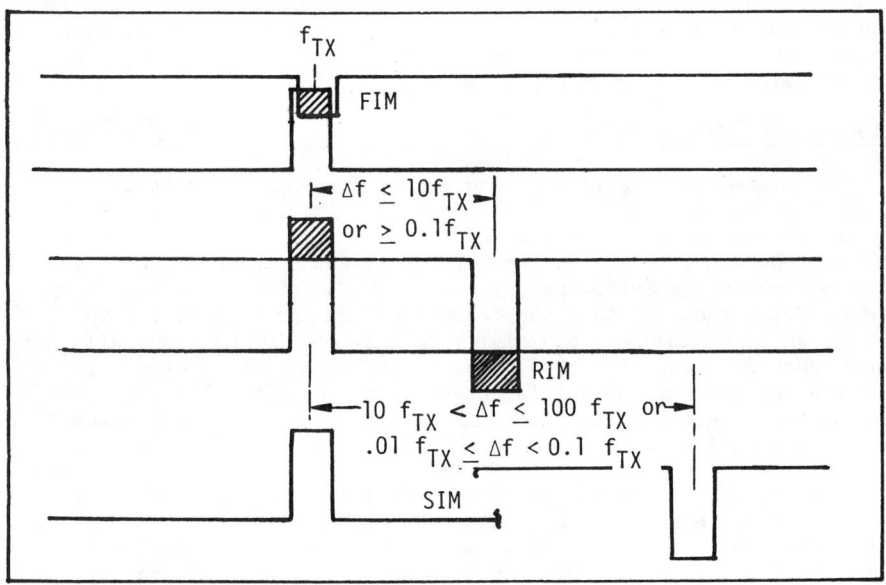

Figure 5.52 - Simplified Math Model of Four Classes of Interference Margins

four situations and Fig. 5.52 shows simplified mathematical models which are used to represent these situations:

(1) *Fundamental Interference Margin (FIM)* is the interference level that would exist if the transmitter fundamental frequency output and receiver fundamental response were aligned in frequency in a manner such that there were no rejection.

(2) *Transmitter Interference Margin (TIM)* is the interference level that would exist if the transmitter fundamental emission were aligned with a receiver spurious response.

(3) *Receiver Interference Margin (RIM)* is the interference level that would exist if the receiver fundamental response were aligned in frequency with a transmitter spurious emission output.

(4) *Spurious Interference Margin (SIM)* is the interference level that would exist if both a transmitter spurious output and a receiver spurious response were aligned in frequency. The spurious interference level is a function of frequency and this must be taken into consideration in the prediction.

The procedure is to first calculate the fundamental interference margin level. If the interference margin is less than the cull level it is not necessary to compute the other three cases since their margins will be still lower. On the other hand, if the interference margin for the fundamental interference level exceeds the cull level, it is next necessary to compute both the transmitter fundamental and the receiver fundamental interference levels. If either of these produce an interference margin that exceeds the cull level, it is also necessary to calculate the spurious interference level.

Illustrative Example 5.20

Calculate the potential EMI between a tropospheric scatter communication transmitter and a radar receiver. Assume that the transmitter operates at 1 GHz (f_{OT}), has a fundamental power output ($P_T(f_{OT})$) of 1 kW (+60 dBm), and all transmitter spurious outputs are at least 60 dB below fundamental output, i.e., spurious outputs are less than 0 dBm. Further assume that the radar receiver operates at 1.2 GHz (f_{OR}), has a fundamental sensitivity $P_R(f_{OR})$ of -100 dBm, and all spurious responses are at least 80 dB above the fundamental response, i.e., spurious sensitivity $P_R(f_{SR})$ is greater than -20 dBm. Thus, calculate the interference margin that will exist for each of the four cases if the transmission loss is 100 dB.

Fig. 5.53 illustrates the functions used in interference prediction and demonstrates the results obtained for this sample problem over a frequency range of $f_R \pm 1$ decade, i.e., 0.1 to 10 GHz. The power available function, P_A, shown in the figure was obtained by applying Eq. (5.41), using simplified notation containing only the frequency variable of this particular problem:

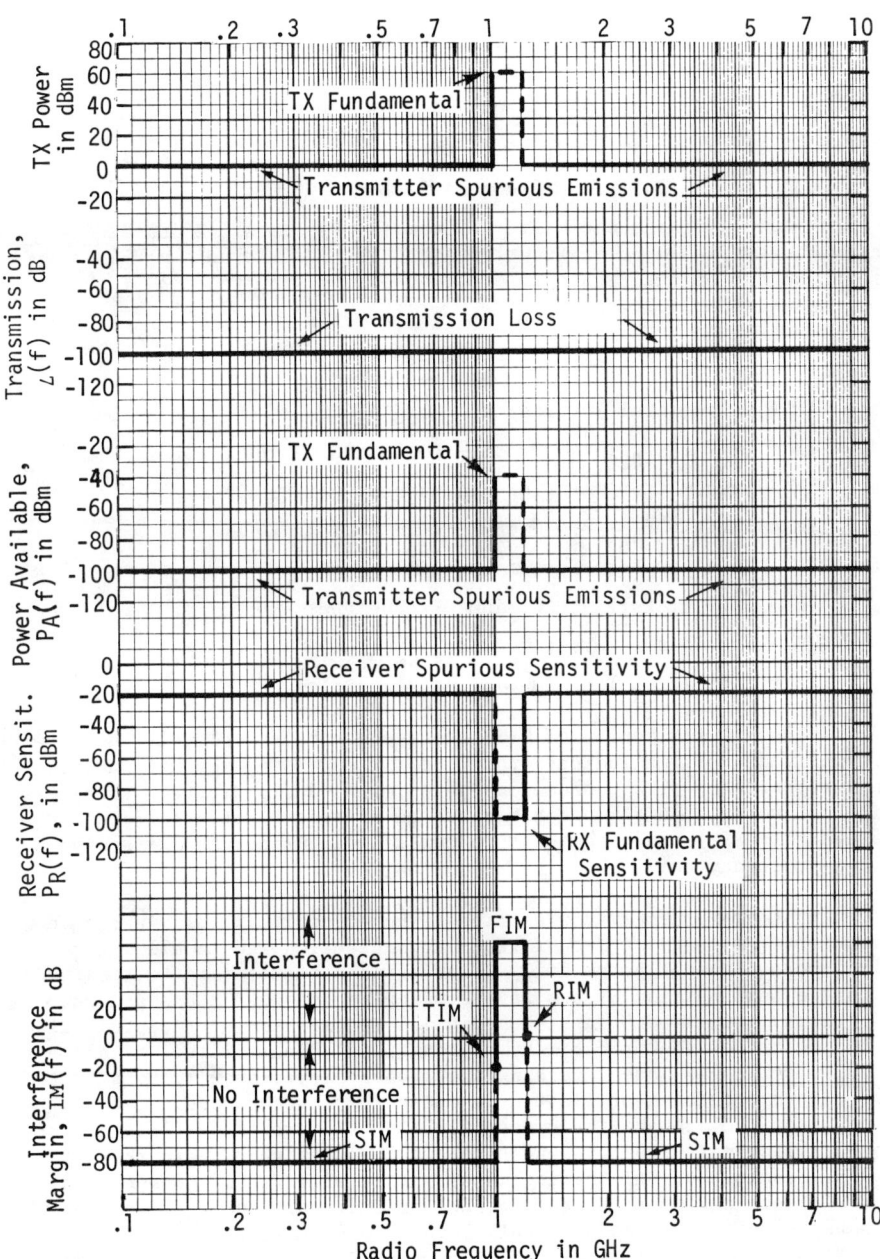

Figure 5.53 - Illustrating Amplitude Culling (See Ex. 2.2)

$$P_A(f) = P_T(f) - L(f) \tag{5.41}$$

For the Transmitter Fundamental Frequency (f_{OT})

$$P_A(f_{OT}) = 60 \text{ dBm } -100 \text{ dB} = -40 \text{ dBm}$$

For Transmitter Spurious Outputs

$$P_A(f_{ST}) = 0 \text{ dBm } -100 \text{ dB} = -100 \text{ dBm}$$

The receiver susceptibility threshold, P_R, is shown immediately below the available power functions, and the interference margin shown at the bottom of the figure is obtained by applying to each of the four cases separately:

$$IM(f) = P_A(f) - P_R(f) \tag{5.42}$$

Case FIM - Fundamental Interference Margin:

$$FIM = IM(f_{OT}, f_{OR}) = -40 \text{ dBm} - (-100 \text{ dBm}) = +60 \text{ dB}$$

Case TIM - Transmitter Interference Margin:

$$TIM = IM(f_{OT}, f_{SR}) = -40 \text{ dBm} - (-20 \text{ dBm}) = -20 \text{ dB}$$

Case RIM - Receiver Interference Margin:

$$RIM = IM(f_{ST}, f_{OR}) = -100 \text{ dBm} - (-100 \text{ dBm}) = 0 \text{ dB}$$

Case SIM - Spurious Interference Margin:

$$SIM = IM(f_{ST}, f_{SR}) = -100 \text{ dBm} - (-20 \text{ dBm}) = -80 \text{ dB}$$

If an EMI I/N margin of 0 dB is used as the cull level, the results of amplitude culling indicate that the TIM and SIM cases will not result in interference. Thus, these may be eliminated from further consideration. However, the FIM and RIM cases do present a potential problem (especially FIM) and hence must be considered further in the second-stage frequency culling.

In order for amplitude culling to be effective in eliminating non-interfering situations from further consideration and to allow for a large percentage of potentially-interfering situations to be retained, a choice must be made of the levels which represent the basic input functions. Since each input function is represented by a distribution which describes the probability that a given level will not be exceeded, a choice is made regarding the probability that will be selected for either retaining or limiting a case from further consideration. Because of the probabilistic nature of the EMI problem it is necessary to establish criteria for amplitude culling such that there results a low probability of eliminating cases with a significant potential of interference during this stage of EMI prediction. Chapter 9 of Vol. 5 discusses this topic and the cull levels in detail.

5.4.1.2 FREQUENCY-CULLING PROCESS

The second step of the multi emitter-receptor, interference prediction and culling process is the frequency-culling stage. This step
modifies the interference margin obtained in the amplitude-culling
stage by accounting for such factors as transmitter bandwidths and
modulation characteristics, bandwidth and selectivity of each receiver
response, and frequency separation between specific transmitter emissions and receiver responses.

Within the selected frequency range determined from the preceding
amplitude cull, frequency culling corrects for bandwidth differences
and frequency separations between transmitter output and receiver response frequencies. Each transmitter output frequency (fundamental,
harmonics, and spurious) is compared with each potential receiver response frequencies (spurious heterodyne, intermodulation, co-channel
and adjacent-channel receiver response frequencies) to determine
specific transmitter emission bands which are capable of producing
EMI. Separation between each mating pair of frequencies is considered
by applying a composite bandwidth, selectivity, and modulation envelope
function at the surviving frequencies. The resulting comparison yields
a correction to the interference margin obtained from amplitude culling.

In amplitude culling, no spectrum spread in power was assumed.
The total power was considered to exist at each emission or response
frequency. No frequency separation (Δf) between a transmitter output
and a receiver response was considered in calculating the interference
margin. In frequency culling, on the other hand, the transmitter modulation envelope and receiver selectivity curve are considered. Taking
into account relative bandwidths, it defines a correction or adjustment factor which is applied to the amplitude-cull interference margin.
In addition, frequency culling considers center frequency separation
between output response pairs and defines a correction factor to the
amplitude cull to account for this separation. Basically, the method
employs superimposing or overlaying (convolution) the receiver selectivity upon the transmitter modulation envelope.

Specifically the frequency cull modifies the amplitude cull result
by the correction factor, $CF(B_T, B_R, \Delta f)$ in Eq. (5.40). If the resulting interference margin still exceeds the cull level, the emission-
response pair is retained for further consideration during the third-
stage detailed prediction. If the resulting interference margin is
less than the cull level, the emission-response pair is eliminated
from further consideration.

A. ON-TUNE CASE, $\Delta f \leq (B_T + B_R)/2$

The basic concept of frequency culling is illustrated by considering various possibilities that may exist between particular output

response pairs as shown in Fig. 5.54. First, if the output and response
occur at the same center frequency (i.e., $\Delta f = 0$), there are two basic
co-channel possibilities that may be considered:

(1) *Receiver bandwidth is either equal to or greater than the*
transmitter bandwidth. $(B_R \geq B_T)$. For this case all the power associated with the transmitter output exists within the receiver bandpass
and no correction is necessary.

(2) *Receiver bandwidth is less than the transmitter bandwidth*
$(B_R < B_T)$. For this case only a portion of the power associated with
the emission output is received and it is necessary to apply a bandwidth correction, CF, to account for the bandwidth differences. The
correction for $\Delta f = 0$ is dependent on the bandwidth ratios and is of
the form:

$$CF(\Delta f = 0) = K \log_{10}(B_R/B_T) \text{ dB, for } B_R < B_T \qquad (5.43)$$

where, B_R = receiver 3-dB bandwidth in Hz

B_T = transmitter 3-dB bandwidth in Hz

K = a constant for a particular emission-response
combination   (5.44)

= 0 for $B_R \geq B_T$ and co-channel frequency alignment

= 10 for noise-like (incoherent) signals for which RMS
levels apply and $B_R < B_T$

= 20 for pulse and transient (coherent) signals for which
peak levels apply and $B_R < B_T$*

Illustrative Example 5.21 - Co-channel Interference ($B_R > B_T$)

Consider a co-channel interference involving a UHF voice transmitter with a 3-dB bandwidth of 10 kHz and a surveillance receiver having
a bandwidth of 1 MHz. Assume that amplitude culling has indicated
that the fundamental interference margin, IM, is +80 dB and both equipments are operating at the same center frequency ($\Delta f = 0$). All the
transmitting power will be received since K = 0 in Eq. (5.44), viz.
$B_R > B_T$. Thus, it is not necessary to apply a frequency-cull correction factor to the results of amplitude culling.

Illustrative Example 5.22 - Co-channel Interference ($B_R < B_T$)

If the previous example were reversed, i.e., the transmitter bandwidth was 1 MHz and the receiver bandwidth was 10 kHz, then only a
portion of the available transmitter power would be received. If the

---

* For pulse signals, peak levels are applicable as long as B_R is
greater than the pulse-repetition rate.

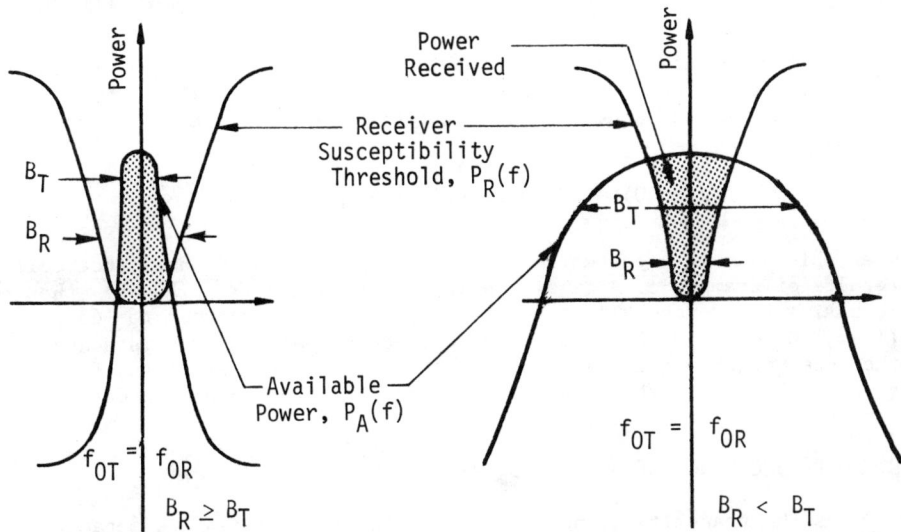

(a) On-Tune Case (Co-channel Frequency Alignment)

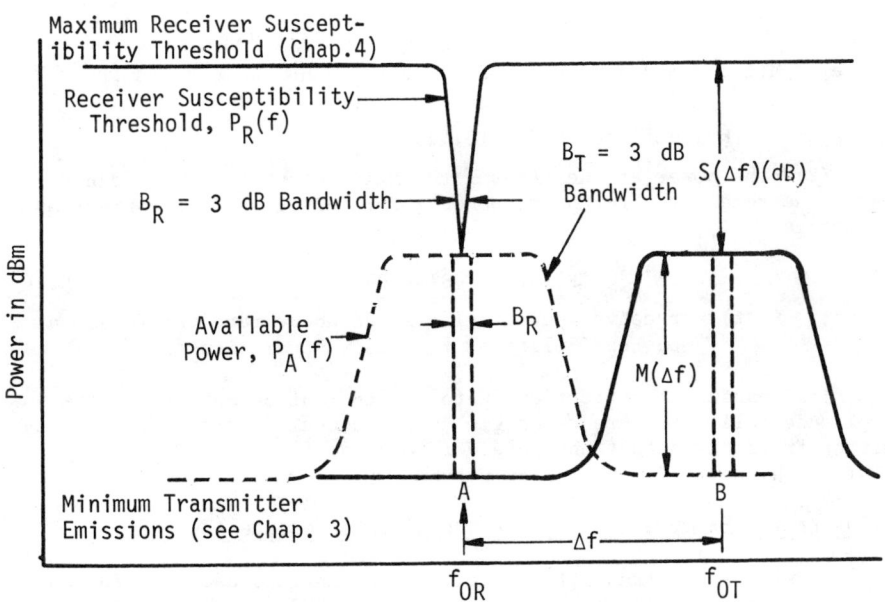

(b) Off-Tune Case (Spurious Frequency Alignment)

Figure 5.54 - Illustration of Frequency Analysis

5.101

transmitter is pulse modulated at 1,000 pps, the received voltage is directly proportional to the bandwidth ratios (the received power is proportional to the square of the ratios; K=20). From Eqs. (5.43) and (5.44):

$$CF(\Delta f = 0) = 20 \log_{10}\left(\frac{10 \text{ kHz}}{1000 \text{ kHz}}\right) = -40 \text{ dB}$$

When this bandwidth correction factor is applied to the amplitude cull results of Example 5.21 (i.e., interference margin IM = +80 dB), the resulting frequency-culling interference margin is reduced to +40 dB (i.e., +80 dB - 40 dB). Thus, where frequency alignment exists and the receiver bandwidth is less than the transmitter bandwidth, frequency culling results in a reduction in the interference potential obtained by amplitude culling.

B. OFF-TUNE CASE, $\Delta f > (B_T + B_R)/2$

As the transmitter and receiver center frequencies are separated, the transmitter power can enter the receiver by either of two other possible means (see Fig. 5.54):

(1) the transmitter emission modulation sidebands can enter the receiver, the main-response frequency, giving a correction:

$$CF_R(\Delta f) = [K \log_{10}(B_R/B_T) + M(\Delta f)] \text{ dB} \qquad (5.45)$$

where, $M(\Delta f)$ = modulation sideband level in dB above transmitter
power at frequency separation Δf

K = defined in Eq. (5.44)

(2) the power at the transmitter main output frequency can enter the receiver off-tune response. For this case, the correction factor is:

$$CF_T(\Delta f) = -S(\Delta f) \text{dB} \qquad (5.46)$$

where, $S(\Delta f)$ = receiver selectivity in dB above receiver fundamental
susceptibility at frequency separation Δf

The final bandwidth correction factor which must be applied to the amplitude-cull interference margin due to non-alignment of the transmitter frequency output and receiver response is either $CF_R(\Delta f)$ or $CF_T(\Delta f)$ whichever is the larger of the two.

Illustrative Example 5.23 - Off-channel Interference ($B_R > B_T$)

Consider the transmitter-receiver combinations used for the co-channel Example 5.21. Assume that the center frequency separation, Δf, is 2 MHz for an AM voice transmitter bandwidth B_T of 10 kHz, and a 1 MHz surveillance receiver bandwidth, B_R. Further, assume that the

transmitter modulation envelope is 90 dB below the fundamental level
($M(\Delta f) = -90$ dB), and the receiver selectivity provides 60 dB of rejection ($S(\Delta f) = 60$ dB) at a 2 MHz frequency separation. The correction factor at R and T frequencies from Eqs. (5.45) and (5.46)
respectively, are (K=10 for voice):

$$CF_R(\Delta f) = 10 \; \log_{10}\left(\frac{1000 \text{ kHz}}{10 \text{ kHz}}\right) -90 \text{ dB} = -70 \text{ dB}$$
$$CF_T(\Delta f) = -S(\Delta f) = -60 \text{ dB}$$

The final frequency-cull correction factor is determined by selecting the algebraically larger of the two correction factors, i.e., it
would be -60 dB. In order to obtain the interference margin resulting
from the additive effects of frequency culling, it is necessary to apply
the frequency-cull correction factor (-60 dB) to the amplitude-cull
interference margin (+80 dB, see Example 5.21). Thus, the resulting
frequency-cull interference margin is reduced to +20 dB.

Illustrative Example 5.24 - Off-channel Interference ($B_R < B_T$)

Consider another example similar to Example 5.23 except that the
transmitter bandwidth and its modulation envelope, and the receiver
bandwidth and selectivity are reversed, i.e.:

> Transmitter Bandwidth, $B_T = 1$ MHz
> Transmitter Modulation Envelope Reduction,
> $M(\Delta f) = -60$ dB
> Receiver Bandwidth, $B_R = 10$ kHz
> Receiver Selectivity Rejection, $S(\Delta f) = 90$ dB

Assume the transmitter is pulsed at 1000 pps (K-factor = 20):

The correction factors at R and T frequencies are obtained by again
applying Eqs. (5.45) and (5.46) respectively.

$$CF_R(\Delta f) = 20 \; \log_{10}\left(\frac{10 \text{ kHz}}{1000 \text{ kHz}}\right) -60 \text{ dB} = -40 \; -60 = -100 \text{ dB}$$
$$CF_T(\Delta f) = S(\Delta f) \text{ dB} = -90 \text{ dB}$$

The frequency cull, correction factor is the larger of the two. In
this example it is -90 dB. When this correction factor is applied to
the value of the amplitude-cull interference margin (i.e., +80 dB,
see Example 5.21), the resulting frequency cull interference margin is
reduced to -10 dB. Thus, by considering both the bandwidth differences
and frequency separations, a considerable reduction in interference
potential is obtained over that previously indicated by amplitude
culling alone.

5.4.1.3   Detailed Prediction

In preceding sections the factors which contribute to interference within a communications-electronics system complex were introduced, and methods of obtaining the interference margin were presented. It is possible to eliminate a large number of non-interfering situations from further prediction consideration by employing amplitude culling and still retain nearly all potential interfering situations. Those cases that remained after amplitude culling were considered further during frequency culling and additional non-EMI situations were eliminated.

The results of frequency culling yield surviving cases that now have a significant potential for producing interference. Each of these remaining cases then must be subjected to still further prediction. The purpose of the detailed prediction is: (1) to include those factors depending on time, distance, and direction which will influence the interference situation between a specific transmitter and receiver and (2) to determine the probability distribution for the resulting interference margin. Some of the more important factors that are considered in the detailed-prediction stage include:

- specific propagation mode(s)
- polarization matching
- near-field antenna gain corrections
- multiple interfering signal effects
- time-dependent statistics, such as resulting from rotating antennas
- probability distribution of interference margin

One of the important steps in the detailed prediction is determining the probability distribution associated with the interference margin. The probability distribution results from considerations presented in Sec. 5.3 and Vol. 5. Probability distributions are associated with the transmitter power, antenna gains, propagation loss, and receiver susceptibility threshold.

5.4.1.4   Performance Prediction

Any signal which hinders or degrades the operation of an electrical or electronic device in accomplishing its intended or design function can be termed interference. However, in order to predict the effects of interference it is necessary to express the results in terms of the impact on operational performance.

The primary problem in expressing results in a form that is useful in evaluating performance is to relate the predicted interference levels to measures of performance. Because of the importance and complexity of this particular aspect of the interference prediction process, a complete chapter is devoted to this subject. The reader is referred to Vol. 5 for a discussion of performance prediction.

5.4.2 SHORT-FORM EMI PREDICTION PROCESS

There are many EMI prediction problems for which only a few transmitter-receiver pairs are involved and the prediction is either performed manually or with the aid of a simple computer program which
may be run on a time-share terminal. This section presents a step-
by-step process for performing a manual EMI prediction using a special
form.* The user simply fills out the prediction form in the same manner as an *income tax form*. References to specific parts of this chapter are provided which may be used to obtain further description of
EMI prediction models or techniques.

A brief description of some of the major EMI prediction considerations that apply to a manual prediction process are now considered. As
previously explained (see Figs. 5.51 and 5.52), there are four possible combinations of transmitter-receiver (TX-RX) emission-responses
for each pair that are accommodated on the form:

● Class FIM - Fundamental Interference Margin- TX fundamental
  emission and RX fundamental response

● Class TIM - Transmitter Interference Margin- TX fundamental
  emission and RX spurious response

● Class RIM - Receiver Interference Margin- TX spurious emission
  and RX fundamental response

● Class SIM - Spurious Interference Margin- TX spurious emission
  and RX spurious response.

5.4.2.1 FACTORS PERTINENT TO THE USE OF THE EMI PREDICTION FORM

The considerations that apply and the calculation procedures used
are different for each of the four categories. The first step in a
prediction process is to determine the frequency limits associated with
each TX-RX pair, as provided for in entries #1 through #7 of the short
form. To determine which emission-response categories apply, the
frequency limits are processed under "Applicability of Four EMI Prediction Cases" in the short form.

After it is determined which of the four cases apply, the next
step is to calculate a preliminary level based on emission and response
amplitudes, and free-space propagation loss from Fig. 5.7. This is
accomplished in lines #8 through #16 on the short form. Line #16 then
shows which, if any, IM cases survive the amplitude cull.

---

*   The form to be presented differs from that of Sec. 5.2.4 in that
it is more comprehensive. However, it does not contain the intended
signal level prediction since I/N ratios (interference margins) are
emphasized here rather than S/I ratios of Sec. 5.2.4.

SHORT FORM EMI PREDICTION

TRANSMITTER AND RECEIVER FREQUENCY LIMITS

1. TX Fundamental Frequency, (f_{OT}) | **220**
2. TX Minimum Spurious Frequency, $(f_{ST})_{min}$ or $.1f_{OT}$ | **22**
3. TX Maximum Spurious Frequency, $(f_{ST})_{max}$ or $10f_{OT}$ | **2200**
4. RX Fundamental Frequency (f_{OR}) | **360**
5. RX Minimum Spurious Frequency, $(f_{SR})_{min}$ or $.1f_{OR}$ | **36**
6. RX Maximum Spurious Frequency, $(f_{SR})_{max}$ or $10f_{OR}$ | **3600**
7. TX-RX Maximum Allowable Frequency Separation for Fundamental EMI, Δf_{max} or $.2f_{OR}$ | **72**

APPLICABILITY OF FOUR EMI PREDICTION CASES:

SIM = TX Harmonic & RX Spurious
Is (2)___**22**___ < (6)**3600**___ ?   ☒ Yes, ☐ No
Is (3)**2200**___ > (5)___**36**___ ?   ☒ Yes, ☐ No
If either is No, there is no EMI Problem-STOP

RIM = TX Harmonic & RX Fundamental
Is (2)___**22**___ < (4)___**360**___ ?   ☒ Yes, ☐ No
Is (3)**2200**___ > (4)___**360**___ ?   ☒ Yes, ☐ No
If either is No, skip RIM and enter N/A on line 38.

TIM = Tx Fundamental & RX Spurious
Is (1)___**220**___ < (6)**3600**___ ?   ☒ Yes, ☐ No
Is (1)___**220**___ > (5)___**36**___ ?   ☒ Yes, ☐ No
If either is No, skip TIM and enter N/A on line 38.
If both RIM and TIM were N/A, skip FIM and enter N/A on line 38.

FIM = TX Fundamental & RX Fundamental
Is |(1) **220**___ -(4)**360**___ |<(7)___**72**___ ?   ☐ Yes, ☒ No
If No, skip FIM and enter N/A on line 38.

Surviving cases ☒ SIM, ☒ RIM, ☒ TIM, ☐ FIM
☐ No cases survived - No EMI problem.

AMPLITUDE CULLING (See Sec. 5.3.1.2, 5.3.2.2, 5.3.3.2)

		FIM	TIM	RIM	SIM	
8.	TX Power, $P_T(f_{OT})$, (peak power if pulsed)	NA	80			dBm
9.	TX Spurious Power Output: $P_T(f_{ST})$ or $P_T(f_{OT})-60dB$		20	20		dBm
10.	TX Antenna Gain in RX Direction: $G_{TR}(f)$ or 0dB	NA	O	O	O	dB
11.	RX Antenna Gain in TX Direction: $G_{RT}(f)$ or 0dB	NA	O	O	O	dB
12.	Propagation Loss, L Using Frequency No.	(1)	(1)	(4)	(2)	
	Loss in dB from Fig. 5.7	-NA	-49	-55	-30	dB
13.	Unintentional Power Available, $P_A(f)$ Add 8 to 12	NA	31	-35	-10	dBm
14.	RX Fundamental Susceptibility, $P_R(f_{OR})$	NA		-100		dBm
15.	RX Spurious Suscept.: $P_R(f_{SR})$ or $P_R(f_{OR})$ + 80dB		-20		-20	dBm
16.	Preliminary EMI Prediction: line 13-14 or 13-15	NA	51	65	10	dB

If EMI margin < -10 dB, EMI Highly Improbable - STOP
If EMI margin > -10 dB, Start Frequency Culling

Sec. 5.4

FREQUENCY CULLING (See Sec. 5.3.1.3 and 5.3.2.3)

Frequency:
☒ kHz; ☐ MHz

BANDWIDTH CORRECTION

17. TX PRF (if pulse) → `100` pps
18. TX Bandwidth, ($B_T = 2/\pi\tau$ if pulse; τ = width) → `64`
19. RX Bandwidth, B_R → `10`
20. Adjustment(from lines 17 to 19; Fig.5.11&5.12) `NA` `-15` `-15` `-15` dB
21. Bandwidth Corrected, EMI Margin = lines 16 + 20 `NA` `36` `56` `-5` dB

IF EMI MARGIN $\leq$ -10 dB, EMI HIGHLY IMPROBABLE - STOP

FREQUENCY CORRECTION

Frequency:
☐ kHz; ☐ MHz

22. RX Local Oscillator Frequency, f_{LO} → `400`
23. RX Intermediate Frequency, f_{IF} → `40`

	FIM	TIM	RIM	SIM	
24. TX-RX Frequency Separation: $\Delta f = \|(1)-(4)\|$	NA				
25. $\Delta f > (B_T + B_R)/2$ (from line 24, use Fig.5.10)	NA				
26. Calculate $(f_{OT} \pm f_{IF})/f_{LO}$ to nearest integer		1			
27. Multiply lines 22 x 26		400			MHz
28. $\Delta f = \|1-23-27\| = \underline{\quad} : \|1+23-27\| = \underline{\quad}$					
29. Select smaller Δf from line 28		140			MHz
30. $\Delta f > (B_T + B_R)/2$(from line 29, use Fig.5.10)		-100			dB
31. Calculate f_{OR}/f_{OT} to nearest integer			2		
32. Multiply lines 1 x 31			440		MHz
33. $\Delta f = \|(4) - (32)\|$			120		MHz
34. $\Delta f > (B_T + B_R)/2$ (from line 33 use Fig.5.10)			-100		dB
35. Calculate Minimum Δf (see Form A)				0	MHz
36. If $\Delta f > (B_T + B_R)/2$ (from line 35 use Fig.5.10)				0	dB

EMI FREQUENCY CORRECTED SUMMARY

	25	30	34	36	
37. Add line 21 to line					
38. Total Here	NA	-49	-29	-5	dB

IF EMI MARGIN < -10 dB, EMI HIGHLY IMPROBABLE

DETAILED CULL (see Sec. 5.3.2.4, 5.3.3.3 and 5.3.4.4)

HARMONIC OR SPURIOUS RESPONSE CORRECTION *

	FIM	TIM	RIM	SIM	
39. Correction from Form B		NA	NA	26	dB
40. Corrected EMI add line 38 to line 39*	NA	NA	NA	21	dB

MODULATION CORRECTION

41. Correction from Form C	NA	NA	NA	0	dB
42. Corrected EMI Add line 40 to line 41	NA	NA	NA	21	dB

POLARIZATION CORRECTION

43. Correction from Form D	NA				dB
44. Corrected EMI Add line 42 to line 44	NA	NA	NA	21	dB

PROPAGATION LOSS CORRECTION

45. Correction from Form E	NA	NA	NA	0	dB
46. Corrected EMI Add line 44 to line 46	NA	NA	NA	21	dB

IF EMI < -10 dB EMI HIGHLY IMPROBABLE
-10 dB $\leq$ EMI $\leq$ 10 dB EMI MARGINAL
EMI > 10 dB EMI PROBABLE

*Apply corrections only if nominal -60dB TX (line 9) and/or +80dB RX (line 15) corrections were used.

FORM A - SIM FREQUENCY SEPARATION CORRECTION in dB

TX Fund(f_{OT}): __220__ ; RX LO(f_{LO}): __400__ ; RX IF (f_{IF}): __40__

N	$(Nf_{OT} + f_{IF})/f_{LO}$	Δp*	$(Nf_{OT} - f_{IF})/f_{LO}$	Δp *
2	1.2	.2	1.0	0
3	1.75	-.25	1.55	-.45
4	2.30	.30	2.10	.10
5	2.85	-.15	2.65	-.35
6	3.40	.40	3.20	.20
7	3.95	-.05	3.75	-.25
8	4.50	.50	4.30	.30
9	5.05	.05	4.85	-.15
10	5.60	-.40	5.40	.40

Select Minimum Value For Δp

Minimum Spurious Frequency Separation = $(\Delta p)_{min} f_{LO}$

---

*Δp is the magnitude of difference between value obtained for $p \pm \Delta p$ = $N(f_{OT} \pm f_{IF})/f_{LO}$ and the nearest integer.

FORM B - HARMONIC AND/OR SPURIOUS RESPONSE CORRECTION IN dB

FIM - No Correction
TIM - Spurious Response Correction use p from line 26
RIM - Harmonic Correction use N from line 31
SIM - Add Spurious and Harmonic Correction use p and N from Form A above.

Category		N or p*								
		2	3	4	5	6	7	8	9	10
HF &	Harmonic	+19	+ 7	- 2	- 9	-14	-19	-23	-27	-30
Below	Response	-13	-17	-20	-22	-24	-26	-27	-29	-30
VHF	Harmonic	+ 6	- 8	-18	-26	-32	-37	-42	-46	-50
	Response	-15	-22	-26	-29	-32	-35	-37	-38	-40
UHF &	Harmonic	+ 5	- 4	-10	-15	-19	-22	-25	-28	-30
Above	Response	+ 8	+ 1	- 4	- 8	-11	-14	-16	-18	-20

*For the receiver image response use a +20 dB correction.

FORM C - MODULATION CORRECTION IN dB*

(Use Δf's from lines 24,29,33, or 35)

Modul Type	B_T line 18	B_R line 19	Modulation Correction in dB
Pulse			0
AM			$20\log(B_T+B_R)/2\Delta f$
FM			$40\log(B_T+B_R)/2\Delta f$

*The combined Frequency Separation Correction from lines 25,30,34,or 36 plus Modulation Correction must not exceed 100 dB

FORM D - ANTENNA POLARIZATION CORRECTION IN dB *

TX Pol: ☐ H; ☐ V; ☐ C.
RX Pol: ☐ H; ☐ V; ☐ C.

	TX	Horizontal		Vertical		Cir
RX		G<10dB	G≥10dB	G<10dB	G>10dB	
Hor	G<10dB	0	0	-16	-16	-3
	G≥10db	0	0	-16	-20	-3
Ver	G<10dB	-16	-16	0	0	-3
	G≥10dB	-16	-20	0	0	-3
Circular		-3	-3	-3	-3	0

*Applies only to intentional radiation region and design frequency

Form E - Propagation Loss Correction in dB

- Sketch Path Profile with TX and RX Antennas on Chart Below

- Check for Line of Sight Between TX and RX

- If Line-of-Sight Exists, Propagation Loss Correction = 0

- If Line-of-Sight is Blocked Calculate Line of Sight Distance, d_{LOS}, from TX to Terrain

- Propagation Loss Correction (dB) = $40 \log_{10} d_{LOS}/d$
  where d = Distance From TX to RX.

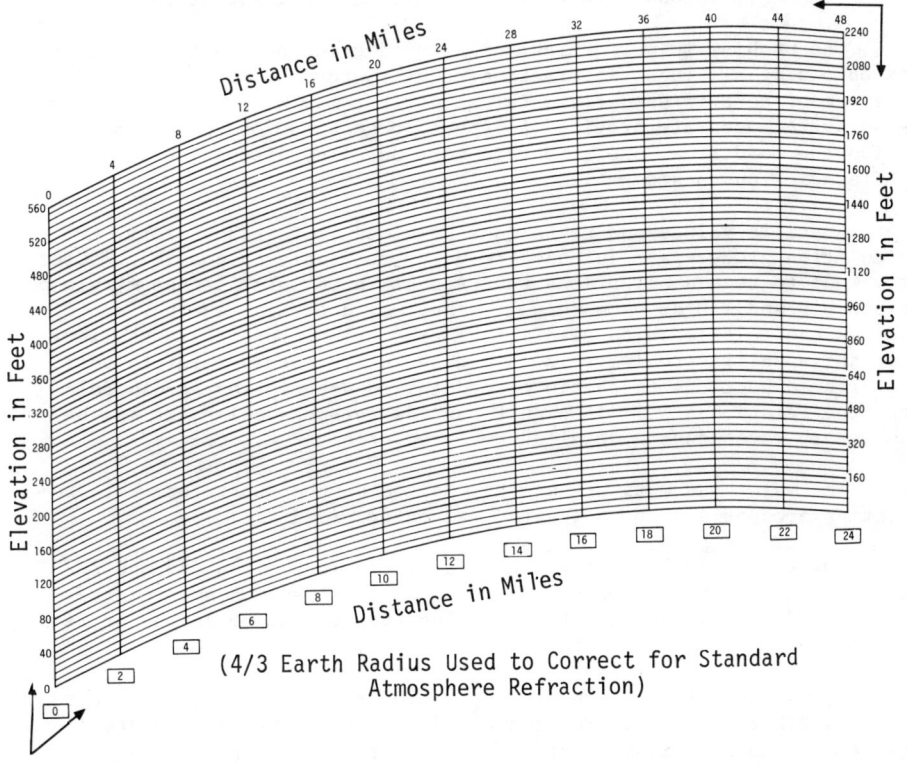

(4/3 Earth Radius Used to Correct for Standard Atmosphere Refraction)

A correction or adjustment is then made as the first step of the frequency-cull process using Figs. 5.11 or 5.12 to account for differences between the transmitter and receiver bandwidths. Entries are made on lines #17 through #21 of the short form. In the next step, a correction is made to account for frequency separation between the particular transmitter output and receiver response. Lines #22 through 38 are used for this purpose. In this step, if specific transmitter and receiver frequencies are not known, calculations may be repeated for different frequencies, if desired. Alternately, the relationship plotted in Fig. 5.10 may be used to determine the frequency separation required for compatibility.

If none of the four EMI prediction cases survive after totaling entries on line #38, then the prediction process ends, and there exists no inter-system EMI problems for the particular TX-RX pair. However, should one or more cases survive (entry on line 38 be algebraically greater than -10dB), then the prediction continues into the third-level of culling, *detail cull*. These entries are listed as lines #39 through 46 on the second page of the short form. The supporting sub-forms, forms A-E, for detail culling are listed on the third and fourth pages of the short form. Details regarding the use of these sub-forms are discussed in Vol. 5 of this handbook series.

One problem in performing an EMI prediction of the type described is obtaining all input information required for the prediction. In the absence of certain specific data values that may be used are provided. Also, references are provided to specific sections of this chapter which the reader may use to obtain more detailed discussion of various EMI prediction considerations.

Specific values and techniques provided in the short-form, EMI prediction process are intended to be as realistic as possible. Users may want to regard situations that result in an interference margin between -10 and +10 dB as marginal. They may determine whether specific values and relationships used in the prediction process are realistic for the specific problem. In this case, users should refer to the referenced sections to obtain more detailed descriptions and models for the relationships. If significant differences exist, it may be necessary to conduct a detailed EMI prediction.

The major assumptions used or implied by the short-form EMI prediction process and the suggested values provided are:

● Frequency limits for transmitter spurious emissions and receiver spurious responses are from 0.1 to 10 times fundamental frequency. This assumes that there are no significant emissions or responses outside these limits.

● Maximum TX-RX frequency separation for fundamental interference is 0.2 times receiver fundamental. This assumes fundamental interference is not significant for larger frequency separations.

● Free space propagation loss is assumed in amplitude culling.

● Levels for transmitter spurious emissions are 60 dB below fundamental emission.

● Levels for receiver spurious susceptibility are 80 dB above fundamental susceptibility.

● An additional 20 dB rejection each is assumed for transmitter and receiver minor emissions and responses.

● Values for antenna gains in unintentional radiation directions and at unintentional frequencies are 0 dB.

● Differences in transmitter and receiver bandwidth are assumed to modify the power available in the manner specified in Table 5.15.

● Frequency separation, Δf, between transmitter emission and receiver response are assumed to reduce the effective power available by an amount given by $40 \log_{10} [(B_T + B_R)/2\Delta f]$.

● A go/no-go cull level of −10 dB is used. Thus, potentially interfering situations are culled only if the mean signal level is less than −10 dB relative to the receiver susceptibility threshold.

Table 5.15 - Bandwidth Corrections in dB

Modulation Type	Bandwidth Conditions	On-Tune $\Delta f \leq (B_T+B_R)/2$	Off-Tune $\Delta f > (B_T+B_R)/2$	Remarks
Noise Like Continuous	$B_R \geq B_T$	No Correction	$10 \log_{10}\left(\dfrac{B_R}{B_T}\right)$	RMS Power Proportional to Bandwidth
	$B_R < B_T$	$10 \log_{10}\left(\dfrac{B_R}{B_T}\right)$		
Pulse	$B_R \geq B_T$	No Correction	$20 \log_{10}\left(\dfrac{B_R}{B_T}\right)$	Peak Voltage Proportional to Bandwidth
	$PRF < B_R < B_T$	$20 \log_{10}\left(\dfrac{B_R}{B_T}\right)$		
	$B_R < PRF$	$20 \log_{10}\left(\dfrac{PRF}{B_T}\right)$	$20 \log_{10}\left(\dfrac{PRF}{B_T}\right)$	Power in $B_R < PRF$

A relatively detailed application of the short-form, EMI prediction process is presented below. It is suggested that the reader review Illustrative Example 5.25 and/or Vol. 5 before attempting to use the process if the use of specific parts of the form are not clear.

Illustrative Example 5.25

Consider the potential EMI which may exist between a P-Band radar transmitting at 220 MHz and UHF AM voice receiver tuned to 360 MHz used for communicating with aircraft. Assume that both the P-Band radar and the UHF receiver are located on the same ship, and their antennas are only 100 ft. apart, but are located with sufficient vertical separation so that the radar antenna main beam does not illuminate the UHF antenna. Assume that the nominal TX-RX characteristics are:

Parameter	P-Band Radar	UHF Receiver
Operating Frequency (f_O)	220 MHz	360 MHz
Peak Power Output	100 kW	NA
Pulse Width (τ)	10 μsec	NA
Pulse Repetition Rate	100 pps	NA
Transmitter Bandwidth ($2/\pi\tau$)	64 kHz	NA
Antenna Gain	23 dB	0 dB
Receiver Bandwidth	NA	10 kHz
Receiver Sensitivity	NA	-100 dBm
Receiver IF	NA	40 MHz
Receiver LO	NA	($f_{OR} + f_{IF}$)

The EMI prediction short form is exercised for this problem and the results are shown written in hand in the preceding pages. Referring to the short form, the spurious (SIM), receiver (RIM), and transmitter (TIM) interfering cases are applicable to this problem. The fundamental (FIM) case is not applicable because the fundamental frequency separation is greater than $0.2 f_{OR}$ = 0.2(360 MHz) = 72 MHz.

As a result of performing the amplitude cull, a positive EMI margin is obtained for the TIM, RIM, and SIM cases as shown on line #16. Hence, it is necessary to continue with the frequency cull. When the frequency cull is applied, the corrected TIM and RIM are both below the -10 dB cull level as shown on line #38. Hence, EMI is highly improbable for these cases and it is not necessary to examine them further in the detailed cull. On the other hand, the frequency-cull corrected SIM case results in an EMI margin that is above the -10 dB cull level (-5 dB on line #38). Hence the SIM case must be examined further during the detailed cull.

When the detailed cull is applied to the SIM case, it is seen that a +26 dB correction is obtained for the harmonic or spurious response correction as shown on line #39. This positive correction results because the SIM case in this particular example involves the transmitter second harmonic (N=2; sub-form A) interfering with the receiver image response. The transmitter second harmonic results in a +6 dB harmonic correction for VHF equipments. This positive correction results because second harmonics tend to exceed the nominal level (i.e., $P_T(f_{OR})$ - 60 dB) used on line 9. Receiver image responses result in a +20 dB correction because receiver tends to be more susceptible to the image than the nominal level (i.e., $P_R(f_{OR})$ + 80 dB) used in line #15. Application of the detailed cull results in a +21 dB EMI margin for the SIM case as shown in lines #40 and 46.

To summarize results of this illustrative example, EMI is probable for the particular conditions considered. Potential EMI results from the transmitter second harmonic existing at the receiver image response frequency of 440 MHz. This potential EMI problem may be avoided by changing the receiver operating frequency so that there is sufficient frequency separation between the transmitter second harmonic and the image response to provide a −31 dB rejection required to reduce the EMI margin below the −10 dB cull level. Referring to Fig. 5.10, if the output-response frequency separation for $(B_T + B_R)/2$ = 37 kHz is greater than 200 kHz, more than 31 dB of rejection will be provided and the potential EMI problem will be eliminated. For example, if the receiver is tuned to a frequency of 360.5 MHz, there should be no EMI problem resulting from the transmitter second harmonic and the receiver image response.

5.5   INTER-SYSTEM EMI CONTROL

Control of EMI should take place at the earliest stage possible of the life cycle of an equipment or system. This is often accomplished in part during the planning and design stages. Yet when an equipment or system, designed and tested for EMI control compliance in accordance with EMI specifications, is put into its operational environment, EMI often develops all over again. This does not mean that the original *intra-system* EMI control program was for naught.* Rather, the number of *inter-system* EMI problems either will be significantly less or their magnitude will be less severe. Inter-system EMI problems are a natural consequence of progress in our modern society.

There are many EMI controls that may be carried out in the design stage to enhance the chances of inter-system EMC. Some of these provide for flexibility and options during the system operational stage of its life cycle. Fig. 5.55 summarizes several EMI control techniques that are available to the system engineer/designer and user. The following discussion summarizes these techniques.

5.5.1   FREQUENCY MANAGEMENT

The first of the principle EMI control techniques shown in Fig. 5.55 is captioned *Frequency Management*. This suggests both transmitter emission control in the electromagnetic spectrum and hardening of receivers against spurious responses. The object here is to design and operationally maintain transmitters so that they occupy the least frequency spectrum possible in order to help control electromagnetic pollution. R-F modulation bandwidth should be no greater than that necessary to accommodate only the transmission base-band intelligence. Ideally, this means that the 100-dB bandwidth of a transmitter's emission spectrum would be very nearly the same as that of the 3-dB bandwidth since emitted energy in the far sidebands do not contribute to system performance. Rather it contaminates the spectrum and generates latent interference problems to either adjacent channel or out-of-band receptors.

When the modulating base-band source is of a pulse type, modulation bandwidth conservation implies that long pulse rise and fall times are used. Practices better than trapezoidal pulses (falloff of 40 dB/decade) are those that employ either gaussian (60 dB/decade falloff) or cosine squared (80 dB/decade falloff) modulations. This represents low pollution to the electromagnetic spectrum. Thus, pulses having small slopes and no discontinuity in the first derivative should be used wherever possible in the design of the base-band modulator.

---

* This may be compared with a family and its relation to its neighborhood. If a family cannot get along with itself, its neighborhood relations are academic. When a family is compatible, its chances of neighborhood compatibility are significantly enhanced.

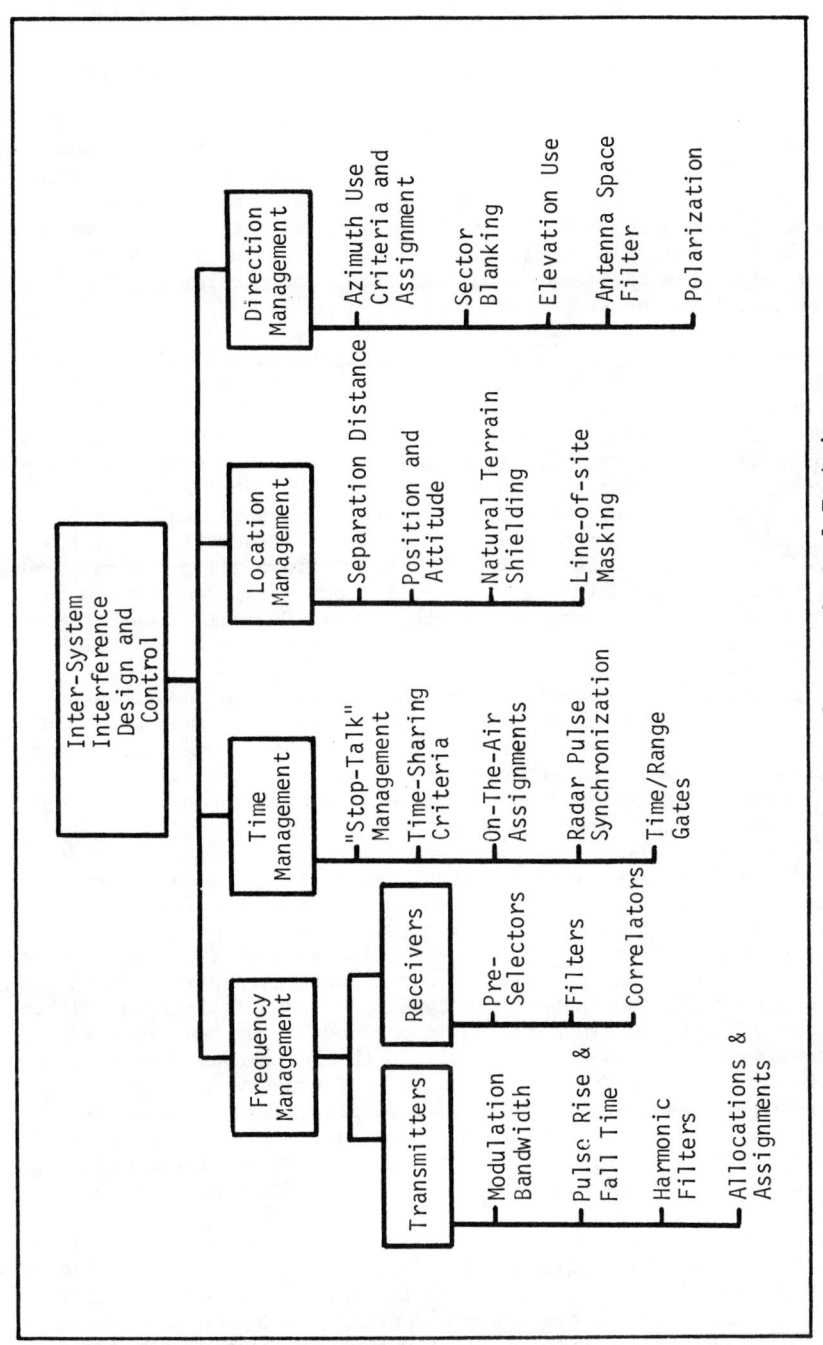

Figure 5.55 – Inter-System Electromagnetic Interference Control Techniques

Since all devices operate to some extent in a non-linear manner,
every transmitter will exhibit some harmonic emissions. The first few
harmonics of typical communications-electronics equipments are approximately 50 to 70 dB down from fundamental emission levels (see Table
5.5). Since these emission sources often constitute major interference
problems, it is logical that their emission levels should be further
reduced in system design whenever practicable. Therefore, one common
practice is to insert a harmonic filter into the output transmission
line which will pass the fundamental emissions and present 30 dB or
more attenuation to the second and higher harmonics.

One of the most important EMI control techniques shown under frequency management in Fig. 5.55 is frequency allocations and assignments. Frequency *allocations* here refers to setting aside of one or
more spectrum bands which are generally dedicated for a particular user
function. These are identified in Chap. 1. Frequency *assignment* refers
to the designation of one or more transmission frequencies (i.e., carrier) to a particular user within a frequency allocated band. For
military and other government functions in the continental U. S., this
is granted by the frequency control officer within a given region. For
non-government users, including industrial, commercial, and consumer,
frequency assignments are granted by the Federal Communications Commission.

The impact on EMI control through both frequency allocations and
assignments has an enormous effect and is the topic of some books and
many articles. It is not discussed further in this volume. Quite often
one of the most convenient, economic and rapid solutions to an EMI
problem in the field, is to change frequency of either the victim receiver or the culprit source.

Under the frequency management topic shown in the figure, are
receivers which represent the victims of electromagnetic interfering
emissions. Since there can be no interference problem without susceptibility of receptors in general or receivers in particular, they
equally contribute to EMI as their transmitting counterparts. Most
receivers are of the superheterodyne type. Accordingly, they are inherently susceptible to intermodulation, cross modulation, and more generally spurious responses developed within the first converter (mixer).
To help curtail some of these problems, one alternative may be not to
use a superheterodyne receiver in the first place, but to use, for
example, a tuned-radio-frequency (TRF) receiver. There is evidence of
a trend in technology away from superhet and to TRF receivers made
possible by high-Q, electronically-tunable and other components.

One classical solution to mitigating intermodulation and spurious
response problem within superheterodyne receivers is to use preselectors.
While not very economical, preselectors limit the tendency for intermodulation in the receiver front end and limit the degree of off-frequency heterodyne mix products within the first converter. Thus,

preselectors are good design measures for other than low selectivity (broadband) receivers.

A sound design practice in receiver front ends is to use a low-pass filter having its cut-off frequency somewhat above the highest frequency to which the receiver is tunable and offering substantial attenuation to harmonics of the receiver's local oscillator. Sometimes a low-pass filter is designed to have a much higher cut-off frequency than the upper band edge. Here the low-pass filter attenuation characteristics take over at those frequencies for which the preselectors performance degenerates.

Many receivers today use one or more types of correlators. While these are primarily intended to improve signal-to-noise ratio for normal applications, correlators also reject other types of unwanted modulations from interfering emitters. Among the various forms of correlators or correlating-like equipments are phase-lock loops, video-enhancement delay lines, pulse-width discriminators, double-threshold detectors, and other anti-jamming devices.

5.5.2 TIME MANAGEMENT

The previous section on frequency management stressed the frequency occupancy and spectrum pollution control aspects of EMI suppression techniques. This section involving time management emphasizes the time domain or temporal aspects of EMI control.

One significant EMI problem in C-E equipments, especially communications types, is that there exists little control on the amount of *chatter* or conversations between the user of a transmitter and its mating receiver. This is shown in Fig. 5.55 as *stop-talk* management. One example of good practice is the abbreviated conversation that takes place between the air-traffic controller and the pilot of an aircraft. Here, conversation is both brief and rather cryptic in that special jargon and nomenclature is used to shorten the average message length. Furthermore, conversation between the controller and the pilot only takes place on an absolutely necessary basis. Thus, idle chatter is eliminated and many more users can use the same channel. In contrast to the aircraft situation many messages between two users of handy talkie radios in a battlefield situation often ramble on and contain unnecessary information. Thus, the amount of time and spectrum occupancy is considerably greater since people in general tend to talk too much especially under hazardous conditions.

As a practical matter it is not likely that *stop-talk* management will become an effective tool for EMI control unless the using organization establishes criteria and discipline with regard to the frequency of use and message duration of its communication system. When messages are of short duration and fairly infrequent per user, such that the average duty cycle is low, it then becomes practical to time share the use of a particular channel or portion of the spectrum. This is evident in land-mobile applications where a dispatcher may be communicating with dozens of vehicles. While the dispatcher's duty cycle may be quite high servicing all users, any one vehicle transmitter is on-the-air a relatively small portion of the time. Thus, by time-sharing techniques it is practical for many users to use this party-line system in a successful manner.

Another aspect of time sharing one or more frequency bands is to make formal assignments of on-the-air criteria. For example, several users having different interests may share a common band with each user assigned particular hours of the day. During any time of the day only one user or using group has authority to transmit. On a 24-hour basis then, several users have an opportunity to use the spectrum. Sometimes this is achieved on a *not-to-interfere basis* in which one user has priority during certain hours of the day but other users may also use that band during this period. This has the provision that if co-channel interference exists, non-assigned users must get off-the-air instantly to make the channel available for the assigned user.

Where two or more co-site radars must operate in and around portions of the same frequency band, one method for curtailing EMI is to have a master oscillator or clock which triggers all radars at the same time. Since the very first part of the pulsed transmission is blanked in radar receivers, none of the radar transmitters can cause interference to any of the receivers. A variation of this scheme is to use either a time or a range gate. Time gating or blanking involves shutting off either the transmitter or receiver for a brief moment when interference otherwise exists from another radar. Range gates, such as employed in fire-control and tracking radar, basically make the receiver mute to interfering sources whose time of emission arrival appears outside of the target range return gate. These are but a few of the many techniques under the topic of time management available to the EMI design and/or applications engineer.

5.5.3 LOCATION MANAGEMENT

Location management here refers to EMI control by the location of the potentially victim receiver with respect to all other emitters in an electromagnetic ambient environment. If the separation is very substantial and/or if there exists protective intervening terrain interference propensity will be greatly mitigated. In this regard, separation distance between transmitters and receivers is one of the most significant forms of control since interfering source emissions are reduced at the rate of 20 dB/decade of separation within line of sight and may approximate 60 dB/decade beyond line of site at VHF and higher frequencies.

The relative position of potentially interfering transmitters to the victim receiver are also significant. If the emitting source and victim receiver are shielded by natural intervening terrain or by a billboard, bulkhead, or other obstacle, the degree of interference would be substantially reduced. Thus, one of the most significant objectives for EMI control is to reduce or eliminate line-of-sight conditions by intervening masking techniques.

5.5.4 Direction Management

Direction management refers to the technique of EMI control by gainfully using the direction and attitude of arrival of electromagnetic signals with respect to the potential victim's receiving antenna. Examples of this include relative azimuth and elevation angles between the different transmitter and receiver antennas. By selecting a receiver at a different elevation angle interference from emitting sources can be reduced. Correspondingly, terrain masking may be achieved in certain azimuth directions from known significant emitting sources. Sector blanking may be used, such as for radars, where over relatively small azimuthal angles the transmitter is turned off in order to protect one or more receivers. This is especially important where transmitters and receivers are co-sited in small real estate areas such as shipboard use in the Navy.

One of the most effective uses of direction management is that of the antenna, per se, in which it may be regarded as a space filter. The antenna offers enhancement to the transmitting signal in its boresight axis and may offer considerable rejection off-axis at either azimuth or elevation angles to either potentially interfering sources or receivers. Thus, if an antenna is properly designed to impede EMI, its off-axis rejection to interfering sources would be at levels considerably below isotropic. In addition to a space filter, the antenna may act also as a frequency filter. Here, the antenna becomes badly defocused for frequencies well removed from the intended operational frequency. Thus, this acts in consonance with the preselectors and low-pass filters of receivers under the frequency-management concept.

Another important aspect of direction management involves the polarization of the arriving electromagnetic wave with respect to the receiving antenna. Pragmatically, cross-polarized signals may offer up to 30 dB rejection by the polarization of the receiving antenna. While this may not sound very significant, it is widely used as a method of EMI control in communications, especially microwave relay.

The foregoing is not intended to be complete but merely provide an overview of some EMI control techniques available to the intersystem designer and user. It is suggested that the reader may want to refer to Vol. 5 of this handbook series for additional information.

5.6   BIBLIOGRAPHY

(1)   Arnold, J. G., "Predicting Spurious Transmitter Signals," *Electronics*, Vol. 34, No. 16, p. 68, Apr. 21, 1961.

(2)   Babcock, W. C., "Intermodulation Interference in Radio Systems," *Bell System Technical Journal*, Vol. 32, No. 1, p. 63, Jan. 1953.

(3)   Baghdady, E. J., *Lectures on Communication System Theory*, New York: McGraw-Hill, 1961.

(4)   Beauchamp, A. J., "A Technique of Intermodulation Interference Determination," *IRE Convention Record*, Part 8, Information Theory, Paper No. 22.5, pp. 26-29, New York, Mar. 1953.

(5)   Crichlow, W.Q., Smith, D.F., Morton, R.N., and Corliss, W.R., "Worldwide Radio Noise Levels Expected in the Frequency Band 10 Kilo-cycles to 100 Megacycles," *National Bureau of Standards Circular 557*, Aug. 25, 1955.

(6)   Egli, J. J., "Radio Propagation Above 40 Mcs Over Irregular Terrain," *Proc. IRE*, Vol. 45, No. 10, pp. 1383-1391, Oct. 1957.

(7)   Firestone, W., MacDonald, A., and Magnuski, H., "Modulation Sideband Splatter of VHF and UHF Transmitters," *Proc. National Electronics Conference*, Oct. 4-6, 1954, Vol. 10, pp. 264-273, Feb. 1955.

(8)   Haber, F., "Study of GSFC Radio Frequency Interference (RFI) Design Guideline for Aerospace Communication Systems," Moore School of Electrical Engineering, University of Pennsylvania, Moore School Report No. 66027 for NASA, pp. 143-147, Apr. 30, 1966.

(9)   Haber, F. and Epstein, B., "The Parameters of Nonlinear Devices from Harmonic Measurements," *IRE Trans. on Electron Devices*, Vol. ED-5, No. 1, pp. 26-28, Jan. 1958.

(10) Heisler, K. G., and Hewitt, H. J., "Interference Notebook," Rome Air Development Center Report No. RADC-TR-66-1, June 1966.

(11) *Interference Reduction Guide for Design Engineers*, Prepared by Filtron Co., Inc., for U. S. Army Electronics Labs., Fort Monmouth, N. J., Vol. 1, pp. 2-116 and 2-128, Aug. 1964, Accession No. AD 619666.

(12) *Interference Reduction Techniques for Receivers*, Quarterly Report No. 4, Contract DA-36-039-AMC-02345(E), Radio Corp. of America, Sept. 30, 1964, AD 455117.

(13) Krstansky, J. J. and Elsner, R. F., "Environment-Generated Intermodulation in Communication Complexes," *Proc. Tenth Tri-Service Conference on Electromagnetic Compatibility*, Chicago, Ill.:   ITT Research Institute, pp. 77-99, Nov. 1964.

(14) Marcus, Robert B., "The Analysis and Synthesis of Radar Emission Spectrums by Digital Computer Techniques," *Proc. Ninth Tri-Service Conference on Electromagnetic Compatibility,* Chicago: Armour Research Foundation, pp. 231-260, Oct. 1963.

(15) McLenon, D., "Measurements of Communication Receiver Interference Vulnerability," *Proc. of the Unclassified Sessions of the Symposium on Electromagnetic Interference*, U. S. Army Signal Corps Research and Development Laboratory, Fort Monmouth, N. J., June 15, 1958.

(16) Otto, J. C. and Garcia, R. R., "Interference Reduction Techniques for Nonlinear Devices," General Electric Co., Final Report on Contract DA 63-039-AMC-02278(E), May 1964.

(17) Schwartz, M., *Information Transmission, Modulation, and Noise*. New York: McGraw-Hill, 1959.

(18) Smith, J. S. and Shepherd, N. H., "The Gaussian Curve-Transmitter Noise Limits Spectrum Utilization," *Proc. of the Unclassified Sessions of the Symposium on Electromagnetic Interference*, U. S. Army Signal Corps Research and Development Laboratory, Fort Monmouth, N. J., pp. 138-144, June 15, 1958.

(19) Tomiyasu, K., "On Spurious Outputs from High Power Pulsed Microwave Tubes and Their Control," *IRE Trans. on Microwave Theory and Techniques*, Vol. MTT9, No. 6, pp. 480-484, Nov. 1961.

(20) Trammell, R. D., Jr., "A Method for Determining Mixer Spurious Response Rejection," *IEEE Trans. on Electromagnetic Compatibility*, Vol. EMC-8, No. 2, pp. 81-89, June 1966.

CHAPTER 6

CABLE WIRING AND HARNESSING

CHAPTER 6
CABLE WIRING AND HARNESSING

This section reviews electromagnetic interference (EMI) situations that result from near-field induction coupling between two or more wires or wire pairs, at least one of which is a potential culprit source and the other(s) the victim wire feeding a susceptible circuit or device. Such wires are generally a common cable harness or located near each other in separate cable runs. Mathematical models of coupling coefficients between these wires as a function of separation distance, height above a ground plane, length, return paths, impedance levels, and frequency are present. Since nearly all wires and cables are terminated in connectors, EMI performance and control techniques regarding the choice and use of connectors are also presented.

Undesired proximity coupling among wires and cables interconnecting networks, chassis, and equipments is one of the principle causes of EMI. Accordingly, this section reviews the subject from the topics of classes of wire and harnesses, magnetic-field coupling, electric-field coupling, cable shields and their grounding, quick-fix de-coupling techniques, and connectors.

6.1 WIRING CLASSES AND HARNESSING

Cables may handle 10 kw (+70 dBm) of 60 Hz, AC power mains supply or 1 Mw (+90 dBm) of VHF/UHF peak R-F transmitter power. On the low-level side, cables may be used to connect an antenna to a receiver-input having a sensitivity of -120 dBm. Thus, this power range covers about 200 dB and presents an enormous EMI coupling threat to low-level circuits from high-level emission sources. It follows, then, that one major aspect of EMI-control is to separate wires and cables into similar classes of power handling and susceptibility levels.

MIL-W-5088C, MIL-STD-461A (Notice 4) and other specifications attempt to classify wiring into four to six groups in order to minimize EMI coupling (maximize isolation between classes). However, the classification seems to be more qualitative and presented in the form of design guides, such as:

Category I - power wiring (AC and DC primary power)

                Category II - secondary power wiring (low voltage and
                                regulated voltage sources)
                Category III - control wiring (transient sources or loads)
                Category IV - sensitive wiring (audio and video circuits)
                Category V   - susceptibile wiring (RF and safety circuits)

    AFSC Design Handbook on EMC* suggests the classification listed
in Table 6.1.

Table 6.1 - Air Force Wiring Classification

Class	Identification	Voltage Current or Power	Frequency
I	DC power circuit DC control circuit	>2 amperes <2 amperes	0 0
II	DC reference circuit AF susceptible circuits	<1 volt or <0.2 amperes	0
III & IV	AC power circuits AC reference circuits AF source circuits	>1 volt or >0.2 amperes >0.2 amperes	<400 Hz <400 Hz <15 kHz
V	RF susceptible circuits		
VI	EMI source circuits	>−45 to −75 dBm >−75 dBm >−75 to −45 dBm >−45 dBm	.15 to 5 MHz 5 to 25 MHz 025 to 1 GHz >1 GHz
VII	Antenna Circuits		

    In order to try to be more quantitative about classification, it
may be more beneficial to classify wiring and cables into levels of
power transmitted or susceptibility of termination as applicable. One
such classification would be achieved by dividing the above 200 dB
power-level range into approximately six equal steps of about 30 dB
each. Table 6.2 is the result of such grouping.

    In Table 6.2, the classification has the advantage that:

        (1) EMI sources and receptors tend to group separately with
a power level dichotomy existing at about −20 dBm.
        (2) Power levels in adjacent wires in a bundle or harness will
not likely exceed a 30 dB spread.

    At first look the classification evidences no particular considera-
tion to either frequency or impedance. As explained later, magnetic
    .

_____

* Air Force Systems Command Design Handbook No. DH 1-4, *Electromagnetic Compatibility.*"

Table 6.2 - Wiring Classification by 30 dB Power-Level Groupings

Class	Power Range	Identification
A	>40 dBm	High Power DC/AC and R-F Sources
B	+10 to +40 dBm	Low Power DC/AC and R-F Sources
C	-20 to +10 dBm	Pulse and Digital Sources Video output circuits
D	-50 to -20 dBm	Audio and sensor susceptible circuits Video input circuits
E	-80 to -50 dBm	RF and IF input circuits Safety Circuits
F	<-80 dBm	Antenna and RF Circuits

field coupling predominates for low-impedance circuits at low frequency while electric-field coupling is greater for higher-impedance
circuits at high-frequency. Yet the above is consistent with this in
that audio and video sources (Class C) and circuits (Class D) have
their own lower frequency classification while I-F and R-F circuits
(Classes E and F) are grouped into different classes at high frequency.
The principle sources of EMI (Classes A and B) are removed from all
susceptible circuits.

Illustrative Example 6.1

A digital computer-controlled surveillance antenna-receiver
system is designed to receive signals in the VHF/UHF spectrum. Providing R-F sensitivities of -90 dBm, the signals are converted to IF in
a double-balanced mixer (no R-F gain). The local-oscillator drive provides 1 mw of R-F power through a 10-dB isolation pad. The receiver
subsystem is instructed by digital words from the computer whose bit
levels are 0 volts (binary 0 or space) and 3 volts (binary 1 or mark)
as obtained from a differential transmitter-receiver line operating at
an impedance level of 120 ohms. Receiver analog outputs are then digi-
tized and sent over similar lines. The AC power mains source is three-
phase, four-wire, 400 Hz. A 1-kw power load is consumed. Finally, all
DC regulated power is generated locally at each "black box" at ± 12
VDC and with current drains varying from 10 ma to 100 ma per box. The
problem is to determine how the interconnecting wiring system cable
harnesses may be grouped.

The following levels result from the above. Some are obtained by
converting voltages and their associated impedances into power. They
may be classified in ascending order in accordance with Table 6.2:

Identification	Power Levels	Wire Class
Antenna-receiver Input Lines	-90 dBm	F
I-F Amplifier Input Lines	-80 dBm	F
Video Input Lines	-30 dBm	D
Video Output Lines to Digitizer	+13 dBm	B
Digital TX-RX Lines	+20 dBm	B
DC Regulated Lines	20 to 30 dBm	B
AC Power Main Lines	+60 dBm	A

Four classes of wire harnessing are indicated. Ordinarily Class F types will not be harnessed together since they are usually located in different areas. Where it is difficult to have more than two or three classes of wiring harnesses, two adjacent classes may be grouped together with a resulting spread of 60 dB between power levels. Never group non-adjacent classes.

6.2 MAGNETIC-FIELD COUPLING

This section reviews electromagnetic interference (EMI) situations that result from near-field magnetic and electric-field induction coupling between two or more wires or wire pairs, at least one of which is a potential culprit source and the other(s) the victim wire(s) feeding a susceptible circuit or device. Such wires may be located in either a common cable harness or near each other in separate cable runs. Mathematical models of coupling coefficients between these wires as a function of separation distance, height above a ground plane, length, return paths, impedance levels, and frequency are presented.

Wires which are close to each other may couple energy from higher level circuits to lower-level circuits by induction-field coupling. This is the physics-coupling mechanism in the near-field* and is divided into magnetic-field and electric-field coupling. This section discusses the former process, presents mathematical models with some empirical data, and describes how the EMI threat may be reduced or eliminated.

6.2.1 COUPLING PHYSICS

Fig. 6.1 shows two nearby wires illustrating the principle means of undesirable magnetic coupling.

Amperes Law states that:

$$\int_0^{2\pi r} H \cdot ds = I \qquad (6.1)$$

where, H is the magnetic-field intensity around a wire carrying a current, I. If the right-hand clutches the wire and the thumb points

---

* See Chap. 5 and Vol. 4, Chap. 3

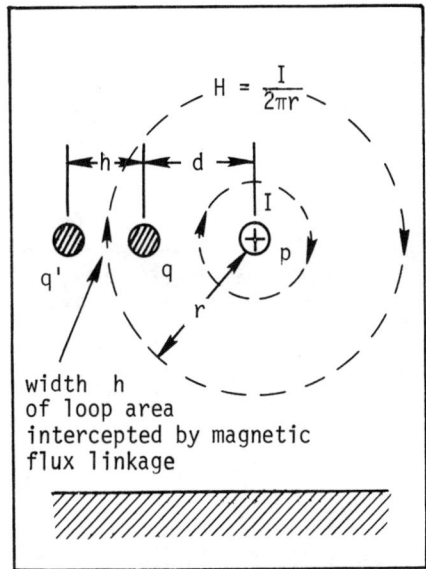

Figure 6.1 - Magnetic Coupling between Source Wire p and Loop Area in Circuit q-q'

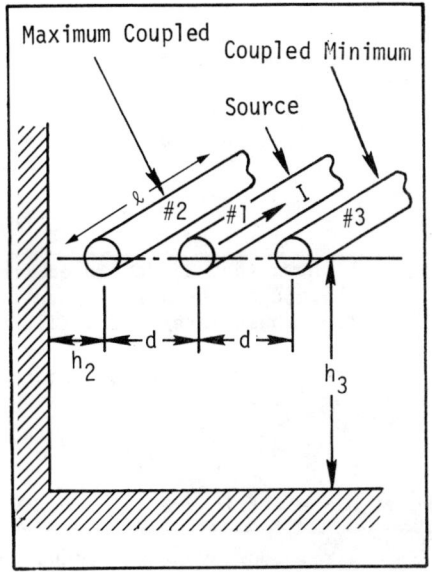

Figure 6.2 - Maximum and Minimum Magnetic Coupling

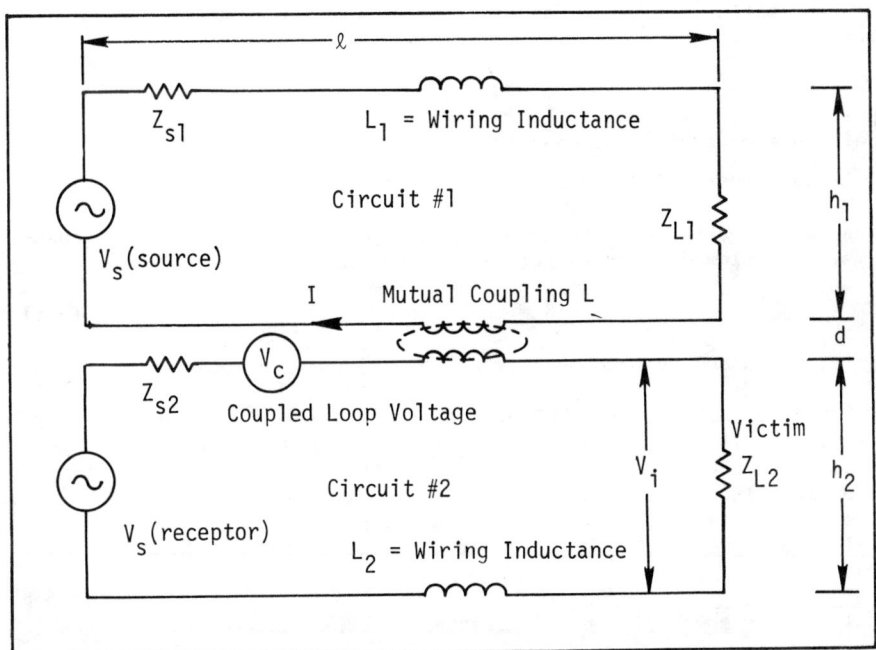

Figure 6.3 - Inductive (Magnetic-Field) Coupling Between Wires

in the direction of current flow, the curled fingers point in the direction of the magnetic-field intensity.

For a circular path, $2\pi r$ (r = any radius), the magnetic-field intensity at a radial distance r in meters is obtained from Eq. (6.1). Thus:

$$H = \frac{I}{2\pi r} \quad \text{Amperes/meter} \tag{6.2}$$

Fig. 6.1 shows that H, developed from wire p encompasses wire q when $r > d$, where d is the wire separation. The ground plane, if not made of a ferrous-base material, will have no effect on H.

A second law, important in understanding the coupling mechanism, is Faraday's law:

$$\int E \cdot ds = -\frac{d\phi}{dt} = -L \frac{dI}{dt} \tag{6.3}$$

where, E = electric-field intensity in Volts/meter

   V = induced voltage around the closed path of the loop,
       formed by outgoing and return wire pair q-q'

   ϕ = magnetic flux passing through or linking the loop.

The magnetic flux is calculated by integrating the flux density over an area, A:

$$\phi = \int B \cdot da \tag{6.4}$$

where, B = magnetic flux density

   da = incremental loop area intercepted by B

Finally, a fourth relation completes the necessary fundamental mechanisms. The flux density is related to the magnetic field:

$$B = \mu H \quad \text{Tesla*} \tag{6.5}$$

where, μ = absolute permeability of the surrounding medium

Fig. 6.2 shows two coupling situations. Circuit #2 is formed by wire #2 and its return path through the left ground plane. The magnetic field from wire #1 is everywhere orthogonal to the plane of the resulting circuit #2 loop area, $A = h_2\ell$. Circuit #3, is formed by wire #3 and its lower ground plane return path. However, the magnetic field from wire #1 is not everywhere orthogonal to the resulting loop

---

* 1 Tesla = 1 weber/m^2 = 10^4 gauss = 240 dBpT

area, $A = h_3\ell$. Only the former situation (worst coupling case) is considered here which is the same as that of Fig. 6.1 when the three wires are in the same plane.

For the situation shown in Fig. 6.2, in which the loop area $A = h_2\ell$, Eq. (6.4) becomes by substitution in Eqs. (6.2) - (6.5):

$$\phi = \int_d^{d+h} B \cdot \ell dr = \mu\ell \int_d^{d+h} H \cdot dr \tag{6.6}$$

$$= \frac{\mu\ell I}{2\pi} \int_d^{d+h} \frac{dr}{r} = \frac{\mu\ell I}{2\pi} \ln_e \left(\frac{d+h}{d}\right) \tag{6.7}$$

where, ℓ = length of wire loop

   h = height or width of wire loop

   d = separation of the two wire circuits

   dr = incremental height of wire loop

Substituting Eq. (6.7) into (6.3) yields:

$$V_c = -\frac{d\phi}{dt} = -\frac{\mu\ell}{2\pi} \ln_e \left(\frac{d+h}{d}\right) \frac{dI}{dt}$$

$$= -\mu\ell f I \ln_e \left(\frac{d+h}{d}\right) \tag{6.8}$$

where, $\left|\frac{dI}{dt}\right| = \left|\frac{d}{dt} (I \cos \omega t)\right| = \omega I = 2\pi f I \tag{6.9}$

Eq. 6.8 indicates that the induced voltage in circuit 2 increases with permeability of media, line length and wire loop height (loop area), culprit source current, I, and frequency. It decreases with separation of wires for $d \gg h_2$. Eq. 6.8 suggests that the magnetic-field induced voltage would become arbitrarily large with an increase in frequency. However, the self inductance in the wire circuits prevents this as illustrated in Fig. 6.3, whereby the EMI voltage, V_i, developed across a victim load, Z_{L2}, is:

$$V_i = \frac{Z_{L2} V_c}{Z_{S2} + Z_{L2} + j\omega L_2} \tag{6.10}$$

$$= \frac{Z_{L2} V_c}{Z_{S2} + Z_{L2}} \quad \text{for } \omega L \ll Z_{S2} + Z_{L2} \tag{6.11}$$

6.7

$$= \frac{Z_{L2}V_c}{2\pi f L_2} \text{ for } \omega L_2 \gg Z_{S2} + Z_{L2} \tag{6.12}$$

Substituting Eq. (6.8) into Eqs. (6.11) and (6.12) yields:

$$Z_T = \frac{V_i}{I} = \frac{Z_{L2}\mu\ell f}{Z_{S2}+Z_{L2}} \ell n_e \left(\frac{d+h}{d}\right) \text{ for } \omega L_2 \ll Z_{S2}+Z_{L2} \tag{6.13}$$

$$= \frac{Z_{L2}\mu\ell}{2\pi L_2} \ell n_e \left(\frac{d+h}{d}\right) \text{ for } \omega L_2 \gg Z_{S2}+Z_{L2} \tag{6.14}$$

where, Z_T = transfer impedance

μ = 0.4π microhenry/meter = 0.38 μh/ft.

L_2 = self inductance of wire. For example, L_2 for #24 (approx #20 to #30) AWG wire $\approx$0.3 μh/ft.

Fig. 6.4 is a plot of Eqs. (6.13) and (6.14) transfer impedance in dB/ohm representing the ratio of induced voltage in the output circuit #2 for an interferring current in circuit #1. Parameter values are Z_{S2} = Z_{L2} = 10, 100 and 1,000 ohms; ℓ = 1 inch, 1 ft., 10 ft., and 100 ft; and (d+h)/d ratios of 1.01, 2.7, and 100 corresponding to d ranging from 0.25" to 10" and h ranging from 0.1 to 5 inches. A series of such mathematical models in Fig. 6.4 are used for computer EMI prediction.

Illustrative Example 6.2

A 60 Hz, 115VAC power supply is driving a 2-kwatt load (I = 17.4 amperes). The power supply cable in a cable tray shares a common length of run of 10 ft. with a 120-ohm digital cable which feeds a computer. The digital cable outgoing and return wires are separated by 0.1" on center and are located 4" from the power cable. A 2-volt level in the digital receiver may cause a "0" to read "1" or vice versa. Determine if a susceptibility problem exists.

From the above, k = (d+h)/d = 4.1"/4" = 1.025. At 60 Hz, Fig. 6.4 shows that a 10 ft. run will exhibit a transfer impedance of about -98 dBΩ. Thus, from Eq. (6.13):

$$V_i = Z_T I \text{ or } V_{dB} = Z_{TdB} + I_{dB} \tag{6.15}$$

$$= -98 \text{ dB/}\Omega + 20 \log_{10} 17.4A = -98 + 25$$

$$= -73 \text{ dBV} = 47 \text{ dB}\mu\text{V} = 140\mu\text{V}$$

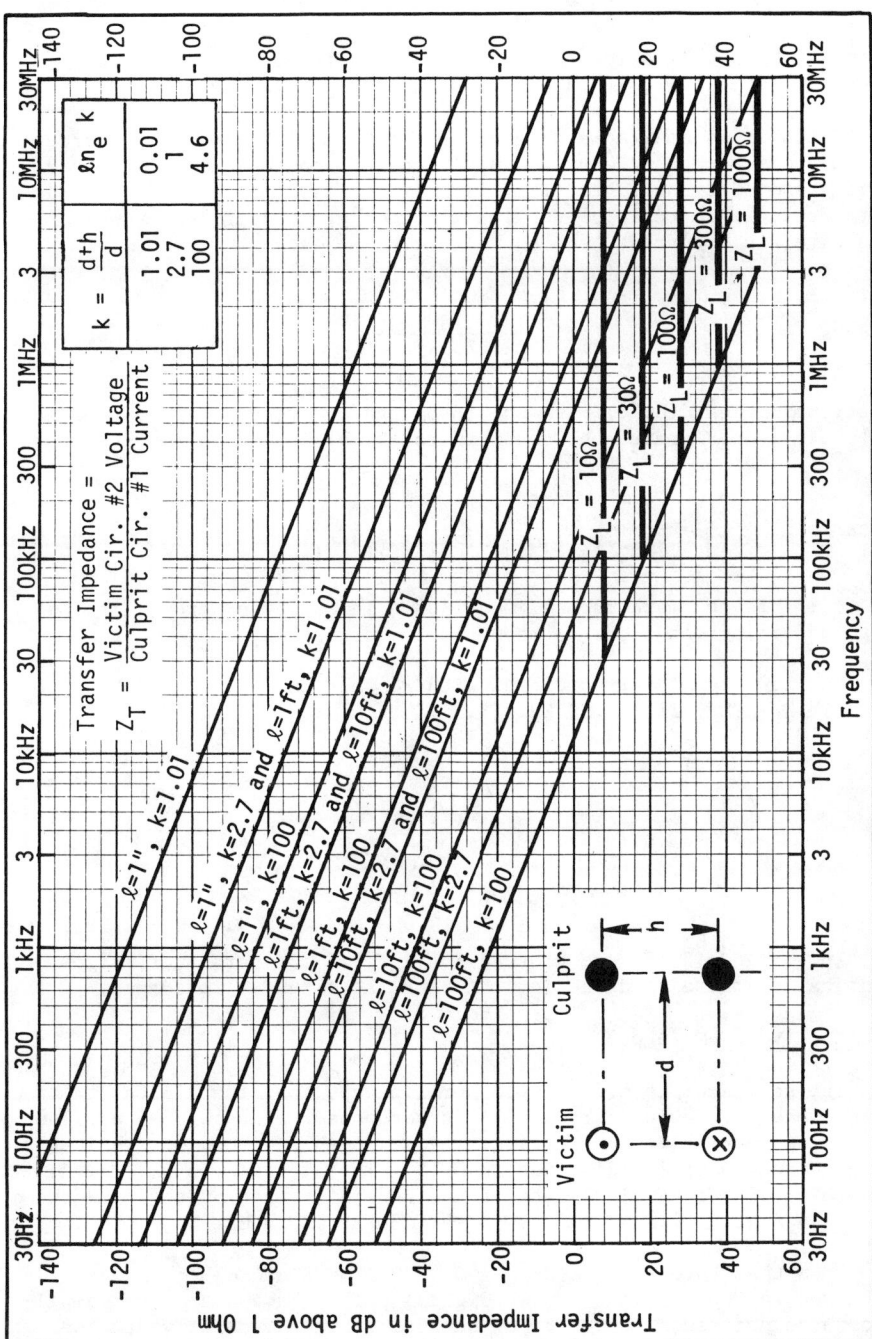

Figure 6.4 – Transfer Impedance between Wires due to Magnetic Coupling

NOTE: (Values for wire pair loops sufficiently remote from a metal ground to neglect ground plane decoupling. If not, values found are slightly conservative)

Since 140µV is well below (by 79 dB) the 2 volt digital receiver, susceptibility level, it is concluded no EMI problem exists.

Often it is desirable to express a potential EMI voltage source, V_S, at the sending end of circuit #1 and to transfer this to the receiving end of circuit #2. This is called cross talk. Since the current flowing in circuit #1 is also limited by the self inductance of it's wire, this current becomes:

$$I = \frac{V_S}{Z_{S1} + Z_{L1} + j\omega L_1} \tag{6.16}$$

Substituting Eqs. (6.16) into Eq. (6.8) and the resulting equation into (16.15), and forming the coupling ratio, V_i/V_S, there results:

$$\text{Cross Talk } Z_1 = \frac{V_i}{V_S} = \frac{Z_{L2}\mu\ell f}{(Z_{S2} + Z_{L2} + j\omega L_2)\ (Z_{S1} + Z_{L1} + j\omega L_1)}\ \ell n_e\left(\frac{d+h}{d}\right) \tag{6.17}$$

Eq. (6.17) is the mathematical model used in computing cross talk between adjoining wires due to magnetic-field coupling.

For low and high frequencies with respect to ωL in which $L_1 = L_2$, and $Z_{S1} \approx Z_{L1} \approx Z_{S2} \approx Z_{L2} = Z$, Eq. (6.17) becomes:

$$\frac{V_i}{V_S} = \frac{\mu\ell f}{4Z}\ \ell n_e\left(\frac{d+h}{d}\right) \text{ for } \omega L \ll 2Z \tag{6.18}$$

$$= \frac{Z\mu\ell}{4\pi^2 fL^2}\ \ell n_e\left(\frac{d+h}{d}\right) \text{ for } \omega L \gg 2Z \tag{6.19}$$

Eqs. (6.18) and (6.19) are plotted in Fig. 6.5 for parameter values identical to those used in Fig. 6.4 for the transfer impedance.

Illustrative Example 6.3

The same Example 6.2 is used here, except that the digital transmitter-receiver cable (circuit #1) is harnessed with a wide-band, low noise video amplifier cable (circuit #2). Both cable types are identical and the separation distance is 0.1". The digital transmitter levels are 4 volts at a clock-rate of 100 kb/sec (5 µsec pulses) and the broadband amplifier sensitivity is -50 dBm at 120 ohms.

From the above, k = (d+h)/d = 0.2"/0.1" = 2 and d/h = 1. As discussed in Chap. 3, the 5 µsec digital pulses have a spectrum amplitude which decreases at 20 dB per decade, i.e., the negative to the

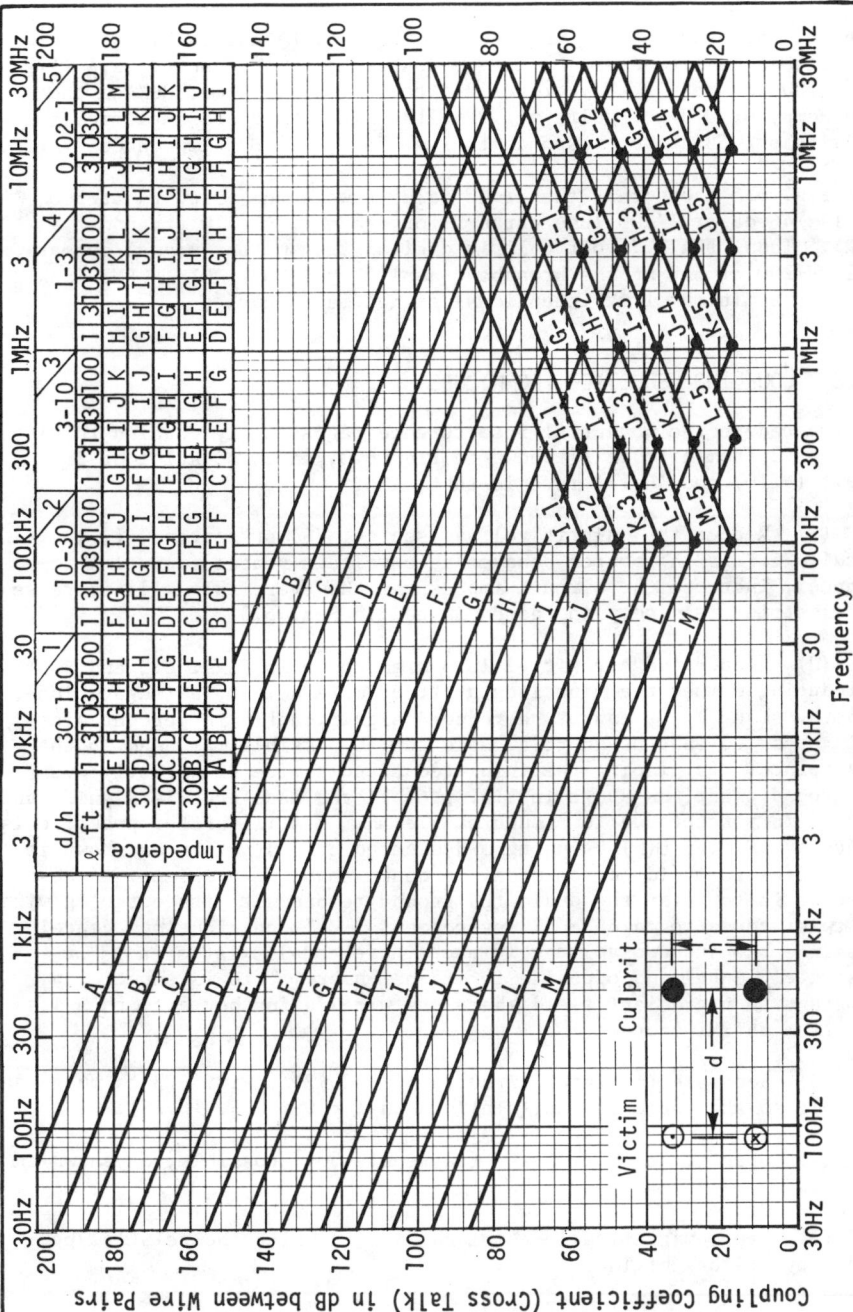

Figure 6.5 – Cross Talk between Wires due to Magnetic (Inductive)Coupling

NOTE: (Values for wire pair loops sufficiently remote from a metal ground to neglect ground plane decoupling. If not, values found are slightly conservative)

slope of increase in coupling shown in Eq. (6.18) and in Fig. 6.5.
The voltage amplitude of the energy contained in the second $(\sin x)/x$
hump (200 kHz to 400 kHz) of the digital signal is about 30% of the
main hump, or 0.3 x 4 volts = 1.2 volts. The broadband amplifier sensitivity of -50 dBm corresponds to 1 mV for a 120 ohm impedance system.

Fig. 6.5 shows that the cross talk between the circuits at 300
kHz is -46 dB or 1/200 (use curve I for d/h = 1, ℓ = 10 ft. and Z =
100Ω). Thus, the 1.2 volt culprit digital transmitter signal appears
as 1.2V/200 or 6mV at the broadband-amplifier input cable. EMI masking of the minimum 1 mV intentional signal may occur.

6.2.2 COUPLING REDUCTION TECHNIQUES

To reduce the potential interference magnetically coupled to wire
circuit #2 from circuit #1 shown in Fig. 6.3, an examination of Eqs.
(6.9), (6.10) and (6.17) reveals that:

(1) Reduce the culprit voltage source, V_S, or current, I, in
circuit #1. If the source behaves as a voltage source (very-low impedance), increase the circuit impedance. However, this will increase
the electric-field coupling as explained in Sec. 6.3.

(2) Reduce the circuit #2 loop area shown in Figs. 6.3 and 6.6(a)
by reducing either the line length, ℓ, or wire height, h (separation),
or both. Usually, ℓ cannot be reduced significantly in percent since
it is presumed that as short a cable run as possible was used in the
first place. By placing the insulated circuit wire #2 directly over
its ground plant, as shown in Fig. 6.6(b), the loop area is significantly
reduced in percent due to a much smaller h. A still better practice is
to achieve Fig. 6.6(b) by using a dedicated ground return as shown in
Fig. 6.6(c) to avoid common-mode impedance coupling problems. A best
practice is to twist the dedicated ground return with its outgoing wire.
The twist tends to make local environmental EMI contributions cancel
out since the induced voltage in each incremental twist area is approximately equal and opposite to its neighbor. This is especially
important in a gradient field where one wire is further away then its
mate from an EMI source.

(3) Separate (isolate) circuits #1 and #2 in Fig. 6.3 as much as
possible so that d>>h and $\ell n_e(d+h)/d \rightarrow \ell n_e 1 = 0$ in Eq. (6.17).

(4) Operate at DC or lower frequencies, if possible, as shown in
Fig. 6.5.

(5) Use a magnetically-shielded wire in which the relative permeability of the shield $\mu_o \gg 1$.

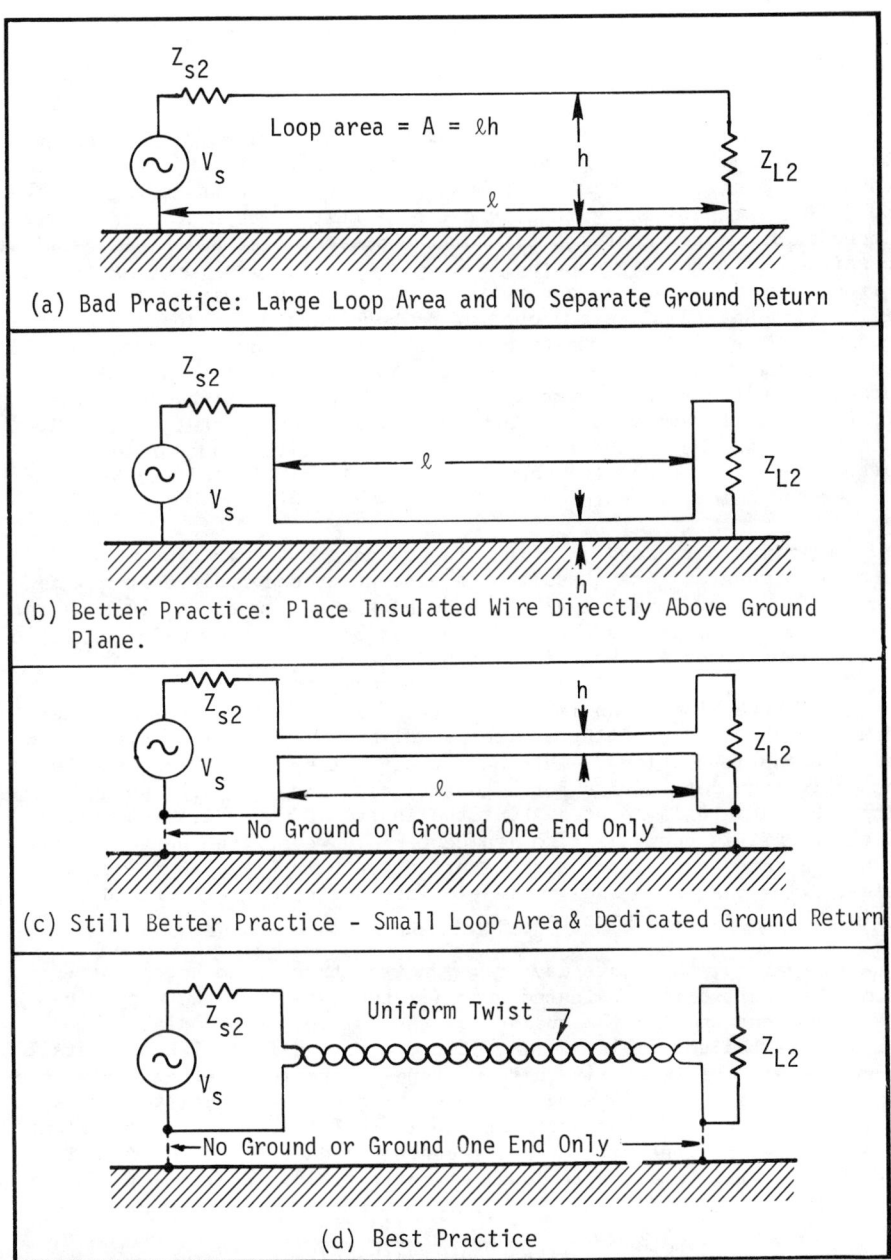

(a) Bad Practice: Large Loop Area and No Separate Ground Return

(b) Better Practice: Place Insulated Wire Directly Above Ground Plane.

(c) Still Better Practice - Small Loop Area & Dedicated Ground Return

(d) Best Practice

Figure 6.6 - Evolution of Reducing Loop Area in Magnetically - Susceptible Circuits Including Circuit Grounding Practice

(6) Operate the sensitive circuit #2 into a differential amplifier so that common-mode, induced currents in both leads will tend to
cancel.

Some additional remarks regarding the above EMI-control techniques
for reducing magnetic coupling in wires and cables are in order. When
a dedicated return wire is used (best practice) as shown in Fig. 6.6(c)
and (d), the topic of how to ground the circuit comes up. This is
discussed in length in Chap. 8. For the present discussion, it is first
questioned if circuit grounding is needed at all. If grounding is required, then it should be accomplished at one point only as shown in
Fig. 6.6(c) and (d). Were grounding of both ends of cable or both
sending and receiving circuits done, then an alternative return path for
the outgoing wire has been introduced. Now, current return will go
through both the return twist and the ground plane. Since the latter
offers a much lower impedance path, nearly all the current will follow
this route and the return twisted wire is effectively out of the circuit. Thus, in effect, the equivalent circuit of Fig. 6.6(a) has been
regenerated all over again.

Sometimes purchased circuits are grounded to their case which
in turn is mounted to a chassis or equipment ground. Since it may be
very difficult or impractical to unground these situations, such as in
a potted package, ground current loops can be avoided by using an isolation transformer as shown in Fig. 6.7.

The above discussion implied that a twisted-wire pair is the best
form of unshielded outgoing and return wire cable (cf. Fig. 6.6(d)) to
reduce magnetic coupling. This may or may not be true. For example,
the equivalent *return wire* for a coaxial cable is its outer sheath.
Since this sheath is located about the inner conductor its geometric
location axis is equivalent to its center. However, the outgoing wire
is also located there, so that the effective h in Fig. 6.6 is zero.
This implies that a coaxial cable provides better magnetic-field isolation than a twisted wire pair.

In reality, the above may or may not be true. The better per-
former of the two is predicated upon (1) the extent of the magnetic-
field gradient and (2) the quality of the twisted vs. coaxial cable.
For a high-gradient field emanating from a wire axially located parallel
to the twist, the twist, if uniform, tends to perform better. Fig. 6.8
shows that the induced voltage in each twist *area* is equal and opposite
to its neighbor. If the coaxial cable is replaced with two equivalent
wires located in line with both the center conductor and the culprit
wire, there will remain a residual magnetic field, ΔH (see Fig. 6.9):

$$\Delta H = \frac{I}{2\pi(R-\Delta R/2)} - \frac{I}{2\pi(R+\Delta R/2)}$$

$$\approx \frac{I\Delta R}{2\pi R^2} \text{ for } R >> \Delta R$$

(6.20)

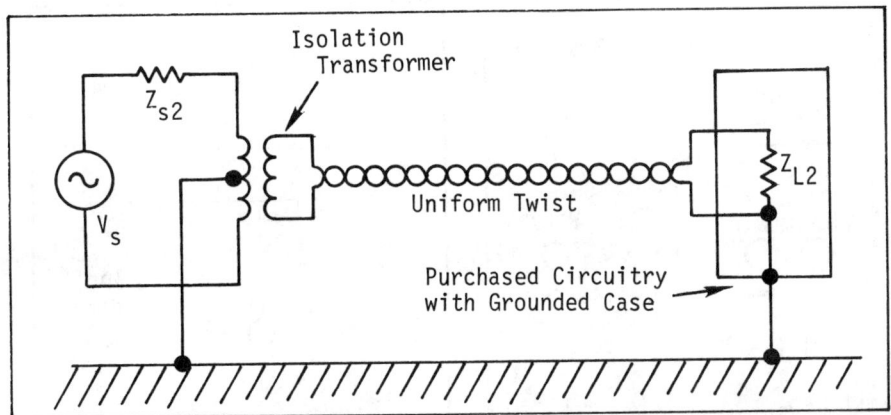

(a) When Source Must Be Grounded and Purchased Circuitry Is
    Grounded and Cannot Be Ungrounded.

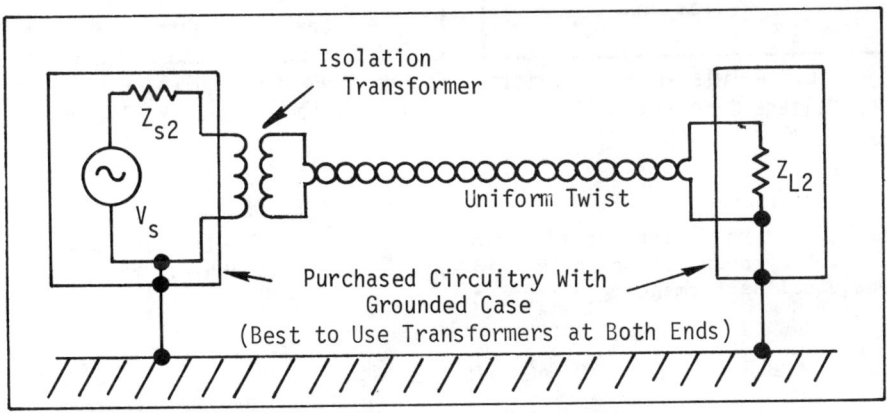

(b) When Both Source and Load Purchased Circuitry Are
    Grounded and Cannot Be Ungrounded

Figure 6.7 - Reducing EMI Common-mode Susceptibility Between Wires In A
             Loop By the Use of Isolation Transformers.

As $\Delta R \to 0$ (small diameter coax) and/or $R \to \infty$ (no gradient or uniform
magnetic field), the induced flux linkage and voltage in the coaxial
line approaches zero.

Regarding the quality of the twist vs. the quality of the coaxial
line, if the former does not have a uniform twist and/or has only a
few twists per unit length, then it will offer less magnetic field
isolation than a coaxial line exhibiting a uniform concentric location
of its center conductor within its sheath. If the reverse conditions

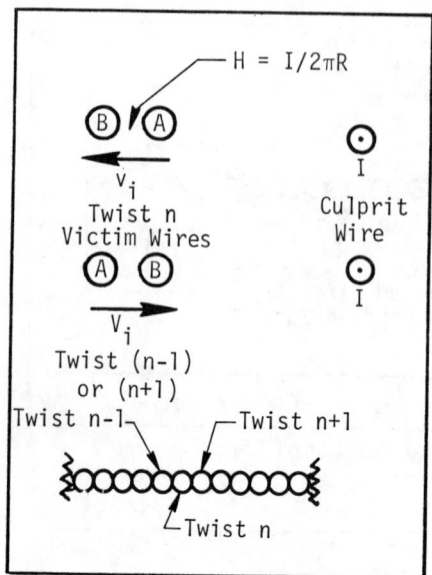

Figure 6.8 - Twist-Loop-to-Twist-Loop Voltage Cancellation

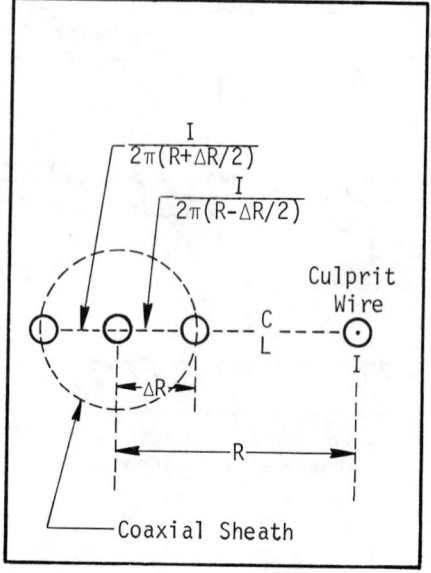

Figure 6.9 - One Equivalent Configuration of a Coaxial Line

apply, then the twist should perform better. Thus, unless the quality of both is known, it may be stated that either will exhibit more or less equal performance.

6.2.3 CABLE SHIELDING TO MAGNETIC FIELDS

Item (5) in the preceding list indicates that further magnetic-field de-coupling can be achieved by using a magnetic shield over the wire. Typical tin-plated, copper-braided shields, used for electric-field de-coupling purposes (shielding) at higher frequencies, will have little-to-no-effect on magnetic-coupling since the shield has a relative permeability, $\mu_o = 1$ - the same as air. However, where magnetic shielding is desired, such as in either secure systems or susceptible equipments operated in the presence of strong magnetic fields, a ferrous-based cable shield may be used. The culprit magnetic field will be reduced at the equivalent victim loop area (Fig. 6.10a) because of the lower reluctance path offered by the field as shown in Fig. 6.10b.

The degree of magnetic shielding, H_{dB}, offered by a permeable cylindrical sheath placed over either a wire pair or a harness is:

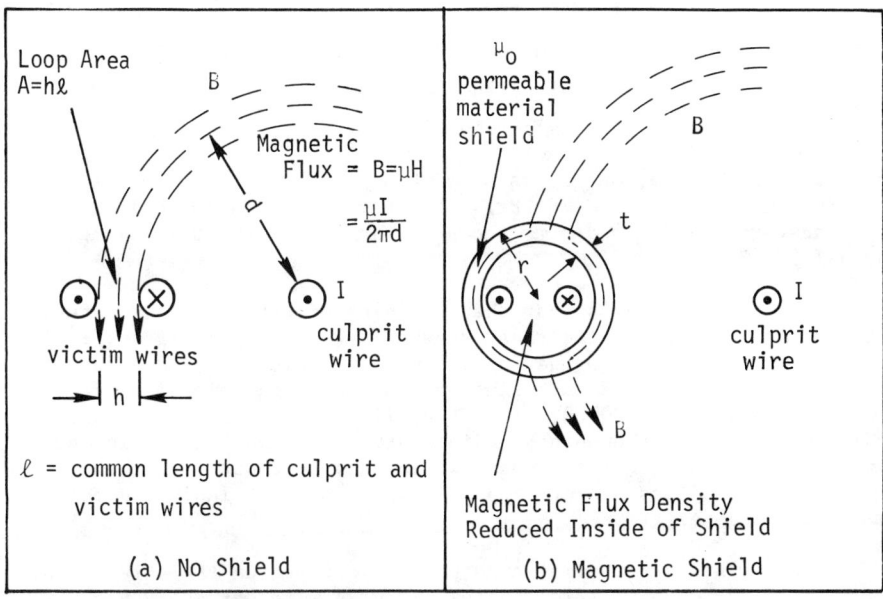

Loop Area
A=hℓ

ℓ = common length of culprit and victim wires

(a) No Shield

Magnetic Flux = B=μH

$$= \frac{\mu I}{2\pi d}$$

Magnetic Flux Density
Reduced Inside of Shield

(b) Magnetic Shield

Figure 6.10   - Reducing Magnetic Flux Loop Area by Using
Permeable Magnetic Shield

Figure 6.11 - Double Knitted Wire
Mesh Tape Shield

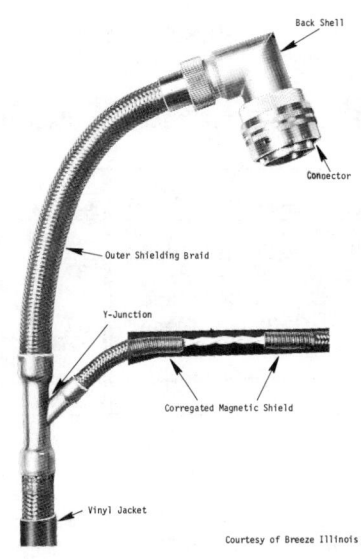

Figure 6.12 - Typical Flexible
Corrugated Shields

$$H_{dB} \approx 20 \log_{10}(\mu_r t/2r) \text{ dB, for } \mu_r t/2r \geq 1 \qquad (6.21)$$

where, μ_r = relative permeability of magnetic sheath

t   = thickness of cylindrical sheath

r   = outer radius of sheath

Eq. (6.21) approximation* holds only when μ_r >> 1 (say, > 100) and for low frequencies (say, f < 10 kHz). Further, the equation only applies for transverse fields, such as shown in Fig. 6.10b - not for longitudinal fields.

Eq. (6.21) indicates that the shielding effectiveness is materially reduced for thin-wall cylindrical sheaths. For example, if a 1/16 inch wall iron pipe (μ = 1,000) having a one-inch radius is used around a large harness in a fixed installation, the magnetic shielding effectiveness is only about 30 dB. If a two mill (0.002") iron tape had been wrapped around a one-inch radius harness with a 50% overlap per pitch, the shielding effectiveness would only be about 3 dB. Thus, higher permeabilities and/or greater thicknesses are indicated for good shielding performance.

There exists a number of high permeable tapes on the market, which can be used for magnetic-field shielding of a wire harness. Fig. 6.11 shows one such tape available in one-inch widths. Based on a tin-coated, copper-clad steel, double-wire mesh with about 50% air space, the equivalent permeability is about 300 per 0.015 inch tape thickness. When wrapped as a bandage, advancing one-half layer per turn, the low-frequency shielding effectiveness for different radii harnesses is shown in Table 6.3.

Table 6.3 - Low-Frequency Magnetic Shielding Effectiveness of Tapes

Magnetic Tape	Radius of Wire Harness				
	r=0.10"	r=0.25"	r=0.5"	r=1"	r=2"
Technit Magnetic Tape	30 dB	21 dB	16 dB	10 dB	4 dB
Netic, 0.004" Tape	32 dB	24 dB	18 dB	12 dB	6 dB
Co-Netic, 0.004" Tape	45 dB	37 dB	31 dB	25 dB	19 dB
Co-Netic, 0.010" Tape	52 dB	44 dB	38 dB	32 dB	26 dB
Hypernom 0.004" Tape	46 dB	38 dB	32 dB	26 dB	20 dB
Corregated Hypernom 8 mil.	NA	49 dB	43 dB	37 dB	31 dB

---

* P. J. Johnson and S. Shenfeld, "Shielding of Cylindrical Tubes at Low Frequencies," 1971 *IEEE International EMC Symposium Record*, 71C29-EMC, pp. 68-76.

Table 6.3 also shows some typical shielding effectiveness expected from other high-permeable, solid-foil tapes when wrapped with a 50% overlap per turn. The permeability is a function of field intensity, degree of annealing, and other parameters. Netic tape offers less shielding than Co-Netic, but because of saturation in high fields, it is used above 5 gauss and Co-Netic is used below 5 gauss. The corregated flexible tube shown in the table is not a tape but is shown because of interest in its high shielding properties. Fig. 6.12 shows a typical corregated flexible harness shield used in EMP environments.

Another form of cable wire harness shield is zipper tubing. It is discussed in Sec. 6.4. When fabricated from tin-coated, copper-clad steel, its magnetic shielding properties are similar to that listed in Table 6.3 for Technit. Zipper tubing distinguishes itself by permitting a rapid installation and removal and is good to check out R & D systems, test arrangements, and quick-fix in the field. It is otherwise relatively heavy, bulky, and expensive.

The problem arises regarding how to ground the shield. Were it not for the need to ground shields as required for electric-field de-coupling discussed in Sec. 6.3, the shield could float from a magnetic-field point-of-view. However, where high-impedance circuits are to be used shield grounding is required. A non-permeable shield can be ungrounded, grounded at either end, or both ends with little effect on the magnetic-field coupling as long as both ends of the *circuit* are not also grounded. No protected circuit current loop is formed by grounding its shield at both ends as long as the shield and return wire are not connected.

6.3 ELECTRIC-FIELD COUPLING

This section is the correlary to the previous section involving magnetic-field coupling in wires and cables, which predominates for lower circuit impedance. Electric-field coupling increases with an increase in circuit impedance and becomes the greater EMI contributer at high frequencies, with the actual frequency of crossover depending upon a number of parameters.

6.3.1 CAPACITIVE (ELECTRIC FIELD) COUPLING PHYSICS

Capacitive coupling predominates when the victim circuit impedance is high relative to 377 ohms. The coupling also increases with frequency and the proximity of the culprit-victim pairs.

Figure 6.13 shows the network involving capacitive coupling between culprit line and victim circuits. The objective is to determine how much of the available culprit source line voltage, V_s, is coupled into the victim load, V_i, for varying conditions. The ratio of the victim-to-culprit voltages then becomes the cable-to-cable coupling, CCC:

$$CCC_{dB} = 20 \log_{10}(V_i/V_s) \qquad (6.22)$$

The negative of CCC_{dB} is often called *cross talk*.

Based on frequently used math models (see Ref.30), the V_i/V_s ratio for capacitive coupling is:

$$\frac{V_i}{V_s} = \left(\frac{1}{1+j/\omega C_c \ Z_z \ell_m} \right) \qquad (6.23)$$

$$\simeq \omega C_c \cdot Z_z \ell_m, \quad \text{for } f \ll 1/2\pi C_c \ Z_z \ell_m \qquad (6.24)$$

where,
$\omega = 2\pi f \times$ frequency in Hz

C_c = two-wire pair coupling capacitance (see Fig 6.13) for D = wire diameter,

$$= \frac{\pi \varepsilon}{\ell n \left(\frac{2d}{D} \right)} \text{farads/m} \quad \text{for } d/D \gg 1* \qquad (6.25)$$

Z_z = parallel combination of victim source and load impedances and wiring capacitance, Z_v, to ground
$= 1/(1/Z_{s2} + 1/Z_{v2} + j\omega C_v)$
ℓ_m = cable length in meters

---

* When d/D is not $\gg$1, $\ell n_e(2d/D)$ must be replaced by $\cosh^{-1}d/D$. For d/D = 1, $\cosh^{-1}$ 1.0 = 0.3.

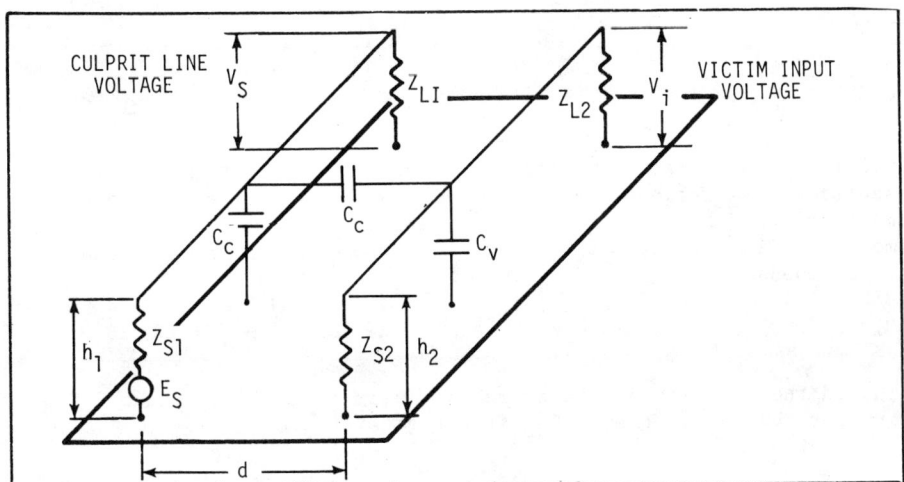

Figure 6.13 - Circuit Representation of Capacitive Coupling Between Parallel Wires Over a Ground Plane.

The C_{cv} term is quite complex (see Ref. 30). It is a function of several variables including:

- Culprit wire radius or AWG number used

- Culprit wire height above ground plane (or source wire pair separation = $2h_1$ by method of images)

- Culprit wire insulation, especially for closely-coupled wire pairs

- Culprit-to-victim separation distance, d

- Victim wire radius or AWG number used

- Victim wire height above ground plane (or victim pair wire separation = $2h_2$ by method of images)

- Victim wire insulation, especially for closely-coupled wire pairs

- When $h_1 \neq h_2$, $h \simeq \sqrt{h_1 h_2}$

When Eq. (6.23) is substituted into Eq. (6.22), CCC may be computed for several parameter values and conditions. Table 6.15 shows CCC for capacitive coupling corresponding to two #22 AWG wire pairs having a common length of 1 meter and an impedance level of 100 ohms. The table parameters are the wire height, h (remember the wire height is 2h for a pair rather than h for a single wire above a ground plane) and the wire pair separation distance, d. Both the h and s terms are in units of mm.

In examining Table 6.15, note that for smaller values (tighter coupling) CCC increases with frequency at 20 dB/decade until the cable approaches an electrical length of $\lambda/2$, whereafter it rolls off. While

CCC is not significantly affected by small changes in h, it is very substantially affected by wire pair separation. In this respect, note that CCC falls off with increasing s at -40 dB/decade. Furthermore, CCC is directly affected by both the cable length, ℓ_m and victim load impedances, Z_2:

$$(CCC)_{dB} = (CCC \text{ in Fig. } 6.15)_{dB} + 20 \log_{10}\left(\frac{Z_v \ell_m}{100}\right) \leq 0 \text{ dB} \quad (6.26)$$

Should the addition of the correction term in Eq.(6.26) make CCC_{dB} greater than zero, then CCC_{dB} is not greater than 0dB.

Illustrative Example 6.1

Compute the capacitive coupling between two wire pairs running 10 meters in a cable tray at a spacing of 3mm. Clock rate is 100Kb/s, τ rise = 500 ns and impedances are 100 ohm, $f_c = 1/\pi\tau_r = 637$ kHz. For h>3mm and S = 3mm, Fig. 6.15 gives (CCC) = -59 dB at 637 kHz. Correcting by Eq. 6.26 gives: -59 dB + 20 dB (length adjustment) = -39 dB. If source pair carries 3.5V logic swing (11 dBB), the result will be: 11 -39 = -28 dBV = 40 mV, much lower than logic noise margin.

6.3.2 COUPLING REDUCTION TECHNIQUES

To reduce the potential interference electrically coupled from wire circuit #1 shown in Fig. 6.13, Eq. (6.23) indicates that:

(1) Reduce the voltage source, V_s, available for coupling in circuit #1 and/or decrease the circuit impedance in circuit #2. The latter will increase the voltage-dividing action across the coupling capacitor, C_c as shown in Fig. 6.13. However, lowering circuit #1 impedance to lower the voltage source will result in increasing the magnetic-field coupling as explained in Sec. 6.2.

(2) Separate (isolate) circuits #1 and #2 as much as possible to reduce mutual coupling circuit capacitance, C_c, so that d>>D and ℓn_e (2d/D) becomes large in Eq.(6.25).

(3) Reduce capacitance C_c by using a shorter length of wire. A long parallel run is also difficult to effectively shield at high frequencies.

(4) Reduce capacitance C_c by interposing a Faraday cage, viz., shield the wires in either or both circuits as explained below.

Culprit: AWG# = 22 . Cable Length = 1 meters. Load: ZS = 100 Ohms.
Victim: AWG# = 22 . Cable Length = 1 meters. Load: ZV = 100 Ohms.

FREQNCY	\=============== h = .5-3 mm ===============							\=============== h = 3-10 mm ===============						
	S=1	S=3	S=10	S=30	S=100	S=300	S=1k	S=1	S=3	S=10	S=30	S=100	S=300	S=1k
10Hz	-145	-162	-181	-201	-221	-241	-261	-144	-155	-167	-184	-204	-224	-244
20Hz	-139	-156	-175	-195	-215	-235	-255	-138	-149	-161	-178	-198	-218	-238
30Hz	-135	-152	-171	-191	-211	-231	-251	-134	-145	-157	-174	-194	-214	-234
50Hz	-131	-148	-167	-187	-207	-227	-247	-130	-141	-153	-170	-190	-210	-230
70Hz	-128	-145	-164	-184	-204	-224	-244	-127	-138	-150	-167	-187	-207	-227
100Hz	-125	-142	-161	-181	-201	-221	-241	-124	-135	-147	-164	-184	-204	-224
200Hz	-119	-136	-155	-175	-195	-215	-235	-118	-129	-141	-158	-178	-198	-218
300Hz	-115	-132	-151	-171	-191	-211	-231	-114	-125	-137	-154	-174	-194	-214
500Hz	-111	-128	-147	-167	-187	-207	-227	-110	-121	-133	-150	-170	-190	-210
700Hz	-108	-125	-144	-164	-184	-204	-224	-107	-118	-130	-147	-167	-187	-207
1kHz	-105	-122	-141	-161	-181	-201	-221	-104	-115	-127	-144	-164	-184	-204
2kHz	-99	-116	-135	-155	-175	-195	-215	-98	-109	-121	-138	-158	-178	-198
3kHz	-95	-112	-131	-151	-171	-191	-211	-94	-105	-117	-134	-154	-174	-194
5kHz	-91	-108	-127	-147	-167	-187	-207	-90	-101	-113	-130	-150	-170	-190
7kHz	-88	-105	-124	-144	-164	-184	-204	-87	-98	-110	-127	-147	-167	-187
10kHz	-85	-102	-121	-141	-161	-181	-201	-84	-95	-107	-124	-144	-164	-184
20kHz	-79	-96	-115	-135	-155	-175	-195	-78	-89	-101	-118	-138	-158	-178
30kHz	-75	-92	-111	-131	-151	-171	-191	-74	-85	-97	-114	-134	-154	-174
50kHz	-71	-88	-107	-127	-147	-167	-187	-70	-81	-93	-110	-130	-150	-170
70kHz	-68	-85	-104	-124	-144	-164	-184	-67	-78	-90	-107	-127	-147	-167
100kHz	-65	-82	-101	-121	-141	-161	-181	-64	-75	-87	-104	-124	-144	-164
200kHz	-59	-76	-95	-115	-135	-155	-175	-58	-69	-81	-98	-118	-138	-158
300kHz	-55	-72	-91	-111	-131	-151	-171	-54	-65	-77	-94	-114	-134	-154
500kHz	-51	-68	-87	-107	-127	-147	-167	-50	-61	-73	-90	-110	-130	-150
700kHz	-48	-65	-84	-104	-124	-144	-164	-47	-58	-70	-87	-107	-127	-147
1MHz	-45	-62	-81	-101	-121	-141	-161	-44	-55	-67	-84	-104	-124	-144
2MHz	-39	-56	-75	-95	-115	-135	-155	-38	-49	-61	-78	-98	-118	-138
3MHz	-36	-52	-71	-91	-111	-131	-151	-34	-45	-57	-74	-94	-114	-134
5MHz	-31	-48	-67	-87	-107	-127	-147	-30	-41	-53	-70	-90	-110	-130
7MHz	-28	-45	-64	-84	-104	-124	-144	-27	-38	-50	-67	-87	-107	-127
10MHz	-26	-42	-61	-81	-101	-121	-141	-24	-35	-47	-64	-84	-104	-124
20MHz	-20	-36	-55	-75	-95	-115	-135	-19	-29	-41	-58	-78	-98	-118
30MHz	-17	-33	-51	-71	-91	-111	-131	-16	-26	-37	-55	-74	-94	-114
50MHz	-13	-28	-47	-67	-87	-107	-127	-13	-22	-33	-50	-70	-90	-110
70MHz	-11	-25	-44	-64	-84	-104	-124	-10	-19	-30	-47	-67	-87	-107
100MHz	-9	-23	-41	-61	-81	-101	-121	-9	-16	-27	-44	-64	-84	-104
200MHz	-6	-17	-35	-55	-75	-95	-115	-6	-12	-22	-38	-58	-78	-98
300MHz	-5	-14	-31	-51	-71	-91	-111	-4	-9	-18	-35	-54	-74	-94
500MHz	-3	-11	-27	-47	-67	-87	-107	-3	-7	-15	-30	-50	-70	-90
700MHz	-3	-9	-24	-44	-64	-84	-104	-3	-6	-12	-28	-47	-67	-87
1GHz	-2	-7	-22	-41	-61	-81	-101	-2	-4	-10	-25	-44	-64	-84
2GHz	-2	-5	-16	-35	-55	-75	-95	-2	-3	-7	-19	-38	-58	-78
3GHz	-1	-4	-14	-31	-51	-71	-91	-1	-2	-5	-16	-34	-54	-74
5GHz	-1	-3	-10	-27	-47	-67	-87	-1	-2	-4	-13	-30	-50	-70
7GHz	-1	-2	-9	-24	-44	-64	-84	-1	-2	-3	-11	-27	-47	-67
10GHz	-1	-2	-7	-21	-41	-61	-81	-1	-1	-3	-9	-24	-44	-64

Figure 6.15 - Cable-to-Cable Capacitive Coupling in dB

Fig. 6.16 shows the two circuit wires above a ground plane together with their mutual capacitance. Wire #2 has a Faraday shield added with the shield grounded to the plane. Thus, the capacitance, C_c, is now between wire #1 and the shield of wire #2, which in essence is ground. Although capacitance C_c^1 is somewhat larger than the previous C_c (because the shield diameter is greater than the wire), the shield is grounded. Also a new capacitance between wire #2 and it's shield is created in the process, but it too is grounded.

Notice that a small impedance, Z_b, is shown for the ground wire to account for its series inductance and resistance. While very low compared with the coupling reactance ($X_c^1 = 1/\omega C_c^1$) at low frequencies, the bond impedance may become significant at high frequencies. The ratio of Z_b/X_c^1 results in the voltage-dividing action improvement offered by the Faraday shield. For example, X_b/X_c^1 may yield 80 dB improvement at 100 kHz, but may approach 0 dB at 30 MHz depending on line length, wiring configuration, and separation. Thus, to reduce the bond impedance, a good high-frequency strap or braid should be used for grounding the shield.

Fig. 6.17 shows wire and shield configurations for shields on both wires. This double-Faraday shield further decouples the mutual wire capacitance as shown in the equivalent circuit. For this situation the voltage-dividing action takes place in two stages yielding an overall decoupling improvement ratio of about $Z_b^2/X_{c1}X_c''$ or approximately twice the dB of the single-wire shield shown in Fig. 6.16.

6.3.3 CABLE-SHIELD GROUNDING

The above underscores the importance of achieving a good ground if the Faraday shield is to perform properly. This also raises the question regarding whether the shield should be grounded at one end, the other end, both ends, the middle, etc. Two considerations govern the practice to be used: (1) do not develop a circuit wire ground-current loop in the process such that low-frequency magnetic-field coupling will become a problem all over again, and (2) do not allow the distance between shield ground point(s) to become a significant fraction of a wavelength at the highest frequency of operation. These considerations will now be examined further.

Fig. 6.18 shows a shield placed over a potentially susceptible twisted-wire pair. Situation (a) corresponds to an intentional circuit wire ground at either or neither end, but not a ground at both ends which would otherwise destroy the effect of magnetic decoupling of the twist as explained earlier. The wire shield, however, may be directly grounded at either or both ends with little difference in magnetic coupling performance as long as $\ell << \lambda$. Situation (b) corresponds to a circuit ground at the sending end; the shield is grounded

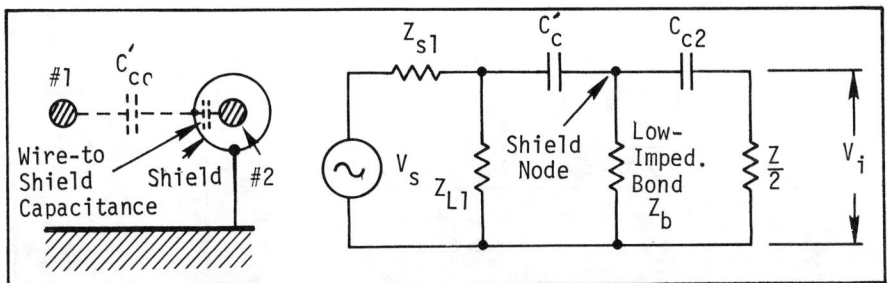

Figure 6.16 - Equivalent Circuit of Wire Shield to Reduce Electric-
Field Coupling (cf. Fig. 6.13)

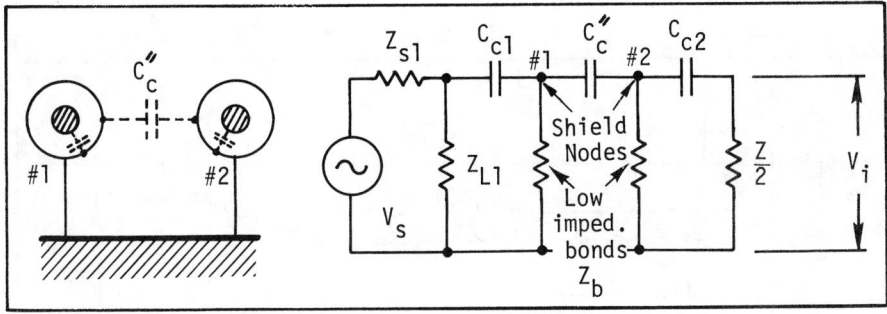

Figure 6.17 -   Equivalent Circuit with both Conductors Shielded

at the receiving end. This performs well as a Faraday shield against
electric-field coupling when $\ell \ll \lambda$. However, the figure also shows
that the return circuit wire is connected to the shield ground. This
error results in a partition of return current with part going through
the return wire and part through the ground plane. Since nearly all
return current flows through the latter because it offers a lower-
impedance path, effective magnetic decoupling is destroyed in the
process. The magnetic circuit is now essentially the same as previously shown in Fig. 6.6(a).

At high frequencies, the shield may have to be grounded at both
ends or more often to continue to perform as a Faraday shield. If the
shield is not well bonded to ground, the voltage-dividing action illustrated in Figs. 6.16 and 6.17 will be compromised. For example,
suppose the shield is grounded at one end as shown in Fig. 6.18(b)
and the shield length is $\ell = \lambda/4$. The shield now acts as a quarter-
wave stub antenna and the sending end of the shield behaves as an
open circuit (high impedance). This is further illustrated on the
Smith Chart shown in Fig. 6.19(a) where the short at one end is converted to an open circuit at the other end. Thus, the shield instead of helping, acts as a pick-up antenna to any interfering signals in the immediate area. By grounding this shield at both ends,

6.25

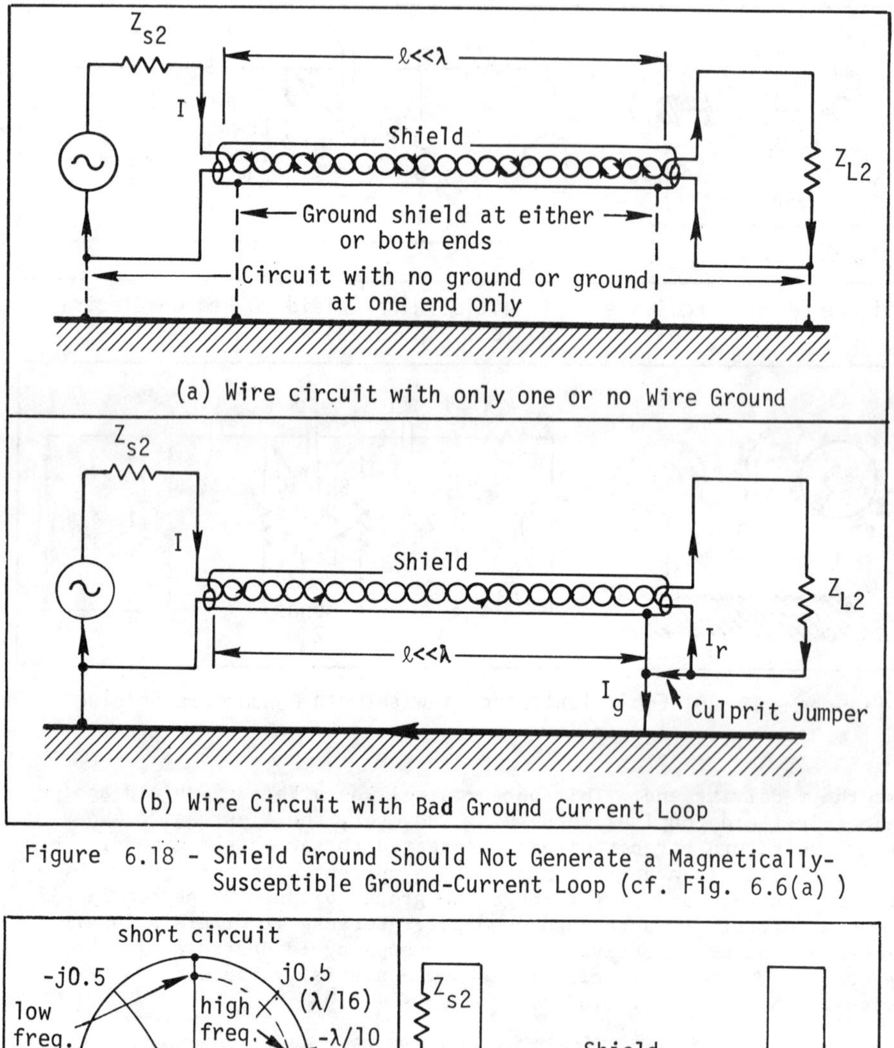

(a) Wire circuit with only one or no Wire Ground

(b) Wire Circuit with Bad Ground Current Loop

Figure 6.18 - Shield Ground Should Not Generate a Magnetically-
Susceptible Ground-Current Loop (cf. Fig. 6.6(a))

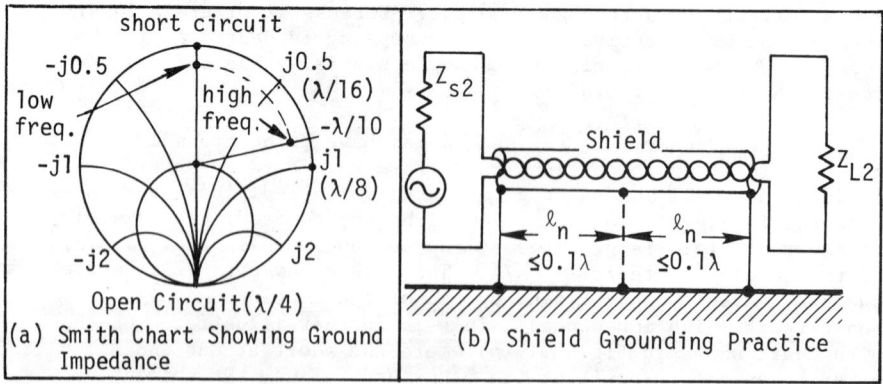

(a) Smith Chart Showing Ground
Impedance

(b) Shield Grounding Practice

Figure 6.19 - Shield Ground Requirements for Effective Faraday Shield

however, shielding performance has been reclaimed; the center of the shield will now be $\lambda/8$ long to either end. Its shielding performance will suffer somewhat near the center since the coupling capacitance C_c^1 per unit length in this region will not be particularly well de-coupled.

There exists no absolute criteria on how often to ground the Faraday shield at high frequencies since the magnitude of a potential EMI situation is a function of many variables. Based on the highest susceptible frequency of operation, one general guideline would be that the shield should be grounded at least every 0.2λ as shown in Fig. 6.19. The bonding impedance, X_b, per se, will then be the prin-ciple element in determinig shielding effectiveness, i.e., electric-field de-coupling.

Illustrative Example 6.4

Two twisted wire-pairs, 3 feet long are located in the same cable harness bundle with each having both sending and receiving impedances of 120 ohms. The wire pair spacing, d, is 3mm and the diameter D is .25mm. The first pair contains a stable local oscillator drive of 10 volts (140 dBμV) at 1 MHz. The second pair feeds a low-noise, wide-band amplifier having a sensitivity of 10μV (20 dBμV). Determine the potential EMI problem, both before and after using shielded wires in which the bonding impedance of each is 1 ohm at 1MHz.

From the Table at Fig. 6.15, the cross talk between wire pairs for ℓ = 1m, Z = 100Ω, and S = 3 for h<3m at 1 MHz is -62 dB. Since the difference between the emission source level (140 dBμV) and the receptor susceptibility level (20 dBμV) is 120 dB, a -62 dB coupling will leave 58 dB of interference margin still remaining. Thus, EMI is highly probable.

Equation 6.25 indicates that the coupling capacitance between wire pairs is approximately 23 pf corresponding to an X_c at 1 MHz of about 7000 ohms. The coupling reduction offered by grounding two added cable shields (now two shielded twisted-pairs) is:

$$\text{Reduction} = Z_b^2/X_c^2$$
$$= 40 \, \log_{10}(Z_b/X_c)$$
$$= 40 \, \log_{10}(1/7000) = 154 \, \text{dB}$$

Thus, the residual 75 dB of interference margin is reduced to 75 dB - 154 dB = -79 dB. EMI will not exist. However, it is cautioned that magnetic coupling must also be calculated to determine if this mode of coupling may now predominate.

The mathematical models discussed above represent a fixed set of conditions in which both emission source and receptor impedances are equal. Certain other simplifications were used. To assure more accurate EMI predictions, families of such models should be used.

6.4   COMPOSITE COUPLING AND SPECIAL CABLES

The previous two sections indicate that magnetic-field coupling between wire pairs is the more significant EMI culprit for low impedances and that electric-field coupling is the more significant for high-impedance circuits. Above about 10 MHz, electric-field coupling will almost always predominate. Between about 100 kHz and 10 MHz, the greater coupling field is a function of many variables and it can be misleading to generalize. However, for many representative situations in this frequency region, there exists more electric-field coupling culprits than magnetic-field.

6.4.1   COMPOSITE MAGNETIC AND ELECTRIC-FIELD COUPLING

For the low-frequency regions (below above 100 kHz, where the lowest notch frequencies exist) either magnetic or electric-field coupling will predominate depending upon both the wiring geometry and impedance levels. As shown below, the wire length and frequency both enter the two coupling relations in the same manner. Thus, the classical question of interest at low frequencies is, for what wiring geometry and circuit impedances will either magnetic or electric-field coupling be the greater culprit? One quantitative answer to this question now follows:

When the magnetic and electric-field cross talk (coupling coefficient) between two wire pairs is set equal for their low-frequency approximations (see Eqs. (6.17) and (6.23) there results.

$$\frac{Z_2 \mu \ell f}{2Z_2 \times 2Z_1} \ln_e\left(\frac{d+h}{d}\right) = Z_z \omega C_c \ell = \frac{Z_2 \pi^2 \varepsilon \ell f}{\ln_e (2d/D)} \tag{6.31}$$

for $Z_{S1} = Z_{L1} = Z_1$; $Z_{S2} = Z_{L2} = Z_2$; and for L-F applications i.e. $Z_2 \ll \frac{1}{C\omega}$

Simplifying Eq. (6.31), there results:

$$Z_1 Z_2 = \left[\frac{\sqrt{\mu/\varepsilon} \times 2.3}{2\pi}\right]^2 \log_{10}\left(\frac{d+h}{d}\right) \log_{10}\left(\frac{2d}{D}\right) \tag{6.32}$$

$$= 19,000 \ \log_{10}\left(\frac{d+h}{d}\right) \log_{10}\left(\frac{2d}{D}\right) \ \text{ohm}^2 \tag{6.33}$$

where, $\sqrt{\mu/\varepsilon}$ = 377 ohms

                2.3 = napierian-to-base-10 logarithm multiplier

In order to gainfully use Eq. (6.33), two-wire pair combinations of the coupling geometry will be illustrated, viz: (1) d (spacing between culprit and victim wires) >> h and D and (2) d h and D. In particular, let d = 10h and 10D, and let d = h and D. Thus, for:

$$d = 10h, (d+h)/d = 1.1 \text{ (culprit-victim far)}$$

$$d = h, (d+h)/d = 2 \text{ (culprit-victim near)}$$

$$d = 20D, 2d/D = 40 \text{ (culprit-victim far)}$$

$$d = 1.1D, 2d/D = 2.2* \text{ (culprit-victim near)}$$

$$d = 1.01D, 2d/D = 2.02* \text{ (culprit-victim very near)}$$

An equivalent geometry matrix is established in terms of the value of Eq. (6.33), viz:

$Z_1 Z_2 =$

(////)	$\dfrac{2d}{D}$	2.02	2.2	4	40
$\dfrac{d+h}{d}$	$\dfrac{d/D}{d/h}$	1.01*	1.1*	2	20
2	1.0	$350\Omega^2$	$1145\Omega^2$	$3440\Omega^2$	$9200\Omega^2$
1.1	10	(//////)	(//////)	(//////)	$1280\Omega^2$

The matrix relations are plotted in Fig. 6.20 for wire circuits #1 and #2 impedances for the indicated wire geometry. The furthest upper-right line corresponds to substantial separation of both wires in each pair and between pairs, whereas the lower left line corresponds to a very close wire pair separation. The central-region lines correspond to more typical conditions existing within a wiring harness or between wires in contiguous harnesses.

Below about 100 kHz, it can be generalized from Fig. 6.20:

(1) When the circuit impedance products, $Z_1 Z_2 \leq 300$ ohms^2, the principle coupling mechanism is magnetic field.

(2) When the circuit impedance products, $Z_1 Z_2 \geq 10k$ ohms^2, the principle coupling mechanism is electric field.

---

*   When d/D is not >>1, $\log_{10}$ (2d/D) must be replaced by 0.434 $\cosh^{-1}$ (d/D). For d/D = 1.1, 0.434 $\cosh^{-1}$ (1.1) = 0.2. Value = 0 when d/D = 1.

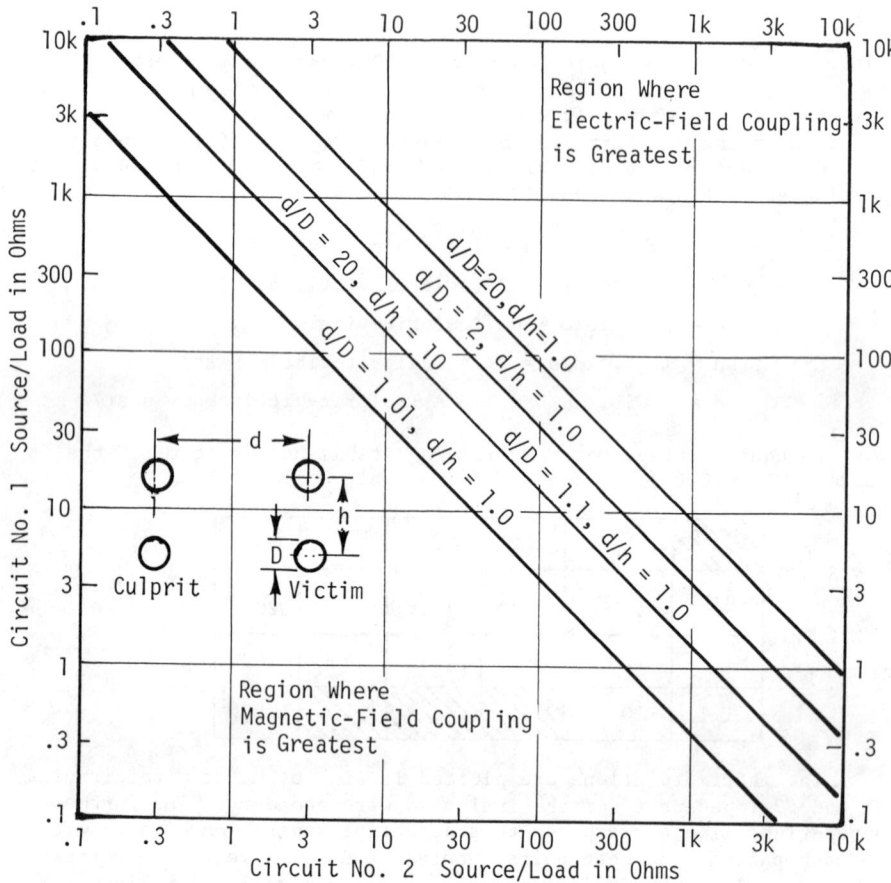

Figure 6.20 - Coupling Mechanism vs. Circuit Impedance for
Low Frequencies

(3) When the circuit impedance products, $300\Omega^2 \leq Z_1 Z_2 \leq 10k\Omega^2$,
the coupling may be either magnetic or electric field depending upon
relative wiring geometry.

Note that the above relations apply only for parallel-wire pairs
as shown in the insert sketch to the figure. If either or both pairs
had used twisted-wire geometry, the magnetic-coupling mode would have
been substantially reduced and the graph lines in Fig. 6.20 would significantly move to the lower left (the H/E-field crossover then might
approximate $1\Omega^2$, more or less). On the other hand, if either or both
parallel-wire pairs had used shields, the electric-field coupling mode
would have been substantially reduced, and the graph lines in the

figure would significantly move to the upper right (the H/E-field crossover then might approximate a circuit impedance product of $1M\Omega^2$, more or less). For twisted-shielded pair wires, the shift in H/E cross-over may move in either direction depending upon the relative de-coupling effectiveness.

6.4.2   Case History of Composite Coupling

To partially illustrate the previous discussion, Fig. 6.21 shows a number of circuit configurations operated at approximately 100 kHz. Unfortunately, both the source and termination impedance are unknown.* Configuration (a) shows a large coupling-loop area consisting of the outgoing wire length and ground-plane return path. Coupling into this circuit is primarily magnetic. The wire shield is grounded at one end and offers essentially no magnetic-field shielding (see Sec. 10.2). Coupling to this circuit is normalized to this condition and is there-fore, called 0-dB reference hereafter.

While circuit (b) uses twisted-wire pair, attempted magnetic-field de-coupling has been defeated since both source and load return wire ends are also grounded to the plane. This makes it perform sub-stantially the same as circuit (a). (Note, the 2-dB improvement is believed to be less than the measurement error and is not regarded significant.) Circuit (c) is similar to circuit (a) except the coax-ial shield return path, which would have resulted in significant mag-netic-field de-coupling, was also defeated by grounding the circuit at both ends to result in a ground-current loop of large area.

Circuit (d) shows a very significant improvement in de-coupling (49 dB) because of the twisted-wire pair (cf. circuit (b)) indicating that the main-coupling mode is indeed magnetic. Here, no circuit loop exists since the right end is floating, i.e., is not grounded. The further improvement of circuit (e) over (d) suggests that the partic-ular coaxial line used offers better magnetic shielding than the twisted pair, which in general may or may not exist (see Sec. 6.2.2). In this case, the coaxial loop area is very small (comparable to or smaller than the twisted pair) because the centroid of the return shield for the coaxial line is effectively located at or near its out-going center wire. The magnetic loop de-coupling integrity is pre-served in both cases by grounding the shield at one end only. It is also possible that some electric-field component may have been removed in circuit (e).

---

*   This sample is shown quite often in the literature, but the geo-metry and impedance levels are not mentioned other than tests were made for the wires at one inch above a ground plane at 100 kHz. Unfortun-ately, the original source of the tests seems to be unknown so that it is difficult to quantatatively verify the previous discussion.

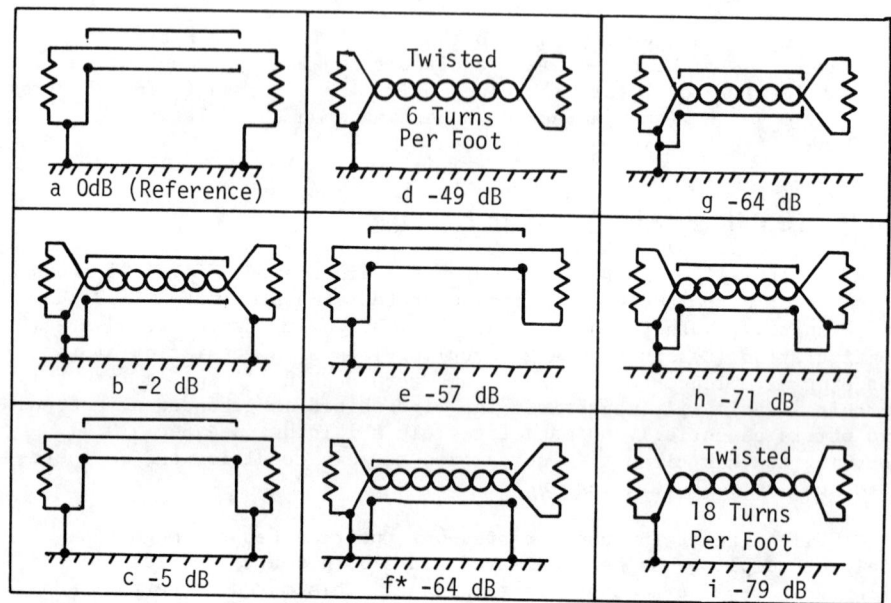

*Preferred Circuit For High Frequencies

Values Given Are For Circuits 1 Inch Above Ground Plane
But Are About The Same For Other Distances From Ground Plane.

Figure 6.21 - Relative Susceptibility of Circuits to Both
            Magnetic and Electric-Field Coupling

     Circuit (f) shows still further improvement by floating the re-
ceiving end of the twisted-pair circuit, previously shown in circuit
(d), to preserve magnetic de-coupling. Since an independent shield is
also used, the further improvement of circuit (f) suggests that some
electric-field coupling is starting to exist after the significant
magnetic-field de-coupling of circuits (d) and (e) have been put into
effect.

     Circuits (f) and (g) in Fig. 6.21 show comparable performance
(-64 dB relative coupling). This indicates that the length of the
wires and associated shields are a very small fraction of a wavelength
(λ=300 meters at 100 kHz) since de-coupling performance of shield
grounding at one end is the same as for both ends.

     It is not clear why circuit (h) should perform better than cir-
cuit (g). One speculation is that circuit (g) further reduces the
equivalent loop area by combining the centroids of the twist and coaxial wire position (cf. circuits (d) and (e)) relative to the outgoing

wire. In other words, the resulting loop area of circuit (h) is less than *either* circuits (d) and (e). Finally, a further twisting of the wire pair in uniform twists per foot shown in circuit (i) gives still better cancellation of the magnetic-field coupling. By twisting the wires in this manner, electric-field coupling also tends to cancel somewhat because capacitively induced currents tend to be common-mode in both wires (i.e., little to no differential-mode signals will couple).

Figs. 6.22* and 6.23 also show some representative electric and magnetic-field coupling situations reported in the literature. Fig. 6.22 shows the electric-field coupling at 60 Hz is increasing with an increase in circuit impedance. This is consistent with both Eq. (6.25) and Figs. 6.15 and 6.20 for high-impedance circuits.

Fig. 6.23 shows the magnetically-coupled induced voltage into the second wire circuit increases at 20 dB/decade as previously discussed in Sec. 6.2. The performance of the unshielded and shielded wires, both having 3/4 inch spacing from the culprit, shows negligible effect of the non-permeable shield below 1 kHz, and negligible effect below 10 kHz when the shielded wire is grounded at one end only. Interestingly enough, the 1K-ohm load impedance (cf. Fig. 6.20) suggests that electric-field coupling is present at 10 kHz, where the 10-foot length of lines corresponds to 10^{-4} wavelengths. It is difficult to explain why such a short length of shielded line ($10^{-4}\lambda$) should be sensitive to shield grounding point(s).*

Finally, Fig. 6.23 shows that about 20 dB of magnetic-field de-coupling resulted from twisting the power-line leads. In light of Fig. 6.21, more de-coupling would have been expected. However, the circuit impedance is relatively high and the twists per foot are unknown.*

6.4.3 Cable Multi-Shields and Grounding

An increasing amount of low-level circuitry is installed or inter-connected and operated near high-emission level, potential EMI sources. Microminiaturization among other circuitry invites this trend. Additionally, many military equipments and systems today are either designed for low-level radiation for security purposes or to be hardened to electromagnetic pickup from an atomic or thermo-nuclear explosion (electromagnetic-pulse (EMP) effect). For example, electric field intensities of about 1,000 V/m may be present 100 miles away from a 10-megaton explosion near the ground. Thus, all unguarded or

---

* This illustrates the danger in trying to develop a rationale for all situations when the reported experiments do not give all of the facts (e.g. Fig. 6.21 with unknown impedances, or Fig. 6.22 with unknown separation).

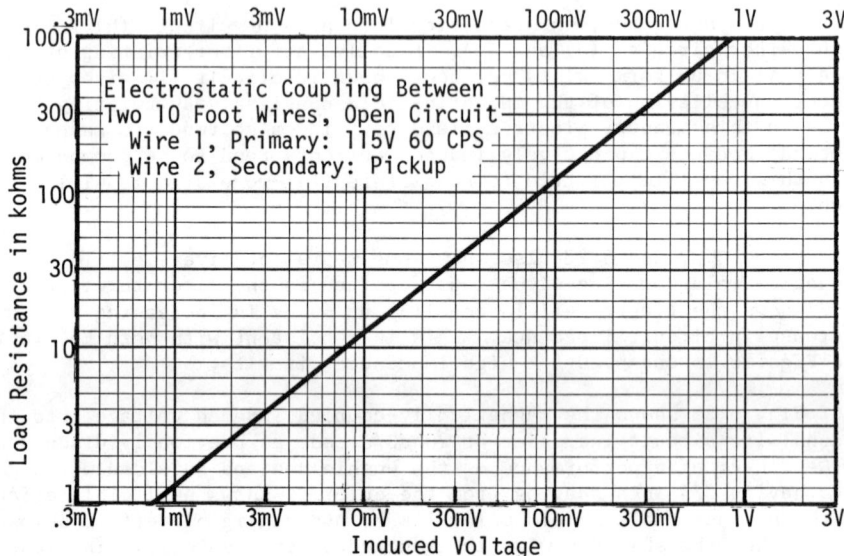

Figure 6.22 - Electrostatic Coupling Between Wires

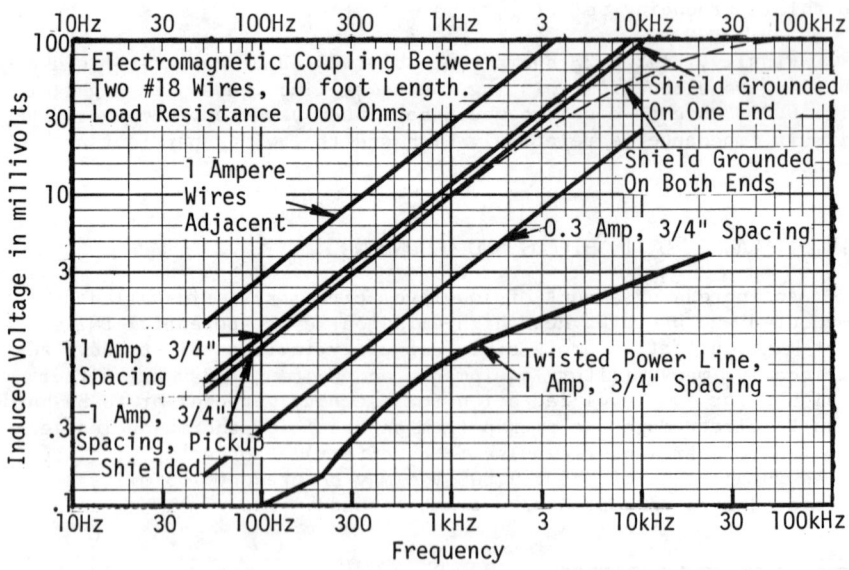

Figure 6.23 - Electromagnetic Coupling Between Wires

unprotected cable and equipment may result in burnout of sensitive components such as transistors, diodes, integrated circuits, receiver
front-end circuitry, sensitive relays, and the like. Accordingly, the
preceding shielding and grounding techniques associated with cables
may prove to be inadequate in severe electromagnetic environments.
This section reviews the more common cables and special cables and
shields including twinax (shielded twisted pair), triax, and quadrax.

6.4.3.1 Coaxial and Triax Cable

In cases of potential EMI, shielded cable should be used to protect against electric fields and in severe situations, against magnetic
fields. Grounded-coax cable installations generally perform well and
are typically used from 20 kHz (sometimes lower) to 10 GHz for many
installations. Coaxial lines, if subjected to strong interference,
may not adequately protect the desired signal. Then, more sophisticated cable and de-coupling techniques must be used dependent upon the
frequency and impedance levels of EMI and how it enters the cable
system. Additional measures taken to reduce noise pickup will conversely reduce outgoing radiation and crosstalk.

Coax cable consists of an inner and cylindrical outer conductor
with *both* conductors carrying the signal currents (source to load and
return). Inasmuch as the outer conductor is usually grounded at the
source, load, bulkheads and other intermediate points, *ground-current
loops* or *common-mode currents* caused by coupling of external noise
sources are also carried on the outer coaxial conductor as shown in
Fig. 6.24. Since both the desired signal and the undesired noise are
carried on the same outer conductor simultaneously, electrical noise
will be introduced into the system. Low-frequency applications (below
about 100 kHz) are particularly susceptible to the ground-current loop
and common-mode interference. In this case, when coaxial cable is
used, the complete coax chain should have a *minimum* number of outer
conductor ground contacts as shown in Fig. 6.25. This requires that
major equipment, relays, switches, connectors, patch panels, etc., be
isolated from ground with the ultimate being one system ground connection at the source. However, all this serves to adversely affect
the Faraday-shield performance at and above HF.

Where strong radiated EMI exists, such as from high-powered radar,
AM and TV broadcast stations, power lines, nearby fluorescent lighting,
and office and industrial machinery, the cable conductors act as receiving antennas or secondary windings of transformers and pick up the
external noise sources. A particularly bad source of EMI pickup is the
crosstalk or induced currents encountered in large multiple-cable installations. To protect against these radiated noise sources, two types
of improved cable are used: triax and twinax cables.

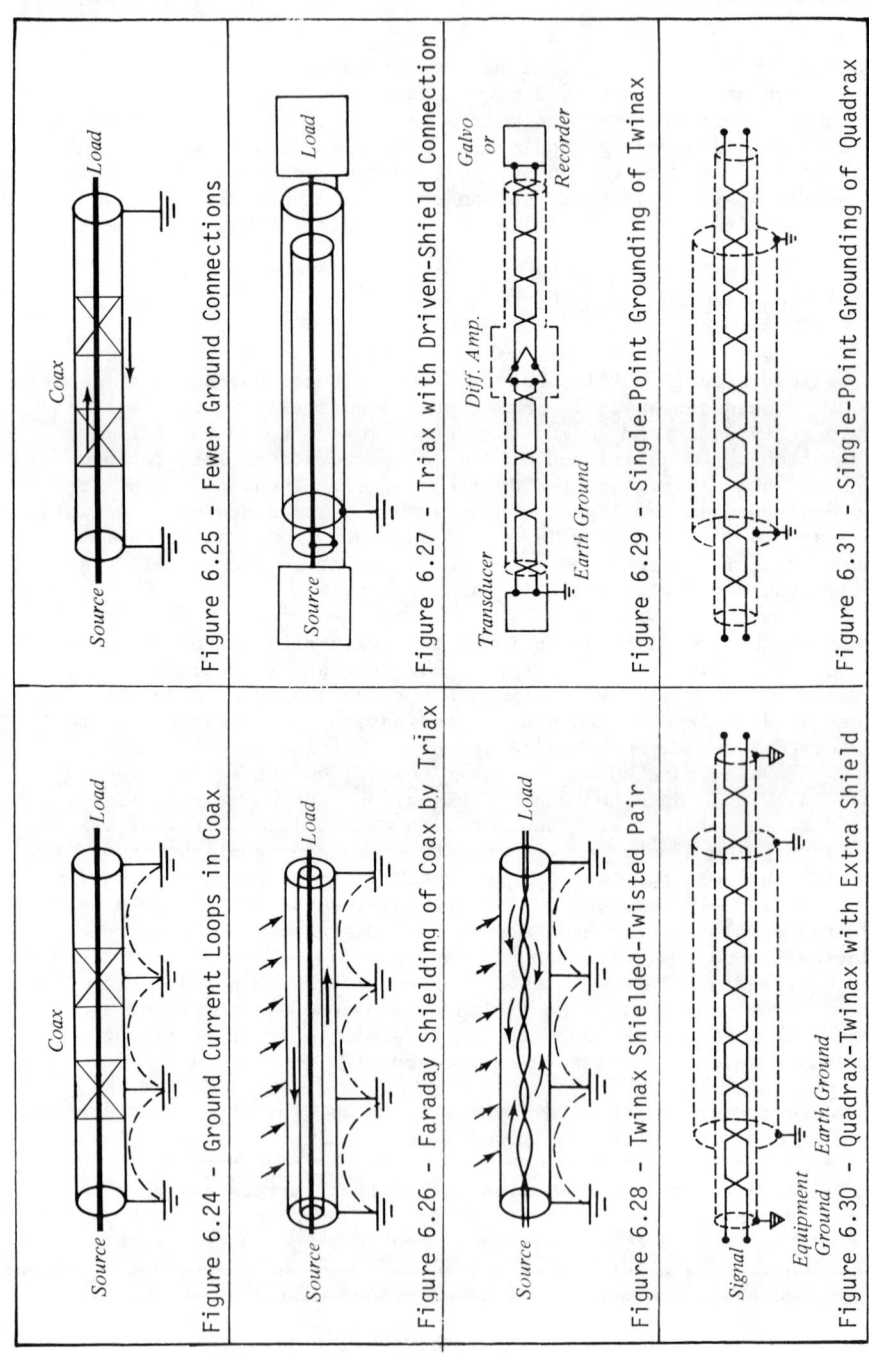

Figure 6.25 - Fewer Ground Connections

Figure 6.27 - Triax with Driven-Shield Connection

Figure 6.29 - Single-Point Grounding of Twinax

Figure 6.31 - Single-Point Grounding of Quadrax

Figure 6.24 - Ground Current Loops in Coax

Figure 6.26 - Faraday Shielding of Coax by Triax

Figure 6.28 - Twinax Shielded-Twisted Pair

Figure 6.30 - Quadrax-Twinax with Extra Shield

Triax cable is coaxial cable with an *additional outer copper braid* insulated from the signal-carrying conductors. It acts as a true Faraday shield and protects the enclosed coaxial conductors. This braid or shield is grounded as shown in Fig. 6.26, and bypasses both *ground-current loop* and *capacitive-field* noise currents away from the signal-carrying coax.

Triax cable may also be used in driven-shield applications where the inner conductor and inner shield are both driven in parallel at the transmitting end and collectively perform with the outer shield as shown in Fig. 6.27. At the receiving end, the inner shield is left floating providing a Faraday shield between the inner conductor and outer shield at frequencies for which $\ell \ll \lambda$. In this way, the cable *distributed capacity* is significantly reduced thereby reducing cable losses and loading. This application, not strictly an EMI situation, is most effective in low-frequency transducer data systems where the distributed capacity in coaxial cables limits the data capability of high-impedance circuits. Another variation of this, is to use only the two outer braids as a low-impedance transmission line (approximately 12 ohms) which can be used to carry high-current pulses to low-impedance laser lamps or electro-explosive ordnance devices. Triax cable and connectors completely *insulated* from ground are available for these applications.*

6.4.3.2 Twinax and Quadrax Cable

Twinax cable is a two-conductor, twisted balanced-wire line having a *specific impedance*, with a grounded shielding braid around both wires as shown in Fig. 6.28. Twisting the two balanced signal-carrying wires provides cancellation against *magnetic* fields as discussed in previous sections. This cable also provides protection against ground-current loops and electric fields, as did triax cable. Twinax cable usefulness, however, is limited to an upper frequency of approximately 10 MHz since it exhibits rather high transmission losses above this frequency. Twinax, having a 124-ohm impedance, is extensively used by the Bell Telephone System for video-TV transmission.

Additional *common-mode rejection* of noise can be obtained in instrumentation systems where low-level transducer information must be remotely recorded. Here, twinax is used with only one ground contact located at the transducer as shown in Fig. 6.29. Special insulated concentric twinax connectors are also available.

---

* A still further variation of the 12Ω impedance line is to use a lossy dielectric between the two braids to absorb HF to UHF EMI and convert it to heat. See Sec. 14.3.2.

For still further EMI decoupling in protected and guarded flex-ible-cable circuits, triax cable with *two* separate and insulated braids, called quadrax, can be used wherein the two braids are inde-pendently connected to system ground and earth ground, respectively (see Chap. 8) as shown in Fig. 6.30. Quadrax cable can also be used to provide additional noise and EMI suppression by connecting both shielding braids to earth ground at one place when a separate equip-ment ground is not available as shown in Fig. 6.31. The inner braid is left floating above ground at all other locations to act as a Faraday shield and provide additional circuit isolation as long as its length is less than about 0.1λ. Coaxial cable with *two* extra in-sulated braids can be used in similar engineering concepts for un-balanced systems.

6.5   QUICK-FIX DE-COUPLING TECHNIQUES

In addition to the preceding magnetic and electric-field de-coupling techniques in wires and cables, there are other EMI-control practices which can be used. While some of them are best carried out in equipment design stages, sometimes an EMI problem will show up only after system installation. Several cable shielding and EMI protection techniques are reviewed, some of which are better applied as quick-fix or retrofit applications.

6.5.1   ZIPPER CABLE SHIELDS

When EMI is induced into a cable harness or bundle which has many wires and connectors, it would appear to be too late to protect the bundle in the normal manner since the harness is already made up (emphasis here is on existing installations). Furthermore, some harnesses are large and complex and represent a substantial previous investment. For these situations, zipper-cable shielding may repre-sent one quick-fix technique. As illustrated in Fig. 6.32, this cable shield can be slipped around a harness and zipped up to close the shield. Where the bundle has one or more branch *break-outs* or transi-tions, special coupling sections may be used as shown in the figure.

Zipper or slide-fastener shields are available from different suppliers* in several shield mesh diameters, densities, and materials. Typical diameter sizes available are 3/4" to 5". Shield braids avail-able include both non-permeable (e.g., tin-plated, copper-wire mesh

---

* Suppliers of EMI zipper-cable shielding include: (1) Metex Corp., 970 New Durham Rd., Edison, N.J. 08817, phone 201-287-0800; (2) Tech-nical Wire Products, Inc., 128 Dermody St., Cranford, N.J., phone 201-272-5500; and (3) Zippertubing Co., 13000 S. Broadway, Los Angeles, Calif., phone 213-321-3901.

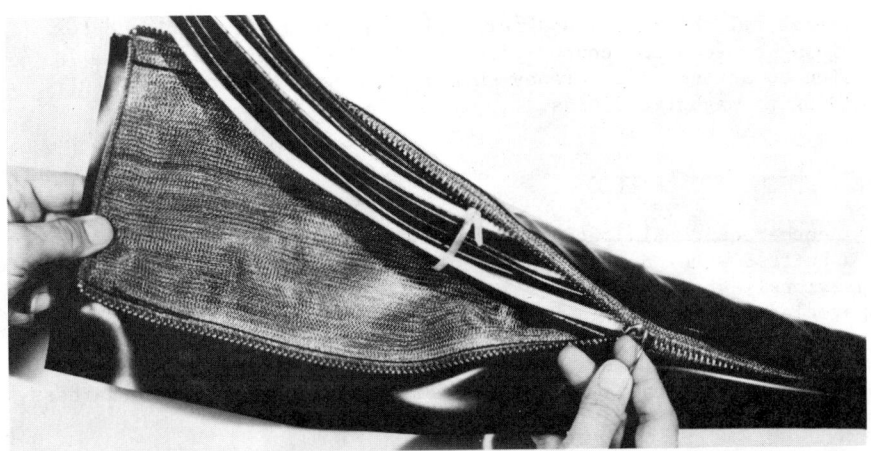

Courtesy of Metex Corp.

(a) Placing Zipper-Cable Shield Over Wire Harness

Split couplers are used to make splices between lengths or between transition components. The loose half of the split coupler is placed in position, the two slide fasteners pulled closed. The splice is completed by securing the ends with standard cable tie-wraps.

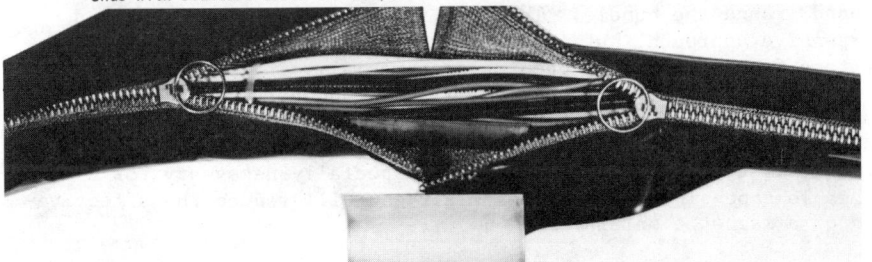

Courtesy of
Metex Corp.

(b) Splices Between Lengths of Shield Use Split Couplers

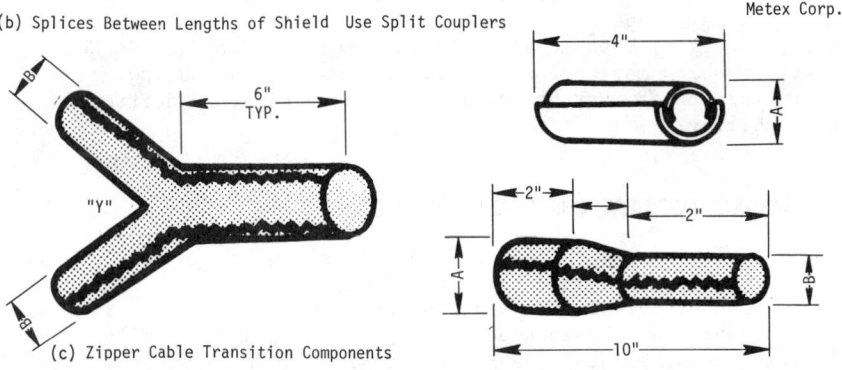

(c) Zipper Cable Transition Components

Figure 6.32 - Quick-Fix Zipper Cable Shields Over Wire Harness

of several mil thickness) and permeable type (e.g., Ferrex by Metex
containing tin-plated, copper-clad, stainless-steel wire mesh). In
addition to acting as a Faraday shield, the permeable braids offer
shielding to magnetic fields.

6.5.2   Mesh-Tape Shields

Another quick-fix technique for shielding a cable harness is to
use a knitted-wire mesh tape applied as a bandage to cover the cable
as previously shown in Fig. 6.11 Typical material is tin-plated,
copper-clad steel, knitted to a width of one inch and thickness of
about 15 mils. The permeability of the tape also offers some magnetic-field suppression. This tape need not be limited to quick-fix
applications as it may also suffice for standard manufactured harness
shields.

6.5.3   Foil-Tape Shields

Another method of covering a cable harness with a shield which
is similar to the above mesh tape, is to use a solid metal tape
wrapped around the bundle. This has the advantage of a quick and
inexpensive approach, but it may present difficulties in preventing
the tape from slipping or opening up especially if the harness is to
be flexed in use. Some tapes come with a conductive adhesive material bonded on one side to prevent this problem. Good electrical continuity is necessary in the tape overlap sections if shielding performance is to be sustained. This is especially necessary for a
permeable tape since a significant air gap will reduce the effectiveness of a magnetic shield.

Magnetic-foil tapes are available in thicknesses from about two
mils to 10 mils. Tape widths vary from about 1/2" to 2". Some tapes
are made of very-high permeable materials such as Netic or Co-Netic*
and hypernom which may have permeabilities as high as 80,000. The
high-permeable tapes must be handled with care to avoid destroying
the permeability.

6.5.4   Conductive Heat-Shrinkable Shielding

Some cable harness shielding material is applied in the form of
a heat-shrinkable tubing and boots as shown in Fig. 6.33. One such
tubing is made of polyolefin whose inside or outside surface is electrically conductive. The conductive coating is usually a silver
system which remains intact even after maximum heat shrinking.

---

* By Perfection Mica Company, 740 Thomas Drive, Bensenville, Ill.,
phone 312-766-7800.

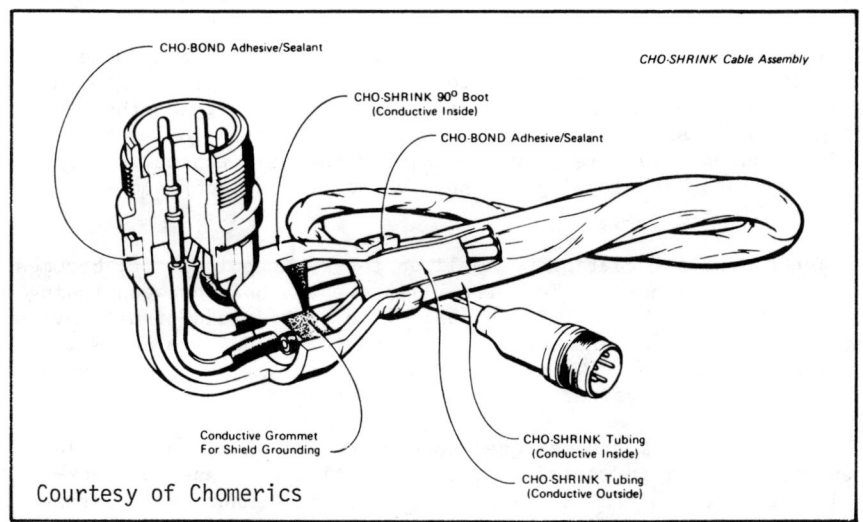

Figure 6.33 - Heat-Shrinkable Shield Tubing and Boots

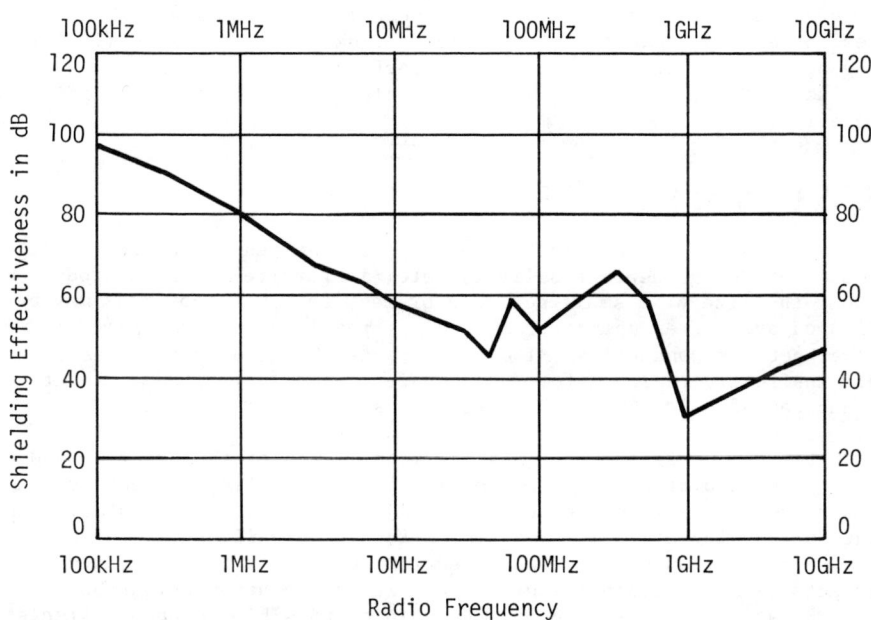

Figure 6.34 - Shielding Effectiveness of Chomerics Heat-Shrinkable
Shields

The heat-shrinkable tubing is available in expanded diameters
from 3/16" to 4", and lengths up to 8 feet. When longer lengths are
required, sections can be spliced together using short lengths of
tubing whose outside surface is conductive. These splice sections
are first shrunk over the joint areas, and the main tubing is then
shrunk in place, overlapping the splices to provide continuous electrical continuity.

The conductive coating, as well as the basic polyolefin, becomes
thicker after shrinking. To further improve the bond between tubing
and the part over which it is shrunk, or to provide additional environmental sealing at the ends, conductive and non-conductive hot-melt
adhesives may be used. These adhesives flow at shrink temperatures
to provide both a physical bond and an environmental seal.

Connector boots are designed to provide EMI shielding, shield
grounding, and strain relief in the termination of connector backshells. Conductive coating, applied to the inside surface of standard
heat-shrinkable polyolefin boots, provides about 40 dB of attenuation
to electric fields at 1 GHz and 100 dB at 10 kHz as shown in Fig.
6.34.

Advantages of boots over R-F backshell adapters are lower weight,
lower cost, and the ability to shrink to a wide range of cable diameters. Boots are supplied with a conductive hot-melt adhesive/
sealant applied to each end. Normal shrink temperatures permit joining and sealing at boot/connector interfaces. Cable transitions,
including T, Y, and other shapes, are also available in polyolefin
material.

6.5.5   FERRITE BEADS AND RODS

Another EMI quick-fix, which has often helped in an electronic
equipment design nearing delivery date, is ferrite beads and rods.
About the size of a small pea, one or many such beads or rods may be
slipped over wires coming from a common regulated power supply or
other network conducting EMI or transients at radio frequencies. They
are ineffective below a few MHz and are typically limited to about
5 amperes of DC or 60/400 Hz power mains.

Ferrite beads and rods have permeabilities of the order of 600.
When slipped over a wire, the resulting magnetic-field intensity con-
centrates in the lower reluctance path of the bead or rod rather than
the surrounding air. This has the effect of significantly increasing
the self inductance of the wire section covered by the bead, i.e.,
it acts as a one-stage filter. Thus, it offers attenuation above a
few MHz as shown in Fig. 6.35. Because of possible resonant effects
exhibited by alternating sections of low inductance (no bead) and high
inductance (bead or rod), EMI attenuation becomes relatively insignificant above about 100 MHz.

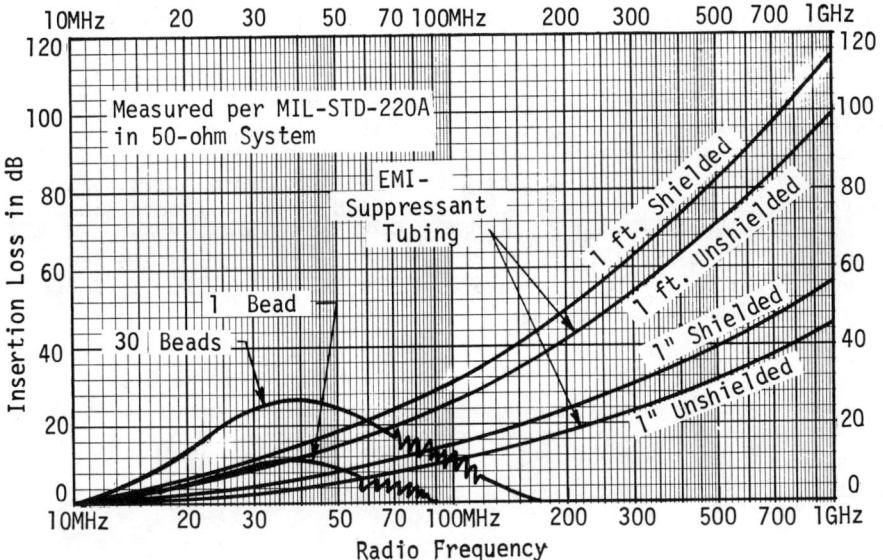

Figure 6.35 - Comparison of Attenuations of Ferrite Beads and Rods with Flexible EMI- Suppressant Tubing

6.5.6   EMI Suppressant Tubing

Another EMI quick-fix technique for wires and cables is based on an extension of some of the advantages of beads and rods and serves to overcome most of the disadvantages. Here, a flexible tubing material may be slipped over an insulated or uninsulated conductor of standard sizes.* Because a lower equivalent permeability of tubing is used compared with beads and rods, little attenuation of EMI is offered below about 5 MHz as shown in Fig. 6.35. On the other hand, no saturation or resonant-frequency properties are exhibited and attenuation above 100 MHz becomes significant.

The principle of operation of the EMI suppressant tubing is similar to that of beads and rods. Having an equivalent permeability of about 10, the self inductance of a wire covered with the tubing is increased so that it acts as a one-stage, distributed filter with series inductance. By avoiding the alternating high and low incremental inductance of beads and rods, tendency to radiate between elements is avoided. The tubing is also available with a shielded

---

* Supplier: Lundy Electronics & Systems, Inc., Glen Head, New York, phone 516-676-1440.

layer of metallized mylar for electric-field shielding at lower
frequencies.

The EMI suppressant tubing, available in one-foot lengths, and
ID's from 0.1 inch to 0.5 inch, may be sliced longitudinally along
its axis and snapped over a cable harness. Similar material in
paste form is used to fill any resulting longitudinal slots, tubing
interfaces, or at cable breakouts. This kind of a cable cover has
an advantage over shielded cables in that it is often difficult to
ground the latter sufficiently often at VHF and UHF to sustain performance, whereas the former converts EMI noise to heat and performs
better at and above VHF.

6.6 BIBLIOGRAPHY

(1) AF/BSD Exhibit 62-87, "Electro-Interference Control Requirements for Minuteman (WS 133B)," 6 December 1962.

(2) Boeing Aircraft Corporation D2-2-2444, "Electro-Interference Control Requirements (Equipment)," 5 March 1959.

(3) Bridges, J. E., and Brueschke, Dr. E. E., "Hazardous Electromagnetic Interaction with Medical Electronics," *Record of the 1970 IEEE International Symposium on Electromagnetic Compatibility*, Vol. 70 C28-EMC, July 14-16, 1970; Anaheim, pp. 173-182.

(4) Cacace, R., and Hassett, R., "Investigation of Measurement Techniques for Transient Magnetic Fields," *Symposium Digest, 7th National Symposium Electromagnetic Compatibility,* June 28-30, 1965; New York.

(5) Clough, L., and Salzetti, J., "Magnetic Induction Susceptibility at Power Frequencies," *Eighth Tri-Service Conference on EMC,* October 30-November 1, 1962; Chicago, Ill., pp. 241-269.

(6) "Designers Guide on Electromagnetic Compatibility, Cabling of Electronic Equipment," *EMC Bulletin*, No. 8, Electronic Industries Association, March, 1965.

(7) Dionne, A. E. J., "Some Practical Approaches in the Control of Interference in Airborne Weapons Systems," *IEEE Third National Symposium on RFI,* Washington, D. C., June, 1961.

(8) Dolle, W. C., and Cory, W. E., "Measurements of the Attenuation of the Electric and Magnetic Fields at Points Close to the Source," *IEEE Transactions on Electromagnetic Compatibility,* Vol. EMC-10, No. 2; September, 1968, pp. 313-319.

(9) Dwight, H. B., "Calculations of Resistance to Ground," *Electrical Engineering,* Vol. 55, pp. 1319-1328, 1936.

(10) Feber, R. R., and Young, F. J., "The Shielding of Electromagnetic Pulses by the Use of Magnetic Materials," *Record of the 1969 IEEE Symposium on Electromagnetic Compatibility,* Vol. 69C3-EMC; June 17-19, 1969; Asbury Park; pp. 73-74.

(11) GM07-59-2617A, "Electro-Interference Control Requirements for Minuteman (WS 133A)," 20 October 1959.

(12) Greenstein, L. J., and Tobin, H. G., "Analysis of Cable Coupled Interference," *IEEE Transactions on Radio Frequency Interference,* Vol. RFI-5, No. 1, pp. 43-55, March, 1962.

(13) Haber, F., "Study of GSFC Radio Frequency Interference (RFI) Design Guideline for Aerospace Communication Systems," Moore School of Electrical Engineering, University of Pennsylvania, Moore School Report No. 66027 for NASA, April 30, 1966.

(14)   Handbook of Instructions for Aerospace Systems Design, Vol. 4, Electromagnetic Compatibility AFSCM 80-9, Code SEG(SEPSM), Wright Patterson Air Force Base, Ohio, 45433, Basic Issue, 20 April 1964.

(15)   Herring, T. H., "A Method for Controlling Airplane Wiring and Equipment Placement to Eliminate AC Magnetic Field Interference," *Proc. Fourth Conference on Radio Interference Reduction,* Chicago: Armour Research Foundation, pp. 412-430, Oct., 1958.

(16)   Herring, T. H., "The Electrical Role of Structure in Large Electronic Systems," *IEEE Fifth National Symposium on Radio Frequency Interference,* New York, June, 1963.

(17)   "Interference Reduction Guide for Design Engineers," Vols. 1 and 2, prepared by Filtron Co., Inc. for U. S. Army Electronics Labs., Fort Monmouth, N. J., Accession AD 619666 and AD 619667.

(18)   Johnson, W. R., et al., "Development of a Space Vehicle Electromagnetic Interference/Compatibility Specification," *NASA*, Contract No. 9-73-5, *TRW*, 08900-60001-T000; June 28, 1968.

(19)   Kaplit, M., "Electromagnetic Coupling Between Coaxial, Single-Wire, Two-Wire, and Shielded Twisted Pair Cables," *Proc. Ninth Tri-Service Conference on Electromagnetic Compatibility*, Chicago; Armour Research Foundation, pp. 183-192, Oct., 1963.

(20)   King, G. J., "Achieving Electromagnetic Compatibility by Control of the Wiring Installations," *Proc. Ninth Tri-Service Conference on Electromagnetic Compatibility*, Chicago; Armour Research Foundation, pp. 193-204, Oct. 1963.

(21)   Martin-Marietta Corp., "Circuit Transient Specification, Apollo Applications Program (AAP) Payload Integration," Martin Document ED-2002-889; August 29, 1969.

(22)   McAdam, W. and Vandeventer, D., "Solving Pickup Problems in Electronic Instrumentation," *ISA Journal*, Vol. 7, No. 4, p. 48, April, 1960.

(23)   McDonald, G. M., and Taylor, G. R., "Shielding Grounding and Circuit-Grounding Effectiveness in Interference Reduction in the 50 Hz to 15 KHz Frequency Region," *IEEE Transactions on Electromagnetic Compatibility*, Vol. EMC-8, No. 1, pp. 8-16, March, 1966.

(24)   Newman, I. M., and Albin, A. L., "An Integrated Approach to Bonding, Grounding, and Cable Selection," *Proceedings Seventh Conference on Radio Interference Reduction and Electromagnetic Compatibility*, Chicago, Ill., Armour Research Foundation, pp. 434-459, November, 1961.

(25) Pearlston, C. B., "Case and Cable Shielding, Bonding, and Grounding Considerations in Electromagnetic Interference," *IRE Transactions on Radio Frequency Interference*, Vol. RFI-4, No. 3, pp. 1-16, October, 1962.

(26) Salati, O. M., Showers, R. M., Haber, F., and Schwartz, R. F., "Training Course in Electromagnetic Compatibility," Moore School of Electrical Engineering, University of Pennsylvania, 1966.

(27) Siegel, N. S., "Near Field Coupling on Aerospace Vehicles," *1970 IEEE EMC Symposium Record*, July 14-16, 1970; Anaheim, California; pp. 211-216.

(28) Weinstock, G. L., "Electromagnetic Interference Control within Aerospace Ground Equipment for the McDonnell Phantom II Aircraft," *IEEE Trans. on Electromagnetic Compatibility*, Vol. EMC-7, No. 2, pp. 85-92, June, 1965.

(29) White, J. V., "Wiring of Data Systems for Minimum Noise," *IEEE Trans. on Radio Frequency Interference*, Vol. RFI-5, No. 1, pp. 77-82, March, 1963.

(30) White, Donald R. J., "EMI Control Methodology and Procedures" Don White Consultants, Inc., Gainesville, Virginia, 1980

CHAPTER 7

CONNECTORS

CHAPTER 7

CONNECTORS

The companion to either the wire cable or harness is the connector. A connector is an assembly of mating contacts which permits quick linking and separation of a cable with either another cable or equipment. The number of wire pins and/or coaxial sheaths making simultaneous contacts may range from two to several hundred. Individual pin contacts are embedded in insulating material to mutually isolate them and to prevent coming in contact with bare hands. In a link or engaged position, the connector should provide a low-impedance path for all internal wires and a low-impedance bond when an outer shell is ued.

This chapter surveys the connector component with emphasis on their EMI control. Thus, the connector back shell, one of the major points of EMI penetration or leakage, is discussed. Other forms of connector problems and EMI control are reviewed together with filter-pin and unbalanced connectors and adaptors.

7.1 CONNECTOR CONTACT PROBLEMS

EMI problems associated with connectors are similar to those manifested by mechanical switches. One principle problem is poor contact which may result directly in arcing or in over heating which eventually leads to arcing. Poor connector contact also invites driven-circuit voltage variations due to contact-impedance modulation of the driving-current source. Common-impedance coupling from outside sources can exist in connector grounding paths. This section discusses the connector contact problems and subsequent sections discuss the other EMI problems and control. Improperly shielded connectors can invite radiated emission penetration or leakage.

7.1.1 CONTACT IMPEDANCE

The geometry of the connector may cause EMI because of discontinuity in its structure. For example, shell discontinuities can create unshielded loops and unwanted localized lumped impedances possibly at sensitive locations in circuits. All connectors become limiting factors in circuits operating at high frequencies. At RF, manufacturers rate connectors in terms of the resulting voltage standing-wave ratio

(VSWR) versus frequency. While a well-designed connector, will not
cause interference, per se, design deficiencies may result in considerable EMI to other circuits due to either radiation coupling or
conduced-path variations. The best design combines low-spring pressure
and good peripheral electrical contact.

Contacts should maintain a low-impedance bond in the link position.
Thus, there exists requirements for testing connectors to detect and
isolate high-impedance contacts:

● Ensure that the applied voltage is about one-tenth the normal
working voltage, so that the normal film and tarnish accumulation is
not broken down by the test voltage.

● Use the highest test frequency to which the driven loads may be
susceptible. It is sufficient to conduct D-C resistance measurements
when the loads are audio and low-frequency circuits.

● Submit the connector to low-level, low-impedance tests, then
perform an environmental test of mechanical forces, wear, and corrosion.

A common cause of faulty connector contacts is damage during mating.
Ensure adequate contact floating to permit insertion without binding and
prevent wedging by correct pin layout. Properly placed guide pins will
reduce bending, gouging, and abrasion due to misalignment. With guides,
alignment occurs without trial-and-error scraping of pins across the
female contact to find the alignment position. Protective coverings
should extend over the male pins to reduce pin damage. Pin overdesign
for an extra low length-to-diameter ratio provides ruggedness. Provide
inexpensive protective plastic caps for use during handling and storage.
Potting the back ends of connectors decreases entry of moisture, fumes,
contaminants, and foreign objects. Clamps prevent wires being pulled and
twisted from their contacts. Ensure good contact pressure over a long
time by use of low fatigue, high-resilience spring materials.

7.1.2 CONTACT FINISHES

Low-impedance contact can occur in either of two ways: (1) simple
contact under pressure in which the pins are pressed together to break
or wipe away film and tarnish, or (2) breakdown contact in which the
pins have a film which is not ruptured by connector mating. Light arcing forms channels of molten metal which provides a low-resistance contact due to reduced cross section. A capacitive effect can exist if
the voltage level between the pins is not enough for the dielectric
oxide film breakdown. High-resistance contact can be produced on some
metal oxides on the surface of the connectors.

To mitigate the above problems, contact plating is used. Plating
yields the advantages of increased tarnish and corrosion resistance.
Gold satisfies both of these parameters. Certain hard-gold alloy platings are preferred from an electrical conductivity corrosion resistance
as well as wearability. For an underplate for the gold alloy, 0.1 mil

of ductile nickel (elongation of not less than 5%) provides the best overall combination from a performance aspect for a sustained period of time.

Nonhermetically sealed connectors contain a copper-based alloy while hermetically sealed connectors usually consist of an iron-nickel alloy. The gold underplating for hermetic seals should be copper over the iron-nickel substrate followed by a plating of nickel. If gold is plated directly to the copper plating, the gold will be diffused into the copper thereby establishing a seal leakage as well as a corrosion mechanism.

The finer the microfinish of the contacts mating surfaces the better the corrosion resistance characteristics and the less the insertion and withdrawal forces. Microfinishes as low as 10 micro-inch are achievable.

7.2 CONNECTOR BACK SHELLS

Connector types may be divided into three classes: (1) low-frequency single and twin-conductor connectors, (2) low-frequency multi-pin connectors, and (3) high-frequency unbalanced-line (coaxial, tri-axial, and quadrax cable) connectors. The dichotomy here between low and high frequency may exist anywhere between 100 kHz and 10 MHz. Co-axial-cable connectors are discussed in a later section. This section discusses the multi-pin connector and shielding of its outer shell to mitigate radiation leakage and penetration.

Multi-pin connectors generally have an external shield which slips over the harness at the connector and secures to the conductor termination or mating shell. The *back shell*, as it is called, serves as a form of strain relief and provides a 360-degree peripheral shielded configuration around the harness assembly at the wire-connector interface. The back shell also serves to terminate (i.e., ground) the shield to either a connecting housing or another mating connector shell assembly. Thus, a good multi-pin connector is one in which the shielding effectiveness of the mated connector equals or exceeds that of an equal length of the interconnecting cable shield.

In an otherwise adequate shield, induced R-F currents that are conducted along cable shields may be coupled to the system wiring at the point of improper cable termination. When a shield is properly terminated, the entire periphery is grounded to a low-impedance reference; this minimizes R-F potentials at the termination (see Sec. 6.3). The use of epoxy or other synthetic-conducting material does not generally result in as good a bond as obtainable in this situation.

Fig. 7.1(a) illustrates a permanent termination of the cable shield to a connector. Here, the outer shield is made continuous with the connector back shell by a soldering or metal-forming bond. Spring

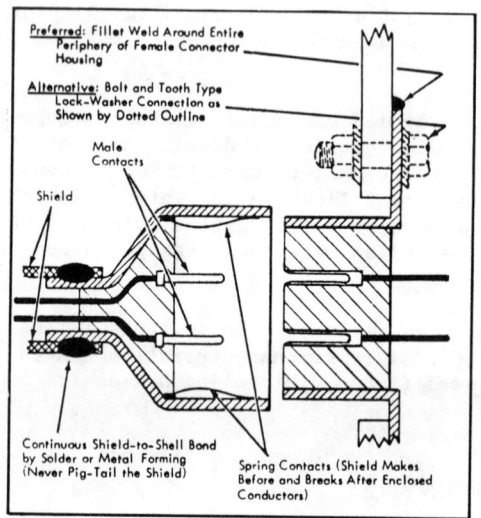

(a) Individual Conductors Are Unshielded

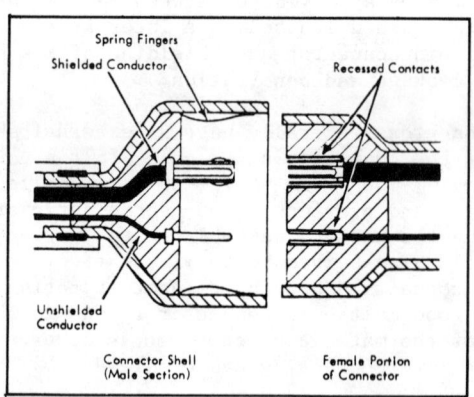

(b) Individual Conductors Are Shielded

(Courtesy AFSC Design Handbook DH 1-4 EMC)

Figure 7.1 - Shield Termination for Electrical Connectors.

fingers are used to carry the shell continuity to the mating connector. The illustration also shows the through path for unshielded individual connectors. When more than one shielded inner conductor must be routed through a single cable and connector, the technique suggested in Fig. 7.1(b) is employed to preserve individual internal-wiring shielding. The internal coaxial shields should never be pulled back, twisted and then bonded to the outer connector sheath, i.e., no portion of the co-axial shield should be broken before it is bonded to the connector shell. Individual shields for connections that are routed through multi-pin coaxial connectors should be terminated individually in the manner illustrated in the figure.

Fig. 7.1 indicated that the cable shield is permanently secured to the connector shell. While offering the best bond, this practice is not particularly cost effective in manufacturing time. Methods of quick mechanical compression bonding of the cable braid to the shell have been developed by the EMC connector manufacturers. Many such connector varieties permit rapid assembly, require no special tools, are field repairable, and permit environmental sealing. They are available in both permeable and non-permeable-base materials to shield against both H and E-fields or E-fields only.

Fig. 7.2 illustrates typical adaptors and backshells used for overall cable shield termination, and Fig. 7.3 shows adaptors for individual termination of shielded wires using the connector shell as ground point.

Courtesy of Glenair, Inc.   Courtesy of Glenair, Inc.

Figure 7.2 - Adaptors and Backshells for Overall Cable Shield Termination

Figure 7.3 - Adaptors for Individual Termination of Shielded Wires Using Connector Shell as Ground

7.3 TERMINATION OF INDIVIDUAL WIRE SHIELDS

When a cable harness contains many individual shielded wires, in which each shield acts as a Faraday cage, continuity through the mating connector interface is obtained via an individual pin for each shield. This suggests that for non-coaxial pin connectors twice as many pin-receptacle contacts are necessary for individual shielded-wire cables. To cut down on the number of extra pin contacts necessary for shield continuity a technique of *daisy-chaining* is sometimes employed where the harness contains many wires carrying signals from DC up to about 1 MHz. In this technique a single dedicated pin is not used for the continuity of *each* individual wire shield in the assembly. Rather one pin may carry up to five individual wire shields connections.

The daisy-chain practice is to peel back the outer braid of each shielded wire and connect these braids in groups of five by an insulated wire looping from outer-shield-to-outer-shield in a daisy-like manner. The final wire from the shield group goes to a separate dedicated feed-thru pin. While this practice compromises Faraday shielding between short lengths of resulting unshielded wires at high frequencies, reliance is made upon the outer cable connector back shell for overall shielded at the cable-connector interface. Cables carrying signals above 1 MHz should not use this practice; in fact, daisy-chaining is regarded obsolete since multi-coaxial pin connectors are now available to give better performance.

One alternative to the daisy-chain technique is the halo-ring technique, in which individual shielded wires in a harness must have a *common* shield ground at the connector. Here a cylindrical conductor (the halo) is used as shown in Fig. 7.4 to ground all applicable shields to ground through one or more connector pins. Where final termination is to exist at an equipment housing, shield halos should be bonded to the ground plane by 1.5 inches or less of 0.25 to 0.5 inch wide, tin-plated, copper strap.

The halo technique is acceptable only when a relative few shielded wires are involved. A preferred method where cost implications become important is to use a collectively crimped peripheral ring as illustrated in Fig. 7.5 for all wire shields exclusive of those intentionally operated as either individual coaxial cables or low-level audio shielded leads. The collective crimping ring uses two ground wires. Connect one wire from the ring to the connector shell where connector design permits. The other wire is carried through the connector. Fig. 7.7 shows what the resulting outer shield grounding configuration would look like.

The best performing method to use, but relatively expensive to manufacture, is called the *interlacing-strap method*, shown in Fig. 7.6. It is used for a common shield ground in multi-shielded wires in harnesses that have a large number of individual internal shields. The

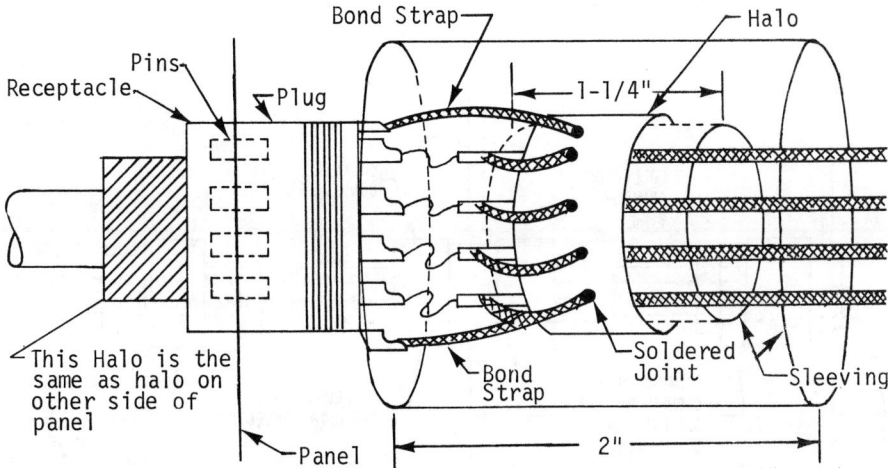

NOTES: 1. Bond Strap may be connected as shown or with 1/4" bond
            strap tied to structure or connector by means of eared
            washer.
         2. Halo is 1/4" to 1/2" wide.

Figure 7.4 - Bonding Ring or Halo at Connector for Terminating
Shields in a Harness

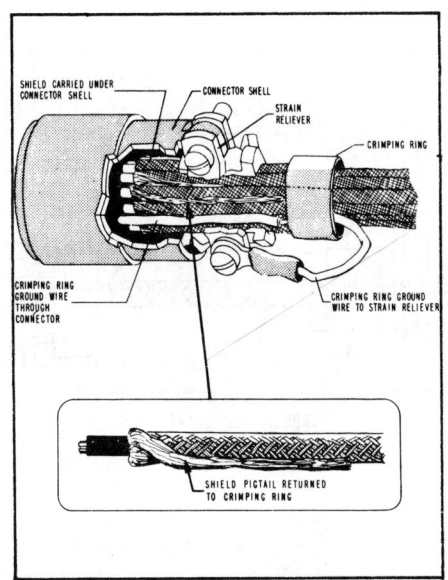

Figure 7.5 - Crimping-Ring
Technique for Terminating Shields

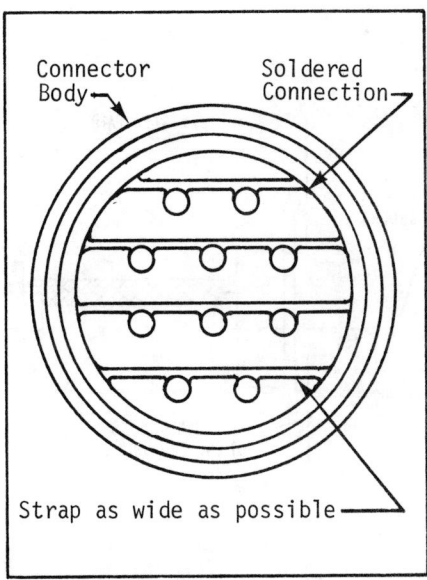

Figure 7.6 - Interlacing Technique
for Terminating Shields

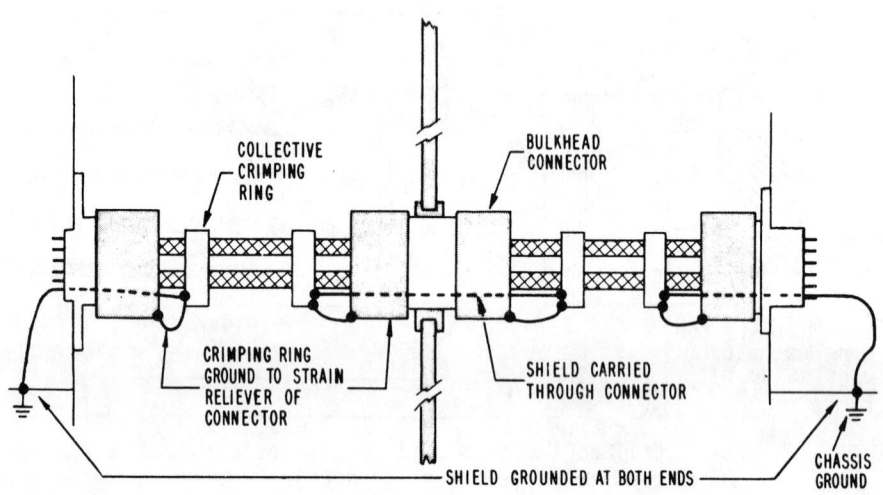

COLLECTIVE CRIMPING RING

BULKHEAD CONNECTOR

CRIMPING RING GROUND TO STRAIN RELIEVER OF CONNECTOR

SHIELD CARRIED THROUGH CONNECTOR

SHIELD GROUNDED AT BOTH ENDS

CHASSIS GROUND

(see Figure 7.5 for Details of Collective Crimping)

Figure 7.7 - Wiring System Termination of Shielded Wires

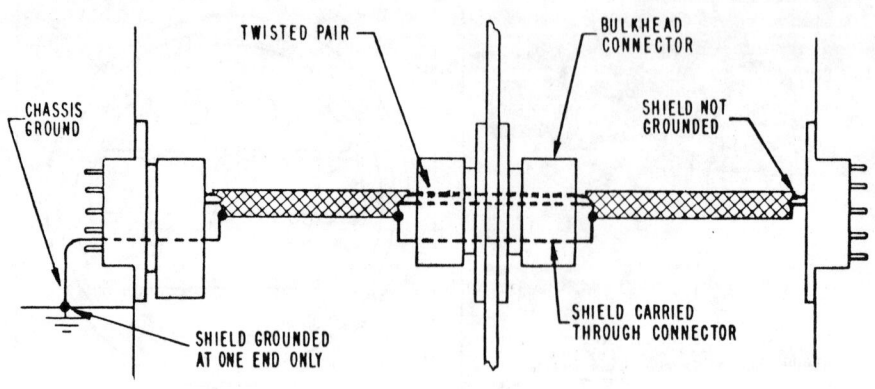

CHASSIS GROUND

TWISTED PAIR

BULKHEAD CONNECTOR

SHIELD NOT GROUNDED

SHIELD CARRIED THROUGH CONNECTOR

SHIELD GROUNDED AT ONE END ONLY

Figure 7.8 - Termination of Shielded Audio Susceptible Wires

interlacing strap should be at least 0.25-inches wide by 10 mils thick
and be bonded securely to the connector as shown in the figure. The
strap follows the wiring system ground as previously shown in Fig. 7.7.

Where multi-shielded wires are to protect audio-susceptible circuits, they should be grounded at one end only as shown in Fig. 7.8.
Individual twisted-wire pair shields should each be insulated from other
pairs to prevent undesired grounding; the shield should never be used
as a signal return.

7.4   FILTER-PIN CONNECTORS

Filters, the topic of Chaps. 13 and 14, offer significant possibilities for controlling conducted interference. Generally, EMI filters are employed as lumped elements in various portions of circuits
and input-output wiring of equipments. In recent years, however, filters have been miniaturized to such small sizes that some can now be
built into the cable-pin assembly. Figure 7.9 illustrates one type of
minature multi-pin connector employing π-type filters in each pin.
Because of the limitation of the obtainable shunt capacitance and series
inductance that can be constructed in the pin, filters of this small
type, typically about 1/8" x 3/8" in size, exhibit little or no attenuation below 1 MHz. Typical attenuation offered by these filter pins
in a 50-ohm system is about 20-dB at 10 MHz and up to 80-dB at 100 MHz.

Another filter-pin connector of a somewhat larger body dimension
is designed to carry 5 amperes. Thus, for low D-C working voltages,
capacitances up to about 1µf are achievable in the larger pins. Fig.
7.10 shows some typical dimensional data of these connectors and associated insertion loss vs frequency. Many of these filters exhibit cut-
off frequencies of the order of 100 kHz when measured in a 50-ohm
system per MIL-STD-220A.

7.5   COAXIAL CONNECTORS

Applications above 10 kHz and more typically above 10 MHz, connectors are of the unbalanced-line, coaxial type in order to mate with
coaxial cables*. Connectors of this type maintain a 360-degree, low-
impedance integrity of the outer shield through the connector interface to the mating connector assembly. A low-impedance shield is
extremely important since this impedance exists in the return-wire path

---

*   Cross talk between wires at high frequencies is due to electric-
field coupling. To provide both a return-wire path and a shield at
and above HF, coaxial lines and connectors are used notwithstanding
some balanced, parallel lines used at HF/VHF.

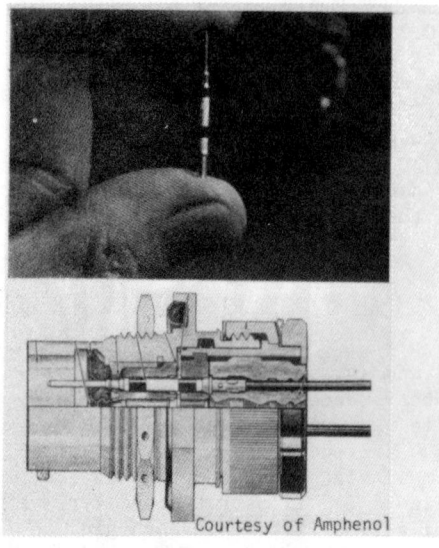

Courtesy of Amphenol

Figure 7.9 - Typical Miniature
Multi-Pin Filter Connector

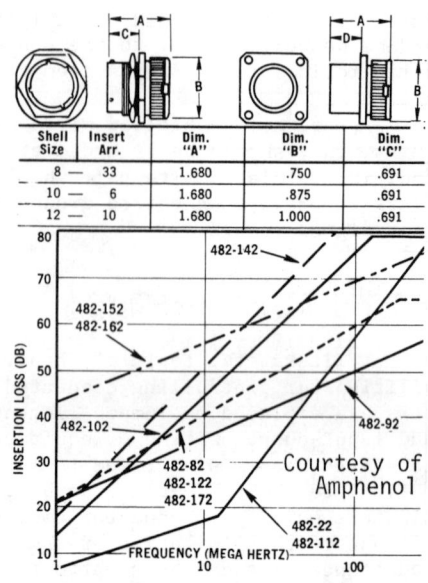

Shell Size	Insert Arr.	Dim. "A"	Dim. "B"	Dim. "C"
8 —	33	1.680	.750	.691
10 —	6	1.680	.875	.691
12 —	10	1.680	1.000	.691

Figure 7.10 - EMI Filtering
Connectors and Insertion Loss

of the associated coaxial cable. Thus, the outer-cable shield impedance of the termination is of paramount consideration in the performance of coaxial connectors.

One significant EMI problem of coaxial connectors is the above impedance mismatch in a 50, 72-ohm or other cable characteristic impedance. Impedance mismatch is rated in terms of maximum voltage standing-wave ratio (VSWR) vs. frequency. The maximum signal amplitude variation as a function of VSWR is the amplitude of the VSWR, per se. For example, depending upon the length of cable, a connector rated with a VSWR of 2:1 at a particular frequency could exhibit a 6-dB peak-to-peak variation in signal or EMI amplitude. Thus, connector VSWR becomes very important especially at frequencies for which an associated cable length approaches or exceeds $\lambda/8$.

Specifying the correct coaxial connector for a specific requirement will simultaneously accomplish most EMI considerations. To do this, Table 7.1 may be used. Work down the characteristics in the left-hand column, making note of the connector series that have the desired characteristics, and progressively eliminate those series that do not. When the checklist is completed, the selection will have been narrowed down to types that will fulfill the requirements.

Table 7.1 - Coaxial Connector Selection Chart*

		S										
	S	M	S	S	T	B		S		U		L
	M	A	M	M	N	N				H	H	
Characteristics	A	A	B	C	C	C	N	C	C	F	N	T
COUPLING — Threaded	X	X		X	X		X	X		X	X	X
Bayonet						X			X			
Quick Disconnect			X									
CABLE ATTACHMENT — Crimp	X	X	X	X	X	X	X	X	X	X		
Clamp	X		X	X	X	X	X	X	X	X	X	X
Solder	X	X	X	X	X				X			
IMPEDANCE — 50 Ohm	X	X	X	X	X	X	X	X	X		X	X
70 Ohm			X	X	X	X	X					
FINISH (BODY)												
Passivated Stainless Steel	X	X	X	X	X		X					
Gold Plated	X	X	X	X								
Silver Plated						X	X	X	X	X	X	X
Nickel Plated						X	X	X	X	X	X	X
SIZE — 0.320" Maximum		X	X	X								
0.400" Maximum	X											
0.625" Maximum					X	X						
0.827" Maximum							X	X	X	X	X	
1.500" Maximum												X
VOLTAGE RATING — 250 VRMS			X	X	X							
500 VRMS	X		X	X	X	X		X				
1500 VRMS							X		X		X	
5000 VRMS												X
SEALING — Moisture	X	X			X	X	X	X	X		X	X
MAXIMUM OPERATING FREQUENCY												
0.3 GHz										X		
2.5 GHz											X	
4.0 GHz			X			X						X
10.0 GHz				X								
11.0 GHz							X	X	X			
12.4 GHz	X	X										
15.0 GHz					X							
18.0 GHz	X	X										
VSWR AT MAXIMUM OPERATING FREQUENCY												
1.25:1		X					X	X				
1.30:1					X	X	X	X				
1.35:1									X	X	X	X
1.40:1	X											
1.50:1			X									
1.60:1				X								
1.70:1				X								
QPL MIL-C-39012	X		X	X	X	X	X	X	X			
BODY MATERIAL												
Beryllium copper	X	X										
Stainless Steel	X	X			X		X					
Brass			X	X	X	X	X	X	X	X	X	X
Aluminum					X		X				X	

*Courtesy of Electronic Products Magazine, May 15, 1972; pp. 148.

CHAPTER 8

GROUNDING AND BONDING

CHAPTER 8

GROUNDING AND BONDING

The subject of grounding appeared in many places in preceding chapters on cables and connectors, such as, where and when should a cable shield be grounded? Other topics dealing with buildings, shielded enclosures, equipments and filters have their own grounding requirements. As it develops, grounding is one of the least understood and is a more significant culprit in the entire intra-system EMI problem.

There are two purposes for grounding devices, cables, equipments, and systems: (1) to prevent a shock hazard in the event that an equipment frame or housing may develop a high voltage due to lightning or an accidental break-down of wiring or components, and (2) to reduce EMI due to electric-field, common-impedance, or other form of interference coupling. Each of these purposes is summarized below although the EMI part of the problem is emphasized in this chapter.

8.1 RATIONALE FOR GROUNDING

This section presents an overview of the reasons why grounding are important.

8.1.1 SHOCK AND LIGHTNING HAZARDS

A-C power distribution in private homes, buildings, hospitals, and industrial sites in the U.S. are governed by local and national codes. The National Fire Protection Association (NFPA) has issued NFPA-STD-70-1971, *National Electric Code*, dealing with standards on wiring and other electrical devices. One requirement is that with each outgoing hot wire (black) and return neutral wire (white) is a reference ground (green) wire. The same applies for three-wire, 115-115-230 VAC systems in which a second hot wire (red) is added as shown in Fig. 8.1.

Theoretically, no return current passes through the green wire reference ground. If the hot side were accidentally shorted to the equipment frame, the latter would become 115 VAC hot to ground. If someone were touching that frame as shown in Fig. 8.2, a path of current would return through his hand and continue through the body and out either (1) the other hand if it were touching some other reference such

8.1

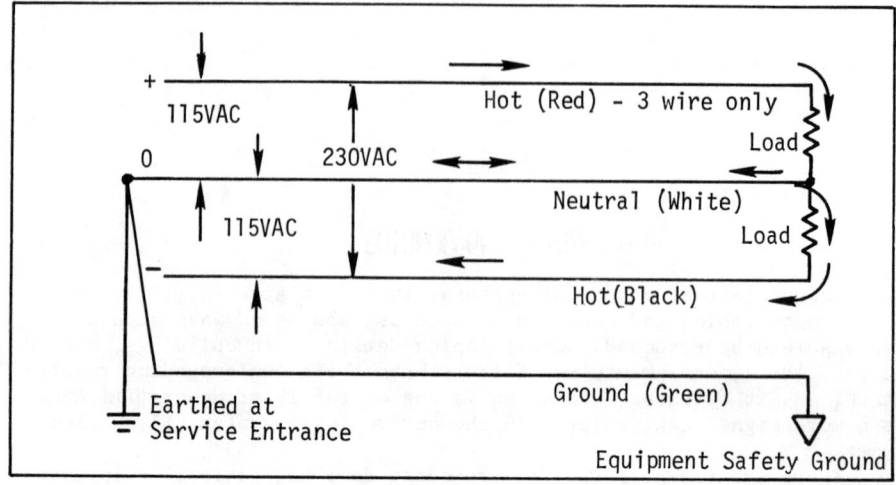

Figure 8.1 - Standard 2 and 3-Wire Electrical Wire Coding

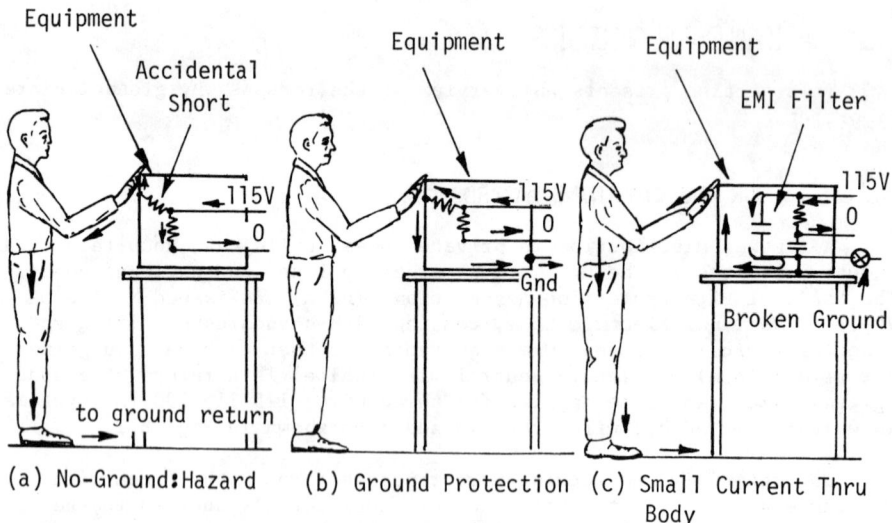

(a) No-Ground:Hazard   (b) Ground Protection   (c) Small Current Thru
                                                         Body

Figure 8.2 - Safety and Shock Hazards with/without Equipment
Grounds and EMI Filters

as ground or (2) the soles of the feet to a concrete floor and thence
to the building ground. Here, 75 mA of current through the body could
be fatal. These are *micro-shock* hazards and can be avoided by grounding
a third wire--the green wire to the equipment frame in a modern A-C
power cord.

To help reduce EMI in equipment of all types, it is common practice
today to use filters in the power-line entry at equipment. Here, either
capacitors or filters are placed from both hot and neutral* lines to
the ground wire to by-pass EMI. The National Electric Code previously
limited the use of such capacitors to 0.1 μf at 60 Hz corresponding to
a leakage reactive current of 5 mA. Thus, if an individual touched an
equipment frame, and the frame was not grounded, the maximum current
through the body would be limited to the 5 mA.

Another kind of shock hazard exists due to either several devices
with EMI filters operated off the same circuit or if one device should
develop a short to frame. Both situations involve the *micro-shock*
hazards in hospitals, clinics, and medical centers in which catheters
via electrodes from EKG, arterial pressure monitors, and similar biophysical instruments are in direct contact with the heart. When a
high-impedance leakage or direct short develops in an equipment, such
as a vacuum cleaner sharing a common ground (green wire) with the medical instruments, a substantial current may flow in the ground wire.
This current will partition with most returning directly to the power
distribution panel and some following another path directly through the
heart as shown in Fig. 8.3.**

The other aspect of grounding involves lightning hazards to buildings and their contents. To protect a structure from damage from
lightning strokes requires that a lower impedance path be provided from
the top of the building to a good building earth ground over that offered by the building, per se. In other words, the lightning stroke
would follow the lower potential gradient of the arrestor-grounding
system rather than that of the protected building. This topic is
covered in length in Chap. 9.

8.1.2   EMI GROUND REFERENCES AND IMPEDANCES

The principle concern about grounding in this chapter is the EMI
problems that develop as a result of wrong or faulty grounds, how to
mathematically model this, and some corrective solutions. Discussions
on grounding a cable or wire shield in previous chapters serve to

---

*   The neutral line is also a source of EMI.
** G.D. Friedlander, "Electricity in Hospitals: Elimination of
Lethal Hazards," *IEEE Spectrum*, Sept. 1971; pp. 40-51.

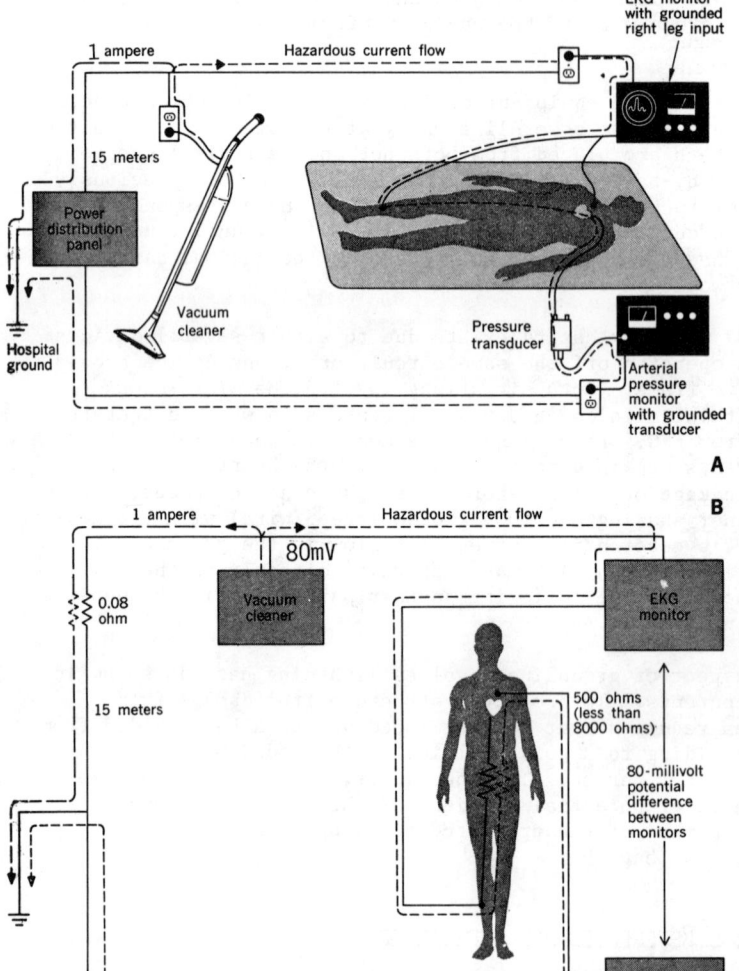

A—Case in which a vacuum cleaner is plugged into a wall outlet on same circuit as the EKG monitor. The cleaner has a three-wire power cord, with the third wire grounding its outer case. But the windings of the motor are exposed to dust (often damp), which provides a good path for an eventual "winding-to-outer-case" short circuit. Because this kind of short circuit makes the case rise to full line voltage, the case is grounded to protect the operator. In this example, the vacuum cleaner has not completely failed, but has developed a fault sufficient to permit 1 ampere to flow down the ground wire and back to the power distribution panel. B—In this analysis of the incident, if the power distribution panel is 15 meters distant and the power wiring is 12 gauge, the 15 meters of ground wire have 0.08 ohm of resistance. The hazard here is that the faulty appliance caused difference in ground potential between two devices and allows a possibly lethal current to flow through the patient.

(Courtesy of IEEE Spectrum)

Fig. 8.3   Microshock Hazard in Hospitals

illustrate this. There, to effectively perform a Farady shield between
conductors, the shield had to be grounded in two or more places when
the cable was longer than 0.1λ at the highest frequency of concern.
While this reduced electric-field coupling at high frequencies, it may
have generated a magnetic-field ground-current loop in the process,
unless certain procedures were followed.

For other than wire-shield situations, different grounding requirements develop. They all involve bonding equipments to a common potential reference to avoid circulating EMI currents because of a difference
in potential between portions of a system or power-mains distribution.
Thus, a concept develops of a ground plane exhibiting zero-potential
difference between any two points thereon. It does not matter what
the absolute potential of a ground plane-to-earth may be since aircraft,
for example, are isolated; a ship is located in the water; and a building
is constructed on earth. All their structures act as a ground-plane
reference.* However, for land structures, such as buildings, the
National Electric Code requires that certain specifications be met such
as connecting the neutral and ground wire together and earthing them
at the utility-service entrance. This reduces the shock hazard but
may develop an undesirable situation from an EMI point-of-view.

8.1.3 ZERO-POTENTIAL GROUND PLANE

Any two points on a metallic structure, whether electrically connected or not, may develop a potential difference at some frequency.
For structural dimensions, ℓ, a potential difference, V, will exist in
the presence of a magnetic or electrical field:

$$V = A \sin(2\pi\ell/\lambda) \qquad\qquad (8.1)$$

$$\sim 2\pi A\ell/\lambda \quad \text{for} \quad \ell < 0.1\lambda \qquad\qquad (8.2)$$

where,   A = Amplitude of induced voltage in the ground plane

λ = wavelength corresponding to frequency

---

*   The low-impedance ground path or ground plane should be the same
as that of a source generator return, or should be connected directly
to the generator reference plane. Thus, aircraft or spacecraft are
connected to earth ground when being serviced by ground-support equipment powered by ground-based power systems. When disconnected from
ground-support equipment, the equipment aboard the craft is referenced
to the vehicle skin, which serves as the ground plane. During flight,
the accumulated excess energy is partially dissipated as an electrostatic discharge in space and partly by conversion to radiant thermal
energy.

Alternatively, the impedance, Z, between two points in a ground plane is:

$$Z = R_{RF}(1 + |\tan 2\pi\ell/\lambda|) \tag{8.3}$$

$$= kR_{DC}(1 + |\tan 2\pi\ell/\lambda|)$$

$$\approx kR_{DC}(1 + 2\pi\ell/\lambda) \text{ for } \ell < \lambda/10 \tag{8.4}$$

$$\approx kR_{DC} \text{ for } \ell < \lambda/20 \tag{8.5}$$

$$\approx 2kR_{DC} \text{ for } \ell \approx \lambda/8, 3\lambda/8$$

$$= \infty \text{ for } \ell = \lambda/4, 3\lambda/4,$$

where, R_{DC} = D-C surface resistance in ohms/square

k = a number greater than one representing the ratio of RF to DC surface resistance, R_{RF}/R_{DC}

and, R_{RF} = $0.26 \times 10^{-6}\sqrt{f}$ ohms/square for copper $\tag{8.6}$

$= 0.26 \times 10^{-6} \sqrt{\mu f/\sigma}$ ohms/square for any conductor $\tag{8.7}$

where, σ = conductivity of material relative to copper

μ = permeability of material relative to copper

Eq. (8.3) indicates that the impedance between two points on a ground plane can become substantial when $\ell > \lambda/8$. Thus, a ground plane offers little equipotential grounding value to two or more equipments which must be grounded thereto at frequencies greater than that corresponding to $\lambda/8$. However, at lower frequencies for which $\ell < \lambda/20$, the impedance between two points in a ground plane is proportional to the R-F impedance in ohms/square as indicated in Eqs. (8.5) and (8.6).

Illustrative Example 8.1

Determine the impedance between two end points in a galvanized steel cable-tray ground plane measuring 6" x 20' at 100 kHz and 10 MHz.

The wavelength λ at 10 MHz is 30 meters $\approx$ 100 ft. For a 20' separation, this corresponds to $\lambda/5$. Thus, Eq. (8.3) in modified form and Eq. (8.7) are:

$$Z = R_{RF} (1 + |\tan 2\pi\ell/\lambda|) \times \text{length/width } \Omega/\text{sq} \tag{8.8}$$

$$= 20'/0.5' \times R_{RF} (1 + |\tan 2\pi/5|)$$

$$= 40R_{RF} (1 + 3.73) = 189R_{RF}$$

where, R_{RF} = 0.26 x $10^{-6}\sqrt{1000 \times 10^7}$.1 = 0.08$\Omega$/sq

Thus,   Z   = 189 x 0.08 = 15 ohms.

The wavelength at 100 kHz is 3 km $\approx 10^4$ feet. Since the 20 ft. separation corresponds to an ℓ = 0.002λ, Eq. (8.5) and (8.7) apply:

$$Z = 40 \times 0.26 \times 10^{-6} \sqrt{1000 \times 10^5}/.1 = 0.32 \text{ ohms}$$

Had the cable tray been made of copper, the impedances end-to-end would have been 150 mohms at 10 MHz and 3.2 mohms at 100 kHz.

The RF impedance between two points in a ground plane only has significance if equipments tied thereto have a potential difference and common circulating currents can flow between them. As explained in the next section, this can cross couple a source EMI voltage to a victim network.

8.2   COMMON-MODE IMPEDANCE COUPLING

Sec. 6.2 described the EMI risks in not using a dedicated return wire with an outgoing conductor, in using a ground plane as return path. Fig. 8.4 illustrates a risky, but not infrequently-found practice in which two different circuits share the same ground plane for their return paths. The voltage, V_c developed across Z, the equivalent impedance of the ground plane, from the potential culprit EMI source, is:

$$V_c = \frac{ZV_1}{R_{g1}+R_{L1}+Z} \approx \frac{ZV_1}{R_{g1}+R_{L1}} \quad \text{for} \quad Z << R_{g1}+R_{L1} \tag{8.9}$$

The resulting voltage, V_i developed across the potential victim load in circuit #2 is:

$$V_i = \frac{R_{L2}V_c}{R_{g2}+R_{L2}} \quad \text{for } Z << R_{g2} + R_{L2} \tag{8.10}$$

Substituting Eq. (8.9) into Eq. (8.10) yields:

$$V_i = \frac{ZR_{L2}V_1}{(R_{g1}+R_{L1})\ (R_{g2}+R_{L2})} \tag{8.11}$$

The cross-talk, CT in dB, between the two circuits due to common-mode impedance coupling, then, may be expressed as a math model:

$$CT_{dB} = 20 \log_{10}[ZR_{L2}/(R_{g1}+R_{L1})\cdot(R_{g2}+R_{L2})] \tag{8.12}$$

where, Z = common impedance appearing in Eq. (8.3).

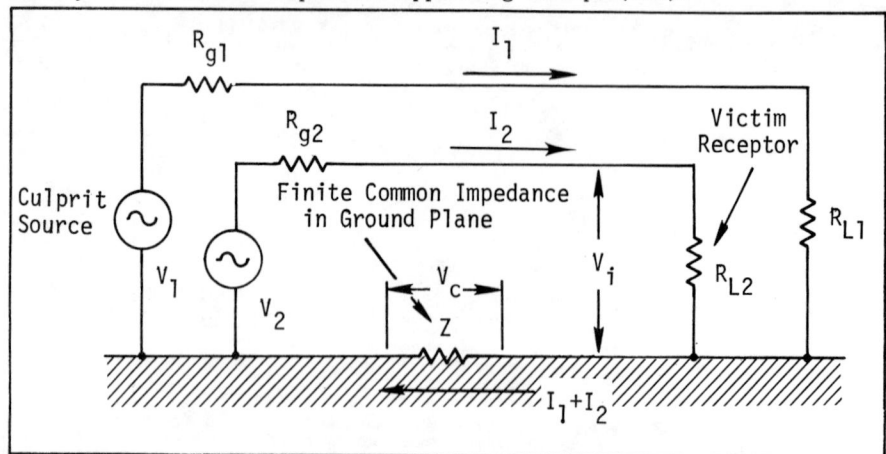

Figure 8.4   - Common-Mode Impedance Coupling Between Circuits

Illustrative Example 8.2

    Two wire conductors are run down the same cable tray described
in Example 8.1 and use the tray as a ground return (a bad practice).
The first circuit is carrying 100 kHz clock pulses at an amplitude of
5 volts. The second circuit is an oscilloscope monitor with a sensor
at one end; the scope has a sensitivity of 1 mV. The source and load
impedances of circuit #1 are 100 ohms. The source and load impedances
of circuit #2 are 100 ohms and 10 Mohms, respectively. Determine if
an EMI problem will exist.

    From the previous example, at 100 kHz, Z = 0.32 ohms. The inter-
ference voltage is determined from Eq. (8.11).

$$V_i = \frac{0.32 \times 10^7 \times 5V}{200 \times 10^7} = 8 \text{ mV}$$

Thus, since V_i = 8 mV > 1 mV sensitivity, EMI due to common-mode
impedance coupling will exist.

8.3   SINGLE VS MULTI-POINT GROUNDING

This section reviews the often controversial topic regarding whether equipments or subsystems should be grounded at one point to a reference, or whether multi-point grounding should be used. The matter of hybrid grounds is also reviewed. It will be shown that all systems are a continuation of each other at higher frequencies, but that certain correct practices should be followed. Conversely, the grounding scheme at lower frequencies is more clearly defined regarding correct and wrong techniques.

8.3.1   SINGLE-POINT GROUNDING

Modern electronic systems seldom have only one ground plane. To mitigate interference, such as due to common-mode impedance coupling, as many separate ground planes as possible are used. Separate ground planes in each subsystem for structural grounds, signal grounds, shield grounds, and A-C prime and secondary power grounds are desirable if economically and logistically practical. Here grouping of ground planes is consistent with a similar technique used for cable classification (see Sec. 6.1). These individual ground planes from each subsystem are finally connected by the shortest route back to the system ground point where they form an overall system potential reference. This method is known as a *single-point ground* and is illustrated in Fig. 8.5.

The single-point or star-type of grounding scheme shown in the figure avoids problems of common-mode impedance coupling discussed in the previous section. The only common path is in the earth ground (for earth-based structures), but this usually consists of a substantial conductor of very-low impedance. Thus, as long as no or low-ground currents flow in any low-impedance common paths, all subsystems or equipments are maintained at essentially the same reference potential.

The problem of implementing the above single-point grounding scheme comes about when: (1) interconnecting cables are used, especially ones having cable shields which have sources and receptors operating over lengths, ℓ, above about $\lambda/20$, and (2) parasitic capacitance exists between subsystem or equipment housings or between subsystems and the grounds of other subsystems. This situation is illustrated in Fig. 8.6. Here, cable shields connect some of the subsystems together so that more than one grounding path from a particular subsystem to the ground point exists. Unless certain precautions are taken, common-impedance ground currents could flow. At high frequencies, the parasitic capacitive reactance represents low-impedance paths, and the bond inductance of a subsystem-to-ground point results in higher impedances. Thus, again common-mode currents may flow or unequal potentials may develop amongst subsystems.

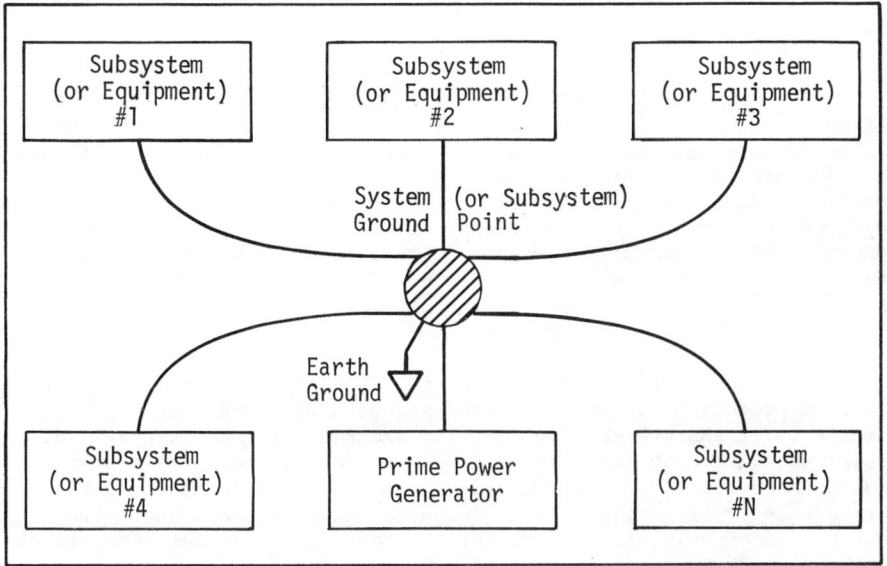

Figure 8.5 - Single-Point or Star Grounding Arrangement

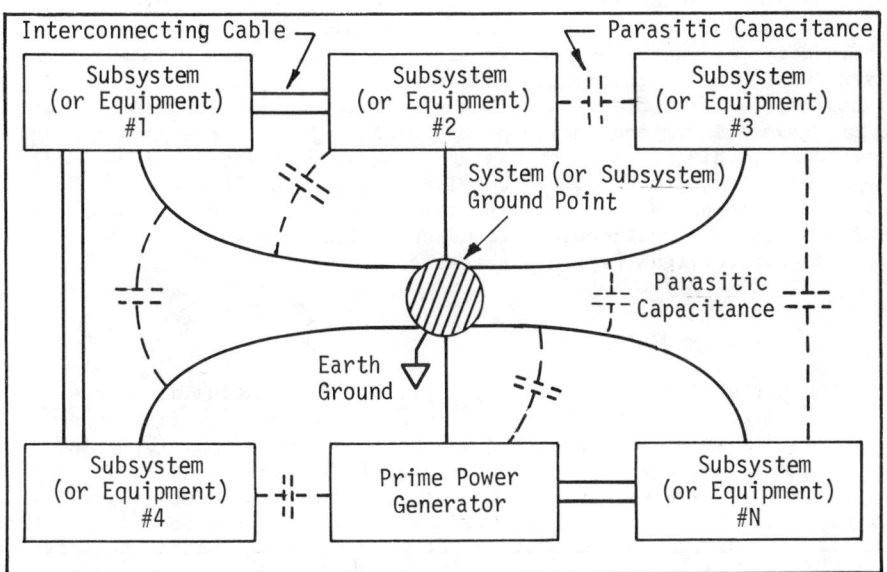

Figure 8.6 - Degeneration of Single-Point Ground of Fig. 8.5
by Interconnecting Cables and Parasitic Capacitance.

8.3.2 MULTI-POINT GROUNDING

Supporters of the multi-point grounding concept argue that pragmatically the situation shown in Fig. 8.6 exists anyway and not that
of the ideal single-point ground shown in Fig. 8.5. Thus, rather than
have an uncontrolled situation as shown in Fig. 8.6, if everything were
heavily bonded to a solid ground conducting plane to form a homogeneous,
low-impedance path, common-mode currents and other EMI problems would
be minimized. An example of such a situation is shown in Fig. 8.7,
where each subsystem or equipment is bonded as directly as possible
to a common low-impedance *equipotential* ground plane. The ground plane
then is earthed for safety purposes.

The facts are that a single-point grounding scheme operates better
at low frequencies and a multi-point ground behaves best at high frequencies. If the over-all system, for example, is a network of audio
equipment, with many low-level sensors and control circuits behaving
as broadband transient noise sources, then the high-frequency performance is irrelevant* since no receptor responds above audio-frequency. Conversely, if the over-all system were a receiver complex of
30 MHz to 1,000 MHz tuners, amplifiers, and displays, then low-level,
low-frequency performance is irrelevant. Here multi-point grounding
applies and interconnecting, unbalanced coaxial lines are used.

The above dichotomy of audio vs VHF/UHF systems make clear the
selection of the correct approach. The problem then narrows down to
one in defining where low and high-frequency cross-over exists for any
given subsystem or equipment. The answer here in part involves the
*highest significant operating frequency of low-level circuits relative
to the physical distance between the furthest located equipments.* In
other words, this *twilight* cross-over frequency region involves: (1)
the magnetic vs electric field coupling problems of Chapter 6, and
(2) the ground-plane impedance problems of Sec. 8.1.2 due to separation.
Hybrid single and multi-point grounding systems are often the best
approach for twilight region applications.

8.3.3 HYBRID GROUNDING

The matter of single-point vs multi-point grounding discussed
in the previous sections is summarized as a general guideline model
in Fig. 8.8. This model is based on the relations of Eq. (8.3) and the
separation criteria of $\ell = \lambda/20$ presented in Eq. (8.5). For low
frequency operation and small dimensions, use single-point grounding.
For high frequency and large dimensions use multi-point grounding. For
transitional situations, one or the other may perform better as shown

---

*   Assuming no H-F spurious or parasitic responses, which is not
always the case.

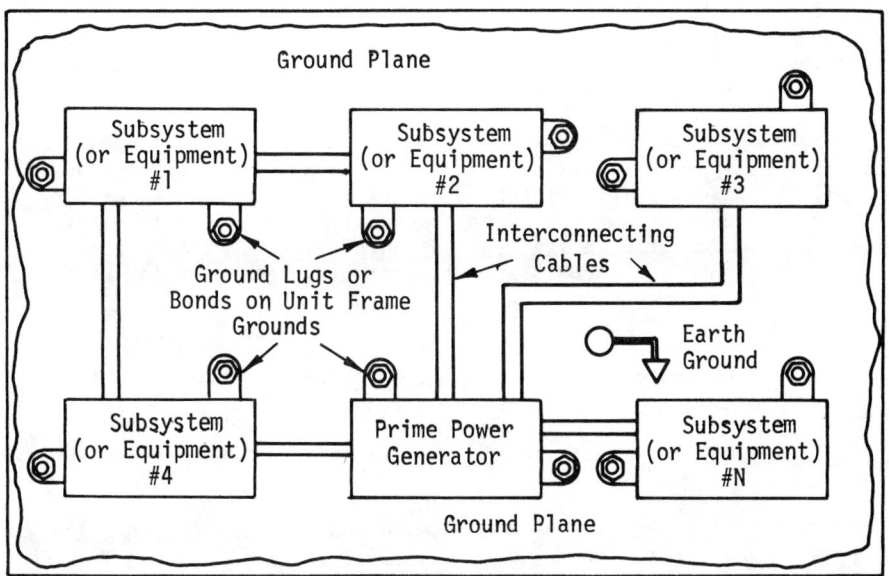

Figure 8.7 - Multi-Point Grounding System

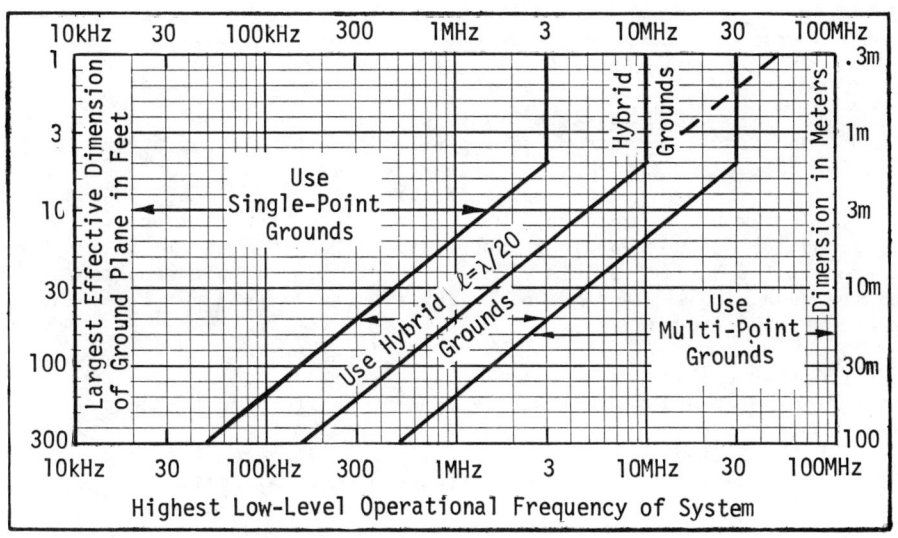

Figure 8.8 - Cross-over Regions of Single-Point vs Multi-Point
Grounding

in the figure. Hybrid grounds perform best in which portions of the
low-frequency systems use single-point grounding while high-frequency
portions use multi-point grounding, all connected in a ground-tree
fashion.

Regarding mathematical modeling and predicting EMI in general due
to grounding, it is extremely difficult to accomplish this for other
than (1) relatively simple system configurations and (2) single-point
grounding. The number of significant parasitic reactances in either
complex systems or at high frequencies and above renders modeling of
a grounding situation unrealistic. Accordingly, the following prediction guidelines are used:

(1) For low-frequency ($\ell < \lambda/20$), single-point grounds, employ the
ground-impedance and common-mode impedance techniques discussed in
Secs. 8.1 and 8.2.

(2) Use cable coupling models for all equipment cables (see Chap.
6).

(3) Use case penetration and leakage models between all applicable
shielding boxes, equipments, and subsystems (see Chaps. 10 and 11).

The term *hybrid grounds* is sometimes used in two somewhat different senses: (1) when a grounding scheme either appears as a
single-point ground at low frequencies and a multi-point ground at
high frequencies, or appears different at both frequencies, and (2)
when a system grounding configuration employs both single-point and
multi-point grounds. Each of these is illustrated below:

Fig. 8.9 shows a low-level video circuit in which both the sensor
and driven circuit chassis must be grounded to the skin of a vehicle
(not by choice) and the coaxial cable shield is grounded to the chassis
at both ends through its mating connectors. A low-frequency ground
current loop would be generated were it not for the capacitor. At
high frequencies the capacitor assures that the cable shield is grounded
to protect the Faraday-shield effect. Thus, this circuit simultaneously
behaves as a single-point ground at low frequencies and a multi-point
ground at high frequencies.

A different kind of an example is shown in Fig. 8.10 in which
all the computer and peripheral frames must be grounded to the power
system *green wire* for safety purposes (shock-hazard protection) pur-
suant to the National Electric Code. Since it is recognized that the
green wire generally contains significant electrical noise trash (cf.
Sec. 9.14), this code conflicts with the desire to float the computer
system ground from the noisy *green wire* ground. Thus, one or more
isolation coils of about 1 mh value are used to provide a low-impedance
(less than 0.4 ohms) safety ground at A-C power line frequencies and
R-F isolation (of the order of 1,000 ohms) in the 50 kHz to 1 MHz
spectrum containing the principle energy of computer pulses. This

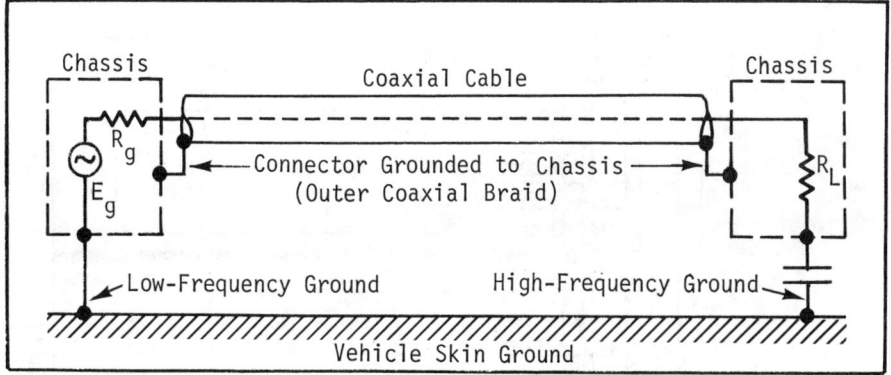

Figure 8.9 - Low-Frequency Ground Current Loop Avoidance with High-Frequency Ground

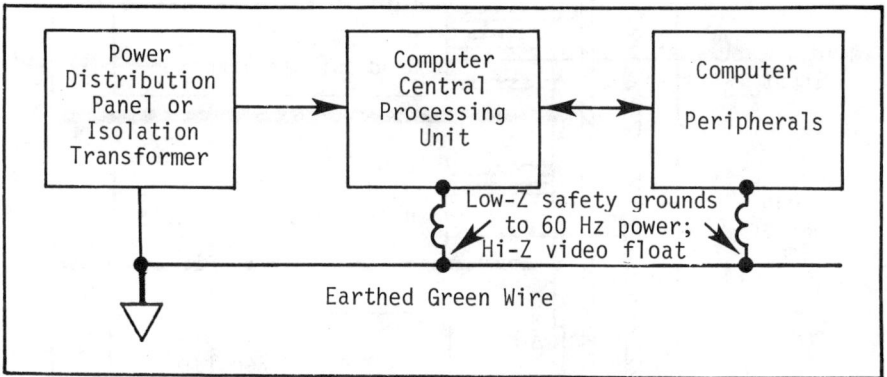

Figure 8.10 - Safety Ground with High Frequency Isolation

inductor helps keep induced transient and EMI noise in the *green wire* off the computer supply voltage logic busses.

   To illustrate the second form of hybrid-ground system, Fig. 8.11 shows a 19-inch cabinet rack containing five separate sliding drawers. Each drawer contains a portion of the system (top to bottom): (1) R-F and I-F preamp circuitry for intercept of microwave signals, (2) I-F and video signal amplifiers, (3) display drivers, displays, and control circuits, (4) low-level, audio circuits and recorders for documenting sensitive multi-channel, hard-line telemetry sensor outputs, and (5) secondary and regulated power supplies. The hybrid aspect results from:

   (1) The R-F and I-F-video drawers are similar. Here, unit-level boxes or stages (interconnecting coaxial cables are grounded at both ends) are multi-point grounded to the drawer-chassis ground plane. The

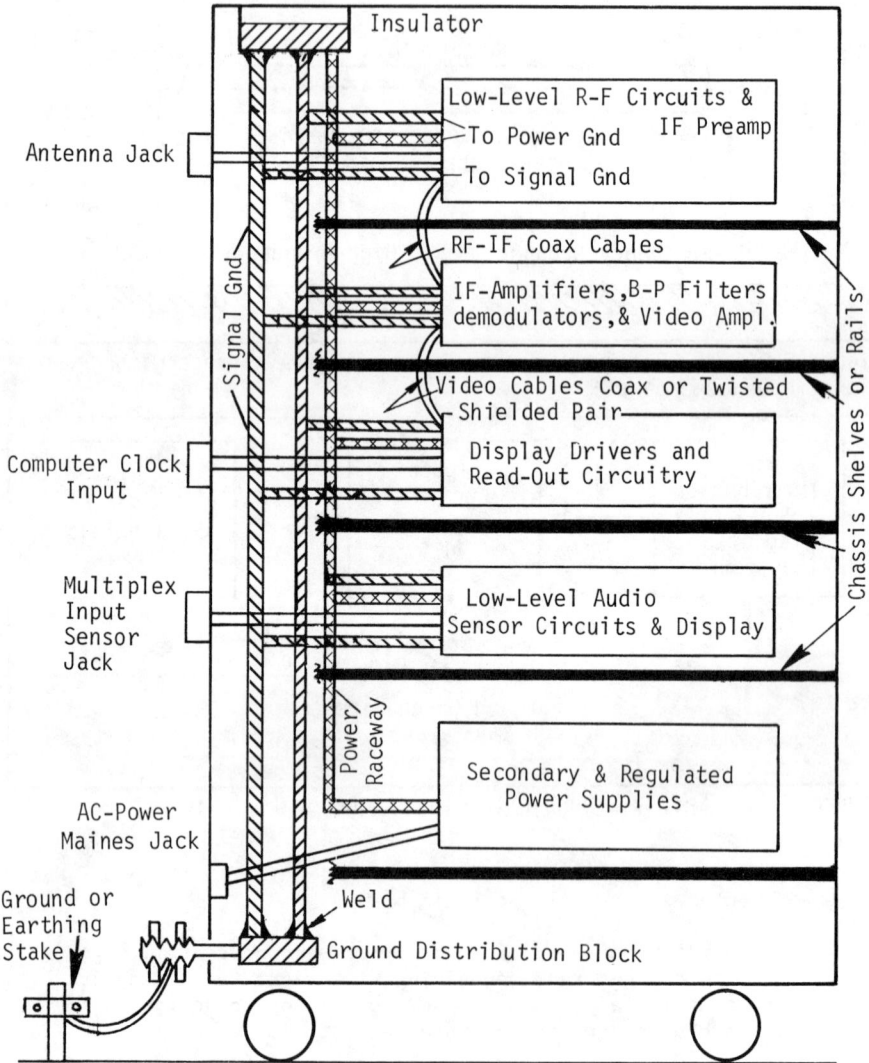

Figure 8.11 - Grounding Arrangment Used in Cabinet Racks

chassis is then grounded to the dagger-pin, chassis-ground bus as
suggested in Fig. 8.12. The power ground to these drawers, on the
other hand, is using a single-point ground from its bus in a manner
identical to the audio drawer.

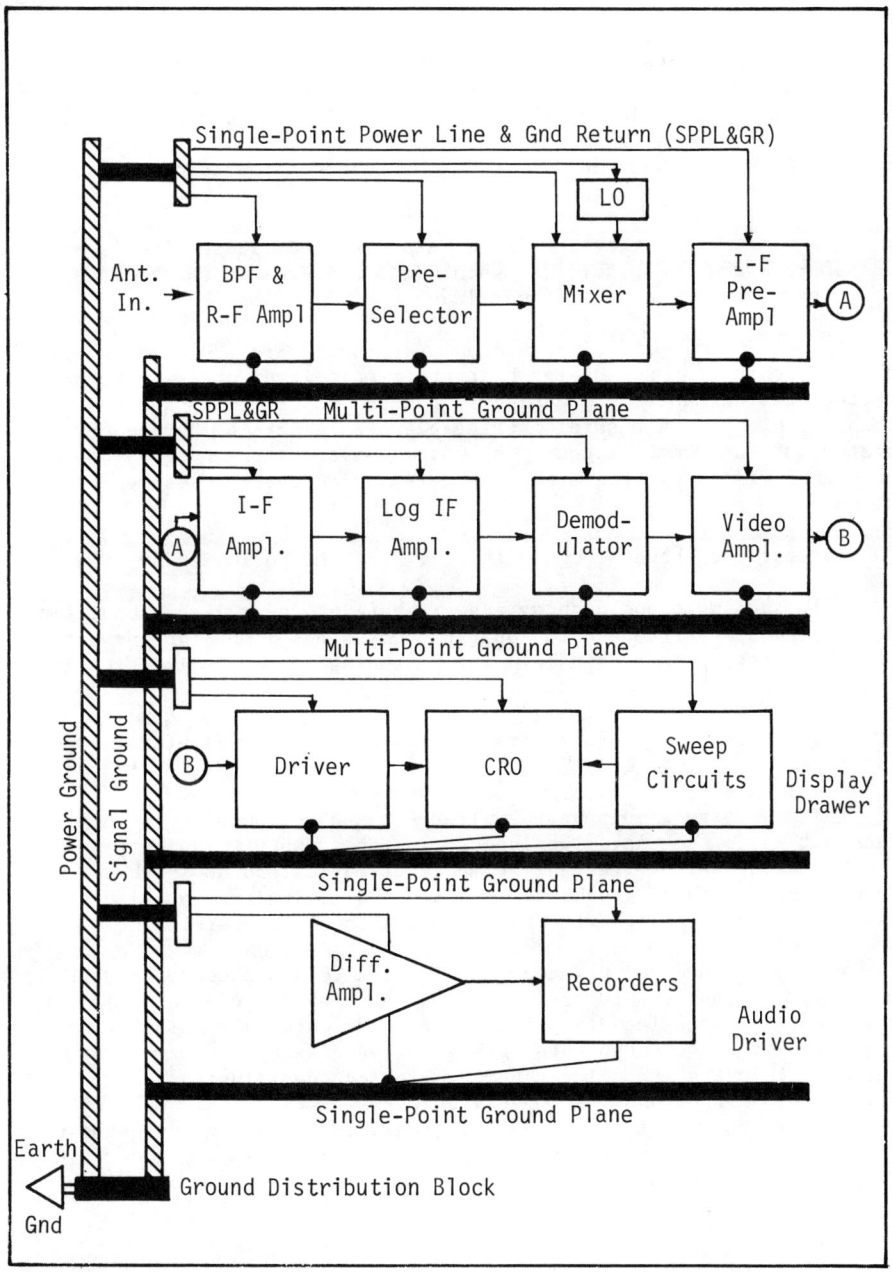

Figure 8.12 - Block Diagram Detail of Fig. 8.11 Hybrid Grounding
Arrangement

(2)   The chassis or signal ground and power ground busses each constitute a multi-point grounding scheme to the drawer level. The individual ground-busses are single-point grounded at the bottom ground distribution block. This avoids circulating common-mode current between chassis or signal ground and power grounds since power-ground current can vary due to transient surges in certain modes of equipment operation.

(3)   Interconnecting cables between different drawer levels are run separate and their shields, when used, are treated in the same grounding manner as at the drawer level.

(4)   The audio and display drawers shown in Fig. 8.12 use single-point grounding throughout for both their unit-level boxes (interconnecting twisted cable is grounded at one end to its unit) and power leads. Cable and unit shield strike-plate holes are all grounded together at the common dagger-pin bus. Similarly, the outgoing power leads and twisted returns are separately bonded on their dagger-pin busses.

To test the above scheme with Fig. 8.8, the following is observed:

(1)   The audio and display drawers have ground runs of about two feet and an upper frequency of operation of about 1 MHz (driver and sweep circuits). Thus, single-point grounding to the strike pins is indicated.

(2)   The R-F and I-F drawers process UHF and 30 MHz signals over a distance of a few feet so that multi-point grounding is indicated.

(3)   The regulated power supplies furnish equipment units having transient surge demands. The longest length is about five feet and significant transient-frequency components may extend up in the H-F region. Here, hybrid grounding is indicated: single-point within a drawer, and multi-point from the power bus to all drawers.

When miniature and printed circuits and IC's are used, network proximity is considerably closer. Thus, multi-point grounding is more economical and practical to produce per card, wafer, or chip. Interconnection of these components through wafer risers, mother boards, etc., should use a grounding scheme following the illustrations of previous paragraphs and the general criteria of Fig. 8.8. This will likely still represent a multi-point or hybrid grounding approach in which any single-point grounding (for hybrid grounds), if used would be to avoid low-frequency ground current loops and/or common-mode impedance coupling.

8.4 ELECTRICAL BONDING

Electrical bonding refers to the process in which components or modules of an assembly, equipments, or subsystems are electrically connected by means of a low-impedance conductor. The purpose is to make the structure homogenous with respect to the flow of R-F currents. This mitigates electrical potential differences which can produce EMI among metallic parts.

An example of the importance of bonding to reduce EMI is shown in Fig. 8.13(a) in which the effectiveness of a filter can be nullified by improper bonding. In that example, the contact resistance of a poor bond does not provide the low-impedance path necessary for shunting interference currents coming from the power mains. This current now flows through the filter capacitors and on to other equipment which was to have been protected. In Fig. 8.13(b), the receiver is not well bonded to a common ground plane reference for both the antenna and the power mains return. Thus, R-F currents appearing on the power mains share a common-impedance path at the bond with R-F signals picked up by the antenna.

8.4.1 EQUIVALENT CIRCUITS OF BONDS

A low-impedance path is possible only when the separation of the bonded members are small compared to a wavelength of the EMI being considered and the bond is a good conductor. This was discussed in Sec. 8.1.3. At high frequencies, structural members behave as transmission lines whose impedance can be inductive or capacitive of varying magnitudes, depending upon geometrical shape and frequency in a manner similar to that explained in connection with Eq. (8.3).

Fig. 8.14 is the equivalent electrical circuit of a bond strap. The circuit contains resistance due to the finite conductance of the strap in series with the self inductance of the bond. Shunt capacitance exists due to the residual capacity of the strap and its mounting. This capacitance and self-inductance form a parallel anti-resonant circuit, resulting in the adverse impedance response shown in the figure.

There is little correlation between the D-C resistance of a bond and its R-F impedance. The measured R-F impedance of artificial bonds, per se, such as jumpers, straps, rivets, etc., is not a reliable indication of the bonding effectiveness in an actual installation. Here, the artificial bond is in parallel with the members to be bonded, and the total impedance includes various parallel paths over which R-F conductive or displacement currents may flow. Thus, a bond strap of low inductance combines with the capacitance of the installation as shown in Figs. 8.14 and 8.15 to form a high impedance anti-resonant circuit at some frequency. The bibliography contain a number of sources presenting math models of various bond configurations.

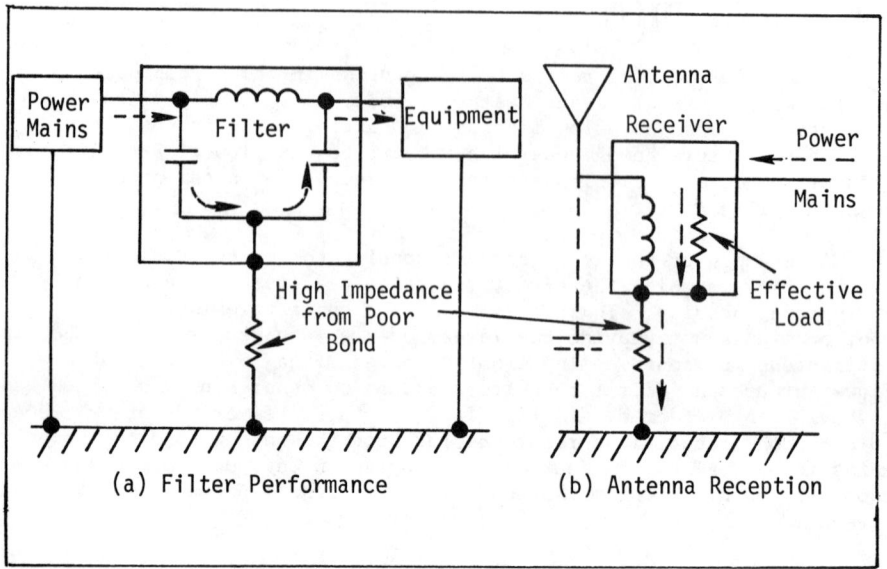

Figure 8.13 - Two Effects of Poor Bonding

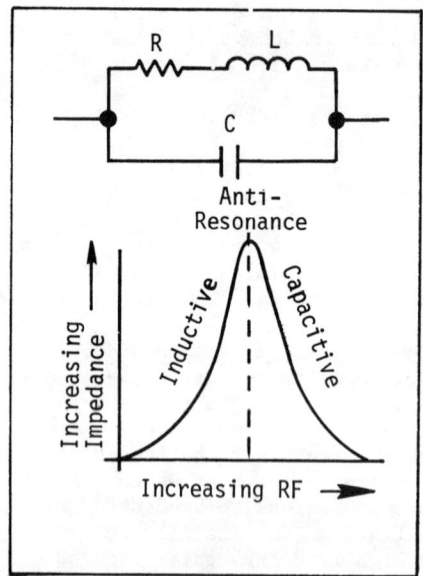

Figure 8.14 - Equivalent Circuit of Bond Strap and Its Impedance

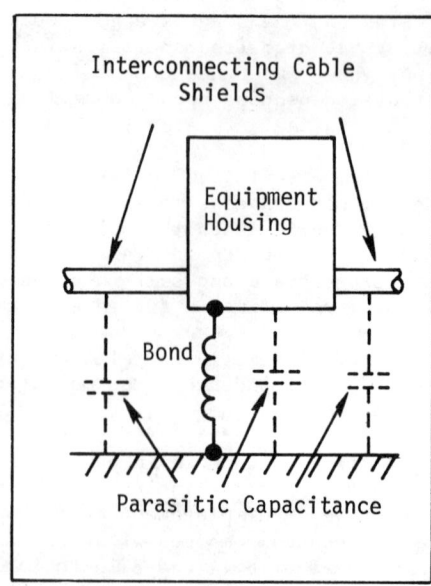

Figure 8.15 - Bond Impedance and Installation Parasitics

8.4.2   Types of Bonds

The best performing electrical bond consists of a permanent, direct, metal-to-metal contact, such as provided by welding, brazing, sweating, or swagging. Though adequate for most purposes, the best soldered joints have appreciable contact resistance and cannot be depended upon for the most satisfactory type of bonding. Semi-permanent joints, such as provided by bolts or rivets can provide effective bonding. However, relative motion of the joined members will likely reduce the bonding effectiveness by introducing a varying impedance.

8.4.3   Bonding Hardware

Star or lock washers should be used with bolt or lock-thread bonding nuts to ensure the continuing tightness of a semi-permanent bonded joint or members. Fig. 8.16(b) shows one recommended arrangement. Star washers are especially effective in cutting through protective or insulating coatings on metal such as anodized aluminum or unintentional oxides or grease films developed during periods between maintenance.

Joints that are press-fitted or joined by screws of the self-tapping or sheet-metal type cannot be relied upon to provide low-impedance R-F paths. Among other considerations, these screws are made on a screw machine in which a jet of coolent oil is used. The threads may thus contain some residual oil in spite of a de-greasing bath. Often there is a need for relative motion between members that should be bonded, as in the case of shock mounts. A flexible metal strap can be used as a bonding agent as shown in Fig. 8.17.

8.4.4   Bonding Jumpers and Bond Straps

Bonding jumpers are short, round, either braided or stranded conductors for application where interference frequencies to be grounded are below about 10 MHz. They are frequently used in low-frequency devices where the development of static charges must be prevented. They are also used to provide good electrical continuity across tubing members and associated clamps such as shown in Fig. 8.16(a). In their application, the clamp should not be relied upon for continuity because of tubing finishes, grease films, and/or oxides.

To provide a low-impedance path at R-F, it is necessary to minimize both the self-inductance and residual capacitance of a bond, in order to maximize the parasitic resonant frequency. Since it is difficult to change the residual capacitance of the strap and mounting, self-inductance becomes the main controllable variable. Thus, bond straps are preferable to round wires of equivalent cross sectional areas.

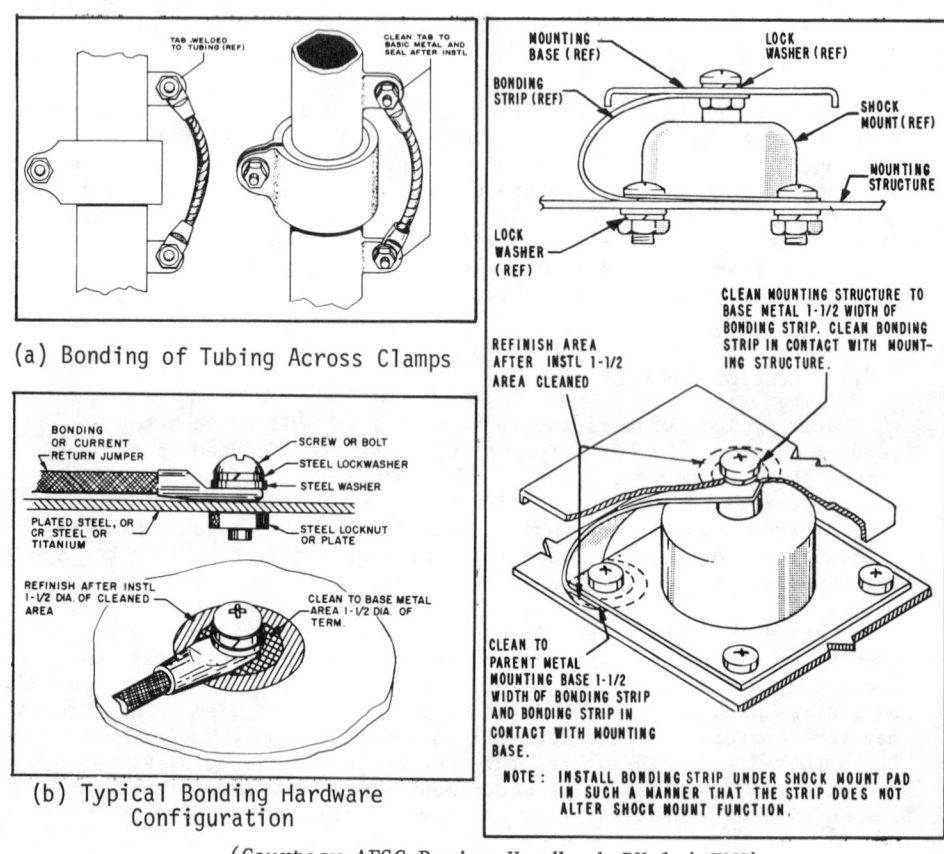

(a) Bonding of Tubing Across Clamps

(b) Typical Bonding Hardware Configuration

(Courtesy AFSC Design Handbook DH 1-4 EMC)

Figure 8.16 - Bonding Connections   Figure 8.17 - Bonding Shock Mounts

Bond straps are either solid, flat, metallic conductors, or a woven braid configuration where many conductors are effectively in parallel. Solid-metal straps are generally preferred for the majority of applications. Braided or stranded bond straps are not generally recommended because of several undesirable characteristics. Oxides may form on each strand of non-protected wire and cause corrosion. Because such corrosion is not uniform, the cross-sectional area of each strand of wire will vary throughout its length. The nonuniform cross-sectional areas (and possible broken strands of wire) may lead to generation of EMI within the cable or strap. Broken strands may act as efficient antennas at high frequencies, and interference may be generated by intermittent contact between strands.

Solid bond straps are also preferable over stranded types because of lower self-inductance. The direct influence of bond-strap construction on R-F impedance is shown in Fig. 8.18, where the impedances of two bonding straps and of No. 12 wire are compared as a function of frequency. The relatively high impedance at high frequencies illustrates that there is no adequate substitute for direct metal-to-metal contact. A rule of thumb for achieving minimum bond strap inductance is that the length-to-width ratio of the strap should be a low value, such as 5:1 or less. This ratio determines the inductance, the major factor in the high-frequency impedance of the strap.

8.4.5   Rack Bonding

Equipment racks (19" or other sizes) provide a convenient means of maintaining electrical continuity between such items as rack-mounted chassis, panels, and ground planes. It also serves as an electrical inter-tie for cable trays. A typical equipment cabinet, with the necessary modifications to provide such bonding, is shown in Fig. 8.19. Bonding between equipment chassis and rack is achieved through equipment front panel and rack right-angle brackets. These brackets are grounded to the unistrut horizontal slide that is welded to the rack frame. The lower surfaces of the rack are treated with a conductive protective finish to facilitate bonding to a ground-plane mat. The ground stud at the top of the rack is used to bond a cable tray, if used, to the rack structure, which is of welded construction. Fig. 8.20 illustrates a typical bonding installation.

Cable trays are bonded together and the cable tray is bonded to the cable chute. The cable chute is bonded to the top of the cabinet; the cabinet is bonded to the flush-mounted grounding insert (which is welded to the ground grid); and the front panel of the equipment is bonded to the rack or cabinet front-panel mounting surface. Nonconductive finishes are removed from the equipment front panel before bonding. The joint between equipment and cabinet may serve a dual purpose: that of achieving a bond and that of preventing interference leakage from the cabinet if the joint is designed to provide shielding. If such shielding is a requirement, conductive gaskets (see Chap. 12) should be used around the joint to ensure that the required metal-to-metal contact is obtained. If equipment is located in a shock-mounted tray, the tray should be bonded across its shock mounts to the rack structure. Connector mounting plates should use conductive gasketing to improve chassis bonding. If chassis removal from the rack structure is required, a one-inch-wide braid with a vinyl sleeving should be used to bond the back of the chassis to the rack. The braid should be long enough to permit withdrawal of the chassis from the rack.

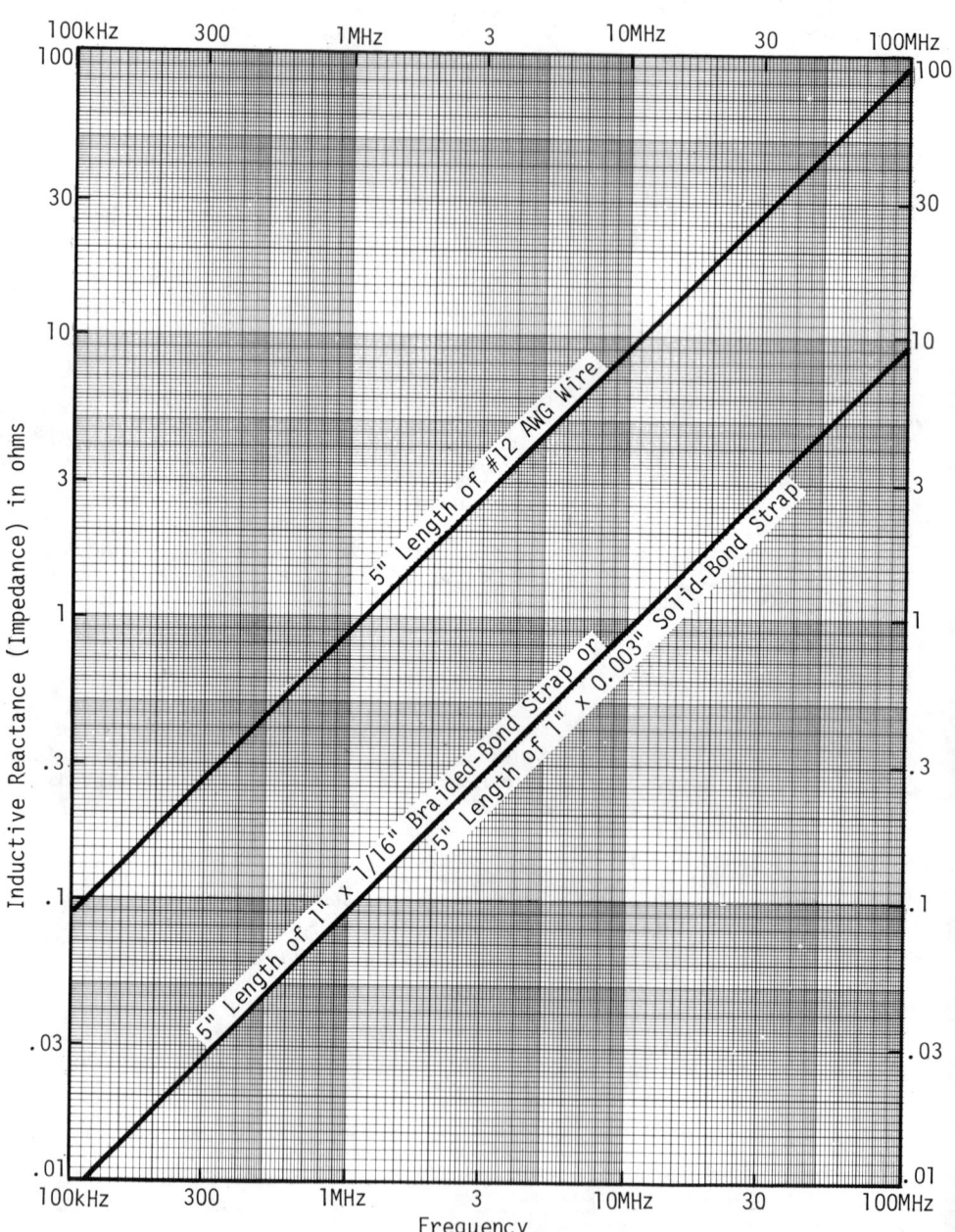

Figure - 8.18 - Impedances of Wire, Braided and Solid-Bond Straps

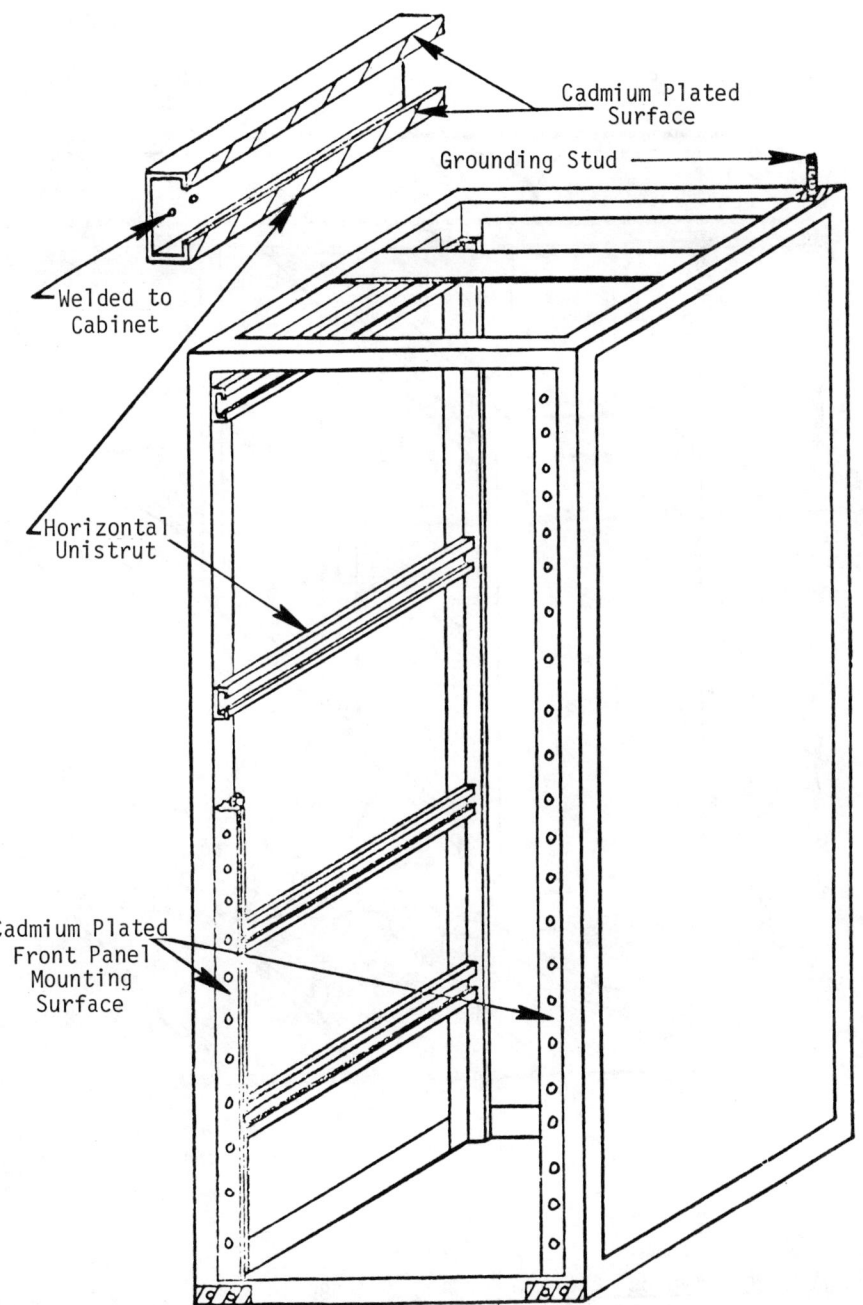

Figure 8.19 - Cabinet Bonding Modifications

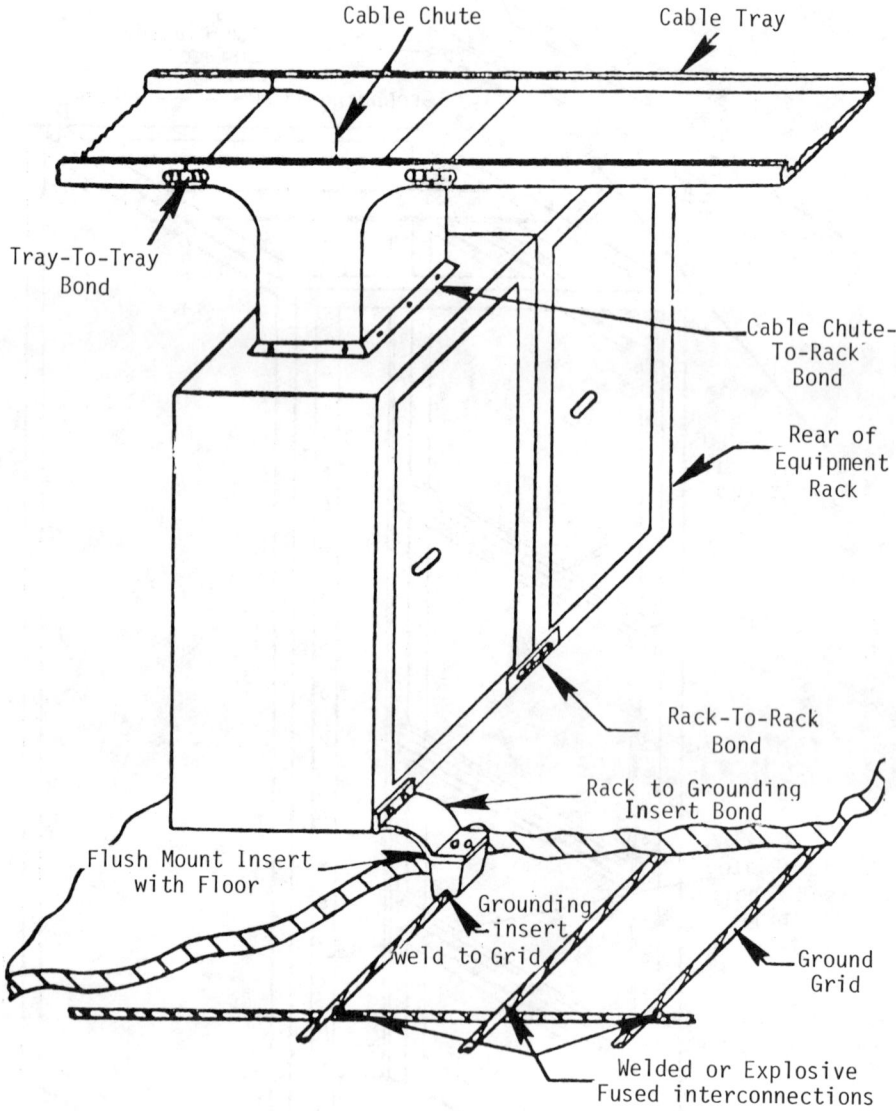

Figure 8.20 - Typical Cabinet Bonding Arrangements

8.5   CORROSION AND CONTROL

This section discusses aspects of bonding dealing with side-effect problems of corrosion and its control.

8.5.1   CORROSION

When two metals are in contact (bonded) in the presence of moisture, corrosion may take place through either of two chemical processes. The first process is termed *galvanic corrosion*, and develops from the formation of a voltaic cell between the metals with moisture acting as an electrolyte. The degree of resultant corrosion depends on the relative positions of the metals in the electro-chemical (sometimes called electromotive) series. This series is shown in Table 8.1, with the metals listed at the top of the table corroding more rapidly than those at the bottom. If the metals differ appreciably in this series, such as aluminum and copper (2.00 volts difference) the resulting electromotive force will cause a continuous ion stream with a significant accompanying decomposition of the more active metal (higher in the series or less noble) as it gradually goes into solution.

Table 8.1 - Electro-Chemical Series

Metal	EMF (Volts)	Metal	EMF (Volts)
Magnesium	+2.37	Lead	+0.13
Magnesium Alloys		Brass	
Beryllium	+1.85	Copper	-0.34
Aluminum	+1.66	Bronze	
Zinc	+0.76	Copper-Nickel Alloys	
Chromium	+0.74	Monel	
Iron or Steel	+0.44	Stainless Steel	
Cast Iron		Silver Solder	
Cadmium	+0.40	Silver	-0.80
Nickel	+0.25	Graphite	
Tin	+0.14	Platinum	-1.20
Lead-Tin Solders		Gold	-1.50

The second chemical corrosion process is termed *electrolytic corrosion*. While this process also requires two metals in contact through an electrolyte, the metals need not have different electrochemical activity, i.e., they can be the same material. In this case, decomposition is attributed to the presence of local electrical currents which may be flowing as a result of using a structure as a power-system ground return.

Since mating bare metal to bare metal is essential for a satisfactory bond, a frequent conflict arises between bonding and finishing specifications. For EMI control, it is preferable to remove the finish

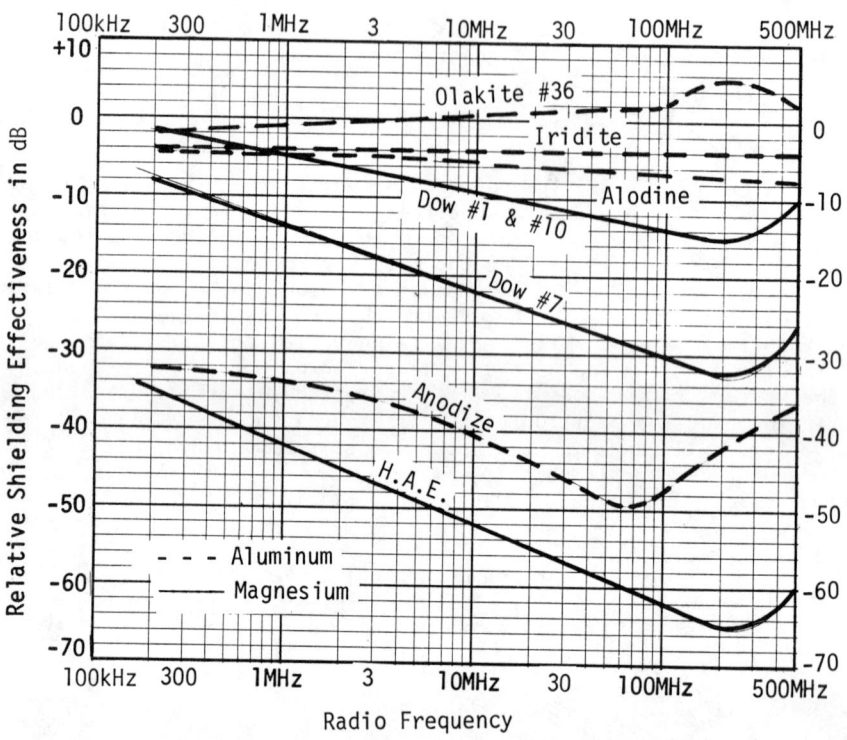

Figure 8.21 - Effect of Various Finishes on Shielding Effectiveness
of Aluminum and Magnesium

where compromising bonding effectiveness would otherwise occur. Certain conductive coatings such as alodine, iridite, and Dow #1, or pro-
tective-metal platings such as cadmium, tin, or silver need not generally be removed. Most other coatings, however, are nonconductive
and destroy the concept of a bond offering a low-impedance R-F path.
For example, anodized aluminum appears to the eye to be a good conductive surface for bonding, but, in reality, it is an insulated
coating. Fig. 8.21 shows the effect of various protective coatings
on the electric-field shielding effectiveness of aluminum and magnesium. The shielding effectiveness is a function of the coating
conductivity (see Chap. 10). The superiority of bare metal over a
12 octave frequency range is evident.

8.5.2   Corrosion Protection

The most effective way to avoid the adverse effects of corrosion is to use metals low in the electro-chemical activity table, such as tin, lead, or copper. This is not generally practical in the design of many structures (e.g., aircraft) due to weight considerations. Consequently, the more active, lighter metals, such as magnesium and aluminum, are employed, although stainless steel has been used in many missile programs.

Joined metals should be close together in the activity series if excessive corrosion is to be avoided. Magnesium and stainless steel form a galvanic couple of high potential (about 3 volts) which tends to a rapid corrosion of the magnesium. Where dissimilar metals must be used, select replacable components for the object of corrosion, such as grounding jumpers, washers, bolts, or clamps, rather than structural members. Thus, the smaller mass should be of the higher potential, such as steel washers for use with brass structures.

When members of the electrolytic couple are widely separated in the activity table, it is often practical to use a plating, such as cadmium or tin, to help reduce the dissimilarity. Sometimes, it is possible to electrically insulate metals with organic and electrolytic finishes and seal the joint against moisture to avoid corrosion. However, this is an unacceptable practice for EMI control. One solution to electrolytic corrosion is to avoid the use of the structure or housing for power-ground return. Any corrosion that is anticipated should be designed to occur in easily replaceable items as previously mentioned. Joints should also be kept tight and well coated after bonding to prevent the entrance of liquids or gases, since a galvanic cell cannot function without moisture.

8.6 BIBLIOGRAPHY

(1)   AF/BSD Exhibit 62-87, "Electro-Interference Control Requirements for Minuteman (WS 133B)," 6 December 1962.

(2)   Armour Research Foundation, "Implementation of bonding practices in existing structures," *Proc. Eighth Tri-Service Conference on Electromagnetic Compatibility,* Chicago; pp. 670-690, Oct. 1962.

(3)   Boeing Aircraft Corporation D2-2-2444, "Electro-Interference Control Requirements (Equipment)," 5 March 1959.

(4)   Craft, Arnold M., "Considerations in the design of bond straps," *IEEE Transactions on Electromagnetic Compatibility,* Vol. EMC-6, No. 3. pp. 58-65; October 1964.

(5)   Denny, H.W., and Byers, K.G., Jr., "A Sweep Frequency Technique for the Measurement of Bonding Impedance Over an Extended Frequency Range," *1967 IEEE EMC Symposium Record,* July 18-20 1967.   Washington, D.C. pp. 195-202.

(6)   Duke, C.A., and Smith, L.E., "The Technique and Instrumentation of Low ImpedanceGround Measurement," *AIEE Conference Paper, No. 58-105,* 1958 (unpublished).

(7)   Dwight, H.B., "Calculations of resistance to ground," *Electrical Engineering,* Vol. 55, pp. 1319-1328, 1936.

(8)   Electronic Instrument Digest, "Grounding low-level instrumentation Systems," Vol. 2, No. 1, p. 16; Jan-Feb 1966.

(9)   Ervin H.W., et al., "The realization of compatible structure grounding systems," *Proceedings Ninth Tri-Service Conference on EMC.,* Chicago; Armour Research Foundation, pp. 155-182, Oct. 1963.

(10) Filtron Co., Inc., "Interference Reduction Guide for Design Engineers," prepared for U.S. Army Electronics Labs., Fort Monmouth, N.J., Vol. 2, pp. 3-89; Aug. 1964.   Accession No. AD 619667.

(11) Foster, J., Buegel, K., and Sowa, C., "Electromagnetic Compatibility Lunar Orbiter," *NASA,* Document No. NAS 1-3800; January 29 1965.

(12) Frederick Research Corp., "Handbook on Radio Frequency Interference," A.H. Sullivan, editor, Vol. 3, pp. 3-62, 1962.

(13) GM07-59-2617A, "Electro-Interference Control Requirements for Minuteman (WS 133A)," 20 October 1959.

(14) Goddard Space Flight Center, Greenblet, Maryland, "Grounding System Requirements for STADAN Stations," GSFC Specification S-533-P-11,

May 1966. *Specification Depository,* Code 252.

(15) "Grounding Low-Level Instrumentation Systems," *Electronic Instrument Digest,* Vol. 2, No. 1, p. 16; January-February 1966.

(16) Haber, F., "Study of GSFC Radio Frequency Interference (RFI) Design Guideline for Aerospace Communication System," Moore School of Electrical Engineering, University of Pennsylvania, *Moore School Report* No. 66027 for NASA, April 30, 1966.

(17) Herring, Thomas H., "The Electrical Role of Structure in Large Electronic Systems," *Fifth National IEEE Symposium on Radio Frequency Interference,* June 1963.

(18) Jakubec, L.G., Jr., "The elimination of Stray RF," CQ, Vol. 16, No. 10; October 1960.

(19) Johnson, W.R., et al, "Development of a Space Vehicle Electromagnetic Interference/Compatibility Specification," *NASA,* Contract No. 9-73-5, *TRW,* 08900-60001-Tooo; June 28 1968.

(20) Marimon, Robert L., "Grounding rules for low-frequency, high-gain amplifiers," *Electronic Design,* Vol. 11, No. 8, p. 56, April 12, 1963.

(21) McAdam, W., and Vandeventer, D., "Solving Pickup Problems in Electronic Instrumentation," *ISA Journal,* Vol. 7, No. 4, P. 48, April 1960.

(22) McDonald, G.M., and Taylor, G.R., "Shielding Grounding and Circuit-Grounding effectiveness in Interference reduction in the 50 Hz to 15 kHz Frequency Region, "*IEEE Transactions on Electromagnetic Compatibility,* Vol. EMC-8, No. 1, pp. 8-16, March 1966.

(23) Mengel, J.T., "Tracking the earth satellite, and data transmission by radio," *Proceedings IRE,* Vol. 44, No. 6, pp. 755-760, June 1956.

(24) Newman, I.M., and Albin, A.L., "An Integrated Approach to Bonding, Grounding, and Cable Selection," *Proceedings Seventh Conference on Radio Interference Reduction and Electromagnetic Compatibility,* Chicago, Ill., Armour Research Foundation, pp. 434-459, November 1961.

(25) Pearlston, C.B., "Case and Cable Shielding, Bonding, and Grounding Considerations in Electromagnetic Interference," *IRE Transactions on Radio Frequency Interference,* Vol. RFI-4, No. 3 pp. 1-16, Oct. 1962.

(26) Robertson, Trevor A., "The role of grounding in eliminating electronic interference," *IEEE Spectrum,* pp. 85-89, Vol. 2, No. 7, July 1965.

(27)   Rudenberg, R., "Grounding Principles and Practices, Part I: Fundamental considerations on ground currents," *Electrical Engineering,* Vol. 64, pp. 1, January 1945.

(28)   Salati, O.M., "Recent Developments in Interference," *IRE Transactions on Radio Frequency Interference,* Vol. 4, No. 2, pp. 24-33, May 1962.

(29)   Salati, O.M., Showers, R.M., Haber, F., Schwartz, R.F., "Training Course in Electromagnetic Compatibility," Moore School of Electrical Engineering, University of Pennsylvania, 1966.

(30)   Schreiber, O.P., "RFI Gasketing," *Electronic Design,* Vol. 8, No. 4, Feb. 1960

(31)   Signal Corps Memo No. M-1301, SCP No. 4144-1085, "Radio Interference Suppression System for 100 KW Engine Generator Unit," June 1950.

(32)   Troup, R.J., and Grubbs, W.C., "A Special Research Paper on Electrical Properties of a Flat-Thin conductive strap for electrical bonding," *Proceedings Tenth Tri-Service Conference on Electromagnetic Compatibility,* Chicago, Ill., Armour Research Foundation, pp. 450-474, November 1964.

(33)   U.S. Department of Commerce, Office of Technical Services, "Bonding Materials, Metallic Mating Surfaces: Low RF Impedance," Report PB 111930, January 1954.

(34)   Weinstock, G.L., "Electromagnetic Interference Control Within Aerospace Ground Equipment for the McDonnell Phantom II Aircraft," *IEEE Transactions on Electromagnetic Compatibility,* Vol. EMC-7, No. 2, pp. 85-92, June 1965.

(35)   Wilson, T.R. and Skene, W.M., "Bonding and Grounding in AerospaceVehicles and Ground Equipment," The Boeing Co., Aerospace Group File No. B-8.

(36)   Wright Patterson Air Force Base, Ohio, "Handbook of Instructions for Aerospace Systems Design," Vol. 4, Electromagnetic Compatibility AFSCM 80-9, Code SEG(SEPSM); Basic Issue, 20 April 1964.

CHAPTER 9

ARCHITECTURAL GROUNDING, WIRING, AND SHIELDING

9.1   STRUCTURAL GROUNDING
      9.1.1   Earth Ground
      9.1.2   Ground Rods and Grids for Power and Safety Grounds
              9.1.2.1   Ground Rods and Earth-Ground Grid Meshes
              9.1.2.2   Design of Grounding System Utilizing Ground Rods
              9.1.2.3   Materials to be Used in Ground Rods and Coatings
              9.1.2.4   Method of Connecting Ground Rod to Structure and Grid Mesh
              9.1.2.5   Design of Grounding System Utilizing Ground-Grid Mesh
              9.1.2.6   Method for Connecting Earth Ground Grid Mesh to Structure
              9.1.2.7   Method for Approximating Combined Ground Resistance of Mesh and Ground Rods
      9.1.3   Soil Impedance
              9.1.3.1   Resistivity Variations as a Function of Soil Type
              9.1.3.2   Resistivity Variations as a Function of Moisture Content
              9.1.3.3   Resistivity Variations as a Function of Temperature
              9.1.3.4   Resistivity Variations Due to Salt Content
              9.1.3.5   Methods for Measuring Soil Resistivity
      9.1.4   Reference Plane Grids and Busses for Instrument Grounds
              9.1.4.1   Bus or Cable Tray Reference Plane
              9.1.4.2   Reference Plane Ground Grids

9.2   BUILDING WIRING AND LIGHTING
      9.2.1   Instrument Power Wiring
      9.2.2   Filtering at Duplex Service
      9.2.3   Conduits and Cable Trays
      9.2.4   Lighting

9.3   SHIELDING
      9.3.1   Typical Building Attenuation
      9.3.2   Exterior Building Materials
              9.3.2.1   Coke-Shell Exteriors
              9.3.2.2   Screen and Conductive Finishes
      9.3.3   Interior Building Materials
              9.3.3.1   Conductive Meshes in Walls, Floors, and Ceilings
              9.3.3.2   Metallized Wallpaper
      9.3.4   Door and Window Design
      9.3.5   Special Shielded Enclosures

9.4   BIBLIOGRAPHY

CHAPTER 9

ARCHITECTURAL GROUNDING, WIRING AND SHIELDING

Many, if not most man-made, electromagnetic noise sources discussed in Chap. 2 emanate from electrical, electromechanical, and electronic devices located in buildings of all types. Consequently, a building represents a plethora of broadband electrical noise culprits which can result in EMI to many receptors also located within the same building. This building *noise-maker* complex also constitutes a threat to receptors in other nearby buildings or areas. While a building offers some natural R-F shielding, if properly designed, it could both protect the inside receptors against outside-world electromagnetic ambients as well as mitigate pollution to this ambient from within.

As explained in the previous chapter, all buildings represent a maze of electrical wiring routed throughout all floors and between floors. This wiring acts as a huge pick-up antenna to both internally developed electrical noise and emissions from without. This antenna complex will conduct EMI to victim receptors as well as re-radiate along the length of the wiring. Further, it is not uncommon for taller buildings to locate an antenna farm on the roof. Here, antennas may be transmitting (e.g.: land mobile, broadcast and microwave relay) as well as receiving both signals and unintentional noise from nearby sources.

This chapter on architectural grounding, wiring, and shielding covers buildings of all types, a few of which are:

- Apartment houses
- Air terminals
- Department stores
- Garages
- Heavy industry
- High rises
- Hospitals
- Houses
- Light industry
- Manufacturing plants
- Medical centers
- Offices
- Petroleum refineries
- Power plants
- Restaurants
- Steel factories
- Theaters
- Warehouses

It is not uncommon for small buildings (e.g.: less than 10,000 square feet of floor space) to have over 1,000 discrete electromagnetic

9.1

emitters and receptors, and for large buildings (over 100,000 square feet of floor space) to have over 10,000 discrete emitters and receptors. For example, every fluorescent lamp is an emitter and every telephone and intercom station is an electromagnetic receptor. Table 9.1 illustrates only a few of the many emitters and receptors existing in buildings. When immersed in the outside world electromagnetic ambient, it is clearly seen that a modern building constitutes both an EMI pollution threat and victim.

This chapter emphasizes EMI control in buildings with particular attention to grounding, wiring and shielding. For convenience of presentation, the topic has been divided into construction above ground and below ground. Among the subjects covered are:

- Artificial shielding
- Cable raceways
- Grounding grids
- Grounding planes
- Intercommunications
- Isolation transformers
- Lightning arresters

- Lighting
- Natural shielding
- Power distribution
- Soil impedance
- Structural shields
- Windows and doors
- Wiring systems

Table 9.1 - A Few Representative Emitters and Receptors Found in Buildings

Emitters and Receptors	Air Terminals	Dept. Stores	Garages	Heavy Industry	High Rises	Hospitals	Houses	Light Industry	Manufacturing Plants	Med. Centers	Offices	Petroleum Refineries	Power Plants	Restaurants & Cafeterias	Steel Factory
Typical Emitters															
Adding machines	X	X	X			X		X	X	X	X	X	X	X	X
Appliances	X				X	X	X	X		X	X			X	
Arc welders				X	X			X	X						
Autos	X		X												
Computers	X			X		X		X	X	X	X	X	X		X
Diathermy						X				X					
Elevators		X			X	X				X	X	X			
Escalators	X	X				X									
Fluorescent lights	X	X	X	X	X	X	X	X	X	X	X	X	X	X	X
Mach. Shop			X	X				X	X	X		X			
Office machines	X			X			X								
Overhead cranes	X		X	X					X			X	X		X
Power tools	X			X		X	X								
Reproduct.		X				X		X	X	X	X	X			
Ultrasonic cleaners			X	X		X		X	X	X					
Typical Receptors															
Automatic controls	X			X		X		X	X			X	X		X
Bio-Medic. instrum.						X				X					
Computers	X		X	X		X		X	X		X	X	X		X
Intercoms	X	X	X	X	X	X	X	X	X	X	X	X	X	X	X
Radios	X	X			X	X	X			X	X				
Recorders/displays	X			X		X				X		X	X		X
Security systems		X		X	X			X	X		X	X	X		X
Telephones	X	X	X	X	X	X	X	X	X	X	X	X	X	X	X
Television	X	X		X		X	X			X			X		X
Test instrum.	X		X	X				X	X			X	X		X

9.1   STRUCTURAL GROUNDING

The subject of structural grounding was discussed at length in Chap. 8. Three purposes for grounding were reviewed: (1) to prevent a shock hazard to personnel in the event that an electrical equipment frame or housing may develop a dangerous voltage to frame due to accidental breakdown (short circuit) of wiring or components, (2) to protect a building and its contents from lightning-stroke damage by providing a very-low impedance path from the top of building to earth, and, (3) to provide a reference voltage (usually called a zero-potential plane) for all electrical and electronic systems and equipments in order to avoid both a shift in operating voltage levels and prevent circulating ground-current loops resulting in common-mode impedance coupling.

A building ground of the neutral and safety (green) wire at the service entrance may exhibit an impedance to earth of the order of 0.5 ohms. During an electrical storm, a typical lightning stroke from a charged-cloud source, having a potential to ground of about 100 MVolts, results in a current of about 30 kAmperes. Assuming a negligible impedance of the lightning arrester and building-girder structure, the neutral *ground* point then is suddenly elevated to 15 kVolts (30 kAmps x 0.5Ω) earth-ground impedance. Were it not for the control of this ground impedance and the line-to-neutral surge spark-gap protectors (diode clamps), many electrical and electronic equipments connected to the black-line-to-neutral would have burned out.

The above kind of problem, i.e., the safety-ground situation and reference-plane grounds for EMI-control of electrical and electronic equipment are the subjects of this section. This section reviews the topic of structural grounding with regard to earth grounds and soil impedance, ground rods and grids, structure frame, power grounds, safety and instrument grounds, and corrosion control in grounding and bonding systems.

9.1.1   EARTH GROUND

The power mains servicing a building in the U.S.A. are governed by local and national codes. The National Fire Protection Association (NFPA) has issued NFPA-STD-70-1971, *National Electric Code*, involving standards on wiring and other electrical devices. One requirement is that both the neutral (white wire) and safety wire (green wire) be earthed at the building power-service entrance. This assures that these lines will be maintained at earth potential in the event the power-input transformer develops a short on the secondary side or that the high-voltage primary of the transformer should short to secondary.

Another aspect of the safety problem involves providing a low-impedance path to earth in order to protect personnel and equipment during lightning strokes in an electrical storm. When the ionized column of lightning strikes a building, it will seek out higher-elevation sharp

surfaces and low-impedance paths to earth. This means that a building's
steel structure and/or internal wiring becomes a natural path for high-
lightning current to follow. If the steel structure is not well earthed,
the current-stroke partition between structure and wiring may follow the
building-wire path. In this case, the wiring could burn out as well as
connected equipment, input filters, transformers, motors, etc. Thus,
the object here is to earth the building structure better than the A-C
power ground, and to connect the structural girders to lightning arresters on top of the building so that lightning surges are kept off the
building wiring.

Another aspect of the earth-ground situation is to recognize that
no local earth region has zero impedance between any two points. Hence,
circulating currents result in potential drops between these points,
which result in common-mode impedance coupling (see Sec. 8.2). Apart
from the soil-impedance problem discussed in a later section, a non-
zero potential earth ground, which should be avoided as a reference for
EMI purposes, results from earth pollution due to both A-C and D-C
currents. Two examples are:

Gas distribution lines are kept at -0.8 volts with respect to
a reference electrode buried in earth. While modern gas lines are insulated in earth and at the meter services, due to nicks in insulation,
called *holidays*, current leakage into earth occurs.

Regarding A-C power distribution, a significant portion of the
60-hZ unbalanced current is returned through earth. Ground currents of
the order of 1,000 amperes have been measured at substations.* While
a typical building ground is not located at a substation, considerably
lesser but significant unbalanced currents can flow through the local
earth to cross modulate 60 Hz and associated R-F contamination upon the
EMI grounding complex.

The above serves to illustrate that lightning and power-safety
earth grounds can conflict with EMI-control reference earth grounds
unless special effort is taken to assure that there exists no significant common-impedance earth grounding path,** viz., that a true single
point ground exists. Sec. 8.3 indicates that single-point grounding,
if accomplished at DC and AC, only applies up to a frequency for which
$\ell = \lambda/20$. Thus, circulating R-F currents associated with 60-Hz building grounds may be effective up to about 200 kHz, more or less.

---

* Two ground rods driven into the earth two feet apart produce a
voltage sufficient to light 100 watt bulbs connected to the ground rods.
** See Secs. 8.2 and 8.3.

9.1.2   GROUND RODS AND GRIDS FOR POWER AND SAFETY GROUNDS

This section discusses techniques in establishing earth grounding
and reference-plane systems that are compatible with requirements of
various structure types and use. Sample problems are presented to
illustrate the use of grounding criteria. Complimentary techniques,
necessary for effective implementation of ground planes, are described
in detail. The following terms are used in subsequent discussions:

Ground Rods: Rods constructed of highly-conductive metals
which are driven into the earth and bonded to metallic or structural
masses above ground to preclude the development of potentials which
may prove hazardous to personnel and equipment.

Earth Ground Grid Meshes: Meshes constructed of highly-conductive materials which are bonded together at all junctions, installed
below the earth's surface, and bonded to metallic or structural masses
above earth to complement ground rods.

Reference Plane Ground Grid Mesh: Highly-conductive mesh construction above earth ground which serves the primary purpose of providing a low-impedance reference plane for electromagnetic shielding
media and electronic equipment users* (see Sec. 9.1.4).

Earth-grounding systems and reference-plane systems serve two distinct functions and, therefore, are presented separately. Earth-grounding systems are not ordinarily sufficient voltage reference planes because of relatively high-earth resistance obtainable from ground rods
and earth-grid meshes. Good results are obtained from reference planes
by connecting such planes to earth ground through a single ground well.
Thus, undesirable earth-loop currents are isolated from the reference
plane by this single-point ground practice.

Underground water and pipelines historically are an excellent media
for connecting structural steel to earth ground due to the large amount
of surface area exposed to the earth and relatively large depths to
prevent freezing which such piping is buried in the earth. It has been
standard practice for many years to bond metallic structures above
ground to water and gas pipes by means of copper bonds or copper-ground
rods. Copper in its various forms creates an undesirable coupling of
dissimilar metals. When in contact with iron or steel, copper acts as
a cathode to accelerate corrosion of the less noble metal (see Sec.
8.5.1). Corrosion between copper bond and the less noble metal will
increase the bond impedance to such an extent that it may act as no
electrical bond at all.

The resultant corrosion of underground and above-ground piping
systems is becoming so expensive to maintain that utility companies

---

*   See Sec. 8.1.3, Zero-Potential Ground Plane

are using coated pipes or non-conductive pipes and couplings which will
eliminate this widely used method of grounding. This presents a significant problem of establishing effective economical grounds so that it
is now necessary to install dedicated grounding systems for nearly all
new buildings.

9.1.2.1   GROUND RODS AND EARTH-GROUND GRID MESHES

Resulting from the need for dedicated grounding systems, it has
become necessary to utilize ground rods and meshes. The following
discussions apply where water and gas pipes are either prohibited for
use as grounds or are not accessible for electrical grounding purposes,
i.e., where ground rods and meshes are to be used in lieu of National
Electrical Code requirements.

The 1972 U.S.A. National Electrical Code reads as follows*:
"Where available on the premises, a metal underground water pipe shall
always be used as the grounding electrode, regardless of its length and
whether supplied by a community or a local underground water piping
system or by a well on the premises. Where the buried portion of the
metal water pipe (including any metal well casings effectively bonded
to the pipe) is less than 10 feet long or where the water pipe is or
is likely to be isolated by insulated sections or joints so that the
effectively grounded portion is less than 10 feet long, it shall be
supplemented by the use of an *additional electrode* specified by Sections 250-82 and 250-83."

National Electrical Code requirements are generally adhered to
for industrial and commercial usage. However, various private con-
cerns and branches of the Armed Services prescribe techniques to be
used to provide earth grounding of structures and equipments using
ground rods and meshes. Copper or copper-clad aluminum ground rods
are commonly used because this metal has long life when buried in the
earth and has the best conductivity of commercially-available metals.

9.1.2.2   DESIGN OF GROUNDING SYSTEM UTILIZING GROUND RODS

Ground resistances are computed either by application of the concepts of field theory or by the average-potential method. Although
the latter method is not exact from the standpoint of applied physics,
it furnishes fairly accurate results and is readily adaptable to problems at hand. In recent years it has been accepted as the only
practical means of solving problems of a more involved nature.

The grounding resistance of many closely-spaced parallel ground
rods is expressed as:

---

*   cf. Sec. 250-81, Water Pipe Electrode, NEC Article 250 - Grounding.

$$R = \frac{\rho}{2\pi nL} \left[\ln_e \left(\frac{4L}{b}\right) - 1 + \frac{2kL}{\sqrt{A}} (\sqrt{n}-1)^2 \right] \text{ ohms} \qquad (9.1)$$

where,   R = grounding resistance in ohms

ρ = soil resistivity in ohm-centimeters

L = length of each rod in cm

b = radius of rods in cm

n = number of equally-spaced rods within area A

A = area of rod coverage in cm^2

k = coefficient explained below

The coefficient, k, in Eq. (9.1) is obtained from the expression $(\rho/\pi) \cdot (k/\sqrt{A})$ for the resistance of a horizontal thin plate. When L approaches infinity, Eq. (9.1) approaches this value. The coefficients k, for square and rectangular plates at each surface, are plotted as curve A in Fig. 9.1. In most practical cases, grids or rod beds are buried to depths much less than $\sqrt{A}$ so that the coefficients k for the surface level hold with sufficient accuracy.

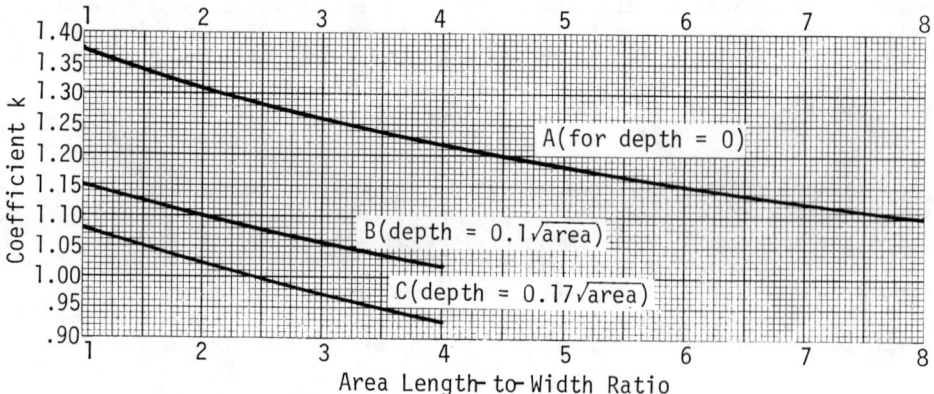

Figure 9.1 - Coefficient k (see text)

The determinations of soil resistivity as a function of locality where measured and depth of ground-rod penetration cannot be practically realized with a high degree of accuracy. Effect of various terrain considerations upon soil resistivity and the accuracy of various methods of measuring soil resistivity are discussed in Sec. 9.1.3.

Rearranging Eq. (9.1) into units more frequently used yields:

$$R = \frac{0.52\rho}{nL} \left[\ln_e\left(\frac{2L}{3b}\right) - 1 + \frac{2kL}{\sqrt{A}} (\sqrt{n} - 1)^2 \right] \quad \text{ohms} \qquad (9.2)$$

where: ρ = soil resistivity in ohm-meter

L = rod or pipe length in feet

b = rod or pipe diameter in inches

n = number of parallel rods

A = area in sq. ft. between rods at farthest outside position

Eq. (9.2) is used to calculate ground resistance as a function of (1) number of evenly-spaced rods, (2) area of coverage, and (3) depth of earth penetration. Resultant data are shown in Fig. 9.2. Calculations have been based upon a value of soil resistivity equal to 50 ohm-meters. Earth resistance resulting from soil resistivities other than this value are:

$$R_1 = R \frac{\rho^1}{\rho} \qquad (9.3)$$

where: R = grounding resistance obtained from Fig. 9.2

ρ = 50 ohm-meters

ρ^1 = measured value of soil resistivity in ohm-meters

Illustrative Example 9.1

A low-resistance earth grounding system is required for a proposed building with foundation dimensions of 150' x 50'. Terrain considerations allow penetration of long ground rods and a soil resistivity of 25 ohm-meters was measured at a depth of 15 feet. The intended use of the building necessitates the requirement that a grounding resistance of no greater than 1 ohm shall exist. Determine the required number of ground rods, depth of penetration, and area of coverage to realize the desired ground resistance.

The foundation area is 150' x 50' = 7500 ft.^{2} The soil resistivity, ρ, is 25 ohm-meters, or one-half the value used for calculating data in Fig. 9.2. Therefore, 2 ohms in Fig. 9.2 is equivalent to the one ohm ground resistance required. The required grounding resistance is obtained from Fig. 9.2 as follows.

(1) Ten, 3/4-inch rods, evenly-spaced over a 7500 ft.^{2} area (or 86^2 ft.^{2}) and driven to a depth of ten feet correspond to $R \approx 1$ ohm.

(2) Three, 3/4-inch rods, evenly-spaced over a 7500 ft.^{2} area and driven to a depth of 30 feet corresponds to $R \approx 1$ ohm.

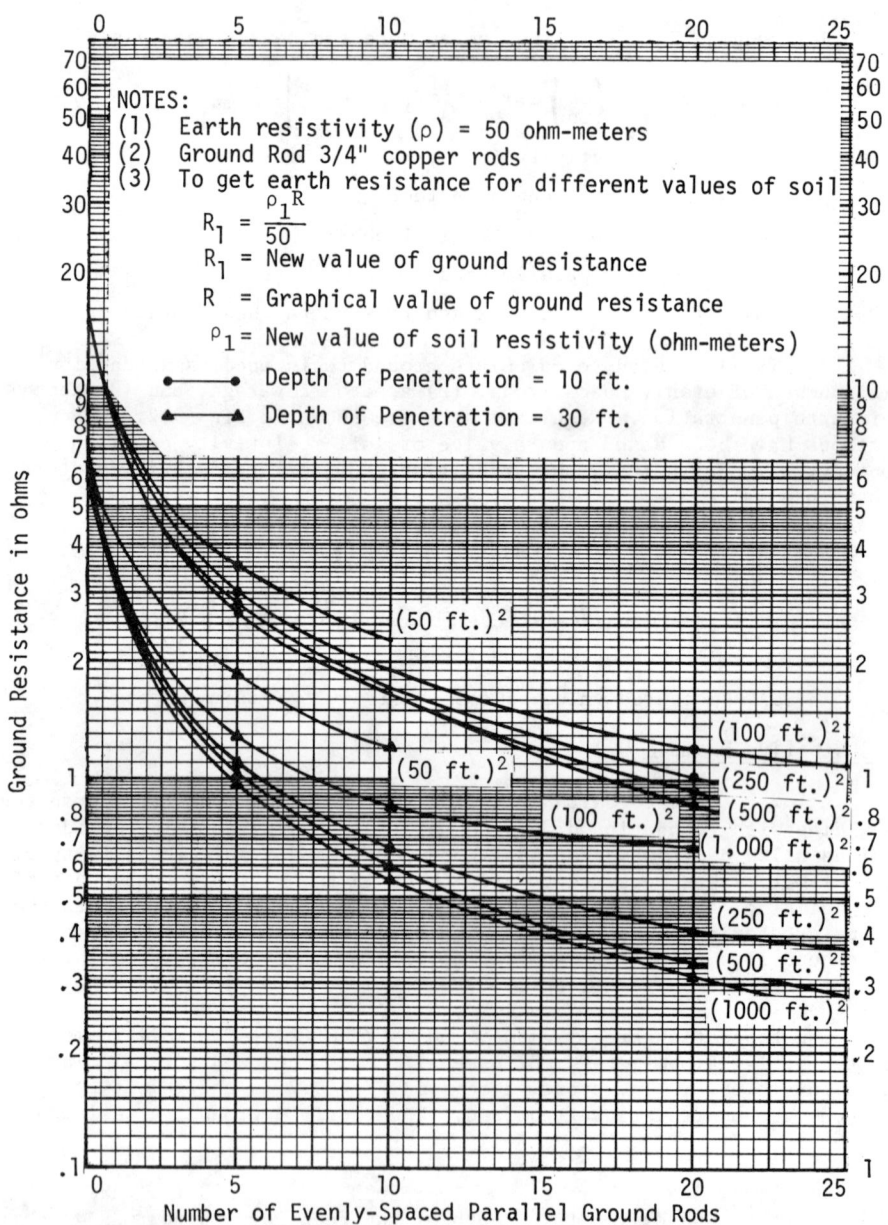

Figure 9.2 - Ground Resistance vs. Number of Evenly-Spaced Parallel
Ground Rods For Various Areas of Coverage and Depths of Earth Penetration

Sufficient tolerances should be added to allow for (1) increase in grounding resistance due to age and corrosion, (2) resistance of rods and tie-cables, (3) bonding resistance resulting from structure to rod and/or tie cables, and (4) variations in soil resistivity. To obtain the decreased resistance, change: (1) depth of penetration, (2) number of rods, and/or (3) area of coverage. A 50% tolerance is often used; therefore the resultant resistance requirement will be .5 ohms. This requirement may be realized by either of the following:

(3) Ten, 3/4-inch rods, evenly-spaced over a 7500 ft.^{2} area and driven to depths of 30 ft. corresponds to $R \approx .5$ ohms.

(4) Seven, 3/4-inch rods, evenly-spaced over a 12,000 ft.^{2} area, and driven to depths of 30 ft. correspond to $R \approx .5$ ohms.

Illustrative Example 9.2

Consider a building identical to Example 9.1, except that the building is intended to house critical instrumentation facilities and the ground resistance is required to be ≤ 0.25 ohms. Such resistance should be practically constant with seasonal change and temperature fluctuations.

Since ground resistance is critical, a tolerance of 50% is again allowed to insure consistent compliance with requirements. Resistance fluctuations might be expected due to factors stipulated in Example 9.1. In order to insure constant ground resistance as a function of climatic changes, rods should be driven to the permanent water level if possible.

The required ground resistance is 0.25 ohms. Thus, use 0.50 x 0.25Ω = 0.125 ohms in Fig. 9.2. Only areas of coverage in excess of 250,000 ft.^{2} utilizing 30 evenly-spaced rods driven to a depth of 30 feet, will result in the specified grounding resistance. This approach is unpractical due to both the extensive area of coverage and large number of ground rods required. Thus, Fig. 9.2 cannot be used to obtain a satisfactory solution to the problem.

Eq. (9.2) is used to calculate the resistances resulting from use of various factors capable of reducing the resistance to specified limits at a reasonable cost. For example, if rods could be driven to the permanent water level where the soil resistivity may be 12.5 ohm-meters, the extra length of rod penetration would reduce the required area of coverage and number of rods to practical values. In lieu of this approach, a ground grid may be used in conjunction with ground rods to obtain the desired ground resistance. A discussion of ground grids is presented in a later section.

9.1.2.3 MATERIALS TO BE USED IN GROUND RODS AND COATINGS

Copper ground rods are commonly used for grounding purposes due to their high conductivity and high-corrosion resistance properties. Fig. 9.3 illustrates the physical makeup of existing ground rods that have proven both effective and capable of being driven to substantial depths. Such rods come in a variety of sizes but the 3/4-inch dia-meter rod appears to be the most compatible with electrical and mechan-ical requirements.

The effects of corrosion must be considered in the selection of compatible ground rods: (1) in most cases, both water and oxygen are necessary for corrosion, (2) the initial rate of corrosion is compara-tively rapid, slowing as protective films form, (3) surface films are important in controlling the rate and distribution of corrosion, (4) increased rate of motion increases corrosion in water, and (5) dis-similar metals in contact accelerate corrosion of the metal higher in the electro-chemical series (see Table 8.1, Sec. 8.5.1).

While copper is reasonably corrosion-resistant, it is anodic (higher in the electro-chemical series) to some metals used under-ground for piping and construction purposes. To reduce copper-rod corrosion, it is therefore desirable to coat copper-grounding rods with a material having the following properties: (1) less anodic to metallic objects located in near vicinity underground, (2) high electrical conductivity, (3) high-corrosion resistance, and (4) galvanically compatible with the base metal.

Tin coatings on copper and copper alloys are normally anodic to the base metal, as indicated in Table 8.1. The tin-copper alloy layer, formed in coating by hot dipping, is cathodic to tin and may be slightly cathodic to copper. Pores in the coating are not usually sites of corrosion attack on copper, and in general the corrosion of tinned copper is essentially corrosion of the tin. The function of tin coatings is usually to provide, between copper and the material in question, a layer which if corroded at all will yield as innocuous a corrosion product as possible.

The advantage obtained from deep-driven ground rods is realized due to the decreased soil resistivity and increased volume of earth associated with deep rods. However, as indicated in Table 9.2, metal corrosivity increases as soil resistivity decreases, which imposes more stringent requirements on the corrosion-resistant properties of ground rods.

Two promising groups of alloys which may be valuable in the prevention of galvanic corrosion, are the austenitic irons (SDTMA-439, Type D2) and austenitic stainless steel of the 18% chromium and 8% nickel variety. However, additional research is required to evaluate the effectiveness of such materials for use in conjunction with grounding.

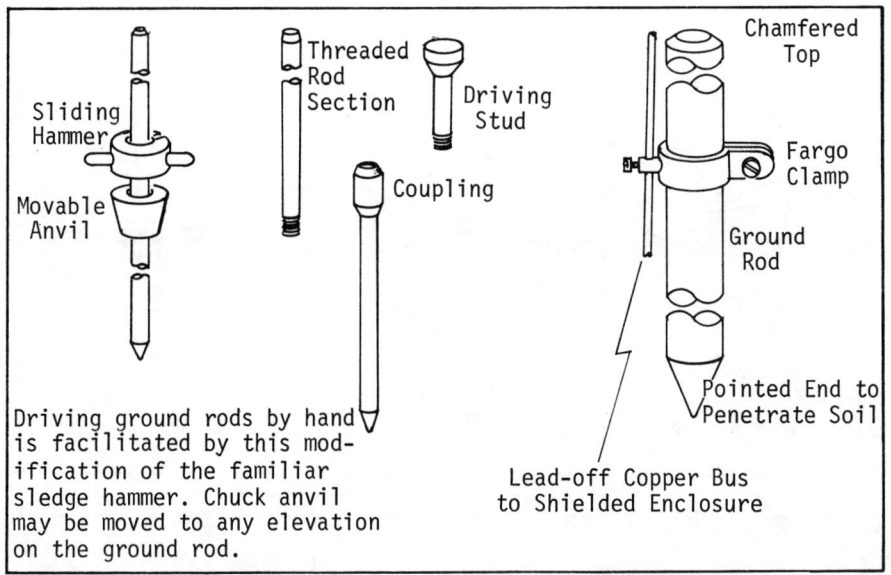

Figure 9.3 - Physical Characteristics of Typical Grounding Rods

Table 9.2 - Metal Corrosivity as a Function of Soil Resistivity

Resistance in Ohms/cm^3	Severity of Galvanic Effects
Less than 400	Extremely severe
400-900	Very severe
900-1,500	Severe
1,500-3,500	Moderate
3,500-8,000	Mild
8,000-20,000	Slight

9.1.2.4 Method of Connecting Ground Rod to Structure and Grid Mesh

Fig. 9.4 illustrates preferred techniques for connecting ground rods to structures and grid meshes. The portion of the ground wire making contact between the Joslyn washer and the base shoe should be tinned to reduce the effects of galvanic corrosion. All bolts and nuts should be securely tightened to prevent bond-impedance deterioration with age and wear. Indicated bonds should be coated with a moisture-proof coating capable of maintaining its physical properties over an extended period of time. The cover which is placed over a

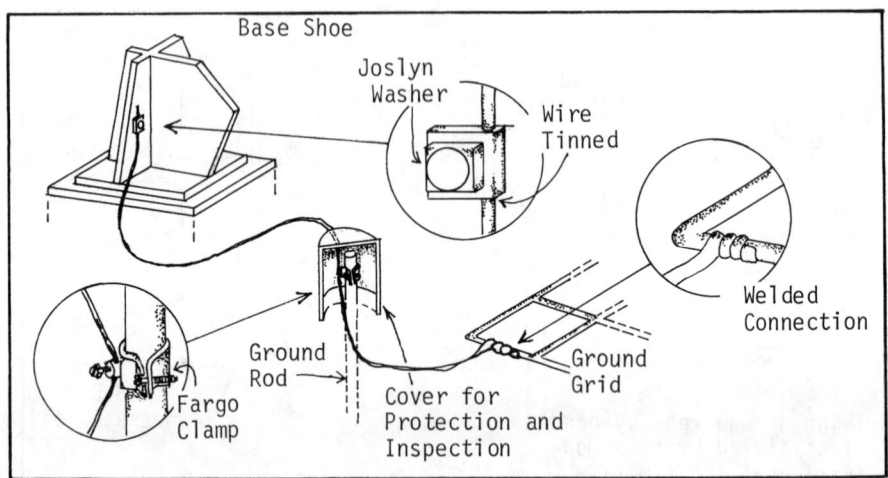

Figure 9.4 - Method for Connecting Ground Rods to Structure and Grid Mesh.

portion of the ground rod extending above the surface of the earth may be of a non-conducting media as long as it remains waterproof over time. Such covers should be removable or provide entrance to facilitate periodic inspection of the ground-rod connection and measurement of earth resistance. The bonding cable can be a 4/0 or solid 1/4" copper wire, or larger sizes. Techniques displayed in Fig. 9.4 can be applied to all types of building structures with steel frames and base shoes.

9.1.2.5 Design of Grounding System Utilizing Ground-Grid Mesh

Ground-grid meshes are often required to compliment rod beds or to be used separately when deep-driven rods are impractical due to soil and terrain considerations. The following formula is used to calculate the ground resistance of an earth ground grid mesh:

$$R = \frac{\rho}{\pi L} \left[\ln_e \left(\frac{2L}{a^1} \right) + k_1 \frac{L}{\sqrt{A}} - k_2 \right] \text{ ohms} \qquad (9.4)$$

where: ρ = soil resistivity in ohm-centimeters

L = total length of all connected conductors in cm

a^1 = a x 2z for conductors buried at a depth of z cm, or

a for conductors on earth surface = conductor radius in cm

A = area covered by conductors, in cm^2

k_1, k_2 = coefficients explained below

9.14

The coefficient k_1 is the same as k used in Eq. (9.2). Coefficient, k_2, has been calculated for loops encircling areas of the same shape and depth as used for calculating k_1 (see Fig. 9.1). Calculated results are shown in Fig. 9.5.

Eq. (9.4) is rearranged into units more frequently used:

$$R = \frac{1.045\rho}{L} \left[\ln_e \left(\frac{2L}{a} \right) + k_1 \frac{L}{\sqrt{A}} - k_2 \right] \text{ ohms} \qquad (9.5)$$

where: a = conductor diameter in inches x depth in feet

L = rod length in feet

ρ = soil resistivity in ohm-meters

Resistance has been calculated as a function of foundation or area of grid coverage and number of grids per side using 4/0 copper cable. The resistivity of the earth was assumed to be 50 ohm-meters for calculation purposes. Coefficients k_1 and k_2 vary only slightly with area and the error introduced by using calculated data for both square and rectangular grids is negligible. Calculated data is presented graphically in Fig. 9.6.

Grounding resistance afforded by buried grid meshes can be reduced significantly by both increasing the number of grids and the area of grid coverage. Data has been extrapolated from Fig. 9.6 and replotted in Fig. 9.7, to show ground resistance as a function of area of grid mesh coverage and single and 30 grids per side.

Various factors are illustrated by such graphical presentations in developing optimum design criteria for earth grid meshes: (1) a far greater reduction in ground resistance is realized by using increased areas of grid coverage up to approximately 90,000 ft.[2]; (2) beyond 90,000 ft.[2], maximum ground resistance reduction is realized by use of number of grids, and (3) the average reduction of ground resistance resulting from additional grids is approximately 0.2 ohms.

Illustrative Example 9.3

An earth ground-grid mesh is to be installed under a building to be located in an area that is not prone to severe climatic fluctuations. The foundation area of the structure is to be approximately 50 foot square (2500 ft.[2]). A grounding resistance of approximately one (1) ohm is required. Measurements indicate that a soil resistivity of 50 ohm-meters exists at a depth of three feet below the earth's surface, and rocky soil precludes measurements at greater depths.

As previously remarked, a ground resistance tolerance of approximately 50% should be allowed in design procedures to account for possible fluctuations in earth resistivity and bond impedance due to age and wear. Therefore, the design resistance should be approximately

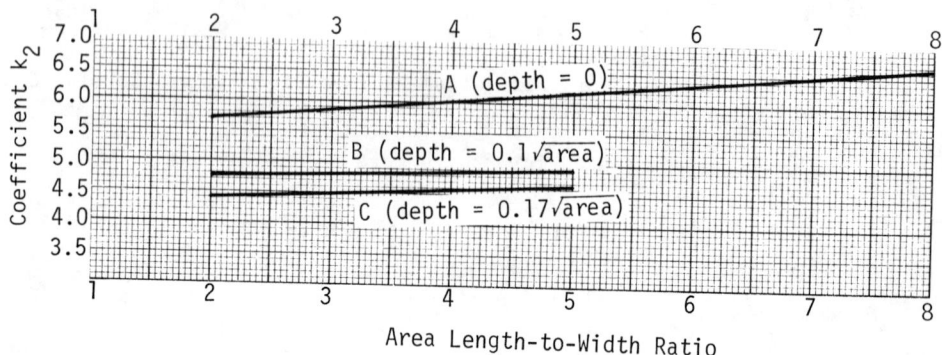

Figure 9.5 – Coefficient k_2 vs. Area Shape and Depth

0.5 ohms. Since the value of soil resistivity is the same as that used for calculating data for Figs. 9.6 and 9.7, graphical values may be used directly. Fig. 9.7 indicates that the required grounding resistance cannot be realized using a number of meshes over an area of 2500 ft.^{2}. An area of coverage of approximately (175 feet)^2 or 30 k ft.^{2} using approximately ten (10) grids per side will yield the required grounding resistance. Fig. 9.6 indicates that either an increased area of coverage or grids per side beyond these amounts will yield a relatively small reduction in ground resistance. Thus, it is recommended that 4/0 copper cable be used to construct a grid mesh 175 foot square having 10 grids per side and buried 3 feet below the surface of the earth. In lieu of the above, the required area of coverage can be reduced considerably, if the earth is artificially treated to reduce the soil resistivity.

9.1.2.6   Method for Connecting Earth Ground Grid Mesh to Structure

Fig. 9.8 illustrates a preferred technique for connecting an earth ground-grid mesh to a structure. The end of the bond cable should be tinned where connection is made with the structure's base shoe in order to reduce effects of galvanic corrosion. A double connection is made between the bond cable and the grid mesh to reduce the possibility of bond deterioration with age and wear. All connections should be wrapped, welded and covered with a protective coating, as indicated. All structure base shoes should be connected in a like manner to the grid mesh. Techniques illustrated in Fig. 9.4 should be followed when grid meshes are used in conjunction with ground rods.

9.1.2.7   Method for Approximating Combined Ground Resistance of Mesh and Ground Rods

In many cases it may be necessary to use a combination of ground rods and a grid mesh below ground to obtain a sufficiently low ground

Assumptions:
(1) 4/0 copper cable
(2) ρ = earth resistivity = 50 ohm-meter
(3) Mesh buried 2 ft. below earth surface

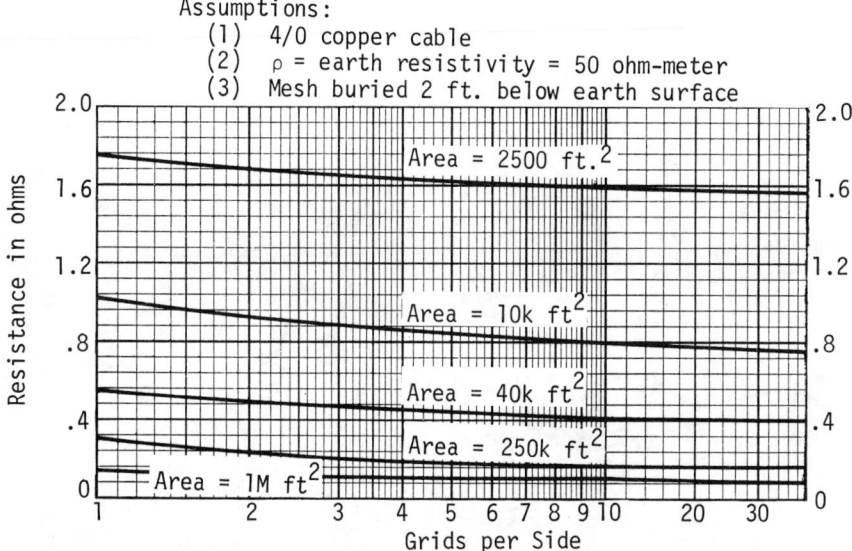

Figure 9.6 - Ground Resistance vs. Number of Grids in Mesh
as a Function of Total Mesh Area

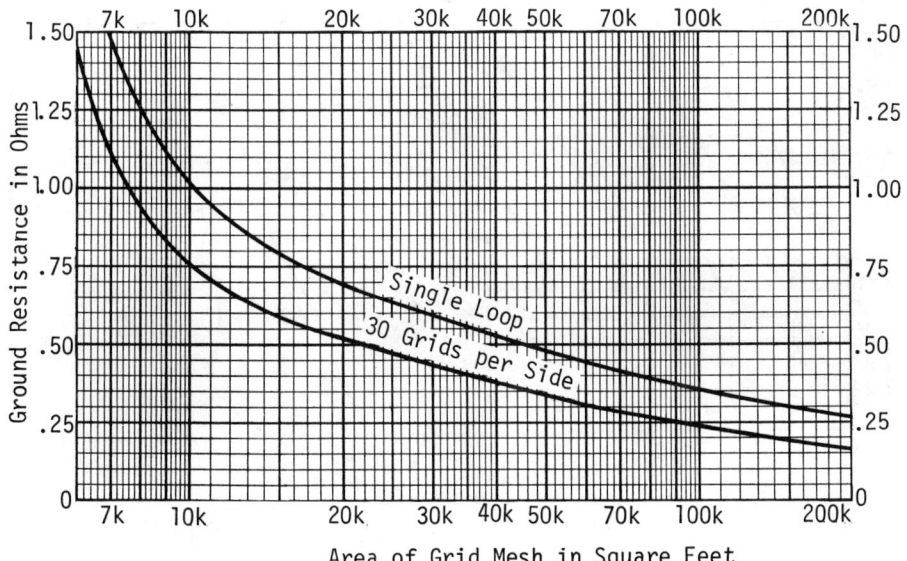

Figure 9.7 - Ground Resistance vs. Grid Mesh

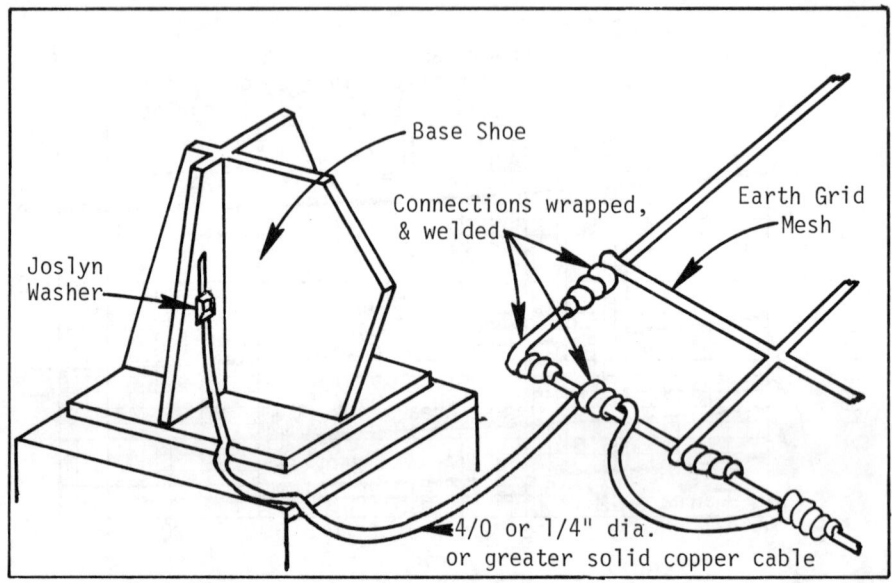

Figure 9.8 - Method for Connecting Earth Ground Grid Mesh to Structure

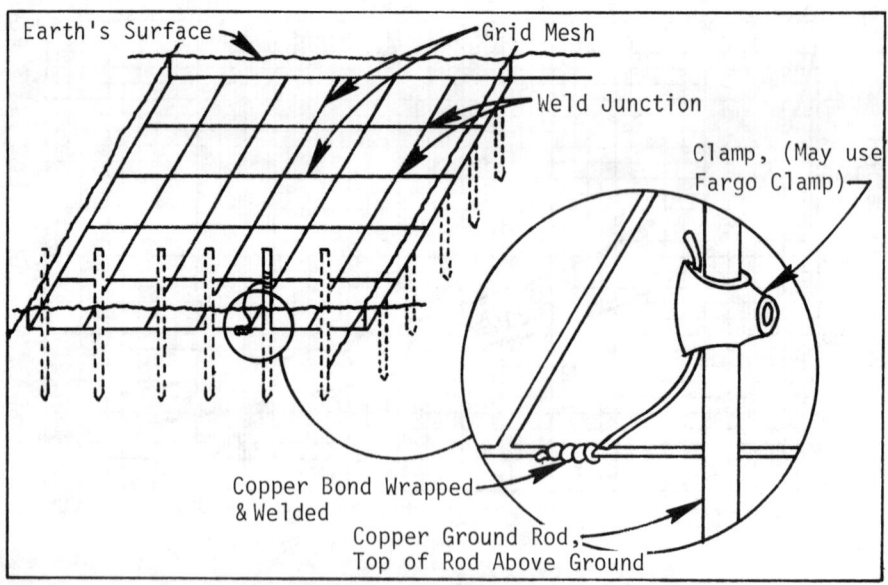

Figure 9.9 - Typical Combination of Ground Rods and Grid Mesh

resistance. Fig. 9.9 illustrates how a combination of a grid mesh
and ground rods might be physically implemented. The mutual resistance between the two grounding systems can be approximated by the
following equation:

$$R_{12} = R_{21} = \frac{\rho}{\pi L} \left[\ln_e \left(\frac{2L}{L_1} \right) + k_1 \frac{L}{\sqrt{A}} - k_2 + 1 \right] \text{ ohms} \qquad (9.6)$$

where: $R_{12} = R_{21}$ = mutual resistance of both systems remaining
                           parameters are equivalent to those used in
                           either equations for rod and mesh resistances.

The combined rod bed and grid resistance is:*

$$R = \frac{R_{11}R_{22} - R_{12}^2}{R_{11} + R_{22} - 2R_{12}} \qquad (9.7)$$

where: R_{11} = resistance of grid alone as presented in Sec. 9.1.2.5

         R_{22} = resistance of rod bed alone as presented in Sec. 9.1.2.2

         R_{12} = mutual resistance between systems

         R = combined resistance.

The reduction in ground resistance achieved by adding rods to a
grid will hardly warrant the extra cost. Yet there are points in favor
of such an arrangement: (1) insure practically constant ground resistance near the earth's surface where soil resistivity may fluctuate due
to extreme climatic conditions, or (2) rods used to provide a reliable
ground source and grid used as a safety measure to equalize fault potentials over the earth's surface.

9.1.3   SOIL IMPEDANCE

Ground rod and grid mesh criteria has been developed in the previous section upon the assumption that a sufficiently low earth resistivity can be realized for effective implementation. In extremely
rocky or frozen soil, deep penetration of ground rods is impractical.
In such cases ground grids might be used but in regions subjected to
extreme climatic variations, earth resistivity will vary considerably
causing resistance changes in shallow buried grid meshes. In various
localities such as dry sandy soils, earth resistivity may be extremely
high regardless of the depth of ground rod penetration. In situations
such as these, other techniques may be utilized to obtain the necessarily low-ground resistance: (1) impregnation of soil with salt solution,

---

* "Analytical Expressions for the Resistance of Grounding Systems,"
by S.J. Schwartz, *Proceedings of AIEE*, August, 1954.

(2) immersion of grid or plate in nearby water sources and connection of such grounding media to structures to be grounded, or (3) utilization of available underground piping systems.

9.1.3.1   RESISTIVITY VARIATIONS AS A FUNCTION OF SOIL TYPE

The ground resistance of any type of electrode that may be used is directly proportional to ρ, the resistivity of the soil. Consider a metallic hemisphere of radius, a, which is buried flush with the surface of the earth as shown in Fig. 9.10. If the resistance of the electrode itself is neglected, the resistance of the earth connection, R, will be that offered to the current flow through the soil volume immediately surround the electrode:

$$R = \rho \int_a^\infty \frac{dx}{2\pi x^2} = \frac{\rho}{2\pi a} \tag{9.8}$$

Earth resistivity is variable as a function of soil type, temperature, and moisture content. Table 9.3 shows data related to soil type. From this table it is noted that a grounding system which is entirely adequate in clay soil will be almost worthless in sandy soil.

9.1.3.2   RESISTIVITY VARIATIONS AS A FUNCTION OF MOISTURE CONTENT

Soils that are relatively good conductors under normal moisture content become good insulators when such content is low. Fig. 9.11 and Table 9.4 show the variation of soil resistivity with moisture content for various soil types. For most soil types, moisture content of 30% will result in a sufficiently low resistivity.

9.1.3.3   RESISTIVITY VARIATIONS AS A FUNCTION OF TEMPERATURE

Soils that have sufficient moisture content to be good conductors at normal temperatures will become ineffective below freezing due to increased resistivity. This is illustrated in Fig. 9.12 and Table 9.5.

9.1.3.4   RESISTIVITY VARIATIONS DUE TO SALT CONTENT

Soil Resistivity varies as a function of the salt content of the soil. Figure 9.13 and Table 9.6 show the effects of salt content upon reduction of the resistivity of various types of soil. Soil can be artificially treated with salt to increase the conductivity. Fig. 9.14 are test data showing resistance variations as a function of time for a specific ground connection.

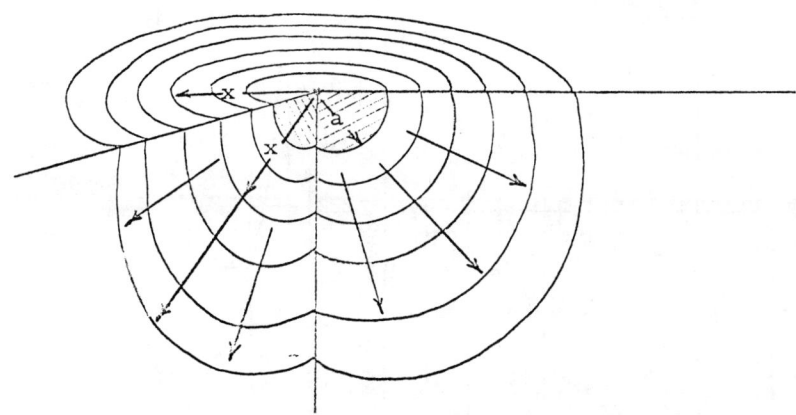

Figure 9.10 - Distribution of Current About A Metallic Hemispherical
Ground Electrode

Table 9.3 - Earth Resistivity of Different Soils

R & ρ Soil Fills	Resistance* Ω 5/8" x 5' rods			Resistivity in Ω/cm^3		
	Avg.	Min.	Max.	Avg.	Min.	Max.
Ashes, Cinders, Brine Waste	14	3.5	41	2,370	500	7,000
Clay, Shale, Gumbo, Loam	24	2	98	4,060	340	16,300
Same, with varying proportions of sand & gravel .	93	6	800	15.8k	1,020	135k
Gravel, sand, stones with little clay or loam	554	35	2,700	94k	59k	458k

*Bureau of Standards Technical Report No. 108

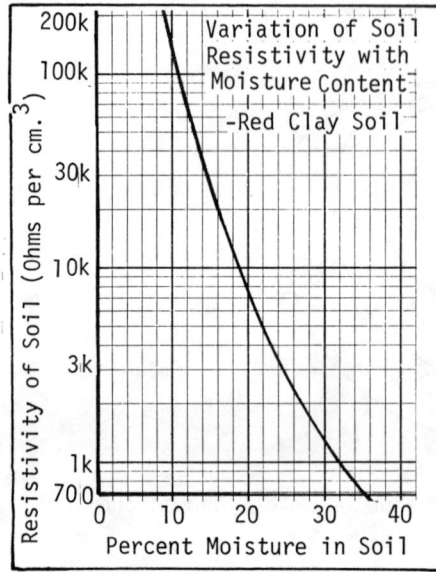

Figure 9.11 - Resistivity of Red Clay Soil with Moisture Content

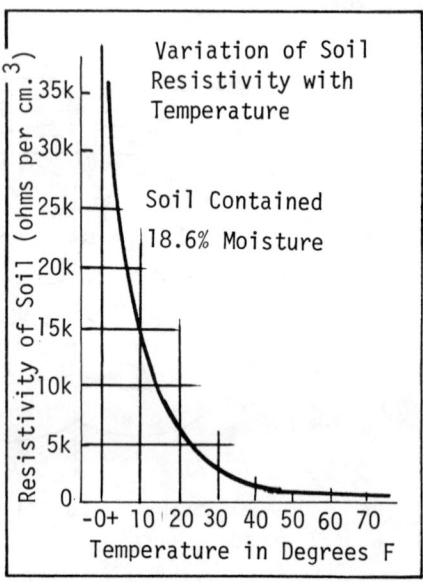

Figure 9.12 - Resistivity of Red Clay Soil With Temperature

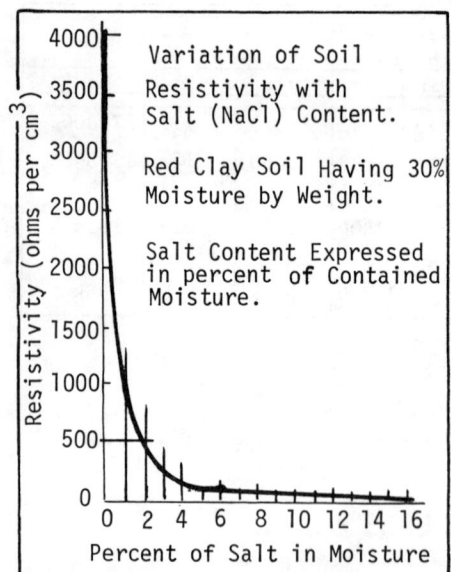

Figure 9.13- Effect of Addition of Salt Upon Resistivity of Red Clay

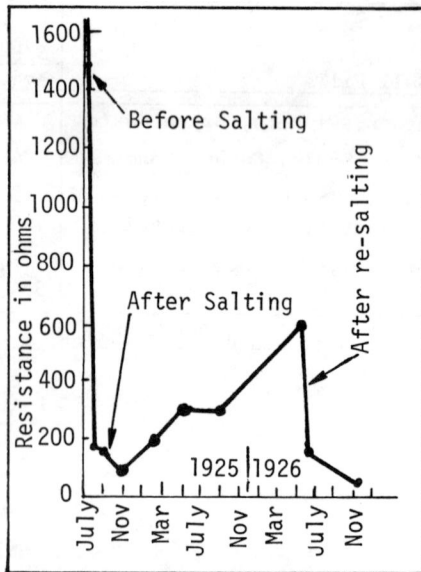

Figure 9.14 - Changes in Resistance of a Ground Connection

Table 9.4 - Effect of Moisture Content on the Resistivity of Soil*

Moisture Content by Wt.	Resistivity in Ω/cm^3	
Soil Type	Top Soil	Sandy Loam
0%	$<1,000 \times 10^6$	$<1,000 \times 10^6$
2.5%	250,000	150,000
5%	165,000	43,000
10%	53,000	18,500
15%	19,000	10,500
20%	12,000	6,300
30%	6,400	4,200

*"An Investigation of Earthing Resistances," by P. J. Higgs, *IEE Journal*, Vol. 68, p. 736, February, 1930.

Table 9.5 - Effect of Temperature on the Resistivity of Soil*
(Sandy Loam 15.2% Moisture)

Temperature		Resistivity
°C	°F	in Ω/cm^3
20	68	7,200
10	50	9,900
0 (water)	32	13,800
0 (ice)		30,000
−5	23	79,000
−15	14	330,000

*"Lightning Arrester Grounds, Parts I, II, and III," by H. M. Towne, *General Electric Review*, Vol. 35, pp. 173, 215, and 280, March, April and May, 1932.

Table 9.6 - The Effect of Salt Content on the Resistivity of Soil*
(Sandy Loam, Moisture Content, 15% by Weight, Temperature 17°C)

Added Salt Weight	Resistivity in Ω/cm^3
0%	10,700
0.1%	1,800
1.0%	460
5%	190
10%	130
20%	100

*"An Investigation of Earthing Resistances," by P. J. Higgs, *IEE Journal*, Vol. 68, p. 736, Feb. 1930.

For all structures requiring a low-resistance earth grounding system which cannot be obtained by use of rods or grids due to soil considerations, the following techniques may be used: (1) artificial salting of soil, (2) immersion in nearby water source of grid or plate connected to the structure to be grounded, (3) or utilization of available underground piping systems or metallic well casing.

9.1.3.5 METHODS FOR MEASURING SOIL RESISTIVITY

It is necessary to know, at least approximately, the value of soil resistivity existing in an area where a structure requiring an earth ground is to be built. Previous discussions and resultant criteria have been based upon the assumption that an approximate value of soil resistivity will be available for design purposes.

After grounding systems have been implemented, techniques must be available for insuring that resultant systems produce the specified grounding resistance. Ground resistance can be expected to fluctuate as a result of climatic changes, age, and wear and must be periodically checked to insure compliance with stipulated specifications.

Ground resistance test requirements vary from situation to situation, and various techniques are available which are compatible with the requirements of all situations. These techniques are identified as (1) Fall-of-Potential-Method, (2) Four-Point Array Method, and (3) Radioactive-Wave Propagation Method. They are referenced in the bibliography.

9.1.4   Reference Plane Grids and Busses for Instrument Grounds

The green safety wire and ground rods and grids of buildings are intended for shock and lightning-hazard control of personnel and equipment damage. However, since nearly all building lighting and equipment (see Table 9.1) share the same green safety wire and building-ground system, they become polluted with both transient and steady-state electrical current flow and noise-voltage drops. For example, a 100 foot run of No. 12 AWG green safety wire exhibits an impedance of 0.160 ohms and may have 50 equipments connected thereto, most of which are not in operation at any one time. However, each may contain two $0.1\mu f$ feed-thru capacitors for electrical-noise suppression between black and green and white and green lines. Thus, 50 equipments x 0.005 amps x 0.16 ohms yields 40 millivolts and 60 Hz and other contamination emissions on the frames of all equipment. Consequently, if Faraday shields of sensitive equipment cables and circuitry are returned to these *grounds*, the common-mode impedance coupling may completely degenerate their EMC performance.

The foregoing situation is shown in Fig. 9.15. The 60 Hz, A-C power source in reality is also a source of R-F noise pollution since it contains pick up from the electromagnetic ambient due to the *antenna effect* of the source-feed system complex.* These source voltages produce drops across the green wire as shown in the equivalent circuit of Fig. 9.16. This serves to common-mode impedance modulated the frame of any equipment k. When the frame of circuit k is also the return side of a sensitive network, a portion of both the 60 Hz and R-F source appear at its input terminals. If the associated frequency-amplitude levels are above the circuit sensitivity, EMI may result. Furthermore, some energy from transients developed in equipment #1 can cross couple to equipment k due to the finite impedances of the black, white and green wires.

Reducing the impedance of the green wire will eliminate the 60 Hz A-C common-mode impedance drop to the frame of equipment k. It will reduce, but not eliminate, the R-F noise from the A-C power mains and other noise making equipments coupling into equipment k. Thus, an argument unfolds for:

(1) Developing a dedicated green wire (either ground plane or bus) for all sensitive equipments.

(2) Developing a dedicated and separate black and white (power feed and distribution) wire system for both housekeeping equipments and sensitive electrical/electronic equipments. This section discusses the former. Sec. 9.2 discusses the building wiring system with emphasis on EMI control.

---

* See Figs. 1.5 and 1.6, Vol. 4.

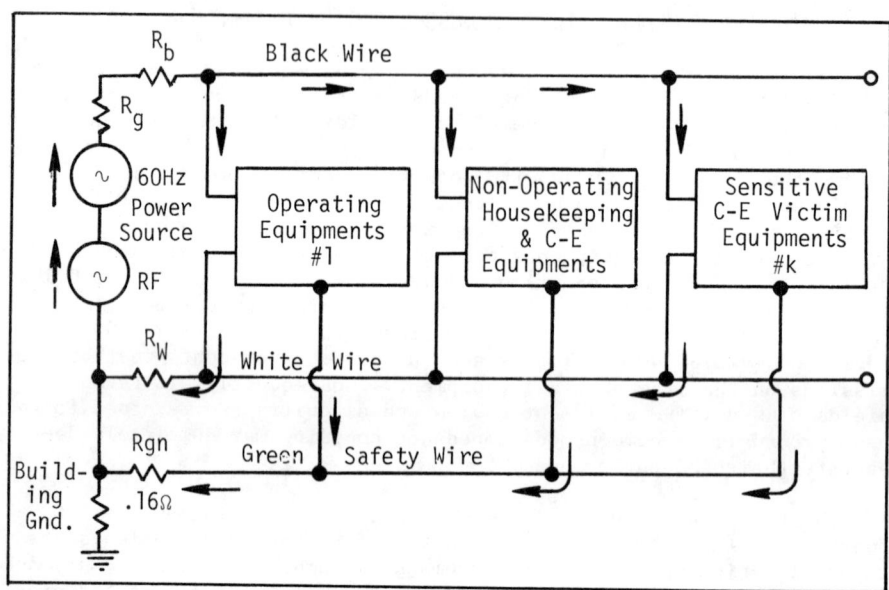

Figure 9.15 - 60-Hz Power and R-F Noise Coupling to Green Safety Wire
and to Chassis/ Frame of Victim Equipments

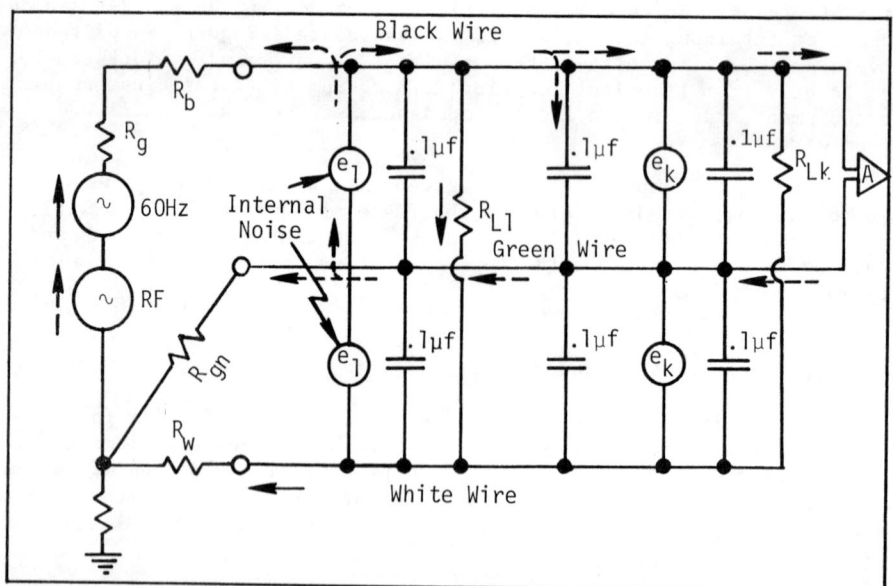

Figure 9.16 - Equivalent Circuit of Fig. 9.15 Showing Green Wire
Common-Impedance Coupling of Both 60Hz & RF Noise

All buildings or structural complexes housing electronic equipments which are susceptible to or capable of generating R-F energy should be supplied with a low-impedance, reference plane ground-grid mesh or green wire bus-bar system as applicable. Conductive shielding media used for the purpose of attenuating R-F energy are dependent upon a low-impedance reference plane for effective attenuation. Shields that are connected to a relatively high-impedance reference plane are ineffective in attenuating R-F electric fields and may radiate energy resulting from potentials induced on the reference plane by extraneous users of the plane. While earth ground is not essential to performance at this function, a single connection to earth is desired for safety protective purposes.

Where sensitive equipment and significant sources of R-F noise are to be located in the same building (almost always the situation), it is necessary to shield cables and equipment housings and to provide a very-low impedance path between these shields and the EMI sources. Fig. 9.17 shows this in which the noise sources may be either cables or equipments which exhibit capacitive coupling to their victims, and both have impedances (Z_1 and Z_2) to some ground reference. From Fig. 9.18, the coupled noise signal, e_s, to the victim voltage, e_v, is:

$$\frac{e_v}{e_s} = \frac{R_v}{R_v + R_s + R_1 + R_2 + R_g + jX_c/2} \tag{9.9}$$

$$\approx \frac{R_v}{R_v + R_s + R_g + X_c/2} \quad \text{for } R_1 \approx R_2 \ll \text{ both other R's and } X_c$$

In an effort to reduce EMI, the Faraday shield (i.e., cable or equipment shield) is placed around the noise source, the victim, or somewhere in between. For illustrative purposes, the latter is shown in Fig. 9.19. From the equivalent circuit of Fig. 9.20, the new emitter-receptor, coupled-signal ratio is:

$$\frac{e_v'}{e_s'} = \frac{R_b}{R_b + R_s + R_1 + R_g/2 + jX_c} \times \frac{R_v}{R_v + R_2 + R_g/2 + jX_c} \tag{9.10}$$

$$\approx \frac{R_b R_v}{R_s R_v + X_c(R_s + R_v) + X_c^2} \quad \text{for } R_1, R_2, R_g \ll \text{ both other R's and } X_c$$

where: R_b = bond impedance to reference of the Faraday shield.

To determine, the de-coupling effectiveness of the shield, SE, the ratio of Eq. (9.10) (after shield) to Eq. (9.9) (before shield) is:

$$SE = \frac{R_b(R_v + R_s + R_g + X_c/2)}{R_s R_v + X_c(R_s + R_v) + X_c^2} \tag{9.11}$$

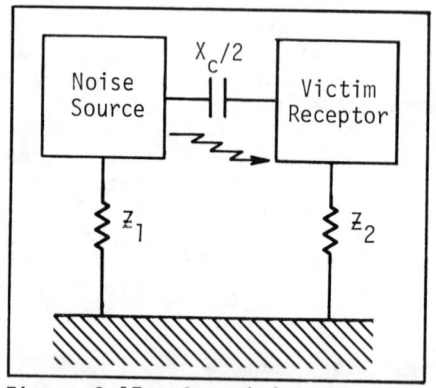

Figure 9.17 - Capacitive Coupling Between Noise Source and Victim

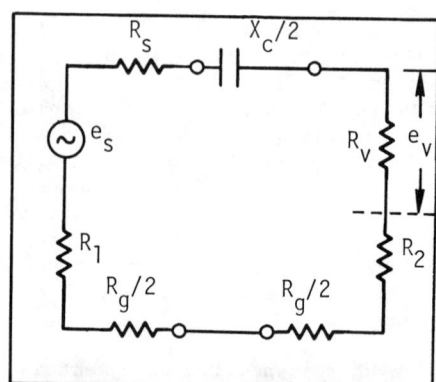

Figure 9.18 - Equivalent Circuit of Fig. 9.17

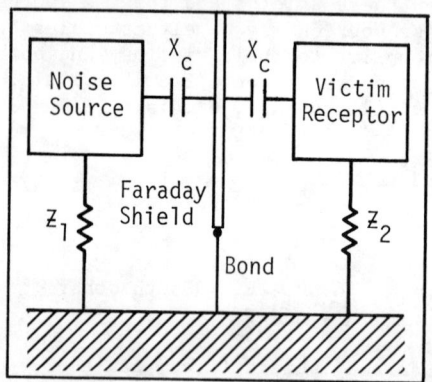

Figure 9.19 - Inserting a Faraday Shield in Fig. 9.17

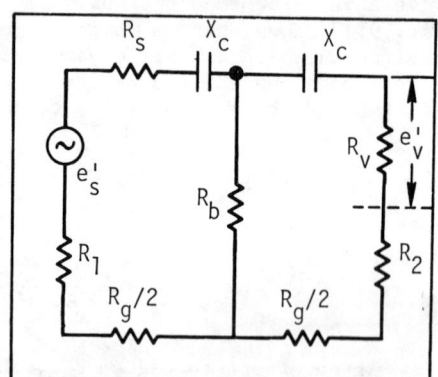

Figure 9.20 - Equivalent Circuit of Fig. 9.19

$$\frac{R_b(R_v+R_s+R_g)}{R_s R_v}, \quad \text{for } X_c \ll R\text{'s at highest}$$ (9.12)
$$\text{frequency of interest}$$

For the shield to perform well in Eq. (9.12):

$$R_b \ll R_s \text{ and } R_v$$ (9.13)

$$R_g \ll R_s \text{ and } R_v$$ (9.14)

$$R_b \text{ and } R_g \ll X_c \text{ at highest}$$ (9.15)
$$\text{frequencies}$$

The bond impedance, R_b, and the reference plane impedance, R_g, both increase with frequency while X_c decreases with frequency. Thus,

9.28

a cross-over frequency will be reached at which the shield is no longer effective and in fact can become *hot and radiate*. Since, R_s, R_v, and X_c can take on any value between any two equipments in a building, there is no unique solution to the problem.

Notwithstanding the above, in an attempt to quantify R_b and R_g, somewhat arbitrary criteria will be selected. R_s and R_v may likely be of the order of 100 ohms. To obtain a 40-dB minimum shielding effectiveness, R_b and $R_g \leq 0.5$ ohms in Eq. (9.12) at the highest frequency for which the building noise source and victim are separated by a distance of $\lambda/20$ (see Eq. (8.5)). Table 9.7 summarizes this:

Table 9.7 - Shield Bond and Reference-Plane Impedance at Maximum
Useful Frequency

Source-Victim Distance in Ft. (m)	Maximum SE Frequency	R_b and R_g at Max SE Freq.
3 ft. (1m)	15 MHz	0.5 Ω
10 ft. (3m)	5 MHz	0.5 Ω
30 ft. (10m)	1.5 MHz	0.5 Ω
100 ft. (30m)	500 kHz	0.5 Ω
300 ft. (100m)	150 kHz	0.5 Ω
1,000 ft. (300m)	50 kHz	0.5 Ω

From Table 9.7, a second criteria can be generated to give the required impedance of both the shield bond and ground-reference plane as a function of building floor length and frequency to yield a 40-dB. This is shown in Fig. 9.21 in which it is presumed that a D-C resistance of 3 milliohm is the least practically obtainable without substantial cost increase. Thus, the R-F impedance becomes asymptotic to this value below the notch frequency.

One of two situations may be presumed to apply to communications electronics equipment located within a building:

(1) They will be located along or near (say within six feet) of power-wiring raceways, or cable trays; or

(2) They may be located anywhere on the floor of a building. The former implies perhaps a few bus lines per floor while the latter implies a reference-plane grid mess. In the limit, the former approaches the latter. Each is now discussed separately.

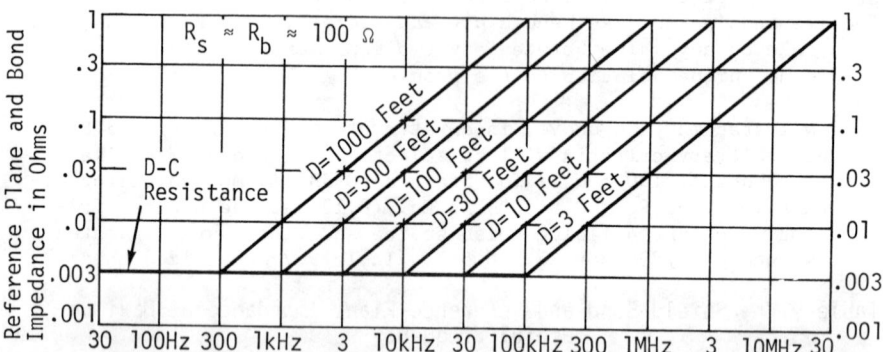

Figure 9.21 - Required Reference Plane and Shield Bond Impedance of Buildings vs. Frequency and Building Floor Length.

9.1.4.1 BUS OR CABLE TRAY REFERENCE PLANE

Based on the criteria used to develop Fig. 9.21, the required impedance is determined and it remains to calculate the ground reference configuration to achieve that impedance. The cross-sectional area of a conductor bus has the following D-C resistance per unit length:

$$R_{dc} = \frac{1}{A\sigma} \quad \text{ohms/meter} \tag{9.16}$$

where: A = cross-sectional area in m^2

σ = conductivity in mhos/m

For copper (σ = 5.8 x 10^7 mhos/m) and aluminum (σ = 3.7 x 10^7 mhos/m), the required cross-sectional area in in^2 for any length, ℓ, in feet to give a 3-mΩ resistance is:

$$A = \frac{(39.4)^2 x \ell_{ft}}{0.003 x \sigma x 3.28} = 473 x 10^3 x \ell_{ft} / \sigma \tag{9.17}$$

$$= 0.0028 \ \ell_{ft} \quad in^2 \text{ for copper} \tag{9.18}$$

$$= 0.0042 \ \ell_{ft} \quad in^2 \text{ for aluminum} \tag{9.19}$$

If aluminum is used as the bus reference frame, Eq. (9.19) shows that the cross-sectional area becomes very large (0.42 sq. in. for 100 feet length of run) when the distance exceeds about 100 feet. This may correspond to a cable tray of 8 inches width, two inch sides and 1/32 inch sheet metal for a 100 foot length. Since a building will ordinarily have more than one cable tray run down a side, the trays may be joined at both the source and far ends with cross cable trays. However, this begins to form the essence of a grid system

(made from trays rather than a bus conductor). It is concluded that
reference planes composed of bus bar stock or cable trays will not
practically provide the required impedance where their lengths exceed
about 100 feet. For buildings having floor lengths in excess of
about 100 feet, a reference-plane grid may have to be used.

9.1.4.2   Reference Plane Ground Grids

Since typically many users will depend upon the reference plane,
design criteria must be compatible with user requirements also. Extraneous user equipment of the reference plane may be susceptible to
relatively small amplitudes of interference. As a result, the following criteria should apply for design purposes of the reference-
plane, ground-grid mesh:

1.   The low-impedance, reference-plane ground-grid mesh should
be designed to comply with the impedance criteria used to develop
Fig. 9.21.

2.   Impedance levels are to be obtained by utilizing wire or bus
size and number of grids to a point where further impedance reduction
is not economically feasible for the percentage impedance reduction
obtained relative to incurred expense.

The initial problem is to determine the number of grids, associated dimensions, and wire size to be used in constructing a reference-
plane grid mesh to obtain the criteria impedance. The maximum impedance (resistance and self-inductance) of the grid is encountered
between diagonally opposite points. The following paragraphs develop
expressions describing variations in impedance between these points
as a function of various parameters.

For a prescribed length and width of a specific reference-plane
grid mesh, the following two factors are working in opposing directions
to change the resistance, R_{AB}, between diagonal points A and B. As
the number of grids is increased, (1) R_{AB} decreases due to the decreasing length of the grid elements, and (2) R_{AB} increases due to
the additional cables added to obtain the increased number of grids.

The maximum resistance offered by a *square grid* mesh is:

$$R_{AB}\big]_n = \frac{R_e]_{n-1}}{2n}\left[\frac{2k]_{n-1}[2(n-1)]}{1+2(n-1)} + 2\right] \qquad (9.20)$$

where: $R_{AB}\big]_n$ = resistance in ohms between points A & B for
                              n grids per side

               n = number of grids per side

           $R_e]_{n-1}$ = resistance in ohms of one element of total length
                              equal to the length of one entire side of the grid

k = multiplication factor to account for additional grids

The resistance of one element of total length equal to the length of one entire side of the grid $(R_e]_{n-1})$ is:

$$R_e]_{a-1} = \rho \frac{\ell}{A} \qquad (9.21)$$

where: ρ = resistivity of conducting media in ohms/circular-mil-foot

ℓ = length of conductor in feet

A = cross-sectional area of conductor in circular-mils

Eq. (9.20) is a recursion type-formula and can only be used by calculating R_{AB} progressively from n=1 to n=n. R_{AB} has been calculated for several values of n grids per side. Such calculations have been performed for grid mesh areas up to $(1000 \text{ ft.})^2$ for 4/0 cable sizes presented graphically in Fig. 9.22.

The distributed self-inductance of the grid mesh must also be considered for development of R-F impedance. The H-F self-inductance of copper cable is:

$$L = 0.609\ell \left[2.303 \log_{10}\left(\frac{4\ell}{d}\right) -1 \right] \mu h \qquad (9.22)$$

where: ℓ = length of conductor in feet

d = diameter of conductor in feet

The H-F self-inductance of a copper-grid mesh consisting of n-grids per side can be computed using Eqs. (9.20) to (9.22).

$$L_{AB}\Big]_n = 0.305\ell \left[(2.303 \log_{10}\left(\frac{4\ell}{d}\right) -1) \right]_n \left[\frac{2k]_{n-1}[2(n-1)]}{1+2(n-1)} +2 \right] \mu h \qquad (9.23)$$

or

$$L_{AB}\Big]_n = L_n k \quad \mu h \qquad (9.24)$$

where: L_n = H-F self-inductance of one element of the grid mesh

$$= .0609\ell \left[2.303 \log_{10}\left(\frac{4\ell}{d}\right) -1 \right] \mu h$$

k = coefficient used in Eq. (9.20) for various numbers of grids in the grid mesh

$$= 0.5 \left(\frac{2k]_{n-1}[2(n-1)]}{1+2(n-1)} +2 \right) \qquad (9.25)$$

L_{AB} has been calculated for values of n (grids per side) up to and including thirty (30). Such calculations have been performed for

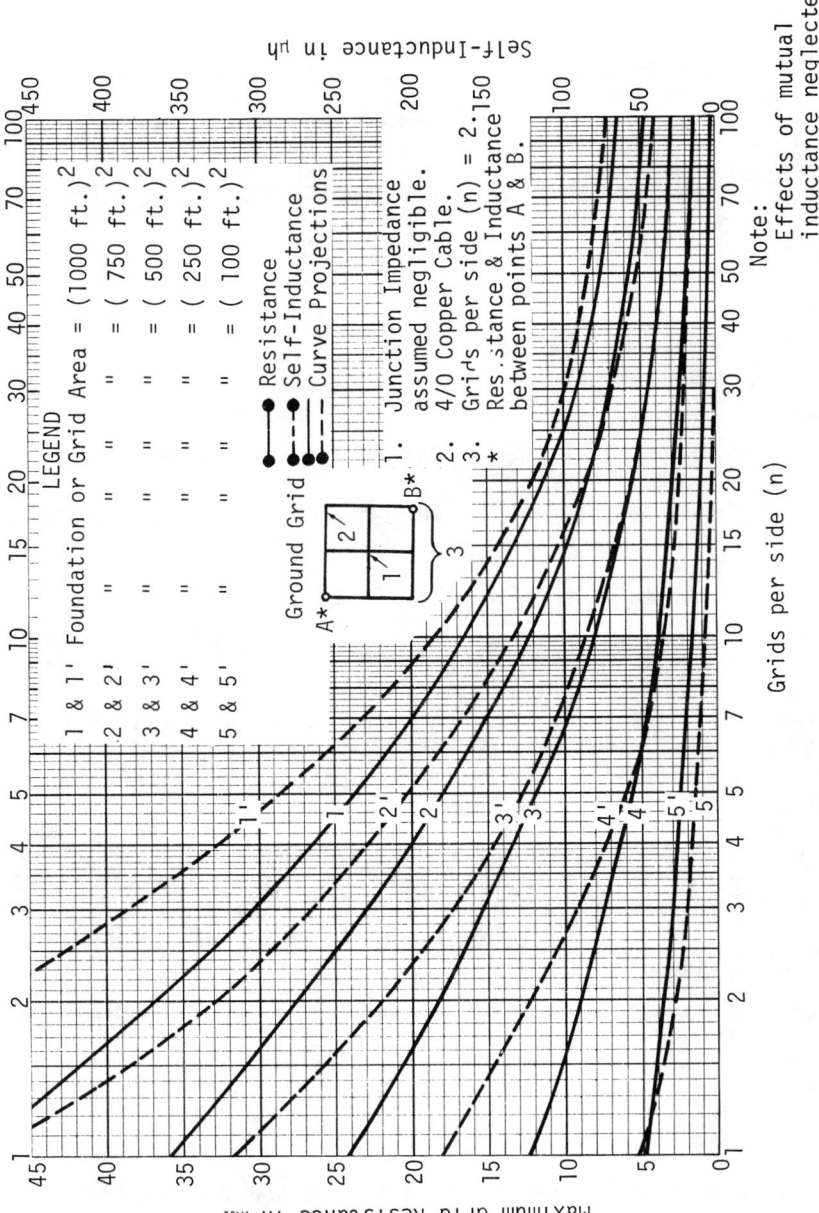

Figure 9.22 - Maximum D-C Grid Resistance and Self Inductance of Grid Mesh
Total Grid Area (4/0 Copper Cable)

grid mesh areas of $(100 \text{ ft.})^2$ to $(1000 \text{ ft.})^2$ and for 4/0 cable size as shown graphically in Fig. 9.22.

The maximum impedance offered by a rectangular-grid mesh is different than the impedance offered by a square-grid mesh with an equivalent number of mesh loops. The following expression has been derived which defines the maximum resistance between points A and B of a rectangular grid mesh:

$$R_{AB}\Big]_{nw} = \frac{R_{e_w}\Big]_{n=1}}{n_w} \left[k + k_n (n_\ell - n_w) \right] \text{ ohms} \qquad (9.26)$$

where: $R_{AB}\Big]_{nw}$ = resistance in ohms of rectangular-grid mesh between points A and B

$R_{e_w}\Big]_{n=1}$ = resistance in ohms of a single conductor of length equal to the width of the grid mesh

n_w = number of grids along the short side or width of the rectangular mesh

k = coefficient computed for square-grid meshes with various numbers of grids per side

k_a = correction factor to account for lengths greater than widths using n_w grids

n = number of grids along the long side or length of the grid mesh

The resistance of a rectangular-grid mesh may be calculated from Eq. (9.26) for length to width ratios of 2 and 3, and for 4/0 cable as a function of foundation or grid mesh area and number of grids on the short side of the rectangular grid mesh. Resultant data is graphically presented in Figs. 9.23 and 9.24. The self-inductance of corresponding grid meshes has been neglected since curves for square-grid meshes indicate that reduction of grid resistance will also result in a reduction of self-inductance.

Illustrative Example 9.4

A reference-plane grid mesh is to be implemented in a structure intended to house sensitive electronic equipment requiring that the maximum D-C resistance offered by such a mesh be no greater than 3 milliohms. The self inductance of a single-floor foundation dimensions 500 ft. by 500 ft., $(500 \text{ ft.})^2$ from the mesh is to be less than 5mΩ at 60 Hz. Determine the number of grids per side.

The impedance obtained from Fig. 9.22 for a square foundation and for R = 5mΩ is:

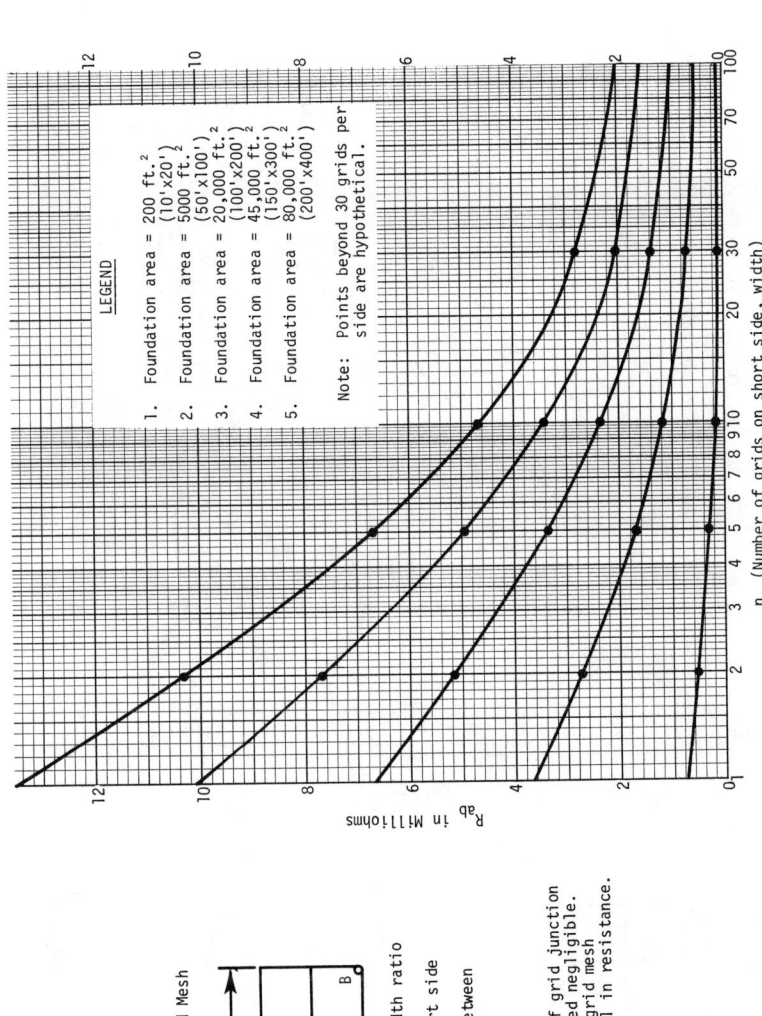

Figure 9.23 - Plot of Maximum Resistance vs. Number of Grids on Short Side of Rectangular Ground Grid Mesh 1/w = 2, 4/0 copper cable

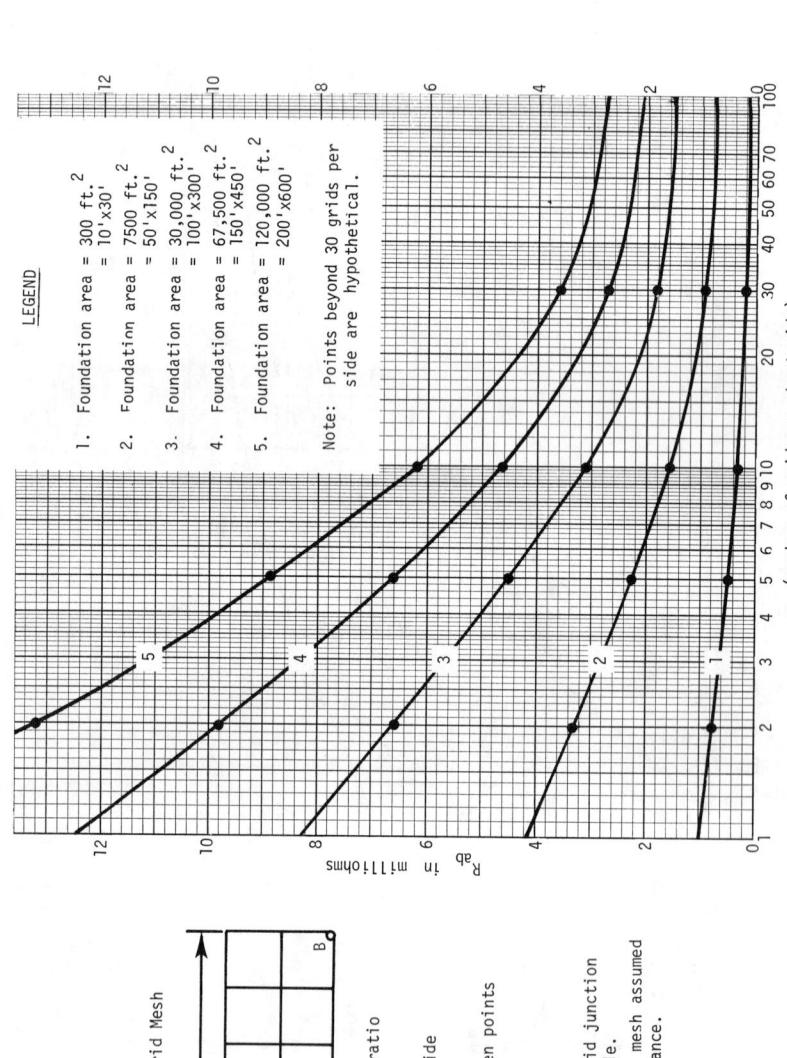

LEGEND

1. Foundation area = 300 ft.^{2}
   = 10'×30'
2. Foundation area = 7500 ft.^{2}
   = 50'×150'
3. Foundation area = 30,000 ft.^{2}
   = 100'×300'
4. Foundation area = 67,500 ft.^{2}
   = 150'×450'
5. Foundation area = 120,000 ft.^{2}
   = 200'×600'

Note: Points beyond 30 grids per side are hypothetical.

n_w (number of grids on short side)

R_{ab} in milliohms

Example Ground Grid Mesh

l/w = length to width ratio
    = 3

n_w = grids on short side
    = 2

R_{ab} = Resistance between points A & B

Assumptions:

1. Resistance of grid junction points negligible.
2. Each leg of grid mesh assumed equal in resistance.

Figure 9.24 - Plot of Maximum Resistance vs. Number of Grids on Short Side of Rectangular Ground Grid Mesh l/w = 3, 4/0 copper cable

4/0 copper cable, 25 grids per side
(crossover every 20 feet)

R = 5mΩ

L = 50μh ≈ 2mΩ at 60 Hz

(A) Materials, Size, Coatings, and Methods of Bonding

Copper cables are recommended for construction of grid reference
planes due to their high electrical conductivity and their corrosion-
resistant properties. Cable size must be selected as a result of resistance requirements and mechanical feasibility of implementation.

Because of the corrosion resistance of copper coating of conductors used in the mesh construction is both impractical and unwar-
ranted. However, in cases where external connection is made to the
reference grid mesh, moisture-proof coatings over bond connections
will reduce both the possibility of corrosion between mating surfaces
and deterioration of welding materials. Such coatings may be in the
form of a paint or plastic material capable of maintaining its moisture-proofing qualities over an extended period of time.

All overlapping joints in the grid construction should be electronically welded or fused together. Such welding techniques will minimize
grid impedance, reduce the possibility of loose connections, and will
result in a more rigid grid construction. All external connections to
the grid mesh should be wrapped and electronically welded.

At no time should an external connection be made to the reference-
plane grid mesh by any metal other than copper or a metal which is
galvanically compatible with copper. Such connections would otherwise result in galvanic corrosion which may deteriorate the effectiveness of the reference plane.

(B) Methods of Physically Implementing Reference-Plane Ground
     Mesh

Several precautions must be followed for effective implementation
of a reference-plane grid mesh:

(1) Adequate provisions must be made to accommodate all prospective users in various parts of the structure.

(2) Tie-on points should be readily accessible for inspection
purposes.

(3) Mesh may be implemented under, between or as an integral
part of flooring materials as long as such materials are non-conductive.

(4) Mesh must not make electrical connection with building structural steel or any other metallic media connected to earth ground, or the advantage of a single-point earth ground will be lost (see Sec. 8.3).

Provisions must be made to accommodate all prospective users in the structure without affecting the primary purpose of the reference plane. If a structure contains more than one floor and if it is conjectured that users on more than one floor will require a low-impedance reference plane, then a grid mesh may be required for each floor. If, however, it is anticipated that users of the reference plane on other floors of the structure, are to be confined to a small area, then satisfactory results can be realized by running a reference plane jumper of relatively large diameter from the main reference plane grid mesh to the required work area. The sum of the maximum reference plane resistance and the total resistance of the jumpers should be less than the resistance specified for the structure.

Fig. 9.25 illustrates a typical application of the above jumpers. Where a multitude of prospective reference-plane users are contemplated on more than one floor, an excessive amount of jumper material would be required and the implementation of extra reference planes would be necessitated. Design and implementation techniques previously discussed should be used. Jumpers should be implemented between the reference planes at various intervals.

(C) Method for Connecting Reference Plane Ground Grid Mesh to
     Earth Ground

The refernce-plane ground-grid mesh should be connected to earth ground for protection of personnel and equipment from hazardous potentials. Various precautions must be taken to insure that such measures are implemented in a manner which will not affect the primary purpose of the reference plane.

The reference plane must be earth grounded at only one point. Building ground can be expected to carry large amounts of power-system fault currents and possibly large currents due to lightning discharges. Such high level circulating loop currents must be isolated from the reference plane.

Figure 9.26 illustrates a method for connecting the reference-plane ground-grid mesh to earth ground. Such procedures can be applied to any building using a reference plane and ground rods. In situations when earth-ground meshes are used in lieu of or to compliment ground rods, techniques indicated in Sections 9.1.2.6 and 9.1.2.4 should be used for making earth-ground connections.

Ground well installations should be made readily accessible for inspection and repair if necessary. Such installations should be

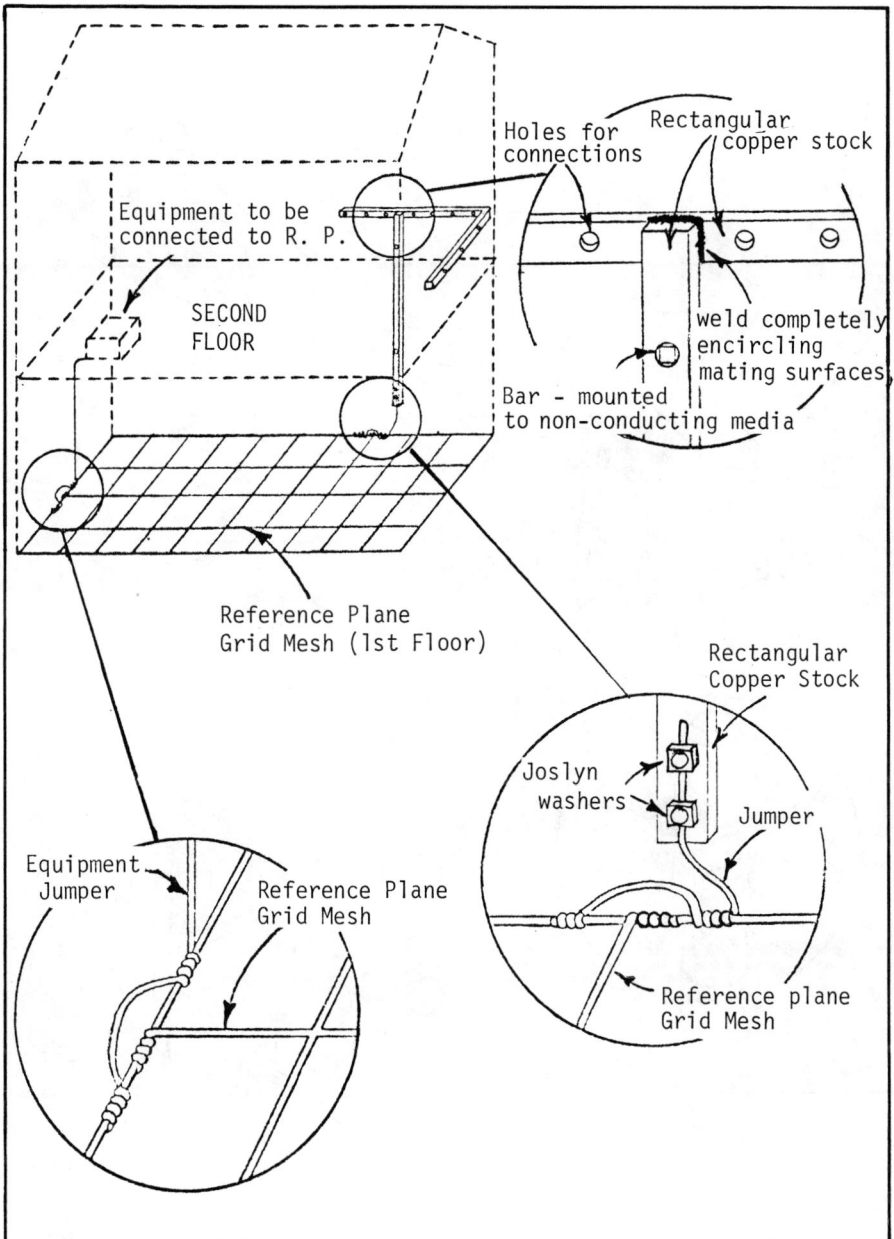

Figure 9.25 - Techniques for Implementing Jumpers to the Reference Plane of a Multi-Story Building.

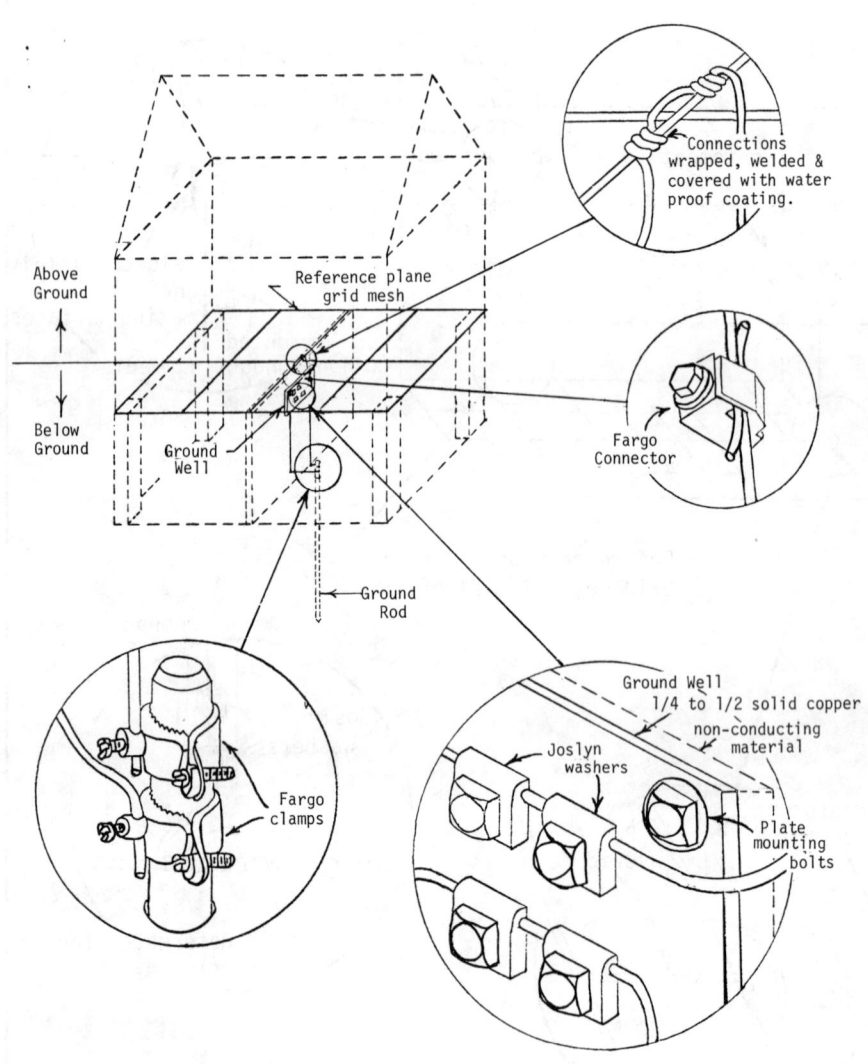

Figure 9.26 — Method for Connecting Reference Plane Ground Grid Mesh to Earth Ground.

installed in basements or in spaces between structure flooring and
ground. More than one contact between the reference plane and the
ground well and between the ground well and earth ground is desirable
to prevent loss of earth ground by default of one contact.

9.2   BUILDING WIRING AND LIGHTING

Since building power wiring distribution and ground planes become
contaminated with EMI, the previous section introduced the need* for
separate ground planes, one earmarked for housekeeping and one for instrument grounds. Housekeeping wiring and ground systems include those
which service building lighting, elevators, escalators, office coffee
pots, accounting and reproduction equipment, cafeteria electrical
appliances, etc. Instrument wiring and grounds include those which
service computers and peripherals, bio-medical instruments, research
equipment, intercoms, security alarms, and the like. Note that the
former emphasizes emitters while the latter stresses receptors. The
grounding portion of this housekeeping-instrument dichotomy was already discussed in Sec. 9.1. This section reviews the power-wiring
implications.

9.2.1   INSTRUMENT POWER WIRING

Based on the dual housekeeping-instrument wiring and ground concept, a three-phase service in a building would be divided such that
two phases are earmarked for housekeeping and lighting and the third
phase would service instrument users. The black, white, and green
wires for the instrument-service phase are separately distributed about
the building with no sharing of these wires with other housekeeping
phases. All duplex outlets for instrument service are marked red or
identified in some other esthetically acceptable, but readily identifiable manner.

The instrument-wiring phase is enclosed in conduit in order to
shield its three wires from radiated pickup. Where conduits terminate
in a circuit-breaker or other utility box, all internal leads are
covered with one foot (more or less) of lossy EMI suppressment tubing.
As described in Chap. 14, this tubing exhibits a cut-off frequency

---

*   After about 50 years of more or less the same wiring practices in
buildings, it may be difficult for architects, electricians, and many
others to imagine a dual-wiring and ground system. However, 50 years
ago there existed virtually no potential EMI susceptible receptors, and
relatively few emitters. If building owners are to be competitive in
the future, they must service the plethora of leasees who have sensitive
receptors of one form or another. Not to be so responsive, will end in
abdicating their fiduciary responsibilities and such buildings will be
flagged by potential lessees as obsolescent.

of about 10 MHz, and offers about 50-dB attenuation at 100 MHz. Thus, the instrument-wiring system pickup of TV, FM, radar, and other signals above about 30 MHz will be attenuated and converted into heat where it can do no EMI damage. The heat rise of the wiring is insignificant. As explained in Sec. 9.2.3, additional EMI protection is needed below 30 MHz if conduit is not used.

An academic argument can develop regarding whether either the instrument-wiring phase containing mostly receptors should be protected, or the electrical-emission housekeeping wire phases should be shielded, or both. It is simple enough to argue that both should be protected, but that is not cost effective. As a general practice, it is best to squelch EMI at the source or as close thereto as possible in order to limit the amount of damage that can result to receptors. However, precedent is more in favor of and pragmatism more strongly indicates protecting the susceptible receptor system. Thus, those users who require an instrument-wiring phase and dedicated grounding system would pay for this additional requirement in a new building offering this option to leasees.

9.2.2   Filtering at Duplex Service

The *red-duplex* boxes for the instrument-phase power distribution would contain two 20-amp, EMI filters, one each for the black and white lead to metal-box ground. The dual EMI filter would have a cut-off frequency of about 10 kHz, per MIL-STD-220A, and would present an attenuation of at least 60-dB at and above 150 kHz. The green wire existing from the duplex would be covered with about 6 inches of the previously mentioned lossy absorptive tubing.

The green wire in the above instrument-ground system is multi-point connected to the reference-plane grid system described in Sec. 9.1.4. These interconnections are also covered with a sleeve of EMI lossy tubing. This assures minimum R-F potential difference between points of the reference-plane ground to all instrument users of the system.

A dual housekeeping and instrument wiring system was put into practice in some buildings by the National Bureau of Standards in Gaithersburg, Maryland, among others. The added measures recommended to suppress EMI picked up by the wiring system acting as an antenna is based on extensive automatic spectrum measurements made on A-C power lines in many different rooms within the Clinical Center of the National Institutes of Health, Bethesda, Maryland. That location is typical suburban and the levels of conducted signals were already causing EMI. Thus, the recommended practice to EMI filter at the duplex box and to use lossy-suppressant tubing to attenuate high ambient signals is a requirement of today.

9.2.3   Conduits and Cable Trays

Previous sections indicated that one phase of the building three-phase wiring system would be dedicated for instrument and receptor users and the wiring would be distributed through conduit to minimize radiated pickup. An alternative to this is not to use conduits, but to route the dedicated, unshielded instrument-phase wiring in cable trays. Since it picks up radiated EMI, this wire system would use about one foot of lossy EMI suppressant tubing every 50 to 100 feet to eliminate conducted EMI above 30 MHz. Since the tubing is ineffective below 10 MHz, it is necessary to either: (1) use a heat-shrinkable shielded wiring cover (see Sec. 6.5), or (2) to cap or cover all open cable trays.

The trade off of either of these two approaches in lieu of conduit would be one predicted on cost. Perhaps one solution would be for the manufacturers of romex wire to develop and mass produce a shielded-foil version in which the plastic dielectric contains an R-F lossy material. This day may come although probably not much before the 1970's have ended.

To be effective, unprotected power-wire distribution for the instrument phase should not be routed in the same cable tray as that for the housekeeping wire unless there exists a metal spectrum divider. If the instrument wiring distribution is protected as outlined in the preceding paragraphs, then it can be routed in a cable tray along with the housekeeping wiring. Under these circumstances, if a cable tray does not exist but ceiling wire hangers are used, then the two wiring systems can also be routed together.

9.2.4   Lighting

Almost all modern buildings use fluorescent lighting throughout. The incandescent lamp is obsolescent. Sec. 2.3 illustrated typical electromagnetic spectrum emissions expected from fluorescent lamps. These emissions would tend to negate much of the purpose behind separating instrument and housekeeping wiring and grounds unless measures are taken to concomitantly suppress fluorescent noise. While this is relatively expensive to do, it involves a shielded fixture housing, conductive plastics to cover the lamps, and a power-line filter.

A more economical initial investment is to install conductive-glass fluorescent bulbs and a shielded fixture containing power-line filters. Actually, the cost is only in the special fixtures since the office leasee can elect to purchase either normal fluorescent tubes or the shielded variety depending upon his receptor susceptibilities. If the rent includes replacing fluorescent lamps at no extra cost, then the landlord can charge an extra incremental rent for shielded lamps if needed by the leasee. A policy of rent additives makes it practical for the building owner to recover most extra costs due to EMI control for users who require these extra provisions. This does not, however, assure that a user will not be an EMI offender within the meaning of the FCC Rules and Regulations, Parts 15 and 18 (see Chap. 6 Vol. 1 of this handbook series).

9.3   SHIELDING

This section shows that most buildings offer relatively little attenuation to outside-world emissions. To protect the contents of buildings against strong ambient electromagnetic environments, it is necessary to use one form or another of screening on the external surfaces as well as the interior walls. Where very sensitive instrumentation is to be used or where very noisy broadband emissions are likely to exist, there appears to be little attenuation to the solid shielded enclosure.

9.3.1   TYPICAL BUILDING ATTENUATION

Two types of building attenuation against electromagnetic ambients is of interest: (1) attenuation as a function of frequency to exterior rooms located around the periphery, and (2) attenuation to deep interior rooms which are protected by at least 100 feet of distance to the exterior walls. Such measurements were made in the Clinical Center of the National Institutes of Health with the rather interesting results shown in Table 9.8.

Table 9.8 - Attenuation of the NIH Clinical Center Building

Frequency Region	External Rooms	Internal Rooms	External To Internal Atten.
AM Broadcast	44 dB	50 dB	6 dB
30 MHz	23 dB	48 dB	25 dB
Low VHF TV/FM	7 dB	48 dB, σ=5 dB	41 dB
Upper VHF-UHF	1 dB	48 dB, σ=4 dB	47 dB

The attenuation of external rooms becomes significantly less as frequency is increased because of the large glass areas compared to a wavelength. The attenuation of ambients to internal rooms, however, seems to show no frequency dependency. This is believed to be due to two somewhat equal but offsetting factors: (1) signals to external rooms are attenuated as shown, and (2) internal attenuation encounters no windows, but drywall, furniture, people and the like which exhibits an increase in attenuation with frequency.

Table 9.8 shows that computers, medical instruments or other sensitive equipment should be located in the center of buildings where substantial natural attenuation protects them from large ERP radiated such as FM and TV broadcast. Yet, esthetically nearly everyone wants the brightness of outside (external) offices where little attenuation is offered to VHF & UHF. Thus, it is not practical to count on the attenuation available to internal rooms. Rather the VHF and UHF attenuation of external rooms will have to be shred up by about 20 dB to assure

TV levels are below 1 V/m in typical surburban areas.

9.3.2   EXTERIOR BUILDING MATERIALS

The exterior of most buildings is either cinder block, concrete, or brick. These materials are nearly transparent to R-F energy at and below VHF (300 MHz). Studies conducted at NBS indicate that the median insertion loss to a single layer of a brick wall (brick veneer) at 518 MHz and 1046 MHz was about the same, viz., only 2.5. dB. Plotted on cumulative probability distribution paper, it was shown that for horizontal polarization only 10% of the measured locations exceeded an insertion loss of 5 dB, while the 90 percentile situation corresponded to about 1 dB. Thus, non-metallic surfaced* buildings are essentially transparent to induced or radiated EMI at and below UHF.

The remaining surfaces of non-metallic buildings are made up of windows and doors. While some doors may be made of metal frames, most have glass. Thus, windows and doors are nearly completely transparent at RF. Most buildings are constructed of I-Beam girders layed out approximately 20 to 40 feet between vertical members and 10 feet between horizontal members. It is postulated that building attenuation on the inside of the girder frame construction will be negligible above those frequencies for which the spacing exceeds 0.1λ. Consequently, even the girders of most buildings would appear to offer little attenuation above about 5 MHz.

One obvious solution to improving the exterior building material R-F attenuation is to use aluminum or steel facing or backing. Screen backs to traditional surface materials and even conductive mastics and plastics have been used. However, they all represent a significant increase in material and installation costs. Thus, a different solution involving the basic building exterior materials is sought. One such solution involves brick, cinder block, and/or concrete made of a mixture of reasonably good grade of high temperature coke.

9.3.2.1   COKE-SHELL EXTERIORS

Before investigating hybrid building materials (coke-embedded brick, etc.), it is useful to summarize some attenuation studies of coke, per se. Coke achieves its shielding properties by virtue of reflection loss (see Chap. 10) at low frequencies in which the volumetric resistivity approximates 0.1 ohm-cm at DC. At high frequencies, both reflection and transmission loss contribute to total attenuation. Thus, the layer thickness of coke chunks is more important at HF. Coke also

---

* Some small buildings are made of corregated steel or aluminum. Some houses use finished aluminum siding which simulates wood panels.

exhibits considerably less insertion loss at LF when it is broken down into small particles or ground into fine dust. Typical values of insertion loss of coke chunks (about 1 inch size) varies from about 60 dB at 100 kHz to 90 dB at 1 GHz per foot thickness for untreated coke vs. about 50 dB at 100 kHz to 70 dB at 1 GHz per foot thickness for coke neutralized with lime slurry.* Powdered coke dust on the other hand offers about 15 dB attenuation per foot at 100 kHz and rises to about 50 dB at 1 GHz.

One approach to the use of coke would be to inject it as a sandwich layer between two surfaces such as wooden siding and brick veneer or between two slabs of concrete or other surfacing and backing. Based on the above data, a two inch sandwich layer of treated coke would expect to exhibit about 10 to 15 dB insertion loss - hardly an impressive figure.

Another approach would be to make brick, concrete, or cinder block of a specified percentage of coke chunks in a manner such that the compression strength would not be significantly less than the normal comparable materials without coke. This tends to result in a relatively low content of coke such that the attenuation of the modified materials approximates 10 dB to 20 dB per foot thickness.

As a result of the above it is concluded that coke, while inexpensive and offering good shielding properties, cannot normally be used in most exterior building materials. Where special structures, such as a blockhouse or bunker building are called for, coke can then gainfully be used.

9.3.2.2 SCREEN AND CONDUCTIVE FINISHES

As discussed in Chap. 11, there are a number of conductive points on the market today. Insertion losses of the order of about 50 dB up to 1 GHz are obtainable, where skin depth becomes the limiting factor. However, they are very expensive in terms of building finishes and would require periodic cleaning unless the appearance of oxides is not objectionable. Thus, conductive paints appear not to be a practical solution for building exteriors in terms of today's costs.

Since building paper and vapor barriers must be used on outside building walls, it would seem that this presents an opportunity to secure relatively inexpensive screen. The screen is attached by staples, tape, or other binding media except at the edges where a one inch overlap of the next screen piece is made. While conductive adhesives or flat conductive caulking could be used, this is both expensive and

---

* Coke has some degree of sulfur content which produces H_2SO_4. Neutralizing coke with lime tends to deteriorate the shielding effectiveness.

unnecessary. Overlap screening can be secured in the same manner as the basic screen provided no layer of electrical insulation (e.g. non-conductive tape) is used. Insertion losses of the screen will approximate 50 dB up to about 1 GHz.

If the above screen is applied, the leaky culprit will then be windows. A typical 3 foot x 6 foot window will be fairly opaque to low frequencies, but will become leaky above about 20 MHz (0.1λ). There is no economical way to RF shield a window without using window screens. Here the sash would have to be metallized (e.g. conductive paint) which would terminate on the building screen discussed in the previous paragraph.

Since all modern buildings are air-conditioned, window screens are passé. They also oxidize and become unsightly after a few years and tend to block light. A far better approach is to use conductive glass (see Chap. 11). While relatively expensive, a light conductive coating can be used, say about 200 ohms/sq. When interior building shielding materials is used (see next section), it is not necessary to use conductive glass on all windows. Only certain floors, sections, or rooms need be so equipped where occupants would be expected to use computers, medical instruments, and the like emitters and/or receptors. Conductive glass windows of the type described would provide about 80 dB or more attenuation below 1 MHz and would gradually degrade to about 20 dB at 1 GHz.

9.3.3   INTERIOR BUILDING MATERIALS

The real opportunity to improve R-F attenuation to radiation appears to more economically stem from the use of interior building materials. Coke aggregate appears not to offer sufficient attenuation per unit thickness as discussed in the preceding section. Other materials, however, offering some promise, are (1) conductive mesh in walls and ceilings, (2) conductive paint, (3) metallic wallpaper, and (4) special windows and doors. This section reviews these topics.

9.3.3.1   CONDUCTIVE MESHES IN WALLS, FLOORS, AND CEILINGS

Table 9.9 and Fig. 9.27 illustrate the magnetic and electric-field shielding effectiveness of various types of screen materials. Number 22 copper screen mesh provides approximately 65-dB attenuation to electric fields and plane waves over a frequency range of 100 Hz to 1 GHz as shown in Table 9.9. Effectiveness decreases as screen aperture size increases and No. 2 galvanized-steel screens provide an efficiency of 24-dB over the same frequency range. Number 60 copper screen provides 32-dB magnetic-field attenuation at 300 kHz, 50 dB at 1 MHz, and 60-dB at and above 20 MHz as shown in Fig. 9.27.

Table 9.9 - Minimum Shielding Effectiveness of Screen
Materials to High-Impedance Electric Fields and Plane Waves

Screen Material	Configuration	Thickness	Attenuation*
Steel	Perforated Sheet	60 mils.	58 dB
Aluminum	1/8" dia, 3/16" ctrs.	60 mils.	48 dB
Aluminum	1/4" dia, 5/16" ctrs.	37 mils	35 dB
30 Mil. Galv. Steel Wire	#2 Screening	NA	24 dB
30 Mil. Galv. Steel Wire	#4 Screening	NA	28 dB
20 Mil. Copper Wire	#12 Screening	NA	50 dB
Copper	#22 Screening	NA	65 dB
20 Mil. Aluminum Wire	#16 Screening	NA	55 dB

*Minimum attenuation in dB from 100 Hz to 1 GHz (cf. Chap. 11)

The above screen mesh can be impregnated in or sandwiched between
drywall and other internal materials with a 1" screen overlap for
seam bonding. Such screening should be electrically continuous
around the shell of the room to be shielded. This necessitates
either soldering of screens or deploying conductive epoxy or caulking
along corner junctions (see Sec. 12.4 and 12.5). A bond of solid
copper cable or sheet metal strap should be connected from the screen
to the reference-plane grid mesh for that floor.

Regarding the ceilings and floor of such rooms, both can be
covered first with similar screen mesh and bonded along the corner
junctions. The ceilings can then be covered with celetex regular,
acoustical tile, or other materials, and the floor is covered with
the normal planned covering materials. An alternative approach is
to use the same screen-impregnated drywall for the ceiling, and to
immerse the concrete floor in screen in which a screen overlap permits
seam bonding at corner junctions.

Other than for doors and windows (discussion later) such a room
enclosure will probably result in about 60 dB of attenuation to
electric and electromagnetic fields over many octaves of the frequency
spectrum. While expensive relative to normal construction, it is considerably cheaper than subsequently installing the least-expensive
shielded enclosure. Further cost lowering will depend upon either:
(1) awaiting the commercial availability of such screen-impregnated
drywall, screen-backed 4 ft. x 8 ft. sheets of 1/8" or 1/4" embossed
and grooved hardboard, panel-veneer finishes for wall coverings, or
(2) a large building might permit the economical purchase in quantity
of special screen-impregnated wallboards.

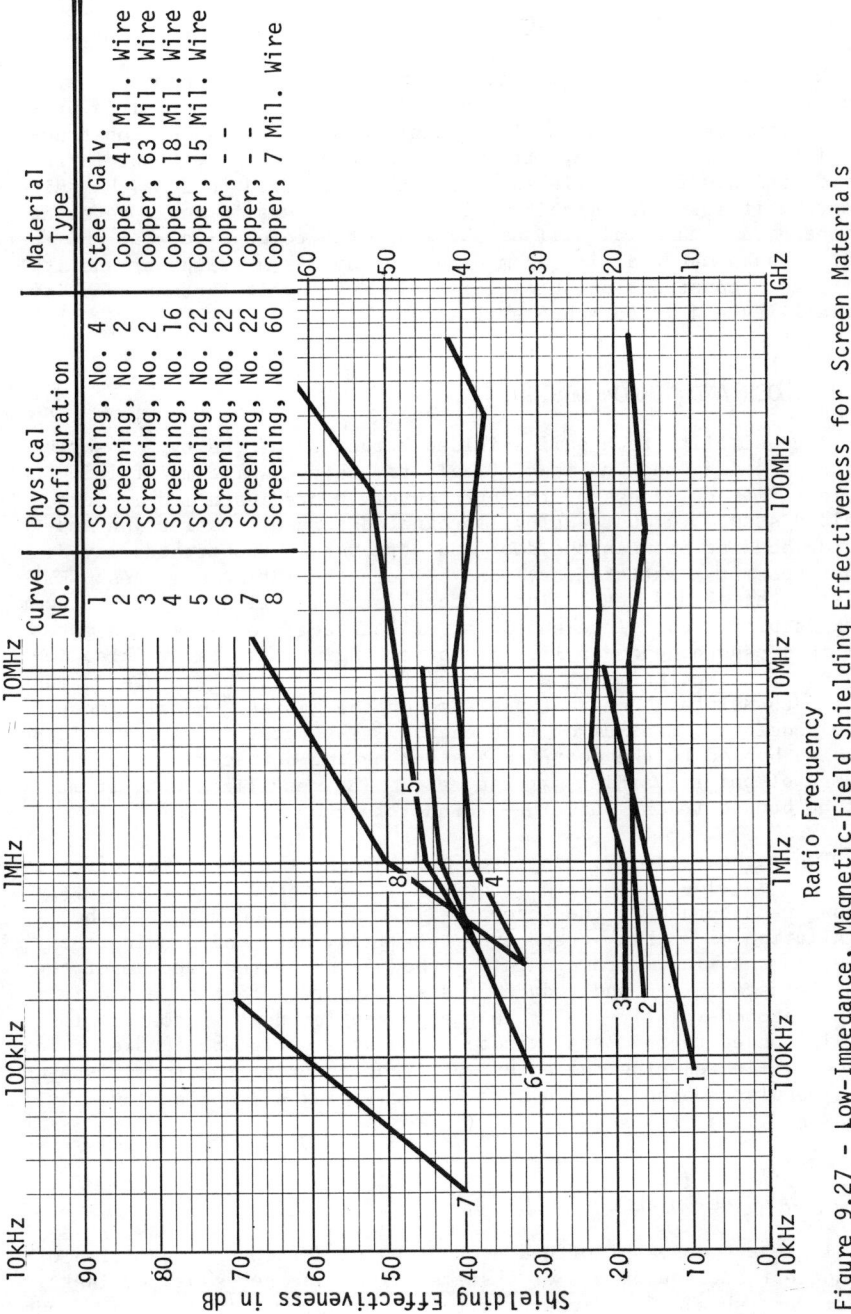

Figure 9.27 - Low-Impedance, Magnetic-Field Shielding Effectiveness for Screen Materials

9.3.3.2 METALLIZED WALLPAPER

Metallized wallpaper is becoming more popular for economic attenuation of electromagnetic fields. Although such material is presently being used as a shielding technique during building construction in a very limited way, its merits of application warrant further consideration. It is believed that such metallized foil may be quite effective if installed *within* walls where moisture is less prevalent and where such material will not be subjected to abusive wear and tear. However, such material is recommended for use as a wallpaper and its merits as a construction shielding media must await further quantity availability and performance evaluation.

9.3.4 DOOR AND WINDOW DESIGN

Accessibility to a semi-shielded structure of the type described in the preceding section is difficult to achieve while maintaining the shielding effectiveness of the structure. Doors installed in such a shielded structure must employ impregnated screen construction and provide bonding between the door and structure. Two methods may be used to permit a satisfactory bond between the mating surfaces: EMI gasket material and metal finger stock (see Sec. 11.2). Metal finger stock is preferred, since shielding effectiveness of gasketing material is dependent upon applied pressure. Aging effects are also important if gasketing material is used for the bond, piano hinges must be used to maintain constant pressure on the gasket. Metal finger stock requires a minimum of maintenance. The metal fingers should be adhesive bonded to the door frame after mating surfaces have been cleaned of paint, varnish, grease, etc. The metallized door frame must be bonded to the building-ground reference plane.

Two methods are available for maintaining the shielding effectiveness of a shielded room with windows. Both methods require a metal window frame bonded to the screen walls. One method requires the installation of a thin screen over the window area and bonding the screen to the window frame. This method is preferred because of the ease of installation and reduced costs. A more esthetic method is to install a commercially-available transparent conductive coated glass that is bonded to the metal window frame by a peripheral gasket. This method, while reducing the amount of light entering the room, is becoming a popular approach where both R-F shielding and optional visibility (see Sec. 11.1) are needed.

9.3.5 SPECIAL SHIELDED ENCLOSURES

In severe cases of offending sources of EMI to the operation of EMI-susceptible equipments or systems, it may be necessary to install one or more shielded enclosures within a building. In this case, one

of two practices is used: (1) if the building is dedicated to a single
purpose such as a hospital, research lab, or the like, the enclosure(s)
can be planned during the building design phase and installed during
construction, or (2) if the building is to be used by many different
commercial, industrial, and/or professional tenants, then let the individual users plan and pay for their own shielded enclosure installation at a later date as needed. The subject of shielded enclosures
is discussed in Sec. 11.2.

9.4 BIBLIOGRAPHY

(1) Cumming, W.R., "Materials for RF Shielded Chambers and Enclosures," *Digest of the Fourth IEEE National Symposium on RFI,* June 1962.

(2) Giehart, G.D., et. al., "Insertion Loss of a Brick Wall at UHF," *NBS Report 7272,* June 26, 1962.

(3) Giehart, G.D., et. al., "Insertion Loss of a Frame Wall at SHF," *NBS Report 7273,* June 26, 1962.

(4) Nicholson, P.F., "Electromagnetic Shielding with Coke," *NRL Memorandum Report 1080,* 18 July 1960.

(5) Sunde, Erling D., "Earth Conductivity Effects in Transmission Systems," D. Van NOstrand Co., Inc., P. 51.

(6) Wenner, Frank, "A Method of Measuring Earth Resistivity," *U.S. Bureau of Standards Bulletin,* Vol. 12, 1916.

(7) White Electromagnetics, Inc., "Architectual Interference Data," Final Report, AF 30(60z)-2691, *RADC Contract,* 20 August 1963.

CHAPTER 10

SHIELDING THEORY AND MATERIALS

CHAPTER 10

SHIELDING THEORY AND MATERIALS

Cable shields, their performance, and methods of grounding have come up many times in previous sections. Cables are not the only receptors where shielding can be effective in reducing radiated EMI. Other components such as relays, solenoids, and amplifiers can also be shielded to prevent radiation to or from other components. In fact, shielding is a major category of EMI control at all levels of EMC, viz., component; chassis or *black box*; equipment; subsystem; system; and entire vehicular or housing structures, such as ships, aircraft, and buildings. Thus, this section presents shielding theory, shielding materials, and some mathematical models of shielding effectiveness.

The performance of shields is a function of whether the source appears as an electric or magnetic field in the near-in induction region or an electromagnetic field in the far-field region. These considerations are a function of both the source and receptor geometry separation, and frequency of operation. Consequently, it is pertinent to first establish criteria for far and near-fields as a function of these parameters.

10.1 FIELD THEORY

The purpose of this section is to present some pragmatic relations about magnetic, electric, and electromagnetic fields as pertinent background to understanding and applying shielding criteria. Since the literature is replete with excellent discussions of Maxwell's equations and field theory, only a few aspects are presented here.

The electric (E_θ, E_r) and magnetic (H_ϕ) fields existing about an oscillating doublet are obtained from applying Maxwell's equations:

$$E_\theta = Z_o k \sin\theta \left[-\left(\frac{\lambda}{2\pi r}\right)^2 \cos\left(\frac{2\pi r}{\lambda} - \omega t\right) - \frac{\lambda}{2\pi r} \sin\left(\frac{2\pi r}{\lambda} - \omega t\right) + \cos\left(\frac{2\pi r}{\lambda} - \omega t\right) \right] \tag{10.1}$$

$$E_r = -2Z_o k \cos\theta \left[\left(\frac{\lambda}{2\pi r}\right) \cos\left(\frac{2\pi r}{\lambda} - \omega t\right) + \frac{\lambda}{2\pi r} \sin\left(\frac{2\pi r}{\lambda} - \omega t\right) \right] \tag{10.2}$$

$$H_\phi = k \sin \theta \left[- \frac{\lambda}{2\pi r} \sin \left(\frac{2\pi r}{\lambda} - \omega t \right) + \cos \left(\frac{2\pi r}{\lambda} - \omega t \right) \right] \qquad (10.3)$$

where, $k = I\ell/2r\lambda$

θ = zenith angle to r

I = current in short wire of length ℓ

$\omega = 2\pi f$ (f = frequency)

λ = wavelength corresponding to f

r = distance from wire to measuring point

t = time

Z_o = free-space impedance $(r \gg \lambda)$

       $= E_\theta / H_\phi = 377\Omega$

Several observations can be made about the near and far fields from Eqs. (10.1) and (10.3)*:

(1) The multiplier, k, contains a 1/r component which also decreases with distance from the source.

(2) When the multiplier, $\lambda/2\pi r$, equals 1 in two of the three E_θ electric-field terms* and the magnetic field term, all coefficients of either the sin or cos are unity and equal. Thus, when $r = \lambda/2\pi$ (about one-sixth wavelength), this corresponds to the transition-field condition or boundary between the near-field (first term of both equations) and far-field (last term).

(3) When $r \gg \lambda/2\pi$ (far-field conditions), only the last term of each equation* is significant. For this condition the wave impedance $Z_o = E_\theta / H_\phi = 377$ ohms. This is called the radiation field (plane waves) and both E_θ and H_ϕ are in time phase although in directional quadrature.

(4) When $r \ll \lambda/2\pi$ (near-field conditions), only the first term of each equation is significant*. For this condition, the wave impedance, $E_\theta / H_\phi = Z_o \lambda / 2\pi r$. Note that the wave impedance is now $\gg Z_o$. This is sometimes called simply an electric-field or a high-impedance field, i.e., high relative to a plane-wave impedance. It is also the induction field and E_θ and H_ϕ are in both time phase and directional quadrature.

(5) Had the oscillating source not been a small straight wire or doublet exhibiting high impedance but rather a small wire loop exhibiting low impedance, the first term appearing in Eq. (10.1) would vanish and a similar first term would appear in Eq. (10.3). For this condition, the wave impedance in the near field, $E_\theta / H_\phi =$

---

* E_r is dropped from further discussion here.

$Z_0 2\pi r/\lambda$. This is sometimes called a magnetic field or a low-impedance field, i.e., low relative to Z_0 the plane-wave (radiation) impedance.

Fig. 10.1 illustrates conceptually the 4th and 5th conditions in the near or induction field. Situation (a) is a monopole or straight wire in which the R-F current is low. Consequently, the source impedance = V/I is a high-impedance. The wave-impedance near in is also high, being made up predominantly of the electric field. The electric field attenuates more rapidly $(1/r^3)$ with an increase in distance than the magnetic field $(1/r^2)$ in the induction region (cf. Eqs. (10.1) and (10.3). Thus, the wave impedance decreases with distance where it asymptotically approaches $Z_0 = 377$ ohms in the far or radiation field. The converse applies for situation (b) wherein a low-impedance source creates a low-impedance wave of predominantly the magnetic-field component. This impedance increases with distance where it asymptotically approaches 377 ohms in the far field. Fig. 10.2 illustrates these impedances of both fields as a function of distance, r.

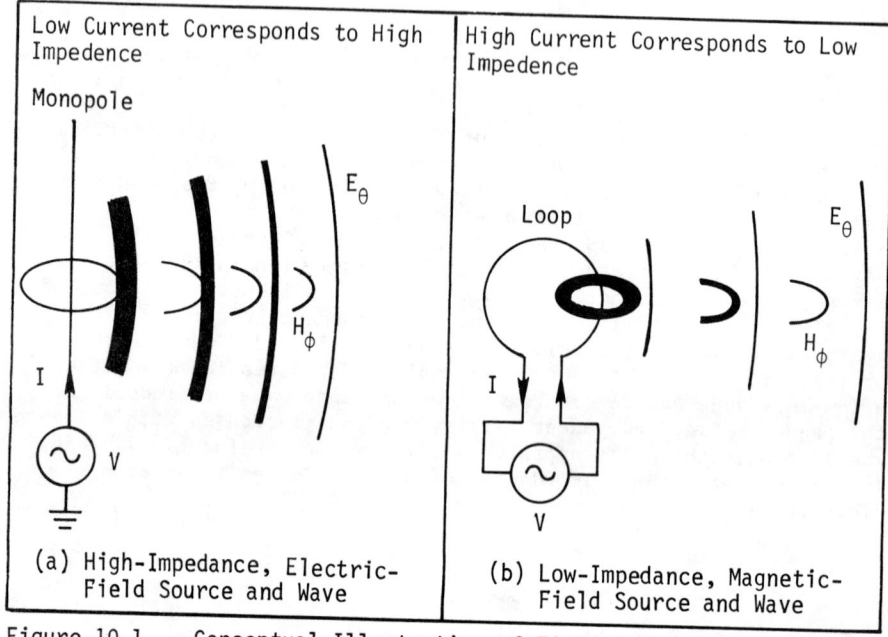

Figure 10.1 - Conceptual Illustration of Field Intensities vs Source
Type and Distance

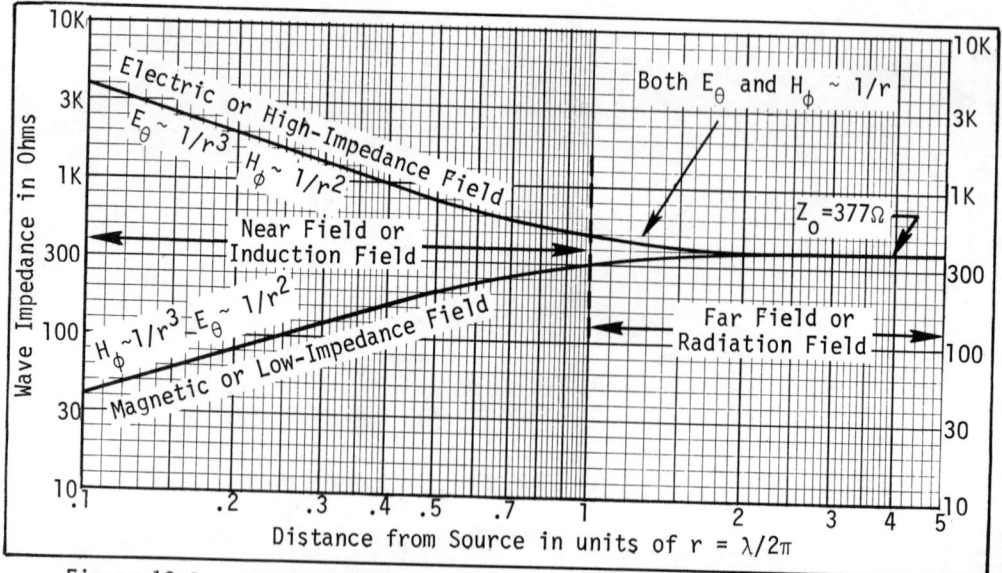

Figure 10.2 - Wave Impedance as a Function of Source Distance

10.2 SHIELDING THEORY

Shielding provided by a metallic barrier can be analyzed from either of two viewpoints: (1) that of field or wave theory or (2) that of circuit theory. In the circuit-theory approach, currents from the interference source induce currents in the shield such that the associated external fields due to both currents are out of phase and tend to cancel. Since the field-theory approach is more widely adopted in the literature, however, it will be used in the remainder of this discussion.

Fig. 10.3 depicts the phenomena of both reflection and transmission that are utilized in removing energy from an incident wave (plane-wave example shown). If an incident plane wave is intercepted by a barrier to its passage, at the region A of the interface, both reflection and transmission occur. The amplitudes of these two portions of the original wave depend on the surface impedance of the barrier material with respect to the impedance of the wave. Since the reflected wave is not proceeding in a direction that contributes to the surviving wave on the far side of the barrier, this is considered a loss mechanism.

The transmitted portion of the incident wave, continuing on in approximately the same direction after penetrating the interface, experiences absorption while traversing the finite thickness of the barrier. At the second barrier interface B of Fig. 10.3, reflection and transmission phenomena again occur. The transmitted portion is the amount of energy that traversed the first interface less the energy absorbed in traversing the barrier and that reflected at B. The second reflection contributes an insignificant amount in the removal of energy and is usually neglected.

At plane-wave (far-field) frequencies, the shielding effectiveness of a barrier in reducing the energy of an electromagnetic field can be readily computed. Each of the contributing factors discussed above is computed separately and then their total contribution is summarized. This is accomplished in the following manner for expressing shielding effectiveness in dB, S_{dB}:

$$S_{dB} = R_{dB} + A_{dB} + B_{dB} \tag{10.4}$$

where,

R_{dB} = reflection loss in dB

A_{dB} = transmission or absorption loss in dB

B_{dB} = internal reflection loss at exiting interface in dB (usually neglected)

The shielding effectiveness to electric or electromagnetic fields may also be measured in terms of the fraction of the impinging field which exists at the other side of the barrier:

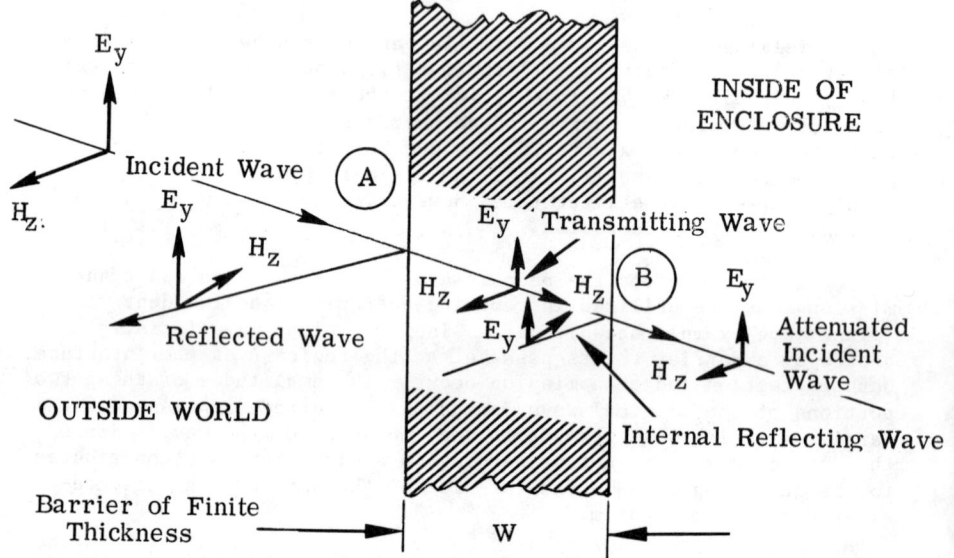

Figure 10.3 - Representation of Shielding Phenomena for Plane Waves

$$S_{dB} = 20 \log_{10}\left(\frac{E_1}{E_2}\right) \qquad (10.5)$$

where, E_1 = impinging field intensity in V/m

E_2 = exiting field intensity in V/m

The individual contributing factors to the shielding effectiveness in Eq. (10.4) are separately computed in the next sections.

10.2.1 ABSORPTION LOSS

The absorption loss, A_{dB}, is independent of the type of wave impinging on the shield and is expressed as follows:

$$A_{dB} = 3.34 \times 10^{-3} \ t\sqrt{fG\mu} = 3.34t\sqrt{f_{MHz}G\mu} \ dB \qquad (10.6)$$

where,   A = attenuation in dB

t = thickness of barrier in mils (unit of 0.001")

f = frequency in Hz

f_{MHz} = frequency in MHz

10.6

G = conductivity, relative to copper (G for Cu = 1)

μ = magnetic permeability of material relative to vacuum
   or copper (μ = 1)

Eq. (10.6) is plotted in Fig. 10.4 for the parameters copper
(G = 1, μ = 1), iron (G = 0.17, μ = 1000), and hypernick (G = 0.06,
μ = 80,000). Absorption loss is the dependent variable and frequency
is the independent variable with thickness in mils as a second parameter. It is noted that the brute-force approach of using a thick
sheet (1/8") of iron at low frequencies (e.g., at 60 Hz) results in
a significant absorption loss (approx. 45 dB). On the other hand, a
thin sheet (e.g., 1 mil) of copper at 1 GHz yields significant
(> 100 dB) absorption loss. This illustrates the difficulty of achieving a significant absorption loss at ELF in contrast to UHF.

The internal reflection loss, B, in Eq. (10.4) is negligible
when A_{dB} is greater than about 4 dB. When A_{dB} is not greater than 4
dB, B_{dB} is negative since it is a coherent term which would have made
E_2 in Eq. (10.5) larger. The value of B_{dB} is shown in the lower right
corner of Fig. 10.4.

10.2.2   REFLECTION LOSS

Reflection loss, R_{dB}, is represented by forming the ratio of the
wave impedance, Z_w, to the surface impedance of the barrier material,
Z_b.

$$R_{dB} = 20 \log_{10} \frac{(K+1)^2}{4K} \approx 20 \log_{10} \left(\frac{Z_W}{4Z_b} \right), \quad \text{for } K>10 \qquad (10.7)$$

Eq. (10.7) indicates that if either the wave impedance is high
(e.g. electric field) and/or the barrier surface impedance is low
(e.g. copper), the loss will be substantial. Conversely, if the wave
impedance is low (e.g. magnetic field) and/or the barrier impedance
is relatively high (e.g. iron), the reflection loss will be significantly less. Each of these situations is now discussed in further
detail.

10.2.3   REFLECTION LOSS TO PLANE WAVES

The reflection loss of a plane wave, R_{dB}, may also be calculated
from:

$$R_{dB} = 108 + 10 \log_{10} (G/\mu f_{MHz}) \text{ dB} \qquad (10.8)$$

Eq. (10.8) is plotted in Fig. 10.5 for copper, iron and hypernick. Compared with absorption loss, the figure indicates that the
reflection loss of plane waves at low-frequencies is the major

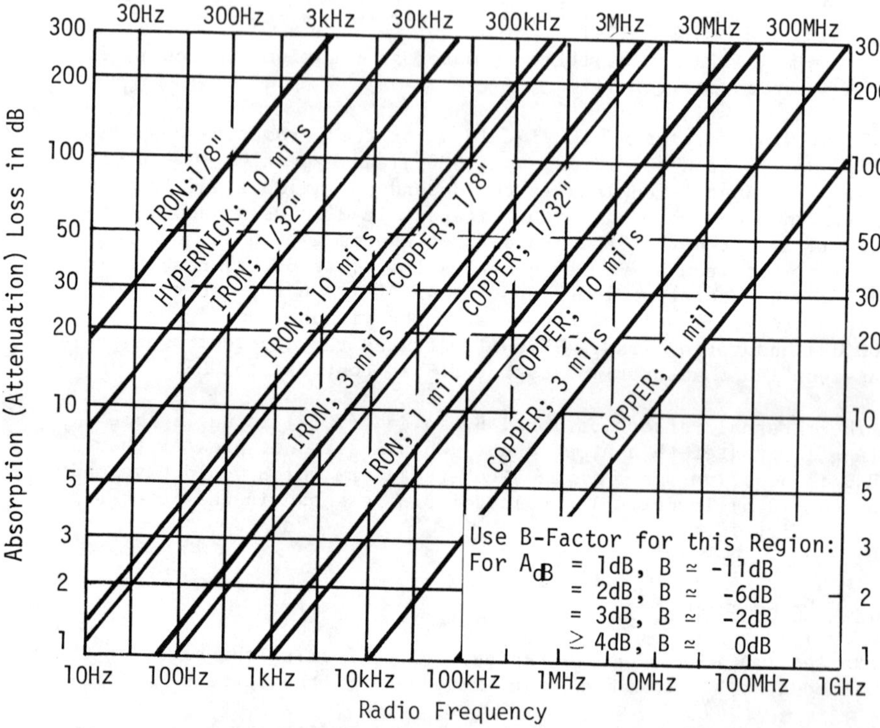

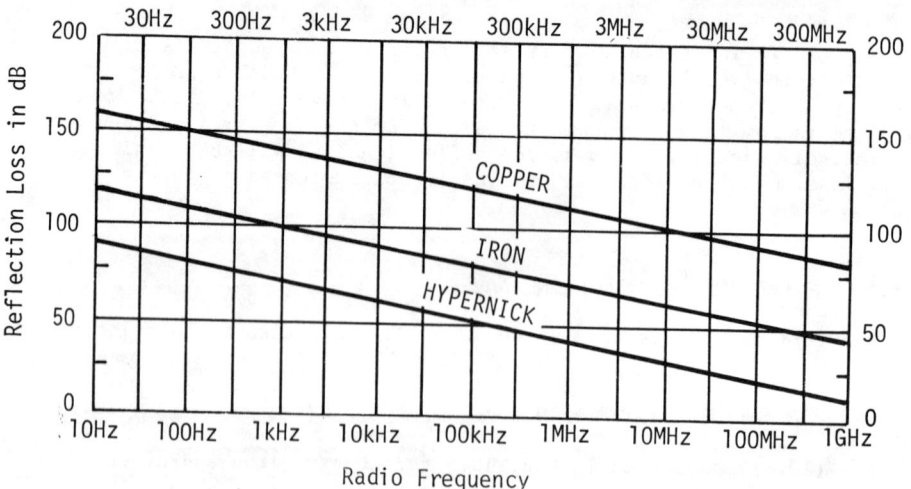

Figure 10.4   - Shielding Absorption (Penetration/Attenuation)
              Loss vs. Radio Frequency, Material, and Thickness

Figure 10.5   - Reflection Loss of Plane Waves vs. Radio Frequency

attenuation mechanism. High conductivity, low permeability material
is more effective in establishing reflection loss, since the barrier
surface impedance is lower with regard to that of a plane wave where
Z_w = 377 ohms and the ratio of the latter to the former (the loss
mechanism) is greater (cf. Eq. (10.7)). At UHF the reflection loss
becomes less effective since the barrier skin depth decreases (surface resistivity increases) and the barrier impedance increases
resulting in a smaller ratio of plane wave to barrier impedance. In
comparing Figs. 10.4 and 10.5, note that at UHF the absorption loss
becomes the more significant loss mechanism of the two.

10.2.4 Reflection Loss to Electric and Magnetic Fields

When there is a substantial difference in the impedance of the
incident wave and the shielding barrier, reflection at the boundary
is significant and good shielding is obtained. The high impedance
wave in the near field is known as an electric-field wave, and its
reflection loss is:

$$R_{dB} = 354 + 10 \log_{10}\left(\frac{G}{f^3 \mu r^2}\right) dB \qquad (10.9)$$

where, r = distance from source to barrier in inches the other
terms are as defined under Eq. (10.6)

Eq. (10.9) is plotted in Fig. 10.6 for the parameters of separation distances, r, of 1 inch, 1 meter (3.3 ft.), and 100 feet
(30 m); and for copper and iron materials. As before, frequency
is the independent variable and reflection loss, R_{dB}, is the dependent variable. The above distance parameter covers a range of 1200
or about 62 dB difference in reflection loss, whereas the G/μ range
for copper to iron is about -38 dB.

Fig. 10.6 shows that the reflection loss of an electric field
decreases with frequency until the separation distance becomes $\lambda/2\pi$,
whence far-field conditions prevail. Thus, Eq. (10.9) applies until
the losses meet that of Eq. (10.8), the plane-wave losses. Thereafter, the two merge. For this reason, the plane wave reflection
losses are also shown as a reference in Fig. 10.6 and are identical
to those previously shown in Fig. 10.5.

For low-impedance or magnetic-field waves, the reflection loss
is:

$$R_{dB} = 20 \log_{10}[(0.462/r)\sqrt{\mu/fG} + 0.136r\sqrt{Gf/\mu} + 0.354] dB \qquad (10.10)$$

Eq. (10.10) is plotted in Fig. 10.7 for the parameters of separation distance, r, of 1 inch, 1 meter (3.3 ft.) and 100 feet (30 m)
and for copper and iron materials. The reflection loss to iron (1

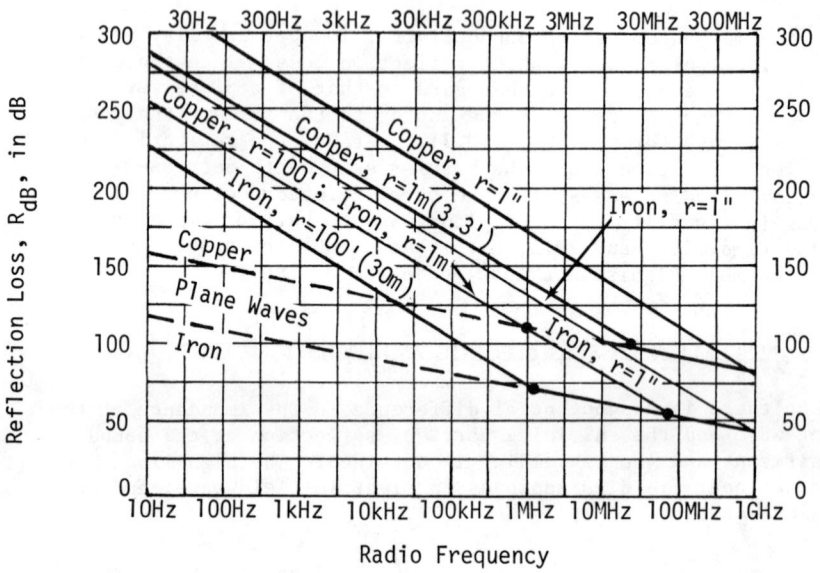

Figure 10.6 - Reflection Loss of Electric Fields
vs. Radio Frequency

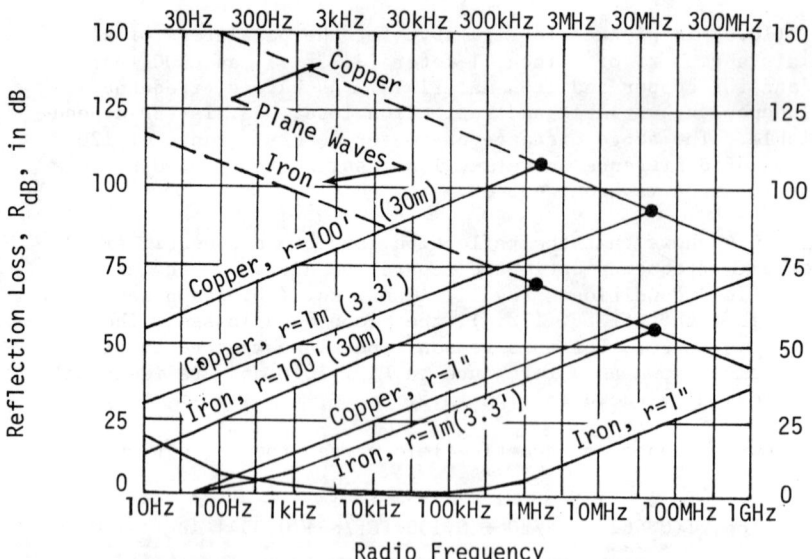

Figure 10.7 - Reflection Loss of Magnetic Fields
vs. Radio Frequency

10.10

inch separation) approaches 0 dB at about 30 kHz when the magnetic-field wave impedance approximates that of the barrier impedance (loss = 0 dB from Eq. (10.7)). Below 30 kHz, the wave impedance is less than the barrier impedance and the loss again increases. The reflection loss of a magnetic field shown in the figure increases with frequency until the source-to-barrier separation distance is about $\lambda/2\pi$, whence the plane-wave losses of Fig. 10.5 again prevail.

In comparing Figs. 10.6 and 10.7, it is noted that reflection-loss shielding for providing a reduction in absolute field intensity to magnetic fields at low frequencies is distinctly different from that for electric fields. Magnetic fields are shielded at DC and ELF only by providing a low reluctance path as an alternative for the incident magnetic-field.

Fig. 10.8 depicts a simple representation of a uniform magnetic field existing in free space. The vertical lines show the direction of the orientation of the magnetic-field vector throughout the two dimensions. Fig. 10.9 shows the effect on the field lines by including a hollow permeable object in this uniform magnetic field. The field-intensity lines enter the object at an angle of 90°to its surface. In the interior of this hollow object the field intensity lines are less intense than in the surrounding free-space medium*. However, the magnetic-field lines in the solid barrier are much more intense than in either the hollow center or the exterior of the barrier. This effect is due to the relative higher reluctances of free space both surrounding the barrier and in the interior, vs that of the barrier itself. The lower reluctance of this barrier divides the field-intensity lines thus reducing the intensity of the absolute magnetic field in the interior of the enclosure to yield a shielding effect. This effect is quite pronounced at DC where shielding effectiveness values in excess of 50 dB have been achieved through the utilization of extremely high permeability materials configured on a double-barrier enclosure.

10.2.5   Composite Absorption and Reflection Loss

When either Eqs. (10.6) to (10.10) or Figs. 10.4 to 10.7 are combined, the overall attenuation or shielding effectiveness given in Eq. (10.4) results. These relationships are plotted in Fig. 10.10. Since there are many variables, the composite curves represent the parameters of copper and iron materials having a thickness of one mil and 1/32 inch; electric and magnetic fields and plane-wave sources; and a source-to-barrier distance of 1 inch and 1 meter (3.3 feet).

---

*   The magnetic field in the inside is about $\mu t/s$ of the value on the outside, where μ is the relative permeability, t is the thickness and s is dimension of one side (see Chap. 6).

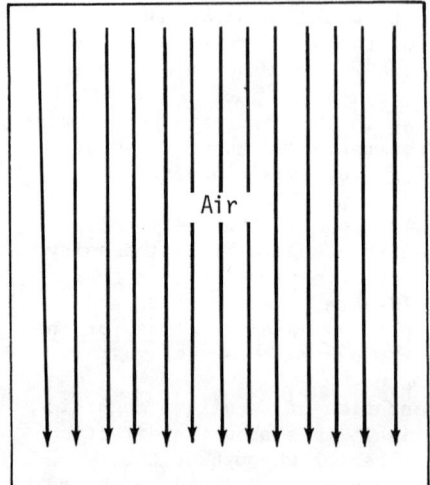

Figure 10.8 - Uniform Magnetic Field

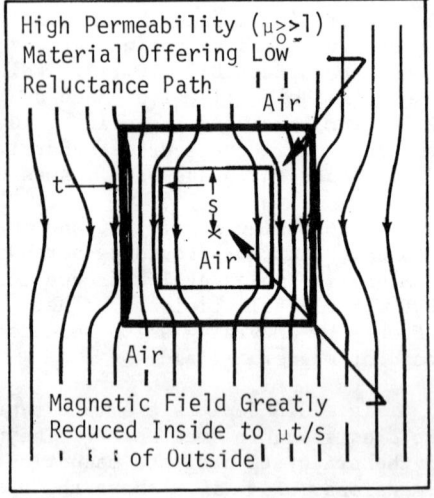

Figure 10.9 - Cross Section of a Hollow Rectangular Solid of High Permeability in Uniform Field

Except for L-F magnetic fields, the figure shows that reflection loss is the principle attenuation mechanism at low frequencies whereas absorption loss is the main mechanism at H-F. Fig. 10.10 is but one of a family of mathematical models which defines shielding attenuation. Other models would reflect different materials, thickness, and emission source distances.

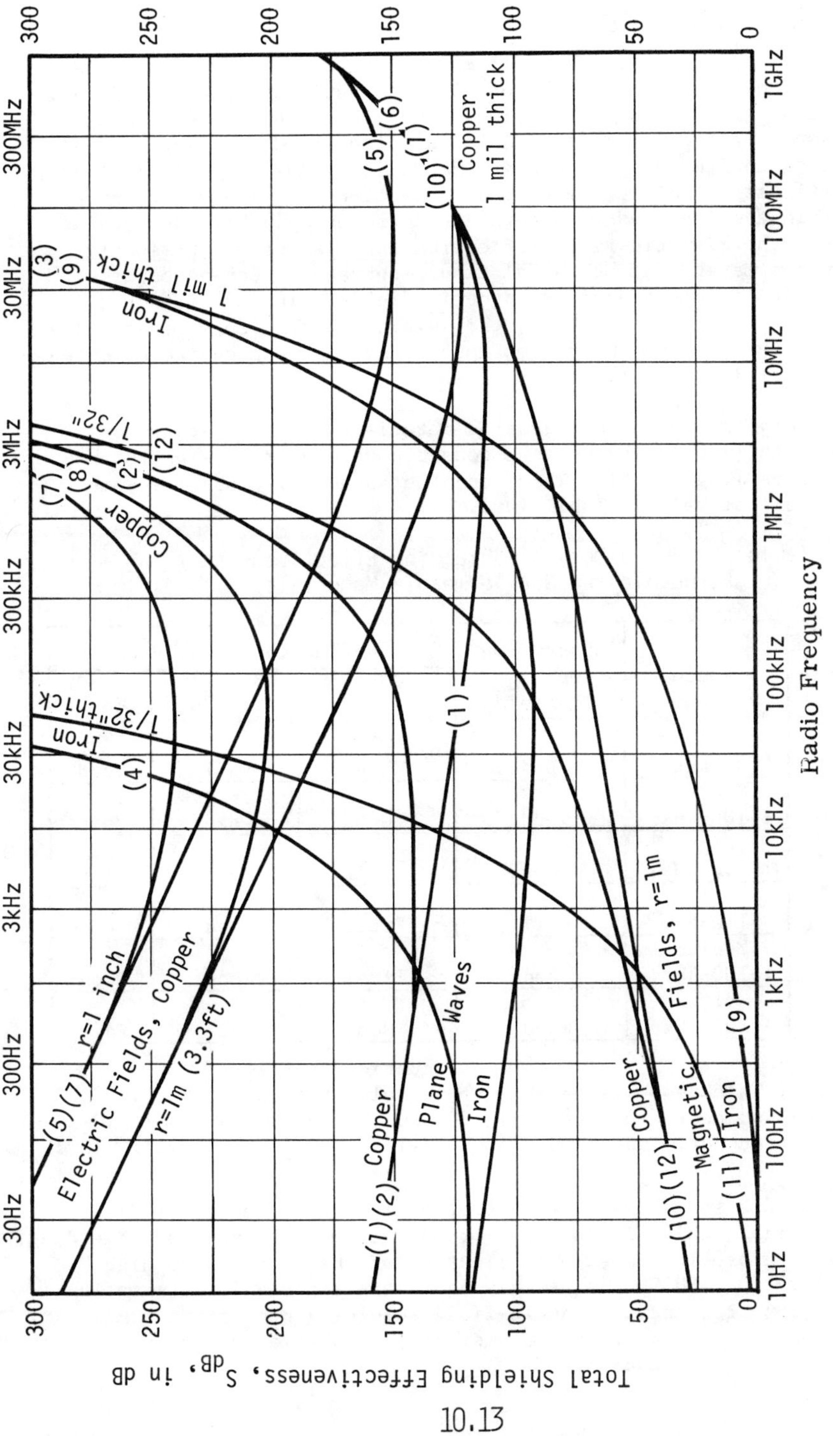

Figure 10.10 - Total Shielding Effectiveness vs Frequency for Electric and Magnetic Fields and Plane Waves

10.3 SHIELDING MATERIALS

Good shielding efficiency for electric (high-impedance) fields is obtained by use of materials of high conductivity, such as copper and aluminum. As shown in Eq. (10.9) and Fig. 10.6, the shielding effectiveness for electric fields is infinite at DC and decreases with an increase in frequency. However, magnetic fields (Eq. (10.10)) are more difficult to shield, since the reflection loss may approach zero for certain combinations of material and frequency. With decreasing frequency, the magnetic field reflection and absorption losses of non-magnetic materials such as aluminum, decrease. Consequently, it is difficult to shield against magnetic fields using non-magnetic materials. At high frequencies, the shielding efficiency is good due to both reflection and absorption losses, so that the choice of materials becomes less important.

Regarding plane waves, magnetic materials provide better absorption loss (Fig. 10.4) whereas good conductors provide better reflection loss (Fig. 10.5). These and the above relations are summarized qualitatively in Table 10.1.

Table 10.1 - Summary of Shielding Effectiveness of
Permeable and Non-Permeable Materials

Permeable Materials	Radio Frequency	Absorption Loss A_{dB} All Fields	Reflection Loss, R_{dB}		
			Electric Fields	Magnetic Fields	Plane Waves
Magnetic ($\mu \geq 1000$)	Low <1 kHz	Bad	Excel.	Fail	Good
	Medium 1–100 kHz	Good	Good	Bad	Fair
	High > 100 kHz	Excel.	Fair	Poor	Fair
Non-Magnetic ($\mu = 1$)	Low < 1 kHz	Fail	Excel.	Bad	Good
	Medium 1–100 kHz	Bad	Excel.	Poor	Good
	High > 100 kHz	Good	Good	Fair	Fair

Assumptions:
    Material Thickness: 1/32"
    Source Distance: 10 feet (3m)
    Radio Frequency: as shown

Attenuation Scores:
    Excellent >150 dB Poor: 30-50 dB
    Good: 100-150 dB Bad: 10-30 dB
    Fair: 50-100 dB Fail: <10 dB

Table 10.2 summarizes the absorption loss of a number of different materials which in one form or another may be used for shielding. The loss is given in dB per mil thickness of the metal. The high permeability (μ 80,000) materials shown are especially interesting for their low-frequency, magnetic-field shielding properties. However, they are prone to saturation at lower field densities and they require careful handling procedures.

Table 10.2 - Characteristics of Metals Used for Shielding

Metal	Conductivity Relative to Copper	Relative Permeability (100 kHz)	Absorption Loss (dB per mil (0.001"))		
			100 Hz	10 kHz	1 MHz
Silver	1.05	1	0.03	0.34	3.40
Copper-Annealed	1.00	1	0.03	0.33	3.33
Copper-Hard Drawn	0.97	1	0.03	0.32	3.25
Gold	0.70	1	0.03	0.28	2.78
Aluminum	0.61	1	0.03	0.26	2.60
Magnesium	0.38	1	0.02	0.20	2.04
Zinc	0.29	1	0.02	0.17	1.70
Brass	0.26	1	0.02	0.17	1.70
Cadmium	0.23	1	0.02	0.16	1.60
Nickel	0.20	1	0.01	0.15	1.49
Bronze	0.18	1	0.01	0.14	1.42
Iron	0.17	1,000	0.44	4.36	43.6
Tin	0.15	1	0.01	0.13	1.29
Steel (SAE 1045)	0.10	1,000	0.33	3.32	33.2
Beryllium	0.10	1	0.01	0.11	1.06
Lead	0.08	1	0.01	0.09	0.93
Hypernom	0.06	80,000	2.28	22.8	228
Monel	0.04	1	0.01	0.07	0.67
Mu-Metal	0.03	80,000	1.63	16.3	163
Permalloy	0.03	80,000	1.63	16.3	163
Stainless Steel	0.02	≈ 1	0.15	1.47	14.7

It is often assumed that most materials which have adequate structural rigidity will also possess sufficient thickness to provide satisfactory shielding efficiency. This is not generally true for equipments operated in the audio-frequency region. At these low frequencies it is necessary to use a high permeability material such as hypernom, mu-metal, permalloy, or Netic or Co-Netic foil to provide satisfactory shielding efficiency to magnetic fields.

While the above equations and figures show a theoretical value of shielding efficiency from magnetic materials which is quite high, in practice such levels are seldom achieved, particularly at low frequencies where the required thickness is substantial. Some of the best results have been obtained by the use of multiple permalloy sheets or the Netic and Co-Netic sandwich foils. These latter products are available in a variety of ready made forms and sizes to fit diverse applications.

Illustrative Example 10.1

A sensitive parallel-T amplifier tuned to 120 Hz is to be located about 1 meter away from a 60 Hz amplidyne. By measurement the magnetic flux density, B, from the amplidyne at a 1 meter distance at its second harmonic is 180 dBpT or 10 gauss (10^{-3} weber/m^2). The cable feeding the tuned amplifier is 16 inches long (0.4 m) and has an equivalent to conductor separation of 0.1 inch (.0025 m). Determine the induced voltage and specify the magnetic shield required to protect the 1 μV amplifier sensitivity, if necessary.

The cable loop area is, A = ℓw = 0.4m x .0025m = 10^{-3}m^2. The magnetic flux, ϕ, crossing the cable loop is BA = 10^{-3} weber/m^2 x 10^{-3}m^2 = 10^{-6} webers. The induced voltage, V, is:

$$V = -\frac{d\phi}{dt} = -\frac{d}{dt} (10^{-6} \text{ webers cos } \omega t)$$

$$= |\omega 10^{-6} \sin \omega t| \text{ volts} = 2\pi \times 120 \text{Hz} \times 10^{-6} = 750 \mu V \quad (58 \text{ dB} \mu V)$$

Since the induced voltage is 58 dB above the 1 μV amplifier sensitivity, about 60 dB of magnetic shielding of the cable is required at 120 Hz. At this frequency, from Fig. 10.10, a 1/32" iron sheet offers about 15 dB attenuation and copper of any thickness offers about 40 dB. Neither will provide the shielding required. Table 10.2 indicates that Hypernom offers 2.3 dB per mil thickness at 100 Hz. Thus, about 26 mils of Hypernom (60 dB attenuation) should adequately shield the twin-T amplifier cable.

The attenuation offered by materials to electric, magnetic, and electromagnetic waves described in the previous sections is achieved theoretically. In practice, however, this attenuation is not often achieved because a shielded enclosure or housing is not completely sealed. In other words, nearly any practical application of shielding has necessary penetrations of one kind or another. The next chapter discusses the loss of such shielding integrity and the practices that may be followed to reclaim the integrity.

10.4 BIBLIOGRAPHY

(1) Adams, W.S., "Graphical presentation of electromagnetic shielding theory," *Proceedings Tenth Tri-Service Conference on Electromagnetic Compatibility,* Chicago: Armour Research Foundation, pp. 421-499, Nov. 1964.

(2) AF/BSD Exhibit 62-87, "Electro-Interference Control Requirements for Minuteman (WS 133B)," 6 December 1962.

(3) Albin, A.L., "Optimum shielding of equipment enclosures," *Electronic Design,* Vol. 8, No. 3, February 3, 1960.

(4) Boeing Aircraft Corporation D2-2-2444, "Electro-Interference Control Requirements (Equipment)," 5 March 1959.

(5) Bridges, J.E., Huenemann, R.F., and Hegner, "Electric-Field Shielding and Measurement," *Record of the 1967 IEEE Symposium on Electromagnetic Compatibility,* Vol. 27C80, July 18-20 1967; Washington, D.C.

(6) Cacace, R., and Hassett, R., "Investigation of Measurement Techniques for Transient Magnetic Fields," *Symposium Digest, 7th National Symposium Electromagnetic Compatibility,* June 28-30, 1965; New York.

(7) Clough, L., and Salzetti, J., "Magnetic Induction Susceptibility at Power Frequencies," *Eighth Tri-Service Conference on EMC,* October 30-November 1, 1962; Chicago, Ill., pp. 241-269.

(8) Cohen, D., "A Shielded Facility for Low-Level Magnetic Measurements," *Journal of Applied Physics,* Vol. 38, No. 3, 1967; pp. 1295-2196.

(9) Cole, N.H., "A Comparison of RF shielding materials," *Electronic Design,* Vol. 10, No. 20; Sept. 27, 1962.

(10) Cowdell, R.B., "Simplified Shielding," *1967 IEEE Electromagnetic Compatibility Symposium Record,* IEEE 27C80; Washington, D.C.; July 18-20, 1967.

(11) Dolle, W.C., and Cory, W.E., "Measurements of the Attenuation of the Electric and Magnetic Fields at Points Close to the Source," *IEEE Transactions on Electromagnetic Compatibility,* Vol. EMC-10, No. 2; September 1968, pp. 313-319.

(12) Dolle, W.C., Van Steenberg, G.N., and Jouffray, O.L., "Effects of Shielded Enclosure Resonances on Measurement Accuracy," *Record of the 1970 IEEE International Symposium on Electromagnetic Compatibility,* Vol. 70C28-EMC, July 14-16, 1970; Anaheim; pp. 417-420.

(13)   Ervin, H.W., "Shield termination prediction method," *Proceedings Eighth Tri-Service Conference on Electromagnetic Compatibility,* Chicago; Armour Research Foundation, Oct. 1962.

(14)   Feber, R.R., and Young, F.J., "The Shielding of Electromagnetic Pulses by the Use of Magnetic Materials," *Record of the 1969 IEEE Symposium on Electromagnetic Compatibility,* Vol. 69C3-EMC; June 17-19, 1969; Asbury Park; pp. 73-74.

(15)   Filtron Company, Inc., "Interference Reduction Guide for Design Engineers," Prepared for U.S. Army Electronics Labs., Fort Monmouth, N.J., Vols. 1 and 2, Accession AD 619666 and AD 619667.

(16)   Foster, J., Buegal, K., and Sowa, C., "Electromagnetic Compatibility Lunar Orbiter," *NASA,* Document No. NAS 1-3800; January 29, 1965.

(17)   Free, W.R., et al, "Compact Chamber for Impedance and Power Testing of VHF Whip Antennas," Final Report on Contract DAAB07-67-C-0575, EES, Georgia Institute of Technology, 1968.

(18)   Free, W.R., et al, "Electromagnetic Interference Measurement Methods-Shielded Enclosure," Final Report on Contract No. DA 28-043 AMC-02381 (E), EES, Georgia Institute of Technology; 1967.

(19)   Free, W.R., "Radiated EMI Measurements in Shielded Enclosure," *Record of the 1967 IEEE Symposium on Electromagnetic Compatibility,* Vol. 27-C80, July 18-20, 1967, Washington, D.C.; pp. 43-53.

(20)   GM07-59-2617A, "Electro-Interference Control Requirements for Minuteman (WS 133A)," 20 October 1959.

(21)   Haber, F., "Study of GSFC Radio Frequency Interference (RFI) Design Guideline for Aerospace Communication Systems," Moore School of Electrical Engineering, University of Pennsylvania, Moore School Report No. 66027 for NASA, April 30, 1966.

(22)   Hollway, D.L., "Screen Rooms and Enclosures," *Proceedings IREA,* Vol. 21, No. 10, October 1960; pp. 660-668.

(23)   Jarva, W., "Shielding efficiency calculation methods for screening waveguide ventilation panels and other perforated electromagnetic shields," *Proceedings Seventh Tri-Service Conference on Radio Interference Reduction and Electronic Compatibility,* Chicago: Armour Research Foundation; pp. 478-498; Nov. 1961.

(24)   Johnson, W.R., et al, "Development of a Space Vehicle Electromagnetic Interference/Compatibility Specification," *NASA,* Contract No. 9-73-5, *TRW,* 08900-60001-T000; June 28, 1968.

(25)   Klouda, J.C., "Practical Aspects in Evaluating Shielded Rooms," *Electro-Technology,* June 1961.

(26)   Kozakoff, D.J., Bolt, A.T., and Howard, F.D., "Shielding Effectiveness of Various Shaped Geometrical Enclosures in Terms of Normalized Parameters," *Record of the 1970 IEEE Regional Electromagnetic Compatibility Symposium Record,* Vol. 70C64-REGEMC; October 6-8, 1970; San Antonio; pp. V-A-1 to V-A-11.

(27)   Lindgren, E.A., "Contemporary RF Enclosures," Erik A. Lindgren and Associates; 1967.

(28)   Lockwood, R.O., "New Technique for the Determination of the Integrity of Shielded Enclosure by the Measurement of the Perpendicular Magnetic Field," *Record of the 1967 IEEE Symposium on Electromagnetic Compatibility,* Vol. 27C80, July 18-20, Washington, D.C.; pp. 61-69.

(29)   McAdam, W., and Vandeventer, D., "Solving Pickup Problems in Electronic Instrumentation," *ISA Journal,* Vol. 7, No. 4, p. 48, April 1960.

(30)   McDonald, G.M., and Taylor, G.R., "Shielding Grounding and Circuit-Grounding effectiveness in Interference reduction in the 50 Hz to 15 kHz Frequency Region," *IEEE Transactions on Electromagnetic Compatibility,* Vol. EMC-8, No. 1, pp. 8-16, March 1966.

(31)   Mendex, H.A., "A New Approach to Electromagnetic Field-Strength Measurements in Shielded Enclosures," *Convention Record of the 1968 WESCON,* Technical Papers Session 19, August 20-23, 1968; Los Angeles.

(32)   Mendez, H.A., "Meaningful EMC Measurements in Shielded Enclosures," *Record of the 1969 IEEE Symposium on Electromagnetic Compatibility,* Vol. 69C3-EMC, June 17-19, 1969, Asbury Park; p. 137.

(33)   MIL-STD 285, "Method of Attenuation Measurements for Enclosures Electromagnetic Shielding, for Electronic Test Purposes."

(34)   Osburn, J.D., and Morris, F.J., "Problems Encountered During the Design and Fabrication of an ELF-VLF-Shielding Enclosure," *Record of the International IEEE Symposium on Electromagnetic Compatibility,* Vol. 70C28-EMC, July 14-16, 1970; Anaheim; pp. 472-478.

(35)   Patton, B.J., and Fitch, J.J., "Design of a Room-Size Magnetic Shield," *Journal of Geophysics Research,* Vol. 67, 1962; pp. 1117-1121.

(36)   Patton, B.J., "Room-Size Enclosure for Geomagnetic Shielding," *Record of the 1970 IEEE International Symposium on Electromagnetic Compatibility,* Vol. 70C28-#MC, July 14-16, 1970, Anaheim; pp. 89-96.

(37) Pearlston, C.B., "Case and Cable Shielding, Bonding and Grounding Considerations in Electromagnetic Interference," *IRE Transactions on Radio Frequency Interference,* Vol. RFI-4, No. 3; pp. 1-6, October 1962.

(38) Pearlston, C.B., "Enclosure shielding in radio interference," *IEEE Third National Symposium on RFI,* Washington, D.C.; June 1961.

(39) Quine, J.P., "Theoretical Formulas for Calculating Shielding Effectiveness of perforated sheets and wire mesh screens," *Proceedings Third Conference on Radio Interference Reduction,* Armour Research Foundation, pp. 315-329; February 1957.

(40) Salati, O.M., "Recent Developments in Interference," *IRE Transactions on Radio Frequency Interference,* Vol. RFI-4, No. 2; pp. 24-33; May 1962.

(41) Schreiber, O.P., "Designing and applying RFI shields and gaskets," *Electronic Design,* Vol. 10, No. 20; Sept. 27, 1962.

(42) Schrieber, O.P., "RF tightness Using Resilient Metallic Gaskets," *Proceedings Second Conference on Radio Interference Reduction;* Chicago, Ill., Armour Research Foundation; pp. 343-359; March 1956.

(43) Schrieber, O.P., "Some useful analogies for RF shielding and gasketing," *IEEE Third National Symposium on Radio Frequency Interference,* Washington, D.C.; June 1961

(44) Schulz, R.B., et al, "Shielding Theory and Practice," *Proceedings Ninth Tri-Service Conference on Electromagnetic Compatibility,* Chicago, Ill., Armour Research Foundation, pp. 597-636; October 1963.

(45) Siegel, N.S., "Near Field Coupling on Aerospace Vehicles," *1970 IEEE EMC Symposium Record,* July 14-16, 1970; Anaheim, California; pp. 211-216.

(46) Stirrat, W.A., "USA ECOM Contribution to Shielding Theory," *IEEE Transactions on Electromagnetic Compatibility,* Vol. EMC-10, No. 1, March 1968, pp. 63-66.

(47) Stuckey, C.W., Free, W.R., and Robertson, D.W., "Preliminary Interpretation of Near-Field Effects on Measurement Accuracy in Shielded Enclosures," *Record of the 1969 IEEE Symposium on Electromagnetic Compatibility,* Vol. 69C3-EMC.

(48) Stuckey, C.W., "The Hooded Antenna - An Approach to Meaningful Field Strength Measurements in Shielded Enclosures," *IEEE Transactions on Electromagnetic Compatibility,* December, 1965; pp. 360-367.

(49) Toler, J.C., and Evans, R., "Shielded Enclosure Specification in Perspective," *1970 Regional Electromagnetic Compatibility Symposium Record,* Vol. 70C64 - REGEMC; October 6-8, 1970; San Antonio; pp. III-A01 to III-A-3.

(50) U.S. Naval Civil Engineering Laboratory, "Proposed Specification for Electromagnetic Shielding of Enclosures and Buildings," Final Project Report, 31 July 1963, Contract NBy-32220, Port Hueneme, Calif. pp. 97-113.

(51) Van Steenberg, G.N, and Willman, J.F., "Shielded Enclosure Measurements Can be Accurate," *21st Annual SWIEECO Record, IEEE,* Catalog No. 69C16-SWIECO; April 1969.

(52) Vaska, C.S., "Problems in shielding electronic equipment," *Proceedings Conference on Radio Interference Reduction,* Chicago, Ill., Armour Research Foundation, pp. 86-103, December 1954.

(53) Weinstock, G.L., "Electromagnetic Interference Control Within Aerospace ground Equipment for the McDonnell Phantom II Aircraft," *IEEE Transactions on Electromagnetic Compatibility,* Vol. EMC-7, No. 2; pp. 85-92, June 1965.

(54) Wright-Patterson Air Force Base, Ohio, "Handbook of Instructions for Aerospace Systems Design," *Vol. 4, Electromagnetic Compatibility AFSCM 80-9, Code SEG (SEPSM),* Bassic Issue, 20 April, 1964.

CHAPTER 11

SHIELDING INTEGRITY PROTECTION

CHAPTER 11

SHIELDING INTEGRITY PROTECTION

The previous chapter discussed the subjects of shielding theory
and materials. With the exception of low-frequency magnetic-field
shielding, it was shown that it is quite simple to obtain more than
100 dB of shielding effectiveness across the entire spectrum from
DC to light for electric and electromagnetic waves. However, since
any practical enclosure has apertures, the theoretical shielding is
never obtained due to loss of integrity. This chapter discusses the
resultant loss of shielding integrity, how the integrity can be reclaimed, and practical applications to shielded boxes, chassis and
equipments, cabinets, rooms and vehicles.

11.1 INTEGRITY OF SHIELDING CONFIGURATIONS

The attenuation offered by materials to electric, magnetic, and
electromagnetic waves described in the previous chapter is achieved
theoretically. In practice, however, this attenuation is not often
achieved because a shielded enclosure or housing is not completely
sealed. In other words, nearly any practical application of shielding
has necessary penetrations and apertures of one kind or another. Some
examples of such penetrations and apertures include:

- Cover plates and access cover members

- Meter windows

- Windows for viewing digital or other displays

- Potentiometer shafts

- Cooling apertures

- Power-line and signal-lead connectors

- Indicator lamps

- Push buttons

- On-off switches

- Fuses

Thus, it is not uncommon to find the plane-wave attenuation of a basic shield material to be 120 dB, for example, while the actual enclosure will exhibit 50 dB in the VHF/UHF portion of the spectrum. Here leakage of one or more of the above types compromises the integrity of the basic shielding material.

From a mathematical modeling point of view, either of two approaches may be used (1) compile a data bank on the shielding effectiveness (attenuation vs frequency) of many equipment materials and configurations and choose that closest to the specimen under examination, or (2) compute the shielding effectiveness based on an inventory of data listed in the above 10 items and use worst case coupling. Since no significant shielding data bank has ever been accumulated and reported on equipments as suggested in the first approach, the approach used here and elsewhere almost always resorts to the second.

11.1.1 BONDING SEAMS AND JOINTS

Loss of R-F shielding integrity across the *interface* of clean mating material members is a main reason why shielding effectiveness is compromised. Here, the conductivity of the interface may be much higher and/or the permeability may be much lower, because of the type of interface bond used. Thus, resulting material interfaces may be classified into two types: physically inhomogenous and physically homogenous.

A physical inhomogenous interface bond results when shielding members are directly connected by screws, rivets, spot welds, and the like. The interface connection is not continuous and there results a bowing or waviness effect between connected members. This in turn develops slits or gaps which leads to radiation or penetration at frequencies approaching 0.01λ. The attenuation A in dB at such a gap follows the waveguide-beyond-cutoff criteria:

$$A_{dB} = 0.0046 \ell f_{MHz} \sqrt{(f_c/f_{MHz})^2 - 1} \text{ dB} \qquad (11.1)$$

where, ℓ = gap depth in inches for overlapping members or the thickness of the material for butting members

f_{MHz} = operating frequency in MHz

f_c = cut-off frequency of gap in MHz

= 5900/g for rectangular gap   (11.2)

= 6920/g for circular gap   (11.3)

g = largest gap transverse dimension in inches

When $f_c \gg f_{MHz}$, Eq. (11.1) becomes:

$$A_{dB} \approx 0.0046 \ell f_c = 27\ell/g \text{ dB for rectangular gap} \qquad (11.4)$$

$$= 32\ell/g \text{ dB for circular gap} \qquad (11.5)$$

Fig. 11.1 is a plot of Eq. (11.1) representing attenuation through a rectangular gap vs frequency as a function of gap dimensions. The figure shows that more than 100 dB attenuation exists over the D-C to 10 GHz spectrum for both g/ℓ ratios greater than about 4 and the largest gap dimension less than 0.2 inch (cut-off frequency of about 30 GHz).

There are a number of techniques available for reducing electromagnetic emission leakage or receptor penetration of a shielded specimen. If members are joined by screws or rivets, Eq. (11.5) shows that A_{dB} may be significantly increased by using more screws or rivets per linear dimension of the interface due to the reduction in the gap, g. Fig. 11.2 shows a joint shielding effectiveness, as a function of screw spacing for the indicated parameters. Also note the improvement due to the application of a typical EMI mesh gasket (see Chap. 12).

Other techniques available for reducing the leakage in a physical inhomogenous mating member bond, involve attempting to eliminate or reduce the inhomogeniety. Fig. 11.3 illustrates some of these approaches. Where members do not have to be disengaged or separated, a continuous seam weld around the periphery of the mating surfaces is preferred. This type of weld is not critical provided it is continuous and has no weld *pin holes*. One exception involves the departure of the weld filler material from the basic shield member material. Hence, either the conductivity or permeability of the weld filler may be much lower resulting in degradation of shielding effectiveness. The seam weld technique is of questionable value when used with the more exotic magnetic materials ($\mu > 1000$; see Table 10.2) which must be annealed before assembly. Here, welding will destroy the specific properties that the annealing produced.

An alternative technique shown in Fig. 11.3 is the overlap seam. All nonconductive material (e.g., paint, rust, coatings, etc.) must be removed from the mating surfaces before they are crimped. Crimping must be performed under sufficient pressure to insure positive contact between all mating surfaces.

Shield members may have to be separated from time to time, such as cover and access plates for equipment alignment or maintenance. Therefore, none of the above techniques is acceptable. A temporary but good bond is required and this is the roll of the R-F gasketing material such as finger stock or resilient mesh. The topic of gaskets is covered in Chap. 12.

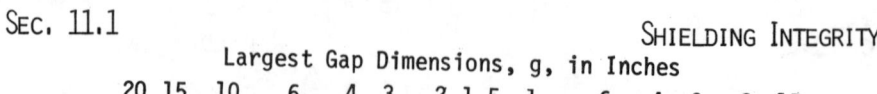

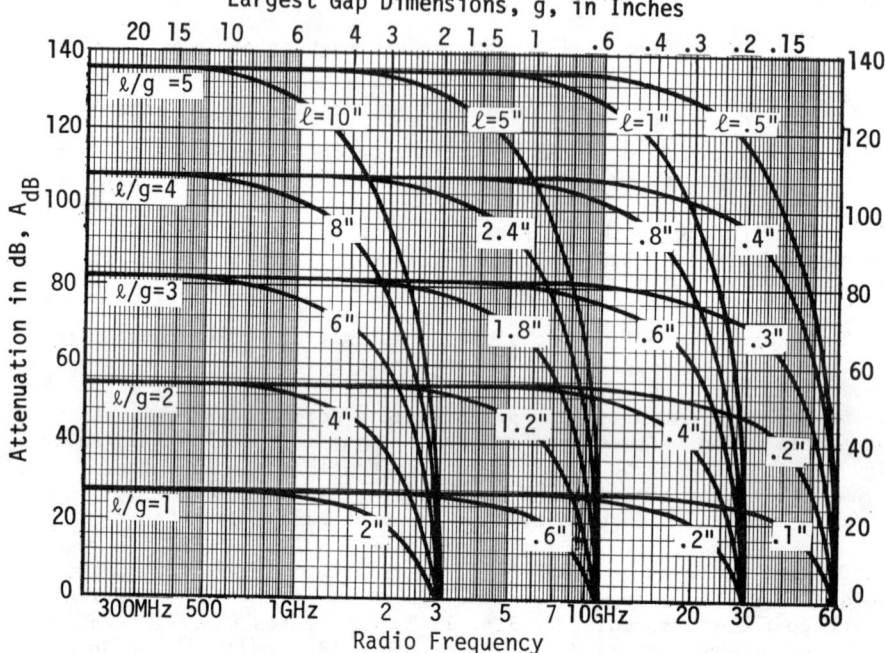

Figure 11.1 - Attenuation Through A Metallic Gap vs Frequency

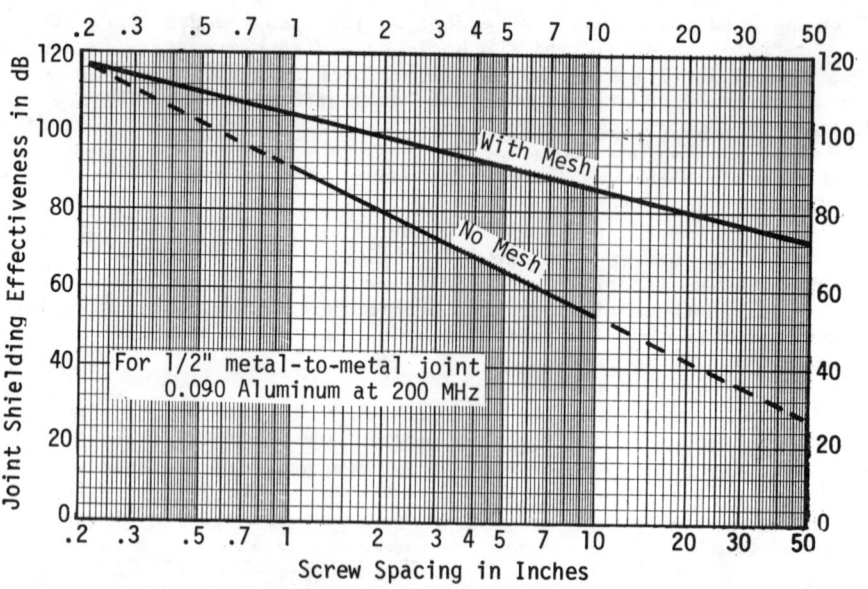

Figure 11.2 - Shielding Effectiveness for Screw-Secured Joints

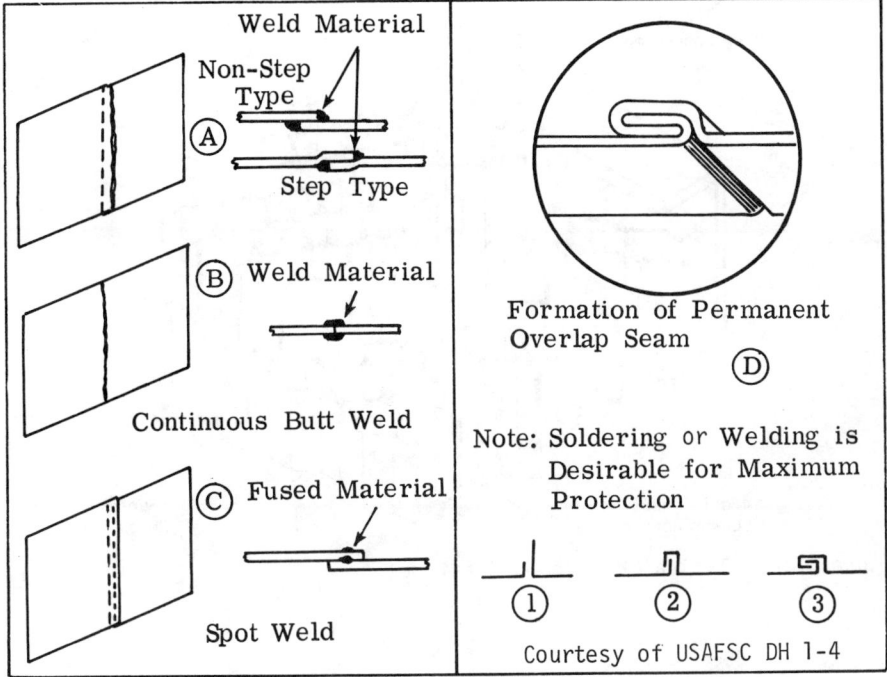

Figure 11.3 - Permanent and Semi-Permanent Shield Seam Configurations

11.1.2 VENTILATION OPENINGS

Most shielding housings or enclosures require either convection or forced-air cooling. Since associated openings will compromise the integrity of the basic shield material, a suitable electromagnetic mask must be sought which will provide substantial attenuation at R-F while not significantly impeding the mechanical flow of air. Two approaches are possible: screened covers and honeycomb aperture covers. As explained in the next section, screens are inexpensive approaches to this problem but are limited in shielding effectiveness and tend to block the flow of air due to turbulance. Thus, a honeycomb material is generally used because it provides higher shielding effectiveness and maintans a streamline flow of air.

In typical honeycomb construction, illustrated in Fig. 11.4, the hexagonal elements use the waveguide-beyond-cut-off technique to accomplish the desired shielding effectiveness. One representative honeycomb configuration is shown in Fig. 11.5. Eq. (11.1) previously indicated the expected attenuation. However, for honeycomb, the shielding effectiveness at frequencies well below cutoff is reduced

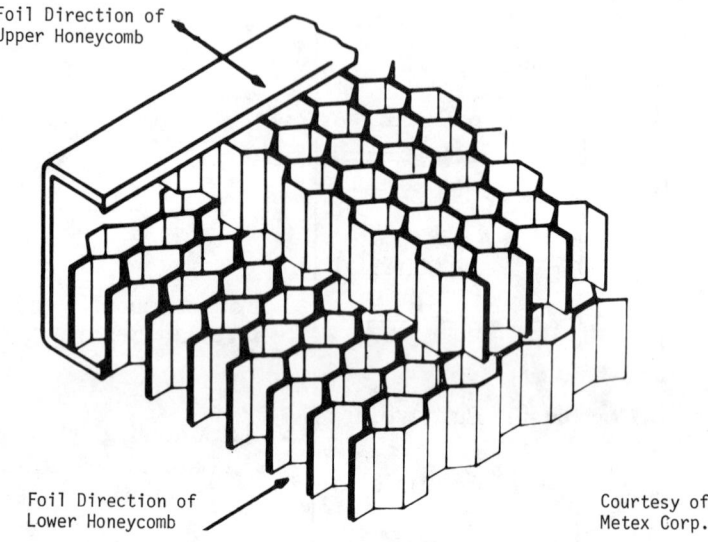

Foil Direction of
Upper Honeycomb

Foil Direction of
Lower Honeycomb

Courtesy of
Metex Corp.

Figure 11.4 - Typical Honeycomb Construction

Courtesy of
Metex Corp.

Figure 11.5 - Representative Honeycomb Configurations

by the number of waveguide elements, N, in the panel since the emerging field from each hexcell coherently combines with its neighbor. Thus, there results for honeycomb ventilation covers:

$$A_{dB} \approx 27\ell/g - 20 \log_{10} N \qquad (11.6)$$

Fig. 11.6 illustrates typical performance of different honeycomb configurations. The L-F magnetic field performance, however, does not follow Eq. (11.6). Rather, the applicable relation is Eq. (10.6).

Sometimes, it is necessary to provide reduction or removal of dust in the ventilation process. Honeycomb construction will not remove dust. Thus, a shield screen is fabricated of a woven-wire mesh. The shielding mesh medium can be either dry (see Fig. 11.7) or wet (to accommodate an oil coating for more dust removal; see Fig. 11.8). Fig. 11.9 shows typical attenuation of shielding mesh covers vs frequency.

When ventilation cover panels are used for convection cooling, it is often common practice to employ a number of perforations in the panel rather than to use honeycomb or screen. Holes are punched out with a die which also cuts the cover panel. For this situation, the shielding effectiveness, A_{dB}, is:

$$A_{dB} = \frac{k\ell}{g} + 20 \log_{10} \left(\frac{C}{D}\right)^2 \qquad (11.7)$$

where,   k = 27 for square perforations (opening holes)

= 32 for circular perforations

ℓ = thickness of cover panel in inches (or cm.)

g = width of square perforations or diameter of circular perforations in inches (or cm.)

C = center-to-center spacing of perforations in inches (or cm.)

D = length of aperture for squares or diameter for circular apertures in inches (or cm.)

If the cover plate perforations are not equally spaced, then C^2 in Eq. (11.7) may be replaced by $C^2 = A/N$, where A = area of aperture = D^2 and N = number of perforations or holes. For this situation, Eq. (11.7) becomes:

$$A_{dB} = \frac{k\ell}{g} - 20 \log_{10} \left(\frac{D^2 N}{A}\right) \qquad (11.8)$$

$$= \frac{k\ell}{g} - 20 \log_{10} N \qquad (11.6)$$

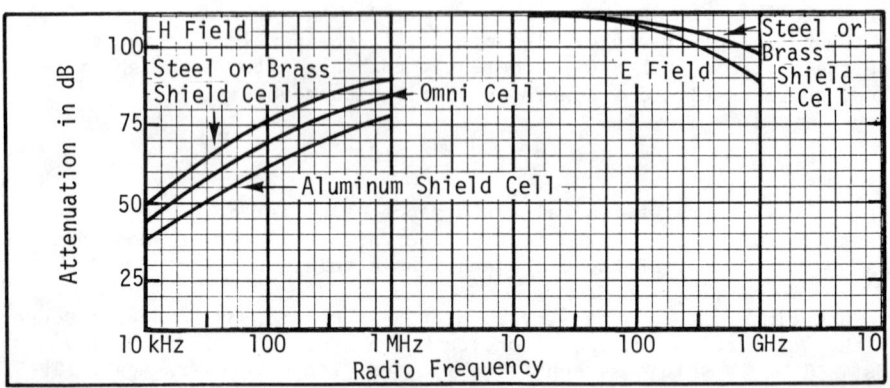

Figure 11.6   - Typical Shielding Effectiveness of Honeycomb
Vent Covers

Both the honeycomb and mesh covers are mounted over the ventilation opening with gasketing material.

11.1.3   Viewing Apertures

Another requirement which compromises the integrity of the basic
shield material is the need for viewing panel meters, digital displays,
scopes, and other types of status monitors or read-out presentations
contained inside the shielded housing or enclosure*. This is accomplished by either a laminated-screen window or a conductive-optical
substrate.

11.1.3.1   Screen Windows

A shield screen window may be used to block R-F penetrations in
which fine knitted wire is laminated between two layers of acrylic or
glass. Fig. 11.10 illustrates this. The wire may be monel with typical sizes of 0.002" diameter (20-25 openings per inch) or 0.0045"
diameter (10-13 openings per inch). This corresponds to a low-shadow
area (15%-20% blockage giving good visibility). Typical shielding
effectiveness is shown in Fig. 11.11. This approach is becoming less
popular to that of the conductive-optical substrate described below
because of the less esthetic aspects of the former. Furthermore, under
some conditions screen window exhibits undesired diffraction-grating
viewing problems.

---

* Such viewing apertures can also be windows in buildings (see Chap.
9), windows in hospital shielded enclosures, canopies in aircraft,
and the like.

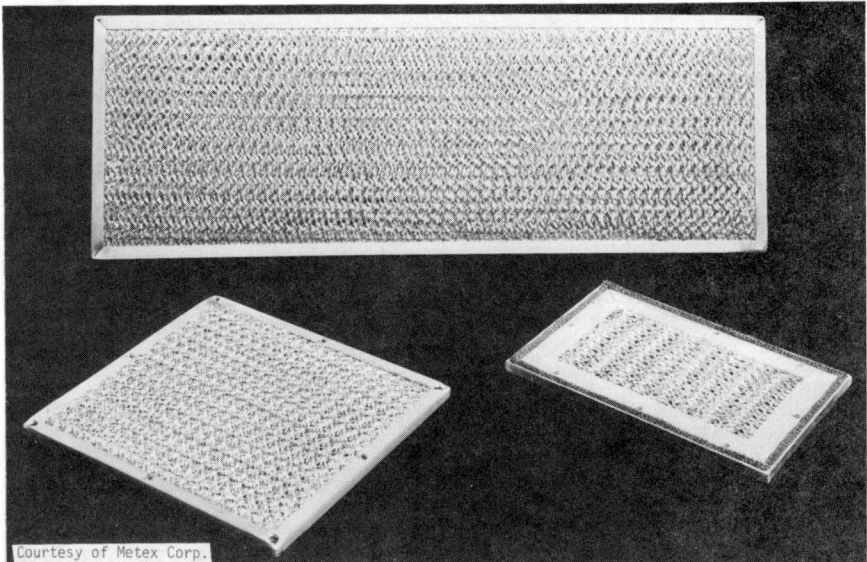

Figure 11.7 - Representative Shield Screen Mesh Ventilation Covers for Air Filtering

Figure 11.8 - Shield Screen Mesh Ventilation Permitting Dust Removal by Oil Impregnation

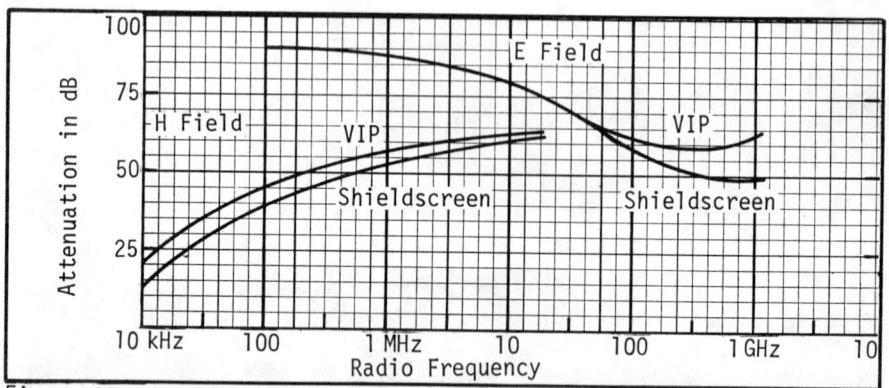

Figure 11.9 - Typical Shielding Effectiveness of Shield Screen
Mesh Vent Covers

11.1.3.2 Conductive Optical Substrate Windows

Another approach is available for providing shielding across apertures through which either optical viewing or the transmission of light
is also necessary. This approach involves the use of a conductive window, a technique in which a thin film of metal is vacuum deposited on
an optical substrate. These conductive window designs such as shown in
Fig. 11.12 are evolved by establishing some or all six basic design
parameters, as applicable:

- Window Material   ● Reticle Requirements
- Conductive Coating   ● EMI Gasketing
- Optical Coating and Finishes   ● Framing and Mounting

Most plastic and glass optical panel materials are suitable as
substrates for the application of conductive-coating.* The commonly
accepted, more standard materials are: glass, acrylic, polycarbonate and
fluorocarbon plastics. The substrates may be clear or colored as required by the application. There are no restrictions on substrate
thickness. Curved or three-dimensional parts can generally be coated.

Most thermosetting and thermoplastic substrates have minute surface scratches produced in their normal manufacture. The application
of the coating will inherently make these more apparent, although
actual user experience indicates no functional problem will arise. The

---

*   Conductive coating can be applied to almost any solid substrate,
making it conductive for use as an EMI shield, switch element, filter
or other active low current carrying device. Acceptable substrates are
those which will not outgas in a high vacuum. A quick test may be made
by checking the substrate for odor. If there is none, it is not likely
to outgas.

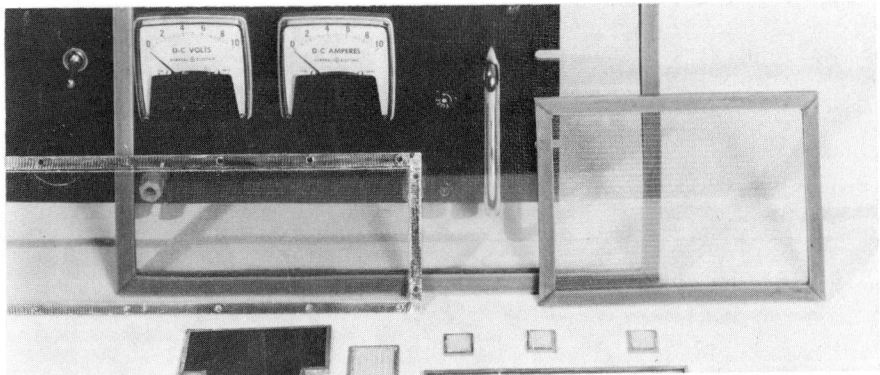

Figure 11.10 - Representative Shield Screen Windows for Viewing

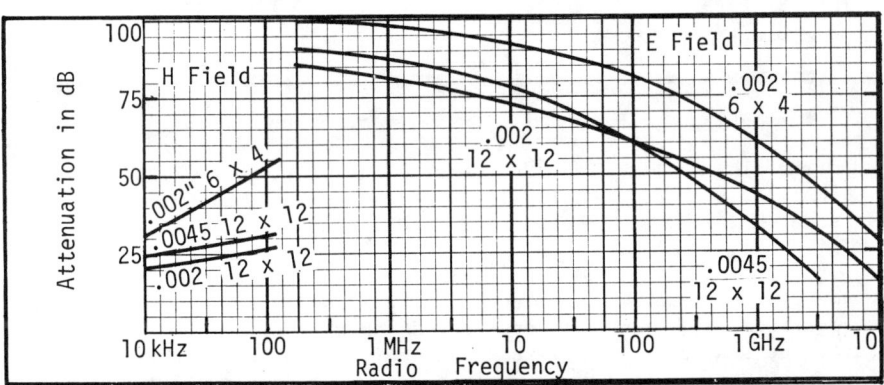

Figure 11.11 - Shielding Effectiveness of Shield Screen Windows

Figure 11.12 - Typical Conductive Optical Viewing Panels

following list illustrates a sample of the large selection of common
substrate materials suitable for conductive coating.

- Glass, Plate
- Glass, single strength
- Glass, float
- Glass, tempered

- Glass laminated, PVB
  film, safety
- Glass, quartz
- Crystals, ruby
- Crystals, quartz
- Vycor[1]
- Pyrex[1]
- Lexan[2]

- Plexiglass, thermoplastic acrylic[3]
- Plexiglass, transparent, colorless
- Plexiglass, frosted, colorless
- Plexiglass, colored: yellow, amber,
  grey, bronze, green, red, blue
- Homalite, thermosetting plastic[4]

- Kapton[5]
- Mylar[5]
- Abcite, coated acrylic[5]
- Polycarbonate
- Self extinguishing plexiglass
- Fluorocarbons

Trademarks of: 1. Corning, 2. General Electric, 3. Rohm & Hass,
4. Homalite, and 5. DuPONT. In the plastic substrate group, the
most scratch resistant materials are ABCITE followed by HOMALITE.

   Polarized filter laminate finishes are available for contrast improvement. Coatings are unaffected by application of laminated circular polarizers. Translucent or frosted finishes, rough in surface
nature, are available. They are best employed on the side opposite
the conductive face. They can only be used for display of rear pro-
jections or where the object is extremely close to the window surface.
Anti-reflective, vacuum-deposited coatings may be applied to windows
before coating.

   Fig. 11.13 illustrates typical shielding effectiveness vs. frequency* for different film coating thicknesses on glass measured in
surface resistance units of ohms/square. Since the film thickness is
deposited in microns, little contribution to attenuation comes from
absorption loss. Accordingly, reflection loss, as previously shown in
Figs. 10.5 and 10.6, is the media of attenuation. Above about the
1 MHz, the loss decreases with an increase in frequency at the rate of
approximately 20-dB per decade and becomes negligible above about
1 GHz.

---

* From Lasitter, Homer A., "Low Frequency Shielding Effectiveness of
Conducted Glass, IEEE Transactions on Electromagnetic Compatibility,"
Vol. EMC-6, No. 2, July, 1964, pp. 17-30.

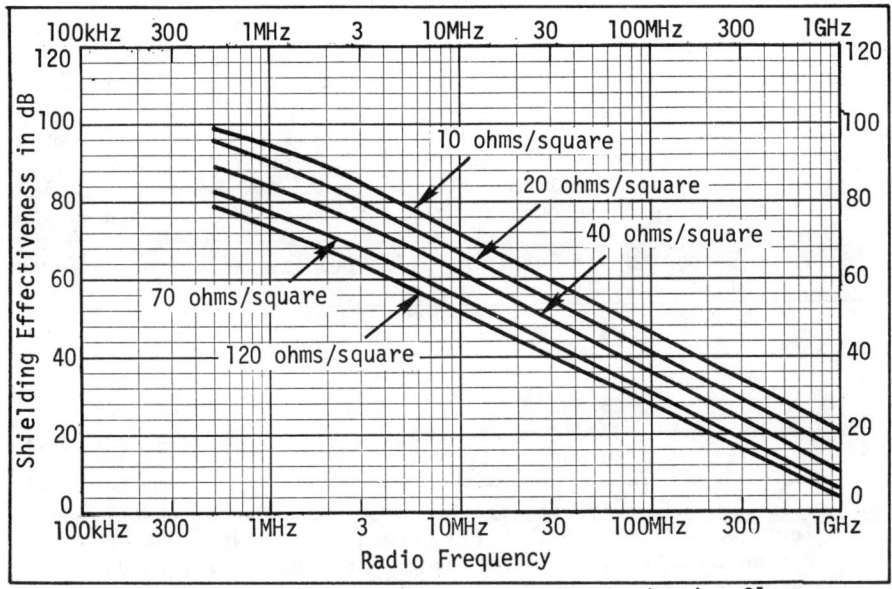

Figure 11.13 - Shielding Effectiveness of Conductive Glass

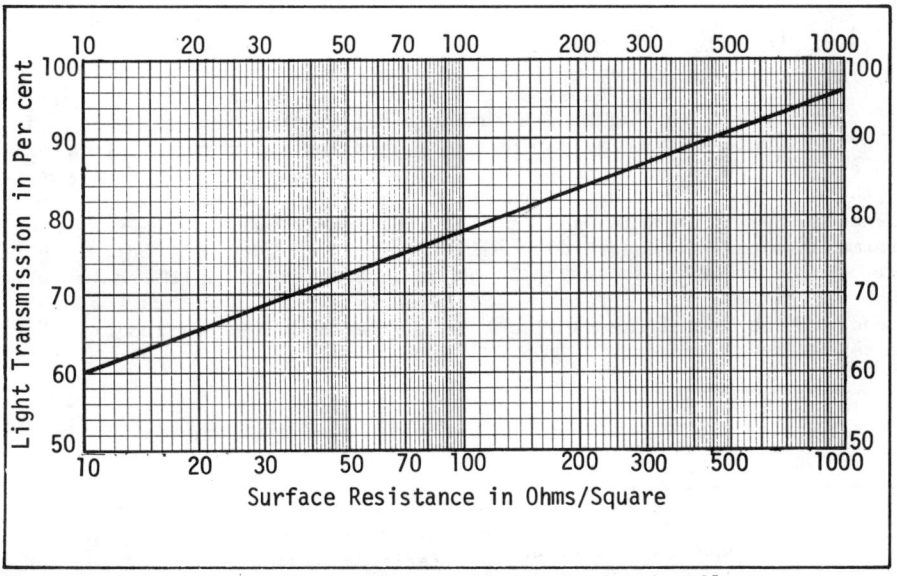

Figure 11.14 - Light Transmission of Conductive Glass

Light transmission vs surface resistance for the above conductive glass is shown in Fig. 11.14. Transmission values of 60-80% correspond to resistances of about 10 to 100 ohms/square. Thus, these values shown in Fig. 11.13 may now be compared with the attenuation data of the shield screen depicted in Fig. 11.11 to determine relative performance for comparable area size specimens. The shield screen is seen to be everywhere superior in shielding effectiveness, as shown in Table 11.1, in which the difference becomes greater with increasing frequency.*

Table 11.1 - Comparison of Shielding Effectiveness of Screen and Conductive Glass Windows

Frequency	Shield Screen	Conductive Glass	Superiority of Shield Screen
1 MHz	98 dB	74-95 dB	3-24 dB
10 MHz	93 dB	52-72 dB	21-41 dB
100 MHz	82 dB	28-46 dB	36-54 dB
1 GHz	60 dB	4-21 dB	39-56 dB

Thus, it is concluded that if significant VHF and UHF attenuation is required for viewing apertures, shield-screen windows should be used. If the esthetics or other considerations do not permit this, conductive glass cannot be relied upon to provide significant R-F attenuation to E-Fields much above 30 MHz.

11.1.4 Control-Shaft Apertures

Another aperture class which compromises the shielding integrity of an equipment housing or instrument panel is that resulting from shafts of potentiometers, tuning dials, and control devices. Generally, an external metallic front panel or housing is either drilled or punched with sufficient clearing tolerance through which the control shaft extends to result in a leaky aperture. The inside wall of the panel hole forms an outer conductor to a coaxially-situated internal control shaft (i.e., the inner conductor). In other words, potential EMI can enter or exit through this effective short-length coaxial line and the extended shaft beyond the panel acts either as a pick-up or radiating antenna.

---

* In reviewing the sales literature of Technical Wire Products, Inc., and other manufacturer's, it is noted that the performance of conductive optical substrates for surface resistivities of the order of 10Ω/sq, more nearly approximates that of the screen. Because measurements were made by different observers, using different test set ups on different specimens variations are expected.

To preserve the shielding integrity of otherwise leaky control-shaft situations, one method of minimizing the degradation of shielding effectiveness is to design a supporting bushing extender to act as a circular waveguide-beyond-cut-off attenuator (cf. Eqs. (11.1) and (11.13)). For 100-dB attenuation in a circular waveguide, the length of the waveguide must be somewhat more than three times its diameter (ℓ/g > in Eq. (11.5)). Fig. 11.15 shows an acceptable use of a metal tube bonded to the wall containing the clearance aperture for control shafts.

If the preceding situation were implemented without regard to the control shaft properties and relations to the added metal tube, little improvement may result for typical metal shafts. This situation corresponds to a low-impedance coaxial line in which an intervening dielectric may result from contaminants such as oil films or oxides. To preclude this from happening, one of two techniques is followed: (1) replace the metallic control shaft with a non-conductive shaft as shown in Fig. 11.15, or (2) use a cylindrical-shim EMI gasket (see Chap. 12) between the shaft and tube. The latter method does not require modification of existing control shafts.

11.1.5   Indicator Buttons and Lamps

Some instruments or equipments require the use of push-buttons, status indicator buttons, and/or indicator lamps. These devices also provide another compromise of shielding integrity by virtue of the required apertures in a front panel or housing. Two techniques are available to mitigate the EMI leakage through such devices:

(1) Encase them in a shielded compartment behind the front panel when they are mounted as shown in Fig. 11.16. Feed-through capacitors or filter-pin conductors are used for hard wiring from outside the compartment to the buttons or indicator lamps since conducted EMI could exist on either side of the barrier.

(2) Use special EMC-designed hardware where such devices are mounted directly to a front panel. Examples of this include wire-mesh indicator lamps (looks like a miniature photo-flash bulb). This mesh serves to reflect and absorb entering or exiting EMI energy.

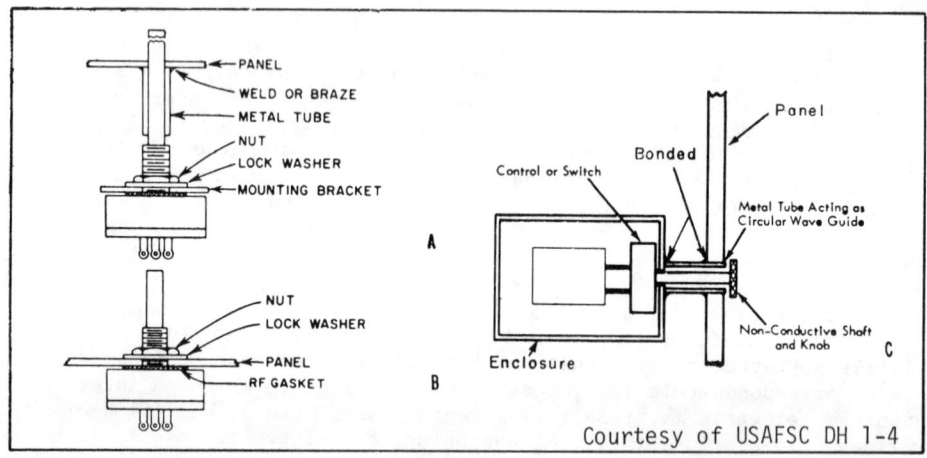

Courtesy of USAFSC DH 1-4

Figure 11.15 - Use of Circular Waveguide in a Permanent
Aperture for Control-Shaft EMI Leakage Control

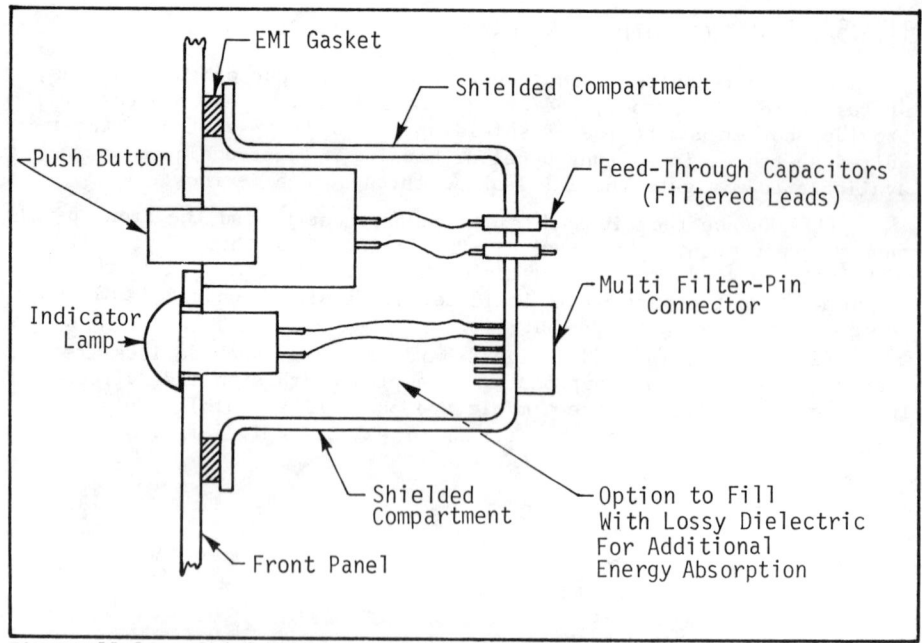

Figure 11.16 - Shielded and Filtered Compartment Technique to Restore
Shielding Integrity of Button and Lamp Apertures

11.2   EMC GASKETS

This section discusses another very important class of techniques used to reinstate loss of shielding integrity at seams and joints where other than permanent fastening methods are permitted.

11.2.1   GASKETING THEORY

Gaskets are employed for either temporary or semi-permanent sealing applications between joints or structures, such as:

Temporary R-F Sealing Applications:

- Securing access doors to enclosures, cabinets, or equipments
- Mounting cover plates or removal panels for equipment maintenance, alignment, or other purposes

Semi-permanent R-F Sealing Applications:

- Mounting either screen or conducted glass windows to housings containing electrical or electronic test equipment
- Mounting honeycomb and other ventilation covers to enclosures, cabinets, or equipment
- Securing parallel members of an equipment housing to a frame structure using machine screws

All gaskets of the non-spring finger stock type, whether they seal EMI, higher-pressure fluid, make a container dunk proof, or simply keep forced ventilating air from escaping at a door-to-cabinet joint, conform to the unavoidable irregularities of the mating surfaces of a joint.   Some examples are:

- The joint between a garden hose and water faucet
- Housing for an emergency radio or beacon to be dropped into the sea
- The joint between the cover and enclosure of a radar pulse modulator

In each example the joint has two relatively rigid mating surfaces, and neither surface is perfectly flat. When the surfaces are mated without a gasket, even high closing forces will not cause the two surfaces to mutually seal. Resultant gaps will allow leaks to exist. A gasket resilient enough to comply to both surfaces under reasonable force, however, will eliminate these leaks. In the garden-hose example, try to prevent a leak by force alone without a gasket. With a gasket placed in the hose fitting against a faucet, even hand torque results in a water tight joint. To try to get the same water-tightness by accurate machining of both surfaces would be prohibitively expensive. Thus, in most cases, the least expensive way to obtain a tight joint (watertight, oil-tight, or EMI tight) is to make the mating surfaces to normal tolerances on flatness, rigidity, and tolerance build-up, and then to add a gasket to compensate for the resulting misfits between the two surfaces.

11.2.1.1   Joint Unevenness

The degree of mis-alignment or misfit of the mating surfaces is commonly called *joint unevenness* and is designated ΔH in Fig. 11.17(a). It is the *maximum* separation between the two surfaces when they are just touching and in the limit becomes the sum of the peak irregularities of both surfaces. If the surfaces are not rigid then the joint unevenness also includes any additional separation between the two surfaces due to joint distortion when pressure is applied.

Fig. 11.17(b) shows the same joint with a gasket installed. The dashed lines indicate the gasket height, H_g, before compression. The compressed minimum gasket height, H_{min}, occurs at the point where the surfaces would touch without a gasket. Compressed maximum gasket height, H_{max}, is at the point of maximum joint separation. Thus, joint unevenness of the mating surface is:

$$\text{Joint Unevenness} = \Delta H = H_{max} - H_{min} \qquad (11.9)$$

11.2.1.2   Required Compression Pressure

Three factors determine the required compression pressure on a gasket, viz, its resiliency, the minimum pressure required for a seal, and the total joint unevenness.

(A) Resiliency

Resiliency is the amount by which a gasket compresses per unit of applied compression pressure. Resiliency is usually expressed in percent of original (uncompressed) gasket height divided by pressure in psi. A soft gasket would compress more than a hard gasket with the same applied pressure. Stated in another way, a soft gasket requires less pressure than a hard gasket to compress the same percentage of gasket height. For example, a sponge neoprene gasket might compress 10% under an applied compression pressure of any 6 psi, but a solid neoprene gasket would require 40 psi for the same 10% deflection as shown in Fig. 11.18.

(B) Minimum Pressure for Seal

A gasket must at least make contact at the point of maximum separation between mating surfaces, i.e., $H_{max} \leq H_g$, in Fig. 11.17. Actually, the pressure at this point must be stated minimum amount in order

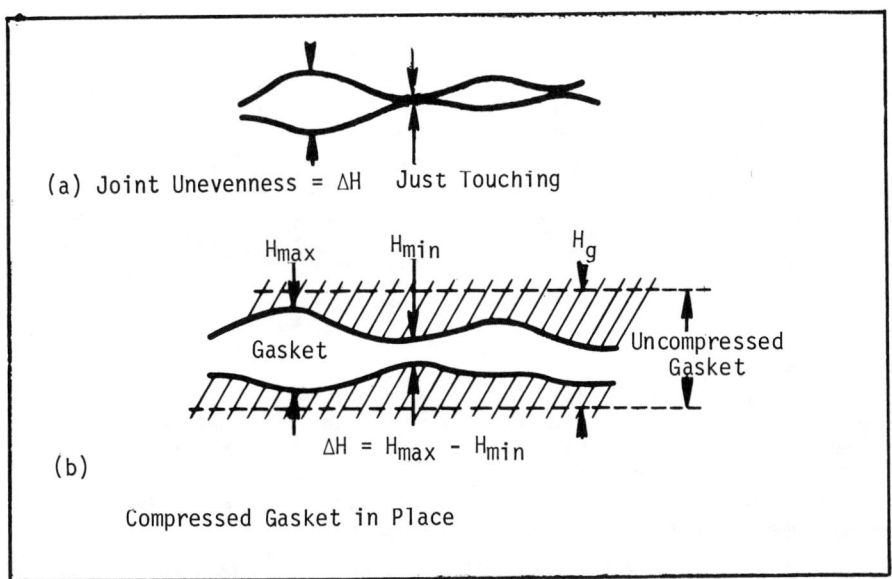

Figure 11.17 - Description of Joint Unevenness

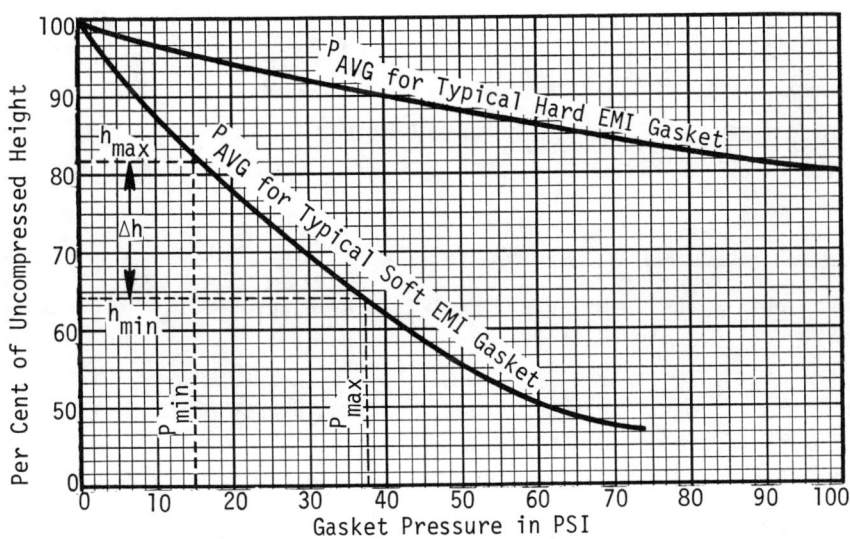

Figure 11.18 - Typical Hard and Soft EMI Gasket Height vs Pressure Relations

to assure an EMI seal. This is easy to understand in the case of a
high-pressure lubricating system. If there is not some required minimum pressure at the point of H_{max}, oil would blow by between the
flanges and the gasketing material. Thus, the pressure at the H_{max}
point must be high enough to prevent blow-by. For EMI gaskets, this
minimum pressure, P_{min}, is determined by the pressure required to
break through corrosion films and to make a suitable low-resistance
contact. P_{min} is typically about 20 psi, but can be as small as 5 psi.

(C) Average Pressure

The average pressure applied to the gasket must also be large
enough to compress the overall gasket so that the difference between
the minimum height and the maximum gasket height (determined by P_{min}
from the previous paragraph) is equal to the joint unevenness, i.e.,
$\Delta H = H_{max} - H_{min}$, as previously presented in Eq. (11.9). In general,
the average pressure should equal or exceed that corresponding to the
average compressed gasket height, H_{avg}:

$$H_{avg} = (H_{max} + H_{min})/2 \qquad\qquad (11.10)$$

and,

$$P_{avg} = (P_{min} + P_{max})/2 \qquad\qquad (11.11)$$

The required compression force, F, in units of pounds, may be
calculated from P_{avg} by determining the surface area of the gasket to
be sandwiched between the mating members:

$$F = P_{avg} A \text{ pounds} \qquad\qquad (11.12)$$

where,   A = gasket area in sq. in.

11.2.1.3 Required Gasket Height

To obtain the required EMI seal from a gasketed joint, the gasket
height must meet these criteria:

The pressure at the point of maximum joint separation
(H_{max}) must correspond to the minimum pressure to obtain
the required EMI seal.

The difference between maximum and minimum compressed
heights of the gasket must equal the joint unevenness
of the mating surfaces.

If the average pressure, available to compress the gasket is P_{avg},
the maximum pressure, P_{max}, is obtained from Eq. (11.11):

$$P_{max} = 2P_{avg} - P_{min} \qquad\qquad (11.13)$$

where, P_{min} is a selected value at least as much as that
          required to obtain a gasket seal.

The percentage of uncompressed height corresponding to P_{min} and
P_{max} in Fig. 11.18 are H_{max} and H_{min}, respectively. To calculate the
required uncompressed gasket height, H_g, as a dimension:

$$H_g = \frac{\Delta H_{inches}}{\Delta H_{decimal}}$$ (11.14)

Thus, the required height is the actual joint unevenness in inches divided by the joint unevenness expressed in decimal equivalent of percent gasket compression (See Fig. 11.18).

Illustrative Example 11.1

Suppose that for a specified soft EMI gasket, the minimum required
pressure for a seal is 10 psi. Picking a value somewhat greater than
this, P_{min} is selected at 15 psi as shown in Fig. 11.18. This corresponds to a maximum uncompressed gasket height of 81%. If the average
pressure available to compress the gasket is 26 psi, the maximum pressure from Eq. (11.13) is:

$$P_{max} = 2 \times 26 \text{ psi} - 15 \text{ psi} = 37 \text{ psi}$$

The corresponding minimum uncompressed gasket height from Fig. 11.18
is 65%. Thus:

$$\Delta H_{decimal} = 81\% - 65\% = 16\% = 0.16$$

The required gasket height for a joint unevenness of 0.032 inches is
obtained from Eq. (11.14):

$$H_g = \frac{0.032}{0.16} = 0.20 \text{ inches}$$

11.2.1.4 COMPRESSION SET

Some gaskets do not return to their original uncompressed height
after release of compression. This is called compression set. It may
be visualized by assuming that the lower curve shown in Fig. 11.18 applies for a particular soft gasket. When compression pressure is removed, the gasket returns to a lesser height whose properties might
look somewhat like the upper curve in Fig. 11.18 (this is exaggerated
for illustrative purposes). The importance of compression set depends
upon how the gasket is to be used; three classes of use are now defined:

Class A, Permanently closed. Compression set is unimportant since
the gasketed component will in all probability never be removed.

Class B, Repeated identical open-close cycles (e.g., hinged door, or symmetrical covers). Here, compression set problems are marginal; further examination of details, however, is indicated.

Class C, Completely interchangeable (complete freedom to reposition gasket on repeat cycles; e.g., round gasket in waveguide). Since the compression-set height at a point of maximum compression may end up being less than minimum compressed height, no contact at all would result between gasket and mating surfaces at this point. For Class C uses, do not reuse gaskets with compression set limits; instead, use a new gasket.

11.2.2   Gasket Types and Materials

There exists a plethora of EMI gasket types, shapes, binders and materials. In fact, the profusion of gaskets is so great that it is likely to be confusing to all but those who specify or use them on some degree of regularity. This is recognized by the suppliers to the point where they have produced creditable application notes and design and order guides.

For convenience of discussion here, EMI gaskets are divided into four types: (1) knitted-wire mesh, (2) oriented immersed wires, (3) conductive plastics and elastomers, and (4) spring-finger stock. The last type is different from the first three types and operates on a significantly different principle. A brief summary of each is presented below followed by a comparison of all four types.

11.2.2.1   Knitted-Wire Mesh Gaskets

Fig. 11.19 shows some examples of knitted-wire mesh gaskets. They are made from resilient, conductive, knitted-wire and somewhat resemble the outer jacket of a coaxial cable. Nearly any metal that can be produced in a fine-wire form can be fabricated into these EMI gaskets. Typical materials used are monel; aluminum; silver-plated brass; and tin-plated, copper-clad steel. These gaskets may employ either an air core or, for maximum resiliency, they may use a spongy neoprene or silicone core. Cross sections may be round, rectangular, or round with fins for mounting. They are generally applied to shielding joints having a periphery of greater than 4" and cross sections between 0.063" and 0.75".

11.2.2.2   Oriented-Immersed Wire Gaskets

Fig. 11.20 shows some examples of oriented immersed-wire gaskets. They are made with a myriad of fine parallel, transverse-conductive wires whose parallel impedance across the gasket interface is very low.

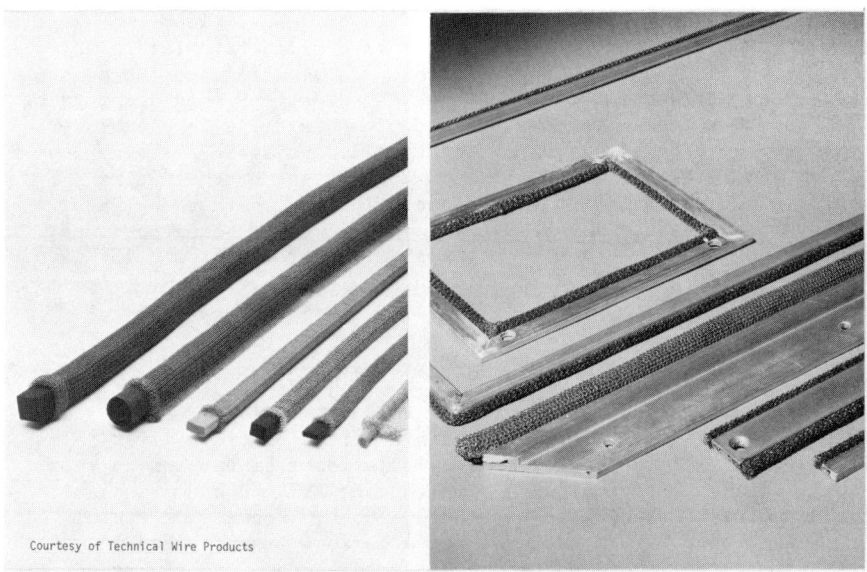

Courtesy of Technical Wire Products

Figure 11.19 - Typical Knitted Wire-Mesh Gaskets

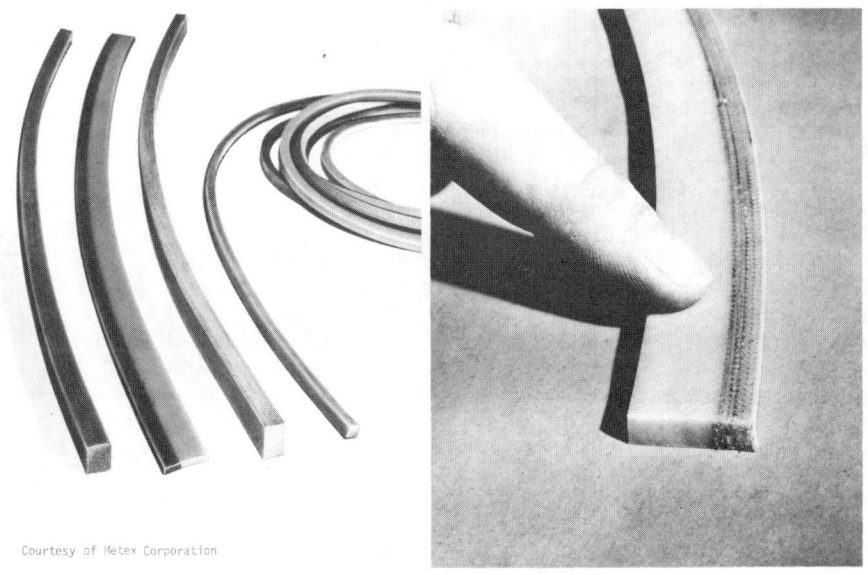

Courtesy of Metex Corporation

Figure 11.20 - Typical Oriented Immersed-Wire Gaskets

Each convulated wire is insulated from its neighbor. They represent
a density of about 1000 wires per sq. in. Typical materials used
are monel or aluminum embedded in either a solid silicone (hard
gasket) or a sponge silicone (soft gasket) elastomer. As such, this
gasket provides a simultaneous EMI and pressure seal. The embedded
wires protrude a few mills on either side to assist in piercing any
residual grease/oil film and oxide on the surface of the mating
members. This characteristic is especially good where aging and
subsequent maintenance may result in a panel member being no longer
clean and degreased. Cross sections available range from 0.125" x
0.125" to 0.625" x 0.500", and come in any lengths.

11.2.2.3 Conductive Plastics and Elastomer Gaskets

Fig. 11.21 shows some examples of conductive plastic and
elastomer gaskets. They are made with a myriad of tiny silver balls
immersed in a silicone rubber or vinyl elastomer binder and carrier.
As such, this gasket provides a simultaneous EMI and hermetic seal.
Offering volume resistivities from 0.001 to 0.01 ohm-m and useful
over a wide range of temperature, these gaskets are provided in
sheets, die cuts, molded parts and extruded shapes. Some versions
are operable down to cryogenic temperatures. They offer low
closing pressures, low compression set and maintenance, and long
life.

11.2.2.4 Spring-Finger Stock Gaskets

Fig. 11.22 shows some examples of beryllium copper, spring-
finger gaskets stamped into different configurations. Basically,
gaskets similar to these were introduced over 30 years ago, and
were the first type of EMI gasket appearing on the market. Since
there existed little elastomer technology in the 1940's, it is
natural that joint unevenness could be accommodated by a series of
individual *fingers*, each capable of flexing a different amount.
Thus, shielded enclosures, cover plates and other heavy-duty
applications used and still use this type of gasket. Recent design
changes, shown in Fig. 11.22, make this type of gasket more competitive with the other gaskets. The new spring-finger contact strips
now offer self-adhesive backing to eliminate older mechanical
fastening methods. They are available in a wide variety of sizes
and shapes. Principal disadvantages are tendency of the fingers to
oxidize and to break-off.

11.2.2.5 Pressure-Sensitive, Foam-Backed Foil Gaskets

Another type of gasket differing from the above is a beryllium-
copper foil backed by a highly-compressible neoprene foam. The

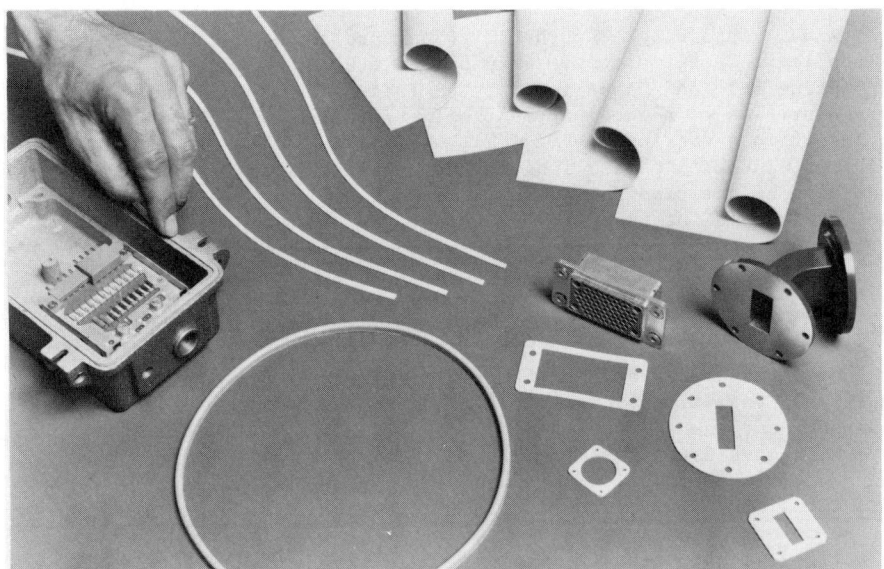

Figure 11.21 - Typical Conductive Elastomer Gaskets

Figure 11.22 - Typical Spring-Finger Strip Gaskets

foam side, containing a synthetic rubber pressure-sensitive adhesive, is applied to cover plates. When placed over an electronics package containing shielded compartments, the foam-backed foil assumes the irregularities of the compartment heights including outside plates to result in a continuous EMI seal. This 1/16" gasket is available in sheet widths to 6" and lengths to 3H or may be die cut. EMI shielding effectiveness of 90 db to electric fields is claimed over the 1 kHz to 10 GHz frequency spectrum. Price range is $6.00 (in 1000 sq. ft. quantity) to $12.00 per sq. ft. (5 sq. ft. quantity).

11.2.2.6 Comparison of Gasket Types and Materials

With the profusion of different gasket types and materials (over 1000 variations), it is confusing to the design or specification engineer tasked with the responsibility of selecting one or more best candidates for his particular application. Accordingly, Table 11.2 is a comparison of some of the principle characteristics of EMI gaskets.* No one type is the best for all applications. For example, those gaskets having relatively low cost tend to have relatively higher volume resistivity resulting in a less impressive shielding effectiveness. Some gaskets are designed to operate down to cryogenic temperatures or up to 500°F, but not both. Since there exists several different methods of mounting, gaskets are available in sheets and strips, die cuts, molded shapes and extruded forms. At the risk of generalizing, conductive plastics and elastomers seem to offer the widest range of applications and price.

To assist the reader in further assessment of EMI gaskets, summary sheets offered by two of the leading gasket manufacturers are presented in Tables 11.3 and 11.4. No attempt has been made to standardize or to audit their data.

---

*   One manufacturer makes a pressure-sensitive foam-backed shielding foil gasket. Sufficient information was not available to include in Table 11.2.

Table 11.2 - Comparison of Gasket Types & Materials

Comparison Factors / Gasket Types			Knitted Wire Mesh	Oriented Immersed Wires	Conductive Plastics and Elastomers	Spring Finger Stock
Available Forms			Strips, Jointless Rings	Strips & Sheets. Jointless Rings Die-cut Shapes	Strips & Sheets. Die-Cut, Molded, Extruded Shapes	Strips
Size	Periphery		>4"			Any
	Cross Section	Min	0.063"			
		Max	0.750"			
Type of Seal	EMI only		Good to Excellent	Good	Also seals Hermetically	Good to Excellent
	EMI plus Hermatic		NA	Fair to Excellent	Good to Excellent	NA
Conductive Material			Silver Plate, Mo-nel, Alum-inum, Steel SN/CU/FE	Monel, Aluminum	Many Tiny Silver Balls	Beryllium-Copper
Binder or Core Material			Rubber, Air Core, Neoprene, Silicone Sponge	Solid & Sponge Si-licone	Silicone or Plastic	NA
Temperature Range			Limited to Core	-70oF to 500oF	-100oF to +400oF	-65°F to 100°F
Available Gasket Heights			.062" to 0.500	.062" to 1.000"	0.020" to 0.160"	.062" to 0.400
Joint Unevenness Accommodations			.020" to 0.160"	.010" to 0.100"	0.003" to 0.030"	.035" to 0.250"
Compression Height Range						7:1
Compression Pressure			5 psi to 100 psi	20 to 100 psi	20 to 100 psi	
EMI Shielding Performance	10 kHz (H)		25-30dB	>45dB	>35dB	>10dB
	10 MHz		>100dB	>100dB	>100dB	>120dB
	1 GHz		>90dB	>90dB	>95dB	>100dB
	10 GHz				>70dB	>100dB

	ECCOSHIELD SV	ECCOSHIELD SV-R	ECCOSHIELD SV-M	ECCOSHIELD SV-P	ECCOSHIELD SV-RS	ECCOSHIELD GLV	ECCOSHIELD SV-RT	ECCOSHIELD SV-MAG	ECCOSHIELD SV-C
CLAIM TO FAME	ORIGINAL PLASTIC RF GASKET SPEC'D IN EVERYWHERE, AVAILABLE IN MANY FORMS	INCREASED TEMP CAPABILITY FOR LIGHT CLOSING PRESSURES	RETAINS HIGH TEMP OF SV-R BUT MUCH LOWER COST, ALSO LOW DUROMETER	LOWEST COST SV-SILICONE, LARGE GASKETS AND LARGE CROSS SECTIONS ECONOMIC	THE ONLY METAL RUBBER FILLED RUBBER WHICH EASILY COMPRESSES	HIGH PERFORMANCE CARBON SYSTEM LOW COST, OF COURSE	CONTINUOUS LENGTHS, MANY FORMS, EXCELLENT PHYSICALS FOR DOORS, CABINETS	SIMPLE WAY TO HOLD COVER OR DOOR IN PLACE WITHOUT HINGES/LATCHES, RF SEAL	POUR YOUR OWN RF GASKETS IN PLACE OR IN SIMPLE MOLDS
BASIC COMPOSITION	SILVER + VINYL ELASTOMER	SILVER + SILICONE RUBBER	SILVER + SILICONE RUBBER	SILVER + FILLER + SILICONE RUBBER	STAINLESS STEEL + SILICONE RUBBER	CARBON + VINYL ELASTOMER	RUBBER SHEATHED IN SILVER-FILLED ELASTOMER	PLASTIC MAGNET SHEATHED IN SILVER-FILLED ELASTOMER	SILVER + SILICONE RUBBER
FORMS AVAILABLE	FLAT SHEET & GASKETS TO, ROD, TUBE, O-RINGS, EXTRUSIONS	FLAT SHEET & GASKETS	FLAT SHEET & GASKETS	FLAT SHEET & GASKETS, O-RINGS, MOLDED GASKETS	FLAT SHEET & GASKETS	LARGE FLAT SHEETS & GASKETS	ROD, TUBE, EXTRUSIONS, MOLDED GASKETS, SPONGE STRIP	STRIPS 1/2" x 1/4" x 2" + FOAM-BACKED VERSION	CASTABLE LIQUID
COLOR	SILVER	SILVER-TAN	SILVER	TAN	GRAY	BLACK	SILVER	SILVER	SILVER-GRAY
DENSITY, g/cc	2.6	3.5	1.1	1.7	1.3	1.4	VARIOUS	(NOT APPLICABLE)	1.8
SERVICE TEMP, °C	-65 to +125	CRYOGENIC TO +250	CRYOGENIC TO +250	CRYOGENIC TO +260	CRYOGENIC TO +260	-65 TO +125	-65 TO +125	-65 TO +125	CRYOGENIC TO +250
DUROMETER, SHORE A	65	40	45	85	75	85	VARIOUS (SOME VERY LOW)	—	75
RELATIVE COMPRESSION SET	GOOD	VERY GOOD	VERY GOOD	EXCELLENT	EXCELLENT	GOOD	EXCELLENT	—	EXCELLENT
VOLUME RESISTIVITY, ohms·cm	2×10^{-3}	1×10^{-3}	$<1 \times 10^{-2}$	$<8 \times 10^{-2}$	$<8 \times 10^{-2}$	5.0	—	—	$<8 \times 10^{-2}$
RESISTANCE THROUGH GASKET, ohms (Note 1)	2×10^{-5}	1×10^{-3}	$<1 \times 10^{-4}$	$<8 \times 10^{-4}$	$<8 \times 10^{-4}$	5×10^{-3}	—	—	$<8 \times 10^{-4}$
BOND TO SELF	ECCOSHIELD VSV	ECCOSHIELD SV-C	ECCOSHIELD SV-C	ECCOSHIELD SV-C	ECCOSHIELD SV-C	ECCOSHIELD VSV	—	—	ECCOSHIELD SV-C
BOND TO METALS	ECCOSHIELD VSM/VCA	ECCOSHIELD RVS	ECCOSHIELD RVS	ECCOSHIELD RVS	ECCOSHIELD RVS	ECCOSHIELD VSM	ECCOSHIELD VCA	ECCOSHIELD VSM	ECCOSHIELD RVS
ELONGATION AT RUPTURE, %	>200	>50	>50	>40	>75	>160	VARIOUS (MOST VERY HIGH)	(NOT APPLICABLE)	>30
TENSILE STRENGTH AT RUPTURE, psi	1000	150	150	250	250	1300	VARIOUS (MOST VERY HIGH)	—	200
THERMAL CONDUCTIVITY BTU/hr/ft²/in	25	30	3	4	6	6	—	—	4
RELATIVE COST	MEDIUM	HIGH	MEDIUM	LOW	MEDIUM	VERY LOW	LOW	MEDIUM	VERY LOW
INSERTION LOSS IMPROVEMENT (Note 2)									
MAGNETIC FIELD, 200 kHz	EXCELLENT	EXCELLENT	EXCELLENT	GOOD	GOOD	NOT RECOMMENDED	EXCELLENT	GOOD	GOOD
ELECTRIC FIELD, 1 MHz	EXCELLENT	EXCELLENT	EXCELLENT	EXCELLENT	GOOD	GOOD	EXCELLENT	EXCELLENT	EXCELLENT
PLANE WAVE, 400 MHz	EXCELLENT	EXCELLENT	GOOD	GOOD	GOOD	GOOD	EXCELLENT	EXCELLENT	EXCELLENT

Note 1: Resistance between surfaces of sheet 1" x 1" x 0.030" Note 2: Insertion loss compared with a non-conductive gasket of the same dimensions, used to seal a metal box (SAE Procedure)

Courtesy Emerson & Cuming, Inc.

Table 11.3 – Typical EMI Gasket Selection Chart (E&C Inc)

PRODUCT TRADE NAME	POLASTRIP	POLASHEET POLASTICK	POLARING	POLA-H	EMI Strips (all metal)	Compressed EMI Gaskets	EMI Strips (Elastomer core)	Combo Strips	Combo Gaskets	Metalex	Perxapine Metalastic	Metalastic	Xecon
Schematic Cross Section													
Construction	Oriented wire in Matrix of Silicone Elastomer (Polastick has pressure sensitive adhesive)				Formed or Compressed Knitted Wire Mesh		Knitted Wire Mesh Over Elastomer Strips	Formed Knitted Wire Strips with Elastomer Strips or Die Cut Elastomer		Formed Knitted Wire with Joined EMI Seal plus Aluminum Extrusions	Expanded Metal in Elastomer	Woven Wire in Elastomer	Silver on Insert Substrate in Elastomer
Available Forms Supplied by Metex	Strips, Gaskets Made by Joining Strips	Sheets, Die Cut Gaskets	Jointless Rings	Solid Metal Paths in Elastomer plus Special Contact Material	Strips, Gaskets Made by Joining Strips	Jointless Rings or Rectangular Gaskets	Strips, Gaskets Made by Joining Strips	Strips, Gaskets Made by Joining Strips	Die Cut Elastomer with Joined EMI Strips	Strips, Fab Lengths, Frames with Joined EMI Strips	Sheets, Die Cut Gaskets	Sheets, Die Cut Gaskets	Sheets, Die Cut Gaskets, Molded Gasket, Strips
Type of Seal — EMI only	Good	Good	Good	Excellent	Good-Excellent	Good-Excellent	Good	Good-Excellent	Good-Excellent	Good-Excellent	Good-Excellent	Fair-Good	Good-Excellent
EMI plus rain tight, drip proof, contain ventilating air	Good	Good-Excellent	Good	Excellent	No	No	Fair-Good	Good-Excellent	Good-Excellent	Excellent	Good-Excellent	Fair-Good	Good-Excellent
EMI plus pressure to 30 psi	Excellent	Good	Excellent	Excellent	No	No	No	Good-Excellent	Good-Excellent	Good	Good	Fair	Excellent
EMI plus pressure over 30 psi	Fair	Fair	Good-Excellent	Fair	No	No	No	Fair	Excellent	Good-Fair	Fair	Poor	Excellent
EMI Rating — 14 kHz (H)	>46 dB	>35 dB	>46 dB	>56 dB	>20->30 dB	>25->30 dB	>25->35 dB	>30 dB	>30 dB	>30 dB	>35 dB	>35 dB	>36 dB
18 MHz (E)	>102 dB	>102 dB	>102 dB	>102 dB	>102 dB	>102 dB	>102 dB	>102 dB	>102 dB	>102 dB	>102 dB	>102 dB	>102 dB
1.0 GHz (P)	>93 dB	>93 dB	>93 dB	>93 dB	83->93 dB	>93 dB	>93 dB	>83->93 dB	>83->93 dB	>93 dB	>85 dB	>40 dB	>93 dB
Maximum Joint Unevenness % of Gasket Height — Class A – Permanently Closed	20%	20%	20%	20%	30-40%	30%	30-50%	30%	30%	30%	15%	10%	25
Class B – Open-Close in same position	17%	17%	17%	17%	25-30%	25%	25-40%	30%	25%	25%	10%	7%	20
Class C – Completely interchangeable	17%	17%	17%	17%	20-25%	20%	20-30%	25%	25%	25%	10%	7%	20
Minimum/Maximum Height	.062"/.625"	.055"/.250"	.062"/.625"	.125"/.625"	.062"/.500"	.040"/.375"	.125"/.625"	.062"/.375"	.062"/.375"	.156"/.250"	.020"/.030"	.016"/.020"	.020"/.250"
Minimum width (greater of actual dim. or portion of height)	.062"/½ H	.125	.062"/½ H	.125"/½ H	.062"/½ H	.062"/½ H	.125"/½ H	.125"/1½ H	.125"/1½ H	.500"	.140"	.125	.090"/H
Recommended Compression Pressure (psi)	20-100	20-100	20-100	20-100	5-100	5-100	5-100	20-100	20-100	5-100	20-100	20-100	20-100
Attachment or Positioning — In slot	Good	Possible	Excellent	Good	Excellent	Excellent	Excellent	Excellent	Excellent	No	No	No	Good
Pressure Sensitive Adhesive	Combo Version Only	Excellent	N/A	Combo Version Only	N/A	N/A	N/A	Excellent	Excellent	N/A	N/A	N/A	N/A
Bond non-EMI Gasket portion (4)	Poor Because Conductive Adhesives Do Not Bond To Silicone				Version with Fins Only (See Note #2)	N/A	Versions with Fins Only (See Note #2)	N/A	N/A	Poor (See Note #3)	N/A	No	Poor
Conductive Adhesive	N/A	N/A	Good If Wide Enough	Excellent	Poor to Good	Poor to Good	Poor to Good	Good-Excellent	Good-Excellent	N/A	N/A	N/A	N/A
Both thru Bolt holes	Excellent	Excellent	Excellent	Excellent	Possible	Good If Wide Enough	Possible Fin Versions	Excellent	Good If Wide Enough	Excellent	Excellent	Excellent	Excellent
Elastomer Temperature Range — Neoprene Version	N/A	N/A	N/A	N/A	N/A	N/A	-30°F to 150°F	-30°F to 150°F	-30°F to 150°F	-30°F to 150°F	N/A	-40°F to 225°F	N/A
Silicone Version	-70°F to 500°F	-80°F to 400°F	-70°F to 500°F	-70°F to 390°F	-80°F to 400°F	-80°F to 400°F	-80°F to 400°F	-80°F to 400°F	-80°F to 400°F	-80°F to 400°F	-80°F to 400°F	-65°F to 500°F	-80°F to 400°F
Standard Metals Available in EMI Portion (others also available)	Monel, Aluminum	Monel, Aluminum	Monel, Aluminum	Tin Plated Copper	Monel, Ferrex™ (1), Aluminum	Monel, Ferrex™ (1), Aluminum	Monel, Ferrex™ (1), Aluminum	Monel, Ferrex™ (1), Aluminum	Monel, Ferrex™ (1), Aluminum	Monel, Ferrex™ (1)	Monel, Aluminum	Aluminum Only	Silver on Insert Substrate

(1) Ferrex™ is the Metex name for tin plated, copper clad steel EMI gasketing. (2) Two versions, ● and ● have fins especially designed for easy attachment. (3) The aluminum extrusion is intended as a convenient attachment method. (4) Most products for which this method is suitable are available from Metex with "dry back" (solvent activated adhesive) adhesive already applied.

Table 11.4 – Typical EMI Gasket Selection Chart (Metex Corp)

11.2.3 Gasket Selection and Mounting

EMI gasket selection involves making suitable matches and trade-offs between (1) available EMI gasket materials and their characteristics (see Tables 11.2 - 11.4), and (2) performance requirements of equipment and design constraints of mating surfaces. Gasket mounting (and hence selection) involves a number of alternatives.

11.2.3.1 Gasket Selection

In selecting one or more suitable EMI gaskets for sealing mating surfaces, gasket characteristics, application requirements and constraints and price are the major considerations. These topics are summarized as follows:

Application Requirements. This is usually stated in the form of equipment performance specifications. They include amount of shielding, pressure sealing, and environmental exposure (e.g., temperature, salt spray, ambient pressure, and corrosive material).

Application Constraints. This is usually imposed by equipment housing design. They include space available, compression force, joint unevenness, contact surface characteristics, and attachment possibilities.

The important matches and trade-offs between application requirements and constraints on one hand and gasket characteristics and price on the other are:

● Gasket height and compressibility must be large enough to compensate for joint unevenness under the available force.

● The gasket must be capable of providing the required EMI sealing and hermetic sealing (when applicable) when compressed by the available force.

● There must be sufficient space for the gasket within the design limitations of the application.

● The gasket must be attached or positioned by a means that fits in with the joint design.

● The metal portion of the EMI gasket must be sufficiently corrosion resistant and compatible with the mating surfaces.

● The EMI gasket must meet the temperature and other environmental needs of the equipment specifications.

Gasket manufacturers and suppliers provide design guide tables to assist the user to select the gasket most nearly meeting the application requirements and constraints. Among the leading EMI gasket manufacturers in U. S. are:

Chomerics, Inc.
77 Dragon Court
Woburn, Mass.   01801
Phone: 617-935-4850

Emerson & Cumming, Inc.
Canton, Mass.   02021
Phone: 617-828-3300

Instrument Specialties Co., Inc.
254 Bergen Street
Little Falls, N. J.   07424
Phone: 201-256-3500

Radcon Corporation
246 Columbus Avenue
Roselle, N. J.   07203
Phone: 201-241-5550

Metex Corporation
970 New Durham Road
Edison, N. J. 08817
Phone: 201-287-0800

Tapecon, Inc.
475 River Street
Rochester, N.Y.   14612
Phone: 716-621-8400

Technical Wire Products, Inc.
128 Dermody Street
Cranford, N. J.   07016
Phone: 201-272-5500

On the topic of price, EMI gaskets cover a wide range depending upon type, size, quantity and performance. However, in order to give some idea of price range, the following remarks are offered. For silver-silicone elastomers having sheet thicknesses of 0.010" to 0.060", prices may range from $17 to $130 per sq. ft. in small quantity. While unit price comes down to about 30-40% in large quantities, a tooling die charge from about $100 to $500 would be spread over the cost of the entire lot. Knitted-wire mesh gaskets will range from about 35¢ to about $4.00 per foot depending upon uncompressed diameter, elastic core if any, and methods of adhesion. Typical custom gaskets of all types about 6"x8" in size will cost about $4-16 each and about 40% of this in production. Several suppliers will provide at no charge a sample of their gaskets. Alternatively, some supply a kit of different sizes and shapes for about $10 to $25.

11.2.3.2   Gasket Mounting

A number of methods is available to position the gasket to a metal mating surface: (1) hold in slot, (2) pressure-sensitive adhesive, (3) bond non-EMI portion of gasket, (4) conductive adhesive, (5) bolt-through bolt holes, and (6) special attachments situations. Each of these methods is summarized below:

Hold in Slot

This method is recommended if the slot can be provided at relatively low cost such as in a die casting. All solid elastomer materials, which embody the gasket material, are essentially incompressible. These products appear to compress because the material flows while it maintains a constant volume. Therefore, when these

products are used in a slot, extra cross-sectional area must be
allowed for the material to flow axially. At least 10% extra volume
and more if possible is recommended such as shown in Fig. 11.23(a).

Pressure-Sensitive Adhesive

This method of mounting is often the least expensive for at-
taching EMI gasket materials. Installation costs are substantially
reduced with only a slight increase in gasket cost over a material
without adhesive backing. Most sponge-elastomer materials are used
for applications which do not require any hermetic sealing. The
adhesive-backed rubber portion of this material serves only as an
inexpensive attachment method for the EMI portion.

Bond Non-EMI Portion of Gasket

Many good non-conductive adhesives are now available to bond
an EMI gasket in position by applying the adhesive to the non-EMI
portion of the gasket. This can be insulated from the mating surfaces by a non-conductive material and is often a good way of mounting EMI gaskets. This method is shown in Fig. 11.23(b).

The designer specifying non-conductive adhesive attachment must
include adequate warnings in applicable drawings and standard procedures for production personnel. These cautions state that adhesive
is to be applied only to the portion of the gasket material not involved with the EMI gasketing function. Experience indicates that
installation workers, either through carelessness or a misguided
desire to do a better job, will apply the non-conductive adhesive
to the *entire* gasket including the EMI gasket portion. It is not
uncommon to hear "This gasket would hold better if I glued all of
it rather than half of it." This occurrence completely degrades
the EMI performance.

Conductive Adhesive

Since good conductive adhesives can provide an adequate electrical contact between the EMI gasket and the mounting surfaces, they
can also be used to mount the gaskets. However, the following cautions should be observed:

• Most conductive adhesives are hard and incompressible. Thus,
if too much adhesive is applied and it is allowed to soak too far
into the EMI gasket material, the compressibility will be destroyed.
Irregularly applied adhesive also has the effect of increasing joint
unevenness.

• The volume resistivity of the adhesive should be .01 ohm-cm
or less, preferably .001 ohm-cm.

• Most conductive adhesives do not bond well to either neoprene
or silicone. This is why all products that have conductive paths in

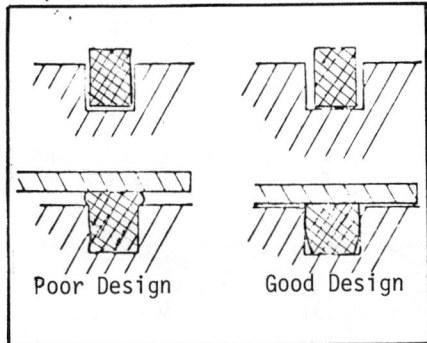

(a) Making Allowance for Solid
    Elastomer Gasket Flow

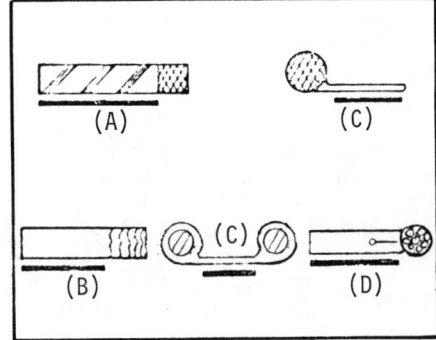

(b) Areas where non-conductive or
    dry-back adhesive can be used

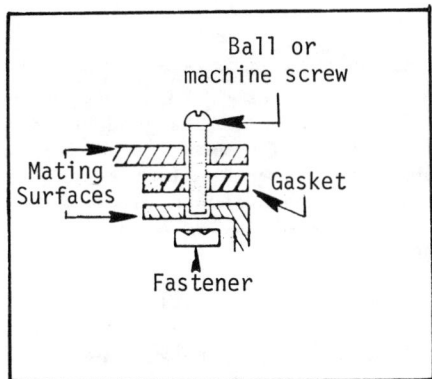

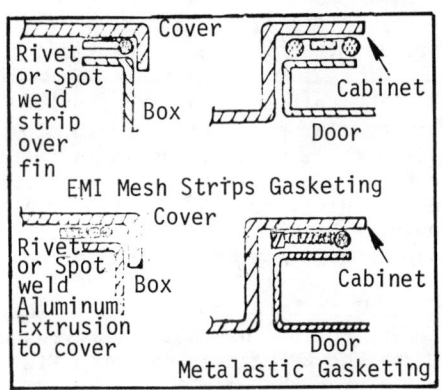

(c) Bolt Through Bolt Holes

(d) Special Mounting Methods

Figure 11.23 - Different Methods of Mounting Gaskets

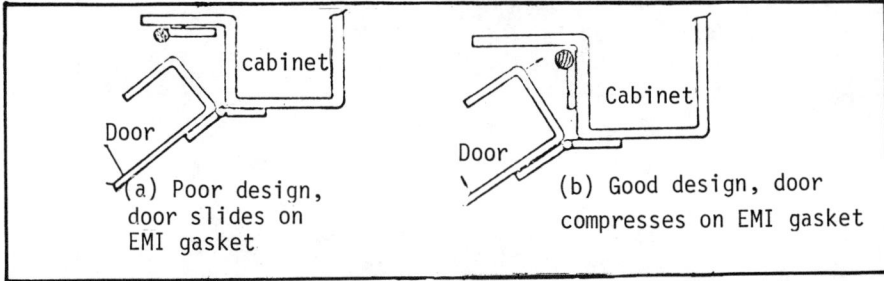

Figure 11.24 - Proper Method of Mounting Gasket in Cabinet Door Well

elastomer are rated *poor* for conductive adhesive bonding by the manufacturers.

- Applying a 1/8" to 1/4" diameter spot of conductive adhesive every 1" to 2" is preferred over a continuous bead.

- Conductive epoxies will attach the gasket permanently. Thus, removal of EMI gasket without destroying it, is almost impossible.

Bolt Through Bolt Holes

This is a very common and inexpensive way to hold gaskets in position as shown in Fig. 11.23(c). For most products, providing bolt holes involves only a small initial tooling charge. There is generally no extra cost for bolt holes in the piece price of the gasket. Bolt holes can be provided in the fin portion of EMI strips or in rectangular cross section EMI strips if they are sufficiently wide, such as over 3/8".

Special Attachment Means Provided

The knitted-mesh fins provided on some versions of EMI strips and the aluminum extrusions in aluminum gasketing were designed to attach these products as shown in Fig. 11.23(d). The mesh fins could be clamped under a strip of metal which is held down by riveting or spot welding, or the mesh fins can be bonded with an adhesive or epoxy. The aluminum extrusions of aluminum gasketing can also be held in position by riveting or bolting.

EMI gaskets should be positioned so they receive little or no sliding motion when being compressed. This is illustrated in Fig. 11.24. The EMI gasket shown in Fig. 11.24(a) is subject to sliding motion when the door is closed. This may cause it to tear loose or to wear out quickly. In Fig. 11.24(b), the gasket is subject to almost pure compression-only forces. This is the preferred position.

11.3 EMC SEALANTS

This section discusses another form of EMC shield integrity protection in the form of conductive epoxies and caulking.

11.3.1 Conductive Epoxies

Conductive epoxies are used to join, bond, and seal two or more metallic mating surfaces. The silver-epoxy resins replace soldering and other bonding techniques and cure at room temperatures. The conductive epoxy adhesive and solder family are used in the following applications:

> Electrical connections to:
> Heat-sensitive components   Capacitor slugs
> Ferrites   Integrated circuits
> Connect electroluminescent panels
> Form bus bars or strips on conductive glass
> Bonding flanges to waveguides
> Bonding waveguide sections
> Bolt holes and fasteners on electronic enclosures
> Joining dissimilar metals
> Sealing I-C packages against moisture and EMI
> Repair of printed circuits
> Interconnecting conductive-metal gaskets
> Field repairs to circuits
> Permanent seam shielding
> Sealing EMI shields

11.3.1.1 Preparation and Curing

The conductive epoxies are easily mixed on a volumetric basis, eliminating much time and equipment that would otherwise be necessary for weighing. Most epoxies can be prepared with either equal volumes or weights of the components. They are formulated with mixed viscosities that produce a light, creamy paste to make application with standard dispensing equipment reasonably easy and fullproof. Typical cure times are one day at room temperature or 30 minutes at 200°F.

11.3.1.2 Typical Properties

Depending upon the type of silver-epoxy resin used, typical volume resistivity will range from 0.001 to 0.02 ohm-cm. Operating temperature range is about -80 to +250°F. Shear strength is about 1200 psi and tensile strength varies with type, but averages about 2500 psi. It exhibits excellent moisture resistance. The cured specific gravity is about two suggesting its relative light weight for many pay-load-limited applications.

11.3.2   Conductive Caulking

Conductive caulking is used to EMI shield and seal two or more metallic mating members mechanically held by other means (cf. Sec.11.3.1, Conductive Epoxies). Silver particles are suspended in resin to provide conductive sealing. Conductive caulking is used in the following applications:

- Caulking EMI-shielded shelter panels
- Caulking EMI-tight cabinets and enclosures
- Improve joint and seam integrity of electronic enclosures
- Protect mating members of shielded conduits
- EMI sealing and grounding bulkhead panel fittings
- Moisture sealing of mating members
- Adhere metal-foil tape to shielded room joints
- Repair damaged conductive gaskets
- Seal ends on wire-mesh gaskets
- Caulking fasteners, panels, and handles

11.3.3   Preparation and Use

The conductive caulking compounds, as in any EMI sealant and bond, requires that the surfaces be thoroughly degreased and cleaned of oxide coatings. The caulking may be applied with conventional caulking guns and dispensing equipment such as small bead-orifice syringes. Hand application with spatula or putty knife may be used. The caulking is free of any corrosive binders. It is used at room temperatures and most caulking will not cure (permanently non-setting). This feature permits easy disassembly of caulked parts for movement or maintenance.

11.3.4   Typical Properties

Depending upon the type of silver resin used, typical volume resistivity will range from 0.005 to 0.02 ohm-cm. Operating temperature is -80 to +400°F. Moisture resistance is excellent. The final specific gravity is about 1.8 suggesting its relative light weight for many payload-limited applications.

11.4 CONDUCTIVE GREASE

Conductive grease is not a member of EMI gaskets and sealants discussed in this chapter. However, it is related in that one of its functions is to provide a low-resistivity contact to mating members. Here, mating members may engage and disengage more often than in most EMI gasketing applications, excepting finger stock used in shielded enclosures.

Conductive grease is a low resistivity, silver-silicone grease which contains no carbon or graphite fillers. The material will maintain its electrical and lubricating properties over a broad environmental range. These conditions include high and low temperatures, resistance to moisture and humidity, inertness to many chemicals, ozone and radiation. Most conductive greases are a viscous paste which can be applied at elevated operating temperatures to vertical or overhead surfaces without dripping or running.

Conductive grease is used on power substation switches and in suspension insulators to reduce EMI noise. It also reduces make-break arcing and pitting of the sliding metal contact surfaces of switches and fills in pitted areas with silver/silicone. In addition, normally-closed switches are prevented from sticking due to corrosion or icing. The grease is effective in maintaining a continuous electrical path between contact surfaces which must be free to move. These include ball and socket connections of power insulators, which if allowed to arc, can generate EMI. Conductive grease is designed to maintain low resistance electrical contact and thereby maintain equipment operating over extended environmental conditions, helping to deliver continuous electrical service.

Conductive grease is used on the contacting surfaces of circuit breakers and knife blade switches. It reduces localized overheating or *hot spots* in turn maintaining the blades spring properties and current rating of the switch or breaker at original equipment level. Lubricating conductively prevents freeze up in operating equipment and permits restoration of marginal or discarded breakers to rated capacity.

Typical volume resistivity is about 0.02 ohms-cm. Operating temperature range is -65° to +450°F. Conductive grease provides excellent moisture resistance and has no corrosion effect on metals. Its pot life is unlimited and unused portions are returned to its container.

11.5 BIBLIOGRAPHY

(1)   Awerkamp, D. R., Evaluation of Radio Frequency Gasket Materials," The Boeing Co., Document No. D2-22832, Feb. 1964.

(2)   Buckley, E. E., "Metal-Foil Shieldings and Conductive Mastics for Inexpensive Shielded Enclosures," *Proceedings of the 5th National Symposium on RFI, IEEE*, June, 1963.

(3)   Cale, N. H., "A Comparison of R-F Shielding Materials," *Electronic Industries*, December, 1962.

(4)   Cowdell, R. B., "Simplified Shielding," *1967 IEEE Electromagnetic Compatibility Symposium Record, IEEE 27C80*, Washington, D.C., July 18-20, 1967.

(5)   "Designing and Applying RFI Shields," *Electrical Design*, September 27, 1962.

(6)   "Electromagnetic Compatibility - A Lecture Series," Volume II, Chapter I, Presented by Armour Research Foundation, 1961, Contract AF33(616)-8507, ARF Project E165.

(7)   "Electromagnetic Shielding Principles," Vol. I, II, Rennsselaer Polytechnic Inst., March, 1956.

(8)   Filtron Company, Inc., "Interference Reduction Guide for Design Engineers," Prepared for U.S. Army Electronics Labs., Fort Monmouth, N. J., Vols. 1 and 2, Accession AD 619666 and AD 619667.

(9)   Gooding, F. H., and Slade, H. B., "Shielding of Communications Cables," *Electrical Engineering*, June, 1955.

(10)   Jarva, W., "Shielding Efficiency Calculation Methods for Screening Waveguide Ventilation Panels and Other Perforated Electromagnetic Shields," *Proceedings Seventh Tri-Service Conference on Radio Interference Reduction and Electronic Compatibility*, Chicago: Armour Research Foundation, pp. 478-498, Nov., 1961.

(11)   "Metal Foil Wallpaper for RF Shielding," *Electrical Design News*, June, 1963.

(12)   Pearlston, C. B., "Case and Cable Shielding, Bonding and Grounding Considerations in Electromagnetic Interference," *IRE Transactions on Radio Frequency Interference*, Vol. RFI-4, No. 3, pp. 1-6, October, 1962.

(13)   Pulsifier, V., "Bonding Materials, Metallic Mating Surfaces: Low R. F. Impedance," Armour Research Foundation, January, 1954.

(14)   Quine, J. P., "Theoretical Formulas for Calculating Shielding Effectiveness of Perforated Sheets and Wire Mesh Screens," *Proceedings Third Conference on Radio Interference Reduction*, Armour Research Foundation, pp. 315-329, February, 1957.

(15)   Schreiber, O. P., "Designing and Applying RFI Shields and Gaskets," *Electronic Design*, Vol. 10, No. 20, September 27, 1962.

(16)   Schreiber, O. P., "Reliable Electrical Contact Theory Applied to RFI Control," *Fourth IRE Symposium on RFI*, June, 1962.

(17)   Schreiber, O. P., "Some Useful Analogies for RF Shielding and Gasketing," *IEEE Third National Symposium on Radio Frequency Interference*, Washington, D. C., June, 1961.

(18)   Schulz, R. B., Plantz, V. C., and Brush, D. R., "Shielding Theory and Practice, *Proceedings of the Ninth Tri-Service Conference on Electromagnetic Compatibility*, Oct. 15-19, 1963, Chicago, Ill., pp. 597-636.

(19)   "Selection of Waveguide Filters for Shielded Compartments," *Radio Engineering*, July, 1961 (translation from Russian).

(20)   Vasaka, C. S., "Shortcuts to R-F Shield Design," Electronic Industries & Tele-Tech., No. 3, p. 72, March, 1957.

(21)   Vogel, S., "RFI Causes, Effects, Cures," *Electronics*, June 21, 1963.

CHAPTER 12

EMI-SHIELDED HOUSINGS

CHAPTER 12

EMI-SHIELDED HOUSINGS

The preceding two chapters covered the subject of shielding. Chap. 10 discussed shielding theory and materials. It was shown that for other than low-frequency magnetic fields, it is easy to obtain more than 100-dB shielding effectiveness across the spectrum for nearly any metal. The shielding problem then develops from the fact that practical enclosures have apertures and penetrations which compromise the effectiveness of the basic shield material. Thus, shielding effectiveness of a housing could be reduced to 60-dB or less because of loss of enclosure integrity.

Chap. 11 reviewed how the shielding integrity of a housing with apertures can be re-instated in either the design or retrofit stage. There, topics involving seams and joints, ventilation covers, access plates, gaskets, and conductive epoxies and caulking were discussed.

It now remains to bring the foregoing material together in the form of practical shielded-housing applications. Consequently, this chapter reviews the subjects of shielded compartments, chassis and equipments, cabinets, shielded rooms and enclosures, huts, and vehicles. Note that the progression of presentation in this chapter is generally one from small physical size (e.g., shielded compartments) to large size (shielded rooms and vehicles). Since Sec. 9.1 presented material on building shielding techniques, this topic is not discussed here.

12.1   SOME BASICS OF SHIELDS AND GROUNDS

In preparation for discussing shielded housings, some basics are
first reviewed. A few were presented in Sec. 6.3 on electric-field
coupling in wire and cable systems. A few rules will be developed
in this section in order to assure consistency of practice.

12.1.1   ELECTRIC-FIELD SHIELDING

Coupling or cross-talk between two members A and B as shown in
Fig. 12.1(a) exists because of induction of charge $-Q_a$ on member B
resulting from the charge $+Q_a$ on member A.* Member A has a charge,
Q_a, because it has both a potential, V_{ag}, and capacitance C_{ag} with
respect to some reference ground, viz:

$$Q_a = C_{ag} V_{ag} \qquad (12.1)$$

If potential V_{ag} were reduced to zero, it's charge would be reduced
to zero from Eq. (12.1), and no coupled charge would exist on member
B.

In a practical system, member A may represent a source of EMI
emission in which case it would have a potential difference of V_{ag}
referenced to some ground pertinent to member B. To decouple its
resultant charge from influencing member B, a completely enclosed
shield is placed around it as shown in Fig. 12.1(b). Here, a charge
of $-Q_a$ is now induced on the shield. However, the net electric flux
flow outside the shield is zero since the two charge sources (member
A and the shield) are equal and opposite. Thus, the induced charge,
Q_b, on member B is zero, whether or not the shield is grounded. To
state this another way, the mutual capacitance, C_{ba} between members
B and A is $C_{ba} = Q_b/V_{ag} = 0$.

If the shield were imperfect in that it has apertures and penetration, such as shown in Fig. 12.1(c), a different situation exists.
Now the electric flux leaving member A no longer terminates entirely
on the shield. Some electric flux leaks out to terminate on member
B resulting in an induced charge $-Q_b$. The mutual capacitance is now
$C_{ba} = -Q_b/V_{ag} \neq 0$. If the flux from member A terminates on both the
shield and member B and not on the reference plane ground, then
$Q_a = |Q_{sh} + Q_b|$. Thus, if a shield contains apertures and penetrations, its internal source of EMI may partially couple to a victim
receptor.

---

*   This develops from basic electrostatics; the proximity of the
reference plane to either members A or B is considered arbitrarily
far removed.

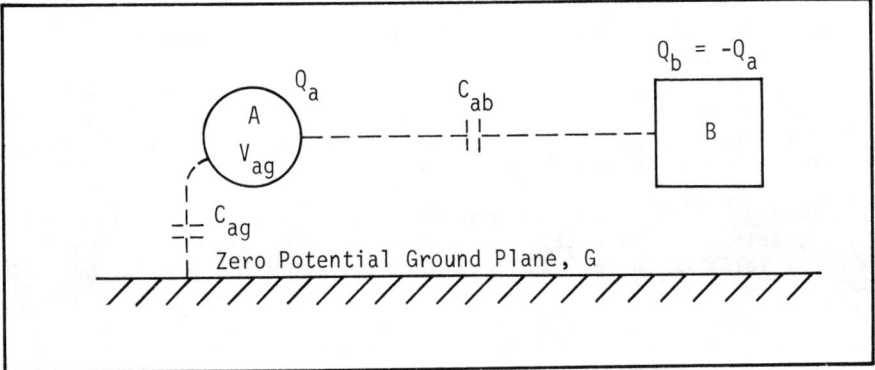

(a)   Two Members in Proximity with No Intervening Shield

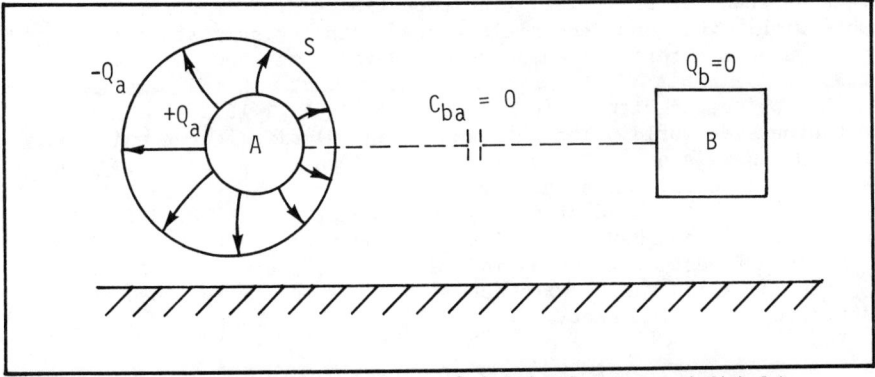

(b)   Reducing the Coupling Between A & B by Interposed Shield

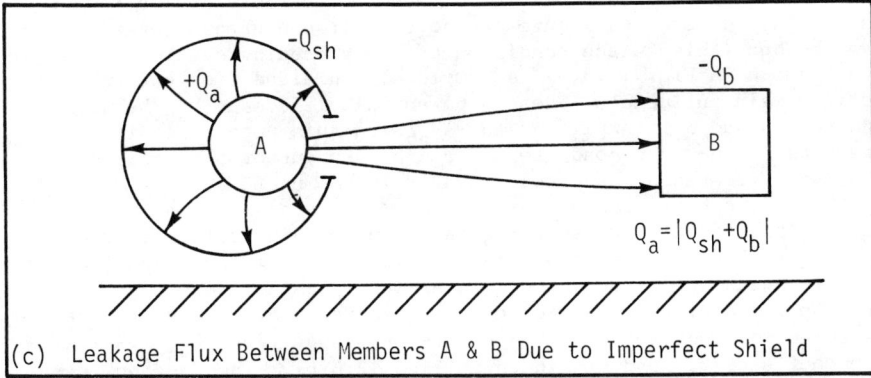

(c)   Leakage Flux Between Members A & B Due to Imperfect Shield

Figure 12.1 - Coupling Between Two Members Due to Charge Induction
and Shielding

Summarizing the foregoing, the mutual-capacitance between conductors within a *complete* shield to conductors outside of a shield is zero. If there exists several conductors within a shielded enclosure the mutual-capacitance amongst these conductors is not zero. The value of these capacitances depends upon the geometric relationships between these inner conductors and the shield conductor itself.

Sec. 6.3 discussed grounding the shield. However, a shield can be at any potential and still provide the same shielding effectiveness since relative changes in the conductor potentials within the shield have no influence on conductors outside of the shield because there is no charge coupling to the outside. Also, changes in potentials of conductors outside of the shield have no effect on the relative potential of conductors within the complete shield. Thus, the shield does not have to be earthed or defined in any way. The only requirement is that the conductors be fully surrounded by a conducting surface as shown in Fig. 12.2. The potential differences for conductors within the grounded-shield case remain the same after the shield has been externally changed to 100 volts as shown.

The conductors within the shield shown in Fig. 12.2 are insulated from the outside world. The internal potential differences exist only because of charges on these conductors. Unless these charges are altered, the potential differences remain the same. However, in most practical cases, the conductors are not completely insulated from the reference plane, but have some direct or indirect tie thereto. Thus, the shield now must be grounded in order not to create an interposed EMI potential which varies with conditions. This is explained more fully in the following section.

12.1.2   ENCLOSURE SHIELDING

When a self-powered amplifier or other electronic device is contained within a metal enclosure and no circuit conductors enter or leave the box (this is the condition of the preceding section), the circuit shown in Fig. 12.3(a) is completely shielded from external electric-field influences. Here a potential difference V_{ab} between conductors a and b is amplified to $-AV_{ab}$ and this potential difference appears as V_{cb} between conductors c and b. Conductor b is called a zero-signal reference conductor since it is common to both V_{ab} and V_{cb}. Parasitic capacitances for the network are also shown in the figure. The effect of these capacitances upon gain is more apparent in the equivalent circuit shown in Fig. 12.3. The mutual capacitances form a feedback loop around the gain element. While these capacitances cannot be avoided, the feedback can be eliminated by bonding the shield enclosured to conductor b. This shorts out capacitance C_{bd} leaving the capacitance C_{ad} and C_{cd} shown in Fig. 12.3(c). The feedback capacitances of Fig. 12.3(b) are related to the geometry of the conductors and shield. The placement of conductors could reduce this feedback but

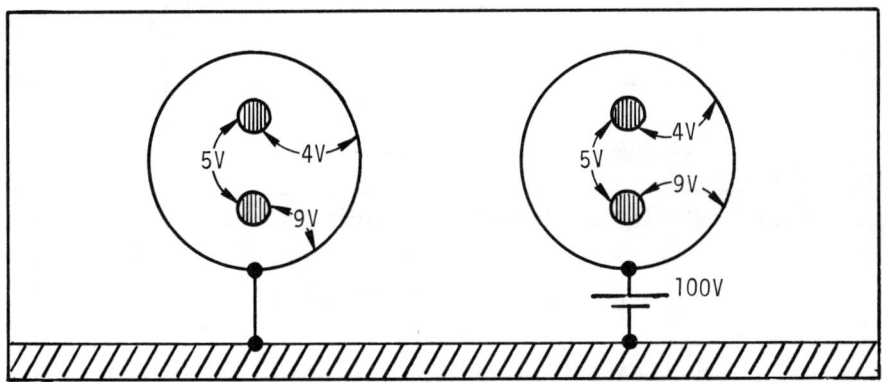

Figure 12.2 - Shielding Effectiveness is Independent of Shield-Ground
Potential

the preferred solution is to ground the common conductor to its
shield.

If conductor b is taken outside of shield d, a new phenomenon
results. Fig. 12.3(d) shows that an external mutual capacitance C_{de}
will now influence the circuit. If conductors b and e have a potential difference, V_{be}, current will flow in capacitance C_{bd} and C_{de} to
close the loop in path bdeb. Since C_{bd} is a feedback element, an
EMI potential difference V_{be} is mixed with signals being processed
by the amplifier gain. This is shown in the equivalent circuit in
Fig. 12.3(e). This coupling is eliminated when C_{bd} is shorted out.
Thus, Rule 1 is now introduced: *A shielded enclosure, to be effective,
must be connected to the zero-signal-reference potential of any circuitry contained within the shield.**

The gain network shown in Fig. 12.3(a) is impractical without
shield penetrating input and output connections. Shielded wires
usually serve the purpose to act as extensions of the shielded enclosure. Such an enclosure is effective when Rule 1 is applied,
which places no restriction on the shield potential relative to the
external environment. This is the key to connecting signal conductors
to a gain element. Since the shield must be at zero-signal-reference
potential, and since the signal is often derived from some reference
point in the external environment, the shield ground is automatically
defined at this external reference potential.

Fig. 12.4(a) shows a gain network inside it's shielded enclosure.
The input and output connections are two-wire shielded conductors.

---

*   See Ralph Morrison, "Grounding and Shielding Techniques in Instrumentation," John Wiley and Sons, 1967, pp. 39.

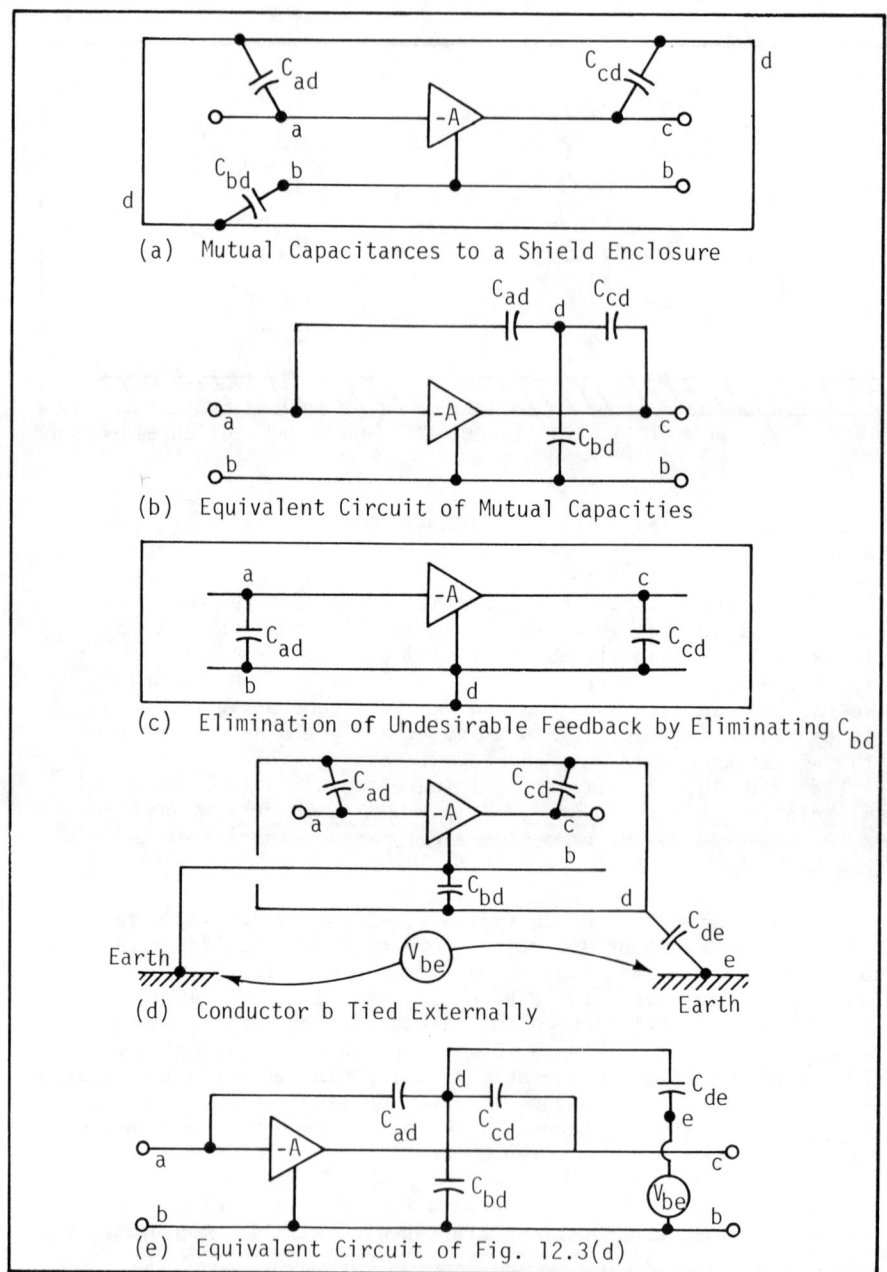

Figure 12.3 - Mutual Capacitances Involving Shields

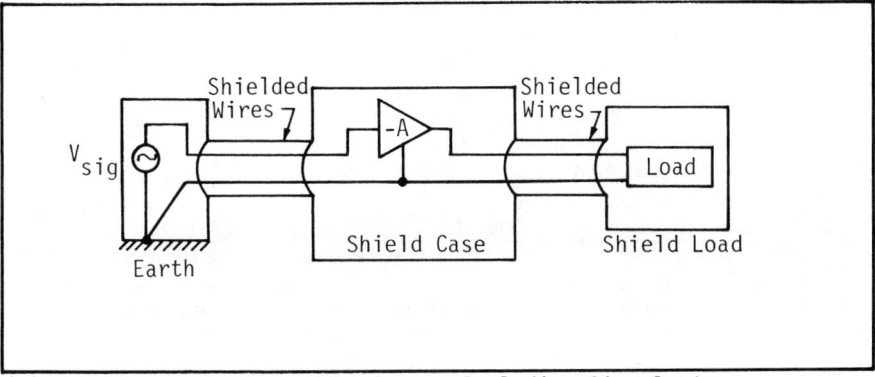

(a)   An Extended Shielded Enclosure Including Signal Lines

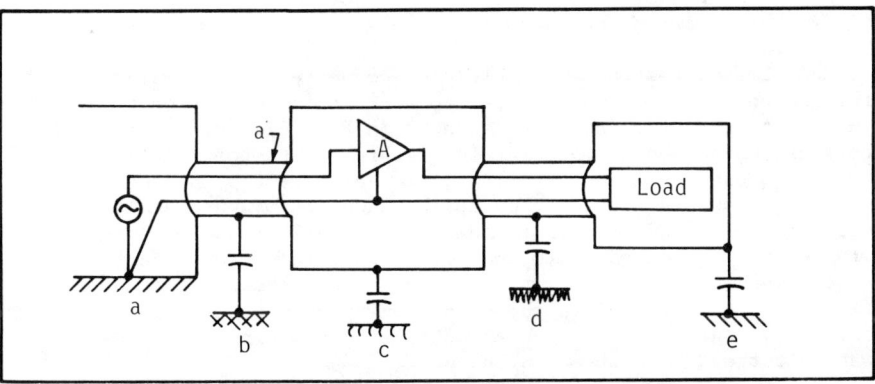

(b)   Mutual Capacitances Between an Enclosure and Other Earths and Grounds

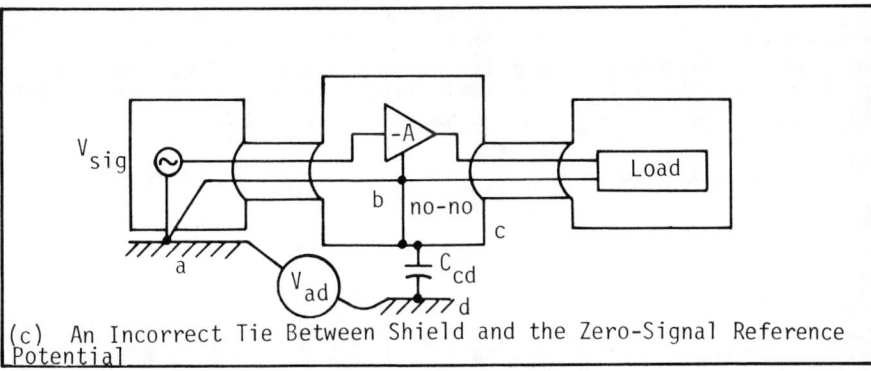

(c)   An Incorrect Tie Between Shield and the Zero-Signal Reference Potential

Figure 12.4 - Shield Element Extensions and Grounding Procedures

The input signal zero is grounded to an earth reference. When the shield is tied to this same earth potential, Rule 1 applies and the system is correctly grounded. Thus, Rule 1 requires that the shield must be connected to zero-signal reference potential. If the signal zero is earthed or grounded the shield makes no sense if the signal source is not also earthed or grounded.*

The enclosures shown in Fig. 12.4(a) often parallel external conductors. For example, long runs of shielded wires are contained in electrical raceways, conduit, floor wells, racks, or along floors. These neighboring conductors are usually at differing potentials, especially at R-F. In particular, they are not the zero-signal-reference-potential of the shield enclosure. Thus, these potentials cause currents to flow in the mutual capacitances between conductors. In Fig. 12.4(b) current flows in loops such as abca or aea. This current flows in shield segments only and not in signal conductors. If it were to flow in the conductors, unwanted pickup will result.

Rule 1 requires that the shield be connected to zero-signal-reference-potential without regard to where this connection should be made. A correct connection is shown in Fig. 12.4(a), whereas an incorrect connection is made in Fig. 12.4(c). The potential difference V_{ad} causes current to flow in capacitance C_{cd} in loop abcda. If current flows in signal conductor b, unwanted pickup results. To avoid this, a second rule is formulated: Rule 2 - *The shield conductor should be connected to the zero-signal-reference potential at the signal-earth connection.* This procedure ensures that parasitic currents will flow in the shield only and not in the signal conductors. The shield can be thought of as a drain path to carry unwanted current back to an earth point.

By Rule 1, the enclosure should be at zero-signal reference potential. If the shield is split in sections Rule 2 places a constraint on the treatment of these segments. The rule requires that the shields be connected in tandem as one continuous conductor and then connected to zero-signal reference potential at the signal-earth point. If the shield segments are individually treated the difficulties exhibited in Fig. 12.4(c) can be expected.

---

*   Note, that HF for which the overall length, ℓ, of the shield configuration is $\ell \geq \lambda/20$, the far end of the shield (right end in Fig. 12.4(a)) is no longer R-F grounded and a whole new set of problems develop - see Sec. 6.3.

12.2   SHIELDED COMPARTMENTS

This section covers one of the physically smallest shielding level requirements, viz., shielded compartments for amplifiers, filters, and the like, where either stage-to-stage or input-output cross-talk is to be kept below a specified amount. Input-output cross-talk must be a greater number of dB than the gain of an overall amplifier or the out-of-band attenuation of a filter. Individual component shields, such as a T-5 can on a relay, are not discussed here since they are covered in Chap. 16.

To illustrate the above, one example of compartment shielding problems involves capacitive coupling from stage-to-stage in which the role of the shield is to provide a Faraday cage to eliminate cross-talk. Fig. 12.5 shows an I-F amplifier and interstage band-pass filter. The need to shield the amplifier from the outside world was recognized by the designer (see Sec. 12.1); hence the outer shield. The need to internally isolate input and output was also recognized; hence the compartments or stage septums. The figure shows a floating, non-grounded, metallic cover plate. Thus, a portion of the high-level output voltage, e_o, is capacitively-coupled back to the amplifiers input voltage, e_i. The amplifier is unstable and may oscillate when:

$$e_o \frac{Z_i}{2/WC_{io}} = .5e_o Z_i WC_{io} > e_i \qquad (12.2)$$

$$\text{or } \frac{e_o}{e_i} = A > \frac{2}{Z_i WC_{io}} \qquad (12.3)$$

where, $.5C_{io}$ = input-output capacity

$\quad\quad\quad W = 2\pi$ x operating frequency

$\quad\quad\quad Z_i$ = input impedance

$\quad\quad\quad A$ = overall amplifier gain

For the situation shown in Fig. 12.5, the ungrounded cover plate serves to intensify the intra-amplifier EMI coupling problem, rather than help.

When the cover plate shown in Fig. 12.5 is grounded to the outer housing, the feedback capacitance is transferred to ground where it can do no harm as shown in Fig. 12.6. However, a small residual capacitance, C_s, between input and output still remains. If Eq. (12.3) is satisfied the amplifier may still oscillate. Thus, compartment shields are added as shown in Fig. 12.7 to further decouple stages by the introduction of additional Faraday cages. To assure the full effectiveness of Fig. 12.7 it is necessary to bond the top of the compartment septums to the cover plate. This is accomplished by gasketing and Fig. 12.8 shows one method used to achieve this. Here a conductive epoxy adhesive is applied to the cover plate over which is

12.9

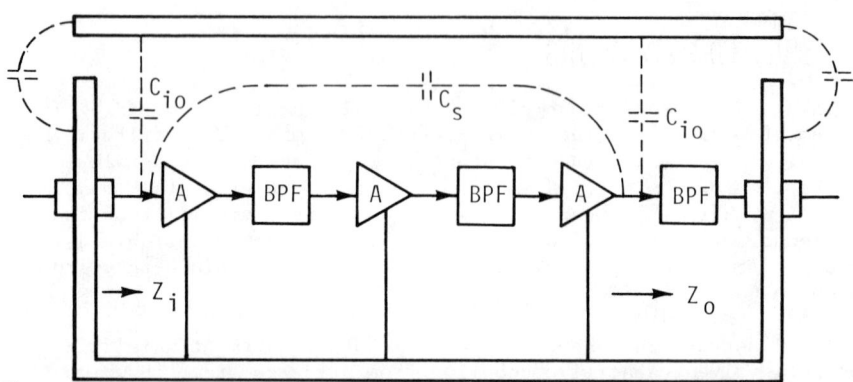

Figure 12.5 - Box Shield for High-Gain Amplifier Showing Two
Paths of Output-Input Feedback

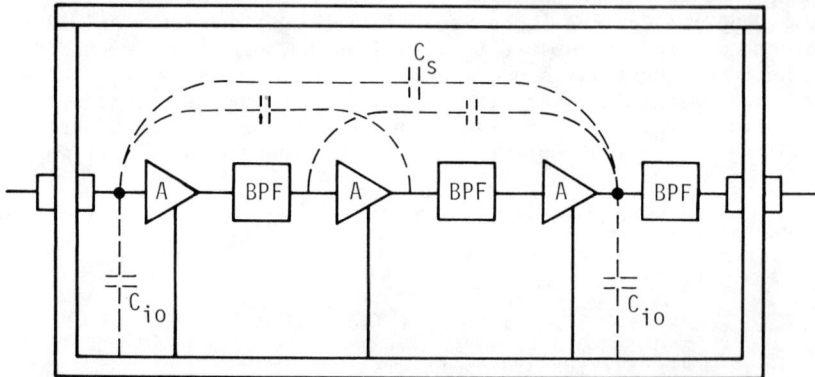

Figure 12.6 - Box Shield Cover Plate in Position to Eliminate
Larger Feedback Path of Fig. 12.1

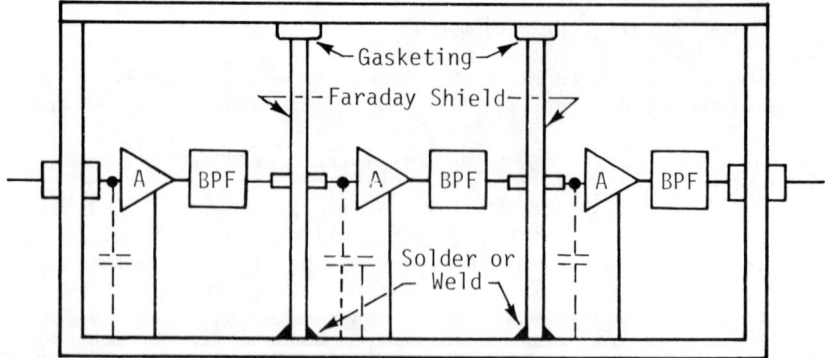

Figure 12.7 - Box Shield With Compartment Septum Shields
to Ground Only Inter-Stage Feedback

Figure 12.8 - Applying Conductive Epoxy to Cover Plate
to Hold Mesh Gasket in Position

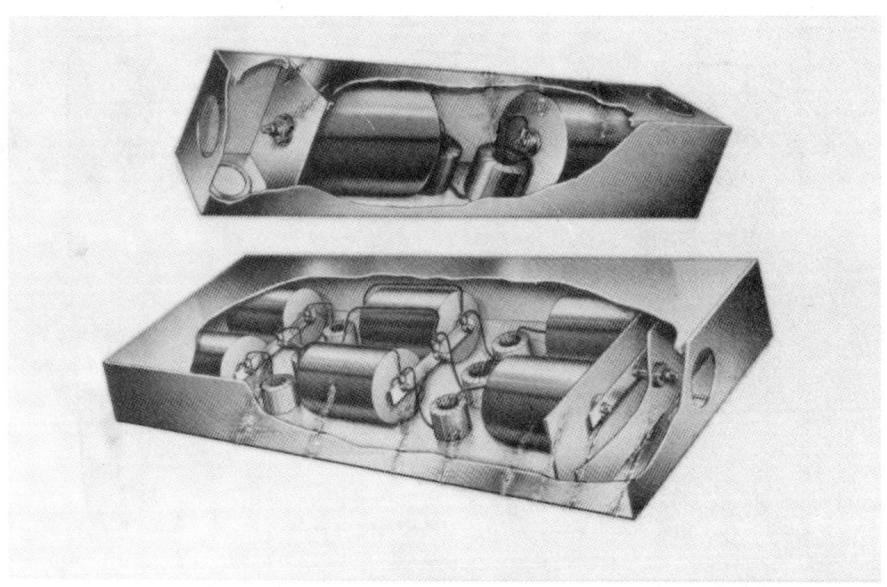

Figure 12.9 - Cutaway View of Typical Shielded Enclosure EMI Filters

placed the EMI mesh gasket (see Chap. 11). With the gasket held in place, it accommodates septum height variations indicated in Fig. 12.7.

Fig. 12.9 shows another shielded compartment used in the construction of power-line filters. The purpose of both the shielded box and the compartments is similar to that discussed in previous paragraphs, viz, to preclude feedback from input to output and between stages by capacitive coupling. However, in this case the metallic configuration is constructed of galvanized steel or other permeable material in order to: (1) prevent magnetic coupling from inter-stage inductors by providing a low-reluctance path to residual magnetic fields from the toroids, and (2) mitigate magnetic coupling (penetration) through the box housing to the outside world. Note the use of separate shields at both ends of the filter to further prevent direct input-output cross-talk and radiation coupling to feed and load wires.

12.3   SHIELDED CHASSIS AND EQUIPMENT

The next higher level up from the shielded compartment is the shielded chassis or equipment. This level may be regarded as an ensemble of many individual electrical and electronic building blocks to result in a single housing. Typical examples of the chassis or equipment-level of shielded housing include electronic test instruments, bio-medical equipment, mobile transceivers, hi-fi amplifiers, and mini computers.

Fig. 12.10 shows the housing construction of an EMI-hardened equipment case. Note the application of several topics on shielding discussed in the previous two chapters and other control techniques:

Base Shielding Material:

> Permeable or non-permeable depending upon EMI environment
> Continuous seam-weld construction

Shielded Integrity Protection:

> Gasketed hinged top
> Gasketed cover plates
> Viewing window screen or conductive glass
> Waveguide attenuators for control shafts
> Screened vent openings
> Unused connector caps

Power-Line Filters

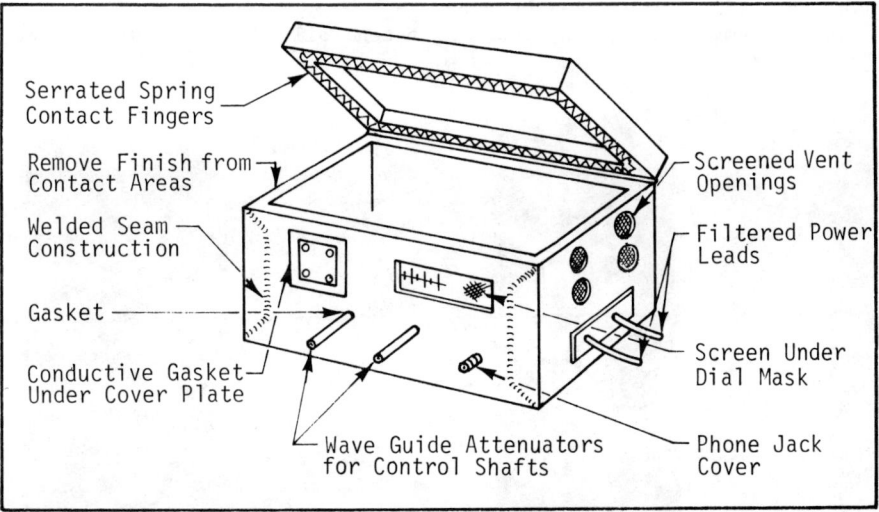

Figure 12.10 - An EMI-Hardened Shielded Equipment Case

12.13

12.4 SHIELDED CABINETS

This section represents a further increase in the physical size of shields from the chassis or box-size equipment of the previous section to the console size here. Shielded cabinets are the topics of this section in which a 19" rack, single or double-bay console, or boxes occupying several cubic feet are the specimen. Examples of typical shielded cabinets are shown in Figs. 12.11 and 12.12. It is noted that these cabinets appear to be similar to most other racks or consoles which have not been EMI hardened. Since both offer shielding then, the question is posed regarding what is the difference?

An EMI shielded cabinet distinguishes itself from other cabinets, typically in the following manner.

(1) The rear access door uses a continuous piano hinge for better door-to-frame bonding by equalizing the pressure along the hinge side.

(2) The rear door is EMI gasketed when in the closed position to provide shielding integrity all around.

(3) The cabinet frame enclosure is usually seam-welded to provide a continuous homogenous bond.

(4) The cabinet is often made of light-gage steel to provide improved magnetic shielding at low frequencies in contrast to that offered by aluminum.

(5) Separate heavy power and signal grounding busses are run vertically up the rear side walls.

(6) Large strike-plate holes on back of drawers and dagger-pin arrangement on back of cabinet grounding busses result in lower common-mode impedance problems. In lieu of this, insulated grounding bus jumper cables to drawers may be used.

(7) Frame bonding and grounding often follows that shown in Fig. 8.20.

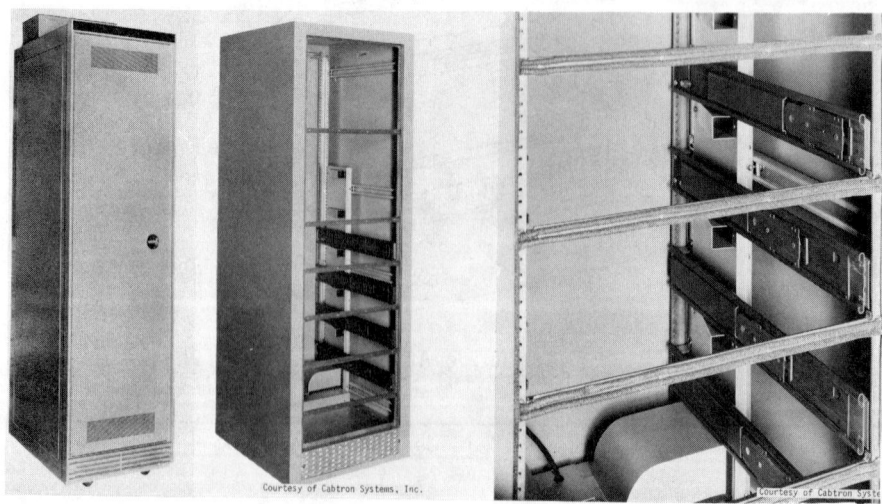

Courtesy of Cabtron Systems, Inc.   Courtesy of Cabtron Syste

12.14

12.5 SHIELDED ROOMS AND ENCLOSURES*

This section continues with shielded housings but for an increase in size to that of the shielded-room configuration. Here sizes may range from an 8 ft. x 8 ft. x 8 ft. small room to a large aircraft hanger.

The shielded room, or shielded enclosure as it is sometimes called,** has been in use for many years for performing electronic measurements where a low-electromagnetic ambient is required, or where potentially-damaging emissions must be contained. Its use has spread to non-measurement applications, such as protecting personnel working near high-power radar sites, containing certain industrial R-F emission sources, and protecting sensitive equipment such as bio-medical instruments and computers.

The main advantage of the shielded enclosure used for performing EMI measurements is that it provides R-F isolation from the outside world. Its use allows meaningful susceptibility measurements to be made, both conducted and radiated, in higher ambient locations where such testing would not ordinarily be possible. However, there still exists some residual electromagnetic ambient inside a shielded room, since the room attenuates rather than eliminates outside-world emissions. In most situations where the outside environment is not abnormally high, a modern shielded room provides attenuation sufficient to reduce all outside emissions to levels below the sensitivity of typical receivers with their normal antennas. Magnetic-field shielding at and below ELF is the exception. For example, the enclosure walls to 60-Hz fields offer little attenuation, and input power lines allow easy emission entry. To reduce 60 magnetic fields and their first few harmonics, it is necessary to use extended frequency-range shielded enclosures.***

12.5.1 ENCLOSURE TYPES AND SIZE

Some electronic systems are physically too large to employ shielded enclosure for testing (see Chap. 2, Vol. 4). The size of shielded enclosure has no theoretical limit and they have been built to enclose aircraft hangers. For example, the Titan ICBM is checked out in a five-story high enclosure with a five-story door. Another example is

---

* See Chap. 2, Vol. 4, this Handbook Series.
** The term *screen room* is still used in some circles. It comes from the early shielded enclosures which were made of copper screen.
*** This is a term which has been adapted by the EMI community for shielded enclosures to imply significant attenuation to magnetic fields at power-line frequencies.

the entire computer facility for an early model Atlas ICBM, located
in a large shielded structure such as shown in Fig. 12.13. Some structures are lined with absorbant material to form an anechoic chamber
(Chap. 2, Vol. 4) as shown in Fig. 12.14.

Shielded enclosures are also constructed in mobile configurations
which are roving electronic laboratories, viz., shielded enclosures
constructed in a trailer or van. This subject is discussed in Sec.
12.6.

One of two metals, either steel or copper, is usually employed
as the basic material in the shielded enclosure. Steel is generally
used in the form of galvanized sheet as shown for the enclosure in
Fig. 12.15 while copper is used in either sheet form or fine-mesh
screening as shown in Fig. 12.16. Which material is to be used depends
on weight restrictions, shielding desired, cost, and other variables.
For equal cost, steel generally provides comparable performance to
copper at radio frequencies down to 150 kHz. Below this, the permeability of steel begins to provide improved magnetic-field shielding.

The first of the modern shielded-enclosure construction methods
was developed a number of years ago at the Naval Air Development
Center, Johnsville, Pennsylvania. This room was made with two layers
of copper screen, separated by a one inch frame. The room was constructed of several panels, called cells, each measuring 4 ft. x 8 ft.
Individual panel edges were butted together and were bolted through
the wood frame that provided the shape and strength for each panel.
One feature of this room was that it could be disassembled, moved,
and reassembled at another location without major modification.

The door for this room was well-braced, framed with wood, and
also covered with copper screen. To the periphery of the door was
attached two sets of spring finger stock, one to provide contact with
the inner edge of the doorjamb, and the other set to push against the
outer edge of the door frame. This second set of spring fingers actually overlapped the door frame opening.

A similar enclosure construction is the cell type which employs
only a single layer of either screening or sheet metal. This enclosure is not widely used since it's shielding performance is less
satisfactory and it is almost as costly as the double-layer, cell-
type. Fig. 12.17 illustrates the three more common types of de-mount-
able shielded enclosure construction.

There appears to be a controversy over the purported advantages
of completely separate layers in the double-shield, isolated-type room
as opposed to the cell-type room shown in Fig. 12.17. Installations
evidence little inherent advantage in either type of construction
where the same total thickness of the same metal is used. Where bolted
seams are used in both, the double-wall isolation construction may be
more effective due to the second opportunity to clamp seams closed.

Figure 12.13 - Example of a Large Shielded Enclosure

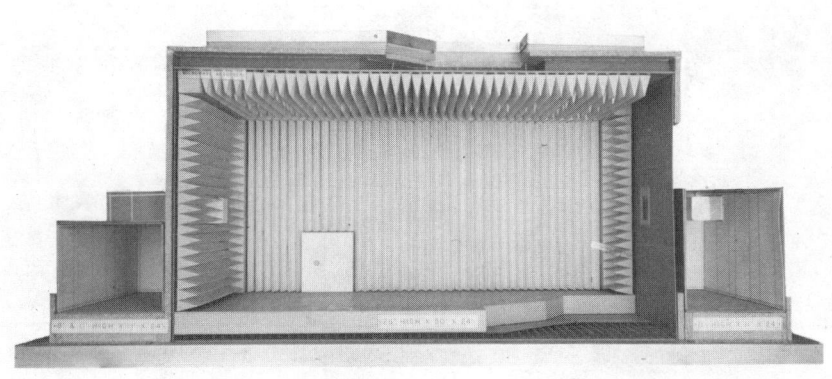

Courtesy of ACE Shielded Products (now Ellis & Watts Co.)

Figure 12.14 - A Large Shielded Anechoic Chamber

Figure 12.15 - Typical Double-Screen Room

Figure 12.16 - Solid, Double-Wall Enclosure

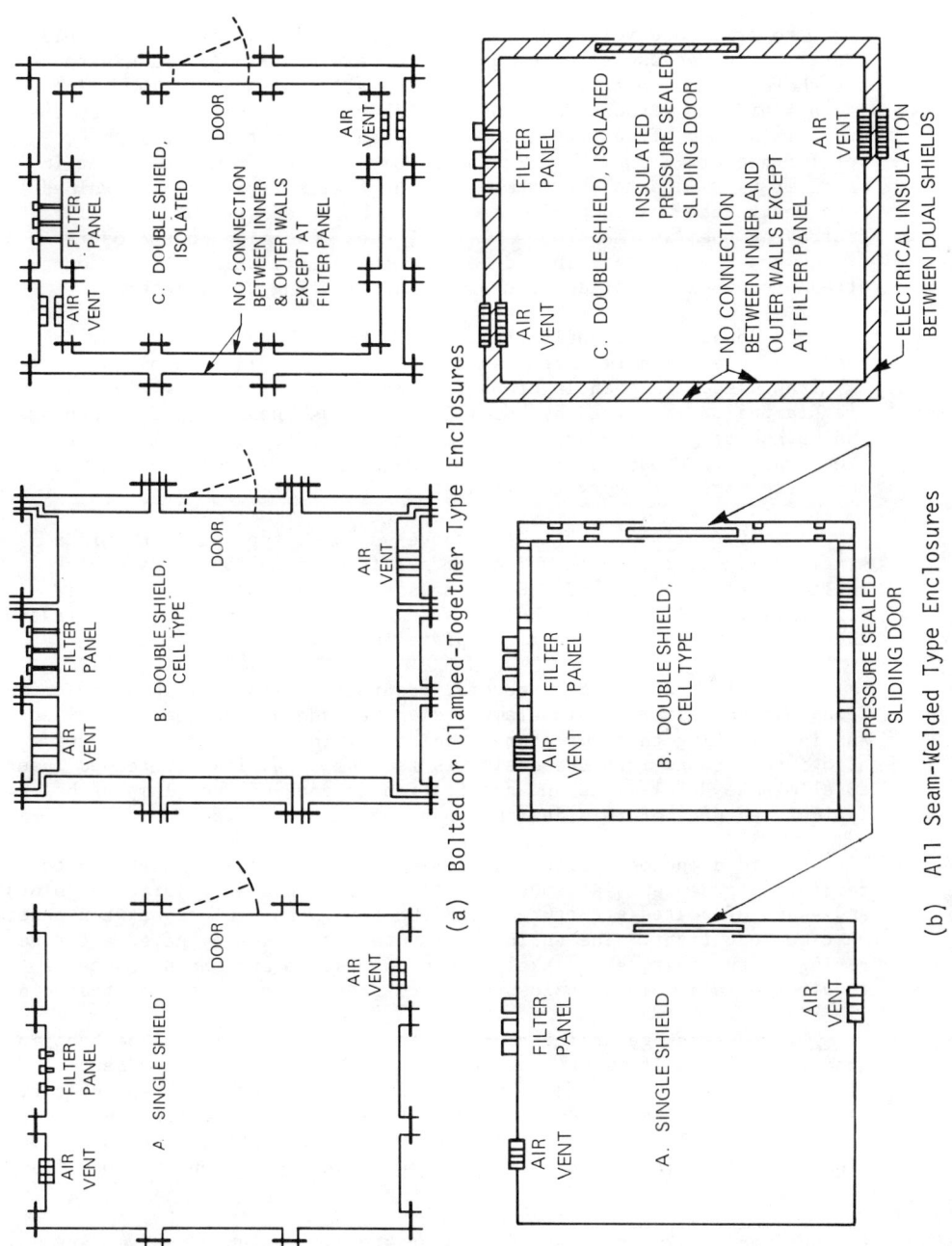

Figure 12.17 - Three Types of Demountable and Seam-Welded Shielded Enclosures

Properly-made welded seams (see Fig. 12.17) favor single-shield construction because electromagnetic leaks are virtually eliminated and there are no possibilities of intra-wall resonances. The trick is to seam weld without developing voids which may not be seen by the eye. This type of construction uses a single sheet of steel on a metal framework. The sheet is under some tension from the manner in which it is welded to the frame, and the frames or panels are welded together. Seam-welded enclosures cannot be moved without literally destroying them in the process. Thus, those ordering an enclosure of the seam-welded type should be sure that there will not be a downstream requirement for demounting it and erecting it elsewhere.

For double-layer enclosures, i.e., the double-shielded room, a sandwich panel with two steel sheets bonded to a 3/4-inch plywood core is often used. The panels are not butted together all the way, but are clamped on each edge by special continuous channels and strapping. The method of joining panels both along the sides of a room and at the corners is shown in Fig. 12.18. Machine screws pull the channel and strap together every few inches.

A different approach to the above sandwich-type enclosure is a single solid-wall type in which the shield material is 0.125" thick-rolled steel which includes U-channels and U-tensioners as shown in Fig. 12.19. All fittings, bolts, screws and spline nuts used to fasten the framing members together are plated steel.

Variations are used by several manufacturers of shielded enclosures. Some newer developments in clamp design include a preassembled welded and interlocking three-way corner which eliminates leakage problems. Also incorporated into the design is the use of a closed threaded insert to eliminate R-F leakage and penetration at each of the clamping bolts. Typical features of this design are shown in Fig. 12.20.

Shielded enclosures require periodic maintenance if they are to retain their designed attenuation. The vulnerable areas are the joints and seams of bolted structures, and the door. Fastenings between panels must be kept tight. The enclosure manufacturer usually gives a torque rating on the fasteners. Exclusive of doors, no such maintenance is required on all-welded enclosures since they use no bolts for fastening.

Finger stock gasketing along the edge of the door must be kept in good condition. If fingers are damaged or broken off, a new section of fingers may be soldered on or bonded* in position as a replacement. As discussed in the preceding chapter these fingers provide a good bond between the enclosure and the door by wiping motion for a short distance along the door frame. To maintain good gasketing action the door frame

---

* See Sec. 11.2.2.4 on spring finger stock in which adhesives are used to hold the stock in place rather than solder.

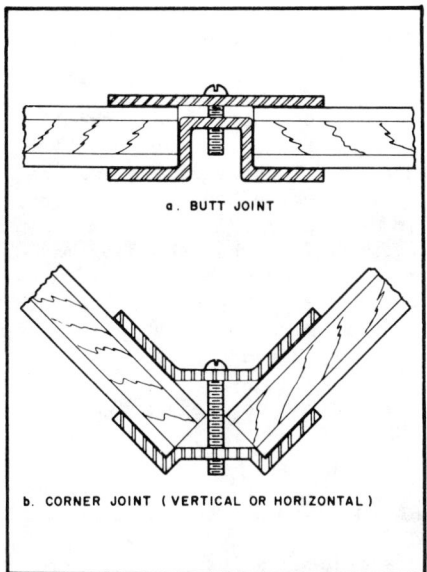

Figure 12.18 - Typical Methods of Joining Sandwich-Type Panels

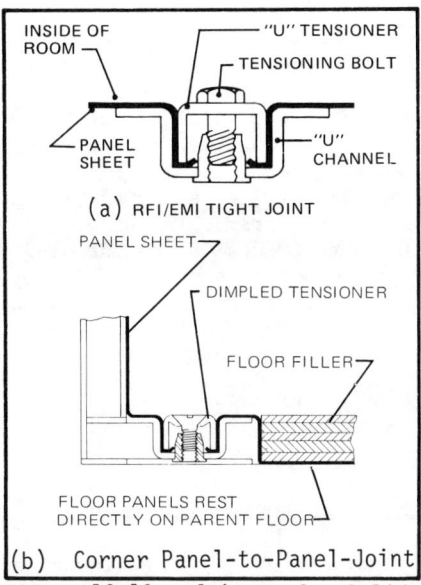

Figure 12.19 - Joiners for Solid Sheet-Stressed Metal Enclosures

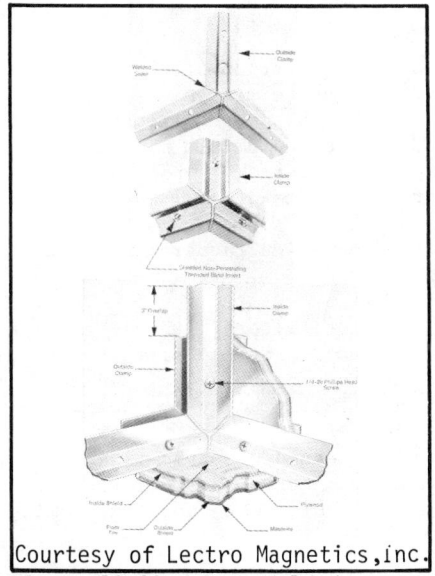

Figure 12.20 - Seam-Welded, Corner-Clamp Design

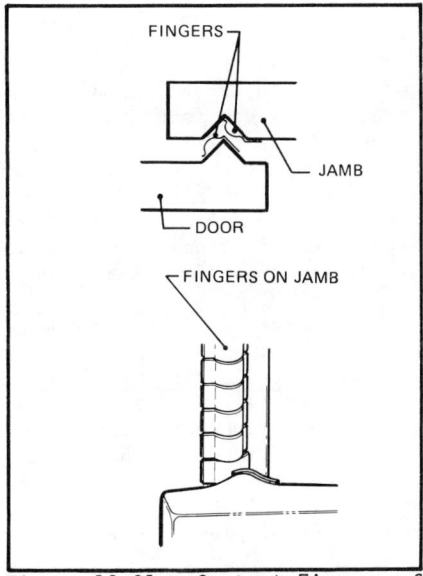

Figure 12.21 - Contact Fingers of Enclosure Door and Jamb

must be kept smooth and clean.* One exception is the pneumatic door, which does not use finger stock.

Some newer shielded enclosure doors are considerably different in concept. They are stronger, with stronger frames to provide better attenuation to R-F emissions around the door. They still use two rows of metallic finger stock or hidden rows to effect a good seal between the door and frame. An example is shown in Fig. 12.21. There are also improvements in the latching arrangements, so that the door may be opened and closed easier. Doors for all-welded shielded enclosures require uniform clamping or sealing pressures well beyond the standard approach. A very effective and reliable high-performance shielded door is the pneumatic-sealed door. This door uses neither finger stock nor R-F gaskets. One version employs a rather complex system that makes use of a set of pressurized air bags to force the door edges against a mating flange.

12.5.2   Enclosure Apertures and Penetrations

Part of providing attenuation by a shielded enclosure is accomplished by the structure including the wall sections, seams, and door. These are of little value, no matter how well designed, if the shielding integrity of the enclosure is comprised by one or more apertures such as suggested in Fig. 12.22. Many of these were discussed in Sec. 11.1.

Other than the access door, the most prominent aperture involves power-service entrance and associated filtering to remove R-F noise on the input lines (see Chap. 2, Vol. 4). The basic requirement for shielded enclosure filters is that they provide a certain minimum attenuation over the useful frequency range of the room, typically from approximately 14 kHz to 10 GHz. One problem is to determine how much attenuation the filters must offer to complement a particular room design or application, since the filter attenuation requirements are only roughly related to the attenuation of the room.

One approach to the problem is to provide filters with attenuation capability somewhat less than that of the room. For example, if the room offers attenuation of 120 dB over most of the frequency range, filters offering 100 dB should prove concomitantly adequate. If the required enclosure performance exceeds the R-F power line filter capability, the filters may be enclosed in a shielded electrical panel to increase low-frequency magnetic isolation. Normal installation practice places the filters outside the enclosure, with the line coming through pipe nipples into the room.

---

*   Some doors use finger stock on the door frame and the door must be kept clean of oxides and grease.

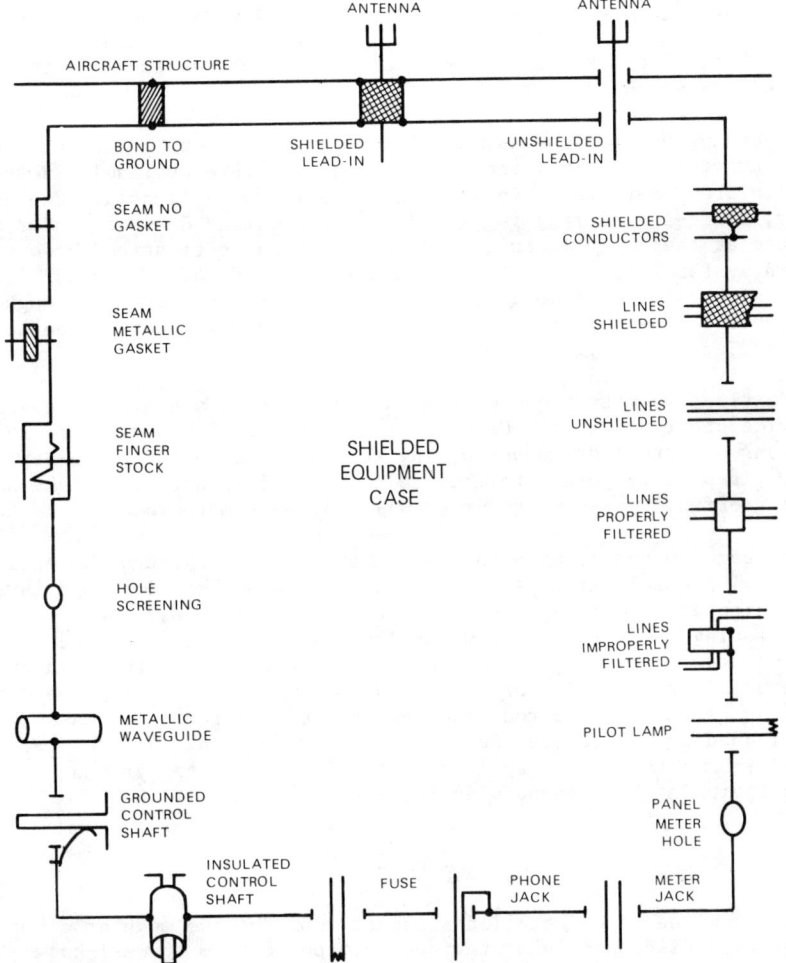

Figure 12.22 - Typical Shielded Enclosure Discontinuities

The mechanical design of high-attenuation filters is an important factor in their performance. To reduce coupling between the input and output of the separate sections, well design compartments are required (see Sec. 12.1).

The enclosure design must provide for lighting, heating and air conditioning. Lighting should be of the incandescent type, since other types usually involve ionization which produces substantial R-F noise (see Secs. 2.3 and 9.2). Special fluorescent fixtures with built-in

filters, shields, and conductive cover windows are available. Although
this type of lighting is effective in lowering the electromagnetic ambient within buildings and laboratories, it is not particularly suitable
for shielded enclosures.

Regarding heating and air conditioning, it is not generally necessary to provide additional facilities since they are available in the
immediate area where the shielded enclosure is to be located. Consequently, force-air ducting is extended from existing ducts to enter the
enclosure at the top. To assure adequate forced ventilation, however,
a return system is needed. This generally consists of an exhaust fan
forcing air to the outside ambient from within the enclosure through
a honeycomb vent which protects the shielding integrity of the enclosure (see Sec. 11.1.2).

Additional penetrations of the enclosure walls are often necessary
to provide other services. Gas, water, and compressed air may be furnished through steel or copper piping which acts as a waveguide-beyond-
cutoff. If pipe is joined to the enclosure wall in a clean, tight bond,
shielding effectiveness of the room will not be compromised.

The same method is used to bring coaxial lines through the enclosure wall. Special fittings are available that are similar to a threaded pipe nipple, except that suitable coaxial fittings are used. Coaxial
cables feeding the fittings from outside the enclousre can reduce the
shielding effectiveness of an enclosure by providing a path of entry for
high-level signals. This form of EMI develops by penetrating the cable
shield whence it is conducted into the enclosure. For this reason,
triax or quadrax cables (see Sec. 6.4) should be used. No cable will
stop EMI from entering the enclosure if it exists in the driving or
loading equipment located outside the room.

12.5.3   ENCLOSURE PERFORMANCE AND CHECKOUT

Well-designed and installed shielded rooms of the modular clamp-
together type discussed in a previous section conform to well-known
shielding requirements of MIL-STD-285, USAF Class I Shielding, or NSA
Spec. 65-6. The shielding effectiveness of some modular rooms, using
two sheets of 24-gauge steel sandwiched on 3/4-inch plywood is shown
in Fig. 12.23. The performance of all-welded rooms is also shown in
the figure for different classes of shielded rooms based on the thickness of the shielding steel used. The performance of all-welded rooms
tends to follow predictable magnetic-field attenuation more closely,
while the lesser slope of magnetic shielding for the modular room is the
result of a derating effect based on the magnetic-seam impedance of the
clamping arrangement.

In order not to create the impression that all cell-type shielded
enclosures are inferior to seam-welded designs in shielding effectiveness, Fig. 12.24 illustrates the performance of a double, electrically-

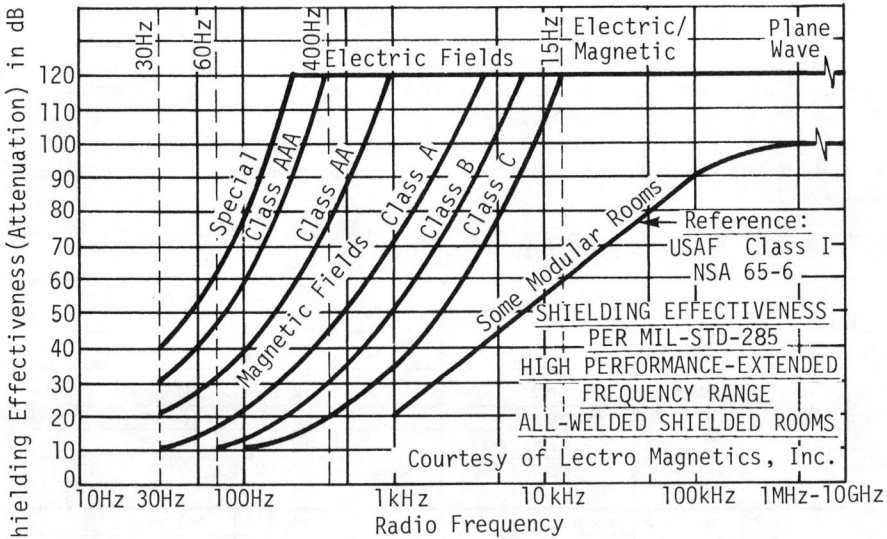

Figure 12.23 - Shielding Effectiveness of Seam-Welded and Some Modular Rooms

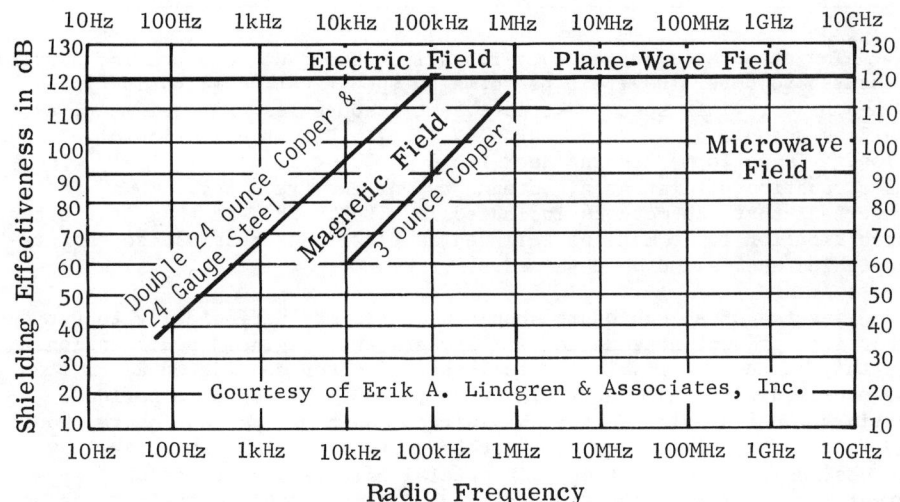

Figure 12.24 - Typical Shielding Effectiveness of a Double Electrically-Isolated Shielded Enclosure

isolated, enclosure. Here the shielding effectiveness is everywhere
superior to that captioned *some modular rooms* shown in Fig. 12.23.
The magnetic-field attenuation of the double 24 oz. copper-steel design
in Fig. 12.24 exceeds that of Classes A, B, and C of Fig. 12.23 up to
70 dB. It is difficult to make such comparisons since the parameters
are often different and measurement details may not be the same. One
attempt to present enclosure attenuation on a comparative basis is shown
in Table 12.1.

Table 12.1 - Summary of Typical Shielded Enclosure Performance

Enclosure Type	Wall Type	Material	Magnetic Fields 60 Hz	15 kHz	Electric & Plane Waves 1 GHz	10 GHz
Double Electrically Isolated	Screen	Copper		68 dB	120 dB	
		Bronze		40 dB	110 dB	
		Galvanized		50 dB	50 dB	
	Solid	24 Ga. Steel	15 dB	82 dB	120 dB	90 dB
		Cu & Steel	18 dB	93 dB	120 dB	105 dB
Cell Type	Screen	Copper		48 dB	90 dB	
		Bronze				
	Solid	24 Ga. Steel		68 dB	100 dB	
		Copper				
Single Shield	Screen	24 Ga. Steel	6 dB		60 dB	
		Bronze				
	Solid	Copper				
		24 Ga. Steel		48 dB	90 dB	

The attenuation of a newly-installed shielded enclosure is, usually,
measured to determine if it performs to specification which is typically
MIL-STD-285. This spec prescribes test frequencies and equipment, as
well as antenna separation distances. Below UHF, the testing of an enclosure is performed in the near-field yielding results may vary widely
as a function of distances, antenna types, and frequency. Thus, it is
important that the methods indicated in MIL-STD-285, if this is the test
specification to be met, be followed as closely as possible so that repeatable results can be obtained.

Testing of an enclosure should be performed periodically to verify
that it's present attenuation still meets the original specification.
In this respect, the shielded enclosure is often considered as an item
of equipment in a laboratory inventory, and is placed on a periodic
calibration schedule. After the initial checkout, the enclosure should
be retested at least once every other year. Interim spot checks may
be desirable in conjunction with special interference tests or in the
event degradation of enclosure attenuation is suspected for any reason.

12.5.4 Cost

Cost is always important and should be resolved early when a new
shielded enclosure is being considered. Cost and shielding effectiveness are interrelated features. If extended-frequency performance is
not required, the modular room is less expensive in its initial cost
for enclosures up to a size of approximately 24' x 20'. The operational
environments and maintenance must be considered to determine the end
cost over a period of years. If performance and/or environments are
paramount, the all-welded room becomes less expensive (see Table 12.2).

One expection to the above involves structures that are significantly larger than 24' x 30', and in particular when such structures
must pass building codes and other environmental factors. Here the
all-welded enclosure is generally less expensive regardless of the
degree of shielding effectiveness required. In such applications,
modular rooms are not self supporting and require considerable framing
whereas all-welded rooms are in accordance with the Uniform Building
Code and are self supporting.

Where tight budgets exist, the purchaser should not overlook the
used market as a source of shielded enclosures of the modular form. The
EMI instrument and rental houses are often aware of such sources. Used
modular enclosures generally can be purchased from 20% to 50% of the
original cost. Prospective purchasers must remember the the added cost
to disassemble, physically move, reassemble, and check out the shielding
effectiveness of such enclosures.

Table 12.2 - Cost vs Attenuation of Solid-Sheet Shielding Materials*

Material	Cost Per Sq. Ft.	Magnetic Field At 10 kHz	Electric Field At 10 kHz	Electric & Plane Wave 10MHz-10GHz
Copper				
8 oz (.018 in)	$0.590	46 dB	> 120 dB	> 120 dB
12 oz (.0162 in)	1.11	48 dB	"	"
16 oz (.0216 in)	1.20	54 dB	"	"
32 oz (.0431 in)	2.08	62 dB	"	"
48 oz (.0647 in)	3.56	68 dB	"	"
Galvanized Cold Rolled Steel				
26 Ga (.0179 in)	$0.18	46 dB	> 120 dB	> 120 dB
22 Ga (.0299 in)	0.29	66 dB	"	"
20 Ga (.0359 in)	0.35	76 dB	"	"
18 Ga (.0478 in)	0.45	92 dB	"	"
14 Ga (.0747 in)	0.67	120 dB	"	"
Cold Rolled Steel				
26 Ga (.0179 in)	$0.16	40 dB	> 120 dB	> 120 dB
22 Ga (.0299 in)	0.26	60 dB	"	"
20 Ga (.0359 in)	0.30	70 dB	"	"
18 Ga (.0478 in)	0.40	86 dB	"	"
14 Ga (.0747 in)	0.63	120 dB	"	"

NOTES:

1. Reference enclosure size is 10' x 20' x 10' high.
2. Prices are as of 1 November 1970, FOB Los Angeles, Calif.
3. Permeability of steel is measured at a µ of 160 and conductivity of 0.17.

12.6 SHIELDED HUTS AND VEHICLES

Shielded huts and vehicles are a special case of shielded enclosures discussed in the previous section. They are characterized
by their transportability which provides a substantial amount of
flexure in transit. Consequently, modular construction is seldom
used and seam-welding is the most popular variety. Techniques similar
to those used in shielded rooms are used to protect shielding integrity
at apertures and penetrations. However, due to structure flexure,
shielding effectiveness of the access door becomes the limiting factor in attenuation performance. Consequently, shielded huts and
vehicles typically exhibit 20 to 30 dB less shielding effectiveness
compared to their stationary counterparts.

Figure 12.25 shows one form of shielded hut. The air vent
flaps are typically backed by gasketed honeycomb. The power supply
is generally a remotely-located diesel engine-driven generator. Some
versions provide a battery bank of 28VDC source for quiet operation
between periods of battery charging. The hut may be transported by
either a flat-bed trailer or by helicopter. When shielded vehicles
are used, an auxiliary power supply, heating, and air conditioning
are furnished external to the driver's cab. To minimize flexure, an
air-bag suspension system is employed.

Courtesy of York Astro

Figure 12.25 - A Typical Transportable Shielded Hut

12.7 BIBLIOGRAPHY

(1) Albin, A. L., "Optimum Shielding of Equipment Enclosures," *Electronic Design*, Vol. 8, No. 3, February 3, 1960.

(2) Butterfield, B., "Polyform - A New Approach to Electromagnetic Shielding Enclosures," *Electrical Design News*, August, 1962.

(3) Cohen, D., "A Shielded Facility for Low-Level Magnetic Measurements," *Journal of Applied Physics*, Vol. 38, No. 3, 1967, pp. 1295-2196.

(4) Dolle, W.C., Van Steenberg, G.N., and Jouffray, O. L., "Effects of Shielded Enclosure Resonances on Measurement Accuracy," *Record of the 1970 IEEE International Symposium on Electromagnetic Compatibility*, Vol. 70C28-EMC, July 14-16, 1970, Anaheim, pp. 417-420.

(5) Free, W. R., et al., "Electromagnetic Interference Measurement Methods-Shielded Enclosure," Final Report on Contract No. DA 28-043 AMC-02381 (E), EES, Georgia Institute of Technology, 1967.

(6) Free, W. R., "Radiated EMI Measurements in Shielded Enclosure," *Record of the 1967 IEEE Symposium on Electromagnetic Compatibility*, Vol. 27-C80, July 18-20, 1967, Washington, D. C., pp. 43-53.

(7) Genistron, Inc. Final Report, "Proposed Specification for Electromagnetic Shielding of Enclosures and Buildings," U.S. Naval Civil Engineering Laboratory Contract NBY-33220.

(8) Hollway, D. L., "Screen Rooms and Enclosures," *Proceedings IREA*, Vol. 21, No. 10, October, 1960, pp. 660-668.

(9) Intrator, A. M., "Using Sheet Steel in the Construction of Shielded Rooms," *Electrical Engineering*, September, 1953.

(10) Jorgenson, C. M., "Shielding in Modern Computer Design," *Data Control*, December, 1958.

(11) Kanellakos, D. P., and Schulz, R. B., "New Techniques for Evaluating the Performance of Shielded Enclosures," Fifth Conference on Radio Interference Reduction and Electronic Compatibility, Armour Research Foundation, Oct., 1959.

(12) Kenny, H. W., and Conard, B. L., "A Practical Approach to R-F Shielded Enclosure Design," Ace Engineering and Machine Co., Inc., 1962. (Also appears as "RFI Shielded Enclosures" in *Electrical Design News*, September, 1962.)

(13) Klouda, J. C., "Practical Aspects of Evaluating Shielded Rooms," *Electro-Technology*, June, 1961.

(14) Kozakoff, D. J., Bolt, A. T., and Howard, F. D., "Shielding Effectiveness of Various Shaped Geometrical Enclosures in Terms of Normalized Parameters," *Record of the 1970 IEEE Regional Electromagnetic Compatibility Symposium Record*, Vol. 70C64-REGEMC, October 6-8, 1970, San Antonio, pp. V-A-1 to V-A-11.

(15) Lindgren, E. A., "Contemporary RF Enclosures," Erik A. Lindgren and Associates, 1967.

(16) Lockwood, R. O., "New Technique for the Determination of the Integrity of Shielded Enclosure by the Measurement of the Perpendicular Magnetic Field," *Record of the 1967 IEEE Symposium on Electromagnetic Compatibility*, Vol. 27C80, July 18-20, Washington, D.C., pp. 61-69.

(17) Mendez, H. A., "A New Approach to Electromagnetic Field-Strength Measurements in Shielded Enclosures," *IEEE Symposium on Electromagnetic Compatibility*, Vol. 69C3-EMC, June 17-19, 1969, Asbury Park, p. 137.

(18) MIL-STD 285, "Method of Attenuation Measurements for Enclosures Electromagnetic Shielding, for Electronic Test Purposes."

(19) Osburn, J. D., and Morris, F. J., "Problems Encountered During the Design and Fabrication of an ELF-VLF-Shielding Enclosure," *Record of the International IEEE Symposium on Electromagnetic Compatibility*, Vol. 70C28-EMC, July 14-16, 1970, Anaheim, pp. 472-478.

(20) Patton, B. J., "Room-Size Enclosure for Geomagnetic Shielding," *Record of the 1970 IEEE International Symposium on Electromagnetic Compatibility*, Vol. 70C28-#MC, July 14-16, 1970, Anaheim, pp. 89-96.

(21) Patton, B. J., and Fitch, J. J., "Design of a Room-Size Magnetic Shield," *Journal of Geophysics Research*, Vol. 67, 1962, pp. 1117-1121.

(22) Pearlston, C. B., "Enclosure Shielding in Radio Interference," *IEEE Third National Symposium on RFI*, Washington, D. C., June, 1961.

(23) Schulz, R. B., and Kanellakos, D. P., "Shielding Enclosure Performance Utilizing New Technqiues," Armour Research Foundation of Illinois Institute of Technology.

(24) Stuckey, C. W., Free, W. R., and Robertson, D. W., "Preliminary Interpretation of Near-Field Effects on Measurement Accuracy in Shielded Enclosures," *Record of the 1969 IEEE Symposium on Electromagnetic Compatibility*, Vol. 69C3-EMC.

(25) Toler, J. C., and Evans, R., "Shielded Enclosure Specification in Perspective," *1970 Regional Electromagnetic Compatibility Symposium Record*, Vol. 70C64 - REGEMC, Oct. 6-8, 1970, San Antonio, pp. III-A01 to III-A-3.

(26) U. S. Naval Civil Engineering Laboratory, "Proposed Specification for Electromagnetic Shielding of Enclosures and Buildings," Final Project Report, 31 July 1963, Contract NBy-32220, Port Hueneme, Calif. pp. 97-113.

(27) Van Steenberg, G. N., and Willman, J. F., "Shielded Enclosure Measurements Can be Accurate," *21st Annual SWIEECO Record, IEEE,* Catalog No. 69C16-SWIECO; April, 1969.

(28) Vasaka, C. S., "Problems in Shielding Electronic Equipment," *Proceedings Conference on Radio Interference Reduction*, Chicago, Ill. Armour Research Founcation, pp. 86-103, December, 1954.

(29) Vasaka, C. S., "Theory, Design and Engineering Evaluation of Radio Frequency Shielded Rooms," Report No. NADC-EL-54129, August, 1956.

CHAPTER 13

COMMUNICATIONS FILTERS

CHAPTER 13

COMMUNICATIONS FILTERS

Electrical filters used to control EMI may be divided into two types: communications or wave filters used in inter-system EMI control, and power-line filters used in intra-system EMI control. This chapter presents a summary of communication filters and filtering techniques and the next chapter surveys EMI power-line filters.

Filters are used to control *inter-system* EMI in one or more of the following ways:

(1) Selectivity in superheterodyne receivers via the I-F bandpass response formed by a combination of a fixed-tuned filter and I-F amplifier.

(2) R-F selectivity in superheterodyne receivers via a tunable pre-selector which both suppresses the image response and provides additional out-of-band rejection to strong unwanted emissions.

(3) R-F selectivity in both TRF and crystal-video receivers via either tunable pre-selectors or fixed-tuned, band-pass filters to reject out-of-band emissions.

(4) Band-rejection or notch filters in receivers to suppress EMI from strong adjacent-channel emissions or for use in wide-band amplifiers.

(5) Low-pass (or high-pass) filters to protect the front-end of receivers and other susceptible circuits or equipments from emissions existing above (or below) the base-band of operation.

(6) High-power, low-pass filters in the output of transmitters to suppress unwanted harmonic radiations.

With few exceptions, these inter-system filters are all characterized by having equal input and output impedances in the pass band of their operational networks. Typical impedance levels are 50 or 72 ohms, but occasionally 600 ohms (or other impedances) may be used at audio

13.1

frequency and 300 ohms may be used at VHF and lower UHF. They are also characterized by protecting low-level susceptible circuits, especially receivers, having sensitivities ranging from -150 dBm to about -30 dBm. One exception is item #6 above in which transmitter outputs may operate with peak-pulse powers from +20 dBm to about +100 dBm. Here filters must have very-low insertion losses in the pass band in order to readily dissipate any absorbed power.

Filters are also used in *intra-system* EMI control in one or more of the following ways but with a different emphasis:

(1) R-F suppression of unwanted signals otherwise entering or exiting from the power lines of AC power mains.

(2) R-F isolation of common-impedance coupled circuitry, such as several networks fed from common power supplies, via low-pass filters.

(3) Conducted broadband noise suppression from power tools, appliances, industrial machinery, office equipment, and other devices developing transients due to arc discharge at the brush-commutator interface of motors.

(4) Conducted broadband noise suppression from non-motor, transient-developing devices such as fluorescent lamps, electric-ignition systems, industrial controls, relays and solenoids, and other switching-action devices.

(5) Protection of susceptible devices such as transducers, computers, and electro-explosive devices.

With some exceptions, intra-system filters are characterized by having *unequal* input and output impedances when installed in their operational environments. For example, impedance sources of power mains are frequently less than one ohm at low frequencies while their loads represent higher impedances. Furthermore, both source and load impedances are frequency dependent. Emphasis for intra-system filtering is to suppress the source rather than protect susceptible circuits such as item #5 above.

As a result of the above distinction between inter- and intra-system filters and filtering techniques, it is convenient to discuss the subject in two separate chapters. This chapter discusses communications filters for inter-system applications.

13.1 MATHEMATICAL FILTER MODELS

There is no way to mathematically model an electrical filter which can be used in EMI prediction when either (1) measured data are unavailable or (2) the equivalent circuit or contents of the filter are unknown. This comes about because filters may range in rate of attenuation beyond cut-off frequency from 20 dB per decade, corresponding to a simple feed-thru capacitor or series inductor to 200 dB (or more) per decade for a 10 (or more)-stage L-C network. This is illustrated in Fig. 13.1 for n = 1 to 20 stages. The abscissa is shown in normalized units of frequency with the 0-dB notch cut-off frequency appearing in the upper left corner.

One danger in using Fig. 13.1 as a general model of low-pass filters for EMI prediction or derived high-pass, band-pass, or band rejection filters discussed later, is that (1) attenuations of more than 100 dB are difficult to achieve due to input-output cross-talk coupling and (2) the filter may completely degenerate in performance a few decades above cut-off due to parasitics. Where open circuitry is used, not involving either connectors or filter shields, it is not uncommon to have direct input-output coupling of the order of 40 to 60 dB, especially in miniature and integrated circuits. Regarding parasitics, unless special precautions are taken in the filter design, and fabrication*, a filter may offer little to no attenuation at two or more decades above cut off.

As a result of the foregoing, it is recommended that installation-measured attenuation performance data always be used for EMI prediction, wherever available. The next best data source to use is either nominal performance data or attenuation measurements of the filter not in its installed position. Finally, if all else fails and the only known data are (1) the filter type and number of stages, (2) the physical size and whether a shield and connectors are used, and (3) the general installation plan, then the maximum performance suggested in Table 13.1 may be used.

Illustrative Example 13.1

An HF (2-30 MHz) receiver is to be designed by its manufacturer who wants to make sure that little chance of interference will exist from high-power radiators operated out-of-band. Accordingly, precautions are being taken to add in front of the preselector at the receiver input terminals a high-pass filter cutting-off at 1.9 MHz and a low-pass filter cutting-off at 32 MHz. Since space is a premium, connectors cannot be used although the miniature filter will be shielded.

---

*   This is the very essence of most EMI filters. They will continue to offer a prescribed amount of attenuation up to 1 GHz, 10 GHz, or whatever the rated value may be.

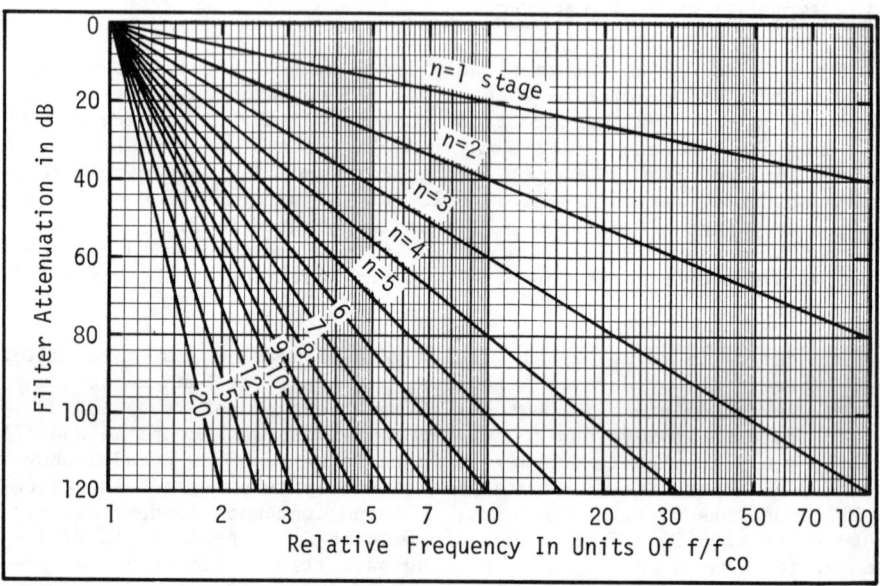

Figure 13.1 - Filter Attenuation vs Frequency

Through the use of the above filter, the manufacturer plans to offer an additional 40 dB of rejection to 54 MHz (Channel 2 TV transmissions) and at least 60 dB to UHF TV transmissions. From Fig. 13.1, for
f/f_{co} = 54 MHz/32 MHz = 1.69 and for A_{dB} = 40 dB, a nine-stage, L-C
filter is required. As a by-product, Fig. 13.1 also shows that over
100 dB attenuation is offered to powerful 5 MWatt UHF TV transmitters
(f/f_{co} = 512 MHz/32 MHz   16). The problem is to determine the more
likely performance in terms of the protection the manufacturer will probably get.

From Table 13.1, the cross talk-parasitic problem will limit the
miniature filter to a 60 dB maximum performance up to VHF. Since this
is greater than the planned 40 dB attenuation at 54 MHz, there is no
problem. For TV Channel 14, however, Table 13.1 indicates that only 40
dB of protection can be expected rather than the planned minimum of 60
dB. Further, to a 3350 MHz radar (f/f_{co} = 105), only 20 dB of protection may be realized.

Summing up, in math modeling interference the procedure to use
where no measured data are available on a filter is (1) calculate the
attenuation from Fig. 13.1, (2) select the applicable maximum attenuation from Table 13.1, and (3) choose the lesser attenuation of the two
figures.

In contrast to power-line filters (see next chapter), one distinguishing feature of communications filters (sometimes called wave filters

Table 13.1 - Model of Maximum Average Attenuation of
Electrical Filters Outside Their Pass Bands

Rejection-Band Frequency Range	Shield & Connectors	Shield Only	No Shield/ Connectors
Microminature or IC Filters			
$f_{co} \leq f \leq 10\ f_{co}$	NA	60 dB	50 dB
$10\ f_{co} \leq f \leq 100\ f_{co}$	NA	40 dB	30 dB
$f > 100\ f_{co}$	NA	20 dB	10 dB
Communication Filters (No Special EMI Precautions)			
$f_{co} \leq f \leq 10\ f_{co}$	80 dB	70 dB	60 dB
$10\ f_{co} \leq f \leq 100\ f_{co}$	60 dB	50 dB	40 dB
$f \geq 100\ f_{co}$	40 dB	30 dB	20 dB
Communications Filters (EMI Hardened)			
$f_{co} \leq f \leq 10\ f_{co}$	90 dB	NA	NA
$10\ f_{co} \leq f \leq 100\ f_{co}$	80 dB	NA	NA
$f > 100\ f_{co}$	70 dB	NA	NA
Power-Line Filters $\leq$ 10 Amps (EMI Type)			
$f_{co} \leq f \leq 10\ f_{co}$	80 dB	NA	NA
$10\ f_{co} \leq f \leq 100\ f_{co}$	80 dB	NA	NA
$f > 100\ f_{co}$	70 dB	NA	NA
Power-Line Filters > 10 Amps (EMI Type)			
$f_{co} \leq f \leq 10\ f_{co}$	100 dB	NA	NA
$10\ f_{co} \leq f \leq 100\ f_{co}$	100 dB	NA	NA
$f > 100\ f_{co}$	90 dB	NA	NA

or electrical filters), is that both the source and load impedance are
either constant over several or more octaves or they are at least definable and predictable. This feature results from the fact that communications filters usually are driven from and terminated by a relatively simple equivalent circuit element.*

Up until about 1950, communication-filter design was still much of
an art because of the clumsy *constant K* and *M-derived* techniques that
were used. The problem involved one of uncertainty in both theoretical
design and physical realizability. By the mid-1950's, however, the use

---

* In contrast, imagine what the equivalent circuit of a power main's
source looks like at RF with its myriad of parasitic elements and the
enormous load variations offered by many devices at RF such as a motor,
incandescent bulb element, transformer, etc.

of modern network synthesis became widespread, and a whole new technique of design had evolved. This, at least removed much uncertainty in theoretical design so that concentration could be directed on physical realizability. Further developments in defining filter capacitive, inductive, and resistive-element parasitics encouraged still further advance in synthesis to pre-adjust element values in order to give a still further predictable performance. Finally, with wide-spread use of digital computers, it became possible to compile an enormous number of design data which account for specific source-to-load impedance ratios, ranges of Q-factors of both capacitors and inductors, etc.

The following material summarizes and excerpts certain portions from A Handbook on *Electrical Filters - Theory and Practice.**

---

* Reprinted by Don White Consultants, 1970, 282 pages.

13.2 PROTOTYPE FILTERS

Modern network synthesis provides a means to derive any low-pass, high pass, band-pass, or band-rejection filter, having any selectivity slope in dB/octave, any cut-off frequency, and any impedance level using a family of low-pass filter prototypes. A prototype filter is defined in terms of a cut-off frequency, f_c, of 1 radian (1 radian = $1/2\pi$ Hz) and an impedance level of 1 ohm as shown in Fig. 13.2 with its dual network.* The spectrum-amplitude response of this prototype is shown in Fig. 13.3 where the 3 dB frequency and 20 dB/decade slope are readily observed.

13.2.1 BANDWIDTH AND IMPEDANCE SCALING

The low-pass filter shown in Fig. 13.2 has no practical value, per se. However, it may be made practical by scaling bandwidth and leveling impedance to any value by using the following rules:

(1) Bandwidth Scaling: divide all reactive (L-C) components by the desired new cut-off angular frequency in radians, $\omega_c = 2\pi f_c$, where f_c is the cut-off frequency in Hz. Thus:

$$L_a = L_b / 2\pi f_c \tag{13.1}$$

$$C_a = C_b / 2\pi f_c \tag{13.2}$$

where,   the subscript a = after and b = before

(2) Impedance Leveling: multiply all resistors and inductors by the desired new impedance level of source and load and divide all capacitors by the new impedance level, Z. Thus:

$$R_a = ZR_b \tag{13.3}$$

$$L_a = ZL_b \tag{13.4}$$

$$C_a = C_b / Z \tag{13.5}$$

(3) Combined Bandwidth and Impedance Scaling: both scalings may be merged into a single operation by combining Eqs. (13.1) through (13.5). Thus:

$$R_a = ZR_b \tag{13.6}$$

$$L_a = ZL_b / 2\pi f_c \tag{13.7}$$

$$C_a = C_b / Z2 \ f_c \tag{13.8}$$

---

* The dual of a network is obtained by replacing any shunt (or series) capacitance with a series (or shunt) inductance and vice versa such that the replaced elements are equal in value to the elements replaced.

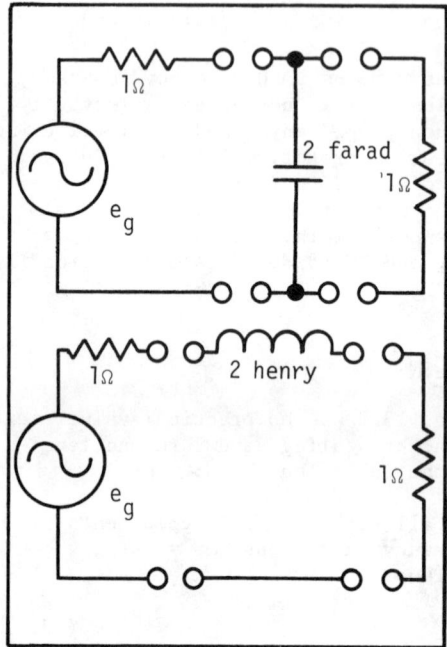

Figure 13.2 - Fundamental Low-Pass
Prototype Filter and Its Dual

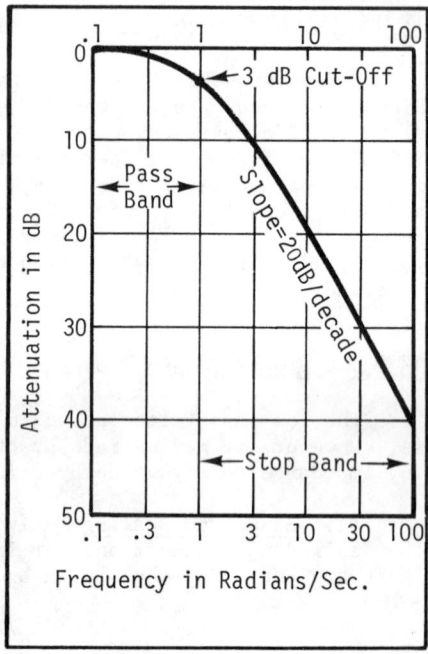

Figure 13.3 - Frequency Response
of Low-Pass Prototype

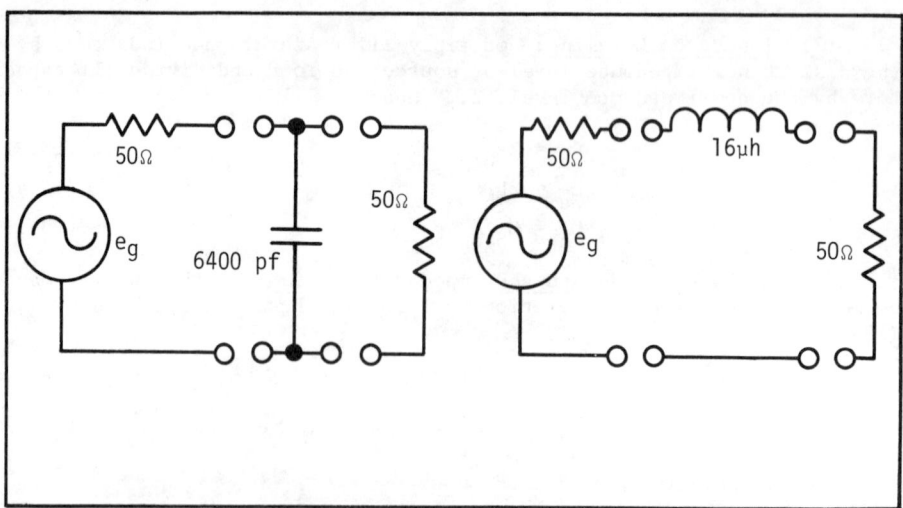

Figure 13.4 - Single-Stage, Low-Pass Filters (Duals) with 1-MHz
Cut-Off and 50-ohm Impedance Source and Load

Illustrative Example 13.2

    Design a single-stage, low-pass filter having a cut-off frequency
of 1 MHz and an input-output impedance level of 50 ohms. Choosing
either of the dual prototypes in Fig. 13.2, the new element values as
obtained from Eqs. (13.6) - (13.8) are:

$$R_a = 50 \text{ x } 1\Omega = 50\Omega$$

$$L_a = 50 \text{ x } 2/2\pi x 10^6 = 16\mu h$$

$$C_a = 2/50 x 2\pi 10^6 = 6400 \text{ pf}$$

Fig. 13.4 shows each 50-ohm filter system having a response identical
to that of Fig. 13.3 except the cut-off frequency is 1 MHz instead of
one radian.

13.2.2 RESPONSE FUNCTIONS

    The preceding filter design was quite simple since it involved a
single reactive element giving 20 dB/decade (6 dB/octave) attenuation
in the stop band. Actually, this is not a sufficient rate of attenuation for most practical applications where a small insertion loss is
desired in the pass band and substantial attenuation may be desired
just outside the cut-off frequency. For these situations multi-stage
filters are required in which the rate of stop-band attenuation beyond
cut-off is 20n dB/decade (6n dB/octave), where n is the number of reactive filter elements or stages.

    There also exists prototype, low-pass filters similar to Fig. 13.2,
but having any number of stages. The value of each reactive element in
such filters is determined from modern network synthesis, and may belong
to one of the following filter response classes:

    (1) Butterworth or maximally-flat amplitude response over the pass
band. The time-domain response, however, has some phase and time-delay
distortion over its pass band especially near the cut-off frequency region. Consequently, its transient response exhibits considerable overshoot. Yet, the Butterworth response is one of the most widely used
functions.

    (2) Tchebycheff or equal-ripple amplitude response over the pass
band. It offers a faster rate of attenuation just outside the stop band
than the Butterworth response. The price paid for this is the ripple in
the pass band which may be designed to have any value, typically from
0.1 dB to 1 dB. The time-delay and phase distortion are greater than
the Butterworth response, although the overshoot of the Tchebycheff response is somewhat less.

(3) Bessel or maximally-flat time delay response. This function exhibits extremely small overshoot. Its rise time, however, is longer than either the Butterworth or Tchebycheff functions.

(4) Butterworth-Thompson response. This exhibits some of the trade-offs of the fast rise time and flat amplitude properties of the Butterworth and the low overshoot of the Bessel function.

(5) Elliptic response, combines some of the advantages of the Tchebycheff amplitude response with a *suck out* (large attenuation) just outside the stop band similar to the old M-derived filters. Its transient response is relatively poor.

13.3   FILTER DESIGN GRAPHS AND TABLES

The above prototype responses are discussed elsewhere in the literature.* For the purpose of the remaining discussion, only the Butterworth response will be presented here. The low-pass prototype of any n-stage filter is shown in Fig. 13.5 for either an odd or even number of stages together with a dual network of the former. Table 13.2 lists the L-C element values of those prototypes for any number of stages from n = 1 to n = 20. Notice that the element values are symmetrical about the center of the filter, with first and last stages being equal, etc.

It remains to establish how the number of stages required can be calculated. The normalized frequency response function for the Butterworth filter was shown earlier in Fig. 13.1 for n = 1 to n = 20 stages. The relative cut-off or notch frequency is 1.0 which appears at the left margin of the figure and the relative stop-band frequency abscissa is extended over two decades. The attenaution (transmission loss) is presented over a 120 dB range. To determine, the number of stages required at twice the cut-off frequency ($f/f_{co}=2$) and to offer an attenuation of 50 dB, for example, Fig. 13.1 shows that n = 8.3 is required. Thus, the next highest integer, n = 9 is chosen.

Illustrative Example 13.3

It is desired to additionally protect the front end of a sensitive 2-30 MHz H-F receiver from nearby VHF TV broadcast transmitters. An analysis of geographic locations, level, and frequencies of channels 4, 5, 7 and 9 indicates that the Channel 4 (66-72 MHz) emission levels will be the most troublesome. Further, analysis indicates that at least 30 dB of additional attenuation is required to prevent intermodulation and other spurious responses in the 72-ohm antenna-receiver system.

---

* Donald R.J. White, "Electrical Filters - Theory and Applications," loc. cit.

To assure that a cut-off frequency of not less than 30 MHz will exist 5% tolerance components, a design cut-off frequency of f_c = 32 MHz is selected. At the lowest culprit emission frequency, f_e = 66 MHz to be protected, the normalized frequency f = f_e/f_c = 66 MHz/32 MHz = 2.06. To achieve at least 30 dB attenuation at 66 MHz, n = 4.9 (use n = 5 stages) is selected from Fig. 13.1. For n = 5 in Table 13.2, the low-pass prototype is shown in Fig. 13.6:

$$R_g = R_\ell = ZR_b = 72 \times 1 = 72 \text{ ohms}$$

Fig. 13.5 shows the final low-pass filter design and Fig. 13.8 shows the frequency response of this filter. Note that 31 dB attenuation exists at 66 MHz which satisfies the 30 dB requirement. The insertion loss is about 2 dB at 30 MHz and is less than 1 dB below 28 MHz.

Other Filter Types

The design of high-pass, band-pass, and band-rejection filters may be obtained directly from the low-pass prototype by a change in the frequency variable. This is discussed in Electrical Filters by White.

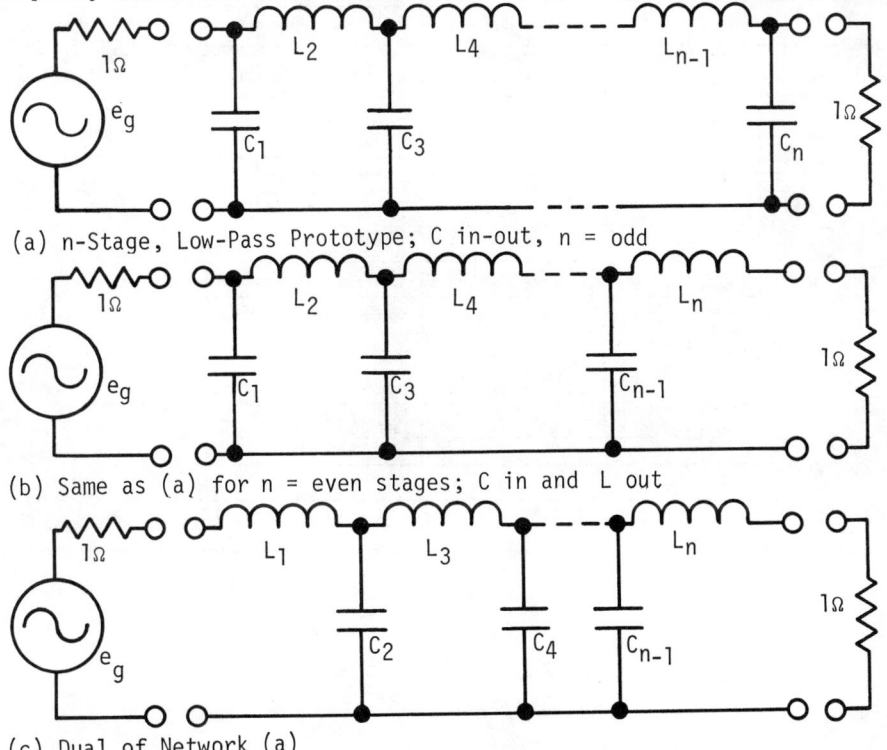

(a) n-Stage, Low-Pass Prototype; C in-out, n = odd

(b) Same as (a) for n = even stages; C in and L out

(c) Dual of Network (a)

Figure 13.5 - Fundamental, n-Stage, Low-Pass Prototype Filters

Table 13.2

ELEMENT VALUES OF BUTTERWORTH LOW-PASS FILTER PROTOTYPES
(Use this table when load and source resistance are within
30% of each other, viz when $0.7 < \overline{R} \le 1.0$)*

n	C_1	L_2	C_3	L_4	C_5	L_6	C_7	L_8	C_9	L_{10}	n
1	2.000										1
2	1.414	1.414									2
3	1.000	2.000	1.000								3
4	0.765	1.848	1.848	0.765							4
5	0.618	1.618	2.000	1.618	0.618						5
6	0.518	1.414	1.932	1.932	1.414	0.518					6
7	0.445	1.247	1.802	2.000	1.802	1.247	0.445				7
8	0.390	1.111	1.663	1.962	1.962	1.663	1.111	0.390			8
9	0.347	1.000	1.532	1.879	2.000	1.879	1.532	1.000	0.347		9
10	0.313	0.908	1.414	1.782	1.975	1.975	1.782	1.414	0.908	0.313	10
11	0.285	0.832	1.319	1.683	1.920	2.000	1.920	1.683	1.319	0.832	11
12	0.261	0.765	1.220	1.591	1.849	1.983	1.983	1.849	1.591	1.220	12
13	0.240	0.707	1.133	1.493	1.768	1.943	2.000	1.943	1.768	1.493	13
14	0.223	0.661	1.066	1.414	1.694	1.889	1.988	1.988	1.889	1.694	14
15	0.209	0.618	1.000	1.338	1.618	1.827	1.956	2.000	1.956	1.827	15
16	0.199	0.581	0.942	1.269	1.545	1.764	1.913	1.990	1.990	1.913	16
17	0.185	0.548	0.892	1.206	1.479	1.699	1.866	1.966	2.000	1.966	17
18	0.174	0.518	0.845	1.147	1.414	1.638	1.813	1.932	1.992	1.992	18
19	0.164	0.491	0.804	1.095	1.354	1.578	1.759	1.891	1.973	2.000	19
20	0.157	0.467	0.765	1.045	1.299	1.521	1.705	1.848	1.945	1.994	20
n	L_1	C_2	L_3	C_4	L_5	C_6	L_7	C_8	L_9	C_{10}	n

n	C_{11}	L_{12}	C_{13}	L_{14}	C_{15}	L_{16}	C_{17}	L_{18}	C_{19}	L_{20}	n
1											1
2											2
3											3
4											4
5					ALL L's in henrys.						5
6					ALL C's in farads.						6
7											7
8											8
9											9
10											10
11	0.285										11
12	0.765	0.261									12
13	1.133	0.707	0.240								13
14	0.414	1.066	0.661	0.223							14
15	1.618	1.338	1.000	0.618	0.209						15
16	1.764	1.545	1.269	0.942	0.581	0.199					16
17	1.866	1.699	1.479	1.206	0.892	0.548	0.185				17
18	1.932	1.813	1.638	1.414	1.147	0.845	0.518	0.174			18
19	1.973	1.891	1.759	1.578	1.354	1.095	0.804	0.491	0.164		19
20	1.994	1.945	1.848	1.705	1.521	1.299	1.045	0.765	0.467	0.157	20
n	L_{11}	C_{12}	L_{13}	C_{14}	L_{15}	C_{16}	L_{17}	C_{18}	L_{19}	C_{20}	n

$$C_1 = C_5 = C_b/Z2\pi f_c$$
$$= 0618/72 \times 2\pi \times 32 \times 10^6 = 43 \text{ pf}$$

$$L_2 = L_4 = ZL_b/2\pi f_c$$
$$= 72 \times 1.618/2\pi \times 32 \times 10^6 = 0.58 \text{ µh}$$

$$C_3 = C_b/Z2\pi f_c$$
$$= 2.00/72 \times 2 \times 32 \times 10^6 = 138 \text{ pf}$$

13.12

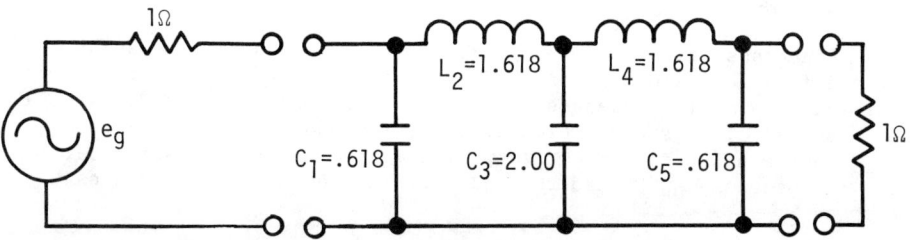

Figure 13.6 5-Stage, Low-Pass Butterworth Prototype

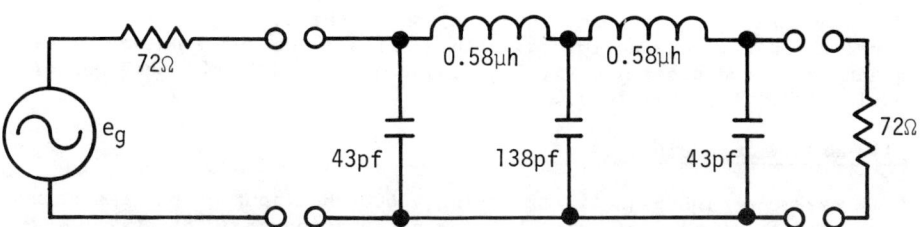

Figure 13.7 - 5 Stage, Low-Pass Filter with a 32 MHz Cut-Off and
72-ohm Impedance Source and Load.

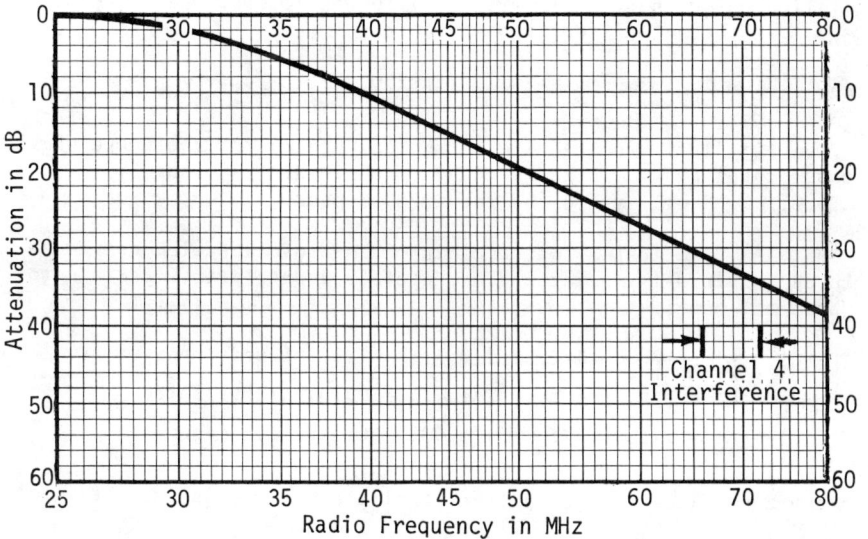

Figure 13.8 - Frequency Response of Filter Shown in Fig. 13.7

13.4 HIGH-PASS FILTERS

High-pass filters having geometric frequency symmetry can be obtained from low-pass filters*. By substituting $\omega_{hp} = 1/\omega_{\ell p}$ in the transfer function of the low-pass prototype, a high-pass prototype is obtained. By use of this transformation, the impedance of an inductance, $L\omega_{\ell p}$, becomes the impedance L/ω_{hp}; the impedance of a capacitance, $1/C\omega_{\ell p}$, becomes ω_{hp}/C; and the value of the resistance(s) remains unchanged.

This transformation is equivalent to replacing all capacitances and inductances with inductances and capacitances respectively, with each taking on the value of the reciprocal of the replaced component. Impedance leveling and bandwidth scaling of the new high-pass prototype, which also has an impedance level of one ohm and a cut-off frequency of 1 rad/sec respectively, are accomplished in the same manner as in Eqs. (13.6) - (13.8).

Illustrative Example 13.4

Assume a high-pass filter having a 600-ohm input-output resistance, a cut-off frequency, ω_c, of 1 MHz, an attenuation of 70 dB at 250 kHz (ω_ℓ), and no ripple (maximally-flat) in the pass band above 1 MHz is desired.

Since the desired attenuation or band-rejection of a high-pass filter lies below the cut-off frequency in contrast to the low-pass filter, form the normalized frequency:

$$\overline{\omega}_{hp} = \omega_c/\omega_\ell = 2\pi \times 10^6/2\pi \times 250 \times 10^3 = 4.0$$

From Fig. 13.1, the required number of stages for the Butterworth low-pass prototype ($\overline{\omega} = 4$ rad/sec and $A_{dB} = 70$ dB) is about 5.9. Therefore, n = 6 will yield the required response.

Table 13.2 indicates that the high-pass prototype element values should be (arbitrarily selecting a capacitor input; primes pertain to high pass and unprimes to low pass):

$$C_1' = L_6' = 1/L_1 = 1/C_6 = 1/0.518 = 1.932 \text{ farads/henrys}$$
$$L_2' = C_5' = 1/C_2 = 1/L_5 = 1/1.414 = 0.707 \text{ henrys/farads}$$
$$C_3' = L_4' = 1/L_3 = 1/C_4 = 1/1.932 = 0.518 \text{ farads/henrys}$$

Employing Eqs. (13.6) - (13.8), the final element values are obtained:

---

* White, "Electrical Filters," loc. cit., Chapter 5.

$$C_1'' = \frac{C_1'}{2\pi R F_c} = \frac{1.932}{2\pi \times 600 \times 10^6} = 512 \text{ pf}$$

$$L_2'' = \frac{RL_2}{2\pi f_c} = \frac{600 \times 0.707}{2\pi \times 10^6} = 67.5 \text{ } \mu h$$

$$C_3'' = \frac{C_3}{2\pi R_c} = \frac{0.518}{2\pi \times 600 \times 10^6} = 137 \text{ pf}$$

$$L_4'' = \frac{RL_4}{2\pi f_c} = \frac{600 \times 0.518}{2\pi \times 10^6} = 49.4 \text{ } \mu h$$

$$C_5'' = \frac{C_5}{2\pi R f_c} = \frac{0.707}{2\pi \times 600 \times 10^6} = 188 \text{ pf}$$

$$L_6'' = \frac{RL_6}{2\pi f_c} = \frac{600 \times 1.932}{2\pi \times 10^6} = 184 \text{ } \mu h$$

The fact that all final filter element values are not symmetrical from the ends to the center has nothing to do with the fact that this is a high-pass filter rather than a low-pass filter. It is only because the design has an even number of **stages** rather than an odd number and the impedance level is other than one ohm. The desired high-pass filter is shown in Fig. 13.9, its dual in Fig. 13.10, and the frequency response of both is depicted in Fig. 13.11. The dual network of the filter in Fig. 13.9, having an even number of elements, is exactly equal to interchanging the source and termination or turning the filter end for end. This illustrates the principle of reciprocity. This equivalence of duality and reciprocity does not apply to a filter having an odd number of elements.

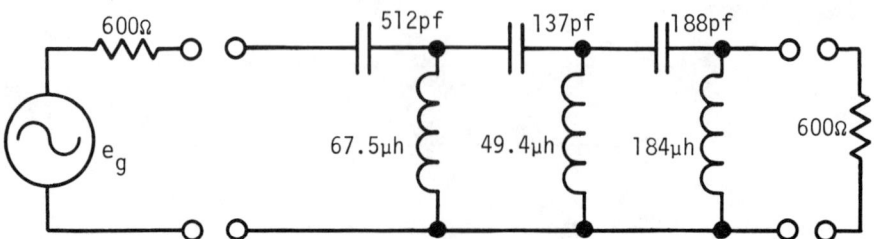

Figure 13.9 - Six-Stage Butterworth High- Pass Filter Having 1MHz
Cut-Off Frequency

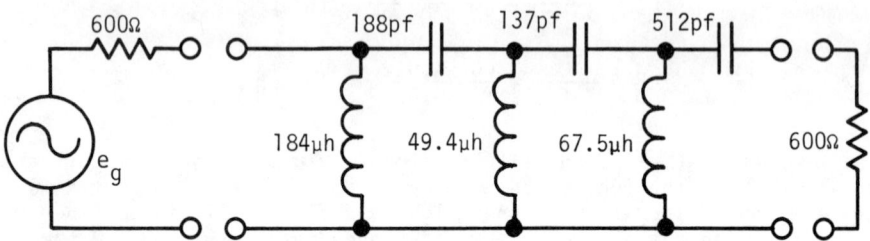

Figure 13.10 - Dual of Network Shown in Fig. 13.9

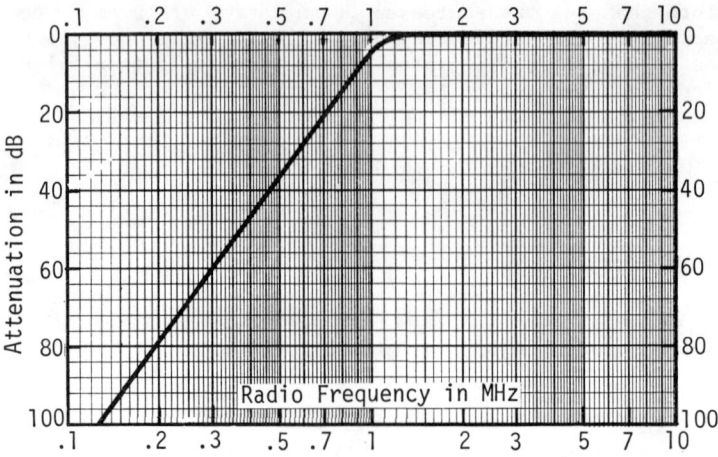

Figure 13.11- Frequency Response of High- Pass Filters Shown in
Figs. 13.9 and 13.10.

13.16

13.5 BAND-PASS AND BAND REJECTION FILTERS

Like the high-pass filter, the design of band-pass filters may also be directly obtained from the low-pass prototype by a change in the frequency variable of the transfer function. The low-pass prototype has a *center-frequency* (in the parlance of band-pass filters) of 0 rad/sec. In order to make a low-pass to band-pass filter transformation, therefore, the frequency variable, ω, must be replaced by a variable displaying a resonance (pole of the transfer function) at $\omega = \omega_o$ rad/sec instead of at 0 rad/sec. Since LC networks can display this resonant effect, the transformed variable will be of the form:

$$\omega_{bp} = \omega - 1/\omega \qquad (13.9)$$

This is equivalent to replacing in the low-pass prototype all shunt capacitances (impedance varies with frequency as $1/\omega$) with parallel-tuned circuits and all series inductances (impedance varies as ω) with series-tuned circuits.

The frequency at which Eq. (13.9) is resonant is:

$$\omega - 1/\omega = 0$$

or,
$$\omega^2 = 1; \quad \omega = \pm 1 \text{ rad/sec} \qquad (13.10)$$

In order for either the impedance of a series-tuned or the admittance of a parallel-tuned network to be reduced to zero (to give band-pass filter action) at $\omega = \omega_0$ rather than at ± 1 rad/sec, the frequency variable in Eq. (13.9) must be normalized to the resonant frequency, ω_0:

$$\omega'_{bp} = \omega_0\left(\frac{\omega}{\omega_0} - \frac{\omega_0}{\omega}\right) \qquad (13.11)$$

where the order of the terms is chosen, as in Eq. (13.9), to correspond to a negative reactance or susceptances for $\omega < \omega_0$, which is the case for tuned circuits.

The change in variable of the low-pass prototype to yield a band-pass network having a center frequency of ω_0, a bandwidth of ω_c, and hence a loaded Q-factor of $QL = \omega_0/\omega_c$, requires bandwidth scaling by dividing the 1-rad/sec cut-off frequency by ω_c; viz,

$$\frac{\omega'_{bp}}{\omega_c} = \frac{\omega_0}{\omega_c}\left(\frac{\omega}{\omega_0} - \frac{\omega_0}{\omega}\right) = QL\left(\frac{\omega}{\omega_0} - \frac{\omega_0}{\omega}\right) \qquad (13.12)$$

$$\frac{\omega^2 - \omega_0^2}{\omega\omega_c} = \frac{\omega}{\omega_c} - \frac{\omega_0^2}{\omega\omega_c} \qquad (13.13)$$

13.17

The right-hand expression of Eq. (13.13) is equivalent to saying that each series inductance in the low-pass prototype (which varies with frequency as ω) can be replaced by:

$$L_s = \frac{L_k}{\omega_c} \tag{13.14}$$

in series with a capacitance (which varies with frequency as $-1/\omega$). This is expressed in the second term of Eq. (13.13) as:

$$C_s = 1/L_k(\omega_o^2/\omega_c) = \omega_c/\omega_o^2 L_k$$

$$= \frac{1}{\omega_o Q_L L_k} \tag{13.15}$$

Similarly, each shunt capacitance is replaced by a capacitance,

$$C_p = \frac{C_k}{\omega_c}, \tag{13.16}$$

in parallel with an inductance, L_p,

$$L_p = \frac{1}{\omega_o Q_L C_k} \tag{13.17}$$

As a check, the resonant frequency of the above elements is:

$$\omega_o^2 = \frac{1}{L_s C_s} = \frac{1}{L_k/\omega_c} \times \frac{\omega_o Q_L L_k}{1} = \omega_c \omega_o Q_L = \omega_c \omega_o \frac{\omega_o}{\omega_c} = \omega_o^2$$

and,

$$\omega_o^2 = \frac{1}{C_k/\omega_c} \times \frac{\omega_o Q_L C_k}{1} = \omega_c \omega_o Q_L = \omega_o^2$$

Finally, as in the cases of the low- and high-pass filters, the impedance level may be changed from one ohm to R ohms by multiplying all resistances and inductances by R ohms and dividing all capacitances by R ohms.

$$L'_{sk} \text{ (new)} = \frac{R L_{sk} \text{ (prototype)}}{\omega_c} \tag{13.18}$$

$$C'_{sk} \text{ (new)}: \omega_o^2 = \frac{1}{L'_{sk} C'_{sk}}; \quad C'_{sk} = \frac{1}{\omega_o^2 L'_{sk}} \tag{13.19}$$

$$= \frac{\omega_c}{\omega_o^2 R L_{sk}} = \frac{1}{\omega_o Q_L R L_{sk}}$$

$$C'_{pk} \text{ (new)} = \frac{C_{pk} \text{ (prototype)}}{R \omega_c} \tag{13.20}$$

13.18

$$L'_{pk} \text{ (new): } \omega_o^2 = \frac{1}{L'_{pk} C'_{pk}}; \quad L'_{pk} = \frac{1}{\omega_o^2 C'_{pk}}$$

$$\text{(13.21)}$$

$$= \frac{R\omega_c}{\omega_o^2 C_{pk}} = \frac{R}{\omega_o Q_L C_{pk}}$$

Illustrative Example 13.5

Assume it is desired to design a 300-ohm, band-pass filter, centered at 15 mc, which will have a 3-db bandwidth of 3 mc ($Q_L = f_o/f_c = 15$ mc/3 mc = 5) and a skirt rejection of at least 30 dB at 3 mc on either side of the 15-mc center frequency. Assume that a maximally-flat (Butterworth) response is desired.

The number of half bandwidths off the center frequency is 3 divided by 3/2, or 2, to yield the normalized frequency, $\omega_{bp} = 2$. Fig. 13.1 shows that a 5-stage Butterworth filter will have the required results.

From Table 13.2 (same source and load resistance), the 5-element prototype values for a 5-stage Butterworth filter may be obtained directly. These values together with the above ω_o, ω_c, and R values are substituted into Eqs.(13.18) through (13.21) as follows:

$$C'_1 = \frac{C_1}{2\pi R f_c} = \frac{0.618}{2\pi \times 300 \times 3 \times 10^6} = 110 \text{ pf}$$

$$L'_1 = \frac{R}{2\pi f_o Q_L C_1} = \frac{300}{2\pi \times 15 \times 10^6 \times 5 \times 0.618} = 1.03 \text{ }\mu h$$

$$L'_2 = \frac{R L_2}{2\pi f_c} = \frac{300 \times 1.618}{2\pi \times 3 \times 10^6} = 25.8 \text{ }\mu h$$

$$C'_2 = C'_4 = \frac{1}{2\pi f_o Q_L R L_2} = \frac{1}{2\pi \times 15 \times 10^6 \times 5 \times 300 \times 1.618} = 4.4 \text{ pf}$$

$$C'_3 = \frac{C_3}{2\pi R f_o} = \frac{2.00}{2\pi \times 300 \times 3 \times 10^6} = 354 \text{ pf}$$

$$L'_3 = \frac{R}{2\pi f_o Q_L C_3} = \frac{300}{2\pi \times 15 \times 10^6 \times 5 \times 2.00} = 0.32 \text{ }\mu h$$

The resulting network is shown in Figs. 13.12 and 13.13.

Like band-pass filters, band rejection filters may also be derived from the low-pass prototype by a change in the frequency variable of the transfer function. This will not be carried out here (cf. White on Filters, loc. cit.).

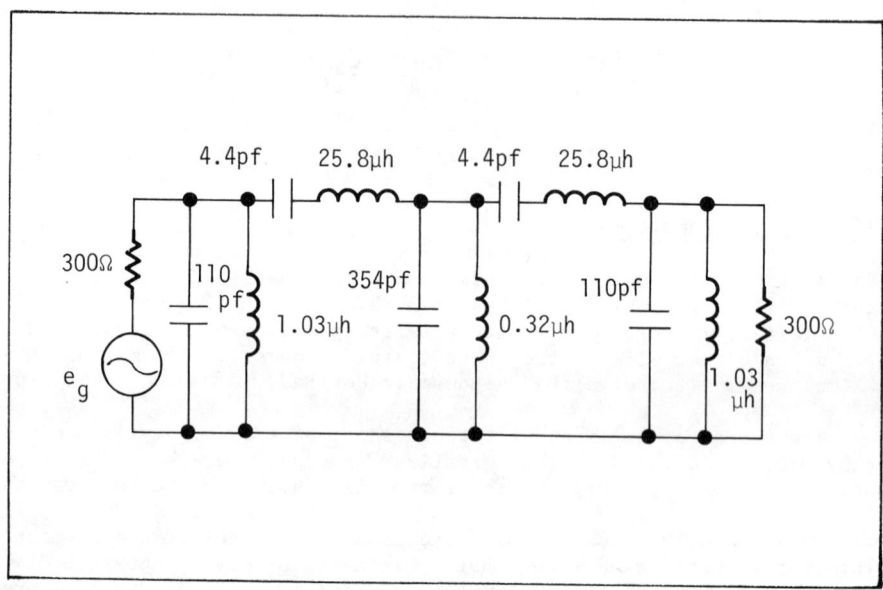

Figure 13.12 - Five-Stage, 15 MHz, Butterworth Band-Pass Filter

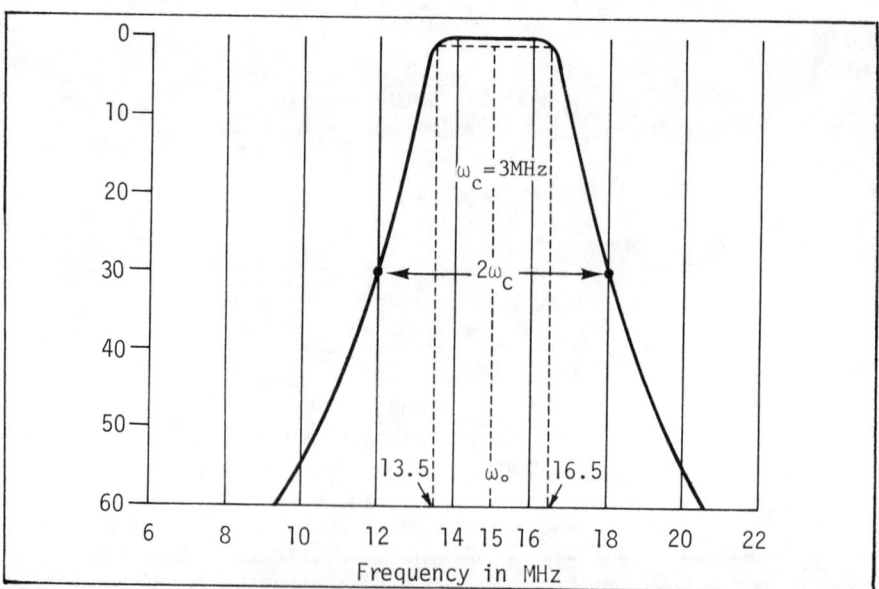

Figure 13.13 - Transmission Response of Five-Stage Filter Depicted in Fig. 13.12.

13.20

13.6 PHYSICAL REALIZABILITY OF FILTERS

Having designed on paper the desired communications filter, it remains to fabricate, align if applicable (band-pass and band-rejection filters), performance test, and install the filter. Fabrication of a filter can be an enormous process and is beyond the scope of this chapter. In general, it may be stated that (1) lumped-element, L-C (or active) filters are constructed below 300 MHz; distributed-element stripline, coaxial, or waveguide filters are fabricated above 1 GHz; and either or hybrids of both techniques may be used between about 300 MHz and 1 GHz.

The following discussion summarizes physical realizability of lumped-passive-element, low and high-pass filters over (1) the cut-off frequency range from 3 Hz to 300 MHz, and (2) the impedance level range from 1 ohm to 50 ohms. For other filter types, and ranges, the references should be used.

The following four definitions of L-C element physical realizability are established without regard to current rating of the inductors or voltage rating of the capacitors other than to assume the filters are used for low-level power transmission of less than one watt. Thus some degree of caution must obviously be exercised in employing the following definitions since certain exceptions may exist.

$$R = \text{Readily realizable (R)} \qquad (13.22)$$
$$1\mu h \leq L \leq 1h$$
$$5pf \leq C \leq 1\mu f$$

$$P = \text{Practical (P):} \qquad (13.23)$$
$$0.2\mu h \leq L \leq 10h$$
$$2pf \leq C \leq 10\mu f$$

$$M = \text{Marginally practical (M):} \qquad (13.24)$$
$$50nh \leq L \leq 100h$$
$$0.5pf \leq C \leq 500\mu f$$

$$I = \text{Impractical (I):} \qquad (13.25)$$

All element values exceeding the bounds of marginal, viz:

$$L < 50nh$$
$$L > 100h$$
$$C < 0.5pf$$
$$C > 500\mu f$$

Table 13.3 lists the physical realizability scores of R, P, M, and I for four different driving and terminating impedance loads

covering eight decades in the frequency spectrum. Note that: (1)
filters are quite realizable at any impedance level for cut-off frequencies in the 3 kHz to 3 MHz frequency range, and (2) an optimum
impedance range appears to be about 100 ohms to 1 kohm. Conversely,
physical realizability of low and high-pass filters are impractical
at the four extreme regions shown in Table 13.3.

Table 13.3 - Physical Realizability of Low-and High-Pass Filters

Impedance Level in ohms	Cut-Off Frequency, f_c							
	3Hz–30Hz	30Hz–300Hz	300Hz–3kHz	3kHz–30kHz	30kHz–300kHz	300kHz–3MHz	3MHz–30MHz	30MHz–300MHz
1Ω – 10Ω	I	M	M	P	R	P	M	I
10Ω – 150Ω	M	M	M	R	R	R	R	M
150Ω – 2500Ω	M	P	R	R	R	R	R	R
2500Ω – $-50k\Omega$	I	M	P	R	R	R	P	I

13.7   BIBLIOGRAPHY

(1)   Frank, J., "Curves Help Determine Envelope Delay of M-Derived Filters," *Electronic Design,* Vol. 9, No. 24; Dec. 6, 1961.

(2)   Geffe, R., "Simplified Modern Filter Design," New York; *John F. Rider;* chap. 1, 1963.

(3)   Herman, F.D., "FM Interference Filter," *Electronics Illustrated,* Vol. 6, No. 1; Jan., 1963.

(4)   Kno, F.F., et. al., "An Aid to the Improvement of Filter Approximation," *IRE Transactions - Circuit Techniques,* Vol. CT-9; Dec., 1962.

(5)   Lauderdale, D.M., "Dual Filter, Phase Detector From Frequency-Discrimination," *Electronic Design,* Vol. 9, No. 23; November 8, 1961.

(6)   RADC-TDR-64-80, AD 600 020, "Communications Interference Reduction Studies."

(7)   Redindel, J., "Practical VHF Filters," Technical Memorandum No. EDL-M-227, AD 236 245, April 12, 1960.

(8)   Strother, G., Jr., "Low-Loss Bandpass Filter Design," *Electrical Design News,* Vol. 7, No. 9, Aug., 1962.

(9)   Warthenen, J., et al., "Control Techniques for Receiver in High RF Fields," RADC-TDR-63-354, AD 418 230.

(10)   White, D.R.J. "A Handbook on Electrical Filters-Synthesis, Design, and Applications, " DWCI Publications, Germantown, Maryland, August 1963.

CHAPTER 14

POWER-LINE FILTERS

CHAPTER 14

POWER-LINE FILTERS

Most conducted forms of intra-system EMI result from equipment or systems sharing the same source of A-C power mains. Here, an electrical noisy source may pollute the power distribution wiring by injecting broadband emissions into wires also feeding other potentially susceptible equipments. Another mechanism involves common impedance coupling in which two or more circuits are fed from a common regulated or unregulated power supply. On the other hand, it frequently develops that a potentially susceptible equipment sharing a common power bus with an EMI source may not be affected thereby. Rather, the power line may have been electromagnetically contaminated to begin with and the mutual connection thereto is academic.

This chapter covers the topic of A-C and D-C power lines and the roll of power-line filters in decoupling noisy EMI sources from potentially susceptible equipments which must share the same power source. Thus, utility supplied power service and the filtering thereof are discussed. Power-line filter considerations, voltage and current rating, and allowable reaction power are reviewed along with filter alternation measurement methods and performance. Finally, a survey is made of available EMI power-line filters including active filters.

14.1 UTILITY-SUPPLIED POWER MAINS

Sec. 2.3 discussed radiation from and conducted noise existing on power transmission lines. This section applies to the power distribution system starting at the utility service poles and ending at the duplex outlet or power output terminals of a building, structure or vehicle.

Most bulk-supplied A-C power is provided by the local power and light company to either a consumer site or a building. Smaller capacity service of the order of 500 amperes or less, is generally furnished as single-phase, three-wire, 60 Hz with neutral grounded at the utility input power-meter panel. Voltage between either of the two lines and neutral is nominally 115 VAC and the voltage between the two non-neutral lines is 230 VAC. Larger capacity service to a facility is generally

furnished as three-phase, four-wire with the neutral grounded at the
input power meter.

14.1.1   Single-Phase Service

Fig. 14.1 illustrates two separate supply service situations having
different capacities. It is noticed that a single-phase output transformer with grounded center tap at the building is taken from one of the
three-phase line pairs of the primary feeder, utility-supplied power
from a nearby power pole or other external distribution arrangement.*

The input power lines are connected to the power meter and thence
to a circuit breaker and distribution box where the neutral terminal is
grounded to earth. 115 VAC is then distributed from both sides through
the building or facility to the various A-C outlet taps. Reasonable
effort is used to balance the two 115 VAC load distributions. In viewing
this arrangement in Fig. 14.1, it is noticed that the primary feed lines,
input service lines, and local power-distribution system throughout the
building act as one extended antenna system with various breaks and
appendages. This *antenna system* picks up radiations from the outside
world as well as electrical noise from internally-located devices within the building or facility.

One of the best ways of demonstrating the magnitude of typical
intercepted radiations is by placing a current probe on one of the
internal power-distribution lines and examining its output. Fig. 14.2
shows a spectrum amplitude measurement plot on an X-Y recorder made
with automatic test equipment. Twenty-six octaves of the spectrum were
examined from 20 Hz to 1 GHz. The power bus contamination is readily
recognized in this X-Y plot by observing the broadcast, shortwave, HF,
TV, FM, and other electromagnetic emission intercepts. The high level
of broadband noise, evidenced below approximately 1 MHz, results from
fluorescent-lamp noise emissions within the building itself for this
particular example. Thus, the utility-furnished power to the building
and its internal distribution system act as a long-wire antenna system
which is replete with electrical noise and signals which may jam sensitive equipment connected thereto.

14.1.2   Three-Phase Service

For service capacities typically greater than about 500 to 1,000
amperes at 60 Hz provided by the public utilities, it is more common
practice to furnish three-phase, four-wire, delta-wye connected to a

---

* In modern residential houses and small buildings in urban areas, the
power lines are burried and the transformer is located on a concrete above
ground.

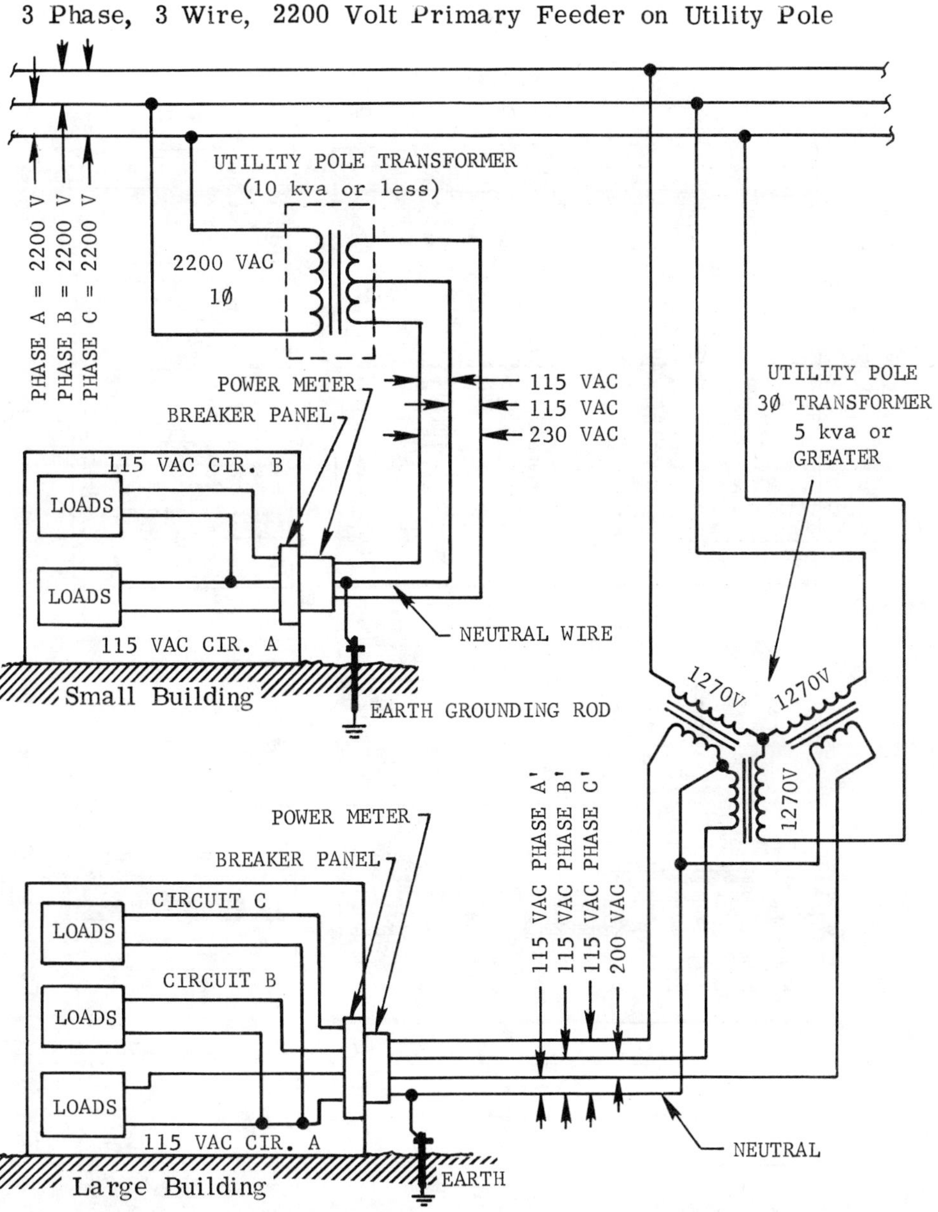

3 Phase, 3 Wire, 2200 Volt Primary Feeder on Utility Pole

Figure 14.1 - Single and Three-Phase Utility Input and Power Distribution
Systems Acting as an Extended Antenna System

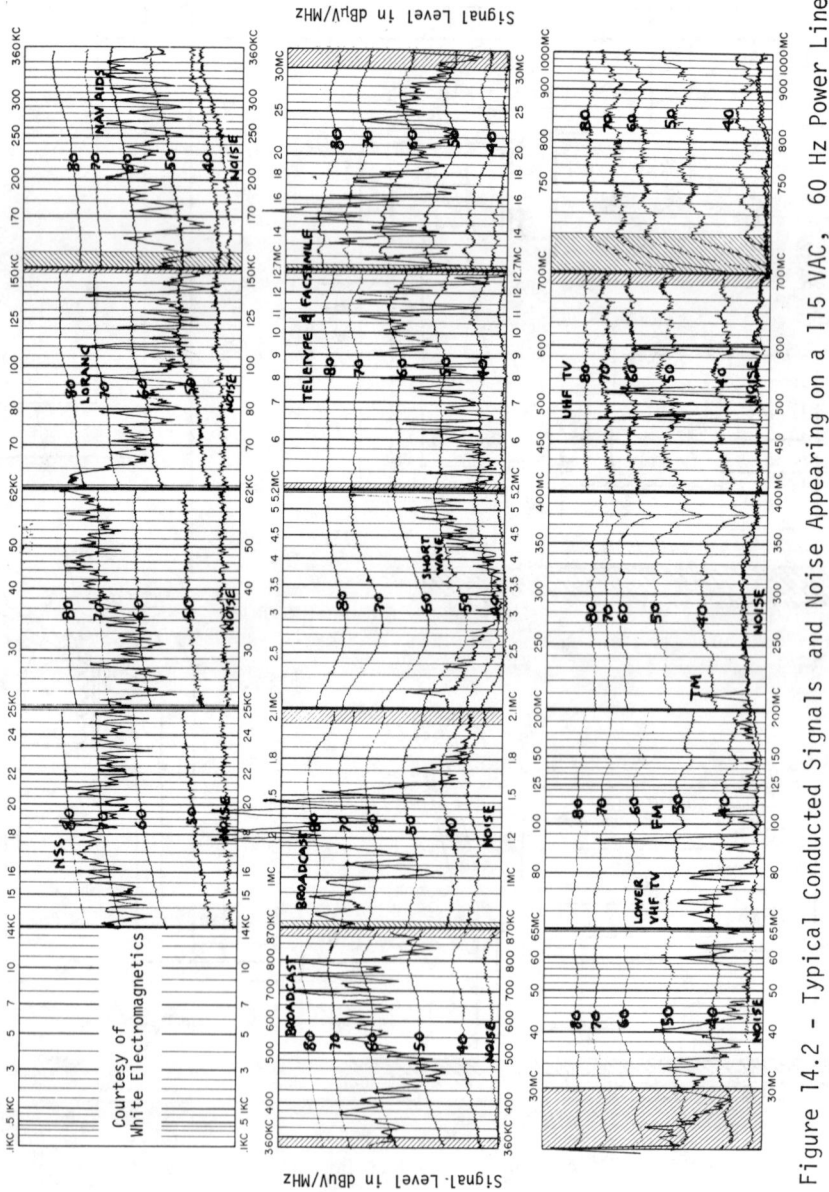

Figure 14.2 - Typical Conducted Signals and Noise Appearing on a 115 VAC, 60 Hz Power Line

building or facility. The purpose for this is to more efficiently
balance up the load distributions on the primary feeders in a given
locale especially where one customer consumes considerably more power
than several other users.

The lower portion of Fig. 14.1 also illustrates a typical three-
phase, four-wire, wye-wye connected service system provided to a facility. The primary voltage reduction transformer exists at a utility pole
or other local distribution as before. Three-phase, four-wire power,
however, is now furnished to the facility rather than single phase,
three-wire as before. The neutral is grounded at the utility input
power meter of the facility. The voltage between any single phase and
neutral is also 115 VAC, 60 Hz. The voltage as measured between any
two phases is approximately 200 VAC, in contrast to the previous 230
VAC. This results from the 120° phase difference between any two of
the three non-neutral lines.

The distribution from a three-phase, four-wire system within a
building or facility, corresponding to larger capacity requirements,
is similar to the single-phase, three-wire distribution. The same
general problems exist with regard to the primary feeder, input service
lines, and internal distribution system acting as an *extended antenna
complex* to pick up both outside world and locally generated electrical
noise. EMI spectrum-amplitude tests performed on typical internal line
distribution systems within a facility will evidence spectrum profile
emissions conducted on the line similar to that previously shown in
Fig. 14.2.

From the foregoing it is concluded that if equipments and systems
are to perform in such an EMI conducted environment, it is necessary
that the conducted emissions previously picked up by the power distribution *antenna system*, must be removed. Accordingly, power-line
filters are used at power-input terminals of many potentially susceptible equipments. They are also used in the power lines of EMI emission sources to prevent further pollution of the power mains. These
filters and their applications are the topic of the next section.

14.2 POWER-LINE FILTER SPECIFICATIONS

As explained in the preceding section, power lines feeding a given
area can act as pick-up antennas for broadcast, shortwave, HF, FM, TV,
communication emissions, radar, etc. across the frequency spectrum.
Further, these lines can conduct wide-band ignition and overhead fluorescent-lamp noise, harmonics from the A-C power mains, nearby office and
machine noise, and virtually any electrical noise which couples to the
input power lines by electric, magnetic, or electromagnetic means.
Since these potentially disturbing noises can cause EMI to sensitive
equipment, it is paramount to filter them out preferably before they
get to user areas. This is accomplished by the use of power-line filters. They must pass the DC, 60 Hz, and/or 400 Hz power-mains frequencies with very little attenuation (e.g., 0.2 dB or less) and provide
perhaps 60 dB or more attenuation from a low frequency such as 10 kHz
to 1 GHz, 10 GHz, or other frequency depending upon the EMI bounds of
potential susceptibility.

14.2.1 FILTER CONSIDERATIONS

If EMI is to be removed from a power line, it remains to establish
if the EMI present is common or differential mode. These modes are defined as follows:

Common-Mode EMI: One in which EMI exists on both black (hot) and
white (neutral) lines, but not on the green (safety) line. Thus, there
exists little to no differential EMI between black and white lines, but
a significant EMI exists between either black or white and green lines.
To be effective, then, filters must be placed between black and green
and/or white and green.

Differential-Mode EMI: One in which EMI exists between the black
and white lines in which one of the two generally contains no EMI. To
be effective, filters must be placed between the black and white, with
or without the green line involved.

As a consequence of the above, EMI filters are generally placed
between both the black and green lines and the white and green lines.
If the EMI is common mode, it will be filtered out to the green line,
whereas if it is differential mode, it will be filtered out between the
black and white lines quite independent of the presence of EMI on the
green line. However, should all three wires contain the same common-
mode, EMI will not be removed by the above filtering action. A differential-mode input circuit (the situation for nearly all networks) will
greatly reduce the impact of nearly equal common-mode inputs.

There are usually only a few specifications of importance in selecting the right power-line filter to accomplish suppression of EMI conducted emission. These specifications include:

(1) Voltage and current rating at the power mains frequency.
This includes allowable voltage drop across the filter under full-load
conditions (e.g., less than 0.2 dB, i.e., 2%, at 30 amps.). This also
includes allowable harmonic distortion of the power-line frequency under
full-load conditions (e.g., all harmonics above 10 kHz are to be more
than 80 dB down from the amplitude of the power mains frequency).

(2) Allowable reactive current at the power-mains frequency (e.g.,
not more than 10% of rated full-load current).

(3) Attenuation expressed in dB over the operational frequency
range for both pass and stop bands for a defined source and load impedance.

(4) Mechanical considerations including size and weight, type of
housing and mounting. Also shielding protection of the filter housing,
per se, to electric, magnetic, and electromagnetic fields.

14.2.2 Voltage and Current Rating

The voltage rating is the important factor in assuring the insulation of the internal capacitor(s) used in the filter will not break down
under maximum supply peak voltage conditions including undesired transients. In order to get the best EMI suppression (attenuation over the designated frequency spectrum) for a defined filter size or weight, more
capacitance per unit volume can be attained with the least voltage rating
possible. Thus, to prevent overdesign and increased cost and weight, the
voltage rating should accommodate the highest peak power supply voltage
expected, and no more. For example, a 150 VAC (RMS rating) filter should
be specified for a 115 VAC (RMS rating) line and not a 250 VAC filter,
provided expected line transients are not more than 20% (see below).

The current rating is important to assure that the internal inductors used in the filter will not saturate or otherwise improperly
perform. Inductors are made using toroids which can saturate and produce power-line harmonics if under-rated. Also, as the load current (I)
is increased, the voltage drop (IZ) across the filter inductor impedance
(Z) will increase. This can result in both (1) poor equivalent voltage
regulation of the power mains and filter combination, and (2) transient
coupling when different loads at the test specimen are turned on or off.
The largest transients are likely to develop when a load is turned off
since many loads appear inductive.

The above considerations are illustrated in Fig. 14.3. The voltage
drop, V_f, across the source generator-filter combination may change by
0.5% (say, 0.5 volt for a 115 VAC supply) due to source regulation when
a significant load is removed (turned off). However, the Q-factor (Q =
$2\pi fL/R$ total) of the inductor-generator-load combination may be of the
order of 20 at some frequencies, so that a momentary transient of voltage amplitude, $V = QV_f \approx 20 \times 0.5$ or about 10 volts can surge to the

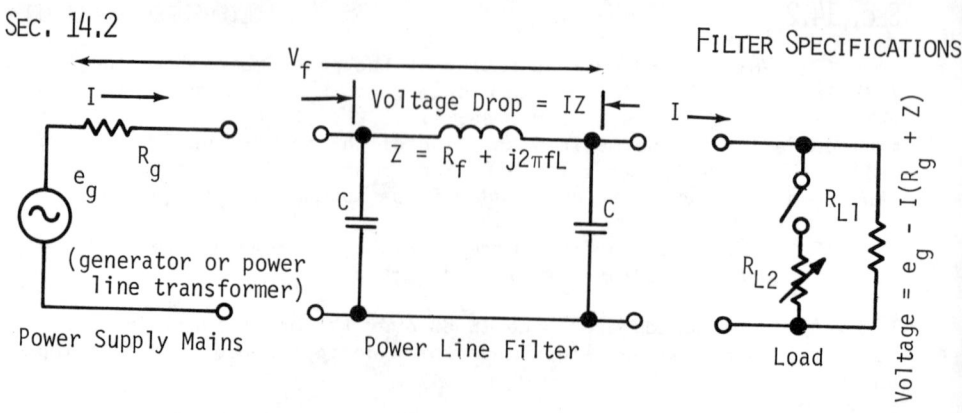

where:

e_g = Power Supply Voltage
I = Current delivered to load from power supply
R_f = Resistance of filter inductor in ohms
R_{L1}= Fixed load resistance

R_{L2} = Variable load resistance
L = Inductance in henrys
C = Capacitance in farads
Z = Impedance in ohms
f = Frequency in Hertz

Figure 14.3 - Simplified Equivalent Circuit of One Power Source, Filter, and Loads

remaining load. The loads, on the other hand may generate transients of several hundred volts, especially if inductive in nature.

14.2.3 ALLOWABLE REACTIVE POWER

For low-frequency performance of power-line filters, capacitors and inductors are generally used (e.f., active filters). Among other parameters, the filter attenuation at any out-of-band frequency increases with an increase either in the capacitor or inductor values. Thus, other than size or economic considerations, the value of the first capacitor, for example, could be made arbitrarily large. Capacitors of 10 µf values, capable of performing up to or beyond 1 GHz, are not uncommon. At a frequency f, of 60 Hz, the value of the resulting capacitive reactance, X_c, is:

$$X_c = \frac{1}{2\pi f C} = \frac{1}{6.28 \times 60 \times 10^{-5}} = 266 \text{ ohms} \tag{14.1}$$

For a 115 VAC supply, the reactive current, I, is I = 115/266 = 0.43 amperes. Should the first capacitor be increased, say to 250 µf, to substantially lower the cut-off frequency, then the reactive current would be about 10 amperes. This not only increases enormously the shock hazard but requires that the source provide the reactive power capacity to furnish this. Stating this another way, if an auxiliary gasoline engine-driven generator power supply were used, a wasted capacity of 10 amps x 115 Volts or more than 1,000 volt-amperes would result.

From the foregoing it is seen that if the power supply capacity is limited, then the user would want to also specify filters which do not draw more than some maximum allowable reactive current or volt-amperes. If a computation is not made, then a figure such as 10% of rated load might be used for maximum allowable filter reactive power. If safety to shock hazard is the governing requirement, then a maximum allowable 5 mA of reactive current corresponds to a 0.01 µf capacitor for a 115 VAC line at 60 Hz.

14.2.4   FILTER ATTENUATION PERFORMANCE

Attenuation over a prescribed frequency range is perhaps the most common way of specifying filter spectrum performance and is also one of the most abused terms in EMI filters. Filter attenuation refers to the ratio of output voltages, before and after filter insertion, as a function of frequency. Attenuation, A, expressed in decibels (dB) is derived in the following manner (see Fig. 14.4):

$$A_{dB} = 10 \ \log_{10} \left(\frac{P_b}{P_a} \right) \qquad (14.2)$$

where, P_b = Power delivered to the load *before*
               the insertion of the filter

P_a = Power delivered to the load *after*
        the insertion of the filter

Since the load impedances, Z, remains unchanged in both the *before* and *after cases*, the respective powers of Eq. (14.2) may be substituted by their voltages and impedances of Eq. (14.2):

$$P = \frac{V^2}{Z_1} \qquad (14.3)$$

Thus, $$A_{dB} = \log_{10} \left(\frac{V_b^{2/Z_1}}{V_a^{2/Z_1}} \right) \qquad (14.4)$$

$$= 10 \ \log_{10} \left(\frac{V_b}{V_a} \right)^2 \quad = 20 \ \log_{10} \left(\frac{V_b}{V_a} \right) \ dB \qquad (14.5)$$

Note that Eq. (14.5) is truly insertion loss expressed in dB and requires that the measurements at any frequency be made by removing and inserting the filter.

Since it is very time consuming to make a series of measurements with the filter both in and out of the installation network at each frequency, a more common method is to permit rapid switching between the two situations such as shown in Fig. 14.5. This is more fully described in MIL-STD-220A (see Volume 1) and Volume 2 of this handbook series on *EMI Test Methods and Procedures*.

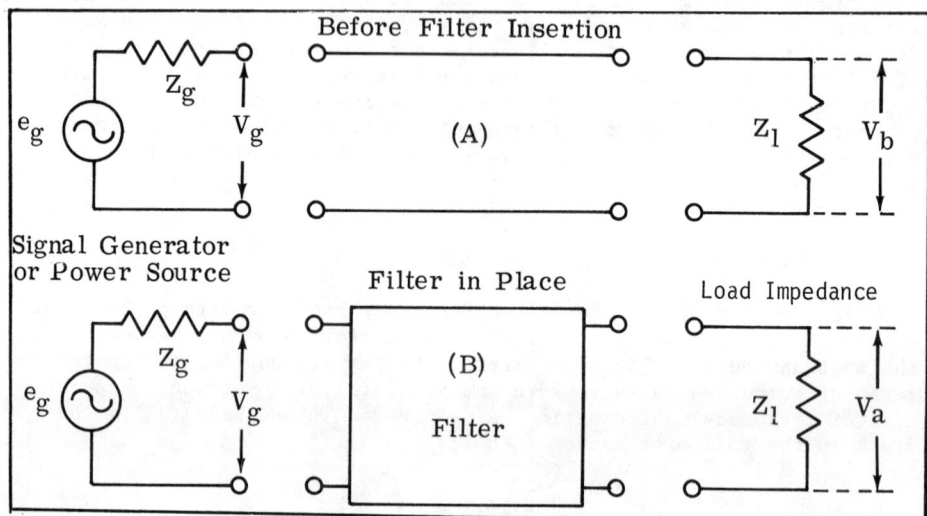

Figure 14.4 - Computing Filter Attenuation by Measuring the Voltage before (V_b) and after (V_a) Filter Insertion (Insertion Loss Method).

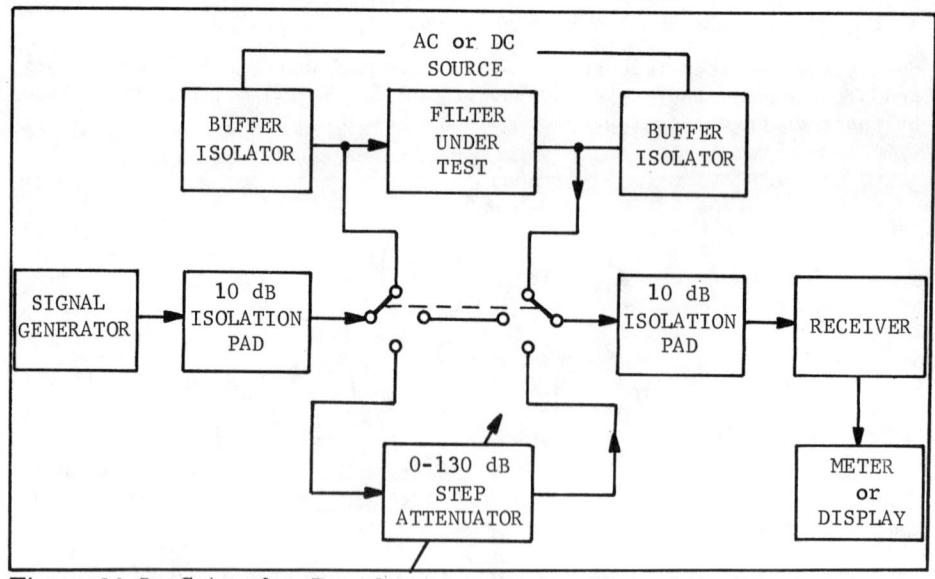

Figure 14.5   Setup for Rapid Measurement of Filter Attenuation (Insertion Loss) in a 50-Ohm Measurement System (per MIL-STD-220A).

14.10

For power filters, however, the above attenuation measurement method is invalid since the impedance of both the source and terminations of real-life installations significantly different from the 50 ohms used in a convenient coaxial measuring system. Since the filter's frequency response behavior is impedance-level dependent,* the test method is nearly meaningless other than some relative basis to compare filters. For example, a 115 VAC power supply mains may provide a 100 ampere service with not more than 5% voltage drop. This corresponds to an impedance source, Z_g, at 60 Hz of (see Fig. 14.3):

$$V_g = e_g - IZ_g = e_g (1 - 0.05)$$

or,

$$Z_g = \frac{0.05 e_g}{I} = \frac{0.05 \times 115V}{100 \text{ amp}} = 0.06\Omega \qquad (14.6)$$

For a 55 ampere load, for example, the termination impedance, Z_L, is:

$$Z_L = \frac{V_L}{I_L} = \frac{110V}{55A} = 2 \text{ ohms} \qquad (14.7)$$

The above example illustrates one typical low-frequency power-source impedance of 60 milliohms and a termination load of 2 ohms at 60 Hz. Now, if a typical single-element filter (e.g., capacitor of inductor) were measured in an impedance system of this amount compared to the 50-ohm system generally used for rating purposes, the results of the spectrum attenuation performance will be significantly different. Fig. 14.6 illustrates this in which either a single shunt capacitor (C = 0.63 μf) or a single series inductor (L = 1.6 mh) results in a cut-off frequency of 10 kHz when measured in a 50 ohm system. Beyond cut-off, the rate of attenuation is 6-dB per octave or 20 dB per decade. Thus, at 10 MHz, for example, the measured attenuation would be 60 dB.

If the above series-inductor or shunt-capacitor filter element were placed in the above $Z_g = 0.06\text{-}\Omega$ and $Z_L = 2\text{-}\Omega$ system, which is more closely related to an actual installation, entirely different results follow. Fig. 14.6 shows that the cut-off frequency of the capacitor filter has increased from 10 kHz to 4.3 MHz and the attenuation thereafter is 52 dB poorer than when measured in a 50-Ω system. In a like manner, Fig. 4.6 shows that the cut-off frequency of the inductor filter has increased to 490 kHz and the attenuation thereafter is 33 dB poorer than when measured in a 50-Ω system. Thus, it is concluded that MIL-STD-220A gives meaningless results regarding attenuation. Both source and load impedance are pertinent to the filters performance (see Sec. 13.2).

---

* See *A Handbook on Electrical Filters - Synthesis, Design and Applications*, by D.R.J. White, 1970, Don White Consultants. Also see Sec. 13.2.1.

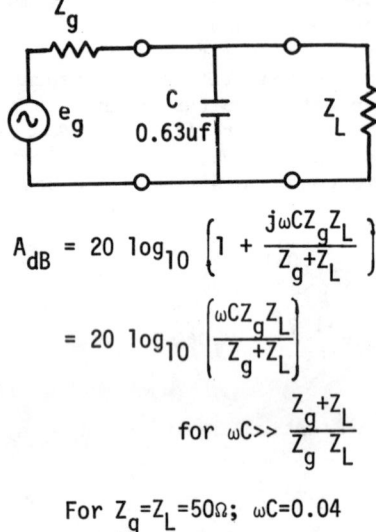

$$A_{dB} = 20 \log_{10} \left[1 + \frac{j\omega C Z_g Z_L}{Z_g + Z_L} \right]$$

$$= 20 \log_{10} \left[\frac{\omega C Z_g Z_L}{Z_g + Z_L} \right]$$

for $\omega C \gg \dfrac{Z_g + Z_L}{Z_g Z_L}$

For $Z_g = Z_L = 50\Omega$; $\omega C = 0.04$
$f_c = 10kHz$; $C = 0.63uf$

$$A_{dB} = 20 \log_{10} \left[1 + \frac{j\omega L}{Z_g + Z_L} \right]$$

$$= 20 \log_{10} \left[\frac{\omega L}{Z_g + Z_L} \right]$$

for $\omega L \gg Z_g + Z_L$

For $Z_g = Z_L = 50\Omega$; $\omega L = 100$
$f_c = 10kHz$; $L = 1.6mh$

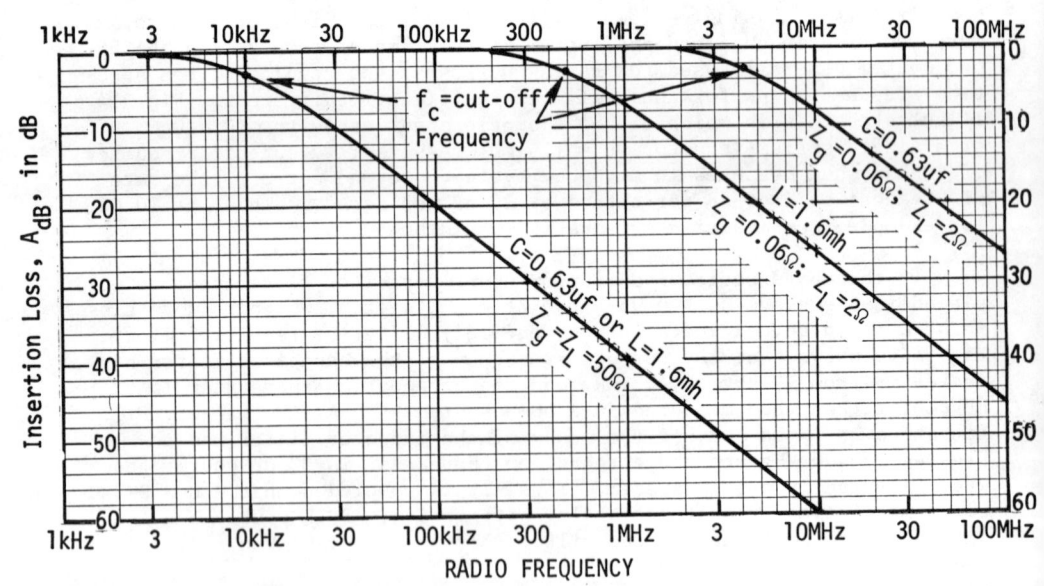

Figure 14.6 - Attenuation (Insertion Loss) of a Single Element Filter in a 50 ohm and in a Low-Impedance Source and Load System

It develops that the actual performance situation for three or more stages of L-C filtering is not quite as bad as implied above for one stage. Depending upon both source and termination impedance, one or more filter stages may be offered as sacrificial elements to establish either a source and/or load impedance. This results in a different number of equivalent filter stages, n, as shown in Table 14.1.

Figs. 14.7-14.10 illustrate different filter configurations from which one may be selected to work into or out of either a high or low, source or load impedance relative to 50 ohms. All filters shown are of the low-pass type (i.e., they use series inductors and shunt capacitors). The philosophy then is to connect either (1) a filter series inductor to a low-impedance source or (2) a shunt capacitor to a high-impedance source such that the impedances of source and filter element are about equal at the desired cut-off frequency. Similarly, a series inductor should face a low-impedance load and a shunt capacitor should face a high-impedance load. This assures optimum use of filter elements and in part compensates for some source and/or load impedances of typical power-mains varying over wide ranges starting about 100 times the power frequencies.

Table 14.1 - Equivalent Number of Filter Stages for Installation Impedance Differing from Design Source and Load Impedances

Installation Impedance		Equivalent Number of Filter Stages*	Cut-Off Frequency for n=1 Stage	
Source	Load		Capacitor	Inductor
Lower	Lower	n	Increase	Decrease
Lower	Same	n-1	Increase	Same
Lower	Higher	n-2	Increase	Increase
Same	Lower	n+1	Same	Decrease
Same	Same	n	Same	Same
Same	Higher	n-1	Same	Increase
Higher	Lower	n+2	Decrease	Decrease
Higher	Same	n+1	Decrease	Same
Higher	Higher	n	Decrease	Increase

14.2.5 FILTER SIZES, WEIGHT AND MOUNTING

As shown in the next section, EMI power-line, low-pass filters may assume any one of a number of configurations. Their size and weight may range up to about six orders of magnitude. The smallest size and weight is about 0.005 in^3 and 10^{-3} lbs; the largest is about 10^4 in^3 and 200 lbs. In general, the filter size and weight will increase with:

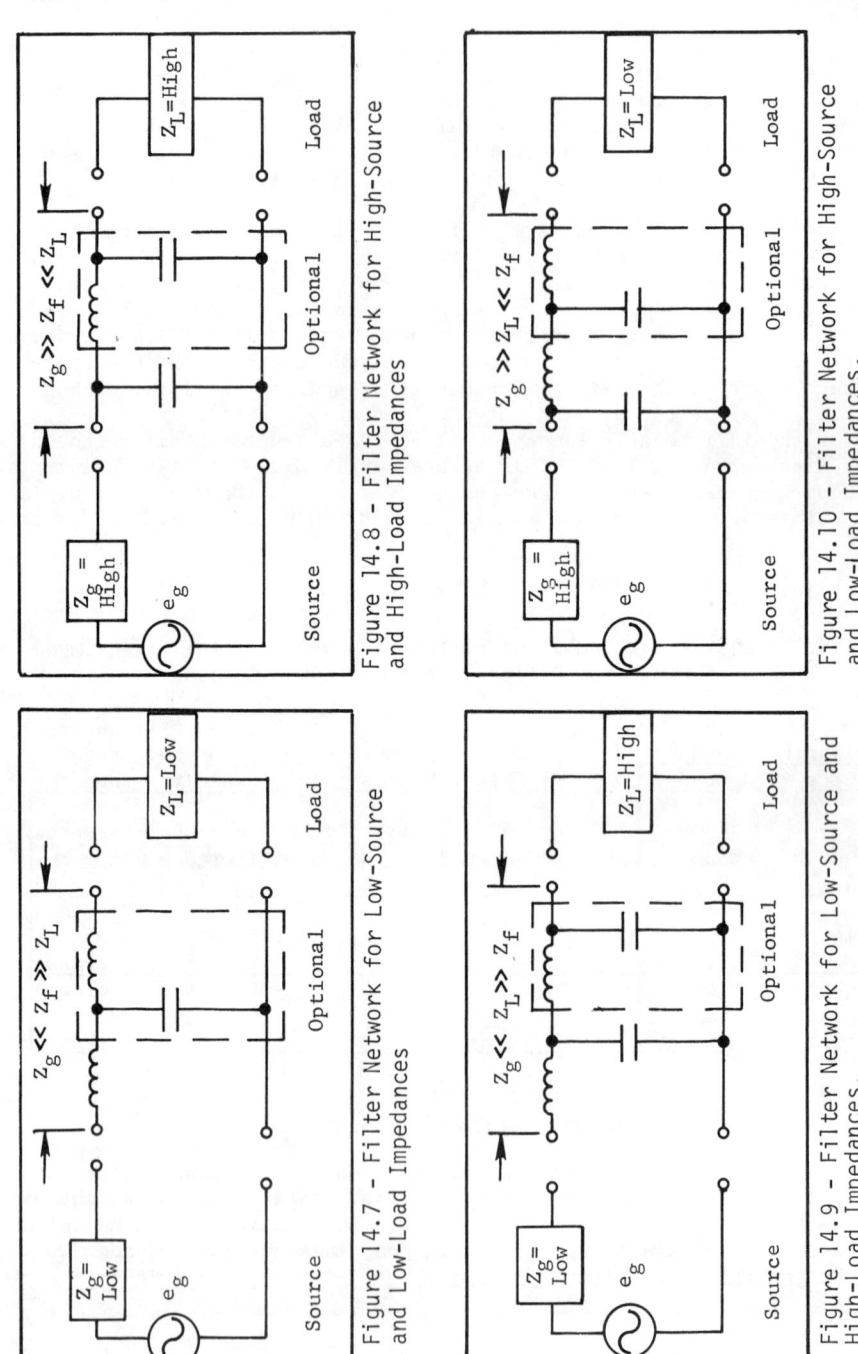

Figure 14.8 - Filter Network for High-Source and High-Load Impedances

Figure 14.10 - Filter Network for High-Source and Low-Load Impedances.

Figure 14.7 - Filter Network for Low-Source and Low-Load Impedances

Figure 14.9 - Filter Network for Low-Source and High-Load Impedances.

- Increase in voltage rating, V volts
- Increase in current rating, I amps
- Decrease in internal voltage drop at rated
  full-load current
- Decrease in cut-off frequency, f_c, in kHz
- Increase in out-of-band attenuation just above
  cut-off frequency (i.e., number of stages)

Cross talk between the filter input and output terminals may be significant (60 dB or even lower, i.e., greater coupling) unless an infinite baffle is used between the two terminal pair. Consequently, when the manufacturer's rate the filter attenuation with frequency, it is understood that the filter is mounted in a suitable bulkhead. For shielded enclosure filters, where attenuation is rated up to 100 dB or more, the entire assembly is shielded in a permeable case. Thus, after filtering, there is reasonable assurance that there will be no cross coupling of magnetic, electric, or electromagnetic fields to the filter output leads which are located inside of the enclosure.

Ambient temperature and shock and vibration specifications for the filter must also be stated unless it is to be operated at room temperature in a fixed location.

14.3   AVAILABLE POWER-LINE FILTERS

This section discusses L-C, lossy-line, and active filters which are available from a number of manufacturer's of power-line filters. The latter two filter types are included since they offer interesting possibilities where passive L-C filters run into limitations.

14.3.1   L-C LOW PASS FILTERS

Nearly every EMI power-line filter on the market is of the low-pass type, passing DC and/or 60 Hz or 400 Hz power mains, and cutting off above these frequencies there are a number of manufacturers of power-line filters which are used either to filter the power mains into an open area, a shielded enclosure, or to an instrument, equipment or device. Among the companies supplying power filters to the EMI/EMC community, are:

1. Aerovox Corp., 740 Belleville Ave., New Bedford, Mass. 02711 (617-994-9661)
2. Allen-Bradley Co., 1201 So. Second Str., Milwaukee, Wisc., 53204 (414-671-2000)
3. American Electronic Labs Inc., P.O. Box 552, Lansdale, Pa., 19446 (215-822-2929)
4. Amphenol Canada Ltd., 44 Metropolitan Rd., Scarborough, Ont., Canada (416-291-4401)
5. Captor Corp., 5040 S. County Rd., 25A, Tipp City, Ohio 45371 (513-667-8484)
6. Cornell-Dubilier Electronics, 150 Ave. L, Newark N.J. 07101 (201-589-7500)
7. Electro Magnetic Filter Co., 231 S. Whisman Rd., Mountain View, Calif. 94040 (415-969-1050)
8. Erie High Frequency Products, Inc., 2206 W. 15th St., Erie, Pa. 16505 (814-456-7084)
9. Erie Technological Products, 5 Fraser Ave., Trenton, Ont., Canada (613-392-9251)
10. Erie Technological Products, 644 W. 12th St., Erie, Pa. 16512 (814-453-5611)
11. Filtron Company, Inc., 131-15 Fowler Ave., Flushing, New York, 11355 (212-445-7000)
12. Genisco Technology Corp., Genistron Div., 18435 Susana Road, Compton, California 90221 (213-774-1850)
13. Gulton Industries, Inc., EMI Devices, Gulton St., Metuchen, N.J. 08849 (201-548-2800)
14. Hopkins Engr. Co., 12900 Foothill Blvd., San Fernando, Calif. 91342 (213-361-8691)
15. I-Tel Inc., 10504 Wheatley St., Kensington, Md. 20795 (301-946-1800)
16. Lectromagnetics, Inc., 6056 W. Jeff. Blvd., L.A., Calif. 90016 (213-870-9383)
17. Lundy Elec. & Systems, Inc., Glen Head, N.Y. 11545 (516-676-1440)

18. Mu-Del Electronics, Inc., 11212 Grandview Ave., Wheaton, Md. 20902 (301-946-3937)
19. Potter Co., The, 500 W. Florence Ave., Inglewood, Calif. 90301 (910-328-6138)
20. R.F. Interonics, A. Div. of KDL NAVCOR, Inc., 100 Pine Aire Dr., Bay Shore, N.Y. 11706 (516-231-6400)
21. Rotron, Inc., P.O. Box 743, Skokie, Ill. 60076 (312-327-4020)
22. Spectrum Control, Inc., 152 E. Main St., Fairview, Pa. 16415 (814-474-5593)
23. Sprague Electric Co., 481 Marshall St., N. Adams, Mass. 01247 (413-664-4411)
24. Standard Electronics Co., 1611 W. 63rd St., Chicago, Ill. 60636 (312-778-4222)
25. U.S. Capacitor Corp. 2151 North Lincoln St., Burbank, Calif. 91504 (213-843-42222)
26. Varian Associates, Microwave Components Div., 611 Hansen Way, Palo Alto, Calif. 94303 (415-326-4000)
27. Welex Electronics, 2431 Linden Lane, Silver Spring, Md. 20910 (301-589-5211)

Fig. 14.11 shows a few typical power-line filters available from some of the above suppliers. Fig. 14.12 illustrates two typical spectrum attenuation plots of available power-line filters used for shielded enclosures. Table 14.2 summarizes some of the power-line filters, characteristics, and approximate prices as of 1972, together with an identification of the suppliers. The method of measure of these insertion loss characteristics was in accordance with MIL-STD-220A procedures.

14.3.2 LOSSY-LINE FILTERS

Lossy-line filters are based on one of two principles of operation: dielectric losses and/or permeable losses. In Fig. 14.13, the dielectric medium intentionally corresponds to a high dissipation factor or loss tangent. The equivalent circuit shown in Fig. 14.14 shows that the dielectric conductive losses convert the R-F energy into heat. In ordinary transmission lines, these losses are negligible, since low dissipation factor dielectrics are used, but for EMI filter design, the high dissipation factor corresponds to a cut-off frequency of about 10 MHz when used in a 50 ohm system. Thus, lumped-element EMI filters which would fail to perform above, say, 100 MHz, perform well to 10GHz when the lossy graphite is used as a potting compound.

The other technique is based on an extension of some of the advantages of ferrite beads and rods and services to eliminate some of the disadvantages. Here a flexible tubing material may be slipped over an insulated or uninsulated conductor of any standard size.as shown in Fig. 14.15. Because a lower permeability of the flexible tubing is used compared with beads and rods little attenuation of EMI is offered below about 10 MHz. On the other hand, no saturation or resonant-frequency properties are exhibited and attenuation above 100 MHz becomes significant.

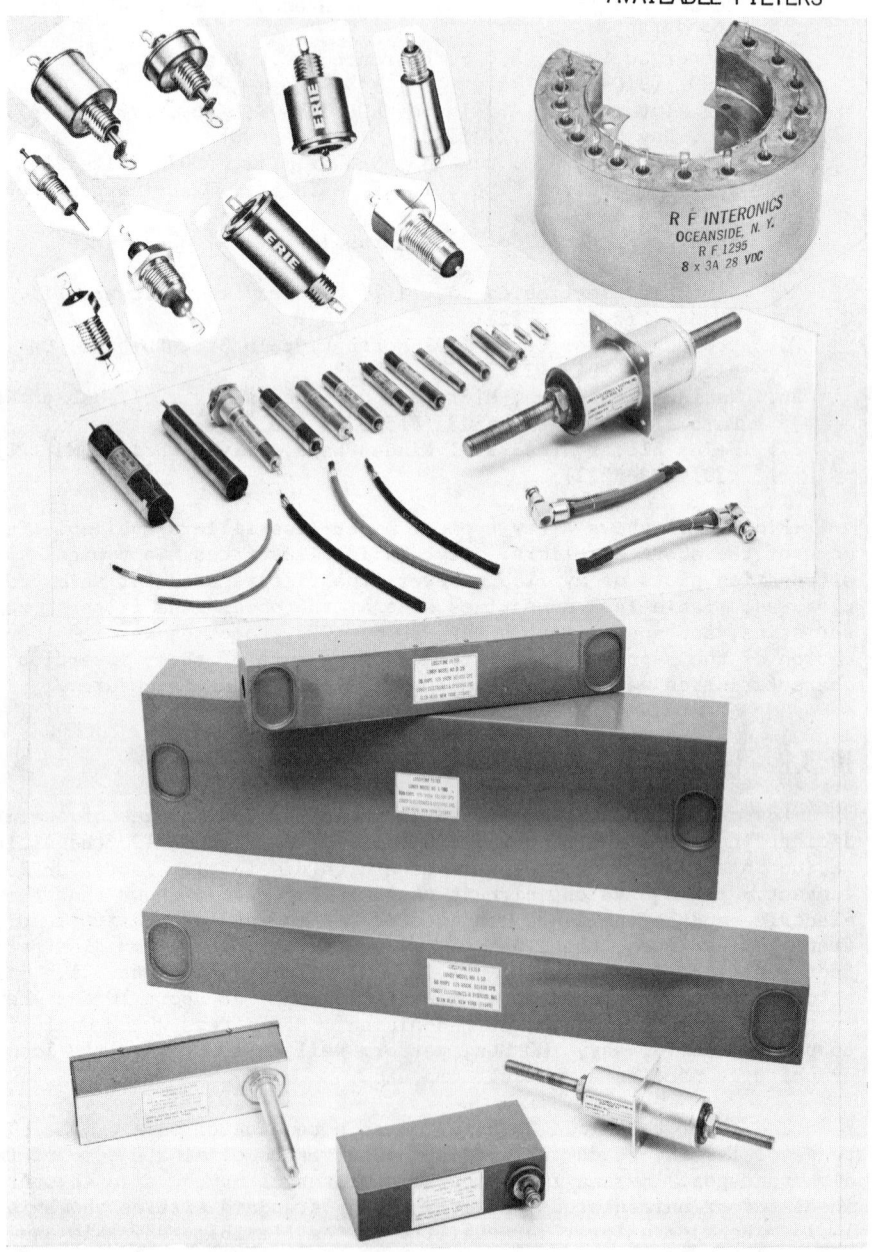

Figure 14.11 - Typical Power-Line Filters

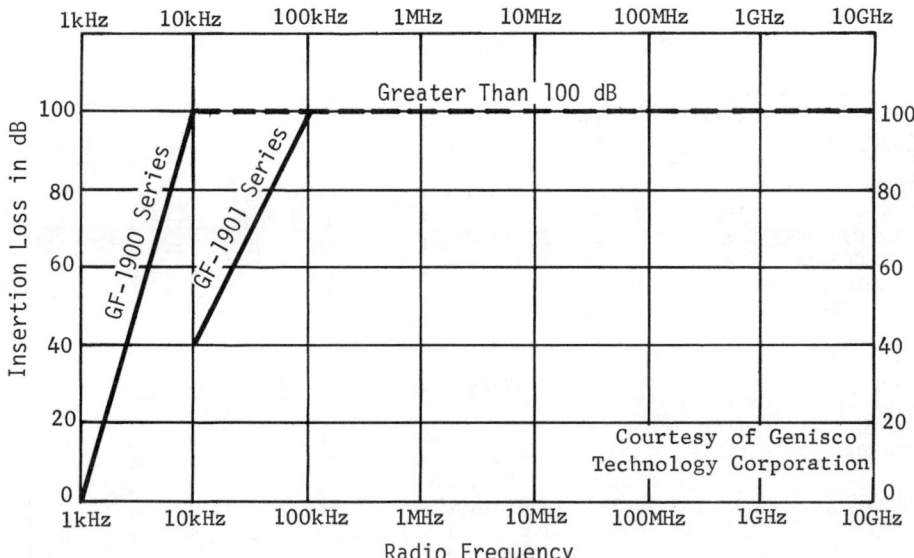

Figure 14.12 - Typical Insertion Loss Characteristics of Power-Line Filters per MIL-STD-220A

The principle of operation of the EMI suppressant tubing is similar to that of beads and rods. Having an equivalent permeability of about 10, the self inductance of a wire covered with the tubing is increased so that it acts as a one-stage distributed filter with series inductance as shown in Fig. 14.16. By avoiding the alternating high and low incremental inductance of beads and rods, tendency to radiate between elements is avoided. The suppressant tubing is also available with a shielded layer of metallized mylar for capacitance shielding (electric field pick up) at lower frequencies. The tubing exhibits no saturation to any DC, 60 Hz or 400 Hz power-line current when shipped over power mains busses for low-pass filter operation. Typical cost is about $5.00 per foot of tubing.

A combination of the above techniques results in a dissipative coaxial-line, ferrite filter, here, the dielectromagnetic loss tangents are very high, the relative permeability is in excess of 10^3 and the relative permitivity is about 10^5. Schiffres has shown that attenuation of the order of 20 dB at 100 kHz and 100 dB at 10 MHz are achievable. However, the use of this kind of a ferrite filter is limited to applications such as squib initiators (electroexplosive devices) where a low D-C resistance between conductors is not objectionable.

Table 14.2 - Some Typical Power-Line Filters* and Prices**

Model No.	Current Rating	Voltage Rating***	Approx. Size	Atten. 150kHz	Atten. 10MHz	Mfg.	** Price
1225-000	1A	50VDC	.4"dia x 1.5"	28dB	64dB	Erie	$ 14
1200-000	10A	50VDC	.4"dia x 1.2"	36dB	70dB	Erie	13
9001-000-1021	1A	185VDC	.4"dia x 1.1"	14dB	80dB	Erie	14
9001-000-1025	5A	185VDC	.4"dia x 1.1"	--	80dB	Erie	15
9930-1000-6000	25A	175VDC	.7"dia x 1.4"	20dB	53dB	Erie	10
9011-000-1000	1A	250VDC	.7"dia x 1.7"	50dB	80dB	Erie	28
9011-000-1007	10A	250	.7"dia x 1.7"	25dB	80dB	Erie	26
9401-000-0003	1A	400VDC	.4"dia x 1.5"	--	60dB	Erie	
9401-000-0007	5A	400VDC	.4"dia x 1.5"	--	31dB	Erie	
RNC-100	0.1A	400VDC	.6"dia x 2.8"	35dB	65dB	Rotron	NA
RNC-104	1A	200VDC	1"dia x 3.5"	80dB	80dB	Rotron	NA
RNC-110	5A	200VDC	.9"dia x 3.1"	55dB	70dB	Rotron	NA
RNC-113	10A	400VDC	1.3"dia x 3.5"	62dB	90dB	Rotron	NA
RNC-115	20A	400VDC	1.5"dia x 3.5"	55dB	85dB	Rotron	NA
5004-7009	1A	200VDC	.7"dia x 2"	40dB	80dB	Potter	$ 9
5004-7015	5A	200VDC	.8"dia x 3"	40dB	80dB	Potter	10
5004-7021	10A	200VDC	1.3"dia x 2.5"	40dB	80dB	Potter	11
5004-7024	20A	200VDC	1.5"dia x 3"	40dB	80dB	Potter	19
5004-7030	50A	200VDC	2"dia x 3"	40dB	80dB	Potter	35
FSR-W-10BN	10A	600VDC	5" x 12" x 29"	100dB	100dB	Filtron	$196
FSR-W-25BN	25A	600VDC	5" x 12" x 37"	100dB	100dB	Filtron	263
FSR-W-50BN	50A	600VDC	5" x 12" x 37"	100dB	100dB	Filtron	316
FSR-W-100BN	100A	600VDC	11" x 17" x 37"	100dB	100dB	Filtron	472
FSR-W-200BN	200A	600VDC	17" x 18" x 40"	100dB	100dB	Filtron	760
GF-1901-25D	25A	600VDC	4" x 4" x 22"	100dB	100dB	Genisco	$ 65
GF-1901-50D	50A	600VDC	4" x 4" x 22"	100dB	100dB	Genisco	70
GF-1901-100D	100A	600VDC	4" x 4" x 22"	100dB	100dB	Genisco	75
GS-1901-200D	200A	600VDC	4" x 4" x 22"	100dB	100dB	Genisco	210
APF 31	20A	250VAC	5" x 4" x 1.5"	20dB	70dB	Cornell-Dubilier	NA
C-7831R	5A	250VAC	3" x 2" x 1.5"	20dB	50dB	TELEC (France)	NA
333-10	10A	250VAC	3" x 2" x 1.2"	20dB	52dB	Schaffner (Switzerland)	NA
20 R6	20A	250VAC	5" x 4" x 1.5"	20dB	70dB	CORCOM	NA

* Only a few manufacturers and a few of their model numbers for DC or 60 Hz power mains are listed here.

** Prices are as of 1973 and rounded off to the nearest dollar. Price based on one unit only; substantial quantity break available. NA = (price) not available for general listing.

*** Rated maximum VDC; VAC RMS = 46% VDC line neutral or 80% VDC line-line.

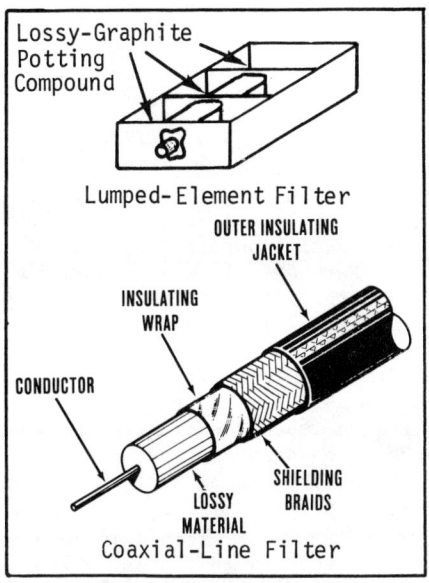

Figure 14.13 - Two Examples of Lossy-Line Filters

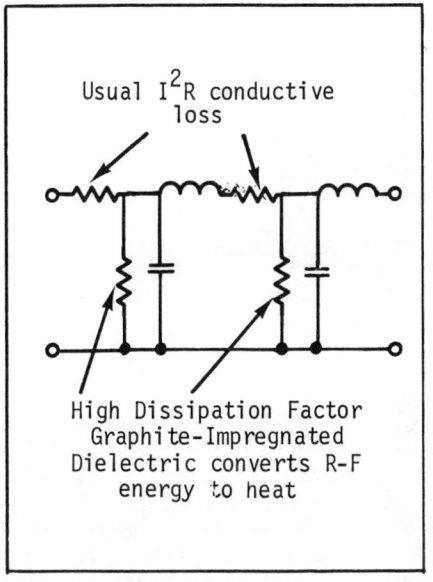

Figure 14.14 - Equivalent Circuit of Fig. 14.13.

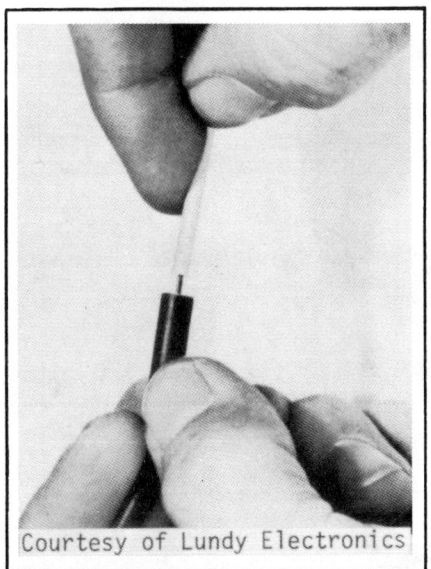

Figure 14.15 - Permeable Flexible Wire Tubing

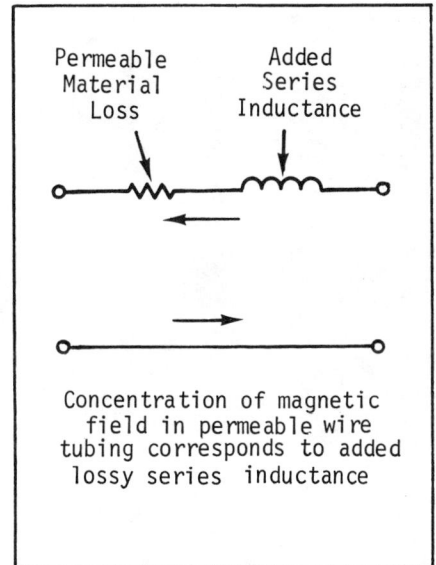

Figure 14.16 - Equivalent Circuit of Fig. 14.15

14.21

14.3.3 ACTIVE FILTERS

Sec. 14.2.5 indicated that passive, low-pass, L-C filters will become very large as the cut-off frequency is lowered into the audio region. While, the capacitor can be chosen with lower values (e.g., 0.01µf), the inductor becomes enormous in size and weight. Thus, an inductorless filter is needed for ELF applications. An active filter is such a device in which operational amplifiers transfer the impedance of an RC network to appear as an inductor. Active parallel or Twin-Tee tuned audio networks are one example. For power-line applications, however, the main power will have to bypass the active filter transistor if large supply currents are to be processed (EMI removed).

When the passband power cannot pass through transistors in active filters, then verters and separators may be used. The term *verter* represents impedance *converter* or *inverter* with impedance transformations having either positive or negative values. The separator passes no energy except for those frequencies for which the feedback of the active elements is made inoperative by low-power filters.

Active low-pass power-line filters have been designed with supply currents up to 100 amperes. In the case of one 24V D-C unit furnishing 30 amps, 40 dB of attenuation existed at 1 Hz and 60 dB at 20 Hz. The physical size compared to a comparable passive L-C network is about 10^{-3} in ratio.

14.4 BIBLIOGRAPHY

(1) "Active Notch Filter," RADC-TDR-64-104, AD 438252.

(2) AF/BSD Exhibit 62-87, "Electro-Interference Control Require-ments for Minuteman (WS 133B)," 6 December 1962.

(3) "Audio Pulse Cancellation Filter," RADC-TDR-64-105, AD 438 133.

(4) Bernstein, S. C., "Insertion Loss Measurements in a 5 ohm System," *Record of the 1969 IEEE Symposium on Electromagnetic Compatibility*, Vol. 69C3-EMC, June 17-19, 1969, Asbury Park, pp. 111-115.

(5) Boeing Aircraft Corporation D2-2-2444, "Electro-Interference Control Requirements (Equipment)," 5 March 1959.

(6) Bridges, J. E., et al., "A Low-Frequency Filter Inherently Free From Spurious Responses," *Proc. Eighth Tri-Service Conference on Electromagnetic Compatibility*, Chicago: Armour Research Foundation, pp. 564-589, Oct., 1962.

(7) Clark, D. B., et al., "Power Filter Insertion-Loss Evaluated in Operational-Type Circuits," *IEEE Transactions on Electromagnetic Compatibility*, Special Filter Issue, June, 1968, pp. 243-255.

(8) Clark, D. B., et al., "Power Filter Insertion-Loss Evaluated in Operational Type Circuits," *NCEL Technical Note N-938*, December, 1967.

(9) "Communications Interference Reduction Studies," RADC-TDR-64-80, AD 600 020.

(10) Cowdell, R. B., "Filters Affect Power Line Voltages," *Record of the 1969 IEEE Symposium on Electromagnetic Compatibility*, Vol. 69C3-EMC, June 17-19, 1969, Asbury Park, pp. 250-256.

(11) Crawford, R. A., "High-Frequency Quartz Crystal Bandpass Filters," *Electrical Design News*, Vol. 7, No. 10, Sept. 1962.

(12) Demetry, J. P., Pasquina, L. N., "Achieving Filter Effective-ness and Insuring Reliability," *Record of the 1969 IEEE Symposium on Electromagnetic Compatibility*, Vol. 69C3-EMC, June 17-19, 1970, Asbury Park, pp. 290-297.

(13) Eisbruck, S. H., and Giordana, F. A., "A Survey of Power-Line Filter Measurement Techniques," *IEEE Transactions on Electromagnetic Compatibility, Special Filter Issue*, June, 1968, p. 238-242.

(14) "Final Report on Interference Analysis of New Components and Circuit Techniques," Report No. 8899-1, RADC-TDR-62-211, AD 285 224.

(15) Fisher, J. F., and Cowdell, R. B., "Dimensions in Measuring Filter Insertion Loss, *Electronic Design News*, May, 1968, pp. 98-103.

(16) Foster, J., Buegel, K., and Sowa, C., "Electromagnetic Compatibility Lunar Orbiter," *NASA*, Document No. NAS 1-3800, January 29, 1965.

(17) Geffe, R., *Simplified Modern Filter Design*, New York: John F. Rider, ch. 1, 1963.

(18) Genistron Division of Genisco Technology Corp., "Development of a Current Injection Probe (CIP) for High Power Level Filter Analysis," Genistron CR 67.013, November, 1966.

(19) GM07-59-2617A, "Electro-Interference Control Requirements for Minuteman (WS 133A)," 20 October 1959.

(20) Grossman, W. K., and Fischer, J. T., "Evaluating Filters in Situ Under Heavy Load Currents and Normal Working Impedances," *1967 IEEE EMC Symposium Record*; July 18-20, 1967, Washington, D.C., pp. 352-364.

(21) Haber, F., "Study of GSFC Radio Frequency Interference (RFI) Design Guideline for Aerospace Communication Systems," Moore School of Electrical Engineering, University of Pennsylvania, Moore School Report No. 66027 for NASA, April 30, 1966.

(22) Hoffart, H. M., "Electromagnetic Interference Reduction Filters," *IEEE Transactions on Electromagnetic Compatibility, Special Filter Issue*, June, 1968, pp. 264-268.

(23) "Interference Reduction Guide for Design Engineers," Vols. 1 and 2, prepared by Filtron Co., Inc. for U. S. Army Electronics Labs., Fort Monmouth, N. J., Accession AD 619666 and AD 619667.

(24) Jobe, D. J., "The Electromagnetic Incompatibility of EMI Filter Test Methods and Test Results," *Record of the 1968 IEEE Symposium on Electromagnetic Compatibility*, Seattle, Washington.

(25) Jobe, D. J., and Jesperson, C. P., "Selection and Test of Power Line Filters for Use in Equipment Designed to Meet Government Electromagnetic Compatibility Specifications," *Record of the 1969 IEEE Symposium on Electromagnetic Compatibility*, Vol. 69C3-EMC, June 17-19, 1969, Asbury Park, pp. 283-289.

(26) Johnson, W. R., et al., "Development of a Space Vehicle Electromagnetic Interference/Compatibility Specification," *NASA*, Contract No. 9-73-5, *TRW*, 08900-60001-T000, June 28, 1968.

(27) Kelkenberg, R. H., "The Effect of Fuel Cell Internal Impedance on Power Supply Noise," *Record of the 1970 IEEE International Symposium on Electromagnetic Compatibility*, Vol. 70C28-EMC, July 14-16, 1970, Anaheim, pp. 417-420.

(28) Lasitter, H. A., "Power Line Impedance Determination Using the '3 Voltmeter' Measurement Method," *Record of the 1969 IEEE Symposium on Electromagnetic Compatibility*, Vol. 69C3-EMC, June 17-19, 1969, Asbury Park, pp. 128-136.

(29) L-STD-220, Method of Insertion Loss Measurement for Radio Frequency Filters.

(30) MIL-STD-C-11693, Capacitors, Feed Through, Radio Interference Reduction, Paper Dielectric, AC and DC (Hermetically Sealed in Metallic Cases).

(31) MIL-STD-F-15733, Filters, Radio Interference, General Specification for.

(32) MIL-STD-220A, "Method of Insertion-Loss Measurement," December 15, 1959.

(33) Milton, J., and Greenwood, E., "Improving the Specification for Power Line Filters," *IEEE Transactions on Electromagnetic Compatibility*, June, 1968, pp. 264-268.

(34) Pender, H., and McIlwain, K. D., *Electrical Engineers Handbook*, New York: John Wiley, 4th ed., pp. 6-33 through 6-62, 1957.

(35) Prye, H. S., and Follett, R. C., "Measuring Filter Insertion Losses Under Rated Load Conditions at Extended Frequencies," *Record of the 1967 IEEE Symposium on Electromagnetic Compatibility*, Vol. 27C80, July 18-20, Washington, D.C., pp. 54-60.

(36) *Reference Data for Radio Engineers*, New York: International Telephone and Telegraph Co., 4th ed., ch. 6 and 7, 1956.

(37) Salati, O. M., "Recent Developments in Interference," *IRE Transactions on Radio Frequency Interference*, Vol. RFI-4, No. 2, pp. 24-33, May, 1962.

(38) Schiffres, Paul, "A Dissipative Coaxial RFI Filter, *IEEE Transactions on Electromagnetic Compatibility*, Vol. EMC-6, January, 1964; pp. 55-61.

(39) Schlicke, H. M., "Effective Broadband Filtering for Interference Elimination in the Frequency Range from 10 Hz to 10 GHz," *IRE Trans. Radio Frequency Interference*, Vol. RFI-4, No. 3, pp. 41-43, Oct., 1962.

(40) Schlicke, H. M., "Theory of Simulated-Skin-Effect Filters," *IEEE Transactions on Electromagnetic Compatibility*, Vol. EMC-6, January, 1964; pp. 47-54.

(41) Schlicke, H. M., and Weidmann, H., "Compatible EMI Filters," *IEEE Spectrum*, October, 1967, pp. 59-68.

(42) Schlicke, H. M., Weidmann, H., and Dadley, H. S., "The Controversial MIL-STD-220A," *Record of the 1969 Symposium on Electromagnetic Compatibility*, Vol. 69C3-EMC, June 17-19, 1969, Asbury Park, pp. 215-226.

(43) Schultz, R. D., "Dissipative Filters for Switching Contacts," *Electronic Design*, Vol. 8, No. 4, Feb. 17, 1960.

(44) Stirrat, W. A., "A General Technique for Interference Filtering," *IRE Trans. Radio Frequency Interference*, Vol. RFI-1, No. 2, pp. 12-17, May, 1960.

(45) Stirrat, W. A., "The Sealed Shield Low-Pass Filter," *IEEE Transactions on Electromagnetic Compatibility*, Vol. EMC-10, No. 2, pp. 233-238, June, 1968.

(46) Warren, W. B., Jr., "Tracking Notch Filter for the Rejection of CW Interference," *The Ninth Tri-Service Conference on EMC;* Chicago, Ill., October 15-17, 1963, pp. 310-319.

(47) Weidmann, H., and McMartin, W. J., "Two Worst-Case Insertion Loss Test Methods for Passive Power-Line Interference Filters," *IEEE Transactions on Electromagnetic Compatibility; Special Filter Issue,* June, 1968, pp. 257-263.

(48) White, D. R. J., "A Handbook on Electrical Filters - Synthesis, Design and Applications," Germantown, Maryland, August, 1963.

CHAPTER 15

POWER-LINE ISOLATION DEVICES

CHAPTER 15

POWER-LINE ISOLATION DEVICES

Many cases of EMI prove to result from electromagnetic pollution of the A-C power mains. Primary contamination comes from the local power distribution lines acting as a huge pick up antenna to all forms of man-made emissions in the general area including within buidlings. This *antenna system* also distributes the EMI to all of its users as well as accepts pollution from the users for redistribution to all others.

The public utilities themselves are not sacrosanct about contaminating their own power distribution system. For example one morning in December 1972, the author counted 18 power interrupt situations in Darnestown, Maryland. It is estimated that these interruptions lasted from a fraction of a second to several seconds with a mean outage of about a second or two. While admittedly this was not typical service, it is one of the worst forms of EMI since both the transients and especially the interrupts can damage equipment, cause loss of memory in computers, effect hospital operations, and the like. With the increasing shortage of power generation and distribution in the future, more service interrupts can be expected in forthcoming years.

Electrical storms in an area also serve to cause brief power outages, sometimes for only a fraction of a 60 Hz cycle. Barely noticeable to occupants of a building, they can result in erroneous readings of biophysical instruments in medical clinics and hospitals, dump computer memory banks and spoil industrial scientific, and medical experiments.

This chapter addresses itself primarily to control of EMI on A-C power lines at the consumer level. While the control of the power-outage problem is not the objective of this chapter, some EMI-control techniques offer the by-product advantage of protection against the consequence of power interruption of service.

Direct insertion power-line filters are not the only means of isolating conducted R-F or electrical noise coming in on the power-line mains from A-C or D-C supply sources. Other methods include using

15.1

a local battery supply with an inverter for auxiliary gasoline or diesel engine-driven generator, or a motor-generator set connected to the power line. Still another method, used either separately or in conjunction with power-line filters, involves using isolation transformers. These devices permit floating the secondary where either grounding one side is undesirable (e.g., to avoid ground current loops), or shock hazards are to be mitigated.

15.1   ISOLATION TRANSFORMERS

Typical transformers isolate one circuit from another while magnetically coupling desired energy from one to the other. Such a transformer is adequate for use with low-gain circuits or insensitive instruments. However, if the transformer must couple into high-gain circuits or sensitive instrumentation, noise potentials between the primary circuit and ground must be prevented from affecting the secondary circuits due to capacitive and resistive coupling between the transformer's windings.

Traditional techniques for keeping electrical noise from reaching the transformer secondary, e.g., the standard Faraday shield (a grounded conducting foil between the windings) will divert most of the primary noise current to ground. Electrical noise still can be coupled into the secondary because of the electrostatic field around the Faraday shield. Unique box-shielding techniques employed in some isolation transformers effectively overcome this problem. There, the impedances between windings are extremely high.

Isolation transformers must be conservatively rated so that they remain cool under full-load conditions and exhibit good voltage regulation, i.e., they do not generate a common-mode impedance coupling problem (see Sec. 8.2). They are especially designed for these important applications:

- For isolating sensitive instrumentation from noisy power lines.
- For maximum common-mode noise rejection
- For isolating noisy equipment from noise-sensitive equipment, both of which share the same power line.
- For minimizing differential-mode noise (noise across winding) resulting from common-mode noise (noise between winding and ground).
- For complete electrostatic blocking. When the complete shield surrounding the transformer secondary is extended around the equipment being guarded, shielding effectiveness is maximized.

15.1.1   COMMON-MODE NOISE REJECTION

Box-shielding techniques employed in the construction of isolation transformers achieve maximum impedance between windings while offering a very-low impedance path for common-mode noise to ground. To accomplish this, the leakage resistance is kept at a high-level (e.g., 10,000 megohms) and the effective interwinding capacitance is maintained at the low values stated below. This provides a significant advantage over the typical isolation transformer using traditional shielding methods as shown in Figs. 15.1 and 15.2.

Three basic levels of quality shielding performance that are commercially available are rated in terms of the interwinding capacitance.

(1)   .005 picofarads (5 x 10^{-15} f)

(2)   .001 picofarads (1 x 15^{-15} f)

(3)   .0005 picofarads (0.5 x 10^{-15} f)

Common-mode noise voltage (V_c) in dB is measured relative to input noise voltage (E), as shown in Fig. 15.3. Measurements are made over the audio-frequency band from 20 Hz to 50 kHz.

For each level of interwinding capacitance, C_x, the common-mode noise rejection is shown in the following table. V_c, is measured across C, a .01 μf capacitor to ground.* Thus:

C_x in pf	$V_c/E = C_x/C$
.005	−126 dB
.001	−140 dB
.0005	−146 dB

15.1.2   Suppression of Differential-Mode Generated By Common-Mode Noise Input

The shielding should minimize differential mode caused by a common-mode noise source. Figs. 15.4 and 15.5 illustrate how the special shielding techniques achieve a reduction of differential-mode voltage greater than 40 dB below the value attainable by the normally best box shielding methods. The ordinary shielded isolation transformer (including a box-shielded type) will itself generate a differential-mode voltage between either primary or secondary terminals due to a common-mode noise voltage appearing between any of these terminals and ground.

In Fig. 15.4, differential-mode appears as a voltage, V_t, across

---

* For noise across capacitances to ground other than C = .01 μf, add −20 dB to above numbers for each factor of 10 *increase* in C, or add +20 dB for each factor of 10 *decrease* in C.

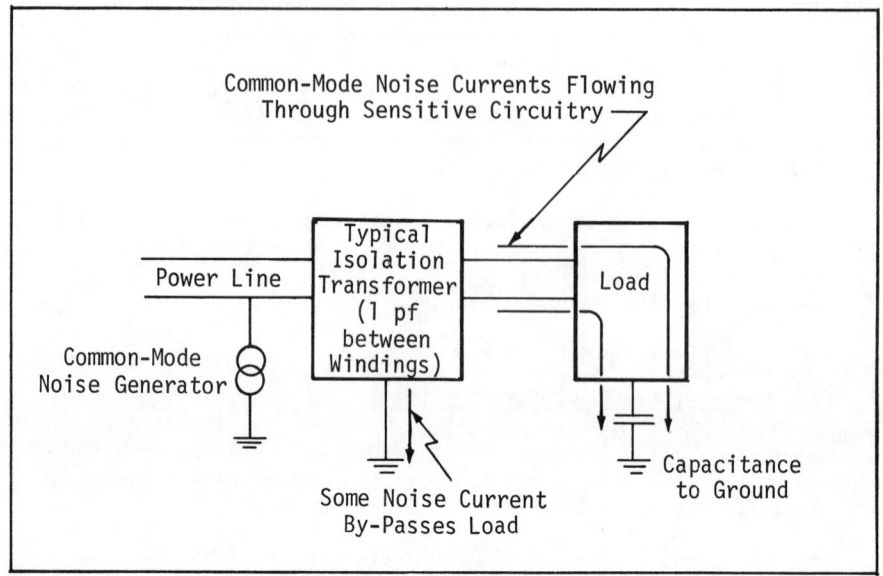

Figure 15.1 - Typical Isolation Transformer

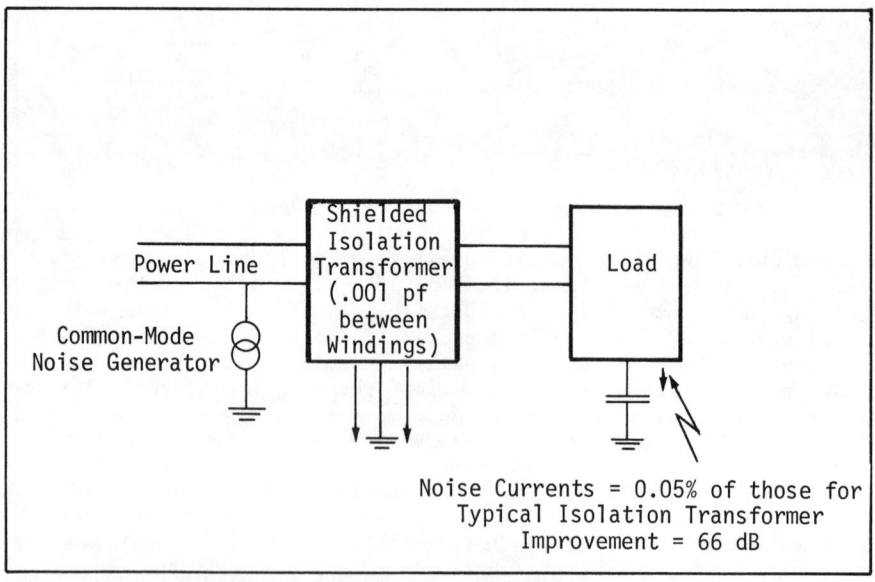

Figure 15.2 - Specially-Shielded Isolation

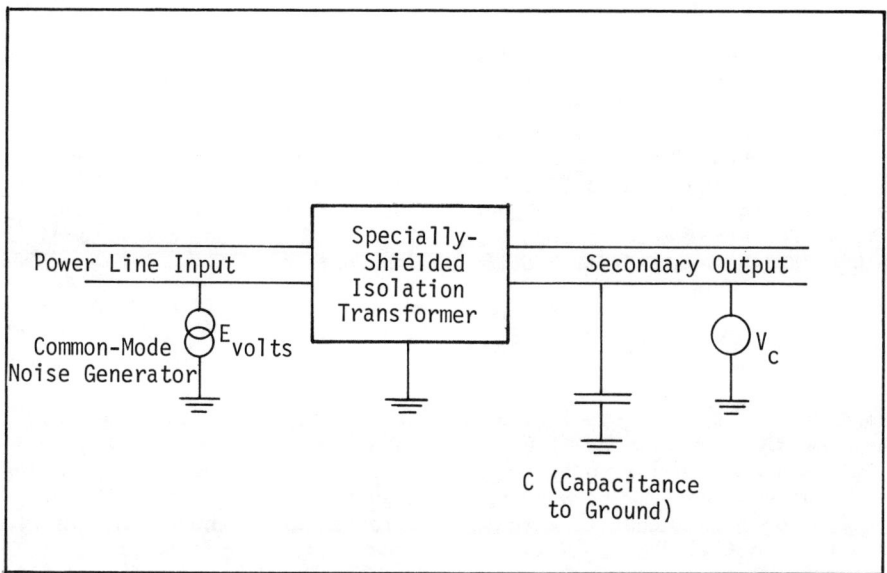

Figure 15.3 - Measuring Common-Mode Noise Voltage

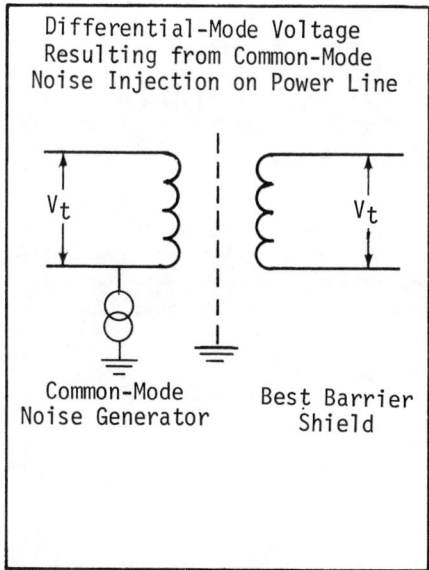

Figure 15.4 - Normal Transverse Voltage from Common-Mode Noise

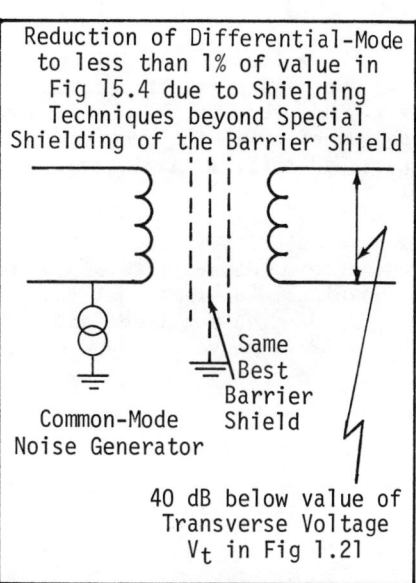

Figure 15.5 - Reduction of Transverse Noise Voltage with Special Shield

both primary and secondary windings of an isolation transformer when
a common-mode noise causes current to flow in the primary winding and
from there to ground via capacitance to a grounded shield. Similarly,
common-mode noise generated by noisy equipment in the secondary environment can be transformed into a differential mode which appears across
the secondary and is magnetically coupled to the primary, thereby contaminating the power line. A good isolation transformer can reduce
this differential-mode to less than 5/1000 of the amount encountered
in standard box-shielded isolation transformers.

15.1.3 MAGNETIC NOISE SUPPRESSION

Often instrumentation in the secondary circuits is sensitive
enough to be affected by electromagnetic noise fields emanating from
the isolation transformer itself. In sound EMI design practices, precautions are taken in isolation transformers to keep these stray fields
to a minimum. For example, a typical magnetic flux density level at
a distance of 18 inches from the geometric center of the transformer
is 0.1 gauss (140 dB_pT). This flux decreases with the cube of the
distance. Thus at a two meter distance, the level would be about 119
$dB_p\pi$.

15.1.4 HIGH-FREQUENCY PERFORMANCE

The performance of EMI-type isolation transformers is in part predicated upon keeping the secondary capacitance to ground to a definable
level. However, it is possible that this capacitance may become self
resonant at some high frequencies such that the secondary impedance is
high. Under these conditions, the common-mode rejection will become
very poor.

One solution is to use EMI filters to suppress EMI at the higher
frequencies. However, these may unbalance the secondary and result in
an undesirable differential mode. A better solution would be to twist
the secondary output lines and cover them with lossy EMI suppressant
tubing. This converts any H-F differential mode to heat.

15.2 BATTERIES AND INVERTERS

Where the required power-mains source due to a light load is of
modest values, say less that 10 kw-hour capacity per day, one or more
12-Volt batteries (or other ratings) connected in series and/or parallel
may be used. Rechargeable batteries, such as nickle-cadmium are useful for capacities less than about 1 kw-hour. For larger load requirements over a typical 8-hour working day, wet cells, such as truck
lead-acid types may be used. Typical ratings are 100 ampere-hours
(1200 watt-hours) each and may be trickle charged over night. Whatever the choice, the use of batteries completely decouples R-F conducted energy on the power mains since the power line is not directly involved.

For A-C power requirements up to about 10 kW-hour per day, batteries may also be used together with an inverter. Inverters come in
many different ratings including both output voltage and power-hand-
ling capacity. It is important that any inverter so used be relatively free from EMI. Since most inverters are DC motor-AC generator sets,
it is likely that one or two sets of brush-commutators are involved.
Thus, both low-frequency radiated magnetic fields and high-frequency
conducted electrical noise can result unless EMI suppressing measures
are taken (see Sec. 16.7.4)

Where loads are less than about 1 kW, battery-driven electronic
inverters may be used to avoid the above brush-commutator noise problem. Most, but not all inverters operate on the principle of the D-C
source driving a 60 or 400 Hz audio oscillator, which is power amplified, and boosted in voltage amplitude by an audio transformer to 115
VAC, 60 or 400 Hz.

For D-C power output requirements at voltage levels differing
from the battery source, static D-C converters may be used. In such
devices, an oscillator is typically used with a squarewave hystersis
core, so that the output waveform to the rectifier is a square wave.
The output before filtering will be rich in harmonics and the rectifiers can cause additional significant H-F electrical noise if they
exhibit reverse recovery problems. Thus, in the selection of a static
D-C converter, it is important to determine if it has been EMI controlled.

15.3 ENGINE-DRIVEN GENERATORS

The engine-driven generator (EDG) may be useful for decoupling
the power mains to sensitive equipment, especially where loads are in
excess of 1 kw. Three disadvantages immediately develop: (1) they
tend to be acoustically noisy, (2) they develop toxic fumes requiring
ventilation, and (3) they develop fuel handling and storage problems
together with an increase in insurance premiums. They also require
periodic maintenance. Not least of the problems, is electrical ignition-system EMI noise.

The gasoline engine must be EMI suppressed since the ignition
system radiates electromagnetically each time a spark plug is fired.
Ignition suppression kits are available on the market and can be purchased at most automotive supply houses. They generally involve replacing the distributor-to-spark plug leads with resistive wire ignition cable. Some of the more elaborate kits include shielding material
or shielded cases for the ignition coil and interrupter-distributor
box. A quick-fix offering some lowering of EMI radiation is obtained
with the use of resistive-type spark plugs such as manufactured by AC
Spark Plug or Champion. It is best, however, if the manufacturer of
the GEDG builds in his own EMI suppression since this can usually be
undertaken at the system level of design layout in which a number of
economical and effective precautions are taken.

Another source of EMI noise from GEDG is from the generator itself, especially if it uses brushes. The collapsing magnetic field in
a generator armature loop, which takes place as a brush leaves the commutator bar, can cause back voltage surges of an order of magnitude or
more than the line voltage. This results in arcing at the brush-commutator interface which in turn develops both radiated and conducted
interference. While conducted EMI can be filtered at the generator
output terminals, radiation is more difficult to suppress. A better
solution to begin with, is to use brushless AC generators such as
synchronous or induction types. Sec. 16.7.4 discusses this topic in
detail.

To avoid some of the above problems, where engine-driven generators must be used, the diesel engine-generator is a better choice.
Since diesel fuel burns - not explodes like gasoline - the resulting
transient in the ignition process is relatively modest. Fuel-handling
hazards are also reduced. Sec. 1.2 of Vol. 4 discusses this topic in
further detail.

15.8

15.4 MOTOR-GENERATOR SETS

Motor-generator (MG) sets provide a fourth means of power-line isolation from conducted R-F emissions. They overcome almost all of the objections listed above in the other three means. They (1) are relatively quiet, (2) do not develop toxic fumes, (3) have no fuel handling and storage problems, (4) are not limited in capacity and do not require recharging, and (5) are relatively maintenance free. MG's are driven directly from the A-C power mains.

MG's rely on the electro-mechanical isolation of the motor and generator to provide power-mains decoupling. However, it does not follow that any MG will accomplish this since some (especially smaller size modern units) use a common stater housing to support both the motor and generator. Here, magnetic-field coupling at low frequencies and parasitic-capacitive coupling at high frequencies can result in relatively little EMI attenuation over a broad portion of the spectrum. While capacitive coupling can be significantly decreased when a Faraday-shield partition is used between the motor and generator (see Sec. 10.2.3), magnetic inductive coupling at low frequencies remains unaffected.

All MG sets properly chosen for EMI power mains isolation will employ separate and distinct motor and generator housings - often with a longer common shaft to further physically isolate them. To prevent a circulating R-F loop, the common shaft may even be conductively isolated by using a non-conductive sleeve bushing or universal joint. For brief power interruptions of a few seconds (see above Darnestown problem), the flywheel effect will smooth out such service outages.

CHAPTER 16

EMI CONTROL IN COMPONENTS

16.1   R-L-C COMPONENTS
        16.1.1   Resistors
        16.1.2   Capacitors
        16.1.3   Inductors

16.2   INSULATORS AND CONDUCTORS
        16.2.1   Insulators
        16.2.2   Conductors

16.3   CONNECTORS

16.4   TRANSIENT CHARACTERISTICS
        16.4.1   Spectrum Occupancy
        16.4.2   Broad and Narrow-Band Emissions

16.5   DIODES

16.6   ACTIVE DEVICES
        16.6.1   Transistors
        16.6.2   Electron Tubes

16.7   INDUCTIVE DEVICES
        16.7.1   Inductors
        16.7.2   Transformers
        16.7.3   Switches and Relays
        16.7.4   Motors and Generators

16.8   LIGHTS
        16.8.1   Incandescent Lamps
        16.8.2   Fluorescent Lamps
        16.8.3   Gas Lamps

16.9   ANTENNAS

16.10 BIBLIOGRAPHY

CHAPTER 16

EMI CONTROL IN COMPONENTS

Previous chapters have discussed EMI prediction and control such as cabling, grounding, shielding, and filtering. These topics represent coupling paths or interfaces between sources and victims of EMI, also discussed in earlier chapters. The next chapter including this chapter examine the very essence of EMC in that EMI-control techniques are carried out on components, circuits, and equipments.

In surveying this chapter on EMI control in components, fundamentals are again stressed. Basic building-block elements of resistors, inductors, capacitors, insulators, conductors, and diodes are reviewed with regard to their EMI problems and control techniques. Equipped with a discussion on transient-noise properties, several sections on transient-producing devices are reviewed. This includes EMI control in transformers, relays, solenoids, motors, and generators. Since these devices also produce significant magnetic fields, flux leakage control provided by shielding is reviewed. The chapter ends with a discussion of EMI problems and control in fluorescent lamps, antennas, and microwave components.

16.1   R-L-C COMPONENTS

This section discusses EMI problems and control techniques that are employed in fabricating and using resistors, inductors, and capacitors. It is shown that these fundamental passive electronic parts in reality do not behave at their stated values at frequencies above about 100 MHz due to parasitics; under certain conditions, their performance even degrades at 1 MHz or below. Thus, filters, for example, often do not perform properly at ten times cut-off frequency, and amplifiers may exhibit out-of-band parasitic oscillation and spurious responses.

16.1.1   RESISTORS

The resistors considered here are: carbon composition, deposited carbon-composition film, pyrolytic carbon film, metal film wirewound, microelectronic, and special purpose. The type of resistor to be used

is determined by considerations of resistance, wattage, cost, compact-
ness, precision, distributed capacitance, distributed inductance, life,
and internal noise. Composition resistors may be of the pellet or filament types. Composition resistors are made of finely-divided carbon
and a binder pressed into a slug with leads imbedded in each end. The
slug is then enclosed in a phenolic or other case and the resistor body
is molded. In some cases, the resistor is enclosed in a ceramic tube
with cement covering both ends. The filament type has the carbon and
binder mixture coated on the outer surface of a glass tube and the
leads are inserted therein. A phenolic tube is then molded around the
resistor body.

16.1.1.1 Resistor Characteristics

Carbon or metal fixed-film resistors are usually made by deposit-
ing a controllable-thickness of resistive material in a continuous film
onto a base. The resistor body is then covered with a plastic or epoxy.
The geometry of film resistors enhances their high-frequency characteristics such that they may be used often up to about 400 MHz.

A microelectronic resistor is a thin layer of silicon on a base or
metal placed over a semiconductor. Close spacing increases capacitance
and coupling leakage. The small size limits available resistances, and
undesired semiconductor junctions may be formed.

Any covering on the resistor body acts as a thermal barrier as well
as protection against moisture. Thus, dissipated energy is conducted
primarily by the leads. Special metal jackets are made to assist heat
energy to leave the resistor body. Bifilar winding of a wirewound resistor reduces internal inductance because adjacent turns carry currents
in opposite directions. However, adjacent turns may exhibit appreciable
shunt capacity. Capacitive currents may have adverse effects on R-F
applications. The Ayrton-Perry winding is preferred, as each resistor
is constructed of two parallel windings in opposite directions; the
turns cross each other at points of no potential difference. A typical
Ayrton-Perry resistor wound on a cylindrical spool exhibits one percent
of the inductance of a conventional spool-wound power resistor.

A composition resistor can exhibit an A-C resistance lower than
its D-C value. This characteristic is known as the *Boella effect*. It
is primarily due to the shunting effect of distributed capacitance that
results from the large number of conducting particles mixed with the
dielectric material. To reduce this effect, resistors with a minimum
of dielectric are used so the dielectric constant and associated loss
factors are minimized. Decreasing the resistor cross-section and increasing the resistor length, as in the filament type of resistor, decreases this problem. Because of the greater amount of dielectric
used, higher values of resistance exhibit greater percentage of change
in value.

Skin effect occuring at high frequencies, causes current flow to be concentrated at the surface with little current flowing in the rest of the cross-section. Because current is not evenly distributed through the entire cross-section of the conductor, skin effect results in an increased effective resistance at R-F over that of the D-C value.

Table 16.1 shows the effect of resistance at different radio frequencies of some general-purpose, axial-lead, carbon-composition resistors of 1MΩ resistance value. The resistor produced by manufacturer No. 1 is described as a *hot molded fixed resistor*. The resistance element is a carbon composition film on a glass body. The manufacturer's published frequency characteristics include both inductance and capacitance effect.

Table 16.1 - Ratio of R-F Resistance to D-C Resistance of a 1MΩ Axial-Lead, Carbon-Composition Resistor

Radio Frequency	Manufacturer No. 1			Manufacturer No. 2		
	1/2 watt	1 watt	2 watt	1/4 watt	1/2 watt	1 watt
10 kHz	1.00	1.00	1.00	1.00	1.00	1.00
100 kHz	0.89	0.85	0.75	1.00	1.00	1.00
1 MHz	0.54	0.46	0.37	0.92	0.89	0.90
10 MHz	0.21	0.15	0.12	0.65	0.60	0.67
100 MHz	0.07	0.04	0.04	0.32	0.28	0.36

16.1.1.2 RESISTOR EQUIVALENT CIRCUITS

The equivalent circuit of a resistor depends upon manufacturing processes, techniques, and raw materials used. One equivalent circuit is shown in Fig. 16.1. The shunt capacitance, C, is of the order of 0.3 pf for a typical 1-watt composition resistor. Fig. 6.2 is the equivalent circuit of a resistor located near the return circuit and operating at a frequency where the capacitance of the return circuit is significant and where distributed capacitance, C_d, is low. For composition resistors, the inductance may be negligible.

Wirewound resistors have a relatively large series inductance and distributed capacitance. They are also affected by skin effect, exhibiting an increase in resistance as the frequency increases. The equivalent circuit of a wirewound resistor is shown in Fig. 16.3.

All resistors generate thermal and current noise. RMS thermal-noise voltage, V_t, is independent of frequency:

$$V_t = \sqrt{4RKTB} \qquad\qquad (16.1)$$

where: R = resistance component affected by thermal agitation

K = Boltzmann's constant = 1.374 x 10^{-23} watts/°K/Hz

T = absolute temperature in degrees Kelvin

B = noise bandwidth in Hz

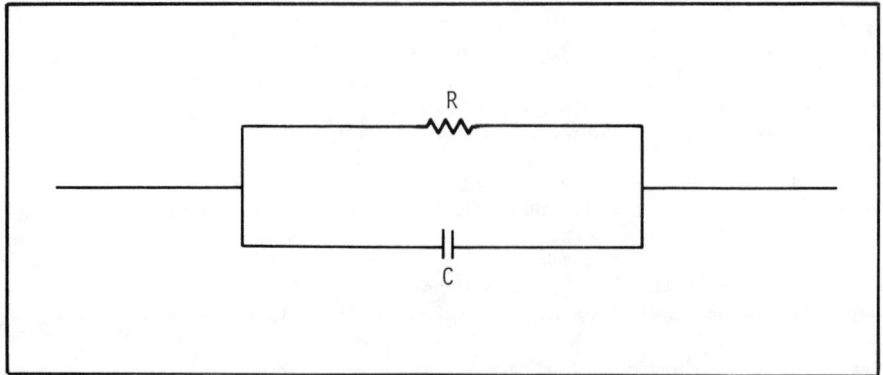

Figure 16.1 - Equivalent Circuit of a Resistor at Low Frequencies

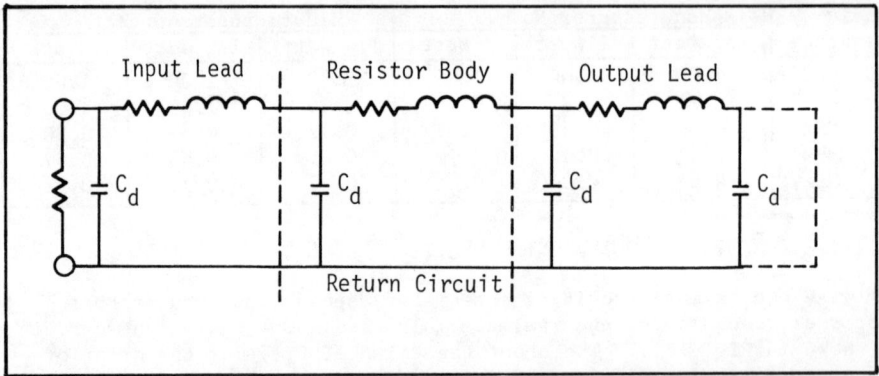

Figure 16.2 - Equivalent Circuit of a Resistor Placed Close to the Return Path

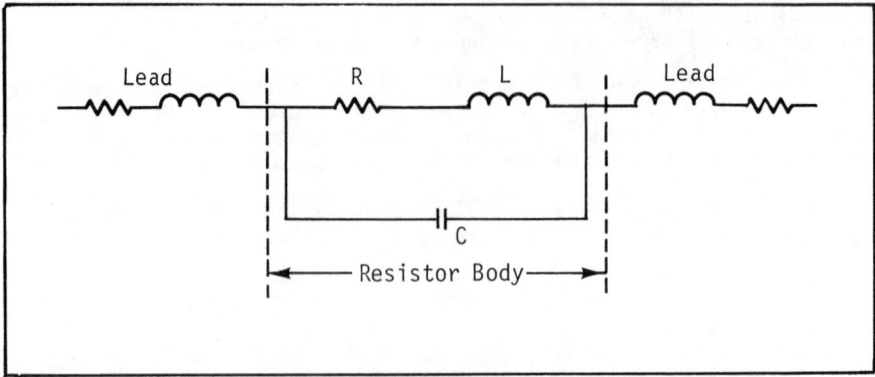

Figure 16.3 - Equivalent Circuit of a Wirewound Resistor

Eq. (16.1) is shown in graphical form in Fig. 16.4 for different values
of resistance and bandwidth.

Current voltage, V_c, in resistors on the other hand, is a function
of frequency and type of resistor:

$$V_c = I\sqrt{k/f} \tag{16.2}$$

where: V_c = RMS voltage/Hz of bandwidth at frequency f

        k = noise quality constant of proportionality

        I = current through the resistor (DC RMS amperes)

        f = frequency in Hz

Current noise varies as shown in Fig. 16.5 where the μV/V scale is
multiplied by the voltage across the resistor. Current noise is generated in molded composition and metalized carbon resistors but usually
not in wirewound and high-stability deposited-carbon resistors.

The RMS addition of the two above noise sources yields:

$$V_{total} = \left(V_t^{\,2} + V_c^{\,2}\right)^{1/2} \tag{16.3}$$

Table 16.2 shows that metal-film and fixed wirewound resistors
generate a lower-noise level than other resistor types, although damage or improper manufacturing processes can result in increased noise.

Table 16.2 - Typical Resistor Noise for 20 Hz to 20 kHz Bandwidth

Resistor Type	μV/V Noise
Metal Film and Wirewound	0.001 to 0.082
Deposited Carbon	0.05 to 0.86
Composition	0.4 to 4.6

Up to approximately 10 MHz, proper spacing and short leads minimize the effects of self and mutual inductance, while various capacitances and dielectric losses are negligible.

A conductor of uniform cross section, diameter d, and distance
D from a return circuit consisting of a conductor of the same dimensions, has an inductance, L, of:

$$L = 2.54 \left[4 \, \ell n_e \left(\frac{D}{d}\right) + 1 \right] \text{ nano henrys/inch} \tag{16.4}$$

The reactance, X_L is:

$$X_L = 16f \left[4 \, \ell n_e \left(\frac{D}{d}\right) + 1 \right] \text{ nano ohms/inch} \tag{16.5}$$

Eqs. (16.4) and (16.5) are applied for a 1-watt, RC 32 resistor (9/16
inch long x 7/32 inch diameter outside dimensions representing one
manufacturer's form of construction) connected with one-half inch coax-

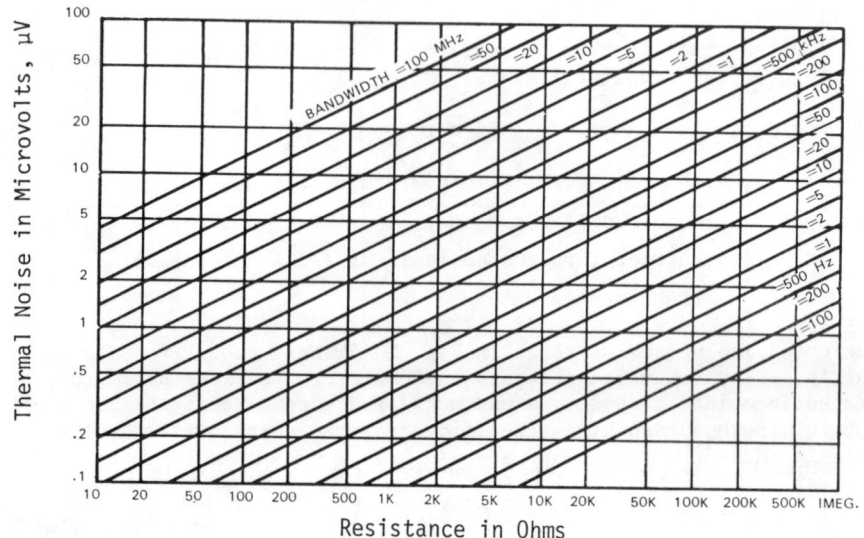

Figure 16.4 - Thermal Noise Voltage of Resistors as a Function of Bandwidth at Room Temperature

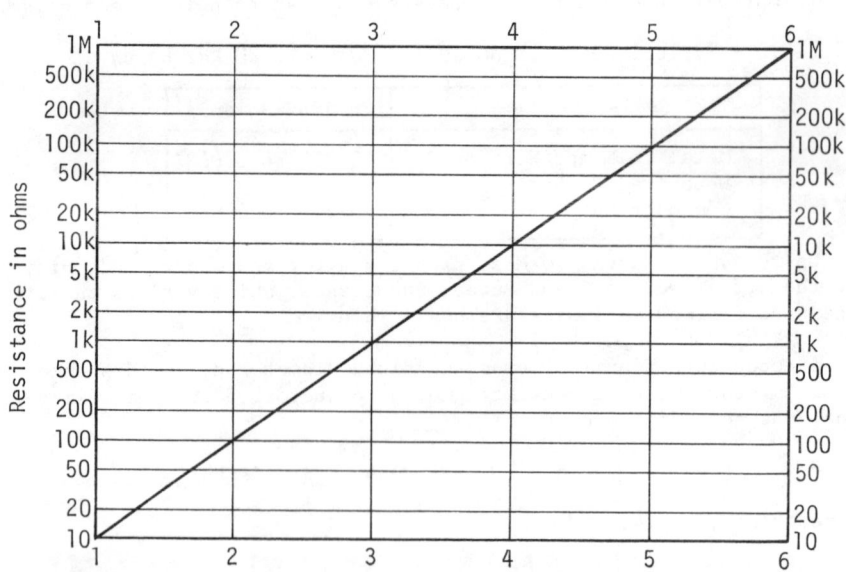

Figure 16.5 - Noise Generated in a Current-Carrying Resistor

ial leads on each end and resistor and leads parallel to a return conductor. Table 16.3 is based upon the actual resistor diameter of 5/32
inch, length 3/8 inch, lead length 1-3/16 inch, and a lead and return
circuit each as 0.40 inch diameter, except for approximating the return conductor as 5/32 inch diameter for the 3/8 inch length opposite
the resistor.

Table 16.3 - Reactance of Resistor at 1 MHz

Spacing D in inches	Resistor Only		Contributed by Leads Only	
	Inductance in nh	Reactance at 1 MHz	Inductance in nH	Reactance at 1 MHz
1/4	2.7	.017Ω	21	.13Ω
1	8.0	.050Ω	35	.22Ω
2	10.7	.067Ω	43	.27Ω

Between the body ends, the capacitance of a 1/2 watt resistor is
about 0.1 to 0.5 pf and the inductance of the leads is effectively in
series with the capacitance. The inductance and capacitance of helical-
form resistors exhibit broadband efforts and parallel resonance at certain frequencies. Therefore, these resistors are often limited to power
frequencies, but proper design can use the characteristics to filter
pulses or reduce undesired frequencies.

Strong electromagnetic fields can affect resistors, usually causing a change in resistance due to heating. While composition resistors
exhibit only the heating effect, spiral-film and ordinary wirewound resistors also behave as inductors which can couple energy into associated
circuits.

Variable-resistor or potentiometer noise may result from several
causes:

● Foreign materials formed on the resistance due to wiper

● Dust particles or chemical contamination

● Formation of oxide films on contract surfaces

● Mechanical untrueness

● Triboelectric effect from the wiper sliding on the element,
   causing a self-generated voltage

● Thermoelectric effect from external or frictional heat.

Noise can be generated when high-current densities exist at the
wiper-to-element interface, such as occurs when a wiper makes contact
with the high spots on the surface of a composition or a ceramic carbon metal (ceramet) element. Due to heating, the contact may change
quickly, often producing arcing and white noise. The wirewound variable resistor wiper can make or break contact with adjacent turns and
cause small arcs. In precision variable resistors, the noise level

is reduced to 100 mV or less. Table 16.4 shows typical noise voltage for resistor currents varying from 10 µA to 1 mA.

Table 16.4 - Variable Resistor Noise Voltage

Resistance in Ohms	Resistor Current		
	1 mA	0.1 mA	0.01 mA
1k	0.01 to 0.03V	0.002V	0.0002V
10k	0.15 to 0.25	0.02	0.002
100k	1.0 to 4.0	0.3	0.035

Table 16.5 illustrates some common resistor EMI situations.

Table 16.5 - Effects of Resistor Characteristics

Resistor Characteristics	Example	Problem
Direct end-to-end capacitance	Attenuator	High-frequency signals fed through to a point not grounded for the signal frequency
Total capacitance (end-to-end and lead-to-ground)	Feedback amplifier plate load resistor	Phase shift to signal components as a function of frequency
Resistance varies with frequency	Some amplifiers, some measurement methods	Boella and skin effects
Inductance	Shunt resistors in attenuators	Change of effective impedance with frequency. Important in low value resistors below 100 MHz.
Inductance	Any resistor	Phase shift, change of effective impedance. Important about 100 MHz.
Susceptibility to R-F fields	Composition and metal film resistors	Can change resistance and overheat
Susceptibility to R-F fields	Ordinary spiral-wound resistors	Induced voltage, proportional to the numbers of turns and field strength, is transferred to circuitry

16.1.2 Capacitors

A capacitance is formed between any two physically separate objects. An intentional capacitance, such as a capacitor, is constructed of two metal plates or foils separated by a controllable dielectric. Its capacitance, C, is:

$$C \approx \frac{A\varepsilon_m}{t} \text{ farads, for } \sqrt{A}/t > 10 \qquad (16.6)$$

where, A = area of each plate or foil in sq. meters

ε_m = permitivity of dielectric medium = $\varepsilon \cdot \varepsilon_o$

ε = dielectric constant relative to air = 1

ε_o = absolute permitivity of air = 8.85×10^{-12} farads/m

t = thickness of medium or separation of plates or foil in meters

Eq. (16.6) applies when the least dimension associated with the area is much greater than the thickness. This corresponds to a negligible capacitance of the fringing field.

In centimeters and inches, Eq. (16.6) becomes:

$$C = \frac{.0885\varepsilon A}{t} \text{ pf (for cm dimensions)} \qquad (16.7)$$

$$= \frac{.225\varepsilon A}{t} \text{ pf (for inch dimensions)} \qquad (16.8)$$

Fig. 16.6 is a plot of Eq. (16.7) for various dielectric constants ranging from 1 to 1,000.

The equivalent circuit of most capacitors over a wide frequency range is shown in Fig. 16.7. Fig. 16.8 is the equivalent circuit of electrolytic capacitors.

In most capacitors, the lead inductance combines with the capacitance to give a terminal impedance, which generally is small at the resonant frequency and increases on either side of this frequency. Above resonance, the capacitor appears as an inductor. Inductance and series resistance limit the rate of change of current during sudden charge or discharge such as for transients. The series resistance affects the dissipation factor and may cause problems in A-C, high-precision, and timing circuits. For most applications, series resistance is considered constant and independent of frequency.

Shunt conductance is caused by current leakage and voltage stress across the dielectric. Current leakage is usually small in solid dielectrics, but may be a problem in both high-precision capacitors and some electrolytic capacitors. Shunt conductance is affected by both

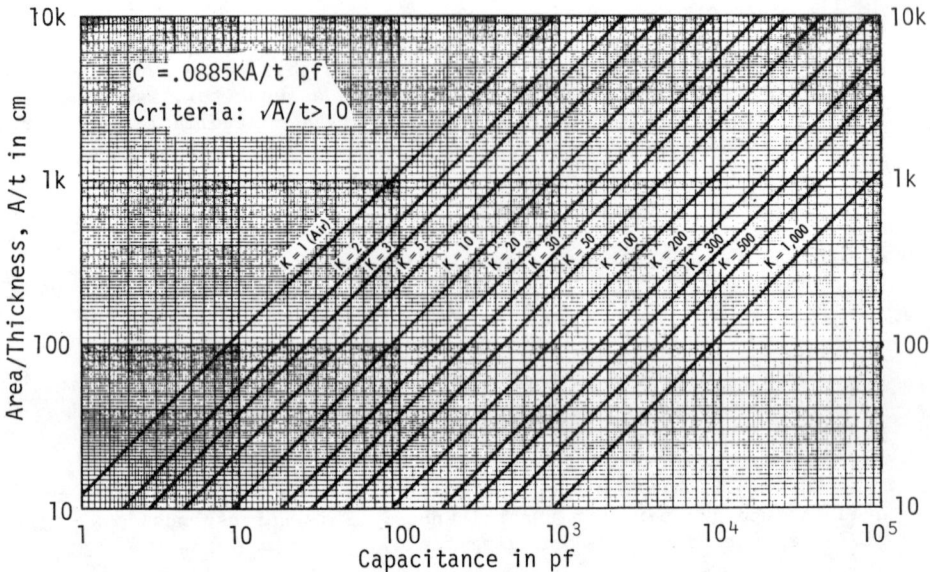

Figure 16.6 - Capacitance of Parallel-Plate Capacitor

the instantaneous and longer-duration application of voltage stress.
The effects are energy loss, heating, and change in power factor. Absorption of energy results in reappearance of voltage on the capacitor
after it has been discharged. Dielectric absorption causes a voltage
stress that is delayed because of the time required to displace charges
from the dielectric.

In high-voltage circuits, a means of discharging capacitors should
be provided to prevent danger to personnel. The self-resonant frequency is determined by many factors, including physical size, dielectric
properties, capacitance, lead inductance, and inductance of the plates.
Fig. 16.9 shows lead-length effects for a 0.5 μf capacitor with 1/4
inch leads. Note the complete degradation of performance above about
5 MHz. Above this frequency, the capacitor behaves as an inductor.
Were this capacitor used as a single-stage filter in a 50 ohm line, the
insertion loss would degenerate above 4 MHz as shown in Fig. 16.10.
This may be contrasted with the performance of a 0.05 μf feed-through
capacitor in which the associated series inductance is very small.
Above 50 MHz, the dielectric and series resistive loss protects the
capacitor's performance as a filter. Three types of resonances can
occur in disc-type capacitors:

● Low-frequency resonance due to long leads

● Medium-high frequency resonance when discs are connected
   in parallel, due to internal leads

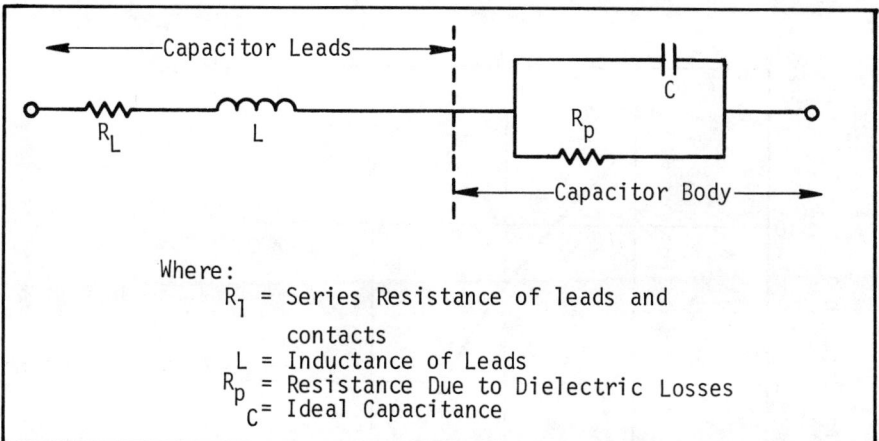

Figure 16.7 - Equivalent Circuit of Most Capacitors Over a Wide
Frequency Range

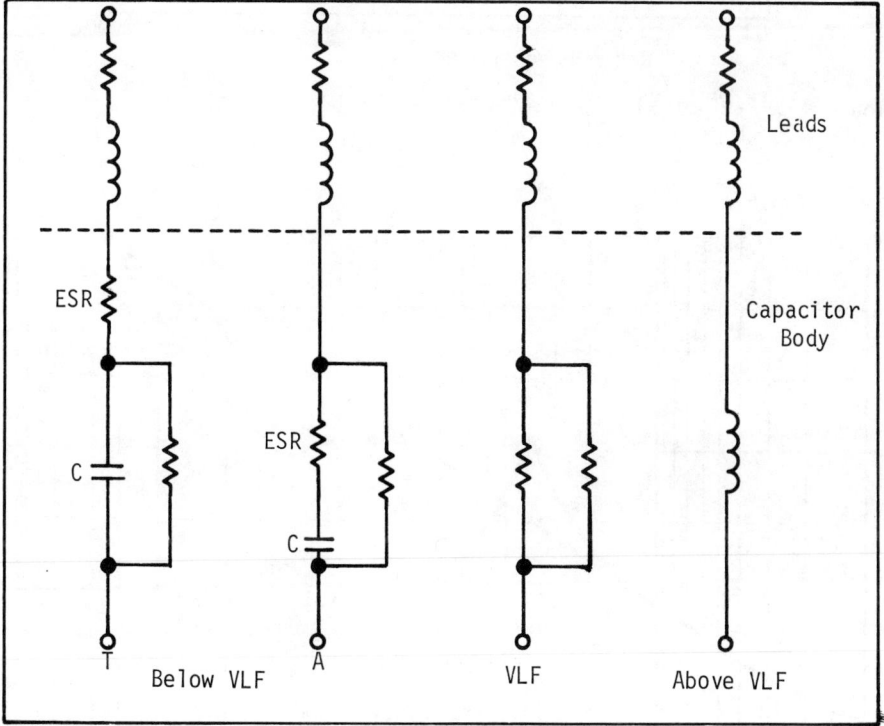

Figure 16.8 - Equivalent Circuit of Many Electrolytic Capacitors

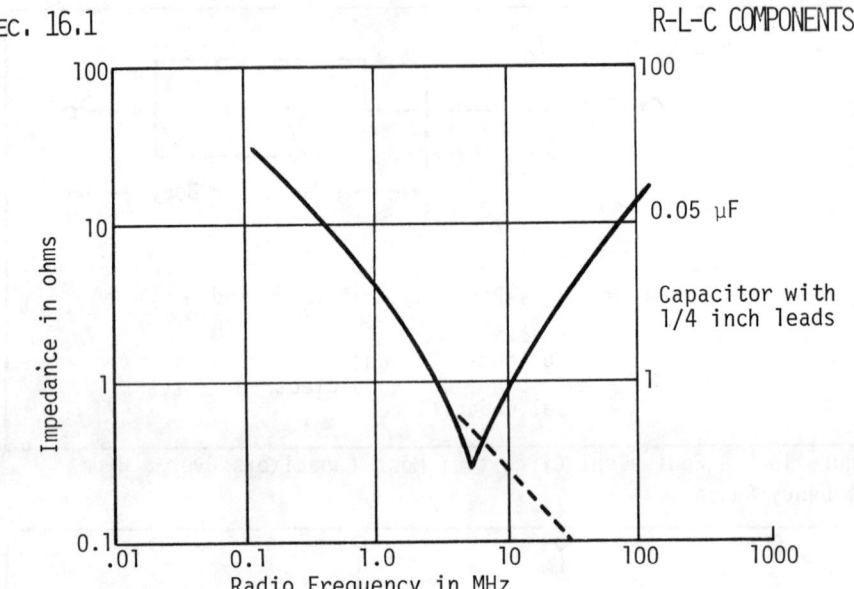

Figure 16.9 - Typical Effect of Lead Length Upon Capacitor

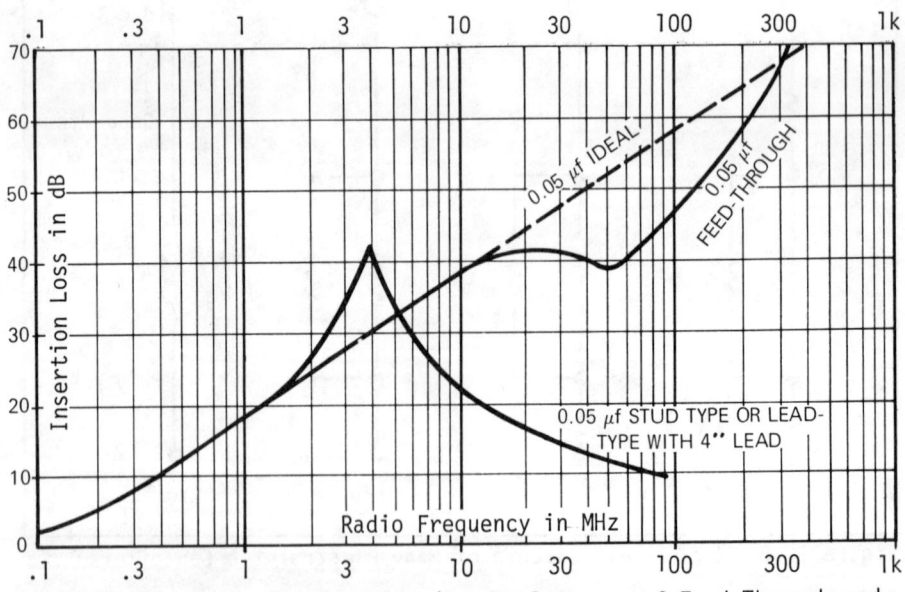

Figure 16.10- Comparison of Filtering Performance of Feed-Through and
Lead-Type Capacitors with Ideal Capacitors

16.12

● High-frequency resonance due to resonant cavity effects
   in high-dielectric capacitors

Ceramic dielectric capacitors and filters using ceramic (barium titanate) dielectric are useful because of their small size and low weight compared with capacitors using more conventional materials such as paper or mica. However, some ceramic capacitors are extremely sensitive to temperature, resulting in a significant degradation in equipment performance when they are used as a bypass operating at a local temperature of -40°C or at an elevated temperature of +100°C. For example, according to one manufacturer's specifications, a nominal 1000 pf ceramic disc feed-through type capacitor, will have a capacity of 330 pf at -40°C. While the operating temperature range is -55° to +85°C, the capacity is rated for a temperature of 20°C ± 1°C. Continuous working voltage rating is 500 volts DC but the capacity is stated for a 0.5 to 5.0 volt range. With 500 volts applied and a feed-through current of 25 amperes, the effective attenuation as a bypass capacitor is reduced by 20 dB at +25°C. Thus, operating conditions must be considered when a capacitor or filter incorporating a ceramic dielectric is used.

Tantalum capacitors also offer attractive space and weight savings under some conditions of operation. Tantalum capacitors come in three types: solid, foil, and wet anode. Solid slug types are made by sintering. This forms a spongy slug of metal that has a large effective surface area and is extremely small for its capacity and voltage rating. Feed-through capacitors of the solid-tantalum type are effective up to 5 GHz.

Foil-type tantalum capacitors can be made in voltage ratings up to 300 volts compared with about 50 volts for the solid and 125 volts for the wet type. The foil-type is limited to audio and low R-F applications because of high internal inductance. Sometimes a tantalum-foil capacitor is shunted with a smaller paper or ceramic capacitor to extend the effective bypass range.

The original wet-type tantalum capacitor is no longer used because of the danger of electrolyte leakage. In the newer wet-type construction, the electrolyte is a gel and the danger of leakage and consequent corrosion is no longer a factor. The wet-type construction offers the smallest size and the largest capacity of all tantalum types.

The tantalum capacitor has a further advantage in that low-temperature performance is greatly superior to the aluminum-electrolytic type. Tantalum capacitors also have longer shelf life and less current leakage especially at high temperatures and after long periods of idle time.

The total inductance of a capacitor is the sum of the internal electrode inductance and the external lead inductance. Table 16.6 lists typical values of internal (electrode) inductance for various types of capacitors.

Table 16.6 - Typical Internal Inductance of Various Capacitors

Capacitor Type	Inductance
Porcelain and ceramic fixed capacitors	1.4 nh
Wet-anode tantalum capacitors	25 nh
Solid tantalum capacitors	20 nh
Foil tantalum, tubular case, with leads	50 nh
Foil tantalum, rectangular case, lug terminals	23 nh

The external (lead) inductance, L, can be determined from the self-inductance of a straight round copper wire:

$$L = 5.08 \; \ell \left[2.303 \; \log_{10} \left(\frac{4\ell}{d} \right) - 0.75 \right] \; nh \qquad (16.9)$$

where, ℓ = lead length in inches

d = lead diameter in inches

For example, for 22 gauge wire (0.025" diameter), the lead inductance is:

3.7 nh for 1/4" total lead length

9.3 nh for 1/2" total lead length

22 nh for 1" total lead length

The internal and external inductances may be added to determine the total inductance in series with the capacitance.

Table 16.7 lists various EMI problems associated with capacitors, and their causes and corrections.

16.1.3   INDUCTORS

An inductor is formed by a portion of a conductor and its return, a single conductor coil, or usually, a multiturn coil. The inductance of different straight wires as a function of length is shown in Fig. 16.11. An intentional inductor exhibits inductance, resistance, capacitance between turns, and capacitance between turns and ground, shields, and other circuits. Fig. 16.12 shows the equivalent circuit of a typical inductor. An aircore inductor may be wound on a non-magnetic metal or insulator core. Magnetic cores are made with steel or iron alloy in sheet, strip, wire, or power form.

The distributed capacitance of an inductor acts as a lumped shunt capacitance, resulting in parallel resonance at some frequency. Other inductor characteristics are power dissipation or loss, saturation, susceptibility to and generation of stray magnetic fields, and instability

Table 16.7 - EMI Problems, Causes, and Corrections in Capacitors

Type of Noise	Causes	Result and Correction
Internal Spike	Combination of temperature and voltage stresses causing breakdown of dielectric	Permanent damage to most capacitors; stress should be decreased by using several capacitors in series. Self-healing capacitors like foil tantalum might help
Random noise in R-F capacitors, e.g., silver-mica, silver ceramic	Flutter or scintillation caused by random and sudden change in capacitance due to improper adhesion of silver to the dielectric, allowing intermittency	Some areas of the capacitor are out of the circuit; proper manufacturer will avoid this
Electrolytic capacitor scintillation	Voltage surges greater than working voltage or exceeding temperature rating	Keep voltage sources and the temperature below the ratings
Noise pulses during charging and discharging at VLF	Release of dielectric stress, particularly with polystyrene and quartz dielectrics	Avoid low frequency use of polystyrene and quartz dielectrics; liquid dielectrics may minimize effect
Noise pulses	Plastic dielectrics such as polyethylene continue to polymerize even after being put in use. Stresses eventually cause noise pulses	Use other dielectrics
Possible noise in polyethylene, quartz and mica dielectric capacitors (particularly small capacitance units	Light, radioactive particles, and X-rays produce ionic action on the dielectric	Shielding or other means
Passing or absorption of energy at certain frequencies	Resonant frequency caused by capacitance, inductance, and physical qualities	Proper design and shielding
Effective inductance of leads is too large even with short leads	Leads of capacitor behave as inductors	Use parallel capacitors connected by short conductors with the return as close as possible to the other conductor; use coax conductors or a flat-strap conductor close to the return conductor

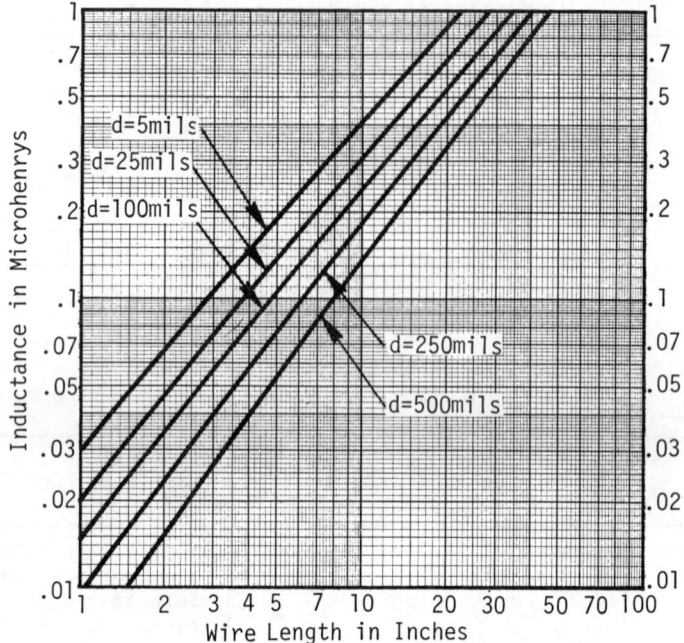

Figure 16.11- Self Inductance of a Straight Round Wire at High
Frequencies

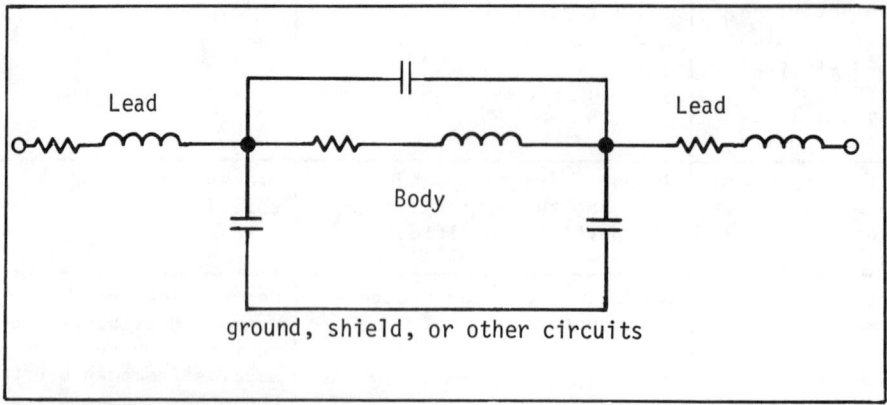

Figure 16.12 - Equivalent Circuit of an Inductor

of characteristics. High-precision circuitry can lose appreciable precision and stability when temperature and humidity cause dielectric
losses from the insulation and support of the inductors.

Current flowing through an inductor coil causes magnetic flux that
produce an EMF proportional to the rate of change of flux linkage. The
self-inductance is:

$$L = N\phi I \qquad\qquad (16.10)$$

where: L = self-inductance in henrys

             N = number of turns, loops, or linkages

             ϕ = magnetic flux or webers (1 Tesla x $1m^2$)

             I = current in amperes

Mutual inductance (inductive coupling) occurs when magnetic flux
lines of one element link with an area of another element. Energy is
thereby coupled from one element to the other and can cause interference to other circuits as suggested in Fig. 16.13. Mutual coupling
decreases with distance, and the least coupling occurs when the inductor axes are at right angles to the area of the victim circuit. Inductors made from toroids have considerably reduced external magnetic
fields since nearly all the field is inside the magnetic toroid material.

Inductors are often shielded to keep their electric and magnetic
fields within a limited space around the inductor. Electrostatic
shields can be made of low electrical resistance material such as copper, aluminum, or zinc. These shields prevent magnetic flux from passing through because voltages are induced that set up eddy currents which
oppose the magnetic fields. A high-permeability material such as Permalloy is used at low frequencies if the flux is undirectional. The
use of high-permeability material results in an induced opposing magnetic field and a magnetic flux short circuit. Thus, there is a negligible magnetic field outside of the shield. Saturation of the shield
reduces its permeability so that the shield no longer behaves as a short
circuit to the flux.

Table 16.8 shows some of the EMC problems encountered with inductors. Usual EMC problems involving inductors are:

EMI Culprit Source	EMI Susceptibility
• Coupling	• Skin Effect
• Counter-EMF when switching	• Core Saturation
• Ringing on Transients	• Resonant Frequencies
	• Mutual Inductance

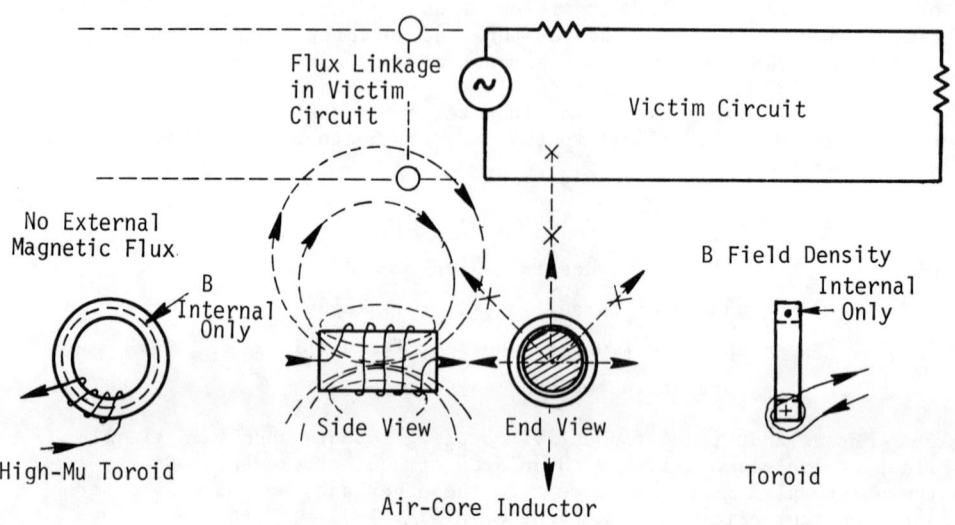

Figure 16.13 - External Magnetic Fields of an Inductor for Air and High Permeability Cores

Table 16.8 - EMI Problems, Causes, and Corrections in Inductors

Problem	Possible Cause	Corrections
Passing or stopping desired signals	Resonance due to inductive and capacitive characteristics	Design and shielding
Power dissipation over the ratings	Skin effect, conductor resistance, eddy current and hysteresis losses	Use of stranded conductors, non-magnetic cores and toroidal geometry winding
Varying effective inductance	Distributed capacitance in coil causes apparent decrease in inductance	Minimize distributed capacitance and/or permit high eddy current effect to reduce effective inductances
Interference by mutual inductance	Proximity of two coils, one coil and a conductor, or any two circuit elements	Separation, shielding and keeping object axes at right angles
Transients	Current applied or cut off to the inductor	Same as for electromagnetic relays

16.2 INSULATORS AND CONDUCTORS

This section reviews the topics of insulators and conductors as potential sources of EMI and some of the control techniques that are employed to suppress interference. Details regarding both the conductor physics and EMI control methods are covered in Chap. 6 on Cable Wiring and Harnessing.

16.2.1 INSULATORS

Insulators permit small leakage currents to flow despite their low conductivity. This may lead to voltage surges and changes in current. The insulation around a cable illustrates this phenomenon. When corona occurs, insulation breaks down and small currents creep along the surface.

Dielectric loss, called loss tangent or power factor, in insulators causes dissipation of power, loss of linearity, and coupling with other circuits. It is especially important in coaxial lines and capacitor construction. This problem is eliminated by selecting proper materials such as polyethelyne or Teflon for the frequency range of interest.

Surface tracking, a condition in which small currents creep across an insulator, is an important source of EMI. It is caused by surface contamination of the insulation by moisture or solid conductive particles, by chemical degradation of the insulation material, or by momentary overvoltage. When a discharge occurs across a low-impedance circuit, significant energy may be released and quick catastrophic degradation occurs. If the circuit impedance is high, a slow discharge occurs over a long time interval. The arc changes paths continuously during this interval, extinguishing in one section but igniting in another. These changes are fast and variable, producing a broad spectrum of interference.

Corona differs from a discharge across insulation, because it is highly concentrated at one point, usually shows a visible glow, and has a humming sound, caused by highly ionized air around the conductor. Special cases of corona may develop in gaseous pockets or voids within solid insulation, usually showing no glow. This also causes EMI, as they develop slowly and lead to insulation failure. Internal breakdown of the insulation produces small changes in current as in surface tracking. Small internal paths are overheated. This causes chemical changes that results in lower local resistivity and accelerated damage. High-voltage gradients sometimes occur across small mechanical voids in an insulator. Gaseous discharges also occur in the voids, with general current changes leading to degradation.

Some typical insulator problems, causes, and remedial measures are shown in Table 16.9.

Table 16.9 - EMI Problems, Causes, and Corrections in Insulators

Problem	Cause	Preventive Measure
Surface tracking causing dI/dt charges which produce wide band of frequencies	Surface contamination, chemical degradation of insulation, momentary overvoltage	Protection from contamination, use of proper material, proper voltage design
Surface tracking resulting in catastrophic degradation	Low impedance discharge circuit releasing high energy	Protection from contamination, use of proper material, proper voltage design
Surface tracking with small but relatively persistent EMI	High impedance discharge circuit which limits energy to circuit component tolerances	Protection from contamination, use of proper material, proper voltage design
Glow discharge which may be visible and audible but causing electrical noise because of dI/dt effects	Corona with high voltage across one conductor to ground or to another conductor	Prevention of voids within insulators, protection from contamination, prevention of sharp points at high potentials, low voltage gradient design

16.2.2 CONDUCTORS

A conductor is any material which readily permits an electric current to flow when subjected to a difference in potential. It is usually desirable to use a low-resistance metal such as copper as the conductor in an electronic circuit. Copper is inexpensive and relatively stable in the ambient temperature range usually encountered. It is easily soldered but will corrode on exposure to the atmosphere. For this reason it is sometimes protected with a plating or coating of tin, silver, or gold.

The selection of a conductor size is generally related to the maximum allowable voltage drop in the conductor or heating effect by power I^2R loss. At RF, skin effect must be considered. Skin effect is the term used to describe an uneven cross-sectional distribution of current density in which there is a concentration of current near the surface or skin of a conductor. This results in a higher conductor resistance at RF. Litzendraht or *litz* wire was designed to reduce skin effect. It is composed of several strands of enamel-coated wire, individually insulated, inter-woven, and connected in parallel at each end. Skin effect combined with the effect of flux linkages between a conductor and its return above about 1 GHz makes conduction within a solid conductor very difficult. Accordingly, about 1 GHz (especially 10 GHz) a transition is often made to waveguide for signal transmission. Wave-guide as a conductor is effective up to about 300 GHz.

Skin effect is also reduced by other means. Since a circular cross-section conductor has the least skin surface per unit of area, it may be advantageous to make the conductor rectangular. Flat, strap or foil-type conductors have inherently lower A-C resistance because of their greater surface per cross-section area. Alternatively, removing the center of the conductor and creating a hollow tube will significantly reduce self-inductance in conductors. For this reason, tubular con-ductors are commonly used in R-F ground systems and in high-power trans-mitters.

Parallel wiring systems may be loaded up to a point where the wire fuses or where corona is probable, whereas arcing limits the power in coaxial and waveguide systems. In general, parallel lines exhibit less loss than equivalent coaxial lines. However, coaxial lines have a less tendency to radiate. Shielded wire is also relatively more expensive. The attenuation offered by shields is due to energy reflection at the boundary and absorption within as external signals pass through. A typical voltage measurement between an external pick-up wire and the central conductor of a coaxial line shows a cross-talk of about 60 dB for the geometry used (see Chap. 6). The use of a double-shield may increase shielding effectiveness by about 25 dB for some geometry, and critical circuits may require even more shielding.

Braided sleeving is sometimes used as a bond across a vibration mount because of its flexibility. However, because of its construction, it has a higher inherent inductance than a strap made of solid copper sheet (see Sec. 8.4). Braided sleeving also makes a convenient cable shield, but it exhibits a limited effectiveness. For low frequency and especially low-impedance applications, an overall insulating cover should be used so that the braid can be grounded at one point and insul-ated from ground over the rest of its length (see Sec. 6.2). As a high-frequency circuit shield it can be bare. However, when it is unprotec-ted the eventual corrosion between intersecting strands will degrade shielding effectiveness.

In many ways teflon is an ideal insulating material for wires in cables. It offers a high insulation with good mechanical protection in extremely thin conductor coatings. This allows a high density of wire within a cable or harness. This in turn may increase circuit-to-circuit coupling within a cable because of closer spacing. Capaci-tive coupling is a function of the distance between conductors, the length of the conductors, and the dielectric constant of the material between conductors. Since capacity is proportional to dielectric con-stant, the capacity will increase by a factor of two when the air be-tween two wires is replaced with teflon. The dielectric constants at 1.0 MHz of common cable dielectrics are listed in Table 16.10. Water is included in the list to illustrate that a small amount of moisture can impact an electronic circuit.

Table 16.10 - Typical Dielectric Constants of Cable Dielectrics

Dielectric	Dielectric Constant
Air	1.0
Polyetheylene	2.26
Polyvinyl chloride (PVC)	3.52
Paper	2.99
Nylon	3.14
Teflon	2.0
Water	78.2

Various methods to minimize EMI in conductors are:

● Multiple-conductor cables that bundle each conductor with its return conductor carrying current in the opposite direction results in magnetic-field cancellation. The amount of cancellation depends upon the relative equality of the currents and the spacing between conductors.

● When two conductors with similar current levels are twisted together, the field generated by one tends to cancel the field of the other if the currents are opposite in direction. The greater the number of uniform twists, the more effective is the cancellation of magnetic fields (see Sec. 6.2).

● A conductor surrounded by a return path shield, as in a coaxial cable, theoretically will not have an external magnetic field. Various factors degrading this are the lack of solidity of the surrounding conductor and its conductivity (see Sec. 6.2.2).

● Skin effect at higher frequencies, has an increasing effect on electric and magnetic fields. At higher frequencies, hollow conductors should be considered to minimize external fields.

● Magnetic shields can be used to minimize magnetic field penetration. Eddy currents that occur in grounded shields create absorption losses that minimize the external magnetic field. The absorption losses are discussed in Sec. 10.2.

● Bare conductors operating at high-voltage potentials can produce large electrostatic fields at sharp corners or points that ionize the surrounding atmosphere, resulting in a broadband, white-noise corona. Minimum bend radius for a high-voltage conductor should be 10 times its outside diameter and connection points and other irregularities should be rounded. Contamination across a component can also lead to corona by reducing the required breakdown potential.

● Conducted interference can be limited by coating conductors with high-permeability material which magnifies skin-effect losses. There is a large effect in the frequency range from about 25 kHz to 50 MHz.

● Ferrite beads are tiny cylindrical beads that may be strung on a wire to increase inductance of the wire and thereby cause attenuation of signals due to filter action. Attenuation by inductive reactance and I^2R losses is a function of frequency because there is no D-C current in the bead. As an example, four ferrite beads can increase the inductance of a two-inch piece of wire by fifteen times and can cause a 6-dB attenuation over a wide frequency range (see Sec. 6.5). The cost is less than that of resistors or RF chokes.

● Lossy flexible sleeving made up from ferrite particles in a suitable binder is a continuation of the ferrite bead and rod concept. Above about 10 MHz conducted EMI is converted to heat and attenuation becomes very significant above about 50 MHz (see Sec. 6.5.4).

● Single-point shielding of conductors must prevent the shield from touching ground at more than one point. A common remedy is to enclose a coaxial conductor in a plastic sheath.

● Ionic atmosphere, created by field gradients in the region of conductance, can promote the deterioration of the insulation or shielding and eventually result in a change from the original characteristics.

16.3 CONNECTORS

Connectors, the interfacing and mating devices between cables and harnesses or between cables and other devices or equipment, were discussed in length in Chap. 7. Thus, the reader is referred to that chapter for details.

Connectors can cause EMI by circuit geometric discontinuities when they:

1. Result in poor mating contacts which develops varying contact potentials resulting in broadband noise.

2. Result in unshielded inductive loops and small capacities that interfere with sensitive circuits.

3. Constitute local circuits of inductance and capacitance having natural resonant and anti-resonant frequencies in the pass band of other circuits.

4. Constitute lumped impedance discontinuities (VSWR's) at certain points in the circuit which cause reflections and standing waves.

5. Provide insertion loss to applied signals. Ideally, connectors should have:

- Negligible resistance
- Chemically inert surfaces
- Resistance to gouging
- Foolproof alignment to minimize contact damage
- Adequate force between contacts
- Little friction to minimize increase in resistance with use
- Contamination-free design
- Provisions for proper connections, including shielding of backshell
- Proper dielectric properties
- Moisture-proofing as required
- Resistance to degradation due to age, wear, maintenance, and repair
- Filter pins incorporated, if necessary
- Compatibility regardless of varying intersystem contractors

There should always be a proper installation, including a good bond between the cable shield(s) and connector shell, as shown in Fig. 16.14. Shields should be bonded completely around the periphery of the connector body. All connectors used as conducting paths for EMI should be

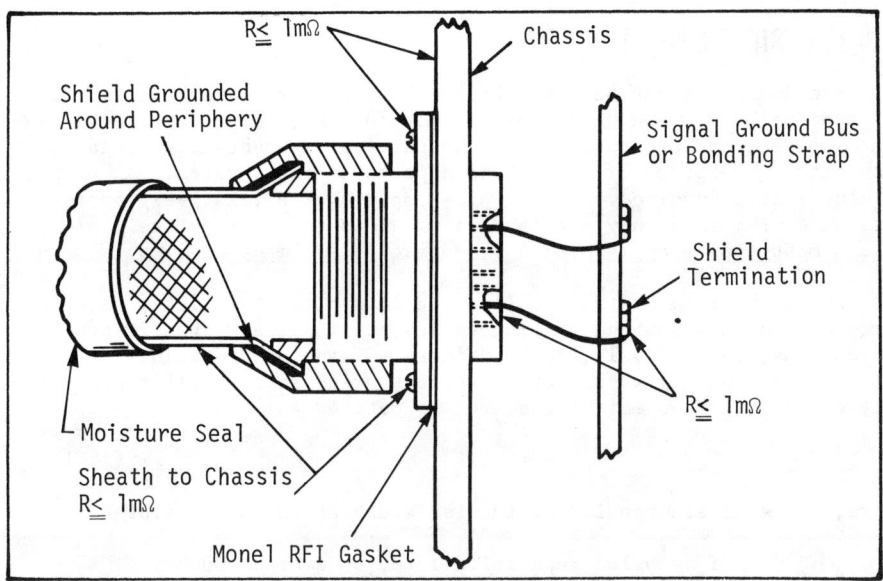

Figure 16.14 Connector Shielding and Grounding

bonded to the static ground with D-C bonding resistance of the order of
1 mΩ or better. Air gaps at the connector-chassis interface should be
eliminated by woven-mesh EMI gasketing means. Other desirable features
are protective coverings that extend over the male pins to reduce pin
damage, the use of caps on unused connectors, the use of clamps to hold
wires steady, contact materials designed for long life and proper pressure, and no loose of faulty contacts that might generate EMI.

Filter pins may be used where interference is in the VHF and UHF
range. Most filter pins are not effective below 1 MHz. Use feed-
through capacitors or filters mounted in a connector box where conducted interference is below 1 MHz.

16.4 TRANSIENT CHARACTERISTICS

Previous sections of this chapter have discussed mathematical models and topics dealing with devices, methods, and techniques which generate couple, and/or control EMI. No particular emphasis was placed upon either the broad or narrowband aspects of the physics, mathematics, or electronics involved. Some topics apply over a relatively narrow portion of the frequency spectrum (i.e., less than one octave), although most topics apply over a broadband (i.e., many octaves).

When either an emitting EMI source or a potential victim receptor performs over a broadband of frequencies, it is likely that it develops or responds, respectively, to transients. Transients distinguish themselves by having a low duty cycle and fast rise and/or fall times. The duty cycle, δ, of an emitting source is defined as:

$$\delta = \tau \times f_r \qquad (16.11)$$

where, τ = equivalent pulse or impulse width at the 50% height

f_r = pulse repetition rate, or average number of pulses/impulses per sec. for random occurrences

Most transients from incidental emitting sources correspond to duty cycles which are very small, i.e., less than 10^{-5}. Table 16.11 lists some approximate duty cycles identified to the nearest order of magnitude corresponding to a few transient sources.

Table 16.11 - Typical Transient Sources

Emitting Transient Source	Repetition Rate	Impulse Width	Duty Cycle
Fluorescent Lamps	100 pps	10^{-7} sec	10^{-5}
Ignition Systems: idle speed	100 pps	10^{-8} sec	10^{-6}
fast speed	10^3 pps	10^{-8} sec	10^{-5}
Relays and Solenoids:			
casual use	10^{-3}	10^{-7} sec	10^{-10}
pin-ball machine	1 pps	10^{-7} sec	10^{-7}
teletype	10 pps	10^{-7} sec	10^{-6}
Brush-Commutator Motor	10^3 pps	10^{-8} sec	10^{-5}
On-Off Switches:			
wall switch	10^{-4} pps	10^{-6} sec	10^{-10}
lathe	10^{-3} pps	10^{-7} sec	10^{-10}
copy machine	10^{-3} pps	10^{-7} sec	10^{-10}

When the duty cycle becomes significantly greater than 10^{-5}, such as 10^{-3} for radar or 0.5 for a computer clock, the emitting source is no longer regarded as a transient although it may still have fast rise times and therefore broadband emissions.

Transient EMI problems before and up to World War II were nearly nonexistent because there were relatively few electrical and electronic devices per capita. With the technology and manufacturing explosion following World War II, the military, industrial, and consumer devices per capita expanded on all fronts together with a gradual transition from steady-state devices to transient devices. Today, the number of repetitive transient devices per capita is estimated by various sources to be between 25 and 100, most of which generate electrical noise rather than respond thereto.

Transients have become a major problem during the early 1970's because so many emitting sources are now operational and because computers, digital and control devices, among others, are especially susceptible. Thus, this section summarizes some of the mathematical models associated with transient noise sources and susceptible equipments.

16.4.1 SPECTRUM OCCUPANCY

A transient is a current or voltage amplitude vs time which is preceded and/or followed by two different level states. Two examples are shown in Fig. 16.15 for an approximation to a transient:

(1) A step function with finite rise or fall time in which two different levels exist on either side of the transition (transient), either of which may take on any amplitude including zero.

(2) A pulse or impulse with finite rise and fall times in which a different level exists with reference to the mid-pulse level of the transition (transient), and in which the preceding and subsequent levels may be any amplitude including zero. Thus, the pulse or impulse is essentially a back-to-back step function.

By using the Fourier integral or Laplace transform, the spectrum occupancy of a transient may be determined in the frequency domain from its structure in the time domain. For the step function shown in Fig. 16.15:

$$\text{Step Function: } \frac{\Delta e}{\tau_r}U(t)t - \frac{\Delta e}{\tau_r}U(t-\tau_r)t \qquad (16.12)$$

$$\text{Laplace Transform: } \frac{\Delta e}{\tau_r s^2}(1-e^{-\tau_r s}) \qquad (16.13)$$

$$\leq \frac{2\Delta e}{\tau_r s^2} = 0.05\ \Delta e/\tau_r f^2, \text{ for } f > 1/\tau_r \qquad (16.14)$$

$$\text{where,} \qquad s = j\omega = j2\pi f \qquad (16.15)$$

Eq. (16.13) is plotted in Fig. 16.16. The envelope of its amplitude, developed in Eq. (16.14), is also shown in Fig. 16.16. The

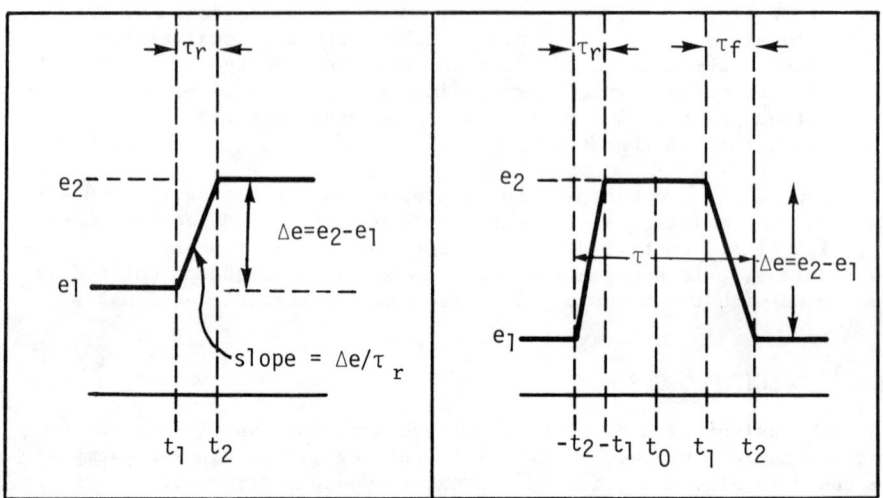

Figure 16.15 - Step Function and Impulse Transients

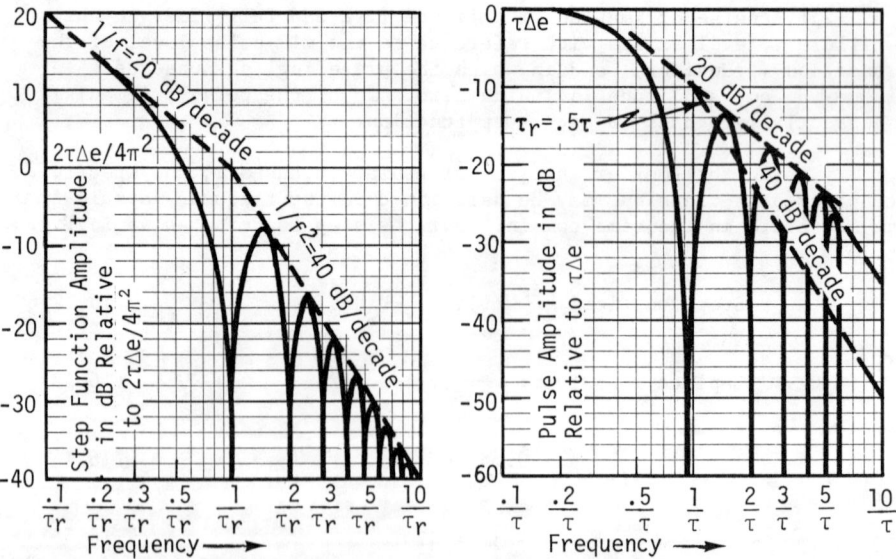

Figure 16.16 - Time Domain Responses of Step and Pulse Functions
With Finite Rise Times

16.28

envelope falls off inversely with frequency squared (40 dB/decade) for a finite rise time and for $f > 1/\tau_r^*$. Should the rise time become arbitrarily large:

$$\lim_{\tau_r \to \infty} \frac{0.05\Delta e}{\tau_r f^2} = 0, \text{ for } f > 1/\tau_r \qquad (16.16)$$

Thus, to reduce the amplitude of a step function in the frequency domain, its transition time (rise or fall time) should be made as large as possible.

Conversely, when the rise time becomes arbitrarily small, Eq. (16.13) becomes:

$$\lim_{\tau_r \to 0} \frac{\Delta e}{\tau_r s^2} (1-e^{-\tau_r s}) = \lim_{\tau_r \to 0} \frac{\Delta e \, e^{-\tau_r s}}{s} = \frac{\Delta e}{s}$$

$$= 0.16\Delta e/f \qquad (16.17)$$

Eq. (16.17) shows that the amplitude of a pure step function (no rise time) in the frequency domain falls off inversely with frequency (20 dB/decade). The amplitude also becomes arbitrarily large as the frequency approaches zero.

The spectral-amplitude distribution shown in Fig. 16.16 is solidly *painted in* during the duration of the transient interface and is zero during all other times. The process repeats for a second step transient and it would be the mirror image (negative going) if the transient has a negative slope. A time series of such transients, random or periodic, increases the probability that a susceptibility problem may occur if the amplitude thresholds are above receptor response limits. As explained later for transient emissions, the potential susceptibility of a receptor is proportional to its bandwidth. Further, should the receptor bandwidth be less than the transient repetition rate for recurring transients, then the response amplitude will be proportional to the square root of bandwidth.

The development of a pulse or impulse-like transient (see Fig. 16.15) follows a nearly identical mathematical procedure to the above. The results of this are:

(1) For finite rise and fall time ($\tau_r = \tau_f < t_2 = \tau/2$):

$$\text{Laplace transform} = \tau\Delta e \, \frac{\sin\pi f(\tau-\tau_r)}{\pi f\tau} \cdot \frac{\sin\pi f\tau_r}{\pi f\tau_r} \qquad (16.18)$$

---

* If the transition is a smooth S-shaped function rather than abrupt, it can be shown that the slope falls off at the rate of 60 dB/decade or faster for $f > 1/\tau_r$.

(2) For zero rise and fall time ($\tau_r = \tau_f = 0$)

$$\text{Laplace transform} = \tau \Delta e \, \frac{\sin \pi f \tau}{\pi f \tau} \tag{16.19}$$

Both equations approach an amplitude of $\tau \Delta e$ as f approaches zero as shown in Fig. 16.16. This is contrasted with the step function which approaches infinity. The envelope of Eq. (16.18) falls off at the rate of 20 dB/decade for $1/\tau < f < 1/\tau_r$, whereafter it falls off at 40 dB/decade due to $1/f^2$ term. The envelope of Eq. (16.19) falls off at 20 dB/decade everywhere beyond the first χ-Axis crossing ($f > 1/\tau$). Thus, to reduce EMI, the rise and fall times, τ_r, should be made as large as possible. Chap. 3 showed that by smoothing the rise and fall time functions, such as a $\cos^2\theta$ term, the emission level will decrease at the rate of 60 dB/decade. This represents a much less EMI threat to out-of-band receptors.

16.4.2 BROAD AND NARROW-BAND EMISSIONS

The previous section implied that there exists some absolute definition to broadband and narrowband emissions. It was stated that spectrum occupancy (e.g., 3-dB bandwidth) of less than one octave was narrowband such as from communications-electronic equipment modulation emissions, while broadband spectral distributions may occupy many octaves, such as from either fluorescent lamps or engine ignition systems. While this is one form of definition of broad and narrowband, it is arbitrary. A more frequent definition use in the EMI Community, as explained below, uses the terms in a *relative* rather than *absolute* sense.

From an EMI prediction and control viewpoint, broad and narrowband emissions may be defined relative to a potentially susceptible victim-receptor bandwidth. It is important to know which type the emission appears to be since different relations are used in either the EMI prediction or control process. For example, it will be shown that when an interfering emission is coherently broadband relative to a receptor bandwidth, such as originating from a transient, the interference-to-noise ratio (I/N) is proportional to the square root of bandwidth. Conversely, when the emission is narrowband relative to a receptor bandwidth, the I/N is *inversely* proportional to the square root of bandwidth. Thus, different approaches regarding receptor bandwidth are used depending upon the relative emission bandwidth and whether the emissions are coherent or appear as white noise.

16.4.2.1 BROADBAND EMISSIONS

Fig. 16.17 shows the spectral intensity distribution of a single pulse emission previously shown in Fig. 16.16. A carrier frequency, f_c, is shown, which when equal to zero, the baseband response only applies with no negative frequencies. The figure also shows the superposition

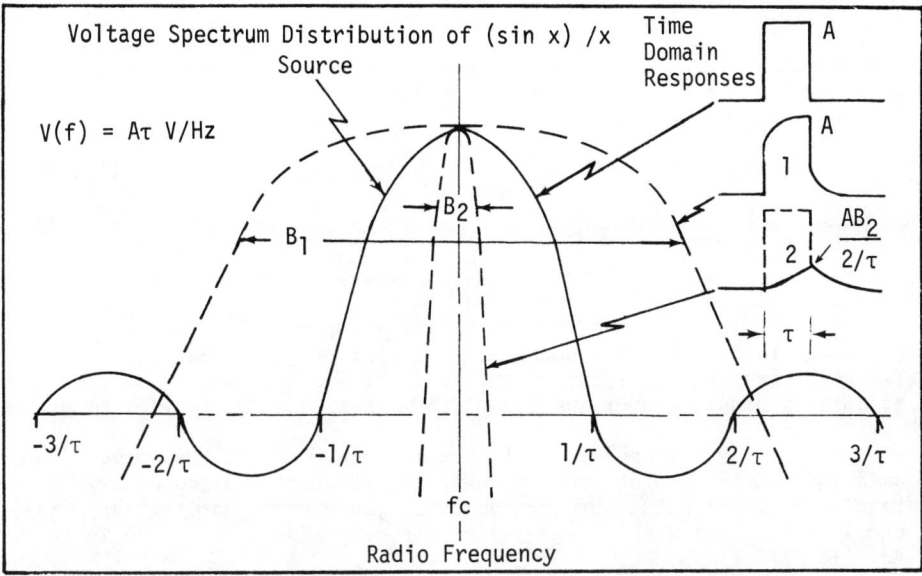

Figure 16.17 - Illustrating Broad and Narrowband Transients

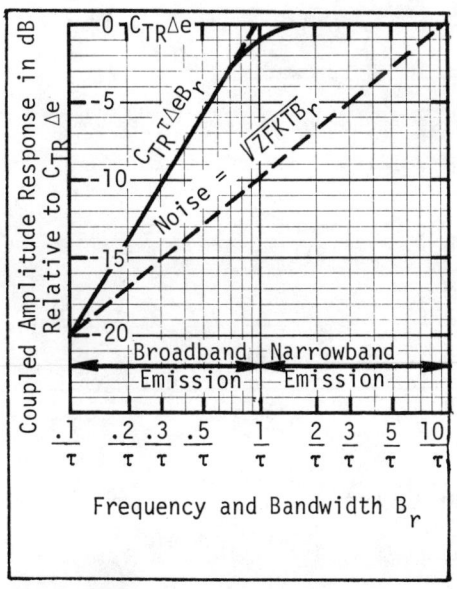

Figure 16.18 - Broad and Narrowband Coupled Emission Levels

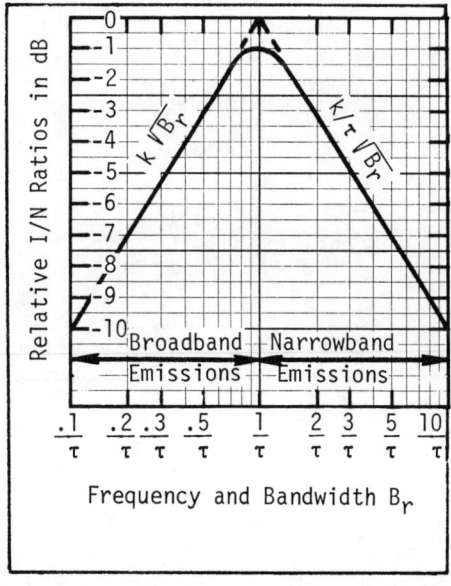

Figure 16.19 - Receptor I/N Ratios vs Emission Source Bandwidth

16.31

of the bandwidth, B_R, of a potentially-susceptible victim receptor. When $B_R \ll 1/\tau$ (case for $B_r = B_2$) the amplitude of the voltage V_{CB}, coupled into the receptor is obtained by the convolution integral, which can be simplified to approximately yield:

$$V_{CB} = C_{TR} \, \tau \Delta e \, \frac{\sin \pi f \tau}{\pi f \tau} \, e^{j \pi f \tau} B_r, \quad \text{for } f \ll 1/\tau \qquad (16.20)$$

$$= C_{TR} \tau \Delta e B_r \qquad (16.21)$$

where, C_{TR} = emission source to receptor coupling coefficient

Eq. 16.21 shows that the potentially-coupled EMI level is proportional to the receptor bandwidth because: (1) the amplitude $(\sin \pi f \tau)/\pi f \tau = 1$ over the bandwidth and (2) the phase term, $e^{j \pi f \tau}$, is relatively constant over the bandwidth. This is shown in Fig. 16.17.

Fig. 16.17 also shows the time-domain response of the pulse emission source as processed through the receptor's narrow bandwidth. Here significant pulse integration takes place such that the shape is nearly triangular with a relative amplitude of TB_r as depicted in Eq. (16.21).

The internal noise power level within the receptors bandwidth, N_r, referred to its input terminals, is:

$$N_r = FKTB_r \qquad (16.22)$$

$$= 4 \times 10^{-21} \, FB_r \text{ watts}$$

where, F = Noise factor of receiver

$KT = 4 \times 10^{-21}$ watts/Hz

For a receptor input impedance, Z, its internal noise voltage, V_r is obtained from Eq. (16.22) (see Fig. 16.18):

$$V_r = \sqrt{ZFKTB_r} \qquad (16.23)$$

Combining Eqs. (16.21) and (16.22), the interference-to-noise ratio (I/N) at the receptor input terminals for broadband emission sources is:

$$\frac{I}{N} = \frac{V_{CB}}{V_r} = \frac{C_{TR} \tau \Delta e B_r}{\sqrt{ZFKTB_r}} = k\sqrt{B_r}, \quad \text{for } B < 1/\tau \qquad (16.24)$$

where, $k = \dfrac{C_{TR} \tau \Delta e}{\sqrt{ZFKT}}$

Thus, increasing the bandwidth makes a receptor more susceptible to transients as long as $B_r < 1/\tau$. This is shown in Fig. 16.19.

When $B_r \to 1/\tau$, most of the pulse energy is contained within the receiver bandwidth (90%) and the pulse amplitude is recovered as shown in Fig. 16.17.

16.4.2.2 NARROWBAND EMISSIONS

Fig. 16.17. also shows the superposition of the bandwidth of a potentially susceptible victim receptor for $B_r >> 1/\tau$ (case for $B_r = B_1$). The amplitude of the voltage, V_{CN}, coupled into the receptor is now the full amount available from the emission source. Here the convolution integral would yield:

$$V_{CN} = C_{TR}\Delta e, \text{ for } B_r >> 1/\tau \qquad (16.25)$$

A further increase in bandwidth will have no effect upon the interference emission level since Eq. (16.25) is independent of bandwidth as long as $B_r >> 1/T$. This is also shown in Fig. 16.17. The I/N ratio, now becomes:

$$\frac{I}{N} = \frac{V_{CN}}{V_r} = \frac{C_{TR}\Delta e}{\sqrt{ZFKTB}} = \frac{k}{\tau\sqrt{B}} \text{ for } B > 1/\tau \qquad (16.26)$$

Eq. (16.26) is shown in Fig. 16.19. Thus, increasing the bandwidth further actually makes a receptor less susceptible to transients provided the I/N ratio is a true measure of its potential susceptibility. This would be a dangerous practice, however, as explained below.

16.4.2.3 SIMULTANEOUS BROAD AND NARROWBAND EMISSIONS

From the foregoing, it should be clear that two independent transient emission sources can act differently upon a receptor. The $1/\tau$ value of one transient may be considerably less than the receptor's bandwidth, B_r, (narrowband emission) while the second source may be considerably greater than B_r (broadband emission). Thus, two emission sources can exhibit simultaneous broad and narrowband properties relative to B_r. They would also be processed differently through the receptor as explained above.

Illustrative Example 16.1

A nearby fluorescent lamp (source #1) radiates with a level of 110 dBμV/m (0.1V/m) measured at the receptor and a equivalent impulse

16.33

width of 0.5μsec. It exhibits a coupling coefficient to the equivalent input terminals of C_{TR} = -30 dB. An engine ignition system
(source #2) radiates with a level of 90 dBμV/m (measured at the
receptor) and equivalent impulse width of 10 nsec. The ignition
source also exhibits a coupling coefficient of -30 dB. The receptor
is a high-speed computer with a basic clock rate of 10 Mbs such that
read amplifier bandwidths of 20 MHz are used. Determine the characteristics of the emission sources and indicate if EMI may be a problem
at the read amplifier where logic sense levels of 1 volt are used.

Recapping:

	τ	$1/\tau$	B_r
Fluorescent noise source #1	5×10^{-7} sec.	2 MHz	--
Ignition noise source #2	10^{-8} sec.	100 MHz	--
Computer Receptor	5×10^{-8} sec.	20 MHz	20 MHz

Regarding sources, the fluorescent lamp appears as a narrowband
emission ($B_r > 1/\tau$) and the computer intercepter levels from Eq. (16.25)
are:

$$V_{CN} = -30 \text{ dB(meter)} + 110 \text{ dBμV/m} = 80 \text{ dBμV} = 10 \text{ mV}$$

The ignition system appears as a broadband emission ($B_r < 1/\tau$) and the
computer coupled levels from Eq. (16.21) are:

$$V_{CB} = -30 \text{ dB(meter)} + 20 \log_{10} \tau B_r + 90 \text{ dBμV/m}$$

$$= -60 \text{ dBμV} + 20 \log_{10} (0.2) = -46 \text{ dBμV} = 200 \text{ μV}$$

Thus, neither source is of sufficient level (1 volt) to trouble the
computer.

16.5 DIODES

Inasmuch as solid-state diodes are also semiconductors, they share many of the characteristics of transistors. Transistors are discussed in Sec. 16.6.1. Diodes act as both sources of EMI and devices which are useful in suppressing EMI. The latter topic is discussed in a subsequent section on relays and switches. This section emphasizes the former properties.

Under conditions of forward bias, a solid-state semiconductor stores a certain amount of charge in the form of minority current carriers in the depletion region. If the diode is then reverse-biased, it conducts heavily in the reverse direction until all of the stored charge has been removed. The resulting conditions are presented in Fig. 16.20. The duration, amplitude, and configuration of the recovery-time pulse (also called switching time or period) is a function of the diode characteristics and circuit parameters. These current spikes generate a broad spectrum of conducted transient emissions.

Rectification involves switching from conduction to cutoff repetitively. This causes high dI/dt values dependent upon the input frequency, minority carrier storage in the diode, and the circuit characteristics. The interference effect can be minimized by one or more of the following measures:

- Placing a bypass capacitor in parallel with each rectifier diode
- Placing a resistor in series with each rectifier diode
- Placing an R-F bypass capacitor to ground from one or both sides of each rectifier diode
- Operating the rectifier diodes well below their rated current capability

The ripple filter that normally follows a rectifier should not be relied upon to filter out the transient emissions. The usual large-value capacitors used for ripple filters exhibit too much series inductance to function effectively as R-F interference filters. The filter capacitors should be shunted with a second capacitor having a value of two or more orders of magnitude of less capacitance.

Diodes are also used to switch at a particular voltage level. The switching action results in high values of dI/dt. Diodes are used as limiters or clippers to cut off input waveforms at a certain level. The greater the amount of limiting, the greater will be the number of spurious frequencies that occur due to the steepening waveform. These switching actions can result in EMI interaction with other components, distributed or actual impedances, and signal discontinuities.

Several design considerations to minimize switching EMI transients are:

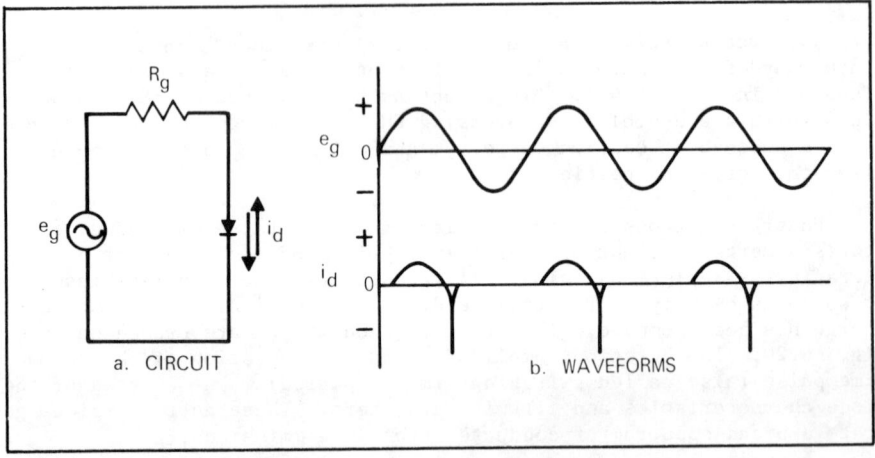

Figure 16.20 - Diode Recovery Periods and Spikes

- Operate at lowest possible voltages and currents
- Anticipate diode-to-diode variation of characteristics
- Use the lowest possible switching rate or rise time and amplitude
- Select diodes with high working and peak inverse voltages
- Use diodes with a slow recovery time (inherent with larger current ratings)

R-F voltage can change the bias of a diode, resulting in improper switching, distortion, or improper output. All diodes are subject to reverse breakdown if they are exposed to R-F voltages greater than their reverse breakdown voltages. Low-power devices (generally those rated 25 mW or less) and small junction devices, such as point-contact diodes, operating in the vicinity of a strong R-F field can absorb sufficient radiated energy to be degraded or burned out. Large junction diodes have a large junction capacitance, of the order of 10 to 15 pf, which will pass high frequencies. If this energy, added to the normal energy, exceeds the thermal limit of the device, damage can occur. Therefore, diodes subjected to R-F fields should be shielded.

When used as an amplifier, a tunnel diode may couple with related circuitry inductance or capacitance to produce parasitic oscillations, usually about 1 MHz. The oscillations should be suppressed by circuit design methods. All zener diodes generate shot and 1/f noise, but the noise level is higher in alloy zeners than in zeners made by a diffusion process. Generally, noise increases with an increase in current but the noise may occur at some point on the zener curve and not at others (this noise is called *spotty*). Most commercial zener diodes exhibit noise levels from 1μV to 1mV.

Diodes with mechanical imperfections may generate noise when physically agitated. Such diodes may not cause trouble if used in a vibration-free environment.

16.6   ACTIVE DEVICES

This section summarizes EMI problems and control in transistors and vacuum tubes.

16.6.1   TRANSISTORS

Transistors have many advantages over vacuum tubes. Some advantages are small size and weight, relatively low-power operation, increased reliability, and no heater or filament power requirement.

Low-level operation can exhibit a problem of sensitivity to interference signals and possible destruction. Inherent semiconductor noise limits the minimum detectable and amplifiable signals. Internal resistance and capacitance or transit time limits the maximum amplifiable frequency. Noise generated by a transistor will appear at the output, but it may be possible to minimize the effects. The three main types of noise generated by a transistor are thermal, shot, and flicker noises.

Thermal noise in transistors is believed to be due to thermal agitation causing random motion of electrons and holes in the material. The equivalent circuit of a thermal noise source is represented by a voltage generator in series with a noiseless resistor as shown in Fig. 16.21(a). The internal noise has a flat-frequency spectrum (white noise) and the amount of output noise power is limited by the circuit bandwidth to which the noise source is connected. The noise spectrum is uniform throughout the frequency range until capacitance causes attenuation due to a reduction in transistor gain. As an example of thermal noise magnitude, 6 μV can be developed across a 500 K resistance corresponding to a bandwidth of 5 kHz at room temperature.

A constant-current generator in parallel with a noiseless resistor represents an equivalent shot-noise source circuit as shown in Fig. 16.21(b). Shot noise is generated by random movement of minority carriers across a junction whenever current is flowing. The current is not uniform due to the random diffusion of minority carriers as well as the random recombination and generation of charges. Shot noise, I_{sn}, has a flat (white noise) spectrum and is proportional to the number of minority carriers, current flow, temperature, and bandwidth.

$$I_{sn}^{2} = 2\ eIB \qquad\qquad (16.27)$$

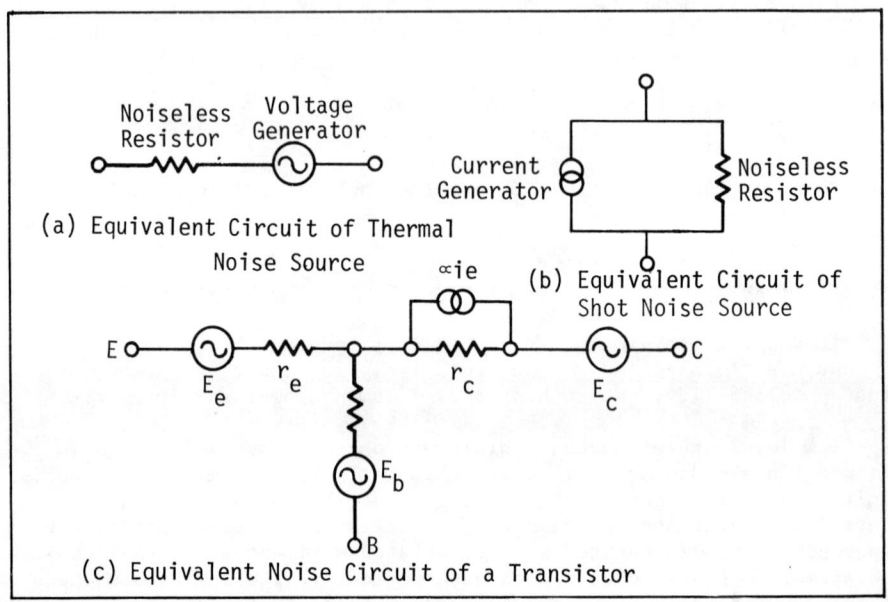

(a) Equivalent Circuit of Thermal
        Noise Source

(b) Equivalent Circuit of
      Shot Noise Source

(c) Equivalent Noise Circuit of a Transistor

Figure 16.21 - Equivalent Noise Circuits

where, I_{sn}^{2} = RMS shot-noise current in amperes

          e = electron charge = 1.6×10^{-19} coulomb

          I = D-C current in amperes

          B = bandwidth in Hz

Flicker noise is called *1/f noise* because it has a power spectrum proportional to 1/f. Thus, for a given bandwidth, the noise decreases inversely with an increase in frequency. It is also called *semiconductor noise* and is believed due to crystal imperfections and surface effects including trapping of carriers by surface charges with associated surface and internal leakage.

As seen in Fig. 16.22 flicker noise (1/f) is usually predominant up to a frequency between 1 and 10 kHz, when thermal and shot (white) noise become predominant. An equivalent circuit of all noise generators in a transistor is shown in Fig. 16.21(c).

In addition to the previous types of noise, amplifying transistors with H-F capabilities (about 100 MHz or greater) may oscillate at spurious frequencies when operated considerably below the device capability. A transistor having usable current gain at 400 MHz, but operated as an amplifier handling frequencies below 80 MHz, tends to react with stray capacitance and inductance within the associated circuitry to oscillate parasitically at unpredictable frequencies. The form is usually that

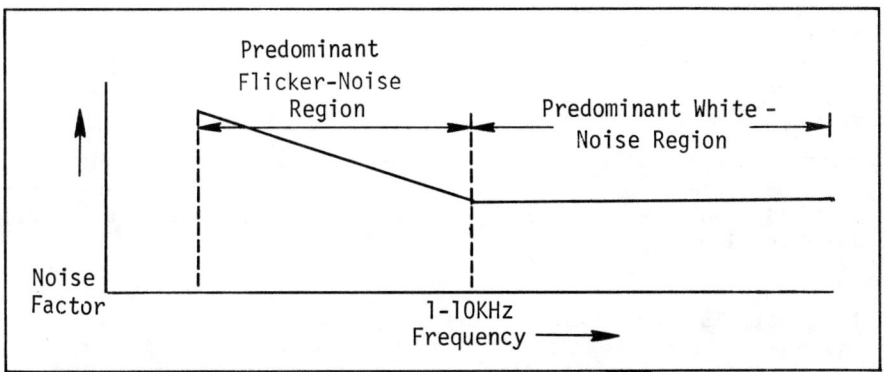

Figure 16.22 - Typical Transistor Noise Versus Frequency

of wideband gaussian noise near or below the operating frequency. The parasitic oscillation is affected by inter-element (junction) capacitances which vary, varacter fashion, over a wide range with changes of element voltage.

Transistors operated in a proportional-control mode usually will not introduce major interference. When used in a switching mode, they may introduce rapid changes in current in the supply and signal lines, and may potentially cause oscillations with fundamental frequencies around 250 kHz to 2 MHz. Inverters may cause serious interference in the range of 15 to 200 MHz with a peak around 60 MHz. Milliampere devices can have switching times in the order of nsec while those of high-amperage rating devices can be a few μsec. Thus, the current gradient can be 10^7 amperes per second. "Soft switching" is a design that controls the rate of change of current, as well as selecting proper dimensions and carrier densities. To minimize interference, it is best to generate required pulses in the required location rather than sending them through long lines, where the number of susceptible components involved could significantly increase. Waveform-control techniques can also be used to make devices responsive to only certain pulse characteristics.

Low-level transistors can be very susceptible to erroneous, or false triggering caused by undesirable signals. The average silicon-controlled-reactifier (SCR) and many other four-layer semi-conductor devices will turn on if a forward voltage with a high dV/dt is applied to the anode; the rapid internal capacitance charging causes a current flow that is similar to a gate turn-on signal. Typical dV/dt characteristic of one 35-amp SCR is 2×10^8 V/sec at an 80°C junction temperature. The dV/dt susceptibility can be minimized by various circuit techniques, as well as temperature limitation.

16.39

Diodes with mechanical imperfections may generate noise when physically agitated. Such diodes may not cause trouble if used in a vibration-free environment.

16.6.2 ELECTRON TUBES

EMI problems associated with electron tubes and their circuits are similar to those with transistors, differing primarily because of power levels and a greater number of elements as well as greater spacing between the elements.

Although ordinary vacuum tubes operating in the proportional-control or switching modes have EMI characteristics similar to those of transistors, gaseous conduction tubes (e.g., thyratron and ignitron) are quite different. The ignitor in an ignitron can be a source of noise because it acts like a spark-producing device. When a gaseous conduction tube is conducting, the space between anode and cathode contains a plasma that can oscillate by itself because of internal instabilities.

The principle types of tube noise are:

- Shot effect
- Partition noise
- Induced noise
- Gas noise
- Secondary emission noise
- Flicker effect

Shot noise (also called Schottky noise or Schot noise) is due to the random fluctuations in the rate of electron emission from the cathode. When the cathode temperature is the current-flow limiting factor, the shot-noise component is:

$$i_{sn}^2 = 2eI_b B \tag{16.28}$$

where, i_{sn} = RMS shot-noise current in amperes

e = electron charge = 1.6×10^{-19} coulomb

I_b = average plate current in amperes

B = bandwidth in Hz

When the plate current is limited by the space charge, many of the fluctuations in the plate current are reduced due to the smoothing effect of the reservoir of electrons in the virtual cathode set up by the space charge. In this case, the following approximations can be used. For diodes:

$$i_{sn}^2 = 4K(0.64)T_c gB \tag{16.29}$$

and for negative-grid triodes:

$$i_{sn}^2 = \frac{4K(0.64T_c g_m mB)}{\sigma}$$ (16.30)

where, K = Boltzmann's constant = 1.38×10^{-23} watts/°Kelvin/Hz

  T_c = cathode temperature in °K

  g = diode-plate conductance in mhos

  g_m = triode transconductance in mhos

  σ = tube parameter (between 0.5 and 1.0)

  B = bandwidth in Hz

The shot-effect noise of a triode can be expressed by considering a resistance which, if applied to the driving grid of a noiseless tube as a source of thermal noise, would produce the same anode-current noise component as is actually present. For triode tubes, the equivalent noise resistance, R_{eq}, is equal to $2.5/g_m$. In amplifiers, the noise voltage generated by R_{eq} is considered to be applied in series with the grid as shown in Fig. 16.23:

$$e^2 = 4KT_c R_{eq} B$$ (16.31)

where, e = RMS noise voltage

  R_{eq} = equivalent grid-noise resistance

Partition noise occurs in multi-collector tubes and is due to fluctuations in the division of current between the electrodes. The noise of a negative-grid pentode amplifier is represented by the equivalent grid resistance (R_{eq}), approximated by:

$$R_{eq} = \frac{I_b}{I_b + I_{c2}} \left(\frac{2.5}{g_m} + \frac{20I_{c2}}{g_m^2} \right)$$ (16.32)

where, I_b = D-C plate current in amperes

  I_{c2} = D-C screen current in amperes

  g_m = transconductance in mhos

The values of R_{eq} of pentodes are usually three to 10 times as great as those of comparable negative-grid triodes. Electron-wave tubes, such as traveling-wave tubes also exhibit partition noise.

At VHF (over 30 MHz), fluctuations in the number of electrons passing a negative grid induce noise currents. Increasing with frequency, the noise currents are introduced into the input circuitry by electronic conductance.

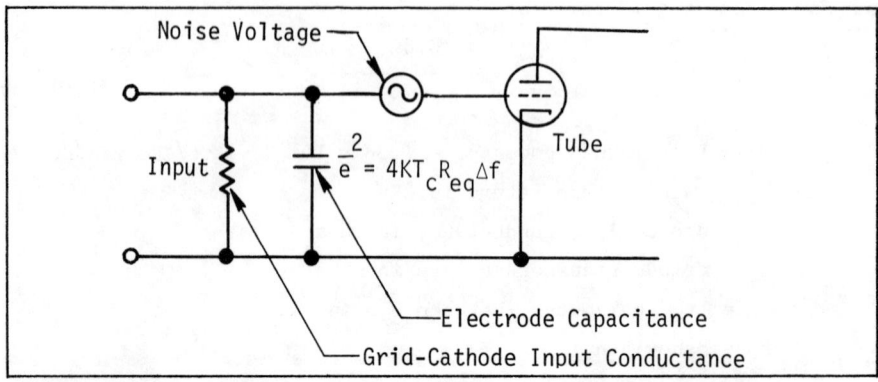

Figure 16.23 - Equivalent Circuit Representing Plate-Current Noise
by A Generator in Series With a Tube Grid

Gas noise is produced by erratic motion of gas molecules in gas or
vacuum tubes. Ionization by collision produces noise when ionized gas
atoms or molecules liberate bursts of electrons when they strike the
cathode.

Secondary emission noise is due to fluctuations in the rate of the
production of secondary electrons. Flicker noise, more common in oxide-
coated cathodes varies as 1/f and is caused by a low-frequency variation in cathode activity.

Some EMI problems involving electron tubes are listed in Table
16.12.

Table 16.12 - EMI Problems, Causes and Corrections in Electron Tubes

Problem	Cause	Corrections
Shot effect	Random fluctuations in the rate of emission of electrons from the cathode	Proper design
Partition noise	Fluctuation in the division of current between electrodes	If important, use tubes with fewer elements
Induced noise	At VHF, fluctuations in the number of electrons passing a negative control grid induces noise currents	Use tube with better grid geometry
Gas noise	Ionization by collision of ionized gas striking the cathode, liberating bursts of electrons	Proper design or use of other electronic components
Flicker noise	Inversely proportional to frequency: due to a low-frequency variation in cathode activity; more common in oxide-coated cathodes	Avoid oxide-coated cathodes
Microphonic effect (vibration of socket with or without sound emanation)	Shock or vibration changes element spacing and damped mechanical oscillation causes changes in plate current, usually setting up vibration through the tube socket or by means of sound waves	Reduce by use of ruggedized tubes
Hum	Use of AC for filament and heater-type tubes	Use DC, use humbucking circuits, use balanced filament supply, bias cathodes negative with respect to filaments
Other tube noise	Leakage from the grid to another electrode, particularly a positive electrode; improper contacts (particularly with low level signals)	Tube replacement or ensure mechanically and electrically good contacts
Large number of unwanted frequencies at amplifier output	Caused by possible feeding in of unwanted frequencies compounded with operation on the non-linear portion of the characteristic curve	Proper design and shielding of the grid circuit
Interfering signals in an R-F field	Pick up of RF signals through tube envelopes	Use tube shields
Affected by nuclear radiation	Causes change in tube characteristics	Use ceramic tubes
Noise current in catcher of klystron amplifier	Electron beam passing through the catcher causes noise current in the shunt impedance of the catcher	Proper design for minimum effect

16.7 INDUCTIVE DEVICES

Sec. 16.4 introduced the topics of characteristics resulting from devices which exhibit fast rise and/or fall times in their amplitudes vs time. This section discusses inductive devices, although not all produce transients. However, these devices are mainly characterized by containing one or more relatively high-current, low-voltage sources which produce predominately magnetic fields. As such, these low-impedance devices produce low-impedance fields which are sources of EMI (see Sec. 10.1). If uncontrolled they may magnetically couple EMI into other low-impedance susceptible circuits, components, and/or wiring (see Sec. 6.2).

The principle inductive devices discussed in this section are inductors, transformers, relays, solenoids, and motors and generators. Each topic is discussed with emphasis placed upon the techniques for controlling EMI. As discussed below the main techniques available for controlling transient-producing devices are the use of diodes and filters, and where the magnetic fields are an EMI problem, shielding is used.

16.7.1 INDUCTORS

Section 16.1.3 discussed inductors as a lumped element whose performance degrades at high frequency due to parasitic capacitance and related effects. Therefore, an inductor may appear as a capacitance above self resonance. If the inductor is used as a filter element, H-F performance may degrade to the point where EMI protection to other circuitry or devices may no longer exist.

An inductor, if not properly constructed or used, may act as a means for coupling undesired EMI into other victim circuits. Fig. 16.13 previously illustrated this in which the magnetic field from the helex is coupling into the wiring of another circuit (cf. Sec. 6.2) when an air-core inductor is used. When wrapped in the shape of a toroid, which has a high permeability (e.g., of the order of 1,000), nearly all the magnetic field is concentrated internally. Thus, little external flux density is available to couple into potential victim circuits. Another advantage of toroids is that their size can be made considerably smaller than air-core inductors because of the greater inductance per inch achievable with high-permeability materials (see Sec. 16.1.3 for more details).

16.7.2 TRANSFORMERS

Transformers are a continuation of the previous paragraphs on inductors but one in which two separate EMI source circuits (not the victim circuit) are to be intentionally coupled with a high efficiency. As in the single inductor, external magnetic fields may couple into victim circuits if special precautions are not taken. Transformers and audio chokes are greater offenders of EMI since they generally carry considerably more power than R-F inductors and toroids.

A transformer or choke coil can be enclosed in a high-conductivity metal enclosure, usually called a shading ring. When the coil's alternating magnetic field cuts the shading ring, the induced voltage in the small resistance causes a large current flow in the ring. This in turn sets up an alternating field that opposes the coil's original magnetic field. The magnetic field outside the ring is thereby reduced. As the resistance is minimized, the field is decreased. This arrangement can be thought of as a transformer with a shorted secondary, viz:

Shading-Ring Concept	Transformer
Coil	Primary Winding
Space around coil and ring	Core
Shading ring	Secondary winding
Magnetic field between coil and ring	Leakage reactance field

The shading ring has no effect on D-C magnetic fields since the fields are not varied. If a conductive box is used, the sides act as a shading ring, with the top and bottom effectively an infinite number of concentric shading rings.

Fig. 16.24 shows the use of a shading ring in a transformer or choke. Television receiver power transformers usually have shading rings to help prevent electron-beam deflection in the CRT by the transformer's magnetic field. Chokes are potentially one of the worst interference sources due to the large harmonic content of power supply current, rectifier switching transients, and the magnetic-circuit gaps that prevent D-C saturation. In the case of this type of choke, the shading ring in Fig. 16.24 must be extended so it covers the gap. The unsymmetric structure of the fields would also require the choke to be enclosed in a high-permeability box of sufficient thickness and distance from the choke to prevent saturation.

Fig. 16.25 shows how a semi-toroidal choke or transformer is made, minimizing the weight and size. The coils act as shading rings as they are placed over a core gap. When the choke is enclosed in a magnetic material housing, external fields can be reduced by approximately 45 dB over a conventional E-lamination transformer.

Although not a reliable practice, another method used to reduce magnetic coupling between transformers and/or chokes or between transformers and circuit wiring is to mount them at right angles to each other. If the magnetic fields of each are confined to a plane, then the coupling will be proportional to the cosine of the angles between their planes with a maximum practical decoupling of about 20 dB being achievable.

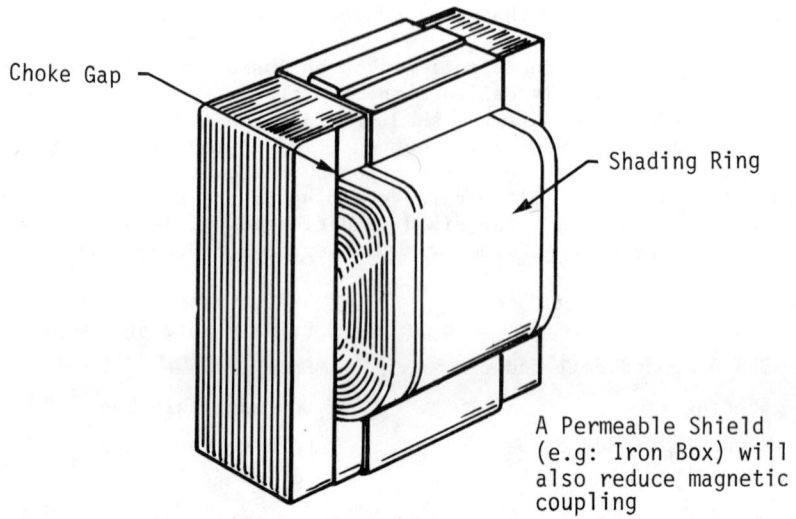

Choke Gap

Shading Ring

A Permeable Shield
(e.g: Iron Box) will
also reduce magnetic
coupling

Figure 16.24 - Shading Ring in a Transformer or Choke

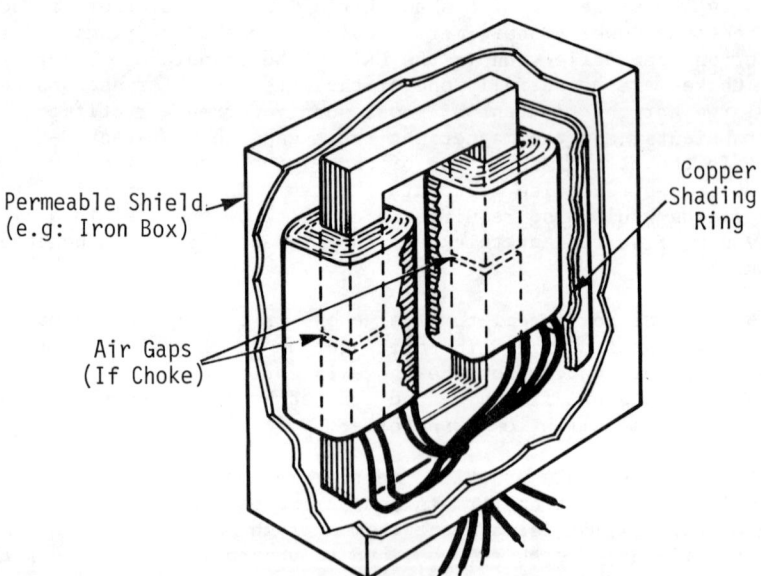

Permeable Shield
(e.g: Iron Box)

Copper
Shading
Ring

Air Gaps
(If Choke)

Figure 16.25 - Magnetically-Shielded Transformer or Choke Having
Semi-Toroidal Construction

This practice, however, is not reliable since fields are generally not confined to a plane and because components might be repositioned during design or maintenance.

16.7.3   SWITCHES AND RELAYS

While not an inductive device, switching action is the purpose for using relays, which are highly inductive. Accordingly, this section reviews switches and relays.

16.7.3.1   SWITCHES

The function of an electrical switch is to interrupt the flow or change the routing of an electrical current. An electrical switch causes the impedance of a circuit to change rapidly between a relatively low value and some large value. The rapid change causes a high dI/dt and/or dV/dt in the switched circuit which, in turn, produces steps in current or voltage waveforms capable of causing interference.

When a power circuit is switched by mechanical contacts, high frequencies capable of causing interference are generated during both closing and opening of the contacts (see Sec. 16.4). As the contacts close, a step function of voltage is applied to the circuit which excites it, causing oscillations in R-L-C reactive elements. EMI is thereby generated that can be transmitted by radiation or conduction. The larger the wattage of the load being switched, the more difficult and expensive it becomes to control the effects of EMI.

Mechanical contacts exhibit the additional problem of contact bounce when closing. Random opening and closing of contacts chops the current, generating H-F oscillations and harmonic components. For inductive circuits, abrupt interruptions can lead to high-induced voltage transients, contact arcing, dielectric breakdown, and associated phenomena. All may cause EMI problems.

The making or breaking of an electrical circuit by a mechanical switch is usually accompanied by the generation of an arc at the switch contacts. Arcing during normal operation of a switch occurs because a highly-ionized gas is substituted for a part of the metallic circuit as the switch contacts move apart. The arc is extinguished when energy stored in the circuit is dissipated, including oscillation with distributed reactances, once the switch contacts exceed the arc-cover distance.

An arc is a phenomenon that dissipates energy that is either supplied to or stored in an electrical circuit. The arcing phenomenon causes deterioration of contact surfaces where the arc is formed, which may destroy the contacts if continued. Arcing phenomenon is also a prime source of interference having a broad-frequency spectrum.

Energy dissipated in an arc depends upon the circuit loads being switched. The most common types of loads to be switched are resistive, lamp, motor, capacitive, and inductive loads. There are two distinct types of loading associated with arcing. One type of loading produces current surges; these occur when a starting transient is greater than the steady operating current. Current surge types of load are lamp, motor and capacitive loads. The second type of loading produces voltage surges; this occurs when the induced voltage is greater than the supply voltage, and is a characteristic of an inductive circuit. A motor, for example, may produce a current surge on start and a voltage surge on stop. A resistive load is subjected to neither a current surge nor a voltage surge because the transient is of the same order of magnitude as the steady-state condition.

Circuits involving energy storage in capacitive or inductive form pose special switching and arcing problems. An arc may strike when the switch contacts first open, clear, then restrike upon discharge of stored energy. Repetitive restriking may occur as the switch contacts move apart.

Arcing during switching of electrical circuits causes problems in reliability because of deterioration of the switching device, and causes an additional problem by generating EMI. Suppression of the arc will reduce these problems.

The usual D-C interruption is caused by making an arc and forcing it into an unstable shape to create instability and eventual collapse. Fast switching minimizes the duration of the arc, although typical inductive D-C circuit arcs last longer than 5 msec. Other elements in the circuit may radiate EMI that was conducted to them.

When AC is interrupted, the duration and magnitude of arcing depends upon the instantaneous voltages at the time of interruption. The interrupter should be designed so that the arc will not reignite after the voltage has passed a zero value. Transients produced upon closing an AC switch also depend upon the instantaneous voltage at the time of switching.

Electrical circuits can be switched by several methods. Arcing is associated with switching by mechanical means such as manual switches, relays, circuit breakers, or thermostats. Arcing generates EMI with a frequency spectrum that extends into the ultra-violet region. Use of solid-state devices may eliminate the arc effect, but solid-state switches exhibit switching transients that may also cause EMI unless special transient-control measures are taken.

In designing equipments containing mechanical switching devices, care must be taken to maintain operational capability of contacts and at the same time suppress EMI generated by switching. Often a gross approach to arc suppression is taken, some standard technique is applied, and a recommended component used without considering all aspects of the

problem. The most direct solution, however, may lead to the generation
of additional problems. Ramifications such as changes in circuit reliability, cost factor, weight, and system effectiveness must be considered.
The system, in fact, may become overdesigned to a point where the generation of attendant problems abrogate basic design objectives. Some parameters such as cost factor become serious and intolerable in large systems. For this reason, recognition of the problems of arc suppression is
especially important in such systems.

Suppression networks can be optimized by considering reliability,
cost, physical size, weight, and other factors. Special networks can be
designed for arc suppression. Such networks may accomplish this by de-
laying or eliminating any transients that occur during the switching
operation. Arc suppression devices should be used as discussed under
relay contacts. When the interrupted currents are large, the only effective means of limiting the interference might be to use short leads,
to filter, and/or shield as much of the switching circuit as necessary.

The dielectric material used in switches must be chosen according
to the intended use of the switch. H-F switches are usually made with
ceramic material for insulation to avoid dielectric breakdown in the
presence of arcing. Some switches are made with a sliding or wiping
action that reduces arcing to a minimum, distributes contact wear, and
reduces contact contamination. The wiping action introduces some interference by the variation of contact resistance for a short time after the
contacts have initially touched or begun separating. A shorting switch
has a *make-before-break* action that introduces only a small amount of
interference because multiple changes in contact impedance do not occur.
Mercury-type switches are excellent because they are quite contamination-
free and exhibit little contact bounce, but their use is limited to stationary system applications with respect to the horizontal plane. Vacuum
switches, in which the contacts are enclosed in a vacuum envelope and
operated by an external solenoid, are useful for aircraft applications.

A high conductance contact occurs when simple contact under pressure
wipes away film and tarnish, or light arcing causes rupturing of non-
conductive oxides or sulfides. There are various types of contact alloys
and platings; the type to be used should be determined by the application.

The thyratron-like latching characteristics of the thyristor make
it ideal for eliminating interference due to contact opening. Reverse-
blocking triode thyristors, commonly called silicon controlled rectifiers (SCR's), and bidirectional triode thyristors, usually referred to
as triacs, are the most common types used in switching and power control.

An SCR can turn off only when the AC current through it naturally
reaches zero in the process of reversing polarity, regardless of the load
power factor. Because it requires no separate and sophisticated turn-
off circuitry, the SCR does not interrupt current abruptly like mechan-

ical contacts. Instead, it opens the circuit as soon as current reaches zero after gate drive has been removed. Circuit disturbances are minimum when current is interrupted at this instant. Turn-on logic can be developed from the opposite half-cycle. The turn-on gate control signal can be applied while the SCR is reverse-biased, but it will not begin to conduct until it begins to be forward-biased.

Because an SCR can conduct in only one direction, full-wave operation calls for two SCR's connected in reverse-parallel with gate control circuitry arranged accordingly. Alternatively a triac, which effectively functions like two SCR's in reverse parallel, can be used.

Thyristors are not entirely free of EMI effects, however. At turn-on time, a voltage spike is produced as forward bias voltage passes through the forward voltage breakover point. At turn-off time, stored carrier charges produce a current spike until a depletion region is established. Furthermore, a thyristor gate element is susceptible to EMI of high dV/dt. Capacitive charging currents can cause the gate to turn on the thyristor even though the magnitude of the gate voltage does not reach the rated triggering level.

16.7.3.2 Relays

A relay is a device that permits one or more circuits to be remotely switched by electrical variations in an independent control circuit. There exists several types of relays: electromagnetic, saturable reactor, bi-metallic or reed, semiconductor, photosensor, and others. There are many designs available, depending upon switching to be performed, power source available, number and type of contacts, and cost.

The most common relay is the electromagnetic solenoid type. EMI problems occur in both the actuator and contact circuits. The electromagnetic solenoid has a large inductance due to the large number of turns and iron mass in the core and armature. When the coil circuit current is interrupted (i.e., de-energized), the collapse of the magnetic field generates a voltage equal to -L(dI/dt). This reverse potential can reach 10 to 20 times the supply voltage in the order of a microsecond, and then decay at a rate determined by the inductance, distributed capacitance, and resistance of the armature winding circuit. The high amplitude voltage surge has a steep wave front that can cause arcing at the point of interruption, along with broadband signals capable of conducting and radiating interference to other circuits. EMI effects of an A-C relay are variable because the voltage and current are continually changing in magnitude, producing results according to the momentary state at the time of switching.

Abrupt changes in circuit current will produce waveforms that cause EMI. Ideal contact operation occurs when the contacts go from fully open to fully closed or vice-versa without arcing. In actual practice,

when contacts close, they bounce and form a closure followed by one or
more openings and reclosures. In small relays, a single bounce may
occur 10 to 50 μsec after the initial closure; final closure may take
10 or more μsec. In large relays, the bounce may be repeated several
times at intervals of a few milliseconds.

Relay arcing can occur when contacts are first opened, continuing
until contact spacing is too large to maintain the arc (depending upon
the surrounding atmosphere and the applied voltage). The make and break
of a contact arc when opening a circuit is proportional to the instantaneous supply voltage, the circuit inductance, and the rate at which the
contacts physically separate. EMI may be worse for contact closure than
for contact opening. Closed contacts can open or vary in contact resistance due to shock, acceleration, or vibration, causing arcing or
at least circuit current changes. If required, contact arc suppression
is used.

When employed in digital systems, contact bouncing and erratic
opening cause data to become in error. Polar or latching relays are
therefore often used in such systems. A polar relay, once actuated in
a given direction, will remain latched on its internal permanent magnet
until its state is reversed by current in the opposite direction.

16.7.3.3   TRANSIENT SUPPRESSION TECHNIQUES

The use of a parallel resistance to suppress EMI in a relay circuit
is shown in Fig. 16.26A. The circuit is not polarity sensitive and provides relatively good suppression depending upon the value of R used.
The effect of relay armature release time (drop-out time) is also a
function of resistance used:

$$V_o = V_r e^{-Lt/(R+R_L)} \qquad (16.33)$$

where: V_o = voltage across relay terminals at any time, t, after
            switch opening (t - 0 at switch opening)

V_r = applied voltage to relay terminals before t = 0

t = time in seconds

L = relay coil inductance in henrys

R = parallel resistance across relay

R_L = D-C resistance of relay coil in ohms

Eq. (16.33) indicates that the dropout time constant (for exponent of
e = -1) is $t_c = (R+R_L)/L$. The disadvantage of using a single shunt resistor is that it consumes power continuously when the relay coil is
energized. Since R affects the drop-out time, it will significantly
affect the time differential when $R > R_L$.

Note: A Resistor and Capacitor in Series May Be Used Directly Across Relay Contact(s)

TYPES OF INDUCTIVE SUPPRESSION	VOLTAGE INPUT	RELAY CONTACTS		REMARKS
		CLOSING	DROPOUT	
A 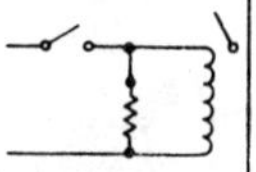 RESISTANCE DAMPING	AC or DC	NO EFFECT	FUNCTION OF RESISTANCE	Increase in power consumption. Resistance should be as low as practicable. Observe power rating E^2/R and heat dissipation.
B 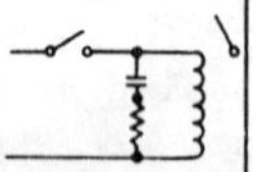 CAPACITANCE SUPPRESSION	DC	SLIGHT EFFECT	SLIGHT EFFECT	Need series resistance of a few ohms. Capacitance value around .01 to $1\mu f$. Capacitance rated 10 times input voltage.
C 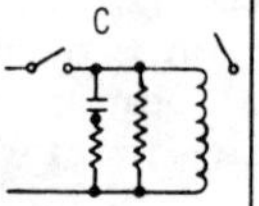 RC SUPPRESSION	DC	SLIGHT EFFECT	FUNCTION OF RESISTANCE	Combination of A and B above.
D 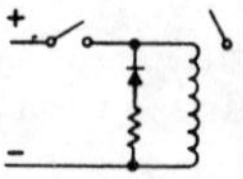 DIODE SUPPRESSION	DC ONLY	NO EFFECT	SLIGHT EFFECT	Polarity critical, diode put in backward or nonconductive direction. PIV should be higher than any transient voltage plus safety factor. Series resistance of a few ohms might be needed to increase inductance life.
E 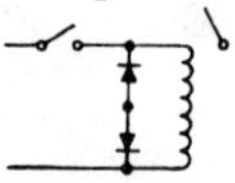 BACK-TO-BACK DIODE SUPPRESSION	AC	NO EFFECT	NO EFFECT	Avalanche voltage should be above input voltage. Power dissipation should be sufficient for transient current. Cost of device is much greater than any of the above divices.

Figure 16.26 - Various Techniques for Suppressing EMI Transients in Relays

In Fig. 16.26B, the R-C EMI suppression circuit limits the transient voltage surge to:

$$V_o = I_{DC}\sqrt{L/C} \qquad (16.34)$$

when the relay resistance R_L is negligible. The capacitor C is typically chosen to be 0.1 to 1.0 µf with a voltage rating of approximately 20 times the maximum D-C input voltage. The use of a capacitor alone will result in a large charging current during relay energizing which may damage the switch contacts or cause transient-noise current surges. Thus, the current-limiting resistor, R, is necessary:

$$R = V_r/I_{max} \qquad (16.35)$$

where, the maximum current, I_{max}, should be limited to typically 10 times the normal coil operating current, I_{coil}:

$$I_{max} = 10\ I_{coil} = 10\ V_r/R_L \qquad (16.36)$$

Substituting this relation into Eq. (16.35) yields:

$$R = \frac{V_r}{10V_r/R_L} = 0.1\ R_L \qquad (16.37)$$

The network will appear resistive at all frequencies if:

$$R_L = R = \sqrt{L/C} \qquad (16.38)$$

This circuit affects contact opening and closing times only slightly.

The series-parallel combination of capacitance and resistance shown in Fig. 16.26C is a combination of the above two circuits. Since it offers no particular advantage over either circuit, it is not often used.

The circuit of Fig. 16.26D results in a polarity-sensitive relay. When the actuating switch is closed, the diode is effectively an open circuit since it is back biased. Thus, it's series resistor is out of the circuit. However, when the switch is opened, the collapsing magnetic field in the relay coil develops a reverse voltage, $V_r = -L\ dI/dt$, and current flows through the diode. The voltage is limited to the forward diode voltage drop due to the internal resistance on the diode and the voltage drop across resistor R. The peak inverse voltage (PIV) rating of the diode should be higher than the maximum applied input voltage or any transient voltages, and should include a sufficient safety factor. The addition of a series diode (an option) to the shunt-diode suppressed coil circuit provides the added advantage of inadvertent polarity reversal protection by the elimination of excessive shunt diode current.

Most germanium diodes exhibit a lower forward resistance and voltage drop which minimizes the magnitude of the interference voltage transient. However, silicon diodes are usually used because of cost, current ratings, and high PIV considerations. When R is greater than the effective forward resistance of the diode (typical case), the resistor becomes the primary transient suppressor and the diode acts as a one-way switch. Thus, the resistor does not consume power when the coil is energized. This circuit provides a small voltage backswing, and there is an increased dropout time:

$$\text{Dropout Time Constant} = \frac{R+R_L}{L} \qquad (16.39)$$

R must be selected to provide the desired voltage backswing. Using a pull-down voltage on the diode will increase the backswing but decrease the dropout time.

The addition of a transistor with the relay coil in the collector circuit drive increases operating sensitivity and provides significant EMI isolation to transients. High-level switching of the contacts are still retained.

Back-to-back zener diodes, as shown in Fig. 16.26E, are effective for both A-C and D-C circuits. When the actuating switch is opened, the high inverse voltage causes one of the diodes to break down due to the zener effect, and the voltage surge is limited. This method is a compromise between the R-C network and the single diode for backswing magnitude and dropout time.

16.7.3.4 OTHER EMI PROBLEMS AND CONTROL TECHNIQUES

EMI suppression is especially required for relays located in areas having susceptible circuits or equipment. Power and signal leads must be isolated, twisted, and/or shielded to avoid magnetic and electric coupling (see Chap. 6). Filters should be used on conductors as necessary at points of entry into an enclosure. Signal circuits may have to be shielded and the shields grounded. Table 16.13 summarizes some of the common EMI problems with relays and their corrective measures.

Where magnetic fields from the relay coil may constitute a potential source of EMI to nearby low-impedance circuits or wiring, the relay should be magnetically shielded. Shielding techniques discussed in Sec. 16.7.2 also apply here.

A. Coaxial Relays

Cross-talk between two or more switched-wire circuits at a relay signal terminal is another EMI problem. While a single relay should not be used to switch both high and low-level circuits due to cross-talk problems, there exists techniques in which it may be used to

Table 16.13 - EMI Problems, Causes, and Correction in Relays

Problem	Cause	Correction
Generation and susceptibility to propagated EMI	Relay circuit characteristics	Twist, shield and filter power leads and enclosing relay in metal
Simultaneous problems in coil and contact circuits	Movement of the armature causes coil magnetic circuit changes and varying contact dI/dt and dV/dt	Proper relay circuitry design, including filtering
Relay contacts generate EMI when they should be quiescent	Vibration	Proper relay design or contact circuit suppression
Noise, misoperation, or permanent damage in semi-conductors	Inductive circuit causes large voltages	Components must be properly rated and/or dI/dt and dV/dt must be diverted or suppressed
Low frequency interference in the signal circuitry	Low frequency spurious voltages	Single-point grounding and twisted leads
High frequency interference in the signal circuitry	High frequency spurious voltages	Multiple-point grounding

select one of two or more circuits to patch to a single input circuit.
One example is coaxial relays with BNC-type connectors in which it is
unusual for isolation to be better than (more than) 35 dB at 1 GHz or
50 dB above 100 MHz. Most coaxial relays of the reed type exhibit
poorer performance. Physically larger, heavier, and more costly type-
N coaxial relays of the non-reed type are available to give better
isolation performance.

A somewhat less-expensive and more reliable technique (i.e., number of operations) is shown in Fig. 16.27 in which several inexpensive
coaxial relays are combined to protect isolation requirements. Here,
three 2-pole relays replace one 4-pole relay. The isolation between
any of the four coaxial lines due to relay contact coupling is twice the
number of dB obtained from any one of the single relays. When connected back-to-back to form a 0-20-40-60 dB remotely-controlled R-F attenuator, Fig. 16.28 results. Up to 100 dB of attenuation has been
realized at 1 GHz using $20 SPDT relays in which individual relay-cir-
cuit isolation is about 30 dB. VSWR padding or impedance-mismatch protection is also preserved in this arrangement over the more expensive
multi-pole relays for other than the 0-dB position.

16.55

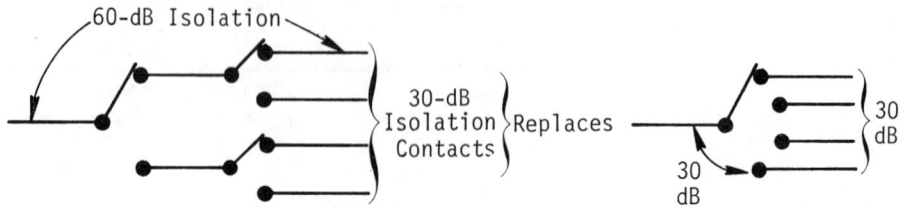

Figure 16.27 - Improvement of Coaxial Relay Circuit Isolation (Cross-Talk) by Combination of SPDT Relays

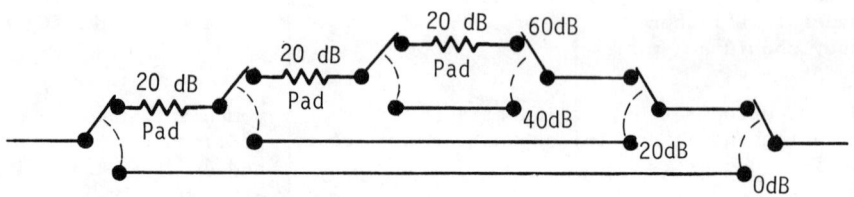

Figure 16.28 - 0-20-40-60 dB Remote-Control R-F Attenuator Offering Substantial Isolation Over Equivalent Two SP4T Relays

B. Solid-State Relays

Solid-state relays are widely used today after being introduced in recent years. For the same reasons that transistors and IC's are preferred by designers, solid-state relays are used because of their small size, reliability, low-power dissipation. Since they have no moving mechanical parts, arcing is inherently prevented to mitigate EMI. This has the added advantage of permitting switching in explosive atmospheres.

Many solid-state A-C relays are designed so that the relay *contacts* are closed at the time of zero-axis crossing of the A-C supply voltage (called zero-voltage switching). Here the transient is significantly reduced. Additionally, many solid-state A-C relays have provision for turning off at a zero-current crossing of the A-C line (called zero-current switching). This eliminates the -L dI/dt arcing problems associated with de-energizing inductive loads.

If the line voltage transients exceeds the maximum contact voltage the relay *contacts* will close until the current waveform passes through zero, or for a maximum of one-half an A-C cycle, then they will reopen. While such transients will not damage the relay, back-to-back zener diodes can be used across the contacts for further suppression. If a dV/dt transient across the *contacts* exceeds a certain rate, this will cause the contacts to close for one-half A-C cycle. Re-closing due to inductive load switching is avoided by an internal R-C network across the *contacts*.

Typical contact resistance (ratios of voltage drop to current through *contacts*) of a 10-Ampere, solid-state relay is about 100 mΩ and rises with a decrease in load current. The off-state leakage is a few mA measured at a D-C voltage equal to the RMS voltage for which the output is designed. The maximum load current is generally rated at room temperature and smoothly decreases with an increase in temperature till it is completely derated to zero at about 100°C.

16.7.4 MOTORS AND GENERATORS

EMI reduction design for electrical machinery is divided into four categories: Interference reduction for large motors and generators, for alternators and synchronous motors, for fractional-horse-power machines, and for special-purpose rotating machines. Any rotating machine with sliding contacts should be regarded as a potential source of EMI because the switching and arcing processes of commutation cause rapid current and voltage changes that limit energy through a wide frequency range.

16.7.4.1 BRUSHES

Brushes and brush leads are the most likely components from which EMI can be radiated or conducted. If a motor or generator is not adequately enclosed, then the brushes and brush leads may require shielding. Provision should be made in the original design of motors or generators for installation of capacitors at the brushes. Brush-generated interference may be reduced by incorporating the following in the design:

Brush Pressure. EMI decreases at all frequencies with increasing brush pressure. Increased brush pressure, however, increases the rate of wear. The necessity for more frequent brush replacement is often a reasonable compromise for decreased interference.

Current Density. EMI decreases with a decrease in current density. As the current density is increased, more heat is generated at the brush surface, sliding on the commutator or slip ring. This heat causes the formation of a thick oxide film on the sliding metal surface. Rapid variations in the sliding contact resistance, resulting from irregularities in this oxide film, cause high-frequency transients that produce interference. To offset the heat increase, a somewhat larger brush-surface area than necessary should be designed. Such a design change will reduce heat and losses due to mechanical friction. On the other hand, if too low a current density is used, nonuniform grooves develop

on the metal surface of the slip ring or commutator, and frequently the
increased friction, due to the wider brush-surface area, sets the brushes into a noisy chatter.

Brush Resistivity. Brush materials of low resistivity are poor
EMI generators and are therefore desirable for use in EMI control.
One example of such a brush is an electrographitic carbon brush with
about two milliohm specific resistance in machines being used at less
than 50 volts. Low-resistance brushes are available with silver copper,
or cadmium impregnated graphite. When used with commutator, the resistance of the brush should match the requirements for good commutation. When used with slip rings, a wide choice of brush material is
available because no switching action is involved.

16.7.4.2   D-C Motors and Generators

Of all rotating machinery, D-C motors and generators are the most
serious offenders in generating EMI because they require commutators
for their operation. Commutation is a switching action that is accompanied by interference-producing transients. When a switch is closed
in an electrical circuit, the input impedance changes from practically
infinity to zero. If the circuit contains inductance and/or capacitance, its voltages and currents cannot return to normal values instan-
taneously because energy stored in the magnetic field of the inductance
(or in the electric field of the capacitance), cannot dissipate in-
stantaneously. Initially, the changing voltages and currents develop
steep wave-fronts which decay as a function of time. Since the bars
of a commutator, sliding rapidly past the contacting brushes, produce
a switching action, this causes extreme variations in impedance, which,
in turn, establish the series of voltage transients, or pulses, that
cause EMI.

Measures can be taken in designing a generator, to minimize the
amount of EMI generated by commutator action. Reduction of commutation transients requires the use of design techniques to provide a
smooth transition from one value of impedance to another within each
armature coil. Interference, produced as a result of commutation, is
reduced by six design techniques: inter-poles, compensating windings,
increased number of armature coils and commutator bars, laminated
brushes, commutator plating, and use of solid-state commutation.

A good way to improve commutation is by adding interpole windings.
Interpoles counterbalance the self-induction of the armature coils
during the commutation period, and also reduce the induced voltage in
the armature coils resulting from the coil-cutting fringing-flux during the commutation period. The use of properly designed interpoles
produces a rapid change in the armature-coil current at the beginning
of the commutating period, reducing the steepness of the transient at
the end of the commutating period.

Compensating windings produce, to a lesser degree, the same effect as interpoles and, in addition, help to prevent field distortion. They also assist in reducing cross flux produced by armature coils. The use of interpoles and compensating windings lessens critical brush positioning requirements with respect to the commutator, and provides EMF in the coils under commutation which oppose the EMF of self and mutual induction in these coils. Increasing the number of coils on the armature (thereby increasing the number of commutator segments, or bars) reduces interference by reducing the current broken per bar and the reactance voltage per coil. The largest number of armature slots in which the coils are uniformly distributed with respect to the commutator bars, should be used, and the armature slots should be as shallow as possible. The use of short-pitch windings reduces interference by reducing the reactance voltage of each coil. The break transients, resulting from the switching action of the commutator, can be smoothed out through the use of laminated brushes. These consist of brush materials of different resistivity, cemented together by nonconducting glue which provides insulation between adjacent brush segments. The ideal operation of laminated brushes is indicated on Fig. 16.29. Having the successive segments of the brush increase in resistance, avoids the sharp current drop after the brush leaves the commutator segment. A more linear coil-current reversal results, thus reducing the break transients. The segments of the laminated brush are insulated from one another by some suitable bonding material, and electrically connected by the brush lead or brush spring. Circulating currents, resulting from the self-inductance of the coil under commutation and from the coil-cutting fringe flux from the pole pieces, must flow through the entire length of two brush laminations. The total resistance of this length is much greater than that presented by a direct path across the face of the brush (as would occur with a solid brush). Circulating currents are, therefore, reduced early in the commutation period, and desirable division of current through the two adjacent commutator bars is achieved.

Good commutation can be achieved over a fairly wide range of brush positions, relative to the magnetic neutral, so that brush positioning becomes less critical and less dependent upon armature current. The design of laminated brushes should include two or, at most, three laminations. The following criteria should be incorporated in the design:

(1)   The thickness of the leading-edge lamination of a two-lamination brush should be about 90 percent of the total thickness, and its resistivity should be as high as allowable for heat dissipation.

(2)   The resistivity of the trailing-edge lamination should be about 15 times that of the leading edge; this lamination should be thick enough to preclude mechanical weakness.

(3)   A thermosetting cement of six-mil thickness should be sufficient to provide electrical insulation between the sections. A cement, that will preclude the formation of a smear of conducting

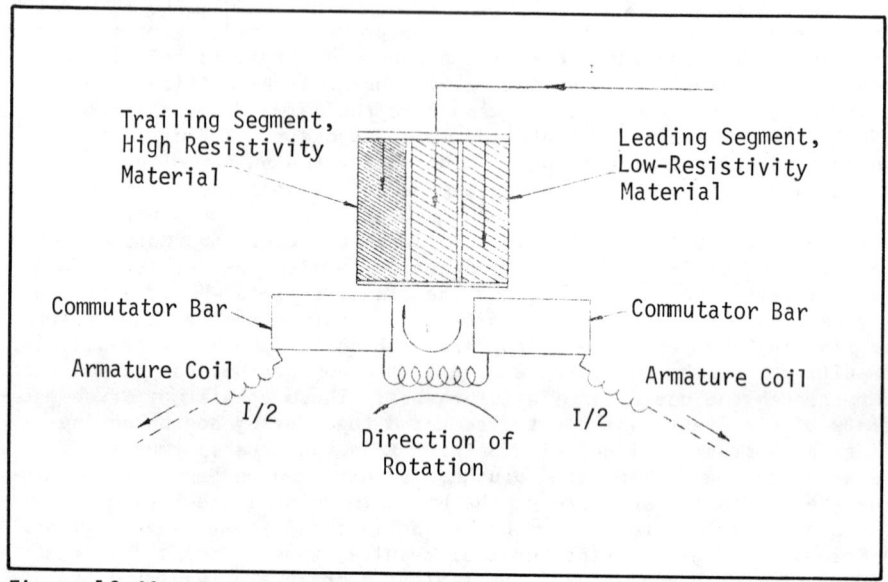

Figure 16.29 - Commutation of an Armature Coil by Laminated Brushes

particles from brush wear on the rubbing edge, should be used; it should have a wear-rate equal to that of the brush.

4.   A brush with varying resistance characteristics from the leading edge to the trailing edge, can be manufactured without the use of insulating separators and will act somewhat like a laminated brush.

A copper commutator, after several hours in contact with a carbon or graphite brush, develops a layer of copper oxide, mixed with carbon particles, from brush wear. This copper-oxide film introduces undirectional electrical properties (polarity effects) as in a copper-oxide rectifier. The oxide layer has a nonlinear resistance of higher value at the brush used as cathode, than at the one used as anode. The cathode brush, as a result, passes current in discontinuous, high current density surges, which cause EMI. Approximately ten times as much interference may result from the cathode brush as from the anode brush. Plating the copper commutator with chromium to a thickness of about one mil will reduce the EMI level from a cathode brush to that of a relatively quiet anode. No adverse effects will result from the platings; the hard chromium surface prevents threading and grooving of the commutator. Wear-rate and sliding friction of many brush materials on chromium are of the same order of magnitude as those for copper.

Design features that improve commutation also reduce EMI. The most effective and economical technique is the installation of capacitors at the brushes. In generators, for example, installing capacitors (e.g., about 0.1µf) at the brushes applies the remedy as close to the interference source as possible. The interference, generated by the commutator and the brushes, will be bypassed to the generator housing. The lead from the brush to the capacitor should be as short as possible, and the capacitor should be bonded to the generator housing to provide a low-impedance path to ground for the EMI currents.

Because of the combined interference-generating characteristics of the commutator and the brushes in a D-C generator, an additional capacitor is installed at the output (armature) terminal. The preferred installation is a feed-through capacitor through the generator housing. The alternate installation is a 0.1 µf bypass capacitor, mounted externally, to maintain electrical contact with the generator housing and minimize the lead length between the terminal and the capacitor. Fig. 16.30 illustrates the mounting of a bypass capacitor at the armature terminal.

Over-all shielding is necessary to prevent the radiation of interference from within the generator. This shielding is afforded by the generator housing, which should be designed to provide maximum shielding effectiveness. Ventilation openings should be screened to prevent radiation of interference into space. No matter how perfectly a generator shield is designed, the shaft provides an exit path for interference because it must penetrate the shield. EMI should be bypassed directly to the generator housing by grounding the shaft through a brush, riding on a special grounding slip ring (or riding directly on the shaft). This grounding will also eliminate bearing interference (bearing static or shaft current). Bearing interference results from a periodic discharge of static electricity that takes place, through the bearing, between the shaft and the housing. Eddy currents, induced in the shaft and the housing by the flux lines in the motor, can cause currents to flow through the bearing. These currents can also be caused by certain combinations of armature segments per pole, air gap and permeability inequalities, rotor accentricities, insulation leakage, or stray electric fields.

Another possible source of leakage from the generator shield is the inspection-band. This band is disadvantageously placed because of its proximity to the EMI-generating brushes and commutator; however, its function of permitting inspection of the brushes and commutator, prevents its being moved to another location. To prevent leakage, the inspection-band should be machined as closely as possible, and should be wide enough to cover the inspection opening adequately, with sufficient overlap to ensure good contact. The band should have machine screws, spaced every two inches, to permit secure tightening. Interference gasketing should be installed around the periphery of the opening. After removal of an inspection-band, all contact surfaces on the band and the generator should be thoroughly cleaned before the band is put back into position.

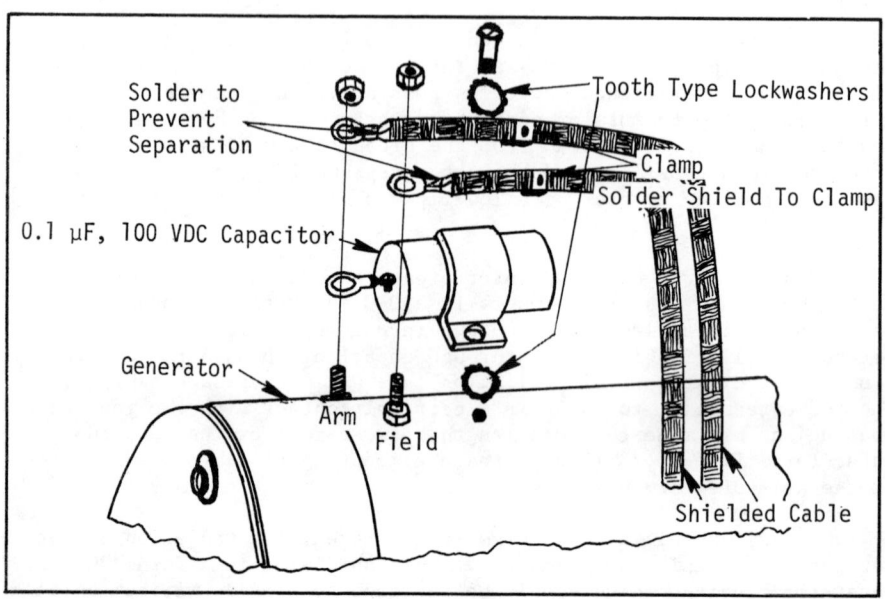

Figure 16.30 - Installation of an EMI Bypass Capacitor on a Generator Armature Terminal

The last shielding consideration for a generator housing is to ensure good contacts and low-impedance paths between the three sections of the generator; the two-end plates, and the main housing. This is accomplished by the bonding and shielding practices discussed in earlier chapters. The design considerations, for minimizing interference generated by brushes and commutation action in D-C generators, also apply to D-C motors. Capacitors, installed to the brushes, bypass the generated interference to ground close to the source, providing an effective and economical means of suppression.

On some D-C motors, an adjustable speed control is included in which the field leads are connected to an externally-mounted rheostat. This arrangement necessitates breaking the shield continuity, and therefore enables interference, generated inside the motor, to be conducted out of the housing. Capacitors installed inside the motor housing connected to these leads, however, will bypass such interference to ground. Fig. 16.31 shows a motor with four installed capacitors; one each for the two brushes, and one each for the two field leads. The feed-through capacitor should be mounted at the positive lead. A less acceptable interference reduction technique for the same motor utilizes bypass capacitors at the brushes.

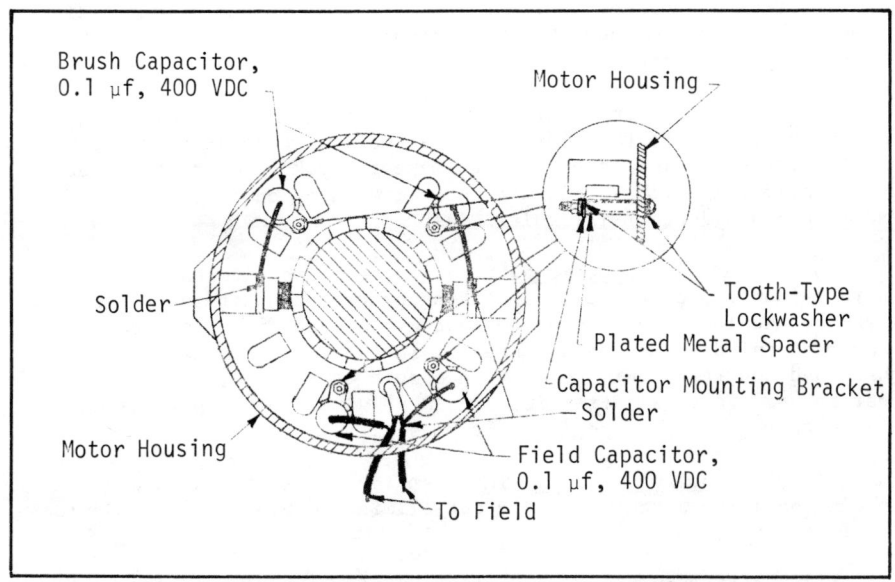

Figure 16.31 - Capacitor Installation at Brushes and Field Leads in a DC Motor

16.7.4.3 ALTERNATORS AND SYNCHRONOUS MOTORS

Alternators and synchronous motors are very similar to D-C generators and motors, except that they supply or use A-C, and therefore have slip rings instead of commutators. Commutator interference is absent in these machines. There is, however, EMI from the brushes and from the generation of harmonics. Brush interference is lessened because most alternators and synchronous motors have stationary armature and rotating fields; heavy power currents need not be supplied to the rotor. Only the much smaller field currents have to be supplied through the brushes. Because commutation is not involved in the selection of brushes, a much wider choice in brush pressure, size, and material is permitted.

In A-C generators, the generation of harmonics and the resonant conditions that create interference can be minimized. Production of as pure a sine wave as possible (an important consideration in the design of alternators) is especially important when EMI reduction techniques are considered. A comparatively small harmonic content may be quite tolerable from all points of view except that of EMI. In the reduction of harmonics, special attention must be given to:

(1) <u>Flux distribution.</u> The most important factor determining the waveform of the generated voltage is the distribution of the

magnetic flux around the periphery of the armature. Sinusoidal distribution, which produces the least amount of interference, may be
achieved by chamfering the pole tips or skewing the pole faces.

(2) Symmetry. In a perfectly symmetrical machine, all even
harmonics disappear; therefore, special care must be exercised to
construct identical pole pieces, to make the yoke and armature perfectly symmetrical, to produce a uniform winding on the armature,
and to avoid all other irregularities.

(3) External connections. In a three-phase alternator, the
third harmonic and its multiplies disappear at the terminals except
when the machine is wye connected and has its neutral grounded. In
this case third harmonics are present in the voltage between any
phase and neutral. This connection should be avoided, or, if it
must be used, special attention should be given to the prevention
of the third harmonic and its multiples.

(4) Distribution factor. The distribution factor should be chosen
to eliminate the lowest harmonic not eliminated by any of the devices
mentioned in (2) or (3).

(5) Tooth ripples. The generation of tooth ripples is greatly
decreased by skewing, through one slot pitch, either the pole shoes
or the armature slots. Tooth ripples may be eliminated altogether by
making the number of armature-slots per pole-pair an odd number. The
chord factors for the harmonics that are contained in the tooth ripples, are then reduced to zero. Slip ring and brush materials should
be such that interference is minimized. The design considerations
applied to brushes and commutator surface materials in D-C machines,
apply equally well here. The effects of brush bounce, due to vibration
or irregularities of armature motion, can be minimized by the use of
two or more brushes per slip ring.

In addition to the interference generated as a result of the brush
action on the slip rings and the harmonics present in the sine-wave
output of an alternator, the exciter is a source of EMI. Because both
the exciter (essentially a D-C generator) and the A-C generator are
installed in a single housing, shielding considerations become a combination problem. Plating of the commutator, the use of proper brushes
and brush pressure, and the application of bypass capacitors are applicable to the exciter; the other design measures can be applied to
the exciter as a separate unit. Although individually designed for
interference reduction, the alternator and exciter each generate some
interference. This residual interference is reduced by shielding and
the use of bypass capacitors installed at terminal outlets.

Shielding of the alternator is incorporated in the design of its
housing. Low-impedance paths between sections of the housing, provisions for bonding, and screening of all ventilating louvres must be
carried out if the overall interference-reduction design is to be

effective. As in D-C generators, no matter how perfect the shield,
a means of escape from the shield for the interference currents is
provided by the alternator shaft which penetrates the shield. The
same procedures for shielding D-C generators therefore apply to alternators.

The alternator terminal outlets provide another means of leakage.
They are prevented from radiating interference by the installation of
capacitors. Bypass capacitors are installed inside the terminal strip
and are connected to the terminal outlet just before the terminal
breaks the shield. This arrangement removes EMI from the lead at the
last possible point, preventing interference from coupling back into
the lead and radiating from the terminals or from their connected
wiring. Another type of installation is to mount feed-through capacitors through the terminal strip.

The problems of interference suppression for alternators also
apply to synchronous motors since they have the same basic components
as alternators. A synchronous motor will operate as an alternator and
vice versa. An induction motor should be used instead of a synchronous
motor whenever possible because of the lower EMI generated by induction
motors.

The primary source of interference within a single-phase induction
motor is the starting device. The starting winding is in series with
a switch (or capacitor and switch) that is closed when power is off.
When the motor reaches approximately 80 percent of its rated speed,
the switch is opened (either by centrifugal force or by a solenoid
coil) and a single pulse of interference is generated. This switch
should be placed in a shielded housing and the leads leaving the housing should be filtered.

16.7.4.4 Portable Fractional-Horsepower Machines

Portable fractional-horsepower machines include such equipment
as portable electric drills and saws. Power is furnished by high-speed,
light weight, A-C/D-C, or A-C electric motors. Such equipment, using
A-C/D-C motors (universal motors), is a major source of EMI because
commutation is essential in its operation. As in D-C motors, an effective, economical method of designing for reduced commutator-brush
interference is by installing capacitors at the brushes. In some
portable A-C/D-C machines, restrictions of size and shape prevent the
installation of capacitors at the brushes, and it is more feasible and
economical to mount the capacitors in other parts of the equipment.
Installing capacitors at the line side of the switch bypasses interference to the unit housing at the last point of exit to the power
lines and prevents the interference from coupling back into an inter-
ference-free lead, and from being conducted by the power lines. If
the mechanical design of the unit prevents the installation of

capacitors on the line side of the switch, it is permissible to install
them on the motor side. Shielding may be used to ensure that no interference couples back into the leads before they leave the unit.

In recent years a new concept in miniature D-C motors has been
introduced, viz., brushless D-C motors featuring solid-state commutation and the associated reduction of conducted and radiated EMI.
The technique involves sensing the exact position of the rotor in
relation to the stator by using two orthogonally located Hall-effect
generators.* If the rotor is constructed of a bipolar cylindrical
magnet, the induced output in the Hall generators will be in phase
quadrature as the rotor rotates.

Fig. 16.32 shows a block diagram of the brushless D-C motor.
The D-C source drives the Hall generators (HG) through the feedback
network and supplies power to the amplifiers. The sensed rotor position signals from the sin and cos HG's are each amplified by a push-
pull amplifier creating four sine waves in quadrature which drive
the coil windings of the stator. The amplifier outputs are rectified
and applied as back EMF's to the feedback network which sets the
value of the reference voltage applied to the HG's through another
set of amplifiers to close the loop. Load, temperature, speed, and
voltage compensation is automatically achieved in the process.

Presently limited in size up to about 7 oz.-in., the brushless
D-C motor should prove to have a great future as a fractional D-C
motor where EMI control, long life (10,000 hours without maintenance,
i.e., brush replacement) and operational parameter compensation are
required.

16.7.4.5 SPECIAL-PURPOSE MACHINES

Special-purpose rotating machinery include rotary inverters, dyna-
motors, motor generators, and generators for electric arc-welding
equipment. The function of conversion is common to most of this equipment: A-C is converted to D-C, or to higher frequency A-C; or D-C is
converted to higher or lower voltage D-C, or to A-C.

A rotary inverter, which converts D-C to A-C, is basically a
D-C motor with added taps on the armature winding; slip rings are connected to these taps to provide the A-C output. Interference is generated by both the A-C and D-C functions, commutator and brush action
in the motor, and by brush action and harmonics in the alternator.

---

*   The Hall generator is a four-terminal device with D-C input, and
an output proportional to the strength and direction of intercepted
magnetic flux. The Hall generator output is also affected by the
strength and direction of the current flow at its input.

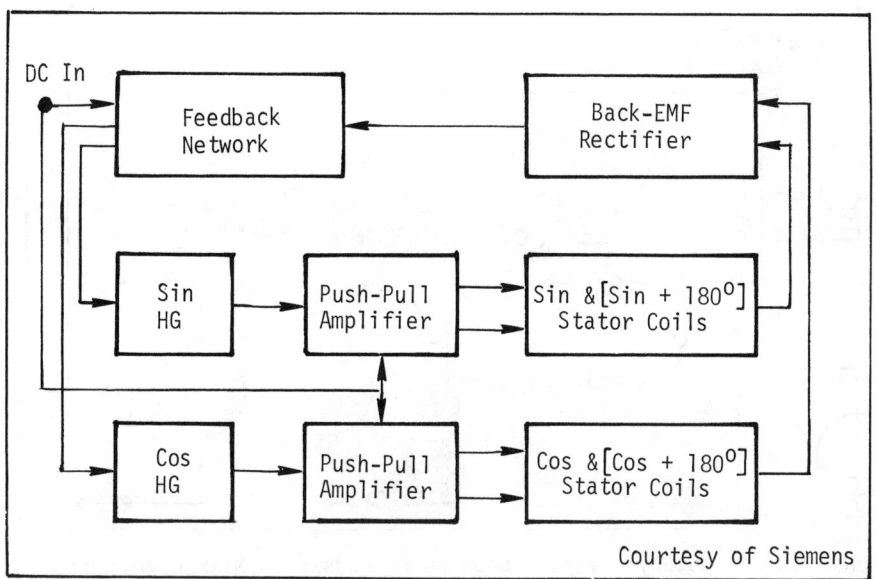

Figure 16.32 - Block Diagram of the Brushless D-C Motor

Fig. 16.33 illustrates an interference reduction design technique for
an inverter. The schematic diagram shows two feed-through capacitors
bypassing EMI from the output leads of the alternator. The D-C lead
is shielded from the feed-through capacitor on the D-C line, a capacitor shield is installed to prevent radiation from the terminal on the
hot side of the capacitor. This shield also provides a ground for
the braid shielding. The A-C output leads do not require shielding
because the EMI generated by the alternator is much less severe than
that generated by the D-C motor. Bypass capacitors, connected to the
brushes in both the motor and the alternator, should be included in
the original design. The housing must adequately shield the unit
with a feed-through capacitor, mounted through the shield for connection to the D-C input lead. The A-C leads may not require suppression in addition to that provided by the capacitors at the brushes.

A dynamotor (a combination D-C motor and generator with a single
magnetic field) has an armature with two separate windings and two
separate commutators, one at each end of the armature. It transforms
low voltage D-C to high voltage D-C, or vice versa. The two commutators make this machine a particularly prolific source of EMI. The
suppression techniques for D-C generators and motors apply to the
dynamotor. Fig. 16.34 illustrates a dynamotor, with feed-through
capacitors bypassing EMI to the housing on both the input and output
leads. Complete shielding of the dynamotor prevents EMI from coupling
through other paths.

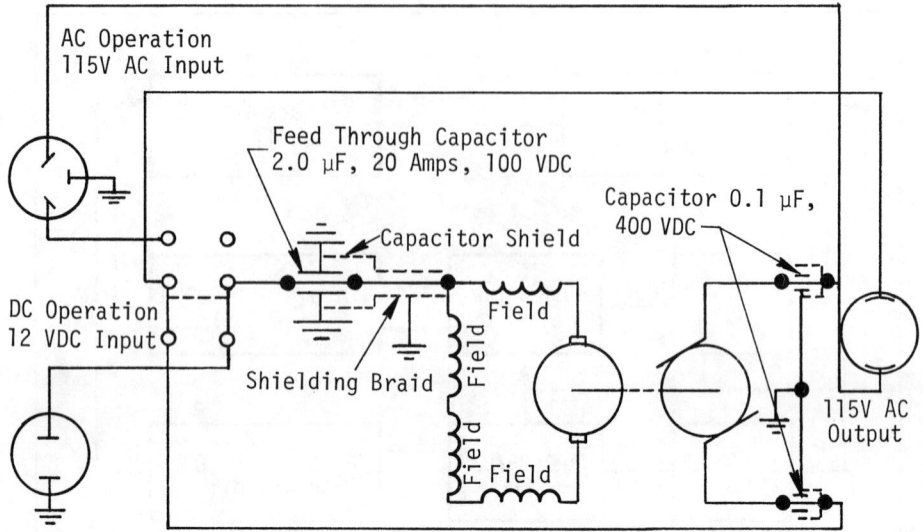

Figure 16.33 - Rotary Inverter Showing Methods for EMI Suppression

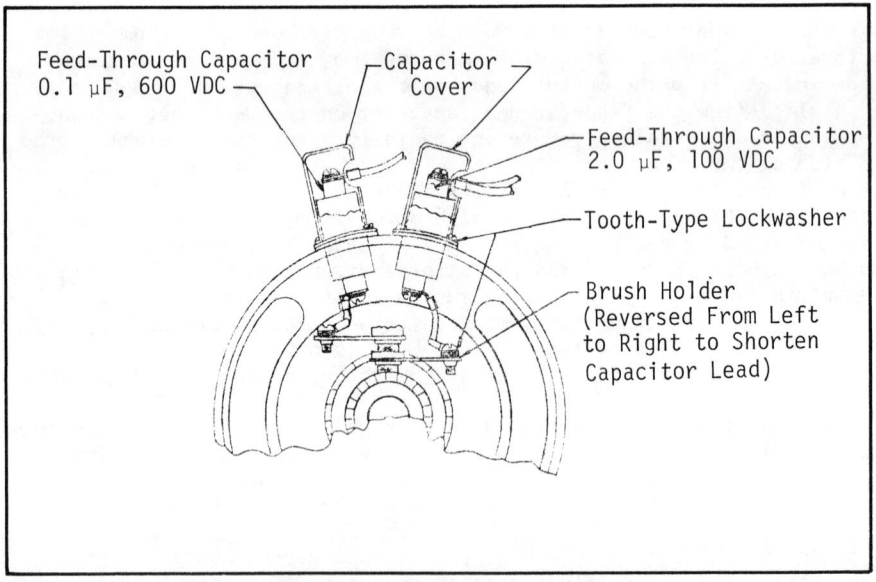

Figure 16.34 - Dynamotor Showing Methods for EMI Suppression

The use of A-C commutator motors should be avoided whenever possible. Universal motors come under this category, as well as repulsion motors and series A-C motors. The performance advantage of these types is their high-starting torque. Their EMI generation, however, is much more severe than that from other types of A-C motors.

High-starting torque with A-C motors can be obtained without increasing EMI by using capacitor-type starting, induction-run motors. These motors use a high-capacitance for starting purposes only. Starting torques of 200 to 350 percent of full load torque are feasible with acceptable starting currents. Ratings from 1/8 to 10 hp are available.

Generators for electric-arc equipment require special attention only when connected to such a severe source of EMI as the electric arc. The generator can be driven either by an A-C or D-C motor to an engine. Little can be done to reduce EMI from the electric arc itself. The equipment should be located away from communication equipment, and in buildings with good shielding characteristics. The leads, from the generator to the welding electrodes, can become very effective EMI radiators and should be adequately shielded.

16.8 LIGHTS

This section describes EMI causes and control in incandescent, fluorescent, and gas lamps.

16.8.1 INCANDESCENT LAMPS

An incandescent lamp, once energized is a fairly stable emitter of infrared and optical energy. Because of its relatively low temperature, comparatively little R-F energy is emitted. Consequently, incandescent lamps generally do not create EMI problems. On rare occasions an incandescent lamp will develop a faulty filament which opens and closes resulting in transient surges. The simplest EMI-control solution is to replace the lamp.

16.8.2 FLUORESCENT LAMPS

Sec. 2.3 discussed emissions from typical fluorescent lamps. It was shown that radiated emission levels are significant up to about 10 MHz and some types exhibit emissions up to 100 MHz or more. R-F radiation from fluorescent lamps comes about as a result of a column of gas being ionized and extinguished 120 times a second for a 60 Hz power mains supply. Thus, these transient surges result in broadband radiation from the bulb as well as conducted emissions back onto the power lines.

EMI is difficult to economically control in fluorescent lamps mounted in their fixtures. One technique is to shield the light-emitting area from the lamp-mount fixture with either conductive glass (this is expensive) or a wire screen (see Sec. 11.1). The A-C lines feeding the fixture are also filtered to keep the 120 pps conducted transients down to controllable levels.

16.8.3 GAS LAMPS

EMI emissions from gas lamps are somewhat smiliar to that of fluorescent lamps. The principle difference is that gas lamps are energized from high-voltage transformers (typically furnishing 10 kV to the lamp) and the gas column remains ionized throughout the A-C cycle. Thus, a steady broadband, non-transient radiation takes place.

While EMI control of conducted and radiated emissions from gas lamps could employ the same techniques used for fluorescent lamps, this is rarely done. Most gas lamps, such as *neon signs*, are too big and cumbersome to shield. Filtering of the A-C power mains may help mitigate some resulting EMI situations. Thus, little is done to contain EMI from gas lamps.

16.9 ANTENNAS

Antennas are a very comprehensive subject and are discussed in length in Vol 5, *EMI Prediction and Analysis*. The principle EMI problems associated with antennas result from either their radiation (for transmitters) or reception (for receivers) in directions other than the main beam and at out-of-band frequencies. In other words, an ideal antenna would act as a complete spatial and frequency filter radiating only within the main beam and only at the designed frequency band of operation. Outside of these directions and frequencies, the ideal antenna would exhibit a large negative gain, such as -50 or -100 dB (or more)/isotropic radiator.

Based on averaging radiation pattern data from many existing and operational antennas classified as low gain, (G<10 dB), medium gain (10 dB<G<25 dB), and high gain (G>10 dB) antennas, the gain in the unintentional radiation regions (outside main antenna pattern) is shown in Table 16.14.

Table 16.14 - Generalized Antenna Models (Unintentional Radiation Regions)

Antenna	Operating Conditions		Mean Gain	Standard
Type	Frequency	Polarization	dB/Isotrope	Dev. in dB
High-Gain G>25 dB	Design Design Non-Design	Design Orthogonal Any	-10 -10 -10	14 14 14
Medium 10 dB<G<25dB	Design Design Non-Design	Design Orthogonal Any	-10 -10 -10	11 13 10
Low-Gain G<10 dB	Design Design Non-Design	Design Orthogonal Any	0 -13 - 3	6 8 6

The gain data listed in Table 16.14 are averaged over 4π steradians (a sphere) less the solid angle of the main beam. Thus, the standard deviations are relatively high since side lobes, back lobes, and diffraction regions are all averaged together. The table shows that only about 10 dB (-10 dB/isotrope) of spatial filtering can be expected for typical antennas (dipoles, monopoles, arrays, rhombic, log periodic, helix, horns, parabolic, etc.).

Low out-of-beam radiation levels are achievable with some antenna design techniques. The common carriers, for example, use special microwave relay horns which exhibit less than -30 dB/isotrope gain in the back hemisphere. Another example involves hooded antennas, a technique involving building an absorptive tunnel or 5-sided box around an antenna at microwave frequencies. For medium gains at SFH, back hemisphere rejection can achieve values of -50 dB/isotropic and 90° off-axis rejection can achieve values of -20 to -30 dB/ isotropic.

It is likely that transmitting antennas designed in the 1980's will be required to meet minimum out-of-beam and/or out-of-band rejection levels to help control electromagnetic pollution. Effective radiated powers (transmitter power x antenna gain) which are 60 dB down (FCC requirements*) from fundamental ERP do not provide sufficient ambient pollution control for high transmitters. Thus, out-of-band rejection levels in excess of 10 dB will likely become a must for antennas.

---

* FCC regulations permit UHF TV transmissions of 5 MW levels to be suppressed by only 60 dB (to 5 watts) to out-of-band emissions.

16.10 BIBLIOGRAPHY

(1) Atkins, J. B., "Worst Case Circuit Design," *IEEE Spectrum*, March, 1965, p. 152.

(2) Clark, Charles M., Jankowski, Herman, Frederick, Carl L., Gauper, Hal, Jr., Goldberg, Leon G., and Hoffart, Henry M., "Electromagnetic Compatibility Principles and Practices," *NASA NHB*, 5320.3.

(3) Cowdell, Hill, Shifman and Skaggs, "Electromagnetic Compatibility Design Guideline for STADAN," TR7-5019, Goddard Space Flight Center, Greenbelt, Maryland.

(4) Drummer and Nordberg, "Fixed and Variable Capacitors," McGraw-Hill Book Co., 1960.

(5) Electromagnetic Compatibility Manual, Naval Air Systems Command, NAVAIR 5335, 1972.

(6) Ficchi, Rocco F., "Practical Design for Electromagnetic Compatibility," Hayden Book Co., New York, 1971.

(7) Henney, Keith and Walsh, Craig, "Electronic Components Handbook," McGraw-Hill Book Co., New York, New York, 1957.

(8) Landee, Davis and Albrecht, "Electronic Designers Handbook," McGraw-Hill Book Co., 1957.

(9) Lansford-Smith, F., "Radiotron Designers Handbook," Radio Corporation of America, Harrison, New Jersey.

(10) "RCA Power Circuits - DC to Microwave," RCA Electronic Components, Harrison, New Jersey.

(11) Robert Pierce Electronic Products, "When is a Capacitor an Inductor," December, 1966.

(12) Siemens, "A New Concept in Miniature DC Motors," *Bulletin 1AD/G*, September, 1969.

(13) Valentino, A. R., "EMC in Digital Integrated Circuits," *ITT Research Institute, 1967 IEEE EMC Symposium Record*.

(14) Wellard, Charles L., "Resistance and Resistors," McGraw-Hill Book Co., 1960.

(15) Welsby, V. G., Wiley, John and Sons, Inc., "The Theory and Design of Inductance Coils," 1960.

CHAPTER 17

EMI CONTROL IN CIRCUITS AND EQUIPMENTS

CHAPTER 17

EMI CONTROL IN CIRCUITS AND EQUIPMENTS

The previous chapter emphasized identifying EMI problems and applying control techniques at the component level. This covered resistors,
capacitors, inductors, diodes, transistors, electron tubes, transformers,
switches, relays, motors, generators, antennas, and some microwave components. This chapter continues with EMI problem identification and control techniques, but at the circuit and equipment level. Included in
this chapter are power supplies, electronic circuits, micro-circuits,
amplifiers, SCR devices, digital circuits and computers, and coding techniques. The chapter is concluded with a bibliography.

The higher the echelon, viz., the greater the ensemble of components
and devices, the more difficult the EMI problems and the control techniques become. The last chapter illustrated many examples at the component level. It is essential that EMI suppression of noise sources,
modes of coupling, and hardening to susceptibility as applicable, be
carried out successfully at the component level since EMI problems at the
next higher level, will be multiplied. Correspondingly, failure to adequately identify EMI problems and effect their solutions at this level
will render good EMI performance at still higher levels more difficult.
These higher levels included are the system and system ensemble levels.
They include buildings and vehicles such as ships, aircraft, spacecraft,
automobile/truck/tank, and the like. These levels which involve inter-
system EMI amongst communications-electronics systems are covered in
Vol. 5 of this handbook series.

17.1 POWER SUPPLIES

EMI problems developing from power supplies originate from one or
more of the following:

- A-C fundamental and harmonics conducted to susceptible circuits

- Transients conducted to susceptible circuits

- Fundamental, harmonics and/or transients radiated to susceptible
circuits or input lines to such circuits

- Poor voltage regulation, especially under transient loads

Chap. 16 discussed the first three topics when the power source is either a D-C or A-C generator. EMI-control by filtering of the source (Chap. 14) and shielding (Chaps 11 and 12) represent other approaches. While related to filtering, this section discusses EMI control when the power source internal impedance becomes a significant percentage of the load impedance of the heaviest using equipment of the power source. Called *voltage regulation* or *common mode source impedance,* this problem can become particularly severe to common users of the power source under transient on-off load conditions.

17.1.1   PRIME POWER IN A COMBAT VEHICLE

To illustrate a problem in voltage regulation, a typical army combat tank uses a hydraulic power pack to slew the turret in azimuth and the gun in evaluation. When the hydraulic pressure reaches some low point, the hydraulic pump motor is connected to the 28 VDC power bus, pulling down a momentary sourge of 1400 amperes, whence it tapers off to about 300 amps in a few seconds before dropping out. Both the resulting on and off transients cause EMI to the fire-control electro-optical sighting and gun-aiming/stabilization subsystems by gun boresight displacement. The problem here is to identify the culprit(s) and fix the problem.

It is academic to argue whether the two electronic subsystem receptors are to blame since they are susceptible, or the hydraulic motor is to blame since it pulls down too heavy a load, or the battery-generator source is to blame because its internal impedance is too high*. It is stated here that the latter is the cause since it is a simple problem of poor voltage regulation in that the battery-generator combination inadequately services several users (the hydraulic motor, the gun-servo drive and electro-optical sights, as well as other subsystems).

Fig. 17.1 illustrates the circuit configuration in which both a 10mΩ load (the 1400 amp hydraulic motor) and the 10mΩ battery-generator source impedance result in a momentary battery bus voltage drop to 14 volts from 28 VDC. This transient has a fall time, $\tau = 1$ μsec as determined by a storage oscilloscope measurement. Thus, using the relation of Sec. 16.4, the spectrum amplitude at the notch frequency of $1/\tau = 1$ MHz is: $2\tau\Delta e/4\pi^2 \approx 0.05 \times 10^{-6} \times 14$ volts = 0.7 μV or .7V/MHz = 117 dBμV/MHz. Since the rate of fall above 1MHz is 40 dB/decade the broadband spectrum amplitude is 77 dBμV/MHz at 10 MHz, 37 dB V/MH, at 100MHz, etc. Thus,

---

* This problem is typical of a hopefully viable vehicular system. In earlier years, it had relatively few or simple electronics subsystems aboard. While the culprit EMI sources were there, there existed few, if any, EMI victims to manifest a problem. As the vehicle became updated and retrofitted with sophisticated electronic subsystems over the years, the latent problems become manifest. This is in part the purpose behind MIL-STD-461A, viz., to preclude or mitigate such eventualities. (Chap. 5, Vol. 1).

this huge transient is a threat to all electronic receptors (especially receivers and amplifiers; see Sec. 3.2) from DC to 100MHz or higher.

To eliminate the transient surge either the generator-battery source impedance must be significantly reduced (not a practical solution here for existing tanks in the field or a *local energy supply* should be created to service the transient demands*. The latter suggests a capacitor, C, to be located near the input terminals of the hydraulic motor as shown in Fig. 17.1. To compute its size:

$$Q = CV \qquad\qquad (17.1)$$

$$I = \frac{dQ}{dt} = \frac{d}{dt}(CV) = C\frac{dV}{dt} \qquad\qquad (17.2)$$

or,
$$C = \frac{I}{dV/dt} \qquad\qquad (17.3)$$

where, C = EMI-suppressing capacitor in farads
dV = Δe = 14 Volts
dt = τ = 1 μsec
I = 1400 Amperes

Making the indicated substitutions into Eq. (17.3) yields:

$$C = \frac{1400}{14/10^{-6}} = 100 \ \mu f$$

A 100 μf capacitor rated at , say 50 VDC, to handle a working voltage of 28 VDC, is not particularly large. However, it must have a self-resonant frequency above 100MHz and preferably up to 1GHz - an impossible situation. To achieve this, two capacitors are used in shunt: (1) the 100 μf capacitor would have a self-resonant frequency somewhat above 1 MHz and (2) a 1 μf feed-through capacitor rated to 1 GHz and capable of carrying a surge current of 1400 amps and an average current of 300 amps for a few seconds with not more than 10 degrees temperature rise. Regarding the latter, the maximum duty cycle would have to be calculated to assure that ratings are not exceeded**

This capacitor arrangement, shown in Fig. 17.1 will service the momentary energy demands of the hydraulic power system during the on-transient condition. However, during the off-transient condition, the hydraulic motor may create a back-EMF inductive transient up to 100 volts.

---

*   A surge inductor would slow down the 1 μsec rise time but the current-carrying capacity would be enormous. Further, the length of line from the battery to load already acts as a small inductor.

** If exceeded, the 1 μf capacitor may have to be substituted by two capacitors, such as: (1) 3 μf by-pass with a self-resonant frequency above 30MHz and (2) a 0.1μf feed-through rated to 1GHz.

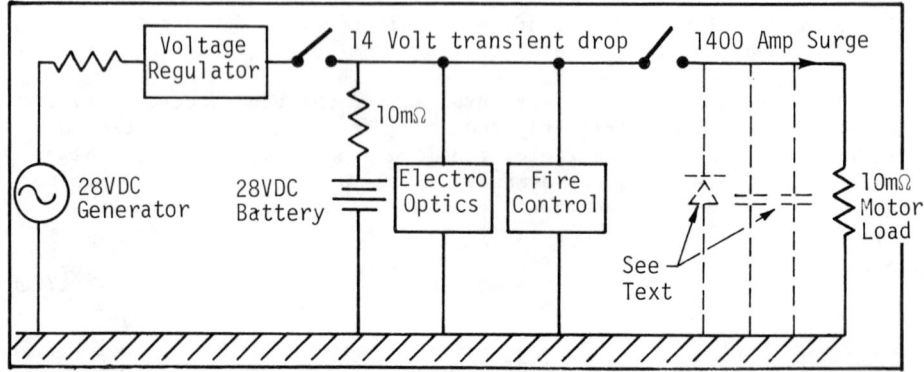

Figure 17.1 - Illustrating Poor Voltage Regulation in a Combat Tank's 28 VDC Battery Bus Supply System.

Since this could rupture the insulation of both capacitors and other electronics packages tied to the battery bus, a lightening-arrestor type surge diode is connected across the line as shown. This will clamp the off-transient surge to manageable values.

17.1.2 REGULATED POWER SUPPLIES

EMI problems resulting from common-mode source impedance is one of the most prevalent culprits of conduced interference. The output voltage of all sources of electric power varies with load and other factors. No source exhibits a zero internal impedance (see 10mΩ in example of previous section). Commercial power systems use regulating equipment in the power plant and on the distribution system to hold the supply voltages within a regulation range of 3 to 7 or more per cent. Such wide limits are entirely unsatisfactory for the output of electronic power supplies used for communication equipment, industrial controls, and test equipment for research and development laboratories.

There are four basic types of units or circuits which are employed to provide close regulation of voltage either directly, jointly, or indirectly via control of other circuit components. These units shown in Fig. 17.2 are (1) reference voltage units, (2) magnetic voltage-control units, and (3) series impedance control, and (4) shunt impedance control. The reference units are typically the zener diode. When connected in series with a suitable resistor across a source of DC, zeners will maintain a constant reference voltage. For very small loads these units alone will often serve to produce satisfactory D-C voltage regulation. For larger current outputs they are bridged across the load circuit to serve as a standard or reference as shown in Fig. 17.2(a). When used in this manner the deviations of the output line voltage appears across the resistor R_x and thus shows the change or error. This error voltage is applied to control units to correct for the deviation. Zener diodes are

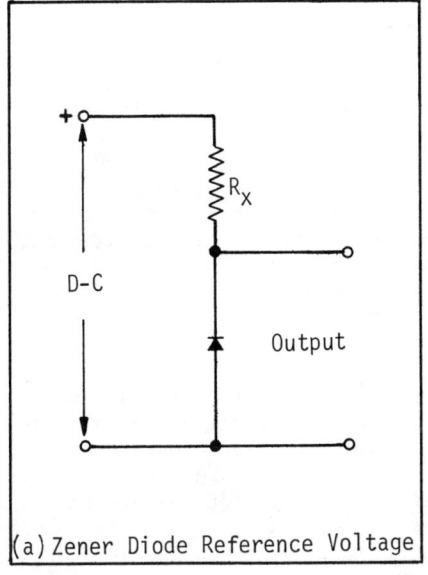

(a) Zener Diode Reference Voltage

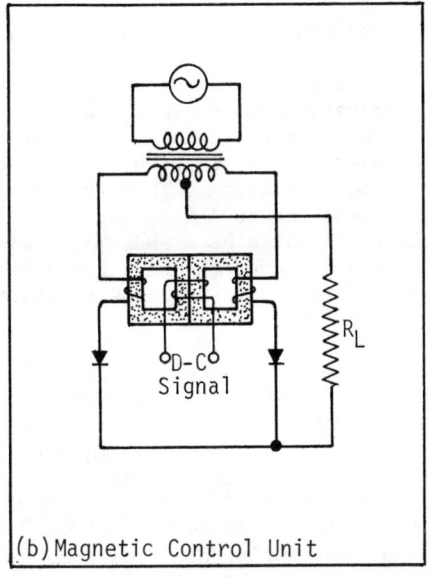

(b) Magnetic Control Unit

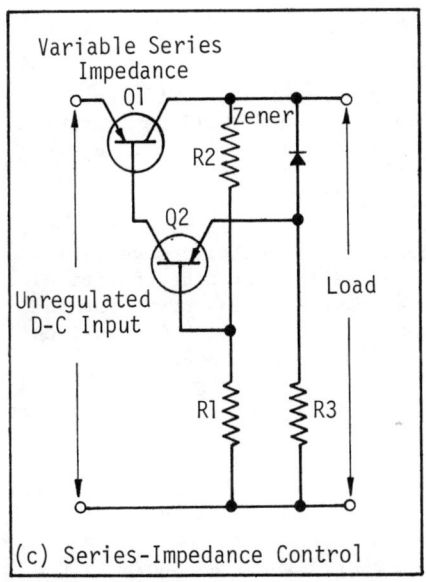

(c) Series-Impedance Control

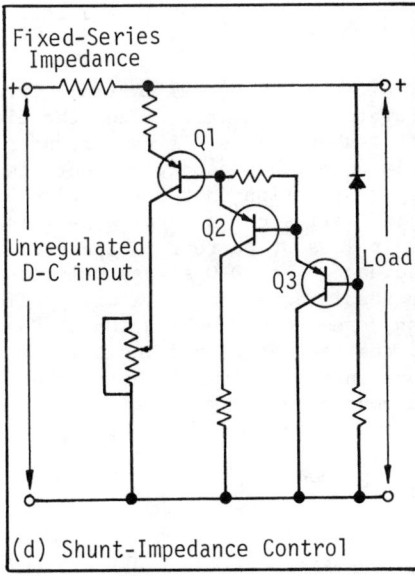

(d) Shunt-Impedance Control

Figure 17.2 - Some Basic Types of Voltage Regulators

available on voltage ratings varying from 2 to 1500 volts.

Magnetic voltage-control units are of three types; constant voltage transformers, voltage regulator units, and magnetic amplifiers. The constant-voltage transformer maintains a nearly constant secondary voltage under varying input primary voltage (e.g. $\pm$ 10%) and varying secondary load conditions. This transformer may be used as a separate unit for regulating an A-C voltage or it may be used in conjunction with D-C rectifiers which have additional control circuits for fine regulation. The magnetic control units are usually a part of a complete AC-DC power supply unit. Both the constant-voltage transformer and the magnetic control units use similar principles for voltage control.

The magnetic amplifier principle is often used to control the magnitude of the D-C voltage and current during the rectification process. One form of the magnetic amplifier as applied to single phase full-wave rectification is shown in Fig. 17.2(b). The amplifier consists of two rectangular iron cores placed adjacent. These cores are constructed of special iron of high permeability having square type of hysteresis loops and they have a cross section which assures saturation for normal operation of the equipment. The rectified current from the outer ends of the center-tapped transformer is led through series coils on the outer legs of the iron cores. These coils tend to produce a high reactance to the flow of the D-C pulses of power. A third coil consisting of many turns of fine wire is placed to surround the inside legs of both magnetic circuits. A variable D-C signal control current is applied to this third coil.

If the D-C signal control current has a magnitude sufficient to saturate the magnetic cores, the flux in these cores will be nearly unaffected by the rectified current pulses in the series coils. Accordingly under this condition there is almost no reactance drop and the only restriction on the rectified current will be the ohmic resistance of the series coils. For a D-C signal current value of zero the rectified pulses of current in the series coils will result in a very-high reactance. For D-C signal values between the two limiting conditions the resulting signal flux will follow the magnetization curve for the iron cores. The combination of the ampere turns of the D-C signal coil and the series rectifying coils will produce a varying degree of reactance in the series circuits. Thus, a very small D-C signal current in the control winding is very effective in controlling the rectified voltage and current on the load R_L.

The D-C output voltage of a rectifier may be controlled by inserting a variable series impedance in the line as suggested in Fig. 17.2(c). This series impedance may ba a magnetic amplifier, the leakage reactance in a magnetic voltage-control circuit or a transistor as shown in the figure.. In this and all other voltage-regulating systems, the input or rectified voltage is always higher than the desired output to permit a sufficient latitude for control. Transistor Q1 in Fig. 17.2(c) is of the

power type and may be replaced by two or more transistors in series if
the voltage deviations and the heat energy to be absorbed in the series
section are sufficiently large. Transistor Q2 is the amplifier for the
voltage sensing unit. The sensing unit consists of a zener diode in
series with resistor R3. This resistor is necessary to assure that the
zener diode always operates on its constant straight-line characteristic.
The zener diode holds the potential at the emitter of Q2 constant, whereas changes in the load potential control the base current and potential
of amplifier Q2 which, in turn, controls the base current of the series
impedance Q1.

In the shunt or parallel-impedance type of voltage regulator, the
voltage drop across the series impedance is maintained constant under
load variations. This is accomplished in Fig. 17.2(d) by varying the
transistor shunt impedance in such a manner that when combined with the
load impedance, both appear as a constant impedance.

For equivalent loop gains, the series-impedance regulator tends to
perform better under input line voltage variations than the shunt type,
whereas the shunt-impedance regulator performs better under output load
variations. Thus, to provide the best under both conditions, it is not
uncommon for a well-regulated supply to provide a combination of series-
shunt regulation with equivalent source impedances of less than 5mΩ, a
0.01% regulation for a 10% line change, a 0.2% regulation for a no-load-
to-full-load change, and a 2mV RMS ripple.

The design of reliable D-C power supplies presents a number of problems. The major problem arises from the transistor's intolerance to
overvoltage. Overvoltages may arise from transients in the load or they
may result from switching and other changes on the input side. Instantaneous voltage *overshoots* or *spikes* can completely destroy a whole bank
of transistors in a matter of microseconds. Conventional fuses and circuit breakers are too slow in operation to offer protection to transistors for transient voltages although they may offer some protection
against overheating with heavy overloads or short circuits. One protection for transistors against voltage spikes is certain semiconductor
diodes which are designed for a rated reverse breakdown voltage, above
which their resistance becomes very low. Such diodes may be placed in
parallel with transistors or other circuit elements to serve as protectors. These diodes act in microseconds.

A second problem in the design of D-C power supplies using transistors is their sensitivity to temperature as it affects their characteristics and life. Thus protection against overloads and overheating
is essential. The best solution lies in the use of inherent self-pro-
tecting regulation which reduces the output current to near zero on overloads and short circuits. This protection may be attained through the
sensing control on a magnetic amplifier or through the transistor regulating circuit.

17.2   SCR POWER CONTROL*

The Silicon Controlled Rectifier (SCR) and related bi-directional solid state switching devices are widely used in power control. They are smaller and lighter than equivalent variable transformers and result in less heat dissipation and power loss than rheostats, but produce high levels of EMI. This section discusses the analysis and prevention of EMI in SCR's.

The A-C power control case is reviewed here. Generation of interference in phase-control applications is analyzed using Fourier techniques, and measured EMI data are presented. An alternate technique, zero-crossover switching, is then analyzed, and measured levels are compared. Techniques of reducing trigger-circuit susceptibility are reviewed, and the application of zero-crossover techniques in proportional control is discussed.

A practical design problem, a self-contained, low-interference light dimmer, is considered, including the effects of visual persistence and beating between a free-running oscillator and line frequency. Related low EMI SCR control applications, including zero-crossover static-control switches, are discussed.

17.2.1   SCR PHASE-CONTROL SWITCHING

In phase-control switching in A-C power control, SCR's are triggered into conduction at a specified phase angle during each half cycle, as shown in Fig. 17.3. Back-to-back SCR's, or a bi-directional switch, may be used to attain this *full-wave* control. Alternatively, one SCR or a rectifier bridge and SCR may be used to provide a D-C biased waveform, but the analysis is similar.

First, a square-wave-truncated sinusoid, shown in Fig. 17.4 is reviewed. The development of the Fourier components proceeds as follows:

The basic square wave, with the Fourier Transform is:

$$\frac{2 \sin \omega T}{\omega} \qquad (17.4)$$

It is shifted to the right and left by an amount D and added to yield:

$$\frac{2 \sin \omega T}{\omega} e^{-j\omega D} + \frac{2 \sin \omega T}{\omega} e^{+j\omega D} = \frac{4 \sin \omega T}{\omega} \cos \omega D \qquad (17.5)$$

---

*   The material for this section was obtained from D. W. Mathias, "Interfence Prevention in SCR Power Control," *IEEE Electromagnetic Compatibility Symposium Record*, Vol. IEEE 69C3-EMC, June 1969, pp. 29-34.

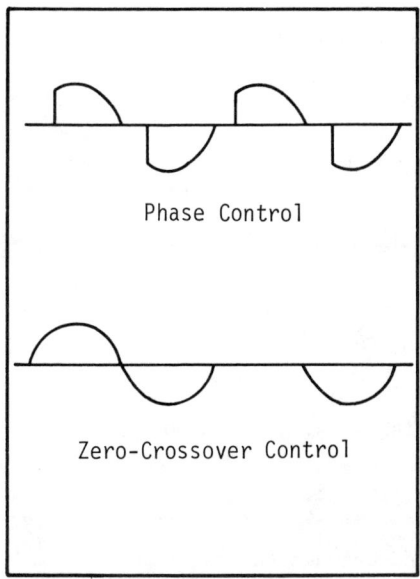

Figure 17.3 - Load Voltage Wave forms

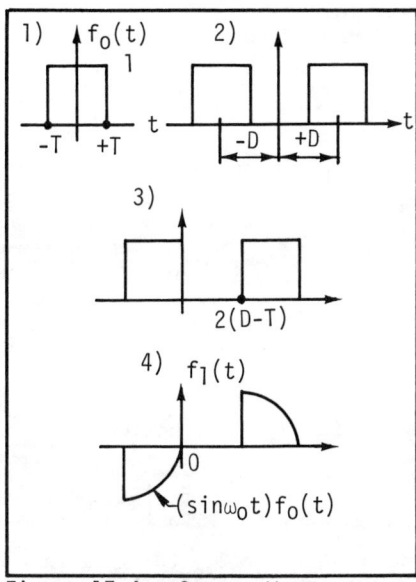

Figure 17.4 - Square-Wave-Truncated Sinusoid

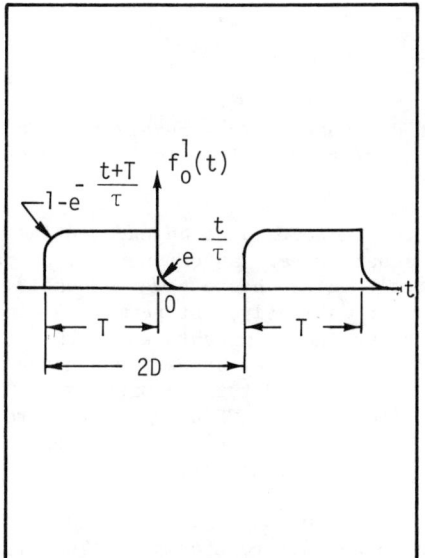

Figure 17.5 - Exponential Pulse Train

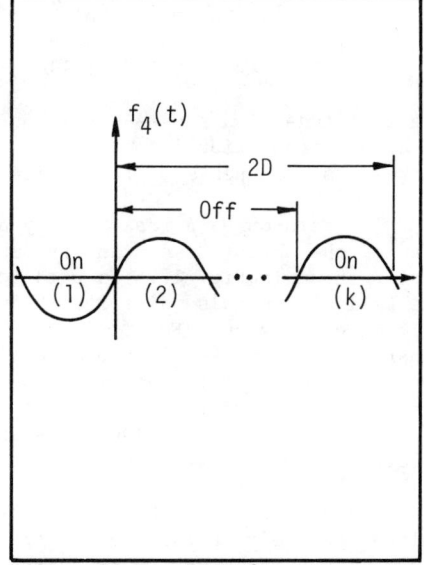

Figure 17.6 - Zero-Crossover Analysis

17.9

and then shifted to the right by an amount D-T to give:

$$\frac{4 \sin \omega T}{\omega} \cos \omega D \cdot e^{-j(D-T)\omega} \qquad (17.6)$$

These square waves are then multiplied by the basic power sinusoid:

$$-j \;\; \frac{2 \sin(\omega-\omega_o)T}{\omega-\omega_o} \cos[(\omega-\omega_o)D]e^{-j(D-T)(\omega-\omega_o)}$$

$$-\frac{2 \sin(\omega+\omega_o)T}{\omega+\omega_o} \cos[(\omega+\omega_o)D]e^{-j(D-T)(\omega+\omega_o)} \qquad (17.7)$$

The magnitude of the harmonic co-efficients is:

$$\frac{2}{T_o} \left| F(j\omega_o n) \right| \quad n = 1, 2, 3 \ldots \qquad (17.8)$$

A phase shift of $2\omega_o$ (D-T) is introduced between the two terms of Eq. (17.7) by the exponentials, and the phase angle of firing thereby affects the magnitude of the sum. For the case of $D = 2T = T_o/4$, the half-power condition, the components are:

$$\frac{1}{\pi} \; [F(j\omega_o, \; 1 \; at \; 3\omega_o, \; 1/3 \; at \; 5\omega_o, \; 1/3 \; at \; 7\omega_o,$$

$$1/5 \; at \; 9\omega_o, \; 1/5 \; at \; 11\omega_o, \; \ldots \qquad (17.9)$$

The amplitude of a square-wave-truncated sinusoid, in general, falls off at a rate proportional to $1/\omega$, or at 20 dB/decade. This is the rate of decrease of the peaks of the basic $(\sin \chi)/\chi$ form.

The turn-on of a real SCR may better be modeled as an exponential than as the abrupt discontinuity in a square wave. An exponential pulse train, shown in Fig. 17.5 is used to truncate the basic power sinusoid. The tail of the pulse distorts the turn-off slightly, but permits a simplified expression. For T<<T, this retains the main features of the square-wave-truncated sinusoid, but introduces a new break frequency at $f \approx 1/2\pi T$, where T is the time constant of the SCR turn-on process. Typical 10% to 90% rise times (t_r) for low current SCR's are on the order of 1 μsec for switching 100 volts and, for a given SCR, are inversely related to voltage switched. The time constant T is (1/2.2)tr.

For high-current SCR's, the rate of rise of current must be held to values typically in the tens of A/μsec to avoid device damage. This is accomplished by inductive or current limiting components in the associated circuitry, and provides a circuitry limited rate of rise rather than a device limited rate of rise.

17.2.2   SCR Zero-Crossover Switching

Fourier analysis of a zero-crossover switched sine wave proceeds in the same manner as that for the square-wave-truncated sinusoid. Eq. (17.7) is valid for this case if:

$$2T = T_o, \quad 2D = \frac{(K-1)}{2} T_o, \tag{17.10}$$

where two out of every k half cycles are allowed to reach the load as shown in Fig. 17.6. This allows calculation of the one-of-i and two-of-i half cycles passed condition where i = 1,2,3 ... Since the phase-shift terms now are integer multiples of the power frequency, the two terms add to produce a sequence proportional to:

$$\frac{1}{n-1} - \frac{1}{n+1} = \frac{2}{n^2-1} \tag{17.11}$$

in which the smoothed envelope of the harmonics decreases in amplitude at a rate proportional to $1/\omega^2$ or at 40 dB/decade.

In this case ω is set equal to harmonics of $2\omega_o/k$, and n is not necessarily an integer. A half-wave rectified sine wave with a period of 2π seconds and amplitude is:

$$\frac{1}{\pi} (1 + \frac{\pi}{2} \text{ sint} + 2/3 \sin 2t - 2/15 \sin 4t \ldots \tag{17.12}$$

Fig. 17.7 shows the familiar unijunction transistor oscillator, phase-control technique. Varying the resistance varies the phase of the output waveform to the SCR gate-pulse transformer. Fig. 17.8 is a simplified block diagram of the zero-crossing circuit. The *disabling means* controls the percentage of half-cycles absent in driving the trigger circuit. Both circuits are powered by the same A-C line voltage as that furnished to the SCR and load.

17.2.3   SCR Filtering

In phase control applications, filtering is often an integral part of SCR circuit design. Device dI/dt ratings for high-frequency devices are given for the SCR shunted by an *RC snubber circuit*. Protection of the device from false triggering due to dV/dt or rapid rise of blocking voltage is required. The most important filtering function is, however, control of current rise time during device turn-on. This limits conducted and radiated electromagnetic interference and protects the device from dI/dt failure.

SCR manufacturers recommend two principle methods of filtering. A single inductor in series with the load-SCR combination may be used with

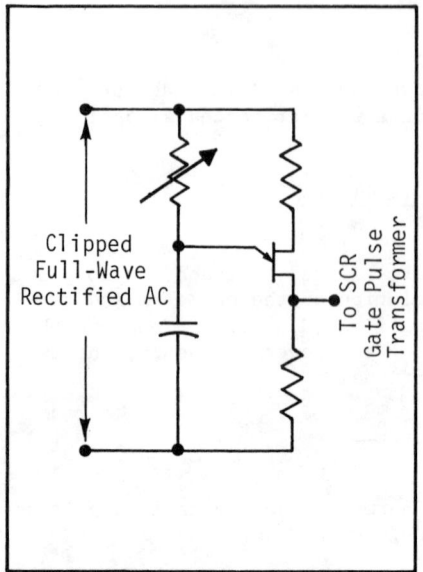

Figure 17.7 - Unijunction Transistor Oscillator Trigger Circuit for Phase Control

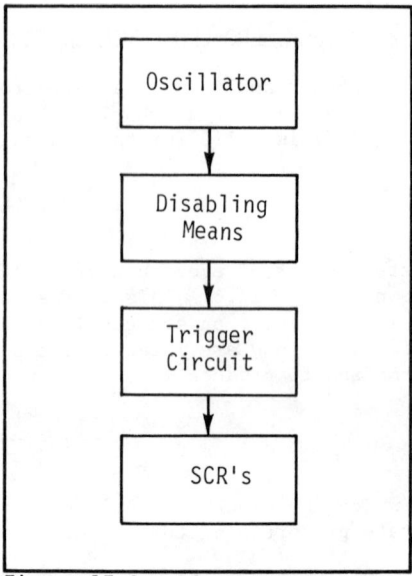

Figure 17.8 - Block Diagram of Zero-Crossover Circuit

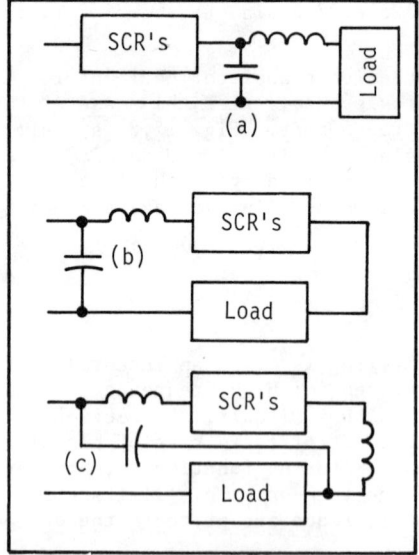

Figure 17.9 - SCR Filtering

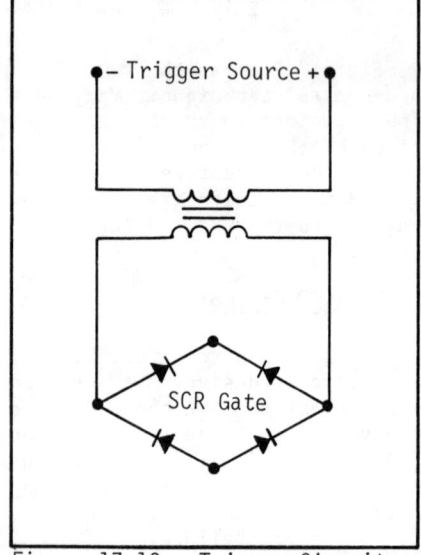

Figure 17.10 - Trigger Circuit Isolation

a parallel capacitance as in Fig. 17.9 (a) and (b) if an R-F ground
is available to terminate the capacitance. Otherwise, the unbalanced
filter creates R-F currents in distributed wiring ground capacity.
Then, a two-inductor balanced filter can be used as in Fig. 17.9(c).
It is often advantageous to use a split-inductor filter with limited
inductance and high saturation current levels.

17.2.4 Trigger-Circuit Susceptibility

Care must be taken to avoid SCR trigger-circuit susceptibility to
either line voltage EMI or to interference in the gate circuit of the
controlled SCR's. Good design practice in regulated supplies for the
trigger circuit, together with R-F filtering, minimize the effects of
power-line fluctuations and transients. Protection of the gate circuit
is more difficult because of the effects of suppression devices which
may adversely affect timing or firing characteristics. For example, in
a pulse transformer coupled-trigger circuit the H-F components of the
trigger-pulse waveform carry a significant portion of the pulse energy.
Rapid application of trigger voltage is desirable to reduce switching
stresses on the device. Thus, filtering is not a practical answer. A
diode bridge, as shown in Fig. 17.10 prevents any reverse interaction
without reducing trigger effectiveness.

17.2.5 Application of the Zero-Crossover Technique

The zero-crossover technique is appropriate in many applications
requiring proportional control. Here, the term *proportional control*
is used in the general sense to describe any requirement for application
of continuous or nearly continuous increments of power to a load rather
than in the special sense of proportional feedback, although the latter
is included. Typical applications are: light dimming, heater control
for industrial processes, aircraft anti-icing or comfort heating, motor-
speed control, and control of electro-chemical processes. When the needs
for low weight and low-power loss coincide with an environment which
is sensitive to EMI, the zero-crossover provides a viable design alternative.

One serious limitation exists. The load must exhibit sufficient
damping to avoid intolerable ripple in the output function. In light-
dimming this requirement can be met, as discussed below, with 400-Hz
power but not with 60-Hz power. In heating applications, the load
thermal inertia is usually sufficient to smooth over variations in average input power over a period of several half cycles of the power waveform. Motor mechanical load systems represent a more critical case, and
again 400-Hz power is preferable, although control is feasible in many
60-Hz applications.

Low-inertia application required dimming 400-Hz aircraft lighting while producing interference levels which were tolerable in an aircraft flight deck area. To prevent annoying flicker-power pulses to the incandescent lamps were required to occur at a rate of at least 50 Hz. The circuit arrangement of Fig. 17.8 was chosen, where the *trigger* block provides zero-crossover triggering in all half cycles if it does not receive a disable signal.

It was found that visual persistence integrates all half cycle pulse trains regardless of their composition if they are repeated at the 50-Hz rate. Thus, a completely unsynchronized oscillator could not be used to produce the disabling signal, since beating between the oscillator and the power frequency often occurred at lower rates regardless of the power frequency. However, use of a free-running oscillator provides completely continuous control of average power. The extreme alternative, selection of a discrete number of pulses during a set period, was rejected because of the resulting discrete power-level increments and complex circuitry. A compromise solution, an oscillator which was loosely coupled to the power waveform, was chosen. Resulting power fractions are closely spaced, but do not cause fluctuations in power at a rate of less than 50 Hz.

17.3 CIRCUITS AND INTERSTAGES

This section reviews EMI problems and control in circuits, amplifiers, amplifier interstages, and micro-circuits. Emphasis is placed on built-in decoupling techniques.

17.3.1 AMPLIFIER DECOUPLING

External radiation from any source is a function of the radiating circuit area, whether it is an electric-field (voltage) or magnetic-field (current) source. It is therefore necessary and prudent to decouple amplifiers and other networks stage-by-stage. Decoupling methods discussed herein emphasize isolation of potential interference voltages and currents.

For high-current stages such as a power-output stage shown in Fig. 17.11, the collector current should be supplied from a low-impedance source. Signal currents that flow through Z_c are supplied by the charge on decoupling capacitor C_d and power supply V_{cc} in series with Z_d. If signal-current flows through the source of V_{cc}, there will be a voltage drop in both the source and in the interconnecting wiring. This voltage will then affect all other circuits connected to that supply and objectionable interaction may result (see Sec. 17.1). In addition, the current through the wiring may induce voltage in other wires. Therefore, a low-impedance path, C_d is provided. A series impedance, Z_d, is provided to raise the impedance of the path and make the power supply a constant-current source. The impedance of C_d and Z_d will force signal currents to flow in the power supply below certain frequencies. The value of C_d and Z_d should be selected so that this frequency is sufficiently low that signal currents do not appear in the power supply and wiring.

The signal current that flows through Z_L is the desired output. This current should also be returned to the emitter with a minimum of disturbance to other circuits. If the emitter and the load are both connected to the chassis or printed circuit board ground, the return current can provide a voltage drop in the ground bus impedance that might interfere with other circuits. A loop is also created that can act as a transformer to couple into other loops. For low-frequency circuits, where the signal system is grounded at only one point, the best return path is a wire twisted with the outgoing wire to Z_L.

At high frequencies, the capacitance of the lead connected to Z_L will cause currents to flow to the chassis along its length and to adjacent wires. This current, too, must return to the emitter. The solution is to provide a shield around the cable to establish a controlled capacitance. The return current for the cable capacitance can then be returned to the emitter by connecting the shield to the emitter. The return current for the load can also be passed through

17.15

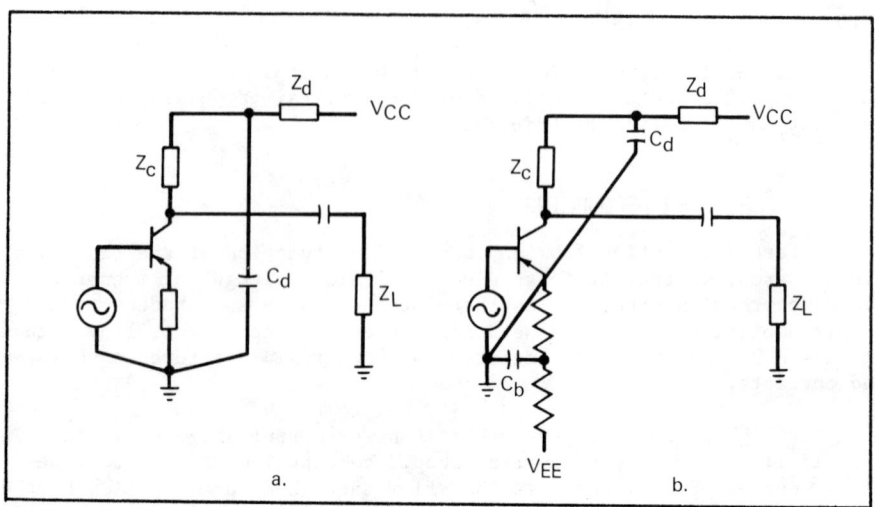

Figure 17.11 - Output Stage Decoupling

the shield, particularly if either the load or the emitter can be
floating. If both emitter and load are grounded, a small percentage
of the return current will pass through interchassis grounds. Twisted
pairs inside shields and transformer coupling decreases this coupling
problem.

Fig. 17.11(b) shows the emitter bypass method used with two-supply
biasing. With the connection shown, the signal current flows through
the emitter bypass capacitor C_b. While it may seem preferable to connect C_d to the emitter side of C_b, so that the collector current does
not pass through the emitter bypass capacitor, that connection allows
disturbances on the power supplied to be coupled into the base emitter
signal loop.

Interference involving turned-output currents is minimized by
connecting the resonating capacitance across the tuning coil rather than
between collector and ground as shown in Fig. 17.12. With the capacitor across the foil, current through the decoupling capacitor at resonance is $e_o/Q\omega L$, where Q is that of the tank circuit, ωL is the impedance of one arm of the tank at resonance, and e_o is the output voltage.
If the capacitor is connected to ground as shown in Fig. 17.12(b), C_d
becomes part of the series resonant circuit, and current through it is
e_o/L, which is higher by a factor Q, which may approximate 100. Thus,
for easier decoupling, the former connection is preferred. In many
cases, most of the current passes through the distributed capacitance
to the chassis, and for which C_d must handle the full tank current.

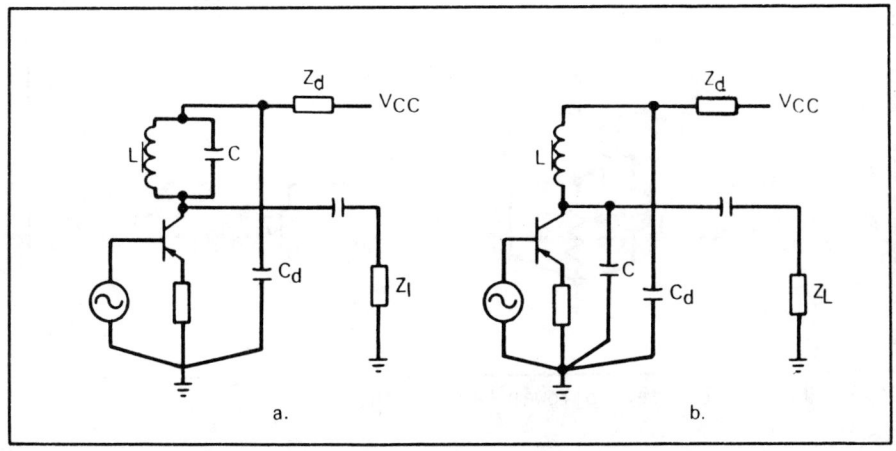

Figure 17.12 - Tuned-Output Stage Decoupling

If the amplifier has signals at a frequency such that $\omega^2 = 1/L$ $(C+C_d)$, then series resonance of Z, C, and C_d as a pi-network can give an objectionable amount of voltage across the decoupling capacitor. The amount of voltage depends on Z_L and Z_d. To approximate the effect, consider the impedance of both to be infinite; then the current through C_d is the short-circuit output current of the transistor. Finite values of Z_c and Z_d reduces this current appreciably.

Emitter-follower stages should be provided with a collector decoupling network to return the collector signal current to the emitter without flowing through the collector supply. Fig. 17.13 shows the collector bypassed to the emitter ground. However, if $Z_L \ll R_e$, which is a typical case when a separate emitter supply is used, current through the chassis can be reduced by returning C_d to the point at which the load current is returned to the chassis.

Interstage coupling of a pair of transistors is shown in Fig. 17.14 (a). The subsequent transistor is represented by its input impedance, Z_i, and its base biasing resistors by R_1 and R_2. The function of this stage is to amplify the input signal represented by e_s and supply maximum current in Z_i and minimum current to the impedances in common with other circuits. At the same time, disturbances on the supplies or in the chassis impedances should supply minimum current to Z_i.

The emitter is shown bypassed to the input signal ground to return the base current signal directly to the driving source without going

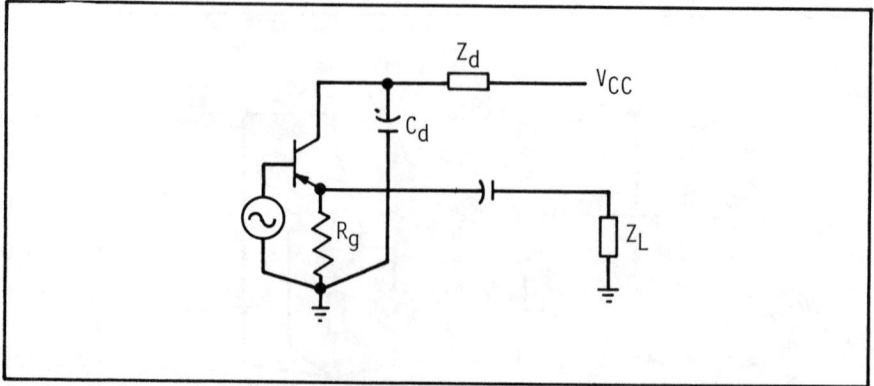

Figure 17.13 - Emitter-Follower Decoupling

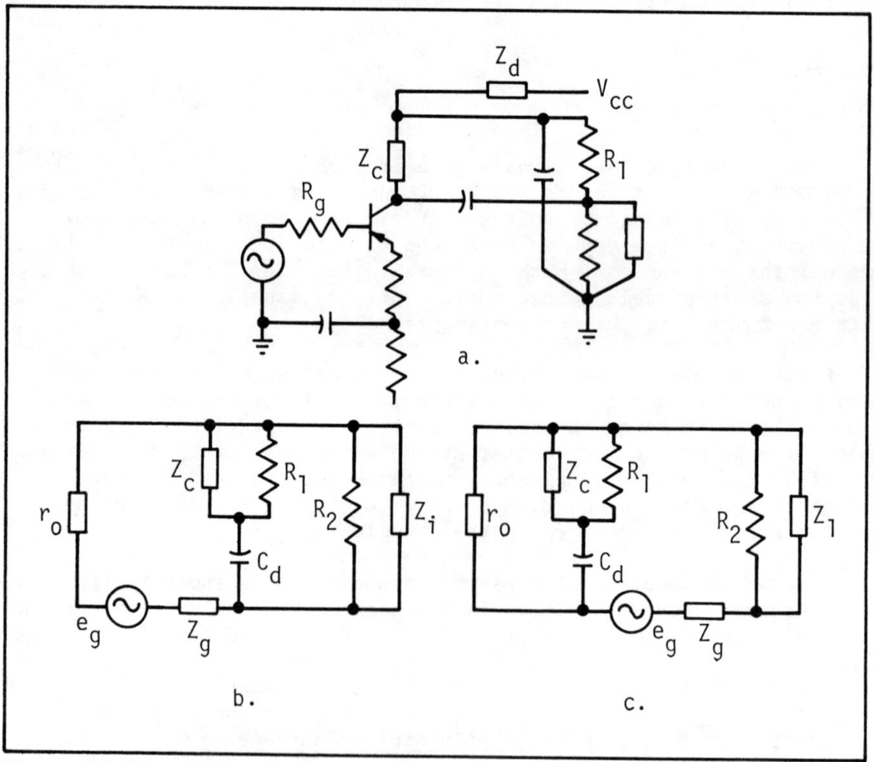

Figure 17.14 - Interstage Decoupling

through the chassis impedance. The ground point of C_d has conflicting
requirements. In the connection shown, all of the transistor current
flows through the chassis so that there may be coupling to other stages.
If C_d is connected back to the first-stage ground, this chassis current
is reduced by the amount of current that flows through Z_c and R_1. If a
current exists in the chassis due to some other source, however, the
configuration of Fig. 17.14(a) minimizes the amplification of this undesired current by the following stages. This is demonstrated in equivalent circuits shown in Fig. 17.14 (b) and (c). The transistor has
been replaced by its output impedance, r_o and e_g and z_g represent the
interference source, the chassis currents. Any current that passes
through z_i must pass through r_o, which is usually a high impedance.
However, in Fig. 17.14(c), which represents the circuit with C_d returned
to the emitter ground, the current through Z_i and flow through R_1 and
Z_c which are usually much lower impedances than r_o.

When flip-flops change state, the current required from the supply
voltages changes momentarily. This pulse is rich in high-frequency components and can couple into wires adjacent to those carrying the supply
current. These transients should be kept on the flip-flop board with
a decoupling network consisting of shunt C and series R or L. When
series L is used, the filter ringing possibilities must be examined.
Where possible, the circuit and board configurations should include
both elements of the flip-flop so that the local current transfer
between these elements necessitates a minimum of energy storage for
the small fraction of time that both elements are simultaneously on
or off.

Some computer circuits use a reference voltage to bias clamping
diodes to obtain constant-level pulses. The resulting pulse currents
in the reference supply and wiring are a source of interfering signals. It is recognized that a series impedance for decoupling can
spoil the reference level. However, a capacitor suitable to the frequency requirements to ground on each board will usually decrease the
amount of high-frequency current in the reference supply and wiring. In
this case, the power supply source impedance at the point of decoupling
must be considered in selecting the size of the decoupling capacitor
(see Sec. 17.1).

17.3.2 AN AUDIO SYSTEM EXAMPLE

An example of the application of EMC techniques to the functional
circuits of an equipment such as the power converter and audio amplifier are shown in Fig. 17.15.

Specifications for the audio amplifier are:

1. Supply voltage: 28 volts

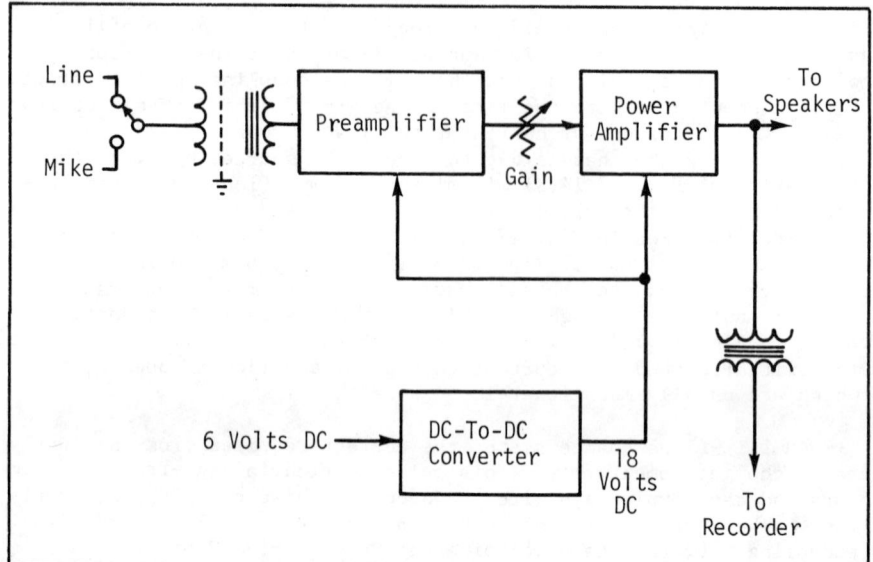

Figure 17.15 - EMC Measures in an Audio Amplifier and Power Supply

2.   Switchable audio-frequency inputs for:

- A long line connected to a remote audio system
- A low level microphone

3.   Audio frequency outputs for:

- Supplying five watts to remote speakers
- Providing a tape recorder output

4.   Quality reproduction of voice is required

5.   Must meet applicable military specifications for conducted and radiation interference over the frequency range of 15 kHz to 400 MHz.

6.   The input power, audio input, and recorder may not be grounded in order to achieve EMI control.

The functional circuitry shown in Fig. 17.15 consists of an audio preamplifier driving an audio-power amplifier, with both powered by a DC-to-DC converter.

A well-shielded input transformer is used to open any ground loops to the distant audio system or microphone. High gain must be used in the preamplifier since the typical microphone has a low-power output.

17.20

Therefore, the input circuitry, including the input transformer, is
very susceptible to EMI. An electrostatic shield is used between the
primary and secondary windings of the transformer to prevent capacitive
coupling of common-mode EMI voltage on the input lines into the preamplifier. Common-mode voltages are those that have the same instantaneous magnitude and polarity on both conductors (see Secs. 6.15 and
8.8). To ensure capacitive and magnetic balance of the transmission
line to the distant audio system or to the microphone, these lines
should be shielded twisted pair. The input transformer should be enclosed in magnetic shields and should preferably be of semi-toroidal
construction. It may be desirable to enclose this transformer in a
simple sheet iron enclosure or provide a shading ring to prevent it from
picking up power supply magnetic interference (see Sec. 16.7).   The
gain control should follow the preamplifier rather than precede it.
Otherwise the noise contribution and interference-pickup of the control
and its wiring will be an appreciable portion of the low-level signal
input.

     Circuit connections to common buses and circuit decoupling from
the common power supply should be designed to minimize common-impedance
coupling of EMI and to eliminate amplifier instability due to feedback
paths. To simplify this, each amplifier stage should be designed for the
minimum permissible audio bandwidth that will result in the specified
overall amplifier bandwidth.

     The DC-to-DC converter portion of Fig. 17.15 is shown in Fig. 17.16.
It has a semi-conductor chopper that supplies a 6 volt, 1000 Hertz
switching rate, square-wave voltage to the power transformer. Switching
in the chopper occurs at points of saturation of the power transformer
core. The secondary of this transformer feeds a semiconductor bridge
rectifier, which, in turn, feeds the output filter that provides the
18 volts DC used to power the preamplifier and power amplifier.

     The chopper, the transformer, and the rectifier cause switching
transients with rise times in the order of 1 μsec. Because the transformer core saturates, it will radiate a strong magnetic field.   To
meet EMC requirements, escape of this interference from the power
supply shield box must be prevented. This is accomplished by providing a completely enclosed shield box of high permeable magnetic
material.

     C_1, C_2, and L form a pi filter on each side of the primary power
input line to prevent the escape of conducted EMI. Filters of this
type present a reactance to the connected circuits in the stop band
and, therefore, reflect EMI. Unless they are carefully made and
grounded, they can also have resonances that permit the transmission
of EMI at one or more resonant frequencies. A type of filter that
absorbs interference usually consists of a length of conductor embedded in a lossy magnetic material. Ideally, there would be a perfect
impedance match to the connected circuits at interference frequencies
so all EMI would be absorbed, and also having infinite attenuation for
interference and no attenuation for the desired frequency passband.

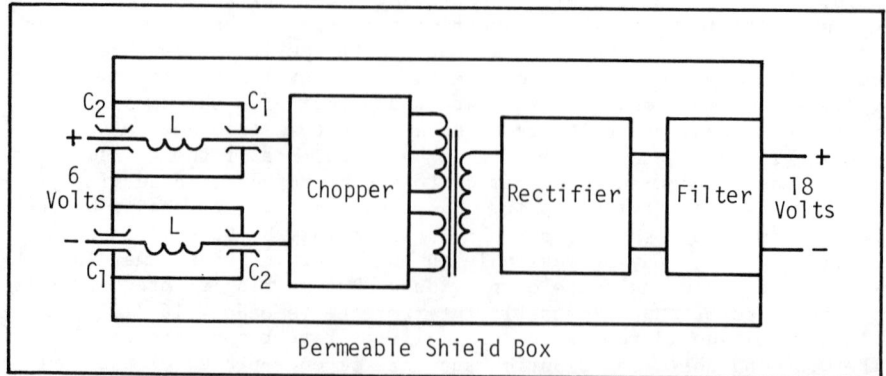

Figure 17.16 - EMC Measures for a DC-to-DC Power Supply

Similar interference filters should be used on the power supply output
leads. Note that separate shield cans are provided on the filter
inductors to prevent either the pickup or radiation of EMI.

To reduce the magnetic field radiated from a power transformer in
this application, it is of toroidal construction or is encased in its
own magnetic shield. Only modest electrostatic shielding is necessary
for the audio-amplifier circuits.

The power supply transformer also fulfills the requirement that
the primary source not be grounded, in order to prevent ground loops
through the power system.

17.3.3 TRANSMITTER EMI CONTROL

All the EMI control measures discussed for the audio amplifier
and its power supply apply to the same type of circuits in the amplitude
modulated (AM) transmitter shown in Fig. 17.17, except that R-F considerations are added. At RF, shielding and decoupling become more
difficult becuase the inductance of even short conductors offers appreciable impedance. The same principles apply but shields and grounding
conductors must make short and direct connections to a sheet of metal
that serves a ground plane rather than the ground tree used in low-
frequency circuits. Circuit layouts on the ground plane are arranged
for minimum radiated and conductive coupling. This is augmented by
the erection of suitable shield walls between the circuits. Circuit
layouts on the ground plane are arranged for minimum radiated and conductive coupling. This is augmented by the erection of suitable shield
walls between the circuits. These walls are soldered or otherwise bonded
directly to the ground plane at many points. R-F coils are oriented with
respect to each other for minimum coupling, are constructed as toroids,
or are enclosed in shield cans bonded directly to the grounded plane.
R-F transmission lines are usually unbalanced coaxial or triax lines
and are run as close to the ground plane as possible.

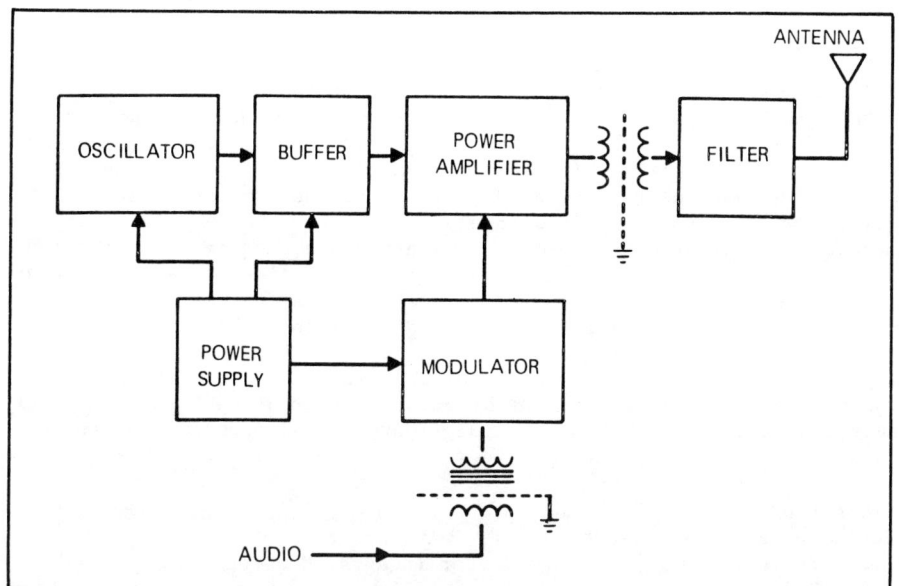

Figure 17.17 - EMC Measures in an A-M Transmitter

    Transmission lines between circuits on different ground planes or
in different equipments are usually unbalanced coaxial lines, because
well-balanced or twisted lines are difficult to achieve at RF. Consequently, the R-F system and all circuit ground planes are bonded to
this with the shortest and largest conductors practicable.

    In RF systems, emphasis is generally on prevention of spurious or
interfering electromagnetic radiation. This usually requires total
enclosure of R-F circuits, with filters or conductors entering those
enclosures designed to prevent the escape of R-F currents. This is
particularly true of the power amplifier stage of a transmitter where
high-power harmonics are produced in normal operation of the amplifier.
A transformer with an electrostatic shield followed by a harmonic
suppression filter before the antenna suppresses the radiation of harmonics to permissible levels. To ensure that the final stage input
is low in harmonic and other spurious R-F energy, a buffer amplifier
is used along with interstage decoupling and shielding.

17.3.4   MICROELECTRONICS*

Application of microelectronics at the system and subsystem level
is progressing rapidly. In order to profit most from advantages in
cost, reliability, weight, volume, etc., a large share of the electronics market will be micro-miniaturized. Early microelectronic applications often took the form of an almost one-to-one correspondence
between conventional circuit parameters and the microelectronic parameters, with slight modification to take advantage of new construction
techniques. In order to achieve maximum technological advantage, future
conversions require more drastic changes, such as the use of digital
instead of analog functions and differential rather than single-ended
amplifiers. Since the introduction of microelectronics has been basically an evolutionary process, considerable insight into anticipated EMI
problems in microelectronics may be gained by comparing the severity of
these problems with those of the more conventional construction techniques.

Although there are some approaches to microelectronics, three
techniques are discussed here: (1) discrete components, (2) thin-
film circuits, and (3) semi-conductor integrated circuits.

17.3.4.1   DISCRETE COMPONENTS

With the use of discrete components, typical approaches to high-
density packaging techniques range from cordwood stacking of conventional components through honeycomb type to flat circuits employing
microsize components, or specifically designed components. Of interest
in the last category are the RCA Micro-Module, the GE TIMM (Thermionic
Integrated Micro-Module) and the Mallory Pellet. For one description,
pellet components consist of composition and film resistors and ceramic or tantalum capacitors of standardized dimensions. These are placed
in a base that accepts the pellets in holes on a standardized matrix.

Of the three major approaches, discrete components offer the greatest flexibility in circuit redesign and the greatest ease of repair.
However, they require the same number of interconnections as conventional circuits, a limiting factor in reliability. In large-scale production they can also be expected to be more expensive than either of
the other two approaches.

---

* Abstracted from "Electrocompability of Microelectronics," by R. B.
Schulz and R. L. Clapsaddle, *IEEE Transactions on EMC*, Vol. EMC-6,
pp. 37-46.

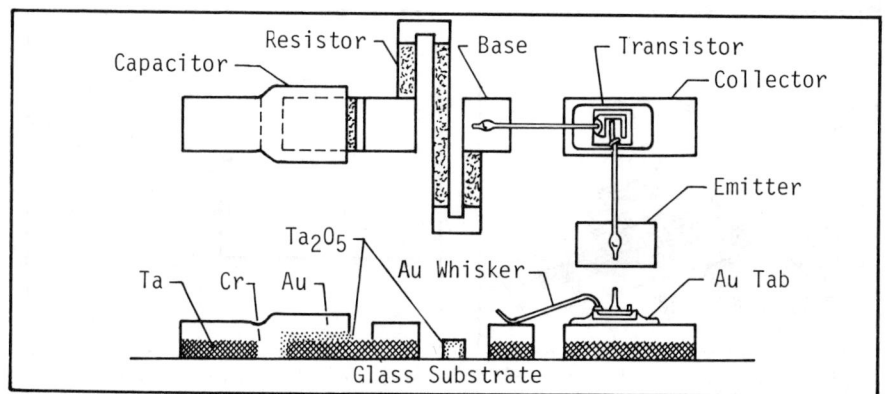

Figure 17.18 - Thin-film Circuit Construction

17.3.4.2 Thin-film Circuits

Thin-film circuits consist of thin films of conductive or insulating material deposited on a substrate, usually glass or ceramic, to form passive components and connection patterns, as shown in Fig. 17.18. They suffer greater limitations on attainable values of resistance, capacitance and inductance than do discrete components. There is, as yet, no sufficiently satisfactory technique for obtaining appreciable inductance. Improvements in techniques and materials offer some promise of overcoming these early problems. The inductance limitation is being overcome at the present time by operating a transistor device in a mode that presents an inductive reactance useful in place of a real inductance and, more often, by simply designing around inductance in circuits. Where this cannot be done, a hybrid approach is sometimes used, i.e., discrete miniaturized inductances are used in conjunction with other circuitry in thin-film form. Thin-film circuits offer potential cost savings over the discrete-component approach. Even more important, they promise higher reliability at reduced cost, weight, and volume.

17.3.4.3 Semiconductor Integrated Circuits

Functional electronic blocks comprise an operational circuit produced by modifying a single block of semiconductor material by diffusion and epitaxial techniques, machining, etc. In practice, semiconductor integrated circuits (SCIC) of the hybrid variety may also utilize auxiliary techniques, such as thin films or even discrete components, when values outside the present range of SCIC techniques are needed. The usual SCIC diffusion approach is to oxidize a silicon wafer, selectively etch through a mask and then diffuse a doping compound into the etched region. A cutaway view of two passive and one active elements is illustrated by Fig. 17.19.

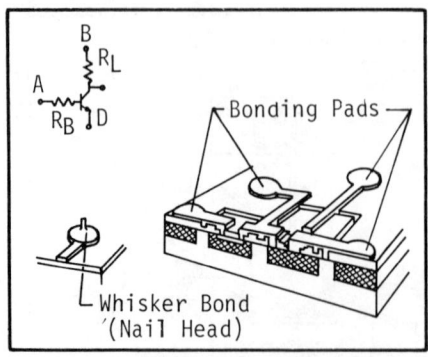

Figure 17.19 - Transistor-resistor
Connected as Inverter

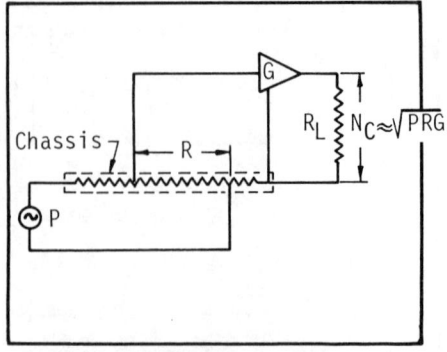

Figure 17.20 - Common-mode
Conductive Coupling

17.3.4.4 Geometric Factors

Reduced size will affect the degree of EMC due to common-mode conductive coupling and due to induction resulting from both magnetic and electric fields. Three induction-field cases must be distinguished: compatibility problems a) within a module, b) between modules and c) between equipments.

In the conducted EMI case, coupling is obtained by means of current flow through a resistance element (such as a chassis) that is common to both the source and the receptor. The response to a conducted interference voltage is shown in Fig. 17.20.

If the very approximate assumption for the discrete-component case (except for cordwood and honeycomb arrangements) is made that a form factor for microelectronics is similar to that for conventional construction, the relative resistance depends only on the relative linear dimensions involved.

For those cases where both the source and the receptor are microminiaturized, it can be expected that any change in relative sensitivity of the receptor will permit a corresponding change in the square root of relative power p of the source. Thus, for construction using similar materials and thicknesses, conducted EMI can be expected to be less severe in proportion to the reduction in linear dimension.

On the other hand, when the same material is used but the thickness is proportional to the linear dimensions, then no significant change can be anticipated in conducted EMI since the coupling resistance will remain constant. The effect of cordwood and honeycomb constructions is generally to shorten the common-mode path to a greater extent than for fixed form factor. EMC due to conductive coupling is further enhanced.

For coupling on a given thin-film wafer, the reduction in common-mode path thickness may be quite severe, with a corresponding increase in coupling resistance. The relative resistance depends not only on relative linear dimensions, but also on relative sheet resistivity. For the SCIC case, coupling on a given substrate may involve radically different materials from the conventional case. The high relative resistivity of the substrate may also increase the severity of the interference problem.

In Intra-Module Compatibility, for the magnetic field, the relative source-antenna efficiency will be proportional to the relative loop areas involved, or proportional to the square of a linear dimension. The relative receptor efficiency will behave similarly. For the close spacings involved within a given module, true induction-field operation may be assumed, with the field strength at the receiver being inversely proportional to the cube of the linear dimension. Intra-module compatibility problems due to magnetic field coupling can be expected to be less severe in proportion to the reduction in linear dimension.

For the electric-field case, the relative antenna efficiency will be proportional to the capacitive effect as will the receptor efficiency. Electric-field effects can thus be anticipated to be more severe than in conventional construction by the inverse ratio of linear size reduction.

For side-by-side modules, inter-module compatibility is essentially the same as for the intramodule case. The only special consideration here is a possible change from source and receptor in separate equipments to source and receptor in adjacent modules, made possible by microelectronics volume reduction.

Inter-equipment compatibility is the problem which includes the source of EMI in one piece of equipment and a receiver of such interference in another piece of equipment, either part of the same system or of different systems. This equipment spacing is determined primarily by construction techniques. Microminiaturization will permit equipments to become closer together, with relationships similar to those developed for the intra-module case.

For the case where equipment location is determined by operational requirements, there will be no change in relative spacing between equipments involved and the interference relationship must be developed. In both magnetic and electric-field EMI cases, the compatibility problem is very much relieved by microelectronics applications.

17.3.4.5 CIRCUIT DESIGN FACTORS

The use of discrete components permits conventional circuit design, provided adequate consideration is given to interelement capacitances. On the other hand, both thin-film and SCIC approaches may require radical circuit redesign.

A number of major problems are inherent in present-day thin-film approaches. Among the more prominent of these is the lack of adequate inductance. Some limited value of inductance may be obtained by utilizing transistor operation in such a way as to present an effective inductive reactance. Also, new techniques are being investigated to obtain adequate inductance. The two chief present-day approaches consist of a hybrid arrangement wherein separate lumped inductances are used, or where circuits are completely redesigned to eliminate the need for inductance. The hybrid approach has little effect upon EMC; the circuit-redesign approach may also have little effect with careful redesign.

Another difficulty of far more importance in EMC considerations is the inability, at present, to maintain close production control over active devices. This lack of good control over the active portions of a circuit leads many designers to turn to digital circuitry in a thin-film version when analog circuitry would have been used in the conventional approach. The digital approach with its high harmonic content in pulse transition can be a source of serious EMI problems. Similar comments with respect to the difficulty of holding tolerances apply to semiconductor integrated circuits. Such difficulty is present because of passive, as well as active, components.

17.3.4.6 EMI-CONTROL

Many EMC measures may be utilized, of which the most important is good basic design. However, it is the purpose of this section to discuss merely two specific approaches, shields and filters.

An undesirable increase in package weight can result from the need to isolate microcircuits from nearby electrostatic and electromagnetic fields. It is possible that, in modules in which heat must be dissipated, the shielding may serve the dual purpose of heat conduction. Fortunately, the smaller circuits of microelectronics require smaller shields. The need for shielding may cause a module that would otherwise be constructed as an integral unit to be broken into multiples, complicating the interconnection problem.

Several shielding methods are available. Where thin-film or SCIC circuitry is arranged on one side of a substrate, a conductive film deposited on the opposite side will provide a shield plate that can be tied to electrical ground. A group of circuits treated this way can be shielded effectively from its neighbor by a ground plate, leaving only the edges unshielded.

One technique that has been used is to coat the outside of a complete encapsulated circuit (with the exception of a small area where leads are brought out) with a conductive coating which is then grounded. Another method makes use of the built-in shielding afforded by mounting microcircuits on standard transistor headers, such as the TO-5 or TO-18 cans.

Regarding using filters for EMI control, the major problem is the lack of large inductances. With present construction techniques, practical values of inductance are limited to approximately 200 μh for pellet-type components. For this reason, high-performance filters are not yet available in microcircuitry. Most filters in use take the form of either high capacitance or resistance-capacitance devices.

The effective input reactance of a transistor might also be utilized as a simulated small-value inductance, but there is limited experience with this approach. Present-day laboratory research is concerned with utilizing some of the more recent materials in combinations which are hoped to produce reasonable values of inductance.

17.4   DIGITAL DATA EQUIPMENT AND COMPUTERS

Recent advances in digital data transmission and the rapid development of computer technology and information theory now make it possible
to tap the potential of electronic communication much more efficiently
than ever before.   Communication systems are no longer restricted to
verbalized message-type information; they are assuming the function of
control systems, involving the exchange of digital data between computers which automatically process, display, and act upon incoming data
from various sources.   Channel efficiency is considerably increased by
setting up a single automated communication system processing time-sharing capability in which all forms of information from all participants
are handled in a uniform binary code.

The present trend is toward the use of digital-data techniques for
communication, data processing, indicator and control systems.   Many
systems use a central data-processing unit that functions as a central
control and interfacing unit for all radio data link, navigation, sta-
bilization, target detection, acquisition, tracking, and weapon control
systems and/or other functions as applicable.   Modern-day requirements
for high data rates lead to systems having short pulse duration and wide
bandwidth, which in turn increases the potential for both emission and
susceptibility problems.

Equipments that use digital techniques are subject to degradation
of data when interference appears in their power, control, or signal
leads.   Digital data equipments, which contain large numbers of solid-
state switching circuits, are also capable of emitting conducted or
radiated EMI.   The emission and susceptibility level of digital data
equipments will vary depending upon design, construction, installation,
and proximity to emission or susceptible circuitry.

17.4.1   EMI EFFECTS

When a digital data equipment is part of a complex system consisting of many electronic units, its operation should not interfere with
other equipments, and it should be unresponsive to extraneous influences in the surrounding electromagnetic environment.   To achieve compatibility, digital data systems should be designed and tested for control of emission and susceptibility characteristics.   Emission and
susceptibility of the equipment can be controlled by isolating the critical circuits and by appropriate circuit design, which includes R-F filtering of interference and selection of circuit logic that minimizes EMI
generation or susceptibility

Digital data systems use a binary system of numbers to perform
their required functions.   Ideally, information in a digital data system
should consist of bistatic circuit conditions in which each circuit element represents either of two all-or-nothing states so that information

is in the form of binary *words* consisting of a series of 1's and 0's.
A binary 1 may be represented by a certain voltage level (positive or
negative), or by a pulse with a defined amplitude, shape, or phase.
A binary 0 may be represented by a voltage level opposite in polarity
to that of a 1, the absence of a level, the absence of a pulse, or by
a pulse of the 1 type but at a different phase. Digital data circuits
are designed to switch states from one condition to the other with minimum transition time.

Binary circuits lend themselves to use in thresholding or comparator
circuits capable of restoring the integrity of pulses or levels that
have become degraded in level or shape by noise, bandwidth limitations,
or other circuit effects. In such a circuit, it is necessary only for
the digital data information level to reach a value sufficient to satisfy
a switching threshold or comparator circuit, and the pulse or level will
be restored thereby to a full-amplitude, relatively noise-free condition.
It is this characteristic, however, that permits EMI to introduce data
errors. If EMI exceeds the switching threshold, or so degrades a signal
that it fails to reach the threshold, the EMI can be quantized into a
full-level bit error.

17.4.1.1 EMI Noise Emission

To achieve high-switching speeds, digital equipments commonly use
low level signal currents or voltages so that transition times between
the states representing 1 and 0 is kept short. High rates of change
produce switching impulse noise. This may require that a trade-off be
made between switching speeds and the EMI effects of noise and switching transients. The predominant interference generated by a digital
equipment can be attributed to the repetitive operation of a multitude
of switching circuits having fast rise times. The interference is
directly related to the organization of the machine, its logic functions,
timing, and basic circuitry. In addition, a solid-state device when conducting produces white noise as a consequence of current flow in the
device. Computers may contain a vast number of circuit elements whose
switching operations are synchronized by clock-timing logic. A computer,
therefore, can be a prolific source of broadband noise (see Sec. 17.5).

17.4.1.2 EMI Susceptibility

An evaluation has been made of the different types of digital integrated circuits as sources of EMI*. Designers of digital equipments or
systems must make technical decisions before selecting the integrated
circuits type that will be used to accomplish a design. When compliance

---

* Richard R. Grim "An Evaluation of the Different Types of Digital Integrated Circuits as sources of EMI," IEEE EMC Symposium Record, IEEE 69C-
EMC, June 1969, pp 43-48.

to an EMI specification is one of the design goals, the EMI performance of the different IC logics enters into decisions regarding logic
selection. An EMI evaluation based on empirical data was established
for some of the major types of integrated circuits. These pertain to
the IC logics as sources of EMl and are not related to the susceptibility performance of the logics.

The integrated circuits evaluated are:

- Fairchild DTL, 940 series (Types 945 and 946)
- Motorola ECL, MFCL-I series (Types 308 and 309)
- Motorola low Power RTL (Types 720 and 710)
- Sylvania TTL, SUHL-I series (Types SF-53 and SG-143)

The logics were those considered to be in general use and representative
of the major logic types employed by most designers.

The choice of the best performing logic from an EMI standpoint will
not eliminate packaging and containment problems, but it will make
decisions regarding packaging and shielding less critical in regards to
achieving compliance. Typical packaging and shielding decisions that
designers are required to make are:

- The types of power buses and wire to use,
- The use of a common ground system or separate signal and chassis
ground,
- The extent of shielding on each digital module,
- The EMI sophistication required in the overall construction of
the equipment cabinetry, and
- The extent of filtering required on leads exiting from the cabi-
net, such as power leads.

Table 17.1 shows the results of the conducted and radiated digital
I-C tests if the relatively small physical size of each digital circuit
is considered, the limited number of integrated circuits used, and the
relatively limited wiring to serve as effective radiators, the conducted and radiated amplitudes generated by all the logics are relatively
high level. It is concluded that all IC logics are potential EMI sources
and that the source problem can be minimized by proper logic selection
but not eliminated.

Since ECL is a high-speed logic and since it performed comparable
to RTL logic but better than the other logics, it is concluded that
selection of this logic in many design situations would be a correct
engineering decision. The table shows that the TTL logic gives more
than 20 dB EMI noise levels than either RTL or ECL logic whereas the
relative excess noise level for DTL logic is more than 10 dB. It would
appear that TTL logic should be avoided where EMI noiseness is critical.

Table 17.1 - Broadband Conducted and Radiated Voltage
Levels from Four Types of Digital IC Logics

Radio Frequency	TTL Logic		DTL Logic		RTL Logic		ECL Logic	
	Cond	Rad	Cond	Rad	Cond	Rad	Cond	Rad
50kHz-1MHz	65	61	53	51	58	57	60	48
1-15MHz	50	40	40	33	40	32	45	29
15-30MHz	38	44	34	46	25	34	20	25
30-50MHz	72	60	58	46	50	36	49	34
50-70MHz	57	60	37	51	(20)	32	31	32
70-90MHz	46	54	37	46	(20)	(20)	(20)	30
90-110MHz	48	60	33	48	(20)	(20)	(20)	31
110-130MHz	44	70	33	54	(20)	(20)	(20)	36
Average	53	56	41	47	32-	29-	34-	33
Overall Avg.	55 dBμV/MHz		44 dBμV/MHz		30 dBμV/MHz		34 dBμV/MHz	

Note: (1) Conducted levels are in units of dBμV/MHz and are measured
            on the circuit logic power supply output bus
       (2) Source is not clear whether radiated levels are in units of
            dBμV/m/MHz (antenna corrected) or dBμV/MHz (antenna uncor-
            rected). The latter is believed to be the case
       (3) Values in parentheses have been inserted to indicate that
            this is the estimated threshold sensitivity of the EMI
            detection system (which was not identified by the source).
            Levels are likely lower than this.
       (4) Where average values show a minus sign behind the number,
            the level is likely lower since data values were limited by
            EMI detection sensitivity.
       (5) Power supply voltages furnished: TTL and DTL circuits:
            +5V; ECL circuits: -5.2V; and RTL circuits: +3.6V.

    The noise margin and noise immunity parameters of gating circuitry
used in digital devices are measures of the gate's susceptibility to
noise signals. Noise margin is defined as the magnitude in volts of
pulse noise, which, when appearing at the input of a digital gate and
riding on the worst-case logic level, will cause a significant reaction
at the output of the gate. D-C noise margin, as in wide, slow rise time
noise pulses, is the difference between the worst-case logic voltage
level and the worst-case switching threshold voltage of the gate. Transients or A-C noise margin, as in fast rise time noise pulses, is, in
most cases, less than the D-C noise margin.

    Noise immunity is a measure of a gate's immunity to noise generated
by neighboring gates. Noise immunity is:

$$\text{Noise immunity} = \frac{\text{worst case noise margin}}{\text{maximum logic voltage swing}} = 100\% \quad (17.13)$$

17.4.1.3 Crosstalk

Crosstalk in a digital circuit may occur in two ways: internally via coupling phenomena within the solid-state device itself, or externally via cables and connectors of associated circuitry (cf. Sec. 17.3.4).

Internal coupling occurs as a consequence of parasitic mutual capacitance and inductance inherent in the semiconductor device. The isolation between input and output of semiconductor devices, is notoriously poor. This inherent characteristic may result in undesirable coupling from one input to another, or from an input to an output. This coupling may also result in crosstalk between two gates in the same integrated circuit package. These problems indicate the necessity for an investigation of the manner in which unused terminals are to be handled. A decision must be made as to whether they can be allowed to float or whether they should be tied to some bias voltage or to ground.

External coupling is due to mutual capacitance and inductance of interconnecting lines and conductors (see Chap. 6). High rates of change of voltage or current on these conductors will cause crosstalk between them. Given the mutual inductance, L_m, and the mutual capacitance, C_m of two interconnecting conductors, the voltage or current crosstalk may be estimated by $L_m \, di/dt$ and $C_m \, dv/dt$ effects.

The effect of crosstalk is to cause noise, usually in the form of differentiated leading and trailing edges of input signals, to appear on the output (or other inputs) of a gate even though the gate inputs did not satisfy the switching logic. A crosstalk signal of sufficient magnitude and width may be amplified and shaped by a string of gates so that it results in a logic signal that causes an error.

17.4.1.4 Magnetic Storage and Electrostatic Effects

Frequent use is made of magnetic materials as information elements in digital data processing systems. Magnetic tapes, discs, or drums may be used as input/output or storage devices, and magnetic toroids are frequently used in shift register, decoder, buffer storage, and memory circuits. Because the magnetic mass and flux density is often quite small, these magnetic materials are easily influenced by extraneous magnetic fields. An A-C EMI field can erase the data stored in magnetic materials. A D-C EMI field can bias the magnetic materials toward an *all ones* or *all zeroes* state. Magnetic transients can produce digital bit errors. The portions of digital data processing systems that use magnetic materials must therefore be protected from internal and external magnetic fields.

Persons accustomed to handling solid-state devices are aware that unless precautions are taken, certain types of diodes and transistors can be degraded by a static charge accumulated on the human hand. The

electrostatic effects become less troublesome after the device is installed, because of the generally low resistance of associated circuitry.
However, cases occur of impaired computer performance attributed to
static charges on a plastic cover over a row of logic cards on digital
data systems.

17.4.1.5 POWER AND GROUND CIRCUIT TRANSIENTS

High rates of change of current in the power and ground terminals
of digital circuits may cause voltage disturbances in the power and
ground distribution systems which may result in the propagation of an
error signal. Because a number of circuits may be packaged to form one
module, a power-supply decoupling capacitor may be necessary within the
module. When the circuits are grouped into modules and each module contains power-supply bypass capacitors, the EMI currents are:

● Ground currents that leave the module and flow into the system
ground

● Signal currents that flow in the bypass capacitor mounted in
the module

High-speed switching circuits may place a large transient current
demand upon power supplies during transition from one logic state to the
other. Knowledge of the rate of change of these currents will allow the
system designer to calculate, approximately, the tolerable impedance of
the bypass capacitor and the system ground. Measurement of the area
under the waveform of the current that passes through the bypass capacitor gives the charge drawn during switching. This measurement allows
the designer to determine the value of the bypass capacitor necessary in
the circuit module.

The path from the module to system ground must be a low impedance.
If the module is a printed circuit card, it is almost always necessary
to use more than one of the connector pins as a ground return. The common-ground return for groups of modules should also be designed for low
impedance.

Susceptibility of a digital circuit to disturbances in the power
and ground systems may be determined by introducing a disturbing pulse
between the ground pin of the module and the system ground, or between
the power pin and the system power. If the logic circuit under test is
one of a string of gates, the width and amplitude of the disturbing pulse
may be varied to determine the minimum pulse that will result in the
propagation of an erroneous logic signal.

17.4.2 DIGITAL DATA ERROR CONTROL

An analog communication or data system can often lose an appreciable

portion of a data word due to EMI without the overall usefulness of the
individual signal transmission and reception being degraded below an acceptable level. For verbalized communication in plain language, a word
with parts missing or in error can often be recognized and restituted
correctly by a human operator. Electromechanical analog systems can
sometimes minimize the effects of EMI by using electrical filters and
mechanical damping. A digital data system, however, operates differently
as a consequence of signal degradation. The introduction of an error bit
into a word (a 1 converted to a 0, or vice versa) converts the entire
word into an error word. The consequences of an error bit in a digital
data word can vary, depending upon the position of the bit in the word.
Yet EMI is as likely to cause an error bit to appear in one place as any
other. A digital data system must therefore be designed to recognize an
error word and to reject the word in its entirety. If the error word
were processed into the data, it possibly could bring about a degradation
far greater in magnitude than that of the original error, for example,
cause dumping of a stored program.

Digital data devices in many applications operate in close proximity to other systems which use high amplitude pulses. Radar, Tacan,
IFF, beacon transponders, tactical data links, and overhead power lines
are typical of potential sources of pulse interference. The pulsed or
bistatic logic used in digital circuits can easily be triggered by pulses
from these systems, or from electrical or magnetic transients on power,
control, or signal leads.

In typical digital-data equipments, some circuits have an inherent
immunity to malfunction under intense electromagnetic influence. Relatively *invulnerable* circuits include high level circuits such as flip-
flops, line drivers, relay drivers, level inverters, and emitter fol-
lowers. On the other hand, low-level circuits whose normal input is within the range of 50mV to ?V peak-to-peak, tend to malfunction when subjected to EMI of moderate field intensities. Circuits in this category
include tape readers and magnetic core memory sensing amplifiers. Digital-data systems employing radio data links are most susceptible to degradation in the radio circuit itself rather than in the logic circuits that
follow the quantization process. The EMI control measures discussed for
receivers (see Chap. 5) in general apply to the digital data receivers.
Other measures are also particularly applicable to digital-data systems.

17.4.2.1 Parity Error Check, Coding and Redundancy

The parity check method involves determining the number of binary
1's in a data word. Circuit logic is arranged to accept either an odd
number or an even number, with a position in the word reserved for the insertion of a parity bit where needed to satisfy the error-checking circuitry. If EMI should cause an extraneous bit to appear, the parity error check would reject the word as containing an error. Usually the circuit logic is designed to try repetitively to read the word before finally

rejecting it. The parity error check method affords no protection
against error words containing two or any even number of error bits.

Error control in data-transmission systems is achieved by incorporating electronic error-control circuits or by adding error-control devices
external to the transmission equipments which perform error-control functions solely. The application of such devices can control only those
errors produced by transmission or equipment deficiencies, and therefore
other techniques must be utilized to prevent human errors. The function
of the transmission system is to transfer data without errors, additions,
or subtractions so that the output of the data link is identical with the
input. It is sometimes sufficient to detect errors and resort to human
intervention to eliminate input of false information in subsequent data
processing. These control methods are known as error-detection methods.
Such methods are wasteful of manpower and time, and the savings in equipment cost can rarely be justified. Error detection and automatic correction methods require more costly equipment, but overall system annual
costs are materially reduced. A number of developments of this latter
type have been completed and are in service.

(A) Redundancy. Basically, all systems performing error detection require transmission redundancy whether automatic correction is employed
or not. There are two transmission techniques utilized in present systems, those systems that transmit only in the forward direction and those
that transmit in both forward and return directions. Both types of systems require the dividing of the input signals (bits or characters) into
word blocks of short enough length to be managed without excessive cost
for storage elements. A forward system requires a transmission channel
in one direction only, whereas a forward and return type of system requires a transmission channel of the two-path type (a duplex channel).
However, to achieve a high degree of error control in a forward system,
it is necessary to resort to coding methods that contain a much higher
order of redundancy, approaching 100 per cent which requires a transmission channel of twice the speed. Two methods are in common use: series
redundancy and parallel redundancy. The series method transmits each
word twice in sequence over a single channel. The parallel method transmits the same word over two or more channels simultaneously. The typical
radio data link may use two independent sideband channels for this purpose. In either case, the two samples of the data word are stored in
buffer registers and compared bit for bit. If the two are not identical,
a reception error is indicated and the word is rejected. Retransmission
is then necessary.

The availability of a duplex channel permits reducing redundancy to
small percentages, and since such a system will permit data to be transmitted in both directions simultaneously, the overall traffic-carrying
capacity of the system remains high, very little time being required for
efficient error control. It therefore becomes a designer's choice as to
which method will be used in an operating system. However, in data links
that are short or between equipment sections that are connected by short

wires, the designer may employ error-control techniques that are quite wasteful of transmission time to lower the overall system cost. For example, in the *loop-check* method, all data bits are transmitted to the distant point and relayed back to the transmitter where the check is made. If an error occurs, the transmitter sends a control signal which causes the last block transmitted to be dumped at the receiver, and the same block is retransmitted. If no error occurs, the transmitter sends a control signal to the receiver which causes it to accept the last transmitted block

(B) ED-AC. The Western Union Error Detection-Automatic Correction System (EDAC) is a highly accurate and low-cost error-control method of error detection and correction for both directions of transmission on a full-duplex channel. The speed of operation can be high on channels of relatively long propagation time, for example, as high as 2,000 bauds on long H-F radio circuits. EDAC transmissions are divided into blocks and since the trunk side is operated synchronously, the same block length is required in each direction. Various block lengths are possible, but one current model operates with blocks of four data characters. These characters are transmitted from nondestructive storage, and 5 check bits derived from a binary summation of the data bits are also included. At the receiving end, each block is stored and checked before delivery. A 3-bit control signal is then returned to the sending end (a corresponding control signal is sent simultaneously on the other side of the channel). When the control signal indicates that an error was made, the preceding four data characters are retransmitted; otherwise these are erased and four new characters are transmitted.

    Mutilation of the control signals does not destroy EDAC accuracy. Retransmission is carried out until acceptance is indicated. Long-period circuit interruptions also have no effect except the lost time involved. If EDAC is operated against a nonsynchronous input or output system, buffers are required.

(C) Beckman ECT. Beckman Instruments, Inc. has developed an Error-Control Transceiver (ECT) to transmit and receive digital data over channels of voice bandwidth. Data are redundantly coded in such a way that the probability the equipment will pass erroneous data is 10^{-14}. At an operating speed of 1,200 bits per second this would be one incorrect block of data in about 2,500 years. Transmission is full-duplex, and any type of input and output can be accomodated by suitable buffers.

(D) SECO. Sequential coding and decoding equipment for the detection correction of errors in high-speed digital transmission has been developed and tested by the Lincoln Laboratory, MIT, Lexington, Mass. Data were transmitted at an average rate of 7,500 bits per second on an 800-mile toll-grade telephone circuit with noteworthy results. This was the first demonstration of an adaptive electronic system for error control on communication channels.

17.4.2.2   Strobing and Gating

Strobing is a process sometimes used in digital systems (including
teletype) to prevent circuit switching transients from degrading the bit
accuracy. Strobing consists of sampling each bit at a time the signal
has reached its best 1 or 0 (mark or space) condition. The strobe sampling time is usually much shorter than the duration of the expected signal.
Circuit switching can cause such effects as switching noise, ringing
due to reactive components of circuits and filters, or slow rise and decay (bias) of pulses due to RC or RL delays. The timing of the strobe
sampling circuit is adjusted so that the EMI effects of a circuit switching sequence has had time to abate before the sample is taken. The
strobed sample is then used to set buffer registers, selector bars, or
other storage devices that will reconstitute the signal into a noise-
free state.

A typical application of the principal of strobing is used with the
waveform that represents the output of a sense amplifier, a circuit that
senses the output of a ferrite core. The ferrite core, the heart of the
memory system, stores 1 or 0 bits as a function of the direction of
magnetization of the core. To read a memory core, the two memory address
conductors at which the core is located are energized simultaneously.
The combined effect of the two currents is to set the addressed core to
its 0 state. If the core were storing a 1, setting it to 0 produces a
magnetic flux change that is inductively coupled to a sense amplifier.
If the core were storing a 0 there would be no flux change and no data
bit output to the sense amplifier.

Other memory cores on the same memory plane and on other planes are
affected by current in only one of the two address lines. The *half-se-
lect* current is insufficient to change the magnetic state of the unad-
dressed cores, but it does cause noise to appear in their sense amplifiers as well as that of the addressed core. Strobing is therefore used
with the sense amplifiers to prevent the noise produced by the current
rise in the address lines from affecting the accuracy of reading memory
cores.

A gated circuit is one that is enabled at a definite time to sense
the input data. The gate is normally opened by a pulse developed from
clock timing circuits and in the absence of the gating pulse, remains
closed and unresponsive to noise or any other signal. For a computer
to malfunction from external interference, the interfering pulse or level
must be of sufficient amplitude to either subtract from or deteriorate
the information pulse, or simulate a data pulse, and at the same time be
in coincidence with the gating pulse.

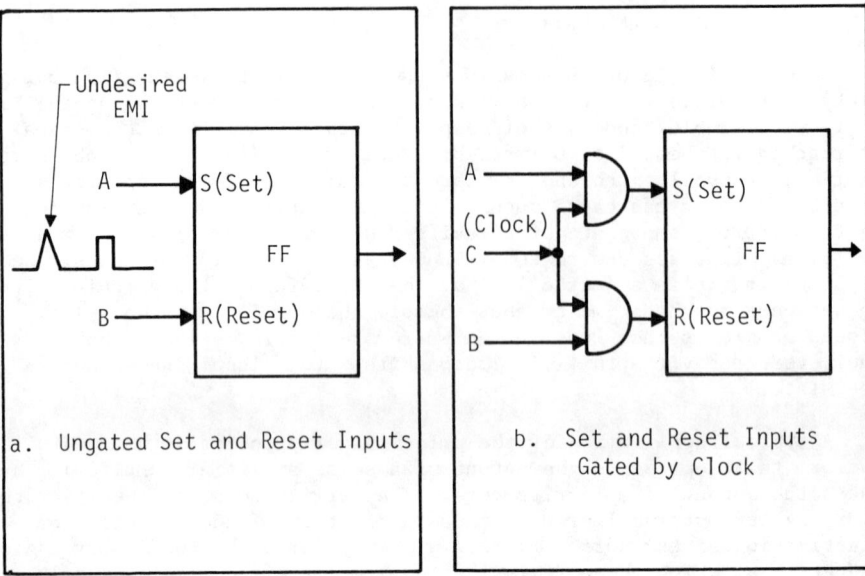

a. Ungated Set and Reset Inputs   b. Set and Reset Inputs
                                              Gated by Clock

Figure 17.21 - Logic Circuit for Interference Suppression

Figure 17.21 illustrates the use of gating to minimize the likeli-
hood of data errors. In Fig. 17.21(a), which does not use gated inputs
to the data register flip-flop, a noise impulse on reset bus B could
cause all data registers to be reset. Similarily, noise on register input A can cause that particular register to be set. The *set* or *reset*
action may take place any time that a noise impulse of sufficient magnitude appears.

Figure 17.21(b) shows the use of gated input logic to lessen the
error rate due to noise. Input A represents the *set* order, which could
be either a pulse or a zero level, and input B represents the *reset* bus.
Input C is the clocked gating pulse. The flip-flop can be neither set
not reset by noise or any other signal unless there is an enabling pulse
from the clock-gating circuitry.

17.5 BIBLIOGRAPHY

(1) Cooper, H. K., "Transient Considerations in Design of Solid State D-C Converters," *Solid-State Design*, Vol. 4, No. 10, 1962.

(2) Egidi, C., "Measurement and Suppression of VHF Radio Interference Caused by Motorcycles and Motor Cars," *IRE Transactions in Radio Frequency Interference*, Vol. RFI-3, No. 1, May, 1961; pp. 30-38.

(3) Evans, G. R., "Effects of Electromagnetic Fields Upon Instrumentation Components and Systems," A Bibliography, Report SB-60-27, AD 243538, *Lockheed Aircraft Corporation*, Sunnyvale, California, July 27, 1960.

(4) Ficchi, Rocco F., "Practical Design for Electromagnetic Compatibility," *Hayden Book Company*, New York, 1971, Chap. 10 - Digital Computer Systems.

(5) Hwei Piao Hsu, "Automotive Ignition Interference," *IEEE Transactions on Electromagnetic Compatibility*, Vol. EMC-6, pp. 15-20.

(6) McCullough, W. M., "Digital Integrated Circuit Susceptibility," *Record of IEEE National Symposium on Electromagnetic Compatibility*, June 28-30, 1965.

(7) Mitchell, J. C., et al., "Electromagnetic Compatibility of Cardiac Pacemakers," *IEEE International Electromagnetic Compatibility Symposium Record*, July 18-20, 1972; pp. 5-10.

(8) Naval Air Systems Command, "Electromagnetic Compatibility Manual," *NAVAIR 5335*, 1972, Chap. 15.

(9) Schulz, R. B., and Clapsaddle, "Electrocompatibility Aspects of Microelectronics," *IEEE Transactions on Electromagnetic Compatibility*, Vol. EMC-6, No. 1, January, 1964, pp. 37-46.

(10) Showers, R. M. and Haber, F., "Studies of Electromagnetic Compatibility of Equipments and Systems," Report No. 65-25, *University of Pennsylvania, Moore School of Electrical Engineering*, May 31, 1965, pp. 24-37.

(11) Spelman, Francis A., "Electrical Interference in Biomedical Systems," *IEEE Transactions on Electromagnetic Compatibility*, Vol. EMC-7, No. 4, December, 1965, pp. 428-436.

(12) Weinstock, G. L., "Electromagnetic Interference Control Within Aerospace Ground Equipments for the McDonnell Phantom II Aircraft, *IEEE Transactions in Electromagnetic Compatibility*, Vol. EMC-7, No. 2, June, 1965, pp. 85-92.

(13) White, J. V., "Wiring of Data Systems for Minimum Noise," *IEEE Transactions on Radio Frequency Interference*, Vol. RFI-5, No. 1, March, 1963; pp. 77-82.

INDEX

Index

B

Index

C

INDEX

INDEX

D

E

INDEX

Index

Index

F

INDEX

INDEX

H

I

Index

Index

N

O

P

Index

Index

R

INDEX

S

INDEX

Index

T

INDEX